PERMUTATION TESTS FOR COMPLEX DATA

Theory, Applications and Software

Fortunato Pesarin • Luigi Salmaso
University of Padua, Italy

WILEY
A John Wiley and Sons, Ltd., Publication

This edition first published 2010

Registered office
John Wiley & Sons Ltd, The Atrium, Southern Gate, Chichester, West Sussex, PO19 8SQ, United Kingdom

For details of our global editorial offices, for customer services and for information about how to apply for permission to reuse the copyright material in this book please see our website at www.wiley.com.

Library of Congress Cataloguing-in-Publication Data
Pesarin, Fortunato.
Permutation tests for complex data : theory, applications, and software / Fortunato Pesarin, Luigi Salmaso.
p. cm.
Includes bibliographical references and index.
ISBN 978-0-470-51641-6 (cloth)
1. Statistical hypothesis testing. 2. Permutations. 3. Multivariate analysis. I. Salmaso, Luigi. II. Title.
QA277.P374 2010
519.5′6 – dc22

2010000776

A catalogue record for this book is available from the British Library.

ISBN: 978-0-470-51641-6

Typeset in 9/11 Times by Laserwords Private Limited, Chennai, India.

To:
Annamaria, Albertina, Annalisa and Alessandro Pesarin
and
Davide, Emanuele, Rosa, Paolina and Serio Salmaso

The obvious is that which is never seen
until someone exposes it simply.
Kahlil Gibran

Contents

Preface

This book deals with the combination-based approach to permutation hypothesis testing in several complex problems frequently encountered in practice. It also deals with a wide range of difficult applications in easy-to-check conditions. The key underlying idea, on which the large majority of testing solutions in multidimensional settings are based, is the nonparametric combination (NPC) of a set of dependent partial tests. This methodology assumes that a testing problem is properly broken down into a set of simpler sub-problems, each provided with a proper permutation solution, and that these sub-problems can be jointly analysed in order to maintain underlying unknown dependence relations.

The first four chapters are devoted to the theory of univariate and multivariate permutation tests, which has been updated. The remaining chapters present real case studies (mainly observational studies) along with recent developments in permutation solutions. Observational studies have enjoyed increasing popularity in recent years for several reasons, including low costs and availability of large data sets, but they differ from experiments because there is no control of the assignment of treatments to subjects. In observational studies the experimenter's main concern is usually to discover an association among variables of interest, possibly indicating one or more causal effects. The robustness of the nonparametric methodology against departures from normality and random sampling are much more relevant in observational studies than in controlled clinical trials. Hence, in this context, the NPC method is particularly suitable. Moreover, given that the NPC method is conditional on a set of sufficient statistics, it shows good general power behaviour, and the Fisher, Liptak or direct combining functions often have power functions which are quite close to the best parametric counterparts, when the latter are applicable, even for moderate sample sizes. Thus NPC tests are relatively efficient and much less demanding in terms of underlying assumptions with respect to parametric competitors and to traditional distribution-free methods based on ranks, which are generally not conditional on sufficient statistics and so rarely present better unconditional power behaviour. One major feature of the NPC with dependent tests, provided that the permutation principle applies, is that we must pay attention to a set of partial tests, each appropriate to the related sub-hypotheses, because the underlying dependence relation structure is nonparametrically and implicitly captured by the combining procedure. In particular, the researcher is not explicitly required to specify the dependence structure on response variables. This aspect is of great importance particularly for non-normal and categorical variables in which dependence relations are generally too difficult to define, and, even when well defined, are hard to cope with. Furthermore, in the presence of a stratification variable, NPC through a multi-phase procedure allows for quite flexible solutions. For instance, we can firstly combine partial tests with respect to variables within each stratum and then combine the combined tests with respect to strata. Alternatively, we can first combine partial tests related to each variable with respect to strata and then combine the combined tests with respect to variables. Moreover, once a global inference is found significant, while controlling for multiplicity it is possible to recover which partial inferences are mostly responsible of that result.

Although dealing with essentially the same methodology as contained in Pesarin (2001), almost all the material included in this book is new, specifically with reference to underlying theory and case studies.

Chapter 1 contains an introduction to general aspects and principles concerning the permutation approach. The main emphasis is on the principles of conditionality, sufficiency and similarity, relationships between conditional and unconditional inferences, why and when conditioning may be necessary, why the permutation approach results from both conditioning with respect to the data set and exchangeability of data in the null hypothesis, etc. Moreover, permutation techniques are discussed along with computational aspects. Basic notation is then introduced. Through a heuristic discussion of simple examples on univariate problems with paired data, two-sample and multi-sample (one-way ANOVA) designs, the practice of permutation testing is introduced. Moreover, discussions on conditional Monte Carlo (CMC) methods for estimating the distribution of a test statistic and some comparisons with parametric and nonparametric counterparts are also presented.

Chapters 2 and 3 formally present: the theory of permutation tests for one-sample and multi-sample problems; proof and related properties of conditional and unconditional unbiasedness; the definition and derivation of conditional and unconditional power functions; confidence intervals for treatment effect δ; the extension of conditional inferences to unconditional counterparts; and a brief discussion on optimal permutation tests and of the permutation central limit theorem.

Chapter 4 presents multivariate permutation testing with the NPC methodology. It includes a discussion on assumptions, properties, sufficient conditions for a complete theory of the NPC of dependent tests, and practical suggestions for making a reasonable selection of the combining function to be used when dealing with practical problems. Also discussed are: the concept of finite-sample consistency, especially useful when the number of observed variables in each subject exceeds that of subjects in the study; the multi-aspect approach; separate testing for two-sided alternatives; testing for multi-sided alternatives; the Behrens–Fisher problem, etc.

Chapter 5 deals with multiple comparisons and multiple testing issues. A brief overview of multiple comparison procedures (MCPs) is presented. The main focus is on closed testing procedures for multiple comparisons and multiple testing. Some hints are also given with reference to weighted methods for controlling family-wise error (FWE) and false discovery rate (FDR), adjustment of stepwise p-values, and optimal subset procedures.

Chapter 6 concerns multivariate permutation approaches for categorical data. A natural multivariate extension of McNemar's test is presented along with the multivariate goodness-of-fit test for ordered variables, the multivariate analysis of variance (MANOVA) test with nominal categorical data, and the issue of stochastic ordering in the presence of multivariate categorical ordinal variables. A permutation approach to test allelic association and genotype-specific effects in the genetic study of a disease is also discussed. An application concerning how to establish whether the distribution of a categorical variable is more heterogeneous (less homogeneous) in one population than in another is presented as well.

Chapter 7 discusses some quite particular problems with repeated measurements and/or missing data. Carry-over effects in repeated measures designs, modelling and inferential issues are treated extensively. Moreover, testing hypothesis problems for repeated measurements and missing data are examined. The rest of the chapter is devoted to permutation testing solutions with missing data.

Chapter 8 refers to permutation approaches for hypothesis testing when a multivariate monotonic stochastic ordering is present (with continuous and/or categorical variables). Umbrella testing problems are also presented. Moreover, two applications are discussed: one concerning the comparison of cancer growth patterns in laboratory animals and the other referring to a functional observational battery study designed to measure the neurotoxicity of perchloroethylene, a solvent used in dry cleaning (Moser, 1989, 1992).

Chapter 9 is concerned with permutation methods for problems of hypothesis testing in the framework of survival analysis.

Chapter 10 deals with statistical shape analysis. Most of the inferential methods known in the shape analysis literature are parametric in nature. They are based on quite stringent assumptions, such as the equality of covariance matrices, the independency of variation within and among landmarks or the multinormality of the model describing landmarks. But, as is well known, the assumption of equal covariance matrices may be unreasonable in certain applications, the multinormal model in the tangent space may be doubted and sometimes there are fewer individuals than landmarks, implying over-dimensioned spaces and loss of power. On the strength of these considerations, an extension of NPC methodology to shape analysis is suggested. Focusing on the case of two independent samples, through an exhaustive comparative simulation study, the behaviour of traditional tests along with nonparametric permutation tests using multi-aspect procedures and domain combinations is evaluated. The case of heterogeneous and dependent variation at each landmark is also analysed, along with the effect of superimposition on the power of NPC tests.

Chapter 11 presents two interesting real case studies in ophthalmology, concerning complex repeated measures problems. For each data set, different analyses have been proposed in order to highlight particular aspects of the data structure itself. In this way we enable the reader to choose the most appropriate analysis for his/her research purposes. The autofluorescence case study concerns a clinical trial in which patients with bilateral age-related macular degeneration were evaluated. In particular, their eyes were observed at several different and fixed positions. Hence, repeated measures issues arise. Five outcome variables were recorded and analysed. The confocal case study concerns a clinical trial with a five-year follow-up period, aiming to evaluate the long-term side-effects of a drug. Fourteen variables and four domains in total were analysed.

Chapter 12 deals with case studies in the field of survival analysis and epidemiology. NPC Test R10 software, SAS, MATLAB® and R codes have been used to perform the analyses. A comparison between logistic regression and NPC methodology in exploratory studies is then provided.

One of the main features of this book is the provision of several different software programs for performing permutation analysis. Various programs have been specifically developed. In particular:

- NPC Lib MATLAB library has been developed by Livio Finos, with consulting team Rosa Arboretti, Francesco Bertoluzzo, Stefano Bonnini, Chiara Brombin, Livio Corain, Fortunato Pesarin, Luigi Salmaso and Aldo Solari. For updates on the NPC Lib MATLAB library we refer to `http://homes.stat.unipd.it/livio`.
- NPC Test Release 10 (R10) standalone software (which is an extended version of the former NPC Test 2.0© produced by Methodologica S.r.l. and designed by Luigi Salmaso) has been updated by Luigi Salmaso, Andrey Pepelyshev, Livio Finos and Livio Corain, with consulting team Rosa Arboretti, Stefano Bonnini, Federico Campigotto and Fortunato Pesarin. For further updates to the NPC Test software we refer to `http://www.gest.unipd.it/~salmaso`.
- R code developed by Dario Basso, with consulting team Stefano Bonnini, Chiara Brombin, Fortunato Pesarin and Luigi Salmaso.
- SAS macros developed by Rosa Arboretti and Luigi Salmaso, with consulting team Stefano Bonnini, Federico Campigotto, Livio Corain and Fortunato Pesarin.

The above software is available from the book's website, `http://www.wiley.com/go/npc`. Raw data for all examples presented in the book, along with corresponding software routines, are also available from the website. Any errata, corrigenda or updates related to theory and software will be posted at `http://www.gest.unipd.it/~salmaso`.

We would like to express our thanks to the members of the Nonparametric Research Group at the University of Padua for their research collaboration on different topics included in this book: Rosa Arboretti, Dario Basso, Stefano Bonnini, Francesco Bertoluzzo, Chiara Brombin, Federico Campigotto, Livio Corain, Francesco Dalla Valle, Livio Finos, Patrizia Furlan, Susanna Ragazzi, Monjed Samuh, Aldo Solari and Francesca Solmi. We also wish to thank Susan Barclay, Richard

Davies, Beth Dufour, Heather Kay, Prachi Sinha-Sahay and the John Wiley & Sons Group in Chichester for their valuable publishing suggestions. Moreover, we owe a debt of thanks to our colleagues in different scientific fields: Ermanno Ancona, PierFrancesco Bassi, Fabio Bellotto, Patrizio Bianchi, Mario Bolzan, Carlo Castoro, Bruno Cozzi, Giovanni Fava, Roberto Filippini, Annarosa Floreani, Alessandro Frigiola, Luca Guarda-Nardini, Franco Grego, Lorenzo Menicanti, Edoardo Midena, Bruno Mozzanega, Virginia Moser, Andrea Peserico, Stefano Piaserico, Alberto Ruol, Luigi Sedea, Luca Spadoni, Tiziano Tempesta, Catherine Tveit, Carla Villanova, and several others with whom we have had stimulating discussions related to complex case studies; some of them are included on the book's website.

We wish to acknowledge Chiara Brombin for her valuable help through all stages of the production of this book and Professors N. Balakrishnan, O. Cucconi, P. Good, S. Kounias, V. Seshadri and J. Stoyanov as well as several colleagues for stimulating us in various ways to do research on multivariate permutation topics and to write the book. We also thank Stefania Galimberti, Ludwig Hothorn and Maria Grazia Valsecchi for revising some chapters.

In addition, we would like to acknowledge the University of Padua (CPDA088513/08) and the Italian Ministry of University and Research (PRIN 2008_2008WKHJPK_002) for providing the financial support for the necessary research and the development of part of the software.

Both authors share full responsibility for any errors or ambiguities, as well as for the ideas expressed throughout the book. A large part of the material presented in the book has been compiled from several publications and real case studies have been fully developed with the proposed different software codes. Although we have tried to detect and correct errors and eliminate ambiguities, there may well be others that have escaped our scrutiny. We take responsibility for and would warmly welcome notification of any that remain.

Finally, the second author (LS) wishes to acknowledge the first author (FP) as an inspiration for his open-mindedness and deep passion for innovative research in permutation methods during the course of our long-lasting collaboration and throughout the writing of this book.

We welcome any suggestions to the improvement of the book and would be very pleased if the book provides users with new insights into the analysis of their data.

Fortunato Pesarin
Department of Statistical Sciences
University of Padua

Luigi Salmaso
Department of Management and Engineering
University of Padua

Padova, January 2010

Notation and Abbreviations

A: an event belonging to the collection $\mathcal{A}$ of events

$\mathcal{A}$: a collection (algebra) of events

$\mathcal{A}_{/A} = \mathcal{A} \bigcap A$: a collection of events conditional on A

ANCOVA: analysis of covariance

ANOVA: analysis of variance

AUC: area under the curve

B: the number of conditional Monte Carlo iterations

$\mathcal{B}n(n, \theta)$: binomial distribution with n trials and probability θ of success in one trial

CDF: cumulative distribution function

CLT: central limit theorem

CMC: conditional Monte Carlo

$\mathbb{C}ov(X, Y) = \mathbb{E}(X \cdot Y) - \mathbb{E}(X) \cdot \mathbb{E}(Y)$: the covariance operator on (X, Y)

CSP: constrained synchronized permutations

$\mathcal{C}y(\eta, \sigma)$: Cauchy distribution with location η and scale σ

d.f.: degrees of freedom

$\delta = \int_{\mathcal{X}} \delta(x) \cdot dF_X(x)$: the fixed treatment effect (same as δ-functional or pseudo-parameter), $\delta \in \Omega$

Δ: stochastic treatment effect

EDF: empirical distribution function: $\hat{F}_{\mathbf{X}}(t) = \hat{F}(t|\mathcal{X}_{/\mathbf{X}}) = \sum_i \mathbb{I}(X_i \le t)/n,\ t \in \mathcal{R}^1$

EPM: empirical probability measure: $\hat{P}_{\mathbf{X}}(A) = \hat{P}(A|\mathcal{X}_{/\mathbf{X}}) = \sum_i \mathbb{I}(X_i \in A)/n,\ A \in \mathcal{A}$

ESF: empirical survival function (same as significance level): $\hat{L}_{\mathbf{X}}(t) = \hat{L}(t|\mathcal{X}_{/\mathbf{X}}) = \sum_i \mathbb{I}(X_i \ge t)/n,\ t \in \mathcal{R}^1$

$\mathbb{E}(X) = \int_{\mathcal{X}} x \cdot dF_X(x)$: the expectation operator (mean value) of X

$\mathbb{E}_A[X)] = \mathbb{E}[X|A] = \int_A x \cdot dF_X(x|A)$: the conditional expectation of X given A

$\stackrel{d}{=}$: equality in distribution: $X \stackrel{d}{=} Y \leftrightarrow F_X(z) = F_Y(z), \forall z \in \mathcal{R}^1$

$\stackrel{d}{>}$: stochastic dominance: $X \stackrel{d}{>} Y \leftrightarrow F_X(z) \le F_Y(z), \forall z$ and $\exists A : F_X(z) < F_Y(z),\ z \in A$, with $\Pr(A) > 0$

$<\neq>$: means '$<$', or '$\neq$', or '$>$'

$\sim$: distributed as: e.g. $X \sim \mathcal{N}0, 1)$ means X follows ths standard normal distribution

$\approx$: permutationally equivalent to

FDR: false discovery rate

FWER: family-wise error rate

$f_P(z) = f(z)$: the density of a variable X, with respect to a dominating measure ξ and related to the probability P

$F_X(z) = F(z) = \Pr\{X \leq z\}$: the CDF of X

$F_{X|A}(z) = \Pr\{X \leq z|A\}$: the conditional CDF of $(X|X \in A)$

$F_T^*(z) = F^*(z) = \Pr\{T^* \leq z|\mathcal{X}_{/\mathbf{X}}\}$: the permutation CDF of T given $\mathbf{X}$

$\mathcal{H}_G(N, \theta, n)$: hypergeometric distribution with N the number of units, $\theta \cdot N$ the number of units of interest, n the sample size

i.i.d.: independent and identically distributed

$\mathbb{I}(A)$: the indicator function, i.e. $\mathbb{I}(A) = 1$ if A is true, and 0 otherwise

ITT: intention-to-treat principle

$\lambda = \Pr\left\{T \geq T^o|\mathcal{X}_{/\mathbf{X}}\right\}$: the attained p-value of test T on data set $\mathbf{X}$

$L_X(t) = L(t) = \Pr\{X \geq t\}$: the significance level function (same as the survival function)

$\boldsymbol{\mu} = \mathbb{E}(\mathbf{X})$: the mean value of vector $\mathbf{X}$

MAD: median of absolute deviations from the median

MANOVA: multivariate analysis of variance

MC: number of Monte Carlo iterations

MCP: multiple comparison procedure

MCAR: missing completely at random

$\mathbb{M}d(X) = \tilde{\mu}$: the median operator on variable X such that $\Pr\{X < \tilde{\mu}\} = \Pr\{X > \tilde{\mu}\}$

$\#(X \in A) = \sum_i \mathbb{I}(X_i \in A)$: number of points X_i belonging to A

n: the (finite) sample size

MNAR: missing not at random

MTP: multiple testing problem

$\mathcal{N}(\mu, \sigma^2)$: Gaussian or normal variable with mean μ and variance σ^2

$\mathcal{N}_V(\boldsymbol{\mu}, \boldsymbol{\Sigma})$: V-dimensional normal variable ($V \geq 1$) with mean vector $\boldsymbol{\mu}$ and covariance matrix $\boldsymbol{\Sigma}$

$O(d_n) = c_n$: given two sequences $\{c_n\}$ and $\{d_n\}$, $O(d_n) = c_n$ if c_n/d_n is bounded as $n \to \infty$

$o(d_n) = c_n$: given two sequences $\{c_n\}$ and $\{d_n\}$, $o(d_n) = c_n$ if $c_n/d_n \to 0$ as $n \to \infty$

Ω: the set of possible values for δ

$\pi(\delta)$: the prior distribution of $\delta \in \Omega$

PCLT: permutation central limit theorem

P: a probability distribution on $(\mathcal{X}, \mathcal{A})$

$\mathcal{P}$: a family of probability distributions

$P(A) = \int_A dP(z)$: the probability of event $A \in \mathcal{A}$ with respect to P

p-FWE: adjusted p-value from a closed testing procedure

$\Pr\{A\}$: a probability statement relative to $A \in \mathcal{A}$

$\mathcal{R}^n$: the set of n-dimensional real numbers

$\mathbb{R}$: the rank operator

$R_i = \mathbb{R}(X_i) = \sum_{1 \leq j \leq n} \mathbb{I}(X_j \leq X_i)$: the rank of X_i within $\{X_1, \ldots, X_n\}$

SLF: sgnificance level function

UMP: uniformly most powerful

UMPS: uniformly most powerful similar

UMPU: uniformly most powerful unbiased

USP: unconstrained synchronized permutations

$\mathbb{V}(X) = \mathbb{E}(X - \mu)^2 = \sigma^2$: the variance operator on variable X

WORE: without replacement random experiment

WRE: with replacement random experiment

X: a univariate or multivariate random variable

$\mathbf{X}$: a sample of n units, $\mathbf{X} = \{X_i, i = 1, \ldots, n\}$

$\mathbf{X}^*$: a permutation of $\mathbf{X}$

$|\mathbf{X}| = \{|X_i|, i = 1, \ldots, n\}$: a vector of absolute values

XOR: exclusive or relationship: (A XOR B) means one but not both

$\mathcal{X}$: the sample space (or support) of variable X

$(\mathcal{X}, \mathcal{A})$: a measurable space

$(\mathcal{X}, \mathcal{A}, P)$: a probability space

$\mathcal{X}_{/\mathbf{X}}$: the orbit or permutation sample space given $\mathbf{X}$

$(\mathcal{X}_{/\mathbf{X}}, \mathcal{A}_{/\mathbf{X}})$: a permutation measurable space

$T \div \mathcal{X} \rightarrow \mathcal{R}^1$: a statistic

$T^o = T(\mathbf{X})$: the observed value of test statistic T evaluated on $\mathbf{X}$

$\mathbf{U}^{*\top}$: the transpose of $\mathbf{U}^*$

$\mathcal{U}(a, b)$: uniform distribution in the interval (a, b)

$\lfloor X \rfloor$: the integer part of X

$\uplus$: the operator for pooling (concatenation) of two data sets: $\mathbf{X} = \mathbf{X}_1 \uplus \mathbf{X}_2$

Z: the unobservable random deviates or errors: $X = \mu + Z$

1

Introduction

1.1 On Permutation Analysis

This book deals with the permutation approach to a variety of univariate and multivariate problems of hypothesis testing in a typical nonparametric framework. A large number of univariate problems may be usefully and effectively solved using traditional parametric or rank-based nonparametric methods as well, although under relatively mild conditions their permutation counterparts are generally asymptotically as good as the best parametric ones (Lehmann, 2009). It should also be noted that permutation methods are essentially of a nonparametric exact nature in a conditional context (see Proposition 2, 3.1.1 and Remarks 1, 2.2.4 and 1, 2.7). In addition, there are a number of parametric tests the distributional behaviour of which is only known asymptotically. Thus, for most sample sizes of practical interest, the relative lack of efficiency of permutation solutions may sometimes be compensated by the lack of approximation of parametric asymptotic counterparts. Even when responses follow the multivariate normal distribution and there are too many nuisance parameters to estimate and remove, due to the fact that each estimate implies a reduction of the degrees of freedom in the overall analysis, it is possible for the permutation solution to be more efficient than its parametric counterpart (note that 'responses', 'variables', 'outcomes', and 'end points' are often used as synonyms). In addition, assumptions regarding the validity of most parametric methods (such as homoscedasticity, normality, regular exponential family, random sampling, etc.) rarely occur in real contexts; so that consequent inferences, when not improper, are necessarily approximated and their approximations are often difficult to assess.

In practice parametric methods reflect a modelling approach and generally require the introduction of a set of stringent assumptions, which are often quite unrealistic, unclear, and difficult to justify. Sometimes these assumptions are merely set on an *ad hoc* basis for specific inferential analyses. Thus they appear to be mostly related to the availability of the methods one wishes to apply rather than to well-discussed necessities obtained from a rational analysis of reality, in accordance with the idea of modifying a problem so that a known method is applicable rather than that of modifying methods in order to properly deal with the problem. For instance, too often and without any justification researchers assume multivariate normality, random sampling from a given population, homoscedasticity of responses also in the alternative, etc., so that it becomes possible to write down a likelihood function and to estimate a variance–covariance matrix and so consequent inferences are without real credibility. In contrast, nonparametric approaches try to keep assumptions at a lower workable level, avoiding those that are difficult to justify or interpret, and possibly without excessive loss of inferential efficiency. Thus, they are based on more realistic foundations for statistical inference. And so they are intrinsically robust and resulting inferences are credible.

Permutation Tests for Complex Data: Theory, Applications and Software Fortunato Pesarin, Luigi Salmaso

However, there are many complex multivariate problems (commonly in areas such as agriculture, biology, business statistics, clinical trials, engineering, the environment, experimental data, finance data, genetics, industrial statistics, marketing, pharmacology, psychology, quality control, social sciences, zoology, etc.) that are difficult to solve outside the conditional framework and, in particular, outside the method of nonparametric combination (NPC) of dependent permutation tests (solutions to several complex problems are discussed in Chapter 4 and beyond).

Moreover, within parametric approaches it is sometimes difficult, if not impossible, to obtain proper solutions even under the assumption of normal errors. Some examples are:

1. problems with paired observations when scale coefficients depend on units;
2. two-sample designs when treatment is effective only on some of the treated subjects, as may occur with drugs having genetic interaction;
3. two-way ANOVA;
4. separate testing in cross-over designs;
5. multivariate tests when the number of observed variables is larger than the sample size;
6. jointly testing for location and scale coefficients in some two-sample experimental problems with positive responses;
7. exact testing for multivariate paired observations when some data are missing, even when not at random;
8. unconditional testing procedures when subjects are randomly assigned to treatments but are obtained by selection-bias sampling from the target population;
9. exact inference in some post-stratification designs;
10. two-sample testing when data are curves or surfaces, i.e. testing with countably many variables.

As regards problem 1, within a parametric framework it is impossible to obtain standard deviation estimates for observed differences on each unit with more than zero degrees of freedom, whereas exact and effective permutation solutions do exist (see Sections 1.9 and 2.6). A similar impossibility also occurs with Wilcoxon's signed rank test. In problem 2, since effects, either random or fixed, behave as if they depend on some unobserved attitudes of the subjects, traditional parametric approaches are not appropriate. Hints as to proper permutation solutions will be provided in Chapters 2, 3 and 4. In problem 3 it is impossible to obtain independent or even uncorrelated separate inferences for main factors and interactions because all related statistics are compared to the same estimate of the error variance (see Remark 8, 2.7). In addition, it is impossible to obtain general parametric solutions in unbalanced designs. We shall see in Example 8, 2.7 and Chapter 11 that, within the permutation approach, it is at least possible to obtain exact, unbiased and uncorrelated separate inferences in both balanced and unbalanced cases. Regarding problem 4, we will see in Remark 5, 2.1.2 that in a typical cross-over problem with paired data ($[A, B]$ in the first group and $[B, A]$ in the second group) two separate hypotheses on treatment effect ($X_B \stackrel{d}{=} X_A$) and on interaction due to treatment administration ($X_{AB} \stackrel{d}{=} X_{BA}$) are tested separately and independently. In problem 5 it is impossible to find estimates of the covariance matrix with more than zero degrees of freedom, whereas the NPC method discussed in Chapter 4 allows for proper solutions which, in addition, are often asymptotically efficient. In problem 6, due to its close analogy with the Behrens–Fisher problem, exact parametric solutions do not exist, whereas, based on concurrent multi-aspect testing, an exact permutation solution does exist, provided that positive data are assumed to be exchangeable in the null hypothesis and the two cumulative distribution functions (CDFs) do not cross in the alternative (see Example 8, 4.6). In problem 7 general exact parametric solutions are impossible unless missing data are missing completely at random and data vectors with at least one missing datum are deleted. In Section 7.10, within the NPC methodology, we will see a general approximate solution and one exact solution even when some of the data are missing not completely at random. In problem 8 any selection-biased mechanism usually produces quite severe modifications to the target population distribution, hence unless the selection mechanism is well defined and the consequent modified distribution is known, no proper parametric inference to the target population is possible; instead, within the permutation approach we may properly extend conditional inferences to unconditional counterparts. Moreover, in cases where the minimal sufficient statistic in the null hypothesis is the whole set of observed data, although the likelihood model would depend on a finite set of parameters, univariate statistics capable of summarizing the necessary information do not exist, so that no parametric method can be claimed to be uniformly better than others; indeed, conditioning on the pooled data set, i.e. considering the permutation counterpart, improves the power behaviour of any test statistics (see Cox and Hinkley, 1974; Lehmann, 1986). However, in order to attenuate the loss of information associated with using one overall statistic, we will find solutions within the so-called multi-aspect methodology based on the NPC of several dependent permutation test statistics, each capable of summarizing information on a specific aspect, so that it takes account of several complementary viewpoints (see Example 3, 4.6) and improves interpretability of results. In problem 9, as far as we know, the exact parametric inference for post-stratification analysis is based on the combination for independent partial tests (one test per stratum), provided that their null continuous distributions are known exactly. In problem 10, as far as can be seen from the literature (see Ramsay and Silverman, 2002; Ferraty and Vieu, 2006), only some regression estimate and predictive problems are solved when data are curves; instead, within the NPC strategy, several testing problems with countably many variables (the coefficients of suitable curve representations) can be efficiently solved.

Although authoritative, we agree only partially with opinions such as that expressed by Kempthorne (1955): 'When one considers the whole problem of experimental inference, that is of tests of significance, estimation of treatment differences and estimation of the errors of estimated differences, there seems little point in the present state of knowledge in using a method of inference other than randomization analysis.'

We agree with the part that stresses the importance for statisticians of referring to conditional procedures of inference and, in particular, to randomization (i.e. permutation) methods. Indeed, there is a wide range of inferential problems which are correctly and effectively solved within a permutation framework; however, there are others which are difficult or perhaps impossible to solve outside it.

We partially disagree, however, because there are very important families of inferential problems, especially connected to unconditional parametric estimation and testing, or to nonparametric prediction, classification, kernel estimation, or more generally within the statistical decision approach, which cannot be dealt with and/or solved in a permutation framework. These are often connected to violations of the so-called exchangeability condition (see Chapter 2) or are related to analysis of too few observed subjects. Moreover, all procedures of exploratory data analysis and all testing methods for which we cannot assume exchangeability of the data with respect to groups (i.e. samples) in the null hypothesis, generally lie outside the permutation approach. In addition, the traditional Bayesian inference (see Remark 4, 3.4, for suggestions on a *permutation Bayesian* approach) also lies outside the permutation approach.

Thus, although we think that permutation methods should be in the tool-kit of every statistician interested in applications, methodology or theory, we disagree because we do not believe that *all* inferential problems of interest for analysis of real problems fall within the permutation approach. In order to apply permutation methods properly, a set of initial conditions must be assumed, and if these conditions are not satisfied, their use may become erroneous.

However, and following remarks made by Berger (2000), these arguments support our decision to develop methods in the area of permutation testing, especially for multivariate complex problems. In this sense, this book attempts to go deeper into the main aspects of conditional methods of inference based on the permutation approach and to systematically study proper solutions to a set of important problems of practical interest. Section 1.4 lists a brief set of circumstances in which conditional testing procedures may be effective or even unavoidable.

1.2 The Permutation Testing Principle

For most problems of hypothesis testing, the observed data set $\mathbf{x} = \{x_1, \ldots, x_n\}$ is usually obtained by a *symbolic* experiment performed n times on a population variable X, and taking values in the sample space $\mathcal{X}$. We sometimes add the word 'symbolic' to names such as experiments, treatments, treatment effects, etc., in order to refer to experimental, pseudo-experimental and observational contexts. For the purposes of analysis, the data set $\mathbf{x}$ is generally partitioned into *groups* or *samples*, according to the so-called *treatment levels* of the symbolic experiment. In the context of the discussion up to and including Section 1.6, we use capital letters for random variables and lower-case for the observed data set. From Section 1.7 onwards, we shall dispense with this distinction, in that only capital letters will be used because the context is always sufficiently clear. Of course, when a data set is observed at its $\mathbf{x}$ value, it is presumed that a sampling experiment on a given underlying population has been performed, so that the resulting sample distribution is related to that of the parent population P. This is, of course, common to any statistical problem, and not peculiar to the permutation framework.

For any general testing problem in the null hypothesis, denoted by H_0, which typically assumes that data come from *only one* (with respect to groups) unknown population distribution P, $H_0 : \{X \sim P \in \mathcal{P}\}$, say, the whole set of observed data $\mathbf{x}$ is considered to be a random sample, taking values in the sample space $\mathcal{X}^n$, where $\mathbf{x}$ is one observation of the n-dimensional sample variable $X^{(n)}$ and where this random sample does not necessarily possess independent and identically distributed (i.i.d.) components (see Chapters 2 and 3 for more details).

We note that the observed data set $\mathbf{x}$ is always a set of sufficient statistics in H_0 for whatever underlying distribution. In order to see this in a simple way, let us assume that H_0 is true and all members of a nonparametric family $\mathcal{P}$ of non-degenerate and distinct distributions are dominated by one *dominating* measure $\xi_{\mathcal{P}}$; moreover, let f_P denote the density of P with respect to $\xi_{\mathcal{P}}$, and $f_P^{(n)}(\mathbf{x})$ denote the density of the sample variable $X^{(n)}$. As the identity $f_P^{(n)}(\mathbf{x}) = f_P^{(n)}(\mathbf{x}) \cdot 1$ is true for all $\mathbf{x} \in \mathcal{X}^n$, except for points such that $f_P^{(n)}(\mathbf{x}) = 0$, due to the well-known factorization theorem, any data set $\mathbf{x}$ is therefore a sufficient set of statistics for whatever member P of the nonparametric family $\mathcal{P}$.

1.2.1 Nonparametric Family of Distributions

Let us consider the following definition.

Definition 1. *A family of distributions $\mathcal{P}$ is said to behave nonparametrically when it is not possible to find a finite-dimensional space Θ such that there is a one-to-one relationship between Θ and $\mathcal{P}$, in the sense that each member P of $\mathcal{P}$ cannot be identified by only one member θ of Θ, and vice versa.*

If of course such a one-to-one relationship exists, θ is called a parameter, Θ is the parameter space, and $\mathcal{P}$ the corresponding parametric family. Families of distributions which are either unspecified or specified except for an infinite number of unknown parameters do satisfy the definition and so are nonparametric. Definition 1 also includes all those situations where the sample size n is smaller than the number of parameters, even though this is finite. All nonparametric families $\mathcal{P}$ which are of interest in permutation analysis are assumed to be sufficiently *rich* in such a way that if x and x' are any two distinct points of $\mathcal{X}$, then $x \neq x'$ implies $f_P(x) \neq f_P(x')$ for at least one $P \in \mathcal{P}$, except for points with null density for P. Also note that the characterization of a family $\mathcal{P}$ as being nonparametric essentially depends on the knowledge we assume about it. When we assume that the underlying family $\mathcal{P}$ contains all continuous distributions, then the data set $\mathbf{x}$ is *complete minimal sufficient*.

Permutation tests are known to be conditional procedures of inference, where conditioning is done with respect to a set of sufficient statistics in the null hypothesis. Thus consequent inferences at least concern the sample data **x** actually observed and the related observed subjects. The act of conditioning on a set of sufficient statistics in H_0, and the assumption of exchangeability *with respect to groups* (samples) for observed data, make permutation tests independent of the underlying likelihood model related to P (see Section 2.1.3). As a consequence, P may be unknown or unspecified, either in some or all of its parameters, or even in its analytic form. We specify this concept in the permutation testing principle.

1.2.2 The Permutation Testing Principle

Let us consider the following definition.

Definition 2. *If two experiments, taking values on the same sample space $\mathcal{X}$ and respectively with underlying distributions P_1 and P_2, both members of $\mathcal{P}$, give the same data set* **x**, *then the two inferences conditional on* **x** *and obtained using the same test statistic must be the same, provided that the exchangeability of data with respect to groups is satisfied in the null hypothesis. Consequently, if two experiments, with underlying distributions P_1 and P_2, give respectively $\mathbf{x}_1$ and $\mathbf{x}_2$, and $\mathbf{x}_1 \neq \mathbf{x}_2$, then the two conditional inferences may be different.*

One of the most important features of the permutation testing principle is that in theory and under a set of mild conditions conditional inferences can be extended unconditionally to all distributions P of $\mathcal{P}$ for which the density with respect to a suitable dominating measure ξ is positive, i.e. $dP(\mathbf{x})/d\xi^n > 0$ (see Section 3.5). It should be emphasized, however, that this feature derives from the sufficiency and conditionality principles of inference (see Cox and Hinkley, 1974; Lehmann, 1986; Berger and Wolpert, 1988), by which inferences are related to all populations sharing the same value of conditioning statistics, particularly those which are sufficient for underlying nuisance entities. For instance, Student's t extends inference to all normal populations which assign positive density to the variance estimate $\hat{\sigma}^2$ and so its inference is for a family of distributions. Therefore, such unconditional extensions should be carried out carefully. Another important feature occurs in multivariate problems, when solved through NPC methods. For these kinds of problems, especially when they are complex and in very mild and easy-to-check conditions (see Chapter 4), it is not necessary to specify or to model the structure of dependence relations for the variables in the underlying population distribution. In this way analysis becomes feasible and results are easy to interpret. For instance, it is known that, for multivariate categorical variables, it is extremely difficult to properly model dependence relations among variables (see Joe, 1997). In practice, therefore, except for very particular cases, only univariate (or component variable by component variable) problems are considered in the literature. From Chapter 4 onwards we will see that, within the permutation testing principle and the NPC of dependent partial tests, a number of rather difficult problems can be effectively and easily solved, provided that partial tests are *marginally unbiased and consistent* (see Section 4.2.1). Also of interest is an application of this principle in the context of the Bayesian permutation approach (see Remark 4, 3.4).

However, the conditioning on sufficient statistics provides permutation tests with good general properties. Among the most important of these, when exchangeability is satisfied in the null hypothesis, is that permutation tests are always exact procedures (see Remark 1, 2.2.4 and Proposition 2, 3.1.1). Another property is that their conditional rejection regions are similar, as intended by Scheffé (1943a, 1943b). The former means that, at least in principle, the null rejection probability can be calculated exactly in all circumstances. The latter means that, if data comes from continuous distributions (where the probability of finding ties in the data set is zero), the null rejection probability is invariant with respect to observed data set **x**, for almost all $\mathbf{x} \in \mathcal{X}^n$, and with respect to the underlying population distribution P (see Chapter 2). As a consequence, conditional rejection regions are similar to the unconditional region. When data comes from non-continuous distributions, unless referring to randomized tests (see Section 2.2), the similarity property is only asymptotically valid. Moreover, if the stochastic dominance condition is satisfied in H_1, permutation tests based on divergence of suitable statistics are *conditionally unbiased* procedures, since the rejection probability of any test T, for all data sets $\mathbf{x} \in \mathcal{X}^n$, satisfies the relation $\Pr\{\lambda(\mathbf{x}(\delta)) \leq \alpha|\mathbf{x}\} = W(\delta, \alpha, T|\mathbf{x}) \geq \alpha$, where $\lambda(\mathbf{x}(\delta))$ indicates the p-value and $W(\delta, \alpha, T|\mathbf{x})$ indicates the *conditional power* of T given $\mathbf{x}$ with fixed treatment effect δ and significance level α (see Section 3.2).

It is worth noting that when exchangeability may be assumed in H_0, the similarity and unbiasedness properties allow for a kind of weak extension of conditional to unconditional inferences, irrespective of the underlying population distribution and the way sample data are collected. Therefore, this weak extension may be made for any sample data, even if they are not collected by well-designed sampling procedures, in which each unit is truly selected at random from a given population and subject to an experiment. Conversely, parametric solutions permit proper extensions only when data comes from well-designed sampling procedures on well-specified parent populations. Specifically, a general situation for unconditional extensions in parametric contexts occurs when likelihood functions are known except for nuisance parameters, and these are removed by invariant statistics or by conditioning on boundedly complete estimates (see Section 3.5 and Remark 2 therein).

For this reason, permutation inferences are proper with most observational data (sometimes called non-experimental), with experimental data, with selection-biased sampling procedures, and with well-designed sampling procedures. However, we must note that well-designed sampling procedures are quite rare even in most experimental problems (see Ludbrook and Dudley, 1998). For instance, if we want to investigate the effects of a drug on rats, the units to be treated are usually not randomly chosen from the population of all rats, but are selected in some way among those available in a laboratory and are *randomly assigned* to the established treatment levels. The same occurs in most clinical trials, where some patients, present in a hospital and that comply with the experiment, are randomly assigned to one of the pre-established treatment levels.

In one sense, the concept of random sampling is rarely achieved in real applications because, for various reasons, real samples are quite often obtained by selection-bias procedures. This implies that most of the forms of unconditional inferences usually associated with parametric tests, being based on the concept of random sampling, are rarely applicable in real situations. In addition, due to the similarity and unbiasedness properties, permutation solutions allow for relaxation of most of the common assumptions needed by parametric counterparts, such as the existence of mean values and variances, and the homoscedasticity of responses in the alternative hypothesis (see also Section 1.4). This is why permutation inferences are so important for both theoretical and application purposes, not only for their potential exactness.

Many authors have emphasized these aspects. A review of the relevant arguments is given in Edgington (1995), Edgington and Onghena (2007), and in Good (2000, 2005). One of these relates to the fact that reference null distributions of ordinary parametric tests are explicitly based on the concept of infinitely repeated and well-designed random sampling from a given well-specified population, the existence of which is often merely virtual. Another argument relates to the fact that, as occurs in many experimental problems, it is often too unrealistic to assume that treatment does not also influence scale coefficients or other distributional aspects, so that traditional parametric solutions may become improper.

Conversely, when exchangeability may be assumed in H_0, reference null distributions of permutation tests always exist because, at least in principle, they are obtained by enumerating all permutations of available data (see Chapter 2). In addition, permutation comparisons of means or of other functionals do not require homoscedasticity in the alternative, provided that underlying CDFs are ordered so that they do not cross each other (see Section 2.1.1). For these reasons, on the one hand permutation inferences generally have a natural interpretation and, on the other, ordinary parametric tests are considered to be rarely applicable to real problems.

1.3 Permutation Approaches

In the literature, three leading approaches to constructing permutation tests are considered: one essentially heuristic, the other two more formal. The heuristic approach, based on intuitive reasoning, is the most commonly adopted, especially for simple problems where common sense may often suffice (see Edgington and Onghena, 2007; Good, 2006; Lunneborg, 1999). But when problems are not simple, this approach can be inadequate. The two formal approaches are much more elegant, effective and precise. One of these is based on the concept of invariance of the reference null distribution under the action of a finite group of transformations (see, for instance, Scheffé, 1943a, 1943b; Lehmann and Stein, 1949; Hoeffding, 1952; Romano, 1990; Nogales et al., 2000). The other, which is as elegant and precise as the invariance approach, is formally based on the concept of conditioning on a set of sufficient statistics in H_0 for the underlying unknown distribution P (Fisher, 1925; Watson, 1957; Lehmann, 1986; Pesarin, 2001). The two formal approaches are substantially equivalent (see Odén and Wedel, 1975).

For very simple problems we often use the heuristic approach, especially if related solutions are essentially clear and no ambiguities arise. We use it in this introductory chapter. From Chapter 2 onwards the concept of conditioning on a set of sufficient statistics is preferred to the invariance approach. The reason for this preference is that the conditioning approach seems to be slightly more natural, easier to work with, and easier to understand than the invariance approach. Moreover, as it can formally characterize the related conditional reference space, it allows for construction of proper solutions to several rather difficult problems, and so it seems apparently more effective. In addition, it allows us to easily establish when and why solutions are exact or approximate (see Remarks 1, 2.2.4 and 1, 2.7 and Proposition 2, 3.1.1, for sufficient conditions leading to exact tests; Chapter 11 for the two-way ANOVA; and Chapter 7 for analysis of some problems with missing data and for problems connected with repeated measurements). It is worth noting, for instance, that the nonparametric analysis of the two-way ANOVA is a very difficult problem when studied using the invariance approach. Hence, with the exception of Pesarin (2001) and Basso et al. (2009a), only heuristic and unsatisfactory solutions have been proposed in the literature and their most important related inferential properties are justified only asymptotically. We shall see, however, that the conditioning approach may be applied for full derivation of exact permutation tests separately for two main factors and their interaction (see Example 8, 2.7). In addition, it is easy to find proper exact solutions for testing with multi-aspect problems (see Example 3, 4.6), cross-over designs (see Remark 5, 2.1.2), multivariate ordered categorical variables (Chapter 6), post-stratification designs (see Remark 4, 2.1.2), etc. The latter are of particular importance in observational study contexts.

1.4 When and Why Conditioning is Appropriate

We know that unconditional parametric testing methods may be available, appropriate and effective when:

1. data sets are obtained by well-defined random sampling designs on well-specified parent populations;
2. population distributions for responses, i.e. the likelihood models, are well defined and treatment effect is just one (and only one) parameter for such a likelihood;

3. with respect to all nuisance entities, well-defined likelihood models are provided with either boundedly complete estimates in H_0 or at least with invariant statistics;
4. at least asymptotically, null sample distributions of suitable test statistics do not depend on any unknown entity.

Therefore, just as there are circumstances in which unconditional parametric testing procedures may be proper from the point of view of interpretation of related inferential results as well as for their efficiency, so there are others in which they may be improper or even impossible to carry out. Conversely, there are circumstances in which conditional testing procedures may become appropriate and at times unavoidable. A brief, incomplete list of these circumstances is as follows:

- Distributional models for responses are nonparametric.
- Distributional models are not well specified.
- Distributional models, although well specified, depend on too many nuisance parameters.
- Treatment effect, even on well-specified models, depends on more than one parameter.
- Even on well-specified models, treatment may have influence on more than one parameter.
- With respect to some nuisance entities, well-specified distributional models do not possess invariant statistics or boundedly complete estimates in H_0.
- Ancillary statistics in well-specified distributional models have a strong influence on inferential results leading to post-stratification analyses.
- Ancillary statistics in well-specified models are confounded with other nuisance entities.
- Asymptotic null sample distributions depend on unknown entities.
- Problems in which the number of nuisance parameters increases with sample size.
- The number of response variables to be analysed is larger than the sample size.
- Sample data come from finite populations or sample sizes are smaller than the number of nuisance parameters.
- In multivariate problems, some variables are categorical and others quantitative.
- Multivariate alternatives are subject to order restrictions.
- In multivariate problems and in view of particular inferences, component variables have different degrees of importance for the analysis (see (f) in Section 4.2.4).
- Data sets contain non-ignorable missing values (see Section 7.10).
- Data sets are obtained by ill-specified selection-bias procedures (see Section 3.5).
- Treatment effects are presumed to act possibly on more than one aspect (a functional or pseudo-parameter, that is, a function of many parameters), so that multi-aspect testing methods are of interest for inferential problems (see Example 3, 4.6).
- Treatment effects may depend on unknown entities, for instance some nuisance parameters.

In addition, we may decide to adopt conditional testing inferences, not only when unconditional counterparts are not possible, but also when we want to lay more importance on the observed data set and related set of observed units than on the population model.

Conditional inferences are also of interest when, for whatever reason, we wish to limit ourselves to conditional methods by explicitly restricting to the actual data set $\mathbf{x}$ (see Greenberg, 1951; Kempthorne, 1966, 1977, 1979, 1982; Basu, 1978, 1980; Thornett, 1982; Greenland, 1991; Celant and Pesarin, 2000, 2001; Pesarin, 2001, 2002). For instance, the latter situation agrees with the idea that, when assessing the reliability of cars, the owner may be mostly interested in his own car, or fleet of cars if he has more than one, because he is responsible for all related reliability maintenance costs, thus giving rise to a *conditional assessment*. Of course, the point of view of the car manufacturer, whose reputation and warranty costs are related to the whole set of similar cars, may be mostly centred on a sort of *average behaviour*, giving rise to a form of *unconditional assessment* related to the car population.

Thus, both conditional and unconditional points of view are important and useful in real problems because there are situations, such as that of the owner, in which we may be interested in conditional inferences, and others, such as that of the manufacturer, in which we may be interested in unconditional inferences. Hence, as both points of view are of research interest, both types of inference are of methodological importance and may often be analysed using the same data set. However, within conditional testing procedures, provided that exchangeability of data with respect to groups is satisfied in the null hypothesis, it should be emphasized that permutation methods play a central role. This is because they allow for reasonably efficient solutions, are useful when dealing with many difficult problems, provide clear interpretations of inferential results, and allow for weak extensions of conditional to unconditional inferences.

1.5 Randomization and Permutation

In most experimental cases units are randomly assigned to treatment levels, i.e. to groups or samples, so that in the null hypothesis observed data appear to have been *randomly assigned* to these levels. Based on this notion of randomization, many authors prefer the term *randomization tests* (Pitman, 1937a, 1937b; Kempthorne, 1977; Edgington, 1995; Good, 2005) or even *re-randomization tests* (Gabriel and Hall, 1983; Lunneborg, 1999) in place of *permutation tests*. In Section 3.2, related to the so-called conditional and post-hoc power functions, a rather more precise notion of re-randomization is used. Although of no great importance, being a mere matter of words, we prefer the term 'permutation' because it is closer to the true state of things and because to some extent it has wider meaning, encompassing the others. Indeed, a sufficient condition for properly applying permutation tests is that the null hypothesis implies that observed data are exchangeable with respect to groups. In Proposition 3, 3.1.1, we shall see an extension to this assumption, leading to testing for composite hypotheses. For instance, in a symbolic experiment where a variable is observed in male and female groups of a given kind of animal, the notion of randomization is difficult to apply exactly because in no way can gender be randomly assigned to units. Instead, the permutation idea is rather more natural because in the null hypothesis, of no distributional difference due to gender, we are led to assume that observed data may be indifferently assigned to either males or females – a notion which justifies exchangeability of data and the permutation of unit labels, but not the randomization of units. The greater emphasis on the notion of randomization by random assignment of units to treatments is because, in the null hypothesis, it is generally easier and more natural to justify the assumption of exchangeability for experimental data than for observational data.

However, when the exchangeability property of data is not satisfied or cannot be assumed in the null hypothesis, both parametric and permutation inferences are generally not exact (for some hints on approximated permutation solutions see Example 8, 4.6). In these cases, especially when even approximate solutions are difficult to obtain, it may be useful to employ bootstrap techniques, which are less demanding in terms of assumptions and may be effective asymptotically or even for exploratory purposes, in spite of the fact that for finite sample sizes related inferences are neither conditional nor unconditional (see Remark 4, 3.8.2).

The conditioning property of permutation tests leads to rather different concepts of related inferences and of power functions. The first two concepts are restricted to the actually observed data set and are called the *conditional inference* and the *conditional or post-hoc power function*, respectively. The second two concepts are related to the parent population and are the *unconditional inference* and the *unconditional power function*. Their definition and determination for some simple problems are discussed in Section 3.2 (an algorithm for evaluating the post-hoc power function is presented in Section 3.2.1). In any case, it should be stressed that, except for two very specific problems (Fisher's exact probability test and McNemar's test) and some cases of asymptotic or approximate calculations, both power functions are generally not expressed in closed form. Of course, being nonparametric, the post-hoc power does not require knowledge of the population distribution P and is the most important for conditional inferences. Instead, the unconditional power implies knowledge of P. Since it measures the probability of a test rejecting the null hypothesis when it is false, it is thus useful for performance comparisons with respect to parametric and non-parametric competitors or in establishing sample sizes so that related unconditional inferences do satisfy assigned requirements. One important feature of both conditional and unconditional powers of permutation tests based on divergence of suitable statistics is that they are monotonically related to the non-centrality functional, that is, the so-called treatment effect or even pseudo-parameter, and that this is true independently of the underlying population distribution P.

1.6 Computational Aspects

One of the major problems associated with permutation tests is that their null distributions, especially for multivariate situations and except for very particular cases (Fisher's exact probability and McNemar tests), are generally impossible to express in closed form and to calculate exactly because they depend to a great extent on the specific data set $\mathbf{x}$, and thus they vary as data vary in the sample space $\mathcal{X}^n$. Furthermore, when sample sizes are not small, direct calculations are practically impossible because of the very large cardinality of the associated permutation sample spaces $\mathcal{X}^n_{/\mathbf{x}}$. Moreover, the approximation of such distributions by means of asymptotic arguments is not always appropriate, unless sample sizes are very large, because their dependence on $\mathbf{x}$ makes it difficult to express and to check the conditions needed to evaluate the degree of approximation in practical cases (see Section 3.7).

Some exact algorithms, not based on the complete enumeration of all possible permutations, have been developed for univariate situations allowing for exact calculation of the permutation distribution in polynomial time (see Pagano and Tritchler, 1983; Zimmerman, 1985a, 1985b; Mehta and Patel, 1980, 1983, 1999; Barabesi, 1998, 2000, 2001). It is also worth noting that there are computer packages (a well-known one is StatXact®) which provide exact calculations in many univariate problems. Approximate calculations in the univariate context are provided, for instance, by Berry and Mielke (1985) (see also Mielke and Berry, 2007), where a suitable distribution sharing the same few moments as the exact one is considered.

For practical reasons, especially for multivariate purposes, in order to obtain appropriate and reliable evaluations of the permutation distributions involved, in this book we suggest using *conditional Monte Carlo* (CMC) procedures.

Although in principle it is always possible to carry out exact calculations by means of specific computing routines based on complete enumeration, in practice the use of conditional simulation algorithms is often required by the excessively large cardinality of permutation spaces $\mathcal{X}^n_{/\mathbf{x}}$ or by the method of NPC of dependent permutation tests, especially for the solution of complex problems. This is because on the one hand complex problems are often multivariate or at least multi-aspect, and on the other the algorithms for exact calculations are generally based on direct calculus of upper tail probabilities, a strategy which may become highly impractical, if not impossible, in multivariate problems because there are no general computing routines useful to identify the critical regions.

We wish to underline the fact that conditional simulations are carried out by means of without-replacement resampling algorithms on the given data set, therefore in place of resampling procedures we prefer to speak of CMC methods. However, it must be emphasized that these procedures are substantially different from the well-known *bootstrap techniques* (see Remarks 3 and 4, 3.8.2). In CMC, in order to obtain random permutations, resampling replicates are done without replacement on the given data set, considered as playing the role of a finite population, provided that sample sizes are finite. Hence, they correspond to a random sampling from the space of data permutations associated with the given data set. In this sense, the name CMC has to some extent the same meaning as *without-replacement resampling*.

Of course, CMC procedures provide good and reliable *statistical estimates* of desired permutation distributions, the accuracy of which depends on the number of iterations. When referring to CMC, some authors use terms such as *approximate permutation tests* or even *modified permutation tests* (see van den Brink and van den Brink, 1990; Dwass, 1957; Vadiveloo, 1983). In this book, the former term is used when they are not exact (see Remark 1, 2.2.4 and Proposition 2, 3.1.1), in the sense that at least one assumption for their exactness, such as the exchangeability of data with respect to groups in the null hypothesis, is violated (some important examples of approximate permutation tests are discussed in Chapters 2–7).

The increasing availability of relatively inexpensive and fast computers has made permutation tests more and more accessible and competitive with respect to their parametric or nonparametric counterparts. In fact, their null distributions can be effectively approximated by easy CMC techniques which, more than numeric evaluations, provide statistical estimations without compromising their desirable statistical properties.

Most well-known statistical packages include specific routines for CMC simulation of permutation distributions, such as MATLAB®, R, SAS, SC©, S-PLUS©, SPSS©, Statistica©, StatXact, etc. Some of these also include routines for exact computations. Most of the examples included in the book are developed in MATLAB® and R codes. Some of them are also developed in SAS. Furthermore, all examples can be analysed using NPC Test R10 standalone software. Data sets, developments in MATLAB®, R, SAS or NPC Test R10 are available from the book's website.

1.7 Basic Notation

Throughout this book, consequences of the main arguments, informal definitions of certain important elements, some relatively important concepts and relevant aspects of analysis are emphasized as 'remarks'. Such remarks are numbered by subsection and, when reference to one of them is necessary, they are cited by number and subsection; for example, Remark 3, 3.8.2 refers to the third remark of subsection 2 in Section 8 of Chapter 3. Examples are numbered by section; for example, Example 2, 2.6 stands for the second example in Section 6 of Chapter 2. Definitions, theorems, lemmas, propositions, etc. are numbered by chapter; for example, Theorem 6, 4.4.2 is the sixth theorem in Chapter 4 and will be found in Section 4.4.2. Figures and tables are numbered by chapter; for example, Table 2.1 means Table 1 in Chapter 2. As a rule, formulae are not numbered.

Generally, theorems from the literature are reported without proof, whereas the most important properties of permutation tests, regarding their conditional and unconditional exactness, unbiasedness, consistency, power function, etc., are explicitly established and proved. Simple proofs of more specific properties as well as extensions of some results are often set as exercises for the reader. Several exercises and problems are given at the end of many sections. A list of references may be found at the end of the volume.

From this point on, unless it is necessary to make reference to specific countable sequences, we suppress the superscript $(^n)$ when referring to $\mathcal{X}^n$, $\mathbf{X}^{(n)}$, $\mathcal{X}^n_{/\mathbf{X}}$, etc., and shall simply write $\mathcal{X}$, $\mathbf{X}$, $\mathcal{X}_{/\mathbf{X}}$, etc. Therefore, we do not distinguish, for instance, between the dimensionalities of response variables, sample spaces of responses, and sample spaces of the experiment, the context generally being sufficient to avoid misunderstandings.

The observed data are assumed to be related to a response variable. Response variables are usually indicated by italic capitals, such as X, Y, if they are univariate and by bold capitals, such as $\mathbf{X}$, $\mathbf{Y}$, if multivariate. Responses are assumed to be observed on statistical units (i.e. individuals or subjects) where units play the role of members of a given population of interest. These units are generally obtained by symbolic sampling experiments carried out on the given population. Similarly to multivariate responses, sample data are also indicated by bold capitals $\mathbf{X}$, $\mathbf{Y}$, etc. The context is generally sufficient to avoid ambiguities. Lower-case letters are generally used to indicate integer numbers, real variables or constants: i, j, h, k, n, t, $\mathbf{x}$, z, etc. The most important exceptions are: A, which represents an event or an experimental factor; B, an experimental factor or the number of iterations of a CMC procedure; MC, the number of ordinary Monte Carlo iterations in simulation studies; and N_i, the cumulative frequencies in contingency tables.

Unless obvious from the context, we usually do not use the notational conventions of linear algebra, so that in general we do not distinguish between column and row vectors, etc. This is because on the one hand the context is always sufficiently clear; on the other it is impractical in linear algebra notation to represent responses that are partly quantitative and partly categorical and which are therefore difficult to work with, especially in unbalanced situations. The only notational convention adopted is that, when necessary, the transpose of a matrix $\mathbf{U}$ is denoted by $\mathbf{U}^\top$. Sometimes $|\mathbf{X}|$ is used to denote the vector of absolute values, as in $|\mathbf{X}| = \{|X_i|, i = 1, \ldots, n\}$.

To indicate sample data sets we often need to use the so-called *unit-by-unit representation*. For instance, in a two-sample univariate design with respectively n_1 and n_2 data, and $n = n_1 + n_2$, we denote the whole data set by $\mathbf{X} = \{X(i), i = 1, \ldots, n; n_1, n_2\}$, where $X(i) = X_i$ is the response related to the ith unit in the list. This notation means that the first n_1 elements in the list belong to the first sample, and the other n_2 elements to the second. This representation is useful in expressing permutation sample points for both categorical and quantitative responses, especially in multivariate situations. The symbol $\biguplus$ is used for pooling (i.e. concatenating) a finite number of data sets into the pooled set, for example $\mathbf{X} = \mathbf{X}_1 \biguplus \mathbf{X}_2$ for pooling two data sets.

In general, we implicitly refer to the population and to the experiment of interest by means of a statistical model such as $(X, \mathcal{X}, \mathcal{A}, P \in \mathcal{P})$, where X represents the response variable, $\mathcal{X}$ is the related sample space, $\mathcal{A}$ is a suitable collection (a σ-algebra) of subsets of $\mathcal{X}$, and P is the underlying parent distribution belonging to the nonparametric family $\mathcal{P}$. Thus, we assume that the response variables take their values on the measurable space $(\mathcal{X}, \mathcal{A})$. Unless necessary, we do not distinguish between random variables and their concrete observations in real sampling experiments.

We often refer to a subsample space such as $\mathcal{X}_{/A}$, the orbit or coset associated with A, that is, the set of points $\mathbf{X}^* \in \mathcal{X}$ sharing condition A, where A is some event of interest belonging to $\mathcal{A}$. The main conditioning set referred to in the context of permutation testing is the pooled set of observed data $\mathbf{X}$ (see Section 1.2). Thus, $\mathbf{X}^* \in \mathcal{X}_{/\mathbf{X}}$ represents a permutation of observed data $\mathbf{X}$ and $\mathcal{X}_{/\mathbf{X}}$ represents the related *permutation sample space*. Moreover, we assume that all statistics of interest are measurable with respect to $(\mathcal{X}, \mathcal{A})$ and, of course, with respect to the conditional or restricted algebra $\mathcal{A}_{/\mathbf{X}} = \mathcal{A} \bigcap \mathcal{X}_{/\mathbf{X}}$, so that conditional probability distributions associated with any $P \in \mathcal{P}$ are well defined on the measurable permutation space $(\mathcal{X}_{/\mathbf{X}}, \mathcal{A}_{/\mathbf{X}})$ induced by conditioning on observed data set $\mathbf{X}$. Note that the measurability assumption with respect to the conditional algebra $\mathcal{A}_{/\mathbf{X}}$ of test statistics T is generally self-evident because all statistics of interest are transformations of the data $\mathbf{X}$ which are required to induce a probability distribution over $(\mathcal{X}_{/\mathbf{X}}, \mathcal{A}_{/\mathbf{X}})$, so that associated conditional inferences have a clear interpretation. In general, unless necessary, we do not indicate the dimensionalities of variables or the cardinalities of sets and spaces, since these are clear from the context. The conditional expectation of X given A is denoted by $\mathbb{E}_A[X]$ or by $\mathbb{E}[X|A]$.

We sometimes need to partition permutation sample spaces $\mathcal{X}_{/\mathbf{X}}$ into sub-orbits (e.g. see Section 7.10.1 and Remarks 4 and 5, 2.1.2) in order to take into consideration restrictions of invariance properties induced by post-stratification arrangements or by some statistics of interest when related to specific problems, so that solutions may become easier to construct.

A test statistic, $T : \mathcal{X} \rightarrow \mathcal{T} \in \mathcal{R}^1$, is a real function of observed data which takes values on a suitable space $\mathcal{T} = T(\mathcal{X}) \subseteq \mathcal{R}^1$ and is usually represented by symbols such as $T = T(\mathbf{X})$ or $T^* = T(\mathbf{X}^*)$, where the second form emphasizes the role of $\mathbf{X}^* \in \mathcal{X}_{/\mathbf{X}}$ as a permutation of the observed data set $\mathbf{X}$. The set $\mathcal{T}$ is also called the support of T, and so the set $\mathcal{T}(\mathbf{X}) = \{T(\mathbf{X}^*), \mathbf{X}^* \in \mathcal{X}_{/\mathbf{X}}\}$ indicates the permutation or conditional support of T associated with the given data set $\mathbf{X}$. In order to emphasize the role of observed data, we use the notation $T^o = T(\mathbf{X})$ to indicate the *observed value* of T on the given data set $\mathbf{X}$. In general the superscript *, as in $\mathbf{X}^*$, F^*, T^*, is used to indicate a variable, a distribution or a statistic related to permutation entities. We also use a hat, as in $\hat{F}$, $\hat{\lambda}$, $\hat{\sigma}$, $\hat{\lambda}$, to indicate an estimate, when referring either to sample estimation or Monte Carlo estimation.

It should be noted that, due to conditioning on the observed data set $\mathbf{X}$, permutations appear to be without-replacement random samples from $\mathbf{X}$, which is then considered to play the role of a finite population, so that estimation of permutation indicators of functionals (or pseudo-parameters) has a close analogy with the sample estimation from finite populations.

1.8 A Problem with Paired Observations

As an initial example, let us consider a testing problem on the effectiveness of training in the reduction of anxiety in a sample of $n = 20$ subjects (Pesarin, 2001). At first glance, the subjects of the experiment are presumed to be 'homogeneous' with respect to the most important experimental conditions, the so-called covariates, such as sex, age and health.

Suppose that anxiety, the variable Y, is measured by means of an Institute for Personality and Ability Testing (IPAT) psychological test, responses to which are quantitative scores corresponding to the sum of sub-responses to a set of different items. Each unit is observed before treatment (occasion 1), also called the *baseline observation*, and one week after a fixed number of training sessions (occasion 2), which are administered with the aim of stochastically reducing baseline values.

Of course, bivariate responses within each specific unit are dependent because they are measured on the same unit on different occasions, whereas the n pairs are assumed to be independent because related to different units. Moreover, due to the assumed homogeneity of all individuals with respect to most common experimental conditions, the set of data pairs $\{(Y_{1i}, Y_{2i}),\ i = 1, \ldots, n\}$ may be viewed as a random sample of n i.i.d. pairs from the bivariate random variable $Y = (Y_1, Y_2)$. Formally, data are represented by a matrix of n pairs $(\mathbf{Y}_1, \mathbf{Y}_2) \in \mathcal{X}$, where $\mathcal{X}$ is the sample space of the experiment.

Observed data are listed in Table 1.1, where the fourth column contains individual differences $X_i = Y_{1i} - Y_{2i}$, $i = 1, \ldots, 20$. The data set and the corresponding software codes are available from the `examples_chapters_1-4` folder on the book's website.

1.8.1 Modelling Responses

The expected treatment effect is that training produces a stochastic reduction of anxiety. Therefore, we can write the hypotheses as

$$H_0 : \{Y_1 \stackrel{d}{=} Y_2\} = \{P_1(t) = P_2(t), \forall t \in \mathcal{R}^1\}$$

against $H_1 : \{Y_1 \stackrel{d}{>} Y_2\}$, where P_1 and P_2 are the marginal distributions of Y_1 and Y_2.

Note that H_0 asserts the distributional (i.e. stochastic) equality of responses and that this is coherent with the hypothesis that training is completely ineffective. Moreover, it should be noted that the *stochastic dominance* of Y_1 with respect to Y_2, stated by H_1 and denoted by the symbol $\stackrel{d}{>}$, may be specified in several ways according to a proper set of side-assumptions. Most common specifications are listed in the following set of additive response models:

- (M.i) With *fixed additive effects*. $Y_{1i} = \mu + Z_{1i}$, $Y_{2i} = \mu - \delta + Z_{2i}$, $i = 1, \ldots, n$, where: μ is a population constant; δ is the treatment effect, assumed to be finite and strictly positive in H_1; Z_{1i} and Z_{2i} are identically distributed *centred random deviates*, the so-called *error components* or error terms, which are assumed to be not necessarily independent within units but are independent between units.

Table 1.1 IPAT data on anxiety in 20 individuals

i	Y_1	Y_2	X
1	19	14	5
2	22	23	−1
3	18	13	5
4	18	17	1
5	24	20	4
6	30	22	8
7	26	30	−4
8	28	21	7
9	15	11	4
10	30	29	1
11	16	17	−1
12	25	20	5
13	22	18	4
14	19	17	2
15	27	22	5
16	23	21	2
17	24	21	3
18	18	15	3
19	28	24	4
20	27	22	5

- (M.ii) With *fixed additive effects but with non-homogeneous units*. $Y_{1i} = \mu + \eta_i + Z_{1i}$, $Y_{2i} = \mu + \eta_i - \delta + Z_{2i}$, $i = 1, \ldots, n$, where η_i are unknown components specific to the ith unit assumed not dependent on treatment levels; all other components have the same meaning as in (M.i).
- (M.iii) With *individually varying additive effects*, i.e. fixed effects specific to each unit. $Y_{1i} = \mu + \eta_i + \sigma_i Z_{1i}$, $Y_{2i} = \mu + \eta_i - \delta_i + \sigma_i Z_{2i}$, $i = 1, \ldots, n$, where σ_i are the scale coefficients and δ_i the treatment effects both specific to the ith unit; in H_1, the δ_i are fixed non-negative finite quantities at least one of which is positive.
- (M.iv) With *generalized stochastic effects*. $Y_{1i} = \mu + \eta_i + \sigma_i Z_{1i}$, $Y_{2i} = \mu + \eta_i + \sigma_i Z_{2i} - \Delta_{2i}$, $i = 1, \ldots, n$, where, in H_1, random effects Δ_{2i}, which may depend in some way on $(\mu, \eta_i, \sigma_i, Z_{1i}, Z_{2i})$, are non-negative stochastic quantities, at least one of which is strictly positive in the alternative.

Model (M.i) is the standard model for homogeneous homoscedastic observations; the other models extend standard conditions. In particular, model (M.ii) assumes homoscedasticity of responses but non-homogeneous units. (M.iii) is consistent with situations in which relevant covariates are not observed, as for instance when some individuals are male and others female with possibly different associated effects and with possibly different scale coefficients. (M.iv) is consistent with any form of stochastic dominance for quantitative responses, in particular with those with fixed or stochastic multiplicative forms. In this chapter we refer mainly to the fixed additive model as in (M.i). We leave the extension of the main results to other models to Chapters 2–4 and some suggested exercises therein.

The null hypothesis may also be written as $H_0 : \{\Pr(Y_1 - Y_2 \leq -t) = \Pr(Y_1 - Y_2 \geq t),\ \forall t \in \mathcal{R}^1\}$, where it is assumed that these probability statements are well defined and are related to distribution P of (Y_1, Y_2). Thus, H_0 is true if and only if the difference $X = Y_1 - Y_2 = \delta + Z_1 - Z_2$

is symmetrically distributed around 0, whereas in the alternative, where in particular we have $\Pr\{X > 0|H_1\} > 1/2$, X is symmetrically distributed around the location, that is, the treatment effect $\delta > 0$. Of course δ corresponds to a suitable indicator, that is, a functional or pseudo-parameter, for the effect, usually the mean, or the trimmed mean, or the median, etc.

Remark 1. When using differences X, models (M.i) and (M.ii) become equivalent. Indeed, both become $\{X_i = Y_{1i} - Y_{2i} = \delta + Z_{1i} - Z_{2i}, i = 1, \ldots, n\}$. This means that when covariates are assumed to influence only individual specific components η_i, differences $\mathbf{X}$ become covariate-free. A nice consequence of this is that when adopting model (M.ii), we do not need units which are necessarily homogeneous with respect to experimental conditions.

1.8.2 Symmetry Induced by Exchangeability

A formal proof of the symmetry property, in H_0, of $X = Z_1 - Z_2$ around 0 may easily be achieved by observing that the two variables Z_1 and Z_2 are exchangeable within units. Exchangeability within units implies both $F_1(t) = F_2(t), \forall t \in \mathcal{R}^1$, and

$$F_{1|t}(z|Z_2 = t) = F_{2|t}(z|Z_1 = t), \quad \forall (t, z) \in \mathcal{R}^2,$$

where F_1, F_2, $F_{1|t}$, and $F_{2|t}$ represent respectively the CDFs of variables Z_1, Z_2, $(Z_1|Z_2 = t)$ and $(Z_2|Z_1 = t)$. Of course, all these CDFs are associated with P, the existence of which is assumed. Hence,

$$\Pr\{(Z_1 - Z_2) \leq z\} = \int_{-\infty}^{+\infty} F_{1|t}(z + t|Z_2 = t) \cdot dF_2(t)$$

and

$$\Pr\{(Z_2 - Z_1) \leq z\} = \int_{-\infty}^{+\infty} F_{2|t}(z + t|Z_1 = t) \cdot dF_1(t),$$

thus $\Pr\{X > z\} = \Pr\{X < -z\}, \forall z \in \mathcal{R}^1$, which is the condition for symmetry of X around 0.

One consequence of this property is that, in H_0, $\Pr\{X < 0\} = \Pr\{X > 0\}$; thus, assuming $\mathbb{E}(Z)$ is finite, X has a null mean value; moreover, assuming the median $\mathbb{M}d(Z)$ has only one value, X has a null median. Instead, in H_1, $\Pr\{X < 0\} > (<) \Pr\{X > 0\}$, according to whether the response variables are such that $Y_1 \overset{d}{<} (\overset{d}{>}) Y_2$. One more consequence is that, in H_0, the vectors of signs $(X_i/|X_i|, i = 1, \ldots, n)$ and of differences $(X_i = Y_{1i} - Y_{2i}, i = 1, \ldots, n)$ are stochastically independent, where when $X_i = 0$, the difference X_i and related sign $X_i/|X_i|$ are excluded from analysis (for proof of this, see Randles and Wolfe, 1979, p. 50).

1.8.3 Further Aspects

Here the probability distribution P of X is assumed to be unknown in some of its parts; that is, in some of its parameters, in its analytic form, or in P as a whole, provided that it belongs to a nonparametric family of non-degenerate distributions $\mathcal{P}$ (see Section 1.2). Moreover, assuming that the mean value $\mathbb{E}(Z)$ is finite, so that we may consider the sample mean $\bar{X} = \sum_i X_i/n$ as a proper indicator of training effect δ, the hypotheses can equally be written as $H_0 : \{\delta = 0\}$ against $H_1 : \{\delta > 0\}$.

In this framework, the vector of pairs $\{(Y_{1i}, Y_{2i}), i = 1, \ldots, n\}$ may be viewed as a random sample of n pairs of observations, where exchangeability is intended within each individual pair (see Remark 3, 2.1.1).

Note that the existence of the mean value $\mathbb{E}(X)$ is necessary only for parametric solutions; for nonparametric permutation solutions we need the existence of a suitable and possibly robust indicator of δ, such as the median $\mathbb{M}d(X)$ or the trimmed mean. Also note that if it were convenient for analysis, we might consider non-degenerate data transformations φ, such as $\varphi(X)$, $\varphi(Y_1) - \varphi(Y_2)$ or more generally $\varphi(Y_1, Y_2) = -\varphi(Y_2, Y_1)$, so that the sample mean of the transformed data is a proper indicator of training effect. These transformations generally modify the distributions of the variables in question and may better fit one of the additive models (M.i)–(M.iv) of Section 1.8.1, so that it is possible to obtain better power behaviour of the resulting test statistics, and even to improve interpretation of the results (see Remarks 1, 2.6, and 2 below).

Remark 1. The one-sample matched pairs problems, where independent units are paired according to some known covariates, are formally equivalent to that of paired observations. The only inessential difference is that error components Z_{1i} and Z_{2i} are now independent, instead of simply exchangeable (see Problem 11, 1.9.5).

Remark 2. The testing problem with paired observations may be solved in several parametric and nonparametric ways, according to explicit assumptions concerning the distribution P of the response variables (Y_1, Y_2). Moreover, in a nonparametric context, the problem of determining the best data transformation φ in order to obtain a best test of the form $\sum_i \varphi_i$ for finite sample sizes is still an open problem (see Runger and Eaton, 1992; see also Section 2.5).

1.8.4 The Student's t-Paired Solution

A first well-known unconditional solution in a parametric framework may be found if the response variable X is assumed to be normally distributed with unknown variance. Accordingly, the response model with fixed additive effects can now be written as $\{Y_{1i} = \mu + \sigma \cdot Z_{1i}, Y_{2i} = \mu - \delta + \sigma \cdot Z_{2i}, i = 1, \ldots, n\}$, where: μ is a population constant; δ is the treatment effect; $\sigma \in \mathcal{R}^+$ is the unknown standard deviation assumed to be independent of units and treatment levels; random errors $Z_{ji} \sim \mathcal{N}(0, 1)$, $j = 1, 2$, are assumed to be normally distributed, with null means and unit variances, and independent with respect to units but not necessarily independent within units.

In this setting, the alternative being one-sided, an optimal unconditional solution (UMP similar) is based on the well-known Student's t test for paired observations, $T = \bar{X} \cdot \sqrt{n}/\hat{\sigma}$, where $\hat{\sigma}^2 = \sum_i (X_i - \bar{X})^2/(n-1)$ and $\bar{X} = \sum_i X_i/n$, because differences are normally distributed: $X_i \sim \mathcal{N}(\delta, \sigma_X^2)$.

In H_0, the unconditional distribution of T is central Student's t with $n-1$ degrees of freedom (d.f.). In H_1, T is distributed as a non-central Student's t with a positive non-central parameter $\sqrt{n}\delta/\sigma_X$, so that large values are significant. Note that the unknown standard deviation σ_X is the only nuisance entity for the problem and T is an invariant statistic with respect to both σ and σ_X. Note also that $\hat{\sigma}$ is a minimal (complete) sufficient statistic for σ_X, either under H_0 or H_1. Using the data of the example we have $T^o = 4.84$ with 19 d.f., which leads to the rejection of H_0 at $\alpha = 0.001$.

One somewhat more efficient parametric solution may be found via covariance analysis, if the pairs (Y_{1i}, Y_{2i}), $i = 1, \ldots, n$, were independent and identically distributed according to a bivariate normal variable and if baseline Y_1 were considered as a covariate for the problem. However, it is worth noting that, although the bivariate normality of (Y_1, Y_2) implies normality of X, the converse is not true. Actually, normality of X is more frequently valid than the bivariate normality of (Y_1, Y_2). In the case of our specific example, the normality of X is an assumption which is difficult to justify because IPAT data are aggregates of a finite number of discrete scores, each related to a specific aspect of anxiety. Of course, the assumption of bivariate normality for the pair (Y_1, Y_2) is even more questionable.

Remark 1. When responses follow model (M.ii) in Section 1.8.1, that is, $Y_{1i} = \mu + \eta_i + \sigma \cdot Z_{1i}$, $Y_{2i} = \mu + \eta_i - \delta + \sigma \cdot Z_{2i}$, $i = 1, \ldots, n$, then the covariance analysis method becomes much more difficult or even impossible. Moreover, when $Y_{1i} = \mu + \eta_i + \sigma_i \cdot Z_{1i}$, $Y_{2i} = \mu + \eta_i - \delta + \sigma_i \cdot Z_{2i}$, $i = 1, \ldots, n$ (i.e. when unknown standard deviations are dependent on units), then no parametric solution can be obtained unless the σ_i and the within-unit correlation coefficients ρ_i, $i = 1, \ldots, n$, are all known.

Remark 2. Student's t can also be applied to response models such as $Y_{1i} = \mu + \eta_i + \sigma_1 \cdot Z_{1i}$, $Y_{2i} = \mu + \eta_i - \delta + \sigma_2 \cdot Z_{2i}$, $i = 1, \ldots, n$, where the two scale coefficients σ_1 and σ_2 may not be equal with respect to measurement occasions but pairs (σ_1, σ_2) are invariant with respect to units, provided that underlying errors are normal and independently distributed. This solution has been used by Scheffé (1943c) in a randomized test for the Behrens–Fisher problem.

1.8.5 The Signed Rank Test Solution

Let us assume that P is completely unknown and X is a continuous variable, so that ties in the observations are assumed to occur with probability zero. In this situation, P as a whole must be considered as a nuisance entity for the testing problem, and a solution must be found either by using invariance arguments or by conditioning on a set of sufficient statistics for P.

Applying invariance arguments and assuming homoscedasticity with respect to units (i.e. the X_i are i.i.d. in P), a suitable solution based on ranks evaluated on absolute values of differences is provided by Wilcoxon's well-known signed rank test (see Randles and Wolfe, 1979; Hollander and Wolfe, 1999; Lehmann, 2006). It is useful to recall that, in this context, we need not assume that $\mathbb{E}(X)$ is finite.

Wilcoxon's test is based on the statistic $W = \sum_i R_i \cdot w_i$, where $R_i = \mathbb{R}(|X_i|) = \sum_{1\leq j\leq n} \mathbb{I}(|X_j| \leq |X_i|)$, in which $\mathbb{I}(\cdot) = 1$ if relation $(\cdot)$ is satisfied and 0 elsewhere, are the ordinary ranks of the absolute values of differences $|X_i|$, $w_i = 1$ if $X_i > 0$ and $w_i = 0$ if $X_i < 0$, $i = 1, \ldots, n$, and $\mathbb{R}$ is the rank operator. If there are no ties, we have $\mathbb{E}(W|H_0) = n(n+1)/4$ and $\mathbb{V}(W|H_0) = n(n+1)(2n+1)/24$ whereas, in H_1, the mean value is larger than $n(n+1)/4$. Moreover, if n is not too small, in H_0 the distribution of $\{W - \mathbb{E}(W|H_0)\}/\sqrt{\mathbb{V}(W|H_0)}$ is well approximated by that of a standard normal variable.

A test statistic which is permutationally equivalent to W is $T = \sum_i R_i \cdot Sg(X_i)$, where $Sg(X_i) = 1$ if $X_i > 0$ and -1 if $X_i < 0$, which in H_0 has mean value $\mathbb{E}(T) = 0$ and variance $\mathbb{V}(T) = \sum_i R_i^2 = n(n+1)(2n+1)/6$ (see Randles and Wolfe, 1979, p. 429).

Unfortunately, the data of the example do not allow direct use of this test because of the excessive number of ties due to the absence of continuity for X. In this case, the permutation distribution of Wilcoxon's signed rank test cannot be approximated by its asymptotic counterpart, and it must be directly evaluated through specific calculations. However, it is worth noting that, assuming continuity for X, the test is distribution-free, and so it is also P-invariant.

Generalized Scores

If, in place of ordinary ranks, a version of the so-called generalized scores is used, $\varphi_i = \varphi(R_i)$, $i = 1, \ldots, n$, we can obtain other nonparametric solutions. Among the many generalized scores, each used for specific problems, those related to the standard normal distribution are the most popular. They consist of replacing ordinary ranks R_i with the related normal scores, $\zeta_i = \Phi^{-1}(R_i/(n+1))$ or $\varsigma_i = \mathbb{E}(Z_{(R_i)})$, $i = 1, \ldots, n$, respectively for the well-known van der Waerden and Fisher–Yates solutions, where Φ is the standard normal CDF and $Z_{(R_i)}$ is the R_ith order statistic of n i.i.d. random elements drawn from a standard normal variable.

1.8.6 The McNemar Solution

If no assumption regarding the continuity of X can be made, we must still consider P as a nuisance entity. With these relaxed assumptions, an invariant solution can be found via the binomial test.

Let $U = \#(X_i > 0) = \sum_i \mathbb{I}(X_i > 0)$ and $\nu = \#(X_i \neq 0)$ be respectively the numbers of positive differences and non-null differences. Thus, in H_0, the statistic U is binomially distributed with parameters ν and $1/2$, $U \sim \mathcal{B}n(\nu, 1/2)$, say. In H_1, U is still binomially distributed, but with parameters ν and $\vartheta = \Pr\{X > 0\} > 1/2$, so that large values of U are significant. This kind of solution essentially corresponds to the one-sided McNemar test, also called the sign test.

With the data from our problem we have $\nu = 20$, $U = 17$ and $\Pr(U \geq 17|\mathbf{X}) = \sum_{i\geq 17}\binom{20}{i}2^{-20} = 0.0013$, which is significant at $\alpha = 0.005$. Note that, when response variables are binary, this test is UMP conditional for one-sided alternatives and UMPU conditional for two-sided alternatives.

It is worth observing that McNemar's solution depends essentially on the number ν of non-null differences, in the sense that removing all units and relative responses presenting null differences from the analysis leads to the same result. One problem, which immediately arises, is concerned with how it is possible to obtain solutions by also including the $n - \nu$ null differences (for suggestions regarding auxiliary randomization procedures, see Lehmann, 1986; Randles, 2001; see also Problems 8, 1.9.5 and 1, 3.4.1 and Remark 2, 3.2.1 for further suggestions).

These two testing solutions do look slightly different in that, in the latter, null differences are assumed to be informative of a substantially null treatment effect, whereas in the former they appear to be totally non-informative. However, it should be noted that this argument is not completely acceptable. On the one hand, if we determine the permutation confidence interval for the treatment effect we see that all null differences play their part in the analysis as well as all other differences (see Remark 2, 3.2.1). On the other, in the multivariate case, all observed unit vectors must be processed in order to maintain underlying dependence relations among the variables (see Sections 4.3.5 and 7.12).

Of course, McNemar's solution can also be used in an obvious way to test for a median in one-sample problems. Indeed, suppose for instance that variable X is continuous, $\mathbf{X} = \{X_i, i = 1, \ldots, n\}$ is the data set, and $H_0 : \{\mathbb{M}d(X) = \tilde{\mu}\}$. Let $\nu = \sum_{i \leq n} \mathbb{I}(X_i \leq \tilde{\mu})$. Thus, in H_0, $\nu \sim \mathcal{B}n(n, 1/2)$ and so the solution is clear. Further extensions are presented in Section 2.6.

Remark 1. McNemar's test may be applied in the case of non-homogeneity in distribution of experimental units: $P_i \neq P_j$, $i \neq j$, that is when the components of $\mathbf{X}$ are independent but not identically distributed. In terms of the response models of Section 1.8.1, this means that the distributions of random errors (Z_{1i}, Z_{2i}) may vary with respect to units and that, in particular, they may have non-constant scale coefficients σ_i, $i = 1, \ldots, n$. This fact allows this test to be used even in some cases where there is a lack of homogeneity of experimental conditions. Therefore, it can be applied when there are censored paired observations (see Good, 1991). Of course, non-constant scale coefficients may have influence on power behaviour. McNemar's solution may also be used when responses are ordered categorical and differences correspond to either positive or negative variations (see Problems 7 and 8, 1.9.5; see also Examples 6–9, 2.6 and Section 6.2).

1.9 The Permutation Solution

1.9.1 General Aspects

Roughly speaking, permutation solutions are conditional on the whole set of observed data which, in H_0, is always a set of sufficient statistics for any kind of underlying non-degenerate distribution P. Let us now examine one solution to our problem under the assumption that P is unknown and that the nonparametric family $\mathcal{P}$ of distributions, to which P belongs, contains only non-degenerate distributions, including discrete, continuous and mixed. Note that because of conditioning and assumed independence of the n units, the multivariate distribution P is $\prod_i P_i$, where P_i is the distribution specific to the ith unit. In terms of the response models of Section 1.8.1, this is consistent with the fact that the distributions of errors (Z_{1i}, Z_{2i}) might be non-invariant with respect to units (see Section 1.8.3). Thus, the conditioning on a set of sufficient statistics allows relaxation of the condition of identical distribution for all units. Moreover, we note that within associative test statistics we need to assume that $\mathbb{E}(X)$ and $\mathbb{E}(Z)$ are finite (see Section 4.5).

1.9.2 The Permutation Sample Space

To proceed with the analysis, let us first observe that the null hypothesis $H_0 : \{Y_1 \stackrel{d}{=} Y_2\}$ implies that the two variables Y_1 and Y_2 are exchangeable within each unit with respect to the two occasions 1 and 2. This means that, in H_0, the two observed values of each unit are considered as if they were randomly assigned to two occasions. In other words, the sign of each difference $X_i, i = 1, \ldots, n$, is considered as if it were randomly assigned with probability 1/2. Thus, one way to solve the testing problem is to refer to a test statistic of the form $T = \sum_i X_i$. Its conditional distribution $F_T(t|\mathbf{X})$, when the observed points $\mathbf{X} = \{X_i, i = 1, \ldots, n\}$ are held fixed, is obtained under assumption that H_0 is true by considering the random attribution in all possible ways of the plus or minus sign to each difference with equal probability (a formal derivation of this statement within the conditional approach is given in Remark 3, 2.1.2). This may be done by referring to the distribution of $T^* = \sum_i X_i^*$, where X_i^* is obtained by attributing the sign + or − to X_i, $i = 1, \ldots, n$, with probability 1/2. Observe that the probability distribution of $\mathbf{X}^* = \{X_i^*, i = 1, \ldots, n\}$, conditional on $\mathbf{X}$, is uniform within the permutation sample space $\mathcal{X}_{/\mathbf{X}}$. That is, all points of $\mathcal{X}_{/\mathbf{X}}$ are equally likely (see Proposition 1, 2.1.3).

The permutation sample space $\mathcal{X}_{/\mathbf{X}}$ of our example then contains $M^{(n)} = 2^\nu$ points because the permutation of signs is ineffective on the $n - \nu$ null differences. Apparently, this solution uses only non-null differences. We shall see in Remark 2, 3.2.1 and Section 3.4 that when determining a conditional power function or a conditional confidence interval for treatment effect δ, null differences enter the process as well as non-null differences, and so they cannot be discarded from the analysis.

Let us denote by $F(t|\mathbf{X}) = \Pr\{T^* \leq t|\mathbf{X}\}$, $t \in \mathcal{R}^1$, the permutation conditional CDF induced by T, given $\mathbf{X}$. Observe that this permutation CDF always exists because, by assumption, $\mathbf{X}$ is a measurable entity with respect to measurable space $(\mathcal{X}, \mathcal{A})$, which is assumed to exist and be well defined.

Remark 1. In H_1, the permutation CDF of T is stochastically larger than that of T in H_0, so that large values of T are significant and the test is unbiased (formal proofs of these ordering properties of $F(t|\mathbf{X};\delta)$ with respect to δ are reported in Sections 3.1–3.5). In practice, by using $T^o = T(\mathbf{X})$ to indicate the observed value of T, if the p-value $\lambda = \Pr\{T^* \geq T^o|\mathbf{X}\}$ is larger than α, for any fixed value of α, then H_0 is accepted, according to traditional testing rules (see Section 2.2.4 for a formal justification of the use of p-values).

Remark 2. When the underlying model is $Y_{1i} = \mu + \eta_i + \sigma_i \cdot Z_{1i}$, $Y_{2i} = \mu + \eta_i - \delta + \sigma_i(\delta) \cdot Z_{2i}$, $i = 1, \ldots, n$, so that location and scale coefficients are both not invariant on units and treatment levels, the permutation solution remains effective (see Problem 11, 3.6.1). On the other hand, when the underlying response model is $Y_{1i} = \mu + \eta_i + \sigma_1 \cdot Z_{1i}$, $Y_{2i} = \mu + \eta_i - \delta + \sigma_2 \cdot Z_{2i}$, $i = 1, \ldots, n$, where the two scale coefficients σ_1 and σ_2 are not equal, due to the lack of exchangeability within units, the permutation solution based on $T^* = \sum_i X_i^*$ is not generally exact. Under this model, when in particular the error terms Z_{1i} and Z_{2i} are both symmetrically distributed around zero, exact permutation solutions do exist. One of these is provided by the sign or McNemar test.

1.9.3 The Conditional Monte Carlo Method

An Algorithm for Inspecting Permutation Sample Spaces

In the case of our example, as well as for all cases where sample sizes are not small, the cardinality $M^{(n)} = \#[\mathbf{X}^* \in \mathcal{X}_{/\mathbf{X}}]$ of the permutation sample space $\mathcal{X}_{/\mathbf{X}}$ (which is finite if n is finite) is too large to enumerate all its points. According to many authors since Dwass (1957), we can inspect this permutation sample space by means of a random sample from it; the idea of random sampling from $\mathcal{X}_{/\mathbf{X}}$ goes back to Eden and Yates (1933). This idea is realized by a simulation of the testing problem conditional on the observed data set $\mathbf{X}$, that is by a *without replacement experiment* (WORE) (see Pesarin, 1992, 2001; see also Chapter 2).

Note that the term *conditional Monte Carlo* (CMC) is used to point out that it is merely an ordinary Monte Carlo simulation carried out on the permutation sample space $\mathcal{X}_{/\mathbf{X}}$, where the set of observed points $\mathbf{X}$ is held fixed. The term *conditional resampling* procedure emphasizes without replacement resampling from the observed data set, considered as a finite population. Of course, in the context of permutation inferences, conditional resampling and CMC have the same meaning.

Essentially, the CMC procedure consists of the following steps:

- (S.a) Calculate, on the given data set $\mathbf{X}$, the observed value T^o of the test statistic T: $T^o = T(\mathbf{X})$.
- (S.b) For each of the n differences in $\mathbf{X}$, consider a random attribution of signs, obtaining a permuted data set $\mathbf{X}^*$.
- (S.c) Calculate $T^* = T(\mathbf{X}^*)$.
- (S.d) Independently, repeat steps (S.b) and (S.c) B times.
- (S.e) The B permutation sets $\mathbf{X}^*$ are a random sample from permutation sample space $\mathcal{X}_{/\mathbf{X}}$.

Thus, the B corresponding values of T^* simulate the null permutation distribution of T. Therefore, they permit the statistical estimation of the permutation CDF $F(t|\mathbf{X})$ and of the significance level (i.e. survival) function $L(t|\mathbf{X}) = \Pr\{T^* \geq t|\mathbf{X}\}$ respectively by their empirical versions: the empirical distribution function (EDF) $\hat{F}^*_B(t) = \sum_{1\leq b\leq B} \mathbb{I}(T^*_b \leq t)/B = \#(T^* \leq t)/B$ and the empirical survival function (ESF) $\hat{L}^*_B(t) = \sum_{1\leq b\leq B} \mathbb{I}(T^*_b \geq t)/B,\ \forall t \in \mathcal{R}^1$.

A Routine for Random Permutations

The algorithm of step (S.b) for random attribution of signs to differences $\mathbf{X}$ may be based on the rule $X^*_i = X_i \cdot S^*_i$, $i = 1, \ldots, n$, where the random variables S^*_i are i.i.d. and each takes the value -1 or $+1$ with equal probability, according to the function:

$$S^* = 2 \cdot \lfloor 2 \cdot \texttt{Rnd} \rfloor - 1,$$

where `Rnd` is a pseudo-random number in the open interval (0, 1) and $\lfloor(\cdot)\rfloor$ the integer part of $(\cdot)$, that is, $S^* = 2\mathcal{B}n(1, 1/2) - 1$.

Estimates of step (S.e) are such that the higher the number B of CMC iterations, the more closely in probability $\hat{F}^*_B(\cdot)$ and $\hat{L}^*_B(\cdot)$ estimate $F(\cdot|\mathbf{X})$ and $L(\cdot|\mathbf{X})$ respectively. In any case, $\hat{F}^*_B(\cdot)$ or $\hat{L}^*_B(\cdot)$ may conveniently be used in place of $F(\cdot|\mathbf{X})$ or $L(\cdot|X)$ for evaluating the agreement of the observed data with H_0. In practice the estimated p-value, which in turn corresponds to the ESF evaluated at the observed value T^o, is given by

$$\hat{\lambda}(\mathbf{X}) = \hat{\lambda} = \hat{L}^*_B(T^o) = \sum\nolimits_b \mathbb{I}(T^*_b \geq T^o)/B.$$

If $\hat{\lambda} \leq \alpha$, we may conclude that the empirical evidence disagrees with H_0, which should be rejected in accordance with traditional rules.

Table 1.2 Conditional Monte Carlo method

$\mathbf{X}$	$\mathbf{X}_1^*$	...	$\mathbf{X}_b^*$	...	$\mathbf{X}_B^*$
T^o	T_1^*	...	T_b^*	...	T_B^*

$\rightarrow \hat{\lambda}(\mathbf{X}) = \sum_{b=1}^{B} \mathbb{I}(T^* \geq T^o) / B.$

Table 1.2 summarizes the CMC procedure: the first line contains the data set $\mathbf{X}$ and the B permutations $\mathbf{X}^*$ randomly chosen from $\mathcal{X}_{/\mathbf{X}}$; the second contains the corresponding values of T.

Remark 1. If, in place of B random permutations, all possible permutations are considered, then the functions $F(t|\mathbf{X})$, $L(t|\mathbf{X})$ and p-value λ are exactly determined. However, due to the well-known Glivenko–Cantelli theorem, the estimated p-value $\hat{\lambda}$, as B tends to infinity, tends almost surely to the true value λ (see Section 2.2.5). Of course, the greater the number B of CMC iterations, the closer in probability the estimate $\hat{\lambda}$ is to its true value λ; and B can be stated in an obvious way so that $\Pr\{|\hat{L}_B^*(t) - L(t|\mathbf{X})| < \varepsilon\} > \eta$, for any t and any suitable choice of $\varepsilon > 0$ and $0 < \eta < 1$. When H_0 is not true, so that $\delta > 0$, the rejection probability, say $\Pr\{\lambda(\mathbf{X}(\delta)) \leq \alpha\} \geq \alpha$, monotonically increases in δ (for a formal proof see Sections 3.1.1 and 3.2), so that T is unbiased.

Analysis of IPAT Data Using R

Let us again refer to the IPAT data set and solve the problem using R code. The variable *anxiety* (Y) is measured before and after a set of training sessions. Let $\mathbf{Y}_1$ be the sample data before treatment (`data$YA`) and $\mathbf{Y}_2$ be the sample data after treatment (`data$YB`).

First of all specify your working directory (say `"C:/path"`), then read the data (type "data" to view). Compute the vector of observed differences $\mathbf{X}$ between the sample data before and after the treatment. Let $B = 10\,000$ be the number of desired permutations. In this example we also set the random number generator seed (by typing `set.seed(101)`(we do this in order to allow readers to obtain the same result. In further examples the generator seed will not be set, so the results might be different).

```
setwd("C:/path")
data<-read.csv("IPAT.csv",header=TRUE)
set.seed(101)
d = data$YA-data$YB
n = dim(data)[1]
B=1000
T<-array(0,dim=c((B+1),1))
T[1] = sum(d)

for(bb in 2:(B+1)){
T[bb] = t(d)%*%(1-2*rbinom(n,1,.5))
}
```

The array $\mathbf{T}$ has dimension $(B + 1) \times 1$ and contains the observed value T^o and the simulated null distribution of the test statistic $T^* = \sum_i [y_{Ai}^* - y_{Bi}^*] S_i^*$, where $S_i^* = 1 - 2\mathcal{B}n(1, 1/2)$, $i = 1, \ldots, n$. In order to obtain a p-value we need to load the function `t2p` that returns an array of $\hat{L}_B^*(t)$ from an array of permutation values of the test statistic.

```
t2p<-function(T){

if(is.null(dim(T))){T<-array(T,dim=c(length(T),1))}
```

```
oth<-seq(1:length(dim(T)))[-1]

B<-dim(T)[1]-1
p<-dim(T)[2]
if(length(dim(T))==3){C<-dim(T)[3]}

rango<-function(x){
r=1-rank(x[-1],ties.method="min")/B+1/B
return(c(mean(x[-1]>=x[1]),r))
}

P=apply(T,oth,rango)
return(P)
}
```

The first element of `P` is the p-value of this analysis:

```
t2p(T)[1]

0.001
```

In the specific example, we obtain $\hat{\lambda} = 0.0003$, which leads to rejection of H_0 at $\alpha = 0.001$. Note that the number of CMC iterations might be smaller than 10 000, for instance 2000 or 1000, without appreciable changes in the conclusions.

The corresponding MATLAB code is given below:

```
D=textimport('IPAT.csv',',',1);
[P T] = NP_1s(D(1).vals-D(1).vals,1000,1);
```

1.9.4 Approximating the Permutation Distribution

If the sample size n is large, a permutation central limit theorem (PCLT; see Section 3.8) may be applied in order to approximate the permutation distribution $F(t|\mathbf{X})$ of T. To this end, according to our experience, we observe the following:

- (a) If n is smaller than about 25, it is possible, by using appropriate computation tools which are easy to implement on desktop computers, to exactly calculate T^* at all points of the permutation sample space and then to calculate $F[t|\mathbf{X}]$ and $L[t|\mathbf{X}]$.
- (b) If n is greater than about 200, σ_X is assumed to be finite and the ratio $(\sum_i X_i^4)/(\sum_i X_i^2)^2$ is small, then $F[t|\mathbf{X}]$ can be approximated by the PCLT. To this end, let us observe that the conditional expectation and variance of S^* are respectively $\mathbb{E}(S^*) = 0$ and $\mathbb{V}(S^*) = 1$. Hence, $\mathbb{E}\left\{\left(\sum_i X_i \cdot S_i^*/n\right)|\mathbf{X}\right\} = 0$ and $\mathbb{V}\left\{\left(\sum_i X_i \cdot S_i^*/n\right)|\mathbf{X}\right\} = \sum_i X_i^2/n^2$, because, conditionally on $\mathbf{X}$, quantities X_i in T^* play the role of fixed quantities. Therefore, the permutation standardized version

$$K^* = \left(\sum_i X_i \cdot S_i^*\right) \Big/ \left(\sum_i X_i^2\right)^{1/2},$$

being the standardized sum of n independent variables, is approximately standard normally distributed (see Chapter 3 for more details on the asymptotic permutation behaviour of test statistics).
- (c) In all other cases, $F[t|\mathbf{X}]$ and $L[t|\mathbf{X}]$ may be approximated, to the desired degree of accuracy, by means of a CMC, performed B times.

Although in this example the sample size n is not sufficiently large for normal approximation, the standardized observed value is $K^o = 3.324$, implying a p-value $\lambda = 0.000\,44$, a value which is very close to the CMC estimate $\hat{\lambda}$.

Remark 1. Observe that test K^* is approximately normally distributed independently of the underlying population distribution P, whereas Student's t, in order to be approximated by the standard normal distribution, requires normality of P.

1.9.5 Problems and Exercises

1) Compare the standardized permutation test statistic K, introduced in Section 1.9.4, to the Student's t appropriate for the same problem and find the main differences. In particular, show that the two tests are asymptotically equivalent.

2) Prove that the two test statistics, K as above and Student's t, are asymptotically equivalent under both H_0 and H_1.

3) Extend the test solution for paired observations to the one-sample problem of testing for symmetry. Note that: (i) X is symmetric with respect to δ if $X - \delta$ is symmetric with respect to 0; (ii) X is symmetric with respect to 0 if and only if $\Pr\{X < -z\} = \Pr\{X > z\}$, $\forall z \in \mathcal{R}^1$; (iii) if X is symmetric with respect to 0, then $\Pr\{X < 0\} = \Pr\{X > 0\}$ (see Section 2.6).

4) Extend the test solution for paired observations when the model for responses is of multiplicative form, $Y_{2i} = \rho \cdot Y_{1i} + \varepsilon_i$, $i = 1, \ldots, n$, so that $H_0 = \{\rho = 1\}$, whereas $H_1 = \{\rho > 1\}$.

5) Discuss the permutation solution for paired observations in the case where $\sigma_1 \neq \sigma_2$.

6) Draw a block diagram for a test of symmetry in a one-sample problem, according to Problem 4.

7) Show that, when there are ties in data in Section 1.9, i.e. the number of zero variations in the categorical responses is positive, a solution not conditional on non-null differences should imply auxiliary randomization (Lehmann, 1986).

8) With reference to Section 1.9 and taking account of Problem 7 above, show that one way to take into consideration the $n - \nu$ null differences is by using auxiliary randomization, according to Lehmann (1986).

9) Prove that the CMC method for testing with paired quantitative observations, illustrated in Section 1.9, may also be used in the case of paired binary ordered categorical observations.

10) Prove that, with reference to the same testing problem for paired observations, the permutation test $T_S^* = \sum_i \left(X_i \cdot S_i^*/|X_i|\right)$, which corresponds to the sum of standardized summands because $\mathbb{V}\{(X_i \cdot S_i^*/|X_i|)|\mathbf{X}\} = 1$, $i = 1, \ldots, n$, and where $X_i \cdot S_i^*/|X_i| = 0$ if $X_i = 0$, coincides with the binomial or McNemar test (note that this solution may be taken into consideration when individual distributions P_i are considerably different from each other).

11) Prove that the matched pairs problem is equivalent to that with paired observations (see Remark 1, 1.8.3).

12) Show that the McNemar test is no more than a test on paired binary observations, either ordered categorical or quantitative.

1.10 A Two-Sample Problem

Let us now discuss, as a second example, a problem (Pesarin, 2001) concerning the comparison of locations of two populations. In a psychological experiment to assess the degree of job satisfaction of two groups of workers, 20 units, assumed to be homogeneous in respect of most important covariates, such as sex, age, general health and social status, were examined in terms of the response variable X, corresponding to the perceived degree of job satisfaction. X was measured by a proper psychological index consisting of a sum of a finite number of items each related to a specific sub-aspect. Before the experiment was carried out, 12 units (group 1) were classified as 'extroverted', X_1, and the remaining 8 units (group 2) were classified as 'introverted', X_2, so that the sample data were $\mathbf{X}_1 = \{X_{1i}, i = 1, \ldots, 12\}$ and $\mathbf{X}_2 = \{X_{2i}, i = 1, \ldots, 8\}$ respectively. The testing problem was to show whether the data conform better to the null hypothesis of no difference in distribution, or to the one-sided (i.e. restricted or dominance) alternative of a difference in favour of 'extroverted'. It is worth noting that since subjects are assigned to symbolic treatment levels (extroverted and introverted) after they were observed, so that subjects were not randomized to treatments, this is a typical observational study where the treatment is merely a post-hoc classification (see Remark 4, 2.1.1 for some related problems). However, since the null hypothesis assumes that there is no distributional difference between two treatment levels, instead of permuting subjects we are allowed to permute observed data (see Section 1.5).

1.10.1 Modelling Responses

The data are given in Table 1.3. Formally, the hypotheses being tested are $H_0 : \{X_1 \stackrel{d}{=} X_2\}$ against $H_1 : \{X_1 \stackrel{d}{>} X_2\}$. Note that H_1 asserts the stochastic dominance of X_1 with respect to X_2. This stochastic dominance may be specified according to several response models, two of which are as follows.

- (M.i) A model with fixed additive effects: $X_{1i} = \mu + \delta + \sigma \cdot Z_{1i}$, $X_{2i} = \mu + \sigma \cdot Z_{2i}$, $i = 1, \ldots, n_j$, $j = 1, 2$, where δ is the treatment effect (note homoscedasticity).
- (M.ii) A model with generalized stochastic effects: $X_{1i} = \mu + \Delta_{1i} + \sigma \cdot Z_{1i}$, $X_{2i} = \mu + \sigma \cdot Z_{2i}$, $i = 1, \ldots, n_j$, $j = 1, 2$, where μ is a population constant, Z_{ji} are exchangeable random errors with null location and unit scale parameter, σ is a scale coefficient not dependent on units or treatment levels, and $\Delta_{1i} \geq 0$ are non-negative random quantities representing individually specific treatment effects, which may depend on (μ, Z_{1i}) but are independent with respect to units, even though not identically distributed.

Model (M.ii) is consistent with any kind of stochastic dominance. In particular, it is consistent with: (a) model (M.i) when, with probability one, $\Delta_{1i} = \delta$; (b) a multiplicative effect model for positive responses, where $X_{1i} = \delta \cdot (\mu + \sigma \cdot Z_{1i})$, with $\delta \geq 1$; (c) an individually varying fixed effect model, where $X_{1i} = \mu + \sigma \cdot Z_{1i} + \delta_{1i}$, when $\Delta_{1i} = \delta_{1i}$ with probability one; (d) a model where the treatment effect may influence both location and scale coefficients, $X_{1i} = \mu + \delta + \sigma(\delta) \cdot Z_{1i}$, where $\sigma(\delta)$ is any monotonic function of δ or of $|\delta|$, provided that the associated CDFs satisfy the stochastic dominance condition $F_1(x) \leq F_2(x)$, $x \in \mathcal{R}^1$ (see Sections 2.1 and 3.1, for further details).

Within this introductory chapter we refer to the fixed effect model (M.i).

Remark 1. The generalized effect model (M.ii) does not imply homoscedasticity of responses in the alternative. We recall that homoscedasticity implies $F_1(x + \delta) = F_2(x)$, $\forall x \in \mathcal{R}^1$, where δ is the

Table 1.3 Job satisfaction of extroverted and introverted groups

X_1 :	66	57	81	62	61	60	73	59	80	55	67	70
X_2 :	64	58	45	43	37	56	44	42				

so-called *location functional* (same as *size effect* or *shift pseudo-parameter*). The terms *functional, size-effect* or *pseudo-parameter* for the treatment effect δ are generally used in place of *parameter* because in the nonparametric context δ is usually a functional expressed in terms of all underlying parameters of the unknown distribution P which, assuming the existence of the mean value $\mathbb{E}(Z)$, in its simplest form is $\delta = \int_{\mathcal{X}} \delta(x)\, dP(x)$.

It is also worth noting that model (M.ii) is consistent with the so-called *placebo effect*. When the placebo, the treatment typically assigned to units in the second group, is supposed to produce an effect, δ_P say, we may model responses (with obvious notation) as $X_{ji} = (\mu + \delta_P) + (W_{Pi} + Z_{ji}) + \Delta_{ji}$, $i = 1, \ldots, n_j$, $j = 1, 2$. From this representation we see that μ becomes $\mu + \delta_P$ and Z changes to $Z + W_P$, so that the placebo effect is included in the population constant and errors.

Furthermore, when the response model becomes $X_{1i} = \mu + \delta + \sigma_1 \cdot Z_{1i}$, $X_{2i} = \mu + \sigma_2 \cdot Z_{2i}$, $i = 1, \ldots, n_j$, $j = 1, 2$, where it is not assumed that $\sigma_1 = \sigma_2$, even in H_0, then we refer to the generalized Behrens–Fisher problem in which, of course, the exchangeability condition in H_0 is violated and so we have to look for approximate solutions (see Example 8, 4.6).

Remark 2. It is common in testing problems to refer to two-sided alternatives, which are usually written as $H_1 : \{X_1 \overset{d}{\neq} X_2\}$. This notation is quite ambiguous. What is usually intended is that it is either $F_1(x) \leq F_2(x)$ or $F_1(x) \geq F_2(x)$, with strict inequality in a set of points of positive probability, and not $F_1(x) \neq F_2(x)$, where two distributions are not equal in a set of points of positive probability. In the former notation it is presumed that the effect, fixed or random, is either positive or negative, but not both. In the latter, instead, it is presumed that the effect can be positive on some subjects and negative on others. Such testing problems are much more intriguing than others and are discussed in Example 5, 4.6, after the introduction of NPC and multi-aspect testing.

1.10.2 The Student t Solution

If we assume that the responses of the two populations are homoscedastic and normally distributed with σ unknown, this problem may be efficiently solved (UMP similar) by a one-sided Student's t for comparing two means. That is,

$$t = \frac{\bar{X}_1 - \bar{X}_2}{[\sum_{ji}(X_{ji} - \bar{X}_j)^2]^{1/2}} \sqrt{\frac{n_1 n_2 (n-2)}{n}},$$

where $\bar{X}_j = \sum_{i \leq n_j} X_{ji}/n_j$, $j = 1, 2$, are the sample means, $n = n_1 + n_2$ is the pooled sample size, and the Student's t has $n - 2$ d.f.

It is worth noting that in the present case the assumption of normality seems rather unnatural because integer numbers are observed and empirical distributions are slightly asymmetric. Thus, we can say that Student's t only provides for an approximate solution and that it is difficult to assess the degree of such approximation. If an underlying continuous model for responses is assumed, then the observed integer values can be considered as if they were truncated, the estimate of population variance obtained is biased downwards, and the resulting inference is anticonservative. However, the results are $t = 4.237$ with 18 d.f. which is significant at $\alpha = 0.001$.

1.10.3 The Permutation Solution

Maintaining the assumption of homoscedasticity in the null hypothesis, with reference to the additive fixed effect model (M.i) we can relax the normality assumption and suppose that the data are distributed according to non-degenerate continuous distributions P_1 and P_2, both from the same nonparametric family $\mathcal{P}$. Accordingly, assuming population means are finite, we can write the hypotheses as $H_0 : \{X_1 \overset{d}{=} X_2\} = \{\delta = 0\}$ against $H_1 : \{X_1 \overset{d}{>} X_2\} = \{\delta > 0\}$.

Remark 1. In the permutation context, in order to apply a test statistic based on comparison of sample means we need only assume that means of involved responses are finite. If by chance we cannot assume population means are finite, we must use a test statistic based, for instance, on comparison of sample medians or trimmed means or EDFs. What is essential is that there is pseudo-parameter (the location functional) δ playing the role of treatment effect and that a proper sampling indicator is available for it (see Section 2.5). Note also that H_0 implies exchangeability of observed data with respect to treatment levels.

The underlying non-degenerate common distribution P is unknown, so we may proceed by conditioning on a set of sufficient statistics for it in H_0. Such a well-known set is $\mathbf{X} = \mathbf{X}_1 \biguplus \mathbf{X}_2 = \{X(i), i = 1, \ldots, n; n_1, n_2\}$, with $\biguplus$ denoting vector concatenation, so that the two vectors are pooled into one; n_1, n_2 and $n = n_1 + n_2$ are the sample sizes of the two groups and pooled sample, respectively. The proof of the sufficiency of the pooled data set $\mathbf{X}$ in H_0 is left as an exercise (see Sections 1.2, 2.1.2 and 2.1.3). The pooled array $\{X(i),\ i = 1, \ldots, n;\ n_1, n_2\}$ is called the *unit-by-unit representation* of the data set $\mathbf{X}$. This assumes that $X(i)$ belongs to group 1 if i satisfies the condition $1 \leq i \leq n_1$, otherwise it belongs to group 2. Remember that we use the same symbol $\mathbf{X}$ to indicate both the set of sample data, regarded as a multivariate random variable, and the pooled vector of observed data, the distinction being clear from the context.

Remark 2. Observe that if in the null hypothesis $\sigma_1 \neq \sigma_2$, then a set of sufficient statistics is the vector of sample data $(\mathbf{X}_1; \mathbf{X}_2)$. Here, it should be emphasized that the data set is partitioned into two subsets $\mathbf{X}_1$ and $\mathbf{X}_2$, so that in this case we are not allowed to exchange data between groups (see Example 8, 4.6, for a discussion in the permutation context).

Conditioning on the whole data set is equivalent to conditioning with respect to the EDF for P, which is also sufficient (note that in H_0, $P_1 = P_2$; see Definition 2, 2.1.3). Hence, in H_0, observed data may be viewed as if they were randomly assigned to two treatment levels. Thus, for this kind of problem, the permutation sample space $\mathcal{X}_{/\mathbf{X}}$ is exactly the set of all permutations of data $\mathbf{X}$, the cardinality of which is $M^{(n)} = n!$ (in the example $n = 12 + 8 = 20$, thus $M^{(n)} = 2.4329 \cdot 10^{18}$). Of course, if we use test statistics such as $T = \sum_i X_{1i}/n_1 - \sum_i X_{2i}/n_2$ or the like, which in turn are differences of two symmetric functions, being invariant with respect to rearrangements of data entry, then this cardinality becomes $M^{(n)} = C_{n,n_1} = \binom{n}{n_1}$, leading to $C_{20,8} = 125\,970$, since in $\mathcal{X}_{/\mathbf{X}}$ there are $n_1! \cdot n_2!$ points sharing the same value of T^* (see Remark 2, 2.3).

If we assume that sample means are proper indicators for the treatment effect δ, a suitable permutation test statistic is $T^*_{1,2} = \bar{X}^*_1 - \bar{X}^*_2$, where $\bar{X}^*_j = \sum_i X^*_{ji}/n_j$, $j = 1, 2$, are the permutation sample means related to the permuted data set $\mathbf{X}^*$. However, $T^*_{1,2}$ is permutationally equivalent to $T^* = \sum_i X^*_{1i}$, because there is an increasing one-to-one relationship between the two statistics since the data set $\mathbf{X}$ is held fixed (see Section 2.4). Indeed, $\sum_{ji} X^*_{ji}$ and $\sum_{ji} X_{ji}$, being equal, are permutation invariant quantities. Thus, T^* and $T^*_{1,2}$ are related by a one-to-one increasing relationship.

In the framework of this problem, it may be seen that T^* is unbiased and consistent (see Section 3.3 and Theorems 5, 3.7.2 and 12, 4.5 for a discussion of these concepts).

Remark 3. In a permutation framework we need not consider standardized forms for the test statistics in question, because standardization is simply an increasing one-to-one relationship. Hence, standardized and non-standardized forms are always permutationally equivalent, provided that the observed data are non-degenerate. Moreover, in order to get unbiasedness of a permutation test, we must assume that two CDFs, F_1 and F_2, do not cross each other (see Sections 2.4 and 3.1 for further reflections).

One way of inspecting $\mathcal{X}_{/\mathbf{X}}$ for a two-sample problem is by modifying step (S.b) of Section 1.9.3 (see also Section 2.2.5) as follows:

- (S.b′) (i) Consider a random permutation $(u_1^*, \ldots, u_n^*)$ of unit labels $(1, \ldots, n)$. (ii) According to unit-by-unit representation, assign the first n_1 corresponding data to group 1 and the other n_2 to group 2, thus obtaining the data permutation $\mathbf{X}^* = \{X(u_i^*), i = 1, \ldots, n; n_1, n_2\}$. (iii) Calculate the test statistic $T^* = T(\mathbf{X}^*)$.

Analysis of Job Data Using R

In the job data set there are two variables: X denoting the degree of job satisfaction and Y denoting the extroverted (1) or introverted (2) group. We can obtain the test statistic $T^* = \bar{X}_1^* - \bar{X}_2^*$ by multiplying the vector of permuted data $\mathbf{X}^*$ for a vector of contrasts `contr`:

```
setwd("C:/path")
data<-read.csv("Job.csv",header=TRUE)
attach(data)
n = table(Y) ; C = length(n);
contr = rep(1/n,n); contr[-c(1:n[1])] = -contr[-c(1:n[1])]
round(contr,digits=3)

 [1]  0.083  0.083  0.083  0.083  0.083  0.083  0.083  0.083  0.083
[10]  0.083  0.083  0.083 -0.125 -0.125 -0.125 -0.125 -0.125 -0.125
[19] -0.125 -0.125

B=1000
T<-array(0,dim=c((B+1),1))
T[1] = X%*%contr
for(bb in 2:(B+1)){
X.star = sample(X)
T[bb] = X.star%*%contr
}
P=t2p(T); P[1]

[1] 2e-04
```

For the data in the above example, we obtain $\hat{\lambda} = 0.0002$, which is significant at $\alpha = 0.001$.

The corresponding MATLAB code is given below:

```
D=textimport('Job.csv',',',1);
reminD(D)
[P T] = NP_2s('X','Y',1000,-1);
```

The data set and the corresponding software codes are available from the `examples_chapters_1-4` folder on the book's website.

Remark 4. As permutations may be considered similar to without replacement random samples from $\mathbf{X}$, when sample sizes are sufficiently large, the permutation CDF $F_T(t|\mathbf{X})$ of T^* may be approximated (see Section 3.8 for conditions allowing this) by that of a normal distribution with mean value $\mathbb{E}(T^*|\mathbf{X}) = n_1 \cdot \sum_{ji} X_{ji}/n$ and variance $\mathbb{V}(T^*|\mathbf{X}) = n_1 \cdot n_2 \cdot \sigma_X^2/(n-1)$, where $\sigma_X^2 = \sum_{ji} X_{ji}^2/n - \left(\sum_{ji} X_{ji}/n\right)^2$ is the variance and $\sum_{ji} X_{ji}/n$ the mean of the pooled data set $\mathbf{X}$ regarded as a finite population.

1.10.4 Rank Solutions

If we assume that P_1 and P_2 are continuous and homoscedastic, the same problem may be solved by the well-known Wilcoxon–Mann–Whitney rank test. This is actually a permutation test based on ranks, $MW = \sum_i R_{1i}$, where $R_{ji} = \mathbb{R}(X_{ji}) = \sum_{gh} \mathbb{I}(X_{gh} \leq X_{ji})$ are the ranks of X_{ji}, $i = 1, \ldots, n_j$, $j = 1, 2$, in the pooled data set $\mathbf{X}$. Alternatively, one of its permutationally equivalent forms may be used. Since the mean value and variance of $\sum_i R_{1i}$ in the null hypothesis are $\mathbb{E}(\sum_i R_{1i}) = n_1(n+1)/2$ and $\mathbb{V}(\sum_i R_{1i}) = n_1 n_2(n+1)/12$ respectively, the standardized version of MW is $T_{MW} = [\sum_i R_{1i} - n_1(n+1)/2]/[n_1 n_2(n+1)/12]^{1/2}$. If sample sizes are not too small, the null distribution of T_{MW} is standard normal. With the data from the example, as $\sum_i R_{1i} = 164$, we have $T_{MW} = 4.146$, which is significant at $\alpha = 0.001$ (this can be compared with the standard normal distribution because sample sizes ($n_1 = 12$, and $n_2 = 8$) are not too small).

Note that, in general, rank transformations are not one-to-one with respect to data $\mathbf{X}$, so they lose the sufficiency property with respect to P, although, under continuity of P, MW is a *maximal invariant test*. This often, but not always, implies some power decay (with regard to the problem of an optimal choice of a test statistic for finite sample sizes, see the discussion in Sections 2.5 and 3.6). Of course, other nonparametric solutions are available, depending on assumptions regarding P.

Furthermore, if assumptions suggest that medians or any other robust statistic are to be preferred in place of mean values as proper indicators for treatment effects, one solution is to use a permutation test statistic of the form $\tilde{T} = \tilde{X}_1 - \tilde{X}_2$, where $\tilde{X}_j$, $j = 1, 2$, is the median or a robust statistic calculated on the jth data group. Alternatively, but not equivalently, another solution is to use the so-called Mood median test. Also, if no assumption regarding the equality of scale parameters with respect to treatment levels or more generally if no assumption of stochastic dominance can be made, another solution is to use a permutation Behrens–Fisher type test (for a discussion within the permutation context, see Example 8, 4.6).

1.10.5 Problems and Exercises

1) Discuss the permutation median test $T_{Md}^* = \tilde{X}_1^* - \tilde{X}_2^*$ for the two-sample problem, where $\tilde{X}_j^* = \mathbb{M}d(\mathbf{X}_j^*) = X_{((n_j+1)/2)}^*$ if n_j is odd and $(X_{(n_j/2)}^* + X_{(1+n_j/2)}^*)/2$ if n_j is even, where $X_{(1)}^* \leq X_{(2)}^* \leq \ldots \leq X_{(n_j)}^*$ are the order statistics of $\mathbf{X}_j^*$, $j = 1, 2$.

2) Prove that in the Behrens–Fisher problem, where the response model is $X_{ji} = \mu + \delta_j + \sigma_j \cdot Z_{ji}$, $i = 1, \ldots, n_j$, $j = 1, 2$, with $\sigma_1 \neq \sigma_2$, and where Z_{ji} are exchangeable random deviates with null mean value, in which the hypotheses are $H_0 : \{\mathbb{E}(X_1) = \mathbb{E}(X_2)\}$ and $H_1 : \{\mathbb{E}(X_1) > \mathbb{E}(X_2)\}$, the dominance of means, i.e. $\mu_1 > \mu_2$, does not imply dominance of responses, $X_1 \stackrel{d}{>} X_2$.

3) Show that, if the response variable is binary, then the test statistic for testing $H_0 : \{X_1 \stackrel{d}{=} X_2\}$ against $H_1 : \{X_1 \stackrel{d}{>} X_2\}$ corresponds to Fisher's exact probability test, which rejects H_0 if $\Pr\{\sum_i X_{1i}^* \geq \sum_i X_{1i}|\mathbf{X}\} \leq \alpha$.

4) With reference to the two-sample testing problem for $H_0 : \{X_1 \stackrel{d}{=} X_2\}$, in which the two data sets are $\mathbf{X}_j = \{X_{ji}, i = 1, \ldots, n_j\}$, $j = 1, 2$, prove that the pooled set $\mathbf{X} = \mathbf{X}_1 \biguplus \mathbf{X}_2$ is a set of sufficient statistics in the null hypothesis for whatever underlying distribution P (see Sections 1.2 and 2.1.2).

1.11 One-Way ANOVA

As a third introductory example, let us consider a one-way ANOVA design. It is a well known that this corresponds to testing for the equality in distribution of $C \geq 2$ groups of data, where C represents the number of treatment levels in a symbolic experiment.

In this framework, units belonging to the jth group, $j = 1, \ldots, C$, are presumed to receive treatment at the jth level. When side-assumptions, specific to the problem, ensure that responses have finite means and are homoscedastic, that is, $\mathbb{E}(|X_j|) < \infty$ and $\mathbb{V}(X_j) = \sigma^2$, $j = 1, \ldots, C$, then the equality of C distributions may be reduced to that of C means.

In order to introduce this problem and justify a permutation solution for it, let us consider the data set in Table 1.4 (from Pollard, 1977, p. 169). The related problem is concerned with the length of worms in three different groups, where the purpose is to test whether the mean length of the worms is the same in all three groups. We may write this formally as $H_0 : \{\mu_1 = \mu_2 = \mu_3\}$ against the alternative H_1 : {at least one equality is false}.

1.11.1 Modelling Responses

Here we consider a fixed effects additive response model, $\mathbf{X} = \{X_{ji} = \mu + \delta_j + \sigma \cdot Z_{ji}, i = 1, \ldots, n_j, j = 1, \ldots, C\}$, where μ is a population constant, δ_j are the fixed treatment effects which satisfy the contrast condition $\sum_j \delta_j = 0$, Z_{ji} are exchangeable random errors with zero mean value and unit scale parameter, σ is a scale coefficient which is assumed to be invariant with respect to groups, and C is the number of groups into which the data are partitioned. Note that, in this model, responses are assumed to be homoscedastic and that scale coefficients are assumed not to be affected by the treatment levels, in particular in the alternative. If data are normally distributed, this problem is solved by Snedecor's well-known F test for the one-way ANOVA layout for $H_0 : \{\delta_1 = \ldots = \delta_C = 0\}$ against $H_1 : \{H_0$ is not true$\}$. That is, with the meaning of the symbols clear,

$$F = \frac{\sum_{j=1}^{C}(\bar{X}_j - \bar{X}_\cdot)^2 n_j}{\sum_{ji}(X_{ji} - \bar{X}_j)^2} \cdot \frac{n - C}{C - 1},$$

the null distribution of which is Fisher's F with $C - 1$ and $n - C$ degrees of freedom for numerator and denominator, respectively.

Observe that, within the homoscedasticity condition, the null hypothesis is equivalent to equality of three distributions: $H_0 : \{X_1 \stackrel{d}{=} X_2 \stackrel{d}{=} X_3\}$. Also observe that this equality implies that the data

Table 1.4 Lengths of worms in three groups

Group		
1	2	3
10.2	12.2	9.2
8.2	10.6	10.5
8.9	9.9	9.2
8.0	13.0	8.7
8.3	8.1	9.0
8.0	10.8	
	11.5	

$\mathbf{X}$ are exchangeable; in particular, they may be viewed as if they were randomly attributed to three groups.

Let us then assume that we can maintain homoscedasticity and the null hypothesis in the form $H_0 : \{X_1 \stackrel{d}{=} X_2 \stackrel{d}{=} X_3\}$, but that we cannot maintain normality. Assuming the existence, in H_0, of a common non-degenerate, continuous, unknown distribution P, the problem may be solved by a rank test such as Kruskal–Wallis, or by any analogous test statistic based on generalized ranks, or by conditioning on a set of sufficient statistics (i.e. using a permutation procedure). Note that because of conditioning, the latter procedure allows for relaxation of continuity for P and for relaxation of finite scale coefficients for responses. It only requires the existence of location coefficients and of proper sampling indicators for them.

The permutation solution also allows for relaxation of some forms of homoscedasticity for responses in H_1 (see Section 2.7 for more details). In fact, the generalized one-way ANOVA model allowing for unbiased permutation solutions assumes that the hypotheses are $H_0 : \{X_1 \stackrel{d}{=} X_2 \stackrel{d}{=} X_3\}$ against $H_1 : \{X_1 \stackrel{d}{\neq} X_2 \stackrel{d}{\neq} X_3\}$, with the restriction that, for every pair $h \neq j$, $h, j = 1, 2, 3$, the corresponding response variables are stochastically ordered (pairwise dominance relationship) according to either $X_h \stackrel{d}{>} X_j$ or $X_h \stackrel{d}{<} X_j$, in such a way that, $\forall t \in \mathcal{R}^1$, the associated CDFs are related according to either $F_h(t) \leq F_j(t)$ or $F_h(t) \geq F_j(t)$.

Remark 1. This dominance assumption may correspond to a model in which treatment may affect both location and scale coefficients, as for instance in $\{X_{ji} = \mu + \delta_j + \sigma(\delta_j) \cdot Z_{ji}, i = 1, \ldots, n_j, j = 1, \ldots, C\}$, where $\sigma(\delta_j)$ are monotonic functions of treatment effects δ_j or of their absolute values $|\delta_j|$, provided that $\sigma(0) = \sigma$ and pairwise stochastic ordering on CDFs is preserved. The latter model is consistent with the notion of randomization (see Section 1.5). Indeed: (i) in the randomization context, units are assumed to be randomly assigned to treatment levels, so that H_0 implies exchangeability of responses; (ii) in the alternative, treatment may jointly affect location and scale coefficients, so that resulting permutation distributions become either stochastically larger or smaller than the null distributions. Also note that the pairwise dominance assumption is consistent with a generalized model with random effects of the form $\{X_{ji} = \mu + \sigma \cdot Z_{ji} + \Delta_{ji}, i = 1, \ldots, n_j, j = 1, \ldots, C\}$, where Δ_{ji} are the stochastic treatment effects which satisfy the (pairwise) ordering condition that for every pair $h \neq j$, $h, j = 1, \ldots, C$, we have either $\Delta_h \stackrel{d}{>} \Delta_j$ or $\Delta_h \stackrel{d}{<} \Delta_j$.

1.11.2 Permutation Solutions

Formalizing the testing problem for a C-sample one-way ANOVA layout, we assume that $\mathbf{X} = \{\mathbf{X}_1, \ldots, \mathbf{X}_C\}$ represents the data set partitioned into C groups, where $\mathbf{X}_j = \{X_{ji}, i = 1, \ldots, n_j\}$, $j = 1, \ldots, C$, are i.i.d. observations from non-degenerate distributions P_j, respectively. It is helpful to use the unit-by-unit representation $\mathbf{X} = \{X(i), i = 1, \ldots, n; n_1, \ldots, n_C\}$, where it is assumed that $X(i) \in \mathbf{X}_1$ if subscript i satisfies the condition $1 \leq i \leq N_1$, $X(i) \in \mathbf{X}_2$ if $N_1 + 1 \leq i \leq N_2$, and so on, where $N_j = \sum_{r \leq j} n_r$, $j = 1, \ldots, C$, are cumulative sample sizes. We also assume that the sample means are proper indicators of treatment effects.

Under the homoscedastic model the hypotheses are

$$H_0 : \{X_1 \stackrel{d}{=} \ldots \stackrel{d}{=} X_C\} = \{\delta_1 = \ldots = \delta_C = 0\}$$

against $H_1 : \{H_0 \text{ is not true}\}$.

If it is suitable for analysis, we may consider a data transformation φ, so that related sample means become proper indicators for treatment effects. According to the CMC procedure, iterations are now done from the pooled data set $\mathbf{X} = \mathbf{X}_1 \biguplus \ldots \biguplus \mathbf{X}_C$, which is still a set of sufficient statistics for the problem in H_0. If symmetric test statistics are used, the related permutation sample space $\mathcal{X}_{/\mathbf{X}}$ contains $n!/(n_1! \cdot \ldots \cdot n_C!)$ distinct points, where $n = \sum_j n_j$ is the total sample size.

According to the above assumptions and Remark 3, 1.10.3, a suitable test statistic based on deviance among sample means is

$$T_C^* = \sum_{j=1}^{C} (\bar{Y}_j^* - \bar{Y}_.)^2 \cdot n_j,$$

where $\bar{Y}_j^* = \sum_i \varphi(X_{ji}^*)/n_j$ and $\bar{Y}_. = \sum_j \bar{Y}_j \cdot n_j/n$. Note that $\bar{Y}_.$ is a permutationally invariant quantity, being based on the sum of all observed data. Hence, statistic T_C^* is permutationally equivalent to $T^* = \sum_{j=1}^{C} n_j \cdot (\bar{Y}_j^*)^2$ (see Example 2, 2.4).

Analysis of Worm Data Using R

Let us define the group variable as Y, and the data are the lengths X of the worms belonging to each group.

```
data<-read.csv("Worms.csv",header=TRUE)
attach(data)
n = table(Y) ; C = length(n); n
Y
1 2 3
6 7 5
```

We can obtain the distribution of the test statistic by computing, at each permutation, the values of $\bar{X}_j^{*2}$, $j = 1, 2, 3$. This is done with the auxiliary dummy variables:

```
I = array(0,dim=c(sum(n),C))
for(i in 1:C){
I[,i]<-ifelse(Y==names(n)[i],1/n[i],0)
}
```

The ith element of the jth column of matrix $\mathbf{I}$ is equal to $1/n_j$ if the ith observation belongs to group j and zero otherwise, $i = 1, \ldots, n$, $n = \sum_j n_j$, $j = 1, 2, 3$. Thus the vector of means by group $\bar{\mathbf{X}} = [\bar{X}_1, \bar{X}_2, \bar{X}_3]$ can easily be derived by multiplying `t(X)` (with dimension $1 \times n$) and $\mathbf{I}$ (with dimension $n \times 3$), and the test statistic can be derived by multiplying the square of $\bar{\mathbf{X}}$ and the vector $\mathbf{n}$. The permutation values of $T^* = \sum_{j=1}^{3} n_j \bar{X}_j^{*2}$ can be obtained similarly by replacing $\mathbf{X}$ with $\mathbf{X}^*$:

```
B=10000
T = array(0,dim=c((B+1),1))
T[1] = ( t(X)%*%I )^2%*%n

for(bb in 2:(B+1)){
X.star=sample(X)
T[bb] = ( t(X.star)%*%I )^2%*%n
}

t2p(T)[1]

[1]  0.0106
```

The corresponding MATLAB code is given below:

```
D=textimport('Worms.csv',',',1);
reminD(D)
[P T] = NP_Cs('X','Y',1000);
```

The data set and the corresponding software codes are available from the `examples_chapters_1-4` folder on the book's website.

The test's p-value is again obtained by applying the `t2p` function to the vector $\mathbf{T}$, and by focusing on the first element of the vector of results. In this case we can conclude that there is a strong evidence against the null hypothesis $\mu_1 = \mu_2 = \mu_3$.

We obtain $\hat{\lambda} = 0.0106$, which leads to the rejection of H_0 at $\alpha = 0.025$. This result fits those obtained by Pollard (1977) by means of the parametric Snedecor F test ($F = 6.30$, with 2 and 15 degrees of freedom) and the Kruskal–Wallis KW rank test ($KW = 7.76$, the null distribution of which is approximated by a central χ^2 with 2 d.f.) both significant at $\alpha = 0.025$. We recall that the Kruskal–Wallis permutation rank test is based on the statistic

$$KW = \left\{ \frac{12}{n(n+1)} \cdot \sum_{j=1}^{C} n_j \cdot \left[\bar{R}_j - \frac{n+1}{2} \right]^2 \right\},$$

where R_{ji} is the rank of X_{ji}, $j = 1, \ldots, C$, $i = 1, \ldots, n_j$, within the pooled data set $\mathbf{X}$, and $\bar{R}_j = \sum_i R_{ji}/n_j$, $j = 1, \ldots, C$, is the jth sample mean rank. For large sample sizes n_j, the null distribution of KW is approximated by that of a central χ^2 with $C - 1$ d.f.

1.11.3 Problems and Exercises

1) Discuss a solution to the one-way ANOVA when, in place of sample means $\bar{X}_j$, sample medians $\tilde{X}_j$ are assumed to be proper indicators for treatment effects.

2) Discuss Mood's median test for the one-way ANOVA and prove that it is a permutation test.

3) Express the rational and heuristic motivations for the choice of the test statistic T in the one-way ANOVA.

4) Compare the permutation solution T above with Snedecor's F based on homoscedastic normal responses, the Kruskal–Wallis KW based on rank transformations, and Mood's test based on frequencies above and below the pooled median. Discuss conditions in which one is better than the others.

5) Prove that the Kruskal–Wallis rank test is permutationally equivalent to $\sum_j n_j \cdot \bar{R}_j^2$.

6) Prove that for the one-way ANOVA problem, the permutation sample space $\mathcal{X}_{/\mathbf{X}}$ associated with $\mathbf{X}$ contains $n!/(n_1! \cdot \ldots \cdot n_C!)$ distinct permutations of $\mathbf{X}$.

2

Theory of One-Dimensional Permutation Tests

2.1 Introduction

2.1.1 Notation and Basic Assumptions

In this chapter we introduce the main terminology, definitions and general theory of permutation tests for some one-dimensional problems. Particular emphasis is given to the two-sample design taken as a guide. Extensions to one-sample and multi-sample designs are generally straightforward. A list of typical one-dimensional testing problems will also be discussed. The analyses for multivariate designs and multi-aspect problems are obtained by the NPC of dependent permutation tests. This is done from Chapter 4 onwards.

The approach used in the previous chapter was essentially heuristic because all test statistics and related results were intuitively justified. In this chapter we use a more formal approach based on conditionality and sufficiency principles (see Cox and Hinkley, 1974; Berger and Wolpert, 1988). Indeed, permutation tests are known to be conditional methods of inference, where the conditioning is done with respect to a set of sufficient statistics in H_0 for the underlying population distribution P (Randles and Wolfe, 1979; Lehmann, 1986; Lehmann and Romano 2005; Pesarin, 2001) and the related conditional reference space is denoted by $\mathcal{X}^n_{/\mathbf{X}}$. Permutation tests can also be derived within the notion that the null distribution of any statistic of interest is invariant with respect to a finite group of transformations (Hoeffding, 1952; Romano, 1990). Two formal approaches, the conditional on sufficient statistics and the group invariant transformations, are essentially equivalent (see Watson, 1957; Odén and Wedel, 1975; Nogales et al., 2000) in that they provide the same solutions. However, we prefer the conditional approach because it is easier to understand, more constructive, more natural to use, test statistics are generally simpler to justify, and inferential conclusions are easier to interpret and explain.

Let us assume that a one-dimensional non-degenerate variable X takes values on the sample space $\mathcal{X}$, and that associated with $(X, \mathcal{X})$ are parent distributions P belonging to a nonparametric family $\mathcal{P}$ (see Definition 1, 1.2). Of course P may belong to any parametric family. In such cases there are parametric counterparts to take into consideration which may often, but not always, perform optimally (see Sections 1.1–1.5). Each P gives the probability measure to events A belonging to a suitable collection (a σ-algebra) $\mathcal{A}$ of events. The family $\mathcal{P}$ may consist of distributions of either quantitative (continuous, discrete and mixed) or categorical (nominal and ordered) kinds of variables. Firstly, we refer to quantitative variables. Specific sections will be devoted to some problems

Permutation Tests for Complex Data: Theory, Applications and Software Fortunato Pesarin, Luigi Salmaso

with nominal and ordered categorical variables. It is assumed that each family $\mathcal{P}$ admits the existence of a common dominating measure $\xi_{\mathcal{P}}$ in respect to which the density (i.e. the Radon–Nikodym derivative), $f_P(X) = dP(X)/d\xi_{\mathcal{P}}$, is well defined. The density $f_P(X)$, as a function of P, may sometimes be regarded as the likelihood of P given by X, the existence of which is admitted by assumption. It is well known that without such an assumption, no statistical problem can be tackled. The density on every observed sample point $X \in \mathcal{X}$ is assumed to satisfy the condition $dP(X)/d\xi_{\mathcal{P}} > 0$ (in what follows we do not distinguish between a variable X and its observed sample points; the distinction will be clear from the context). For quantitative variables defined on the real line the probability measure P is equivalent to the CDF $F_P(x) = \int_{t \leq x} dP(t)$, $x \in \mathcal{R}^1$. The notation $(X, \mathcal{X}, \mathcal{A}, P \in \mathcal{P})$ summarizes the statistical model associated with the problem at hand.

Let $\mathbf{X}_j = \{X_{ji}, i = 1, \ldots, n_j\} \in \mathcal{X}^{n_j}$ be the i.i.d. sample data of size n_j from the model $(X, \mathcal{X}, \mathcal{A}, P_j \in \mathcal{P})$, $j = 1, 2$, respectively. For data sets with two independent samples we may write $\mathbf{X} = \{X_{11}, \ldots, X_{1n_1}, X_{21}, \ldots, X_{2n_2}\} \in \mathcal{X}^n$, whose related model is $(\mathbf{X}, \mathcal{X}^n, \mathcal{A}^{(n)}, P^{(n)} \in \mathcal{P}^{(n)})$, where $n = n_1 + n_2$ and $P^{(n)} = P_1^{n_1} \cdot P_2^{n_2}$ (and where the meaning of the symbols is otherwise clear). To denote data sets in the permutation context it may be convenient to use the unit-by-unit representation $\mathbf{X} = \mathbf{X}^{(n)} = (\mathbf{X}_1, \mathbf{X}_2) = \{X(i), i = 1, \ldots, n;\ n_1, n_2\}$, where it is intended that the first n_1 data in the list belong to the first sample and the rest to the second. In practice, with $\mathbf{u}^* = (u_1^*, \ldots, u_n^*)$ denoting a permutation of unit labels $\mathbf{u} = (1, \ldots, n)$, $\mathbf{X}^* = \{X^*(i) = X(u_i^*), i = 1, \ldots, n; n_1, n_2\}$ is the related permutation of $\mathbf{X}$. And so, $\mathbf{X}_1^* = \{X_{1i}^* = X(u_i^*), i = 1, \ldots, n_1\}$ and $\mathbf{X}_2^* = \{X_{2i}^* = X(u_i^*), i = n_1 + 1, \ldots, n\}$ are the two permuted samples respectively. We normally also use the same symbol $\mathbf{X}$ to denote the pooled data set as obtained by $\mathbf{X} = \mathbf{X}_1 \biguplus \mathbf{X}_2$, where $\biguplus$ is the symbol for concatenating two data files. It is worth noting that, more than with respect to units, permutations operate with respect to data associated with unit labels i and u_i^*, $1 \leq i \leq n$, and not with respect to individuals (see Section 1.5). This is of particular importance for multivariate problems where data vectors are permuted, as well as for observational studies where individuals cannot generally be permuted.

Here we discuss testing problems for stochastic dominance (i.e. one-sided) alternatives as generated by symbolic treatments with non-negative random shift effects Δ. In particular, the alternative assumes that treatments produce effects Δ_1 and Δ_2 respectively, and that $\Delta_1 \overset{d}{>} \Delta_2$, where $\overset{d}{>}$ stands for distributional (i.e. stochastic) dominance. Thus, the hypotheses are $H_0 : \{X_1 \overset{d}{=} X_2 \overset{d}{=} X\} \equiv \{P_1 = P_2\}$, and $H_1 : \{(X_1 + \Delta_1) \overset{d}{>} (X_2 + \Delta_2)\}$, respectively. Specifying this for quantitative variables, we can write $H_0 : \{F_1(t) = F_2(t), \forall t \in \mathcal{R}^1\}$ and $H_1 : \{F_1(\cdot) \leq F_2(\cdot)\}$, where the inequality is strict in a set of positive probability with respect to both distributions. Note that data of two samples are exchangeable in H_0, in accordance with the notion that responses behave as they were randomized to treatments; also note that two CDFs do not cross in H_1. Without loss of generality, we assume that effects in H_1 are such that $\Delta_1 = \Delta \overset{d}{>} 0$ and $\Pr\{\Delta_2 = 0\} = 1$. The latter condition agrees with the notion that an 'active treatment' is only assigned to subjects of the first sample and a 'placebo' to those of the second. The condition $\Delta \overset{d}{>} 0$ can be specified as $\{\Delta_i \geq 0$, $i = 1, \ldots, n_1\}$, with strict inequality for $\nu \geq 1$ subjects, where $0 < \nu/n_1 \leq 1$ and $\nu \to \infty$ almost surely as $n_1 \to \infty$. Moreover, we may let Δ depend on subjects and on related null responses, so that pairs (X_{1i}, Δ_i), $i = 1, \ldots, n_1$, satisfy the relation $(X_{1i} + \Delta_i) \overset{d}{\geq} X_{1i}$ with ν strict inequalities (see Example 3, 4.6). In this situation, since effects Δ may depend on null responses X_1, stochastic dominance $(X_1 + \Delta) \overset{d}{>} X_2 = X$ is compatible with non-homoscedastic situations in the alternative. This gives a considerable, useful advantage to the permutation approach over the traditional parametric methodology. Thus, the null hypothesis can also be written as $H_0 : \{\Delta = 0\}$ and the alternative as $H_1 : \{\Delta > 0\}$. Other than measurability, no further distributional assumption on random effects Δ is required. In particular, existence of moments of any positive order is not required. To emphasize the roles of sample sizes and effects, we sometimes use $\mathbf{X}^{(n)}(\Delta) = \{X_{11} + \Delta_1, \ldots, X_{1n_1} + \Delta_{n_1}, X_{21}, \ldots, X_{2n_2}\}$ to denote data sets; and so $\mathbf{X}^{(n)}(0)$ denotes data in H_0. Extensions to non-positive and two-sided alternatives are straightforward. Also of importance is the extension to composite null hypotheses such as $H_0 : \{\Delta \leq 0\}$; this extension is made in Proposition 3, 3.1.1.

Remark 1. In what follows we suppress the superscript $^{(n)}$ from $\mathbf{X}^{(n)}$ and $\mathbf{Z}^{(n)}$ unless it is necessary to avoid misunderstandings as when considering sequences of samples. By the additive model (M.ii) in Section 1.10.1, we may write that $X_{ji} = \mu + \Delta_{ji} + Z_{ji}$, and so the data set is $\mathbf{X}(\Delta) = \{\mu + \Delta_{1i} + Z_{1i}, i = 1, \ldots, n_1;\ \mu + Z_{2i}, i = 1, \ldots, n_2\}$. Note that the additive model for the data set can equivalently be written as $\mathbf{X}(\Delta) = (\mathbf{Z}_1 + \Delta, \mathbf{Z}_2)$. In fact, without loss of generality we can put $\mu = 0$ because it is a nuisance quantity common to all units and thus is not essential for comparing X_1 to X_2, in that the test statistic $T(\mathbf{Z} + \mu)$ is permutationally equivalent to $T(\mathbf{Z})$ since they lead to exactly the same inference (see Problem 16, 2.4).

It is also worth noting that the alternative $H_1 : \Pr\{\Delta > 0\} > 0$ may imply that $\Pr\{\Delta = 0\} > 0$. The latter can be useful for interpreting experiments in which the treatment is not active on all treated individuals, as may occur for some drugs with genetic efficacy where only those individuals with a specific genetic configuration are really affected. The case of $\Pr\{\Delta < 0\} > 0$ AND $\Pr\{\Delta > 0\} > 0$, called the *multi-sided alternative*, which may interpret situations where treatment can produce negative effects on some individuals, can be ineffective on others, and can produce positive effects on the rest, will be discussed in Example 5, 4.6, in connection with the so-called *multi-aspect* testing.

2.1.2 The Conditional Reference Space

It was shown in Section 1.2 that the data set $\mathbf{X}$ is always and trivially a set of sufficient statistics for the underlying distribution P in the null hypothesis and that inferences we wish to obtain are conditional on such a set of sufficient statistics. However, as for any given testing problem we may condition on different sets of sufficient statistics, it seems reasonable to condition on a minimal sufficient set. In this way we make the best use of the available information. This requirement is met in general by referring to the set $\mathbf{X}$ itself because in the nonparametric setting it is only known that the underlying distribution P belongs to the nonparametric family $\mathcal{P}$, and so no further reduction of dimensionality is possible without loss of information. Besides, in cases where it is known that P belongs to a parametric family for which the minimal sufficient statistic is the whole data set $\mathbf{X}$ and there are nuisance entities to remove, then there is no possibility of avoiding the conditioning with respect to $\mathbf{X}$, that is, acting outside the permutation approach. However, since all test statistics are functions T mapping $\mathcal{X}^n$ into $\mathcal{R}^1$, which implies a maximal reduction of dimensionality, there is a weakness with the nonparametric paradigm because in general a test statistic T cannot be selected among all possible statistics according to unconditionally optimal criteria (see Sections 2.5 and 3.6 for a discussion). However, it is worth noting that the same weakness also occurs for all other parametric and nonparametric methods when $\mathbf{X}$ is minimal sufficient because in no way it is possible to use all the information carried by $\mathbf{X}$ by considering only one statistic T. In order to attenuate the loss of information associated with using one overall statistic expressed in closed form, we will find solutions within the multi-aspect methodology by combining a set of tests for specific complementary viewpoints (e.g. Examples 3–8, 4.6).

Let $\mathbf{X}$ be the actual data set according to the model $(\mathbf{X}, \mathcal{X}^n, \mathcal{A}^{(n)}, P^{(n)} \in \mathcal{P}^{(n)})$. To deal with the conditional approach, let us define the conditional reference space $\mathcal{X}^n_{/\mathbf{X}}$ associated with $\mathbf{X}$ under the assumption that H_0 is true. We have the following:

Definition 1. (The conditional reference space). *Essentially, the conditional reference space is the set of points of the sample space $\mathcal{X}^n$ which are equivalent to $\mathbf{X}$ in terms of information carried by the associated underlying likelihood. It is indicated by the symbol $\mathcal{X}^n_{/\mathbf{X}}$.*

Thus, $\mathcal{X}^n_{/\mathbf{X}}$ contains all points $\mathbf{X}^*$ such that the likelihood ratio $dP^{(n)}(\mathbf{X})/dP^{(n)}(\mathbf{X}^*)$ is independent of P, and so corresponds to the *orbit* (or *coset*) of equivalent points associated with $\mathbf{X}$. Given that in H_0 the density $dP^{(n)}(\mathbf{X})/d\xi_{\mathcal{P}}^{(n)} = \prod_{ji} dP(X_{ji})/d\xi_{\mathcal{P}}$ is by assumption exchangeable in its arguments because $dP^{(n)}(\mathbf{X}) = dP^{(n)}(\mathbf{X}^*)$ for every permutation $\mathbf{X}^*$ of $\mathbf{X}$, then $\mathcal{X}^n_{/\mathbf{X}}$, or simply $\mathcal{X}_{/\mathbf{X}}$ if no ambiguity arises by suppressing superscript n, contains all permutations of $\mathbf{X}$. That is, $\mathcal{X}_{/\mathbf{X}} = \{\bigcup_{\mathbf{u}^*}[X(u_i^*), i = 1, \ldots, n]\}$, in which $\mathbf{u}^*$ is any permutation of unit labels $(1, \ldots, n)$. Therefore, every element $\mathbf{X}^* \in \mathcal{X}_{/\mathbf{X}}$ is a set of sufficient statistics for P in H_0. One consequence of this is that sample space $\mathcal{X}^n$ is partitioned into orbits $\mathcal{X}_{/\mathbf{X}}$, in that any point $\mathbf{X} \in \mathcal{X}^n$ such that $dP^{(n)}(\mathbf{X})/d\xi_{\mathcal{P}}^{(n)} > 0$ belongs to one and only one of such orbits, and so $\mathbf{X}^\dagger \in \mathcal{X}_{/\mathbf{X}}$ implies that $\mathcal{X}_{/\mathbf{X}^\dagger} = \mathcal{X}_{/\mathbf{X}}$, a condition which emphasizes the invariance of conditional reference spaces with respect to data permutations. Indeed, using $\mathcal{X}\backslash\mathcal{X}_{/\mathbf{X}}$ to denote such a partition, if $\mathbf{X}_1$ and $\mathbf{X}_2$ are two distinct points of $\mathcal{X}\backslash\mathcal{X}_{/\mathbf{X}}$, then no point of $\mathcal{X}_{/\mathbf{X}_1}$ can also be a point of $\mathcal{X}_{/\mathbf{X}_2}$, because the intersection of two orbits is empty: $\mathcal{X}_{/\mathbf{X}_1} \bigcap \mathcal{X}_{/\mathbf{X}_2} = \varnothing$, say. Therefore, we can write $\mathcal{X} = \bigcup_{\mathcal{X}\backslash\mathcal{X}_{/\mathbf{X}}} \mathcal{X}_{/\mathbf{X}}$. Conditional reference spaces $\mathcal{X}_{/\mathbf{X}}$ are also called *permutation sample spaces*. Characterizations of $\mathcal{X}_{/\mathbf{X}}$ for one-sample paired data, stratified and cross-over designs are given in Remarks 3, 4 and 5, respectively.

Remark 1. Suppose that the statistical model $(X, \mathcal{X}, \mathcal{A}, P \in \mathcal{P})$ is V-dimensional, and so the variable $\mathbf{X} = (X_1, \ldots, X_V)$ is defined over a V-dimensional sample space $\mathcal{X}$, where X_h, $h = 1, \ldots, V$, is the hth component of $\mathbf{X}$. In this way the observed data set is $\mathbf{X} = \{X_{hji}, i = 1, \ldots, n_j, j = 1, 2, h = 1, \ldots, V\} = \{\mathbf{X}(i), i = 1, \ldots, n; n_1, n_2\}$. In such a case, permutations of V-dimensional vectors are to be taken into consideration since $dP^{(n)}(\mathbf{X})/d\xi_{\mathcal{P}}^{(n)} = \prod_{ji} dP(X_{1ji}, \ldots, X_{Vji})/d\xi_{\mathcal{P}}$, where $\mathcal{P}$ and $\xi_{\mathcal{P}}$ are the V-dimensional family of parent distributions and the V-dimensional dominating measure, respectively. Thus, $\mathbf{X}^* = \{\mathbf{X}(u_i^*), i = 1, \ldots, n; n_1, n_2\}$, in which $\mathbf{X}(u_i^*) = (X_{1u_i^*}, \ldots, X_{Vu_i^*})$ is the V-dimensional vector associated with the u_i^*th unit label in the list.

Remark 2. What is really required in defining the permutation sample space $\mathcal{X}_{/\mathbf{X}}$ is that $dP^{(n)}(\mathbf{X})/d\xi_{\mathcal{P}}^{(n)}$ is invariant over rearrangements of data vectors, so that data are exchangeable. There are several ways to obtain exchangeable but not independent data. Five are as follows:

(i) A very typical situation occurs when rank transformations are used. Actually, the rank or also the generalized rank of any datum, say $R(X_{ji}) = \sum_{r=1}^{2} \sum_{s=1}^{n_r} \mathbb{I}(X_{rs} \leq X_{ji})$, is always a permutation invariant quantity although ranks, due to the linear relation $\sum_{ji} R(X_{ji}) = constant$, are not independent variables. Indeed, all rank tests are nothing other than permutation tests based on ranks. This is also true for randomized ranks, according to proposals made by Bell and Doksum (1967) and Fassò and Pesarin (1986), provided that they are assigned to data prior to permutation analysis.

(ii) Suppose that in the one-dimensional situation with quantitative data we consider the empirical deviates, that is, the transformations $Y_{ji} = X_{ji} - \bar{X}$, where $\bar{X} = \sum_{ji} X_{ji}/n$ is the pooled sample mean. As a consequence the Y_{ji}, due to the linear relation $\sum_{ji} Y_{ji} = 0$, are not independent but are exchangeable since $\bar{X}$ is a permutation invariant quantity. The same result occurs when deviates from any pooled statistics are used, for example $Y_{ji} = X_{ji} - \tilde{X}$, where $\tilde{X} = \mathbb{M}d(\mathbf{X})$ is the pooled median and $\mathbb{M}d(\cdot)$ the median operator, since $\tilde{X}$ is a permutation invariant quantity. Of course, transformations such as $Y_{ji} = X_{ji} - \bar{X}_j$, where $\bar{X}_j = \sum_{i \leq n_j} X_{ji}/n_j$, do not satisfy exchangeability other than within samples, since the pair $(\bar{X}_1, \bar{X}_2)$ is not permutation invariant. Indeed the conditional probability for a random permutation $(\bar{X}_1^*, \bar{X}_2^*)$ to be different from $(\bar{X}_1, \bar{X}_2)$, given the observed data set $\mathbf{X}$, is greater than zero. This fact implies at least that related inferences, provided that they are asymptotically exact, are at most approximate (see Example 8, 4.6).

(iii) Suppose that empirical data standardization (studentization) is considered: $\hat{Y}_{ji} = (X_{ji} - \bar{X})/\hat{\sigma}$, where $\hat{\sigma} = \{\sum_{ji}(X_{ji} - \bar{X})^2/n\}^{1/2}$ is a permutation invariant quantity which 'estimates' the population standard deviation σ in the null hypothesis. Of course the Y_{ji} are exchangeable in the null hypothesis without being independent. A similar result occurs when any scale coefficient is evaluated from the pooled data, for example, the median of absolute deviations from the median, $MAD(\mathbf{X}) = \mathbb{M}d[|\mathbf{X} - \tilde{X}|]$. It is, however, to be emphasized that the standardized data $\hat{Y}_{ji}$, the empirical residuals $Y_{ji} = X_{ji} - \bar{X}$, and the untransformed data X_{ji} always lead to the same conclusion and so they are permutationally equivalent (see Section 2.4). Thus, in one-dimensional problems data standardization is generally not required. It may be required in some multivariate analyses when component variables have different marginal distributions.

Particular attention will be devoted to the so-called scale-invariant transformations. Suppose that X_j have locations μ_j and scale coefficients σ_j, $j = 1, 2$, respectively, with σ_1 not necessarily equal to σ_2, and that the null hypothesis is $H_0 : \{\mu_1 = \mu_2\}$. In such a case, the group-dependent rescaled deviates $\hat{Y}_{ji} = (X_{ji} - \bar{X})/\hat{\sigma}_j$ are exchangeable only within groups but not between groups. Known as the Behrens–Fisher permutation problem, this leads to approximate solutions (see Example 8, 4.6 and Pesarin, 2001). This lack of exchangeability comes out when group-dependent data transformations are used before analysis. However, if under H_0 it is possible to assume that $\sigma_1 = \sigma_2$ and so data are exchangeable, then based on $Y_{ji} = (X_{ji} - \bar{X})$, which are exchangeable deviates, we have exact solutions as in (ii). It is worth observing that the fact that $\sigma_1 = \sigma_2$ in H_0 by assumption can be justified by design when subjects are randomly assigned to treatments.

(iv) As a further example let us consider a bivariate model $\{(X_{ji}, Y_{ji}), i = 1, \ldots, n_j; j = 1, 2\}$, where the X_{ji} are taken with the role of covariates for responses Y_{ji}. First, let us observe that the residuals (empirical deviates), such as $\hat{Y}_{ji} = [Y_{ji} - \hat{\beta}(X_{ji})]$, where $\hat{\beta}$ is any suitable estimate of the regression function β evaluated on the pooled set of pairs $(\mathbf{Y}, \mathbf{X})$, are exchangeable in the null hypothesis as $\hat{\beta}$ is a permutation invariant quantity. Thus the permutation strategy can be applied to the residuals $\hat{Y}_{ji}$. This strategy can be useful in some observational problems where covariates X cannot be assumed exchangeable in H_0 and it is required to eliminate their contribution to responses Y (see Example 11, 2.7). Of course, it is assumed that β is estimable (for linear regression functions within the least squares approach it is required that the dimensionality of covariates $\mathbf{X}$ is smaller than sample size n and that the associated variance–covariance estimated matrix has full rank).

(v) Suppose that in the bivariate model $\{(X_{ji}, Y_{ji}), i = 1, \ldots, n_j; j = 1, 2\}$ as in (iv), the covariate X is nominal categorical with $K > 1$ distinct categories $A_1, \ldots, A_K$. We can then stratify (or partition) the given problem into K sub-problems: $\{(Y_{kji}|X = A_k), k = 1, \ldots, K, i = 1, \ldots, n_j; j = 1, 2\}$. Data within each stratum are independent, and so exchangeable in H_0, and strata are also independent, so that analyses can be carried out separately within each stratum and then, to obtain one overall inference, partial results can be suitably combined. This strategy will be used in Chapter 4 as a particular case of NPC of a set of partial permutation tests; it is also used in several application problems discussed from Chapter 5 onwards (see also Remark 4 below).

Remark 3. It is worth noting that in the univariate paired data design (see Section 1.8), since the differences of any two individual observations in the null hypothesis are symmetrically distributed around 0, the set of absolute values of differences $|\mathbf{X}| = \{|X_i|, i = 1, \ldots, n\}$ is a set of sufficient statistics for P. For a simple proof of this, let us make the following assumptions:

(i) Let $f_P(t) = dP(t)/d\xi$ indicate the density corresponding to P with respect to the dominating measure ξ.
(ii) In H_0, f_P is symmetric with respect to 0 (see Section 1.8.2), so that $f_P(t) = f_P(-t)$, $\forall t \in \mathcal{R}^1$.

Let $\mathbf{t}$ and $\mathbf{t}'$ be any two points from sample space $\mathcal{X}^n$. The two points $\mathbf{t} = (t_1, \ldots, t_n)$ and $\mathbf{t}' = (t'_1, \ldots, t'_n)$ lie in the same orbit of a sufficient statistic for any density f_P symmetric around 0 if and only if the likelihood ratio

$$\frac{f_P(t_1) \cdot \ldots \cdot f_P(t_n)}{f_P(t'_1) \cdot \ldots \cdot f_P(t'_n)} = \rho_f(\mathbf{t}, \mathbf{t}')$$

does not depend on f_P. Now, due to the assumed symmetry of f_P, the ratio $\rho_f(\mathbf{t}, \mathbf{t}')$ is f_P-independent if $f_P(t_i) = f_P(t'_i)$, $i = 1, \ldots, n$; that is, if $t_i = \pm t'_i$. This implies that the set of points $\mathcal{X}_{/\mathbf{X}} \in \mathcal{R}^n$ which contains the same amount of information with respect to P as that contained in $\mathbf{X}$ is obtained by giving signs in all possible ways to the elements of $\mathbf{X}$ (see also the intuitive reasoning of Section 1.9.1): $\mathcal{X}_{/\mathbf{X}} = \{X_i S_i^*, i = 1, \ldots, n, S_i^* = 1 - 2\mathcal{B}n(1, 1/2)\}$. And so the set of absolute values $|\mathbf{X}|$, as well as any of its permutations (including $\mathbf{X}$ itself) is sufficient for P in H_0. Of course, in the alternative when differences are symmetric around δ, $|\mathbf{X} - \delta|$ and $\mathbf{X} - \delta$ are both sufficient. Note that the intuitive permutation solution in Section 1.9 properly is a conditional solution. Of course, the non-null permutation distribution of any statistic is not independent of the underlying population distribution P (see Problem 13, 2.6.2, for an extension to the situation in which units have different distributions).

Observe that the set of pairs $\{(Y_{1i}, Y_{2i}),\ i = 1, \ldots, n\}$ is also sufficient. The latter allows for easier intuitive interpretation of how the permutation principle operates. In fact, random assignment of signs to differences $X_i = Y_{1i} - Y_{2i}$ in the null hypothesis is equivalent to the assumption of exchangeability of paired observations (Y_{1i}, Y_{2i}) within each unit and independently with respect to units.

If, instead of only two, we are concerned with R observations for each unit, that is, the data set is $\mathbf{X} = \{X_{ir},\ r = 1, \ldots, R,\ i = 1, \ldots, n\}$, and the null hypothesis is that there is no effect due to the rth observation, that is, $H_0 : \{X_{i1} \overset{d}{=} \ldots \overset{d}{=} X_{iR},\ i = 1, \ldots, n\}$, then the permutation sample space becomes $\mathcal{X}_{/\mathbf{X}} = \prod_i \mathcal{X}_{/\mathbf{X}_i}$, where $\mathbf{X}_i = (X_{i1}, \ldots, X_{iR})$. $\mathcal{X}_{/\mathbf{X}}$ is therefore the cartesian product of individual permutation spaces $\mathcal{X}_{/\mathbf{X}_i}$; its cardinality is then $M^{(n)} = [R!]^n$ (see Example 9, 2.7).

Remark 4. Suppose we are given a stratified two-sample design in which data are $\mathbf{X} = \{X_{sji},\ i = 1, \ldots, n_{sj},\ j = 1, 2,\ s = 1, \ldots S\}$, where $S \geq 2$ is the number of strata. Suppose also that the hypotheses are $H_0 : \{\mathbf{X}_1 \overset{d}{=} \mathbf{X}_2\} = \{X_{s1} \overset{d}{=} X_{s2}, s = 1, \ldots, S\} = \{\bigcap_{s \leq S}(X_{s1} \overset{d}{=} X_{s2})\}$ against some alternative H_1. In such a situation, again the pooled data set $\mathbf{X}$ is a set of sufficient statistics for the underlying distribution P in H_0. However, supposing that strata influence responses, in the sense that P_s is not equal to $P_{s'}$ for some $s \neq s'$, the set of points for which the likelihood ratio is not dependent on P is now characterized by $dP^{(n)}(\mathbf{X})/dP^{(n)}(\mathbf{X}^*) = \prod_{s \leq S} dP_s^{(n)}(\mathbf{X}_s)/dP_s^{(n)}(\mathbf{X}_s^*)$, where it is emphasized that permutations are admitted only within strata. Thus, the resulting permutation sample space becomes the cartesian product of stratified permutation sample spaces $\mathcal{X}_{/\mathbf{X}_s}$, that is, $\mathcal{X}_{/\mathbf{X}} = \prod_{s \leq S} \mathcal{X}_{/\mathbf{X}_s}$. On the one hand this implies that, since strata are generally required not to be designed prior to data collection, this model is also valid for post-stratification designs, which are so useful in observational studies; on the other hand separate inferences on $H_{0s} : \{X_{s1} \overset{d}{=} X_{s2}\}$, $s = 1, \ldots S$, each against an appropriate alternative H_{1s}, are independent conditionally, and so related statistics are independent with respect to strata. It is worth noting here that the latter result is much more difficult to obtain within the invariance under a finite group of transformations approach.

Remark 5. Consider a simple *cross-over design* in which n units are randomly partitioned in two groups of size n_1 and n_2, with $n = n_1 + n_2$. Units of group $j = 1$ have assigned treatment A before treatment B, those in group 2 have assigned B before A. In this kind of cross-over experiment, where typically B is an active treatment and A a placebo, two rather different testing problems are of interest: (i) whether or not treatment is effective, that is, response Y_A is equal to Y_B; (ii) whether or not the two methods of treatment administration are equivalent (looking for a kind of interaction of group j and treatment k), that is, response Y_{AB} is equal to Y_{BA}.

Assume that the response data behave according to the model $Y_{jki} = \mu + \gamma_{ji} + \delta_{jk} + Z_{jki}, i = 1, \ldots, n_j, k = 1, 2, j = 1, 2$, where: μ is an unknown population constant; γ_{ji} are the so-called individual or block effects, which are generally considered as nuisance entities, and so are not of interest for analysis; and satisfy the side condition $\sum_{ji} \gamma_{ji} = 0$; δ_{jk} are effects due to treatment k on the jth group, and Z_{jki} are random errors assumed to be independent with respect to units and exchangeable within units. Within-unit differences of paired observations behave according to $X_{j \cdot i} = Y_{j1i} - Y_{j2i} = \delta_{j1} - \delta_{j2} + Z_{j1i} - Z_{j2i} = \delta_j + \sigma(\delta_j) \cdot Z_{ji}$, where: individual nuisance effects γ as well as population constant μ are compensated; the δ_j represent incremental effects specific to the jth group; Z_{ji} are the error components symmetrically distributed around 0 (see Section 1.8.2); and $\sigma(\delta_j)$ are scale coefficients, which may depend on main effects but are assumed homoscedastic with respect to units, that is, i-independent. The two separate sets of hypotheses of interest are $H_0' : \{\delta_1 = \delta_2 = 0\}$ against $H_1' : \{(\delta_1 \neq 0) \bigcup (\delta_2 \neq 0)\}$, that in at least one group there is a non-null effect, *and* $H_0'' : \{X_1 \overset{d}{=} X_2\} \equiv \{\delta_1 = \delta_2\}$ against $H_1'' : \{\delta_1 \neq \delta_2\}$, that the two effects are different. Hence, we must act separately and jointly for both sub-problems.

Due to the within-units exchangeability, the underlying density of each pair of responses in H_0' is such that $f_{ji}(Y_{j1i}, Y_{j2i}) = f_{ji}(Y_{j2i}, Y_{j1i}), i = 1, \ldots, n_j, j = 1, 2$, where f depends on the unknown population distribution P. In order to characterize the reference sample space, let us consider the likelihood ratio

$$dP^{(n)}(\mathbf{Y})/dP^{(n)}(\mathbf{Y}') = \prod\nolimits_{ji} f_{ji}(Y_{j1i}, Y_{j2i}) / \prod\nolimits_{ji} f_{ji}(Y'_{j1i}, Y'_{j2i}).$$

This does not depend on f if Y'_{j1i} is Y_{j1i} or Y_{j2i} with probability 1/2 each *and* if n pairs (Y_{j1i}, Y_{j2i}) are permuted with respect to groups. Thus $\mathcal{X}_{/\mathbf{X}} = \{\bigcup_{\mathbf{S}^*}[X_i S_i^*, i = 1, \ldots, n]\} \times \{\bigcup_{\mathbf{u}^*}[X(u_i^*), i = 1, \ldots, n]\}$, in which $\mathbf{S}^* = \{S_i^* = 1 - 2 \cdot \mathcal{B}n(1, 1/2), i = 1, \ldots, n\}$ are n i.i.d. elementary binomials, and $\mathbf{u}^*$ are permutations of unit labels $(1, \ldots, n)$. Thus the reference space is the cartesian product of within-unit exchanges *times* between-group exchanges of paired observations and so contains $2^n \cdot n!$ points. Therefore two null hypotheses can be tested separately and independently, in the sense that truth of H_0'' does not depend on truth of H_0', and vice versa. This implies that the two statistics for H_0' and H_0'' are conditionally independent. It is also worth noting that the cardinality of $\mathcal{X}_{/\mathbf{X}}$ is smaller than that of the corresponding full randomized design with $2n$ subjects; actually $2^n n! < (2n)! \simeq (2n)^{2n} e^{-2n} \sqrt{4\pi n}$.

2.1.3 *Conditioning on a Set of Sufficient Statistics*

Sufficiency in H_0 of $\mathcal{X}_{/\mathbf{X}}$ for P implies that the null conditional probability of every event $A^{(n)} \in \mathcal{A}^{(n)}$, given $\mathcal{X}_{/\mathbf{X}}$, is independent of P; that is, $\Pr\{\mathbf{X}^* \in A^{(n)}; P|\mathcal{X}_{/\mathbf{X}}\} = \Pr\{\mathbf{X}^* \in A^{(n)}|\mathcal{X}_{/\mathbf{X}}\}$, where the meaning of the symbols is clear. Thus, the permutation distribution induced by any statistic $T : \mathcal{X}^n \rightarrow \mathcal{R}^1$, namely $F_T(t|\mathcal{X}_{/\mathbf{X}}) = F_T^*(t) = \Pr\{T^* \leq t|\mathcal{X}_{/\mathbf{X}}\}$, is P-invariant. For this reason, T^* is also said to be a *P-invariant statistic*. Hence, any related conditional inference is distribution-free and nonparametric. Moreover, since for finite sample sizes the number $M^{(n)}$ of points in $\mathcal{X}_{/\mathbf{X}}$ is finite, where $M^{(n)} = \sum_{\mathcal{X}_{/\mathbf{X}}} \mathbb{I}(\mathbf{X}^* \in \mathcal{X}_{/\mathbf{X}}) < \infty$, a relevant consequence of both independence on P and finiteness of $M^{(n)}$ is that the permutation (conditional) probability of every $A^{(n)} \in \mathcal{A}^{(n)}$ is defined and calculated as

$$\Pr\{\mathbf{X}^* \in A^{(n)}|\mathcal{X}_{/\mathbf{X}}\} = \frac{\sum_{\mathbf{X}^* \in A^{(n)}} dP^{(n)}(\mathbf{X}^*)}{\sum_{\mathbf{X}^* \in \mathcal{X}_{/\mathbf{X}}} dP^{(n)}(\mathbf{X}^*)} = \frac{\sum_{\mathcal{X}_{/\mathbf{X}}} \mathbb{I}(\mathbf{X}^* \in A^{(n)})}{M^{(n)}},$$

because by definition $dP^{(n)}(\mathbf{X}) = dP^{(n)}(\mathbf{X}^*)$ for every permutation $\mathbf{X}^* \in \mathcal{X}_{/\mathbf{X}}$. Therefore, the restriction (i.e. projection) of the collection of events $\mathcal{A}$ over the permutation sample space $\mathcal{X}_{/\mathbf{X}}$, that is, $\mathcal{A} \bigcap \mathcal{X}_{/\mathbf{X}} = \mathcal{A}_{/\mathbf{X}}$ consisting of conditional events given $\mathbf{X}$, defines the *permutation measurable space* $(\mathcal{X}_{/\mathbf{X}}, \mathcal{A}_{/\mathbf{X}})$ on which the permutation probability $\Pr\{A|\mathcal{X}_{/\mathbf{X}}\}$ is defined.

As a consequence, we clearly have the following proposition.

Proposition 1. *In* $H_0 : \{X_1 \overset{d}{=} X_2\} \equiv \{P_1 = P_2 = P\}$, *provided that in* $\mathcal{X}_{/\mathbf{X}}$ *there are no multiple points, that is, when* $\sum_{\mathcal{X}_{/\mathbf{X}}} \mathbb{I}(\mathbf{X}^* = \mathbf{x}) = 1$ *if* $\mathbf{x} \in \mathcal{X}_{/\mathbf{X}}$, *and 0 elsewhere, permutations* $\mathbf{X}^*$ *are equally likely:*

$$\Pr\{\mathbf{X} = \mathbf{x}|\mathcal{X}_{/\mathbf{X}}\} = \Pr\{\mathbf{X}^* = \mathbf{x}|\mathcal{X}_{/\mathbf{X}}\} = \begin{cases} 1/M^{(n)} & \text{if } \mathbf{x} \in \mathcal{X}_{/\mathbf{X}}, \\ 0 & \text{if } \mathbf{x} \notin \mathcal{X}_{/\mathbf{X}}, \end{cases}$$

and so the observed data set $\mathbf{X}$ *as well as any of its permutations* $\mathbf{X}^*$ *are uniformly distributed over* $\mathcal{X}_{/\mathbf{X}}$ *conditionally.*

This property, which can be extended in a straightforward way to one-sample, multi-sample, and multidimensional contexts, shows that the null permutation distribution $\Pr\{\mathbf{X}^* = \mathbf{x}|\mathcal{X}_{/\mathbf{X}}\}$ only depends on the observed data set $\mathbf{X}$. From this point of view, the data set $\mathbf{X}$ can be viewed as the n-dimensional parameter for the related permutation CDF $F_T^*(t|\mathcal{X}_{/\mathbf{X}})$. Of course, in stratified designs (see Remark 4, 2.1.2) permutations are equally likely within strata.

In H_1, where it is assumed that there exists $A \in \mathcal{A}$ such that $0 < P_1(A) \neq P_2(A) > 0$, a set of sufficient statistics is the pair $(\mathbf{X}_1, \mathbf{X}_2)$. Consequently, the data are exchangeable within but not between samples, and so the observed data set $\mathbf{X}$ is not uniformly distributed over $\mathcal{X}_{/\mathbf{X}}$ conditionally. Hence, if we are able to find statistics sensitive to such a non-uniform distribution, we are then able to construct permutation tests. However, the problem of establishing the best test when P is unknown remains open (Lehmann, 1986; Lehmann and Romano, 2005; Braun and Feng, 2001; see also Section 2.5).

Definition 2. *For each permutation* $\mathbf{X}^* \in \mathcal{X}_{/\mathbf{X}}$ *the empirical probability measure (EPM) of any event* $A \in \mathcal{A}$ *is defined as*

$$\hat{P}_{\mathbf{X}^*}(A) = \sum_{i \leq n} \mathbb{I}(X_i^* \in A)/n = \sum_{i \leq n} \mathbb{I}(X_i \in A)/n = \hat{P}_{\mathbf{X}}(A),$$

which then is a permutation invariant function given $\mathbf{X}$. *Similarly, for quantitative variables the empirical distribution function (EDF) is such that for every* $t \in \mathcal{R}^1$ *and every* $\mathbf{X}^* \in \mathcal{X}_{/\mathbf{X}}$,

$$\hat{F}_{\mathbf{X}^*}(t) = \sum_{i \leq n} \mathbb{I}(X_i^* \leq t)/n = \sum_{i \leq n} \mathbb{I}(X_i \leq t)/n = \hat{F}_{\mathbf{X}}(t),$$

which also is a permutation invariant function given $\mathbf{X}$.

Accordingly, we may state the following proposition.

Proposition 2. *For any given data set* $\mathbf{X} \in \mathcal{X}^n$, *the EPM* $\hat{P}_{\mathbf{X}}$ *and, when defined, the EDF* $\hat{F}_{\mathbf{X}}$ *are permutation invariant functions which characterize* $\mathcal{X}_{/\mathbf{X}}$, *and so* $\mathcal{X}_{/\mathbf{X}}$ *can equivalently be defined as the set of points of* $\mathcal{X}^n$ *sharing the same EPM* $\hat{P}_{\mathbf{X}}$, *or the same EDF* $\hat{F}_{\mathbf{X}}$. *Accordingly we may also write* $\hat{P}_{\mathbf{X}}(A) = \hat{P}(A|\mathcal{X}_{/\mathbf{X}})$, *etc.*

One consequence of Proposition 2 is that the EPM $\hat{P}_{\mathbf{X}}$ (or the EDF $\hat{F}_{\mathbf{X}}$, when defined) is a sufficient statistical function for P in H_0. Hence, conditioning on $\mathcal{X}_{/\mathbf{X}}$ is equivalent to conditioning on $\hat{P}_{\mathbf{X}}$, or on $\hat{F}_{\mathbf{X}}$. One more consequence is that, for any statistic $T : \mathcal{X}^n \to \mathcal{R}^1$ and all $t \in \mathcal{R}^1$, $\Pr\{T(\mathbf{X}^*) \leq t|\mathcal{X}_{/\mathbf{X}}\} = \Pr\{T(\mathbf{X}^*) \leq t|\hat{P}_{\mathbf{X}}\} = \Pr\{T(\mathbf{X}^*) \leq t|\hat{P}_{\mathbf{X}^\dagger}\}$, $\mathbf{X}^\dagger \in \mathcal{X}_{/\mathbf{X}}$. From now on, we prefer to use symbols such as $\Pr\{T(\mathbf{X}^*) \leq t|\mathcal{X}_{/\mathbf{X}}\}$ to emphasize the conditional status of permutation distributions. A further consequence is that the permutation null distribution of T can be interpreted as if it were generated by a *without replacement random experiment* (WORE) with constant probability on a finite population whose distribution is $\hat{P}_{\mathbf{X}}$, or $\hat{F}_{\mathbf{X}}$. Indeed, any random permutation $\mathbf{X}^*$ can be seen as a simple random sample from $\mathbf{X}$, which in turn plays the role of a finite population with equally likely elements (see Section 1.9.3). We recall that with replacement random experiments (WREs) by i.i.d. resamplings from $\hat{P}_{\mathbf{X}}$, or $\hat{F}_{\mathbf{X}}$, give rise to bootstrap methods. The latter, however, are not proper conditional procedures and so, at least for finite sample sizes, their inferential interpretations may not be entirely clear (see Remark 4, 3.8.2).

Remark 1. Quite a general result, in very mild conditions, is obtained for large sample sizes (Hoeffding, 1952; Romano, 1989, 1990; Janssen and Pauls, 2003). This essentially states that for sufficiently large n_1 and n_2, the null permutation distribution $\Pr\{T(\mathbf{X}^{(n)}) \leq t|\hat{P}_{\mathbf{X}^{(n)}}\}$ of any regular (continuous non-degenerate) statistic T approximates its population (or unconditional) counterpart $\Pr\{T(\mathbf{X}^{(n)}) \leq t;\, P\}$. Indeed, based on the Glivenko–Cantelli theorem, which states that the EPM $\hat{P}_{\mathbf{X}^{(n)}}$ strongly converges to P, and on the theorem for continuous functions of empirical processes (Shorack and Wellner, 1986), as sample sizes diverge it can be proved that $\Pr\{T(\mathbf{X}^{(n)}) \leq t|\hat{P}_{\mathbf{X}^{(n)}}\}$ converges in probability to $\Pr\{T(\mathbf{X}^{(n)}) \leq t;\, P\}$ for all real t.

Remark 2. In stratified problems (see Remark 4, 2.1.2), since the EPM (or the EDF) is a mixture of partial EPMs, it is difficult to express it in simple way. Henceforth we use only notational conventions such as $\{\cdot|\mathcal{X}_{/\mathbf{X}}\}$ to denote relations conditional on sets of sufficient statistics. Also note that this convention is somewhat more precise than $\{\cdot|\mathbf{X}\}$ used in Chapter 1.

Remark 3. Note that the null permutation distribution $F_T^*(t|\mathcal{X}_{/\mathbf{X}})$, induced by the test statistic T given the data set $\mathbf{X}$, essentially depends only on exchangeable errors $\mathbf{Z}$ which define the observed responses $\mathbf{X}$ because all nuisance quantities, such as μ and σ, are irrelevant (see Problem 1, 3.6.1). Moreover, if sample size goes to infinity and the CDF $F_T^*(t|\mathcal{X}_{/\mathbf{X}})$ becomes continuous, the integral transformation $U^* = F_T^*(T^*|\mathcal{X}_{/\mathbf{X}})$ becomes uniformly distributed on the unit interval.

2.2 Definition of Permutation Tests

2.2.1 General Aspects

A test statistic is a non-degenerate measurable function T, mapping $\mathcal{X}^n$ into $\mathcal{R}^1$, which satisfies properties suitable for inference. Suppose then that $T : \mathcal{X}^n \to \mathcal{R}^1$ is such an appropriate test statistic (see Section 2.5) for which, without loss of generality, we assume that large values are evidence against H_0. We define the *permutation support* induced by the pair $(T, \mathbf{X})$ as the set $\mathcal{T}_{\mathbf{X}} = \{T^* = T(\mathbf{X}^*) : \mathbf{X}^* \in \mathcal{X}_{/\mathbf{X}}\}$ containing all possible values assumed by T as $\mathbf{X}^*$ varies in $\mathcal{X}_{/\mathbf{X}}$. Of course, when more than one aspect is of interest for the analysis, a test can be associated with a vector of statistics, $\mathbf{T} = (T_1, \ldots, T_K) : \mathcal{X}^n \to \mathcal{R}^K$, where $1 \leq K$ is the finite (or at most countable, as in Section 4.5) number of aspects under consideration. The theory of multidimensional permutation tests, including the NPC of several dependent tests, is developed in Chapter 4. Several applications to complex problems are discussed in subsequent chapters.

Let us suppose that H_0 is true so that, according to Proposition 1, 2.1.3, $\mathbf{X}^*$ is uniformly distributed over $\mathcal{X}_{/\mathbf{X}}$, and let us put the $M^{(n)}$ members of $\mathcal{T}_{\mathbf{X}}$ in non-decreasing order, $T^*_{(1)} \leq T^*_{(2)} \leq \ldots \leq T^*_{(M^{(n)})}$. In this way for each value of $\alpha \in (0, 1)$, $T_\alpha(\mathbf{X}) = T_\alpha = T^*_{(M^{(n)}_\alpha)}$ defines the permutation critical value associated with the pair $(T, \mathbf{X})$, where $M_\alpha^{(n)} = \sum_{\mathcal{X}_{/\mathbf{X}}} \mathbb{I}[T(\mathbf{X}^*) < T_\alpha]$ is the number of permutation values T^* that are strictly less than T_α. Note that when X is a continuous variable and T is a regular function, the elements of $\mathcal{T}_\mathbf{X}$ are distinct with probability one because multiple points have a probability of zero. Therefore, they can be set in increasing order: $T^*_{(1)} < T^*_{(2)} < \ldots < T^*_{(M^{(n)})}$. Also note that permutation critical values T_α depend on $\mathcal{X}_{/\mathbf{X}}$ and not merely on $\mathbf{X}$. Indeed, the relation $T_\alpha = T_\alpha(\mathbf{X}) = T_\alpha(\mathbf{X}^\dagger)$ is satisfied by all $\mathbf{X}^\dagger \in \mathcal{X}_{/\mathbf{X}}$ because two associated orbits $\mathcal{X}_{/\mathbf{X}}$ and $\mathcal{X}_{/\mathbf{X}^\dagger}$ coincide, that is, $\mathcal{X}_{/\mathbf{X}} = \mathcal{X}_{/\mathbf{X}^\dagger}$. This implies that, for each fixed $\alpha \in (0, 1)$, T_α is a fixed value in the permutation support $\mathcal{T}_\mathbf{X}$ which in turn may vary as $\mathbf{X}$ varies in $\mathcal{X}^n$. The latter gives rise to obvious difficulties in expressing the critical value T_α and the conditional and unconditional power functions of T in closed forms.

2.2.2 *Randomized Permutation Tests*

The randomized version of the permutation test ϕ_R associated with the pair $(T, \mathbf{X})$ is defined as

$$\phi_R = \begin{cases} 1 & \text{if} \quad T^o > T_\alpha, \\ \gamma & \text{if} \quad T^o = T_\alpha, \\ 0 & \text{if} \quad T^o < T_\alpha, \end{cases}$$

where $T^o = T(\mathbf{X})$ is the value of T calculated on observed data $\mathbf{X}$ and the probability γ, when $T^o = T_\alpha$, is given by

$$\gamma = \left[\alpha - \Pr\left\{T^o > T_\alpha | \mathcal{X}_{/\mathbf{X}}\right\}\right] / \Pr\left\{T^o = T_\alpha | \mathcal{X}_{/\mathbf{X}}\right\}.$$

It is worth observing that to apply ϕ_R it is usual to make use of a result of an auxiliary random experiment independent of the data set $\mathbf{X}$. For instance, this can be realized by rejecting H_0 if $U \geq \gamma$, for $T^o = T_\alpha$, where U is a random value from the uniform variable $\mathcal{U}(0, 1)$.

It can be immediately proven that, for all data sets $\mathbf{X} \in \mathcal{X}^n$, the conditional expectation in H_0 of ϕ_R, for any $\alpha \in (0, 1)$, satisfies

$$\mathbb{E}\{\phi_R(\mathbf{X}) | \mathcal{X}_{/\mathbf{X}}\} = \Pr\left\{T^o(\mathbf{X}) > T_\alpha(\mathbf{X}) | \mathcal{X}_{/\mathbf{X}}\right\} + \gamma \cdot \Pr\left\{T^o(\mathbf{X}) = T_\alpha(\mathbf{X}) | \mathcal{X}_{/\mathbf{X}}\right\} = \alpha,$$

where the role of the data set $\mathbf{X}$ is well emphasized.

Consequently, randomized permutation tests are of exact size α. Moreover, they are provided with the so-called *uniform* (or strong) *similarity* property, as stated in the following proposition.

Proposition 3. (Uniform similarity of randomized permutation tests). *Assume that the exchangeability condition on data* $\mathbf{X}$ *is satisfied. Then for all underlying distributions* P *and uniformly for all data sets* $\mathbf{X} \in \mathcal{X}^n$, *the conditional rejection probability of* ϕ_R *is* $\mathbf{X}$-P*-invariant in* H_0. (Scheffé, 1943b; Lehmann and Scheffé, 1950, 1955; Watson, 1957; Lehmann, 1986; Lehmann and Romano, 2005; Pesarin, 2001).

Note that the uniform similarity is formally valid also for degenerate distributions, when all values in any data set $\mathbf{X}$ are coincident with probability one. In such a case, the permutation support $\mathcal{T}(\mathbf{X})$ associated with $(T, \mathbf{X})$, containing only one point, is also degenerate, and so the uniform similarity property is essentially ineffective because tests ϕ_R become purely randomized, since the rejection of H_0 only depends on the result of an auxiliary random experiment. For this reason we generally do not consider degenerate distributions, which in any case are of no practical interest.

2.2.3 *Non-randomized Permutation Tests*

In application contexts the so-called simple or non-randomized version is generally preferred. This is defined as

$$\phi = \begin{cases} 1 & \text{if} \quad T^o \geq T_\alpha, \\ 0 & \text{if} \quad T^o < T_\alpha, \end{cases}$$

in which the associated type I error rate, in H_0, is $\mathbb{E}\{\phi(\mathbf{X})|\mathcal{X}_{/\mathbf{X}}\} = \Pr\left\{T^o \geq T_\alpha|\mathcal{X}_{/\mathbf{X}}\right\} = \sum_{\mathcal{X}_{/\mathbf{X}}} \mathbb{I}[T(\mathbf{X}^*) \geq T_\alpha]/M^{(n)} = \alpha_a \geq \alpha$, where α_a are the so-called *attainable α-values* associated with $(T, \mathbf{X})$.

For any given pair $(T, \mathbf{X})$, the associated attainable α-values belong to the set $\Lambda_{\mathbf{X}}^{(n)} = \{L_{\mathbf{X}}(t) : dL_{\mathbf{X}}(t) > 0\}$ of step points of the significance level function $L_{\mathbf{X}}(t) = \Pr[T^* \geq t|\mathcal{X}_{/\mathbf{X}}]$. $\Lambda_{\mathbf{X}}^{(n)}$ is always a discrete set, the elements of which depend on the pair $(T, \mathbf{X})$, and so on n. Therefore, for non-randomized permutation tests not all values of type I error rates are possible in practice. It is worth noting that, especially when the response variable X is discrete or when there are ties in $\mathbf{X}$, the attainable α-values depend on the observed data set $\mathbf{X}$, hence the set $\Lambda_{\mathbf{X}}^{(n)}$ is not $\mathbf{X}$-invariant. Because of this, sometimes the power function of non-randomized tests may apparently decrease for increasing sample sizes (see Remark 1 below). However, when there are no multiple points in $\mathcal{T}_{\mathbf{X}}$, that is, for continuous variables X and regular test statistics T, or when these points have constant multiplicity, that is, $\exists c : \forall t \in \mathcal{T}_{\mathbf{X}}, \sum_{\mathcal{X}_{/\mathbf{X}}} \mathbb{I}(T^* = t) = c \geq 1$, $\Lambda_{\mathbf{X}}^{(n)}$ becomes $\mathbf{X}$-invariant, although dependent on n. In this case $\Lambda_{\mathbf{X}}^{(n)} = \Lambda^{(n)} = \{mc/M^{(n)}, m = 1, \ldots, M^{(n)}/c\}$, and so α_a-values have constant jumps of $c/M^{(n)}$. For instance, in the two-sample problem with quantitative variables, associated with a statistic such as the comparison of sample means $\bar{X}_1^* - \bar{X}_2^*$, we get $c = n_1!n_2!$. Naturally, when n is not too small and if there are constant multiple points in $\mathcal{T}_{\mathbf{X}}$, in practice there is no substantial difference between randomized ϕ_R and non-randomized ϕ. For instance, in a two-sample problem with $(n_1, n_2) = (10, 10)$ the minimum attainable α-value is $\min \alpha_a = 1/184\,756$, with $(n_1, n_2) = (20, 20)$ it is $\min \alpha_a = 1.179714 \cdot 10^{-12}$, and with $(n_1, n_2) = (50, 50)$ it is $\min \alpha_a = 9.911\,630 \cdot 10^{-30}$.

It is worth noting that, if we wish to test at a desired type I error rate of α_d and choose $\alpha_d \geq \alpha_a \in \Lambda_{\mathbf{X}}^{(n)}$, then non-randomized permutation tests become conservative. Of course, if the desired α-value $\alpha_d \in \Lambda_{\mathbf{X}}^{(n)}$, then $\alpha_d = \alpha_a$. Henceforth we refer to the non-randomized version ϕ and the attainable type I error rates are indicated with the usual symbol α.

Remark 1. To show that power at nominal $\alpha = 0.05$ may apparently decrease by increasing sample sizes, let us consider two situations related to two-sample designs for one-sided alternatives where sample sizes are: (a) 3 and 3; (b) 4 and 3. In situation (a) the attainable α is exactly 0.05, whereas in situation (b) the closest value not exceeding 0.05 is 0.02857 (the next is 0.05714). So if the effect δ is not large it is naturally possible for the apparent power in the first situation to be larger than that in the second. It is not so, for instance, with sample sizes (a′) 8 and 8 and (b′) 9 and 8. In (a′) the closest attainable α is 0.04996, whereas in (b′) it is 0.04998. In such a case the power in (b′) may increase with respect to (a′) because sample size and attainable α both increase (for a discussion on power behaviour, see Section 3.2).

2.2.4 *The p-Value*

It is known that determining the critical values T_α of a test statistic T, given the observed data set $\mathbf{X}$, in practice presents obvious difficulties. Therefore, it is common to make reference to the so-called p-value associated with $(T, \mathbf{X})$. This is defined as $\lambda = \lambda_T(\mathbf{X}) = L_{\mathbf{X}}(T^o) = \Pr\{T^* \geq T^o|\mathcal{X}_{/\mathbf{X}}\}$, which can be determined exactly by complete enumeration of $\mathcal{T}_{\mathbf{X}}$ or estimated, to the desired degree of accuracy, by a conditional Monte Carlo algorithm based on a random sampling from $\mathcal{X}_{/\mathbf{X}}$ (see the algorithm in Section 2.2.5). For quite simple problems it can be evaluated exactly by efficient computing routines such as those described in Pagano and Tritchler (1983) and Mehta and Patel (1980, 1983); moreover, according to Berry and Mielke (1985) and Mielke and Berry (2007) it can be evaluated approximately by using a suitable approximating distribution, for example within Pearson's system of distributions, sharing the same few moments of the exact permutation distribution, when these are known in closed form in terms of actual data $\mathbf{X}$.

Note that the p-value λ is a non-increasing function of T^o and that it has a one-to-one relationship with the attainable α-value of ϕ, in the sense that $\lambda_T(\mathbf{X}) > \alpha$ implies $T^o < T_\alpha$, and vice versa. Hence, the non-randomized version can be equivalently stated as

$$\phi = \begin{cases} 1 & \text{if} \quad \lambda_T(\mathbf{X}) \leq \alpha, \\ 0 & \text{if} \quad \lambda_T(\mathbf{X}) > \alpha. \end{cases}$$

It is to be emphasized that attainable α-values play the role of critical values, in the sense that *α is the exact critical value for $\lambda_T(\mathbf{X})$*. In this sense, the p-value $\lambda_T(\mathbf{X})$ itself can be used as a test statistic. Moreover, in H_0 we have that $\mathbb{E}\{\phi(\mathbf{X})|\mathcal{X}_{/\mathbf{X}}\} = \Pr\{\lambda_T(\mathbf{X}) \leq \alpha|\mathcal{X}_{/\mathbf{X}}\} = \alpha$ for every $\alpha \in \Lambda_{\mathbf{X}}^{(n)}$. It is worth noting that for the practical determination of λ we need not know whether H_0 is true or not when working with the data set $\mathbf{X}$. This because, by imposed exchangeability, permutations $\mathbf{X}^*$ are constructed so as to be equally likely on $\mathcal{X}_{/\mathbf{X}}$, and so the alternative may be considered to be active only for the observed value T^o (see the CMC algorithm).

If X is continuous and non-degenerate and T is a regular function so that attainable α-values belong to $\Lambda^{(n)}$ for almost all data sets $\mathbf{X} \in \mathcal{X}^n$, then non-randomized permutation tests ϕ are provided with the similarity property in the *almost sure* form. This is essentially due to the fact that for continuous variables the probability of finding ties in the data sets is zero and so with probability one $\Lambda_{\mathbf{X}}^{(n)} = \Lambda^{(n)}$. This property is stated in the following proposition.

Proposition 4. (Almost sure similarity of ϕ in the continuous case). *If X is a continuous variable and T is a regular continuous non-degenerate function, then the attainable α-values of ϕ are $\mathbf{X}$-P-invariant for almost all $\mathbf{X} \in \mathcal{X}^n$ and with probability one with respect to the underlying population distribution P.*

It is worth noting that, since for discrete or mixed variables, in which ties have positive probability, the attainable α-values belong to $\Lambda_{\mathbf{X}}^{(n)}$, which depends on the observed data $\mathbf{X}$, non-randomized permutation tests lose the similarity property for finite sample sizes. However, this property is always asymptotically satisfied, so we can say that it is satisfied at least approximately.

Proposition 5. (Uniform null distribution of p-values). *Based on Proposition* 1, 2.1.3, *if X is a continuous variable and T is a regular continuous non-degenerate function, then p-values $\lambda_T(\mathbf{X})$ are uniformly distributed in its support $\Lambda(\mathbf{X})$.*

Remark 1. In accordance with Propositions 5 above and 1, 2.1.3, in the null hypothesis the elements of the permutation support $\mathcal{T}_{\mathbf{X}}$ are equally likely provided that they are distinct. This essentially represents a characterization of exactness of a permutation test, that is, *a test T is exact if the null distribution of $\lambda_T(\mathbf{X})$ over $\Lambda(\mathbf{X})$ is uniform.*

2.2.5 A CMC Algorithm for Estimating the p-Value

A CMC algorithm for evaluating the p-value λ of a test statistic T on a data set $\mathbf{X}(\Delta) = \{X(i; \Delta), i = 1, \ldots, n; n_1, n_2\}$ includes the following steps:

Table 2.1 The CMC algorithm for the p-value λ

$\mathbf{X}$	$\mathbf{X}_1^*$	$\cdots$	$\mathbf{X}_b^*$	$\cdots$	$\mathbf{X}_B^*$
T^o	T_1^*	$\cdots$	T_b^*	$\cdots$	T_B^*

$\rightarrow \hat{\lambda} = \sum_{b=1}^{B} \mathbb{I}(T_b^* \geq T^o) / B.$

1. Calculate, on the given data set $\mathbf{X}(\Delta)$, the observed value T^o of the test statistic T, i.e. $T^o(\Delta) = T[\mathbf{X}(\Delta)]$.
2. Take a random permutation $\mathbf{X}^*(\Delta)$ of $\mathbf{X}(\Delta)$. This is obtained by considering a random permutation $\mathbf{u}^* = (u_1^*, \ldots, u_n^*)$ of unit labels $\mathbf{u} = (1, \ldots, n)$ and so $\mathbf{X}^*(\Delta) = \{X(u_i^*; \Delta), i = 1, \ldots, n; n_1, n_2\}$.
3. Calculate $T^*(\Delta) = T(\mathbf{X}^*(\Delta))$.
4. Independently repeat steps 2 and 3 B times.
5. The set $\{\mathbf{X}_b^*(\Delta), b = 1, \ldots, B\}$ of B permutations is a random sample from the permutation sample space $\mathcal{X}_{/\mathbf{X}}$, and so the corresponding values $\{T_b^*(\Delta), b = 1 \ldots, B\}$ simulate the null permutation distribution of T. Therefore, the p-value is estimated as $\hat{\lambda}(\mathbf{X}(\Delta)) = \sum_{1 \leq b \leq B} \mathbb{I}[T_b^*(\Delta) \geq T^o(\Delta)]/B$, that is, the proportion of permutation values not smaller than the observed one.

The CMC algorithm is summarized in Table 2.1.

It is worth noting that if algorithms for obtaining all possible permutations were available, as for instance with the StatXact® software, instead of steps 2 and 3 it is straightforward to enumerate the whole permutation sample space and the corresponding true distribution. In such a case we would have $\lambda(\mathbf{X}(\Delta)) = \sum_{\mathcal{X}_{/\mathbf{X}(\Delta)}} \mathbb{I}[T^*(\Delta) \geq T^o(\Delta)]/M^{(n)}$. We also observe that the CMC values $\{T_b^*(\Delta), b = 1 \ldots, B\}$ allow us to calculate, $\forall t \in \mathcal{R}^1$, the permutation EDF of T as $\hat{F}_B^*(t) = \sum_{1 \leq b \leq B} \mathbb{I}[T_b^*(\Delta) \leq t]/B$ and the ESF as $\hat{L}_B^*(t) = \sum_{1 \leq b \leq B} \mathbb{I}[T_b^*(\Delta) \geq t]/B$. Therefore, $\hat{\lambda} = \hat{L}_B^*[T^o(\Delta)]$. Of course, if B diverges to infinity, due to the well-known Glivenko–Cantelli theorem, the p-value $\hat{\lambda}$, the EDF $\hat{F}_B^*$, and the ESF $\hat{L}_B^*$ strongly converge to their respective true values.

To estimate the p-value λ some authors (e.g. Edgington and Onghena, 2007) use expressions such as $\sum_{0 \leq b \leq B} \mathbb{I}[T_b^* \geq T^o]/(B+1)$, where they set the observed value $T^o = T_0^*$ as if it were obtained by one of the random permutations. We think this expression is incorrect because it provides biased estimates in both H_0 and H_1, although the bias vanishes as B diverges. Indeed, since for each b the conditional mean value of $\mathbb{I}(T_b^* \geq T^o)$ in H_0, due to Proposition 5, 2.2.4, is $\mathbb{E}_{\mathcal{X}_{/\mathbf{X}}}[\mathbb{I}(T_b^* \geq T^o)] = \lambda$, we generally have that $\mathbb{E}_{\mathcal{X}_{/\mathbf{X}}}\left\{\sum_{0 \leq b \leq B} \mathbb{I}[T_b^* \geq T^o]/(B+1)\right\} = (1 + B\lambda)/(B+1)$, which for $B = 1$ gives $\mathbb{E}_{\mathcal{X}_{/\mathbf{X}}}\left\{\mathbb{I}[T_0^* \geq T^o]/2 + \mathbb{I}[T_1^* \geq T^o]/2\right\} = (1+\lambda)/2$. These expressions show biasedness. Furthermore, it is to be emphasized that in H_1 the distribution of the observed value T^o depends on the effect Δ, whereas, due to the imposed exchangeability leading to equally likely permutations $\mathbf{X}^*$, all values T_b^*, $b = 1, \ldots, B$, are calculated as if the null hypothesis were true. Thus, since T^o has the same distribution of T_b^* only when H_0 is true, that expression has no statistical meaning.

2.3 Some Useful Test Statistics

There are a number of test statistics for two-sample designs used to deal with practical problems. With obvious meaning of the symbols, a brief, incomplete list of those most commonly used is as follows:

1. $\bar{X}_1^* - \bar{X}_2^* = \sum_i X_{1i}^*/n_1 - \sum_i X_{2i}^*/n_2$, for comparison of sample means.
2. $\bar{G}_1^* - \bar{G}_2^* = \exp\left\{\sum_i \log(X_{1i}^*)/n_1 - \sum_i \log(X_{2i}^*)/n_2\right\}$, for comparison of sample geometric means, provided that data are positive: $X \overset{p}{>} 0$.

3. $\bar{A}_1^* - \bar{A}_2^* = \left(\sum_i n_1/X_{1i}^*\right)^{-1} - \left(\sum_i n_2/X_{2i}^*\right)^{-1}$, for comparison of sample harmonic means, provided that data are positive: $X \overset{p}{>} 0$.
4. $\mu_{k1}^* - \mu_{k2}^* = \left\{\left[\sum_i (X_{1i}^*)^k/n_1\right]^{1/k} - \left[\sum_i (X_{2i}^*)^k\right]^{1/k}\right\} \cdot S_g(k)$, for comparison of sample kth moments, provided that $X \overset{p}{>} 0$ and where the sign $S_g(k)$ is $+1$ or -1 according to whether $k > 0$ or $k < 0$. It is worth noting that (4) includes: (1) for $k = 1$, (2) for $\lim k \to 0$, and a function of (3) for $k = -1$. Moreover, if $\Pr\{X \leq 0\} > 0$, i.e. when data may assume negative values with positive probability, then we need $k > 0$, i.e. only positive moments can be considered.
5. $\max[0, \bar{X}_1^* - \bar{X}_2^*]$ or $\max[0, \bar{X}_2^* - \bar{X}_1^*]$, the first suitable for alternatives $X_1 \overset{d}{>} X_2$, and the second for $X_1 \overset{d}{<} X_2$, to take into account only permutations coherent with the active alternatives (this is used in Example 5, 4.6, for dealing with a complex problem where it is supposed that a treatment may have a positive effect with some subjects and negative with some others).
6. $\bar{R}_1^* - \bar{R}_2^* = \sum_i R_{1i}^*/n_1 - \sum_i R_{2i}^*/n_2$, for comparison of sample means of ordinary ranks, leading to a version of the Wilcoxon–Mann–Whitney test, where $\mathbb{R}(X_{ji}^*) = R_{ji}^* = \sum_{r=1}^2 \sum_{s \leq n_r} \mathbb{I}(X_{rs}^* \leq X_{ji}^*)$.
7. $\bar{\psi}_1^* - \bar{\psi}_2^* = \sum_i \psi_{1i}^*/n_1 - \sum_i \psi_{2i}^*/n_2$ for comparison of sample means of generalized ranks: $\psi_{ji}^* = \Psi[R_{ji}^*/(n+1)]$, where Ψ is the inverse CDF of any suitable distribution, such as the standard normal; $\psi_{ji}^* = \Phi^{-1}[R_{ji}^*/(n+1)]$, leading to the van der Waerden test; or the exponential, $\psi_{ji}^* = -\log[R_{ji}^*/(n+1)]$, leading to a form of log-rank test; or the logistic, $\psi_{ji}^* = \log[R_{ji}^*/(n+1-R_{ji}^*)]$; or the chi-squared with given degrees of freedom, etc. Ψ can also be the mean value of the ith out of n order statistics of any suitable variable W: $\psi_{ji}^* = \psi(X_{j(i)}^*) = \mathbb{I}(X_{(i,n)} \in \mathbf{X}_j^*)\mathbb{E}(W_{(i,n)})$, $j = 1, 2$, where $W_{(i,n)}$ is the ith order statistic among n random data from variable W, i.e. $W_{(1,n)} \leq W_{(2,n)} \leq \ldots \leq W_{(n,n)}$; when W is standard normal, this leads to a version of the Fisher–Yates test.
8. When variable X is numeric or ordered categorical and $\varphi(X)$ is any non-decreasing score function, $\bar{\varphi}_1^* - \bar{\varphi}_2^* = \sum_i \varphi(X_{1i}^*)/n_1 - \sum_i \varphi(X_{2i}^*)/n_2$ is the comparison of sample means of φ-scores.
9. $\hat{p}_1^* - \hat{p}_2^* = \sum_i \mathbb{I}(X_{1i}^* \leq \ddot{x})/n_1 - \sum_i \mathbb{I}(X_{2i}^* \leq \ddot{x})/n_2$, for comparison of sample percentages of data not larger than a preset value $\ddot{x}$. When $\ddot{x}$ is the pooled median: $\widetilde{X} = \mathbb{M}d(\mathbf{X}) = X_{((n+1)/2)}$ if n is odd and $\widetilde{X} = (X_{(n/2)} + X_{(1+n/2)})/2$ if n is even, where $X_{(1)} \leq X_{(2)} \leq \ldots \leq X_{(n)}$ are the observed order statistics. This corresponds to a version of the Brown–Mood median test.
10. If, instead of only one, $K \geq 2$ preset ordered values ($\ddot{x}_1 < \ldots < \ddot{x}_K$) are considered, then $T_P^* = \sum_{k=1}^K (\hat{p}_{1k}^* - \hat{p}_{2k}^*)$ or $T_{PS}^* = \sum_{k=1}^K (\hat{p}_{1k}^* - \hat{p}_{2k}^*) / [\hat{p}_k(1-\hat{p}_k)]^{1/2}$, where $\hat{p}_{jk}^* = \sum_i \mathbb{I}(X_{ji}^* \leq \ddot{x}_k)/n_j$ and $\hat{p}_k = \sum_{ji} \mathbb{I}(X_{ji} \leq \ddot{x}_k)/n$ are the proportion of values not larger than $\ddot{x}_k$ in the jth sample, $j = 1, 2$, and in the pooled sample, respectively. T_P^* corresponds to divergences of cumulative frequencies on the set of preset points and involves taking account of data in fixed cumulative intervals (similar to the Cramér–von Mises test), whereas T_{PS}^*, provided that $\hat{p}_1 > 0$ and $\hat{p}_K < 1$, corresponds to divergence of standardized frequencies (similar to the Anderson–Darling test). T_P^* and T_{PS}^* are extensions of the Brown–Mood test; moreover, they may be seen as a way, alternative to (8), to analyse ordered categorical variables.
11. $T_{CM}^* = \sum_i [\hat{F}_2^*(X_i) - \hat{F}_1^*(X_i)]$, for comparison of sample EDFs: $\hat{F}_j^*(t) = \sum_i \mathbb{I}(X_{ji}^* \leq t)/n_j$, $t \in \mathcal{R}^1$, $j = 1, 2$, leading to the Cramér–von Mises divergence test. This may be seen as a generalization of T_P^* in (9) in which $\ddot{x}_k = X_{(k)}$ and, if there are no ties, $K = n$.
12. $T_{AD}^* = \sum_i [\hat{F}_2^*(X_i) - \hat{F}_1^*(X_i)] / \{\bar{F}(X_i)[1 - \bar{F}(X_i)]\}^{1/2}$, where $\bar{F}(t) = \sum_{ji} \mathbb{I}(X_{ji} \leq t)/n$, $t \in \mathcal{R}^1$, and 0 is assigned to summands of the form 0/0, which leads to the Anderson–Darling EDF divergence test. This may be seen as a generalization of T_{PS}^* in (10).
13. $T_{KS}^* = \max_i \left\{\hat{F}_2^*(X_i) - \hat{F}_1^*(X_i)\right\}$, leading to a version of the Kolmogorov–Smirnov EDF divergence test.

14. $T^*_{KD} = \max_i \left\{[\hat{F}^*_2(X_i) - \hat{F}^*_1(X_i)] / \{\bar{F}(X_i)[1 - \bar{F}(X_i)]\}^{1/2}\right\}$, leading to a version of the Kolmogorov–Smirnov–Anderson–Darling EDF divergence test.
15. $T^*_\Psi = \sum_i\{\Psi[\hat{F}^*_2(X_i)] - \Psi[\hat{F}^*_1(X_i)]\}$ for a generalized divergence of EDFs, where Ψ is the inverse CDF of any suitable and completely specified distribution. For instance, when $\Psi^*_{ji} = \Phi^{-1}[(n_j\hat{F}^*_j(X_i) + 0.5)/(n_j + 1)]$, $j = 1, 2$, i.e. Ψ is the inverse CDF of a standard normal distribution, this gives a version of the Liptak divergence test (similar to the van der Waerden test); when $\Psi^*_{ji} = -\log[(n_j(1 - \hat{F}^*_j(X_i)) + 0.5)/(n_j + 1)]$, it gives a version of Fisher divergence test (similar to the log-rank test); when $\Psi^*_{ji} = n\,\bar{F}(X^*_{ji})$ it is similar to the Wilcoxon–Mann–Whitney test; and so on.
16. As a special case of (15), one can also use statistics based on divergence of cumulative EDF $T^*_\Delta = \sum_{k\leq n}\{\sum_{i\leq k}[\hat{F}^*_2(X_i) - \hat{F}^*_1(X_i)] \cdot \varphi_i\}$, where φ_i are permutationally invariant weights.
17. Divergence of sample medians $T^*_{Md} = \tilde{X}^*_1 - \tilde{X}^*_2$, where $\tilde{X}^*_j = \mathbb{M}d[X^*_{ji}, i = 1, \ldots n_j]$, $j = 1, 2$.
18. Divergence of preset sample π-quantiles $T^*_\pi = \tilde{X}^*_{\pi 1} - \tilde{X}^*_{\pi 2}$, where $\tilde{X}^*_{\pi j} = [\inf\{t : \hat{F}^*_j(t) \geq \pi\} + \sup\{t : \hat{F}^*_j(t) \leq \pi\}]/2$ with $0 < \pi < 1$, in which the EDFs are $\hat{F}^*_j(t) = \sum_i \mathbb{I}[X^*_{ji} \leq t]/n_j$, $t \in \mathcal{R}^1$, $j = 1, 2$ (of course, for $\pi = 1/2$ we get the median, $\tilde{X}^*_{\pi j} = \tilde{X}^*_j$, say). Also of interest is a sort of multi-quantile test: supposing $0 < \pi_1 < \pi_2 < \ldots < \pi_k < 1$ are $k > 1$ suitable preset quantities, then $\{T^*_{\pi_h} = \check{Y}^*_{\pi_h 1} - \check{Y}^*_{\pi_h 2}, h = 1, \ldots, k\}$ is a k-dimensional test statistic (see Chapter 4 for multivariate testing).
19. Divergence of sample trimmed means of order r, $T^*_r = \tilde{X}^*_{r1} - \tilde{X}^*_{r2}$, where $0 \leq r < \min[n_1, n_2]/2$, and $\tilde{X}^*_{rj} = \sum_{i=r+1}^{n_j-r} X^*_{j(i)}/(n_j - 2r)$, $X^*_{j(i)}$ being the ith ordered value in the jth group, $j = 1, 2$.

Remark 1. Sometimes instead of the EDF $\hat{F}(t)$ we may use the so-called *normalized EDF* introduced by Ruymgaart (1980) and defined as $\ddot{F}(t) = \frac{1}{n}[\sum_i \mathbb{I}(X_i < t) + \frac{1}{2}\mathbb{I}(X_i = t)]$. The reason for this is that its behaviour is slightly more regular than that of $\hat{F}$, especially for discrete variables. Its transformations $n\ddot{F}(X_{ji})$ give rise to the notion of midranks (see Remark 2, 2.8.2).

Remark 2. It is worth noting that all statistics (1)–(19) share the common form $T(\mathbf{X}) = S_1(\mathbf{X}_1) - S_2(\mathbf{X}_2)$ where S_j, $j = 1, 2$, are symmetric functions, that is, invariant with respect to data entry, and sharing the same analytic form so that are indicators of the same quantity. For this reason, the number of distinct points of related permutation support $\mathcal{T}(\mathbf{X})$ is $M^{(n)} = C_{n,n_1} = n!/(n_1!n_2!)$, as each point has multiplicity $n_1!n_2!$.

Of course, every test statistic can be expressed in one of its permutationally equivalent forms (see Section 2.4). Moreover, it is straightforward to extend most of them to two-sided alternatives, and to one-sample and multi-sample cases. Observe also that statistics corresponding to (5), (13), (14) and (17)–(19) in the list are typically non-associative statistics.

2.4 Equivalence of Permutation Statistics

This section briefly discusses the concept of permutationally equivalent statistics. This is useful in simplifying computations and sometimes in facilitating the establishment of the asymptotic equivalence of permutation solutions with respect to some of their parametric counterparts.

Let $\mathbf{X} \in \mathcal{X}$ be the given data set. We have the following definition of the permutation equivalence of two statistics:

Definition 3. *Two statistics T_1 and T_2, both mapping $\mathcal{X}$ into $\mathcal{R}^1$, are said to be permutationally equivalent when, for all points $\mathbf{X} \in \mathcal{X}$ and $\mathbf{X}^* \in \mathcal{X}_{/\mathbf{X}}$, the relationship $\{T_1(\mathbf{X}^*) \leq T_1(\mathbf{X})\}$ is true if* *and only if* $\{T_2(\mathbf{X}^*) \leq T_2(\mathbf{X})\}$ *is true, where* $\mathbf{X}^*$ *indicates any permutation of* $\mathbf{X}$ *and* $\mathcal{X}_{/\mathbf{X}}$ *indicates the associated permutation sample space. This permutation equivalence relation is indicated by* $T_1 \approx T_2$.

With reference to this definition we have the following theorem and corollaries.

Theorem 1. *If between the two statistics* T_1 *and* T_2 *there is a one-to-one increasing relationship, then they are permutationally equivalent and* $\Pr\{T_1(\mathbf{X}^*) \leq T_1(\mathbf{X})|\mathcal{X}_{/\mathbf{X}}\} = \Pr\{T_2(\mathbf{X}^*) \leq T_2(\mathbf{X})|\mathcal{X}_{/\mathbf{X}}\}$, *where these probabilities are evaluated with respect to permutation distribution induced by the sampling experiment and defined on the permutation measurable space* $(\mathcal{X}_{/\mathbf{X}}, \mathcal{A}_{/\mathbf{X}})$.

Proof. Let us consider any one-to-one increasing relationship ψ, $\psi : \mathcal{R}^1 \leftrightarrow \mathcal{R}^1$, and assume $T_2 = \psi(T_1)$. Then

$$\{T_1(\mathbf{X}^*) \leq T_1(\mathbf{X})\} \rightarrow \{\psi[T_1(\mathbf{X}^*)] \leq \psi[T_1(\mathbf{X})]\} = \{T_2(\mathbf{X}^*) \leq T_2(\mathbf{X})\}$$

and

$$\{T_2(\mathbf{X}^*) \leq T_2(\mathbf{X})\} \rightarrow \{\psi^{-1}[T_2(\mathbf{X}^*)] \leq \psi^{-1}[T_2(\mathbf{X})]\} = \{T_1(\mathbf{X}^*) \leq T_1(\mathbf{X})\};$$

thus the relationship $\Pr\{T_1(\mathbf{X}^*) \leq T_1(\mathbf{X})|\mathcal{X}_{/\mathbf{X}}\} = \Pr\{T_2(\mathbf{X}^*) \leq T_2(\mathbf{X})|\mathcal{X}_{/\mathbf{X}}\}$ is straightforward.

Remark 1. The equivalence established in Definition 1 and Theorem 1 above, being valid for all $\mathbf{X} \in \mathcal{X}$, is certain, not merely in probability. However, it is common to analyse permutation equivalence of T_1 and T_2 for a given data set $\mathbf{X}$ by only considering points of the permutation space $\mathcal{X}_{/\mathbf{X}}$. It should then be emphasized that what we are asking for with the notion of equivalent tests is that this is at least valid for almost all data sets $\mathbf{X} \in \mathcal{X}$, that is, valid except for a subset of sample points of null probability with respect to the underlying population distribution P.

Corollary 1. *If* T_1 *and* T_2 *are related by a decreasing one-to-one relationship, then they are permutationally equivalent in the sense that* $\{T_1(\mathbf{X}^*) \leq T_1(\mathbf{X})\} \leftrightarrow \{T_2(\mathbf{X}^*) \geq T_2(\mathbf{X})\}$ *for all* $\mathbf{X} \in \mathcal{X}$ *and* $\mathbf{X}^* \in \mathcal{X}_{/\mathbf{X}}$.

Corollary 2. *The permutation equivalence relation is reflexive*: $T_1 \approx T_1$.

Corollary 3. *The permutation equivalence relation is transitive: if* $T_1 \approx T_2$ *and* $T_2 \approx T_3$, *then* $T_1 \approx T_3$.

Proofs of the above corollaries are straightforward and are left as exercises (see Problem 1, 2.4.2).

Corollary 4. *If two statistics* T_1 *and* T_2 *are one-to-one related given the data set* $\mathbf{X}$ *they are permutationally equivalent*: $T_1 \approx T_2$.

The proof of this corollary is trivial. In fact, for every permutation $\mathbf{X}^* \in \mathcal{X}_{/\mathbf{X}}$ it is obvious that the relation $[(T_1(\mathbf{X}^*) \geq T_1(\mathbf{X}))|\mathcal{X}_{/\mathbf{X}}]$ is true if and only if $[(T_2(\mathbf{X}^*) \geq T_2(\mathbf{X}))|\mathcal{X}_{/\mathbf{X}}]$ is true. This corollary in practice is equivalent to Theorem 1, in the sense that if T_1 and T_2 are equivalent for any given $\mathbf{X}$, they are also unconditionally equivalent.

Remark 2. Corollary 4 is perhaps the most frequently used when proving the permutation equivalence of test statistics. A typical example is the equivalence of $T_1^* = \bar{X}_1^* - \bar{X}_2^*$ and $T_2^* = \sum_i X_{1i}^*$ in a two-sample design. Actually, as $\sum_{ji} X_{ji}^* = \sum_{ji} X_{ji} = SX$, we have $T_1^* = \sum_i X_{1i}^*/n_1 - [SX - \sum_i X_{1i}^*]/n_2 = \sum_i X_{1i}^*[\frac{n}{n_1 n_2}] - \frac{SX}{n_2} \approx \sum_i X_{1i}^*$ because $[\frac{n}{n_1 n_2}]$ and SX/n_2 are permutation constant quantities given $\mathbf{X}$. Thus T_1 and T_2 are one-to-one related, and so equivalent given $\mathbf{X}$.

2.4.1 Some Examples

Example 1. Let us first prove permutation equivalence between the Mann–Whitney rank test (Section 1.10.4) $T^*_{MW} = [\sum_i R^*_{1i} - n_1(n+1)/2]/[n_1 n_2(n+1)/12]^{1/2}$ and the permutation test $T^* = \sum_{i=1}^{n_1} \bar{F}_n(X^*_{1i})$, where $\bar{F}_n(t) = \sum_{i=1}^{n} \mathbb{I}(X_i \leq t)/n$ is the pooled EDF. Indeed: (i) T^*_{MW} is permutationally equivalent to $\sum_i R^*_{1i}$ because they are linearly related except for a positive coefficient; (ii) by definition $R^*_{1i} = \sum_{j=1}^{n} \mathbb{I}(X_j \leq X^*_{1i})$, so that $R^*_{1i} = n\bar{F}_n(X^*_{1i})$. Thus, as between T^*_{MW} and T^* there is a one-to-one increasing relationship, their permutation equivalence is proved.

Example 2. Consider the Snedecor F statistic for the one-way ANOVA (see Section 1.11.1). With clear notation, we prove that $F^* = \frac{n-C}{C-1}\sum_{j=1}^{C} n_j(\bar{X}^*_j - \bar{X}_.)^2 n_j / \sum_{ji}(X^*_{ji} - \bar{X}^*_j)^2$ is permutationally equivalent to $\sum_{j=1}^{C} n_j(\bar{X}^*_j)^2$, an expression much simpler for computation. Indeed, let us note that: (i) $\bar{X}_. = \bar{X}^*_.$ because both sum all the data; (b) $\sum_{ji}(X^*_{ji} - \bar{X}_.)^2 = \sum_{ji}(X^*_{ji} - \bar{X}^*_j)^2 + \sum_j n_j(\bar{X}^*_j - \bar{X}_.)^2 = \sum_{ji}(X_{ji} - \bar{X}_.)^2 = \sum_{ji}(X_{ji} - \bar{X}_j)^2 + \sum_j n_j(\bar{X}_j - \bar{X}_.)^2$. Therefore, the following chain of equivalence relations clearly holds:

$$\Pr\left\{\frac{\sum_j n_j(\bar{X}^*_j - \bar{X}_.)^2}{\sum_{ji}(X^*_{ji} - \bar{X}^*_j)^2} \geq \frac{\sum_j n_j(\bar{X}_j - \bar{X}_.)^2}{\sum_{ji}(X_{ji} - \bar{X}_j)^2} |\mathcal{X}_{/\mathbf{X}}\right\}$$

$$= \Pr\left\{\frac{\sum_{ji}(X^*_{ji} - \bar{X}^*_j)^2}{\sum_j n_j(\bar{X}^*_j - \bar{X}_.)^2} + 1 \leq \frac{\sum_{ji}(X_{ji} - \bar{X}_j)^2}{\sum_j n_j(\bar{X}_j - \bar{X}_.)^2} + 1 |\mathcal{X}_{/\mathbf{X}}\right\}$$

$$= \Pr\left\{\frac{\sum_{ji}(X^*_{ji} - \bar{X}_.)^2}{\sum_j n_j(\bar{X}^*_j - \bar{X}_.)^2} \leq \frac{\sum_{ji}(X_{ji} - \bar{X}_.)^2}{\sum_j n_j(\bar{X}_j - \bar{X}_.)^2} |\mathcal{X}_{/\mathbf{X}}\right\}$$

$$= \Pr\left\{\sum_j n_j(\bar{X}^*_j - \bar{X}_.)^2 \geq \sum_j n_j(\bar{X}_j - \bar{X}_.)^2 |\mathcal{X}_{/\mathbf{X}}\right\}$$

$$= \Pr\left\{\sum_j n_j(\bar{X}^*_j)^2 \geq \sum_j n_j(\bar{X}_j)^2 |\mathcal{X}_{/\mathbf{X}}\right\}.$$

Example 3. As a particular case of Snedecor's F, the permutation equivalence of Student's t statistic in a two-sample design for two-sided alternatives with $T^* = \sum_j n_j(\bar{X}^*_j)^2$ clearly holds. Instead, by defining $S^*_g = 1$ if $\bar{X}^*_1 - \bar{X}^*_2 > 0$ and $S^*_g = -1$ if $\bar{X}^*_1 - \bar{X}^*_2 < 0$, for one-sided alternatives the following chain of equivalence relations occurs:

$$\Pr\left\{\frac{\bar{X}^*_1 - \bar{X}^*_2}{\sqrt{\sum_{ji}(X^*_{ji} - \bar{X}^*_j)^2}} \geq \frac{\bar{X}_1 - \bar{X}_2}{\sqrt{\sum_{ji}(X_{ji} - \bar{X}_j)^2}} |\mathcal{X}_{/\mathbf{X}}\right\}$$

$$= \Pr\left\{\frac{S^*_g \cdot (\bar{X}^*_1 - \bar{X}^*_2)^2}{\sum_{ji}(X^*_{ji} - \bar{X}^*_j)^2} \geq \frac{S_g \cdot (\bar{X}_1 - \bar{X}_2)^2}{\sum_{ji}(X_{ji} - \bar{X}_j)^2} |\mathcal{X}_{/\mathbf{X}}\right\}$$

$$= \Pr\left\{\frac{\sum_{ji}(X^*_{ji} - \bar{X}^*_j)^2}{S^*_g \cdot (\bar{X}^*_1 - \bar{X}^*_2)^2} + 1 \leq \frac{\sum_{ji}(X_{ji} - \bar{X}_j)^2}{S_g \cdot (\bar{X}_1 - \bar{X}_2)^2} + 1 |\mathcal{X}_{/\mathbf{X}}\right\}$$

$$= \Pr\left\{\frac{\sum_{ji}(X^*_{ji} - \bar{X}_{..})^2}{S^*_g \cdot (\bar{X}^*_1 - \bar{X}^*_2)^2} \leq \frac{\sum_{ji}(X_{ji} - \bar{X}_{..})^2}{S_g \cdot (\bar{X}_1 - \bar{X}_2)^2} |\mathcal{X}_{/\mathbf{X}}\right\}$$

$$= \Pr\left\{S_g^* \cdot (\bar{X}_1^* - \bar{X}_2^*)^2 \geq S_g \cdot (\bar{X}_1 - \bar{X}_2)^2 \,|\mathcal{X}_{/\mathbf{X}}\right\}$$
$$= \Pr\left\{\bar{X}_1^* - \bar{X}_2^* \geq \bar{X}_1 - \bar{X}_2 \,|\mathcal{X}_{/\mathbf{X}}\right\}$$
$$= \Pr\left\{\bar{X}_1^* \geq \bar{X}_1 \,|\mathcal{X}_{/\mathbf{X}}\right\}.$$

Thus, the one-sided test statistic for $\{X_1 \overset{d}{>} X_2\}$ becomes $T^* = \sum_i X_{1i}^*$ and for $\{X_1 \overset{d}{<} X_2\}$ it is $T^* = \sum_i X_{2i}^*$.

Remark 1. It should be noted, however, that if instead of Student's t statistic, the standardized statistic $K^* = (\bar{X}_1^* - \bar{X}_2^*)/\sigma_E$ is used, where the permutation constant σ_E is the permutation standard error of the numerator or even its unconditional (population) counterpart, then, based on Remark 2, 2.4, $K^* \approx \bar{X}_1^*$ for $\{X_1 \overset{d}{>} X_2\}$ alternatives and $K^* \approx \bar{X}_2^*$ for $\{X_1 \overset{d}{<} X_2\}$.

2.4.2 Problems and Exercises

1) Prove Corollaries 1–3 of Section 2.4.

2) Prove that, in the paired data design, the test statistic $T^* = \sum_i X_i S_i^*$ in Section 1.9 is not permutationally equivalent to $T_Y^* = \sum_i Y_{1i} S_i^*$ (remember that $T^* = \sum_i (Y_{1i} - Y_{2i}) \cdot S_i^*$).

3) With reference to the two-sample design, show that two tests $T^* = \sum_i X_{1i}^*$ and $T_\varphi^* = \sum_i \varphi(X_{1i}^*)$ are not permutationally equivalent unless φ is a non-degenerate linear transformation.

4) Establish that when responses are binary, e.g. (0,1) or $(-1, +1)$, then the binomial or one-sided McNemar test of Section 1.8.6 is not equivalent to the permutation solution based on the test statistic $T_Y = \sum_i Y_{1i}$.

5) Show that, if the response variable is binary, then two-sided permutation tests $T_1^* = (f_{11}^* - f_{21}^*)^2$ and $T_2^* = (f_{11}^*/n_1 - f_{21}^*/n_2)^2$, respectively related to comparisons of absolute and relative frequencies, are not permutationally equivalent.

6) Prove that Mood's median test for the two-sample problem, comparing frequencies below the pooled median $\tilde{X}$ of $\mathbf{X}$, is permutationally equivalent neither to $T = \bar{X}_1$ nor to $\tilde{T} = \tilde{X}_1$.

7) Prove that, for testing $H_0 : \{\delta = 0\}$ against the set of two-sided alternatives $H_1 : \{\delta \neq 0\}$ in a two-sample problem, test statistics $|\bar{X}_1 - \bar{X}_2|$, $(\bar{X}_1 - \bar{X}_2)^2$ and $\sum_{j=1}^2 n_j \cdot \bar{X}_j^2$ are permutationally equivalent.

8) Prove that, for the two-sample problem for testing $H_0 : \{X_1 \overset{d}{=} X_2\}$ against $H_1 : \{X_1 \overset{d}{>} X_2\}$, the two tests $T_1 = \sum_i X_{1i}$ and $T_2 = -\sum_i X_{2i}$ are permutationally equivalent.

9) Prove that, in the two-sample problem for testing $H_0 : \{X_1 \overset{d}{=} X_2\}$ against $H_1 : \{X_1 \overset{d}{>} X_2\}$, the four tests $T_1 = \sum_i X_{1i}$, $T_2 = -\sum_i X_{2i}$, $T_3 = 1/\sum_i X_{2i}$ and $T_4 = \sum_i X_{1i}/\sum_i X_{2i}$ are permutationally equivalent, provided that the response variable X is positive.

10) With reference to point (iii) in Remark 2, 2.1.2, prove that tests T based on the standardized data $\hat{Y}_{ji} = (X_{ji} - \bar{X})/\hat{\sigma}$ with $\hat{\sigma}$ a function of the pooled data $\mathbf{X}$, the empirical residuals $Y_{ji} = X_{ji} - \bar{X}$, and the untransformed data X_{ji} are permutationally equivalent.

11) With reference to point (iv) in Remark 2, 2.1.2, prove that tests T based on the empirical residuals $\hat{Y}_{ji} = [Y_{ji} - \hat{\beta}(X_{ji})]$ and the marginal data Y_{ji} are not permutationally equivalent (where $\hat{\beta}$ is any estimate of the regression function β evaluated on the pooled set of pairs $(\mathbf{Y}, \mathbf{X})$, provided that β is estimable).

12) Prove that Mood's median test for the one-way ANOVA is not permutationally equivalent to the test T^* in Section 1.11.2.

13) Show that the permutation test T^* for the one-way ANOVA is invariant with respect to linear increasing transformations of data $\mathbf{X}$.

14) Show that the permutation test T^* above is not invariant with respect to nonlinear one-to-one transformations φ applied to data $\mathbf{X}$.

15) With reference to Example 1, 2.4.1, show that in a two-sample design the Cramér–von Mises test, i.e. $T^*_{CM} = \sum_i [\hat{F}^*_2(X_i) - \hat{F}^*_1(X_i)]$ as in (11) in Section 2.3, and Wilcoxon–Mann–Whitney test are not permutationally equivalent.

16) With reference to Remark 1, 2.1.1, prove that if φ is any non-degenerate and non-decreasing data transformation, then two test statistics $T^* = \sum_i \varphi(X^*_{1i})$ and $T^*_\mu = \sum_i \varphi(X^*_{1i} + \mu)$, where μ is a constant, are permutationally equivalent. (Hint: for each $i = 1, \ldots, n_1$, $\varphi(X^*_{1i}) \geq \varphi(X_{1i})$ implies $\varphi(X^*_{1i} + \mu) \geq \varphi(X_{1i} + \mu)$ and vice versa.)

2.5 Arguments for Selecting Permutation Tests

The suggested permutation solution for two-sample designs in continuous situations, also partially discussed in Chapter 1, is conditional on a set of minimal sufficient statistics for the unknown distribution P. Note that the Wilcoxon–Mann–Whitney rank sum test T_{MW} is essentially based on the permutation distribution of T calculated on the rank transformation of data $\mathbf{X}$. It is well known that, in general, the rank transformation is not one-to-one with the data, so that the vector of ranks $\mathbf{R}$ is not yet a set of sufficient statistics for P, it is only a maximal invariant transformation and so T_{MW} is distribution-free and nonparametric. Thus, unless the data transformation φ leading to a best unconditional test statistic corresponds to the rank transformation (see Nikitin, 1995, for some examples of population distributions admitting the Wilcoxon–Mann–Whitney test as asymptotically optimal), in most cases any rank test is no more powerful than its permutation counterpart. Instead, if data are i.i.d. from normally distributed populations, the permutation test is generally less powerful than the Student's t counterpart because in this case $\mathbf{X}$ is not minimal sufficient, although, except for very small sample sizes, their power difference is negligible and vanishes asymptotically. However, it is worth noting that in the multidimensional case 'best' parametric solutions are not always better than their permutation counterparts.

On the one hand, permutation tests may be regarded as precursors of rank tests (see the review by Bell and Sen, 1984). In rank tests, test statistics are generally expressed in terms of linear functions of ranks $\mathbf{R}$ associated with the original sample $\mathbf{X}$. However, permutation tests are generally preferable to rank tests, because the former are genuinely conditional on a set of sufficient statistics. It is well known that this generally implies desirable properties for a testing procedure (see Lehmann, 1986, Chapters 3–6). Rank tests are more commonly applied especially in univariate situations, because asymptotic distributions are easier to achieve. Well-known introductory books on rank tests, among others, are those by Fraser (1957), Puri and Sen (1971), Randles and Wolfe (1979), Hollander and Wolfe (1999), and Hettmansperger (1984); the formal theory is well developed in books by Hájek and Šidák (1967) and Hájek et al. (1999).

Conditioning on a set of sufficient statistics for P confers the distribution-free property on permutation tests (see Sections 1.2 and 2.1.3). However, when sample sizes are not sufficiently large, and especially in cases where data come from discrete populations, we must focus on the fact that permutation distributions of any statistics of interest are essentially discrete, so that it may be difficult to attain exactly every desired value for type I error rate α. In these cases, we have to tolerate approximations, use the auxiliary randomization technique, or adopt the so-called attainable α-values (see Section 2.2.3).

Among the general arguments related to permutation tests, there are two important questions we wish to discuss:

(a) How can we motivate our preference for a test statistic based, for instance, on divergence of sample means such as $T^* = \sum_i X^*_{1i}$?

(b) Why is its one-sided critical region a half-line, say $T \geq T_\alpha > 0$, so that large values are significant?

The main arguments behind these questions are as follows.

(a) The conditions for the optimal selection of a permutation test statistic T are reported in Lehmann (1986) (see also Hoeffding, 1951a; and Section 3.6). These conditions are so restrictive that they almost never occur in the nonparametric context. Essentially, they are related to the fact that a best test statistic is expressed as a function of the population distribution P. But, in the context of permutation testing, on the one hand the population distribution P is assumed unknown; on the other the permutation distribution depends on the pair $(T, \mathbf{X})$. Thus, a best permutation test statistic for all data sets $\mathbf{X}$ and for all population distributions P does not generally exist because, depending on P, it cannot be uniquely determined. Hence, in a general nonparametric framework, permutation tests are heuristically established on the basis of intuitive and reasonable arguments, both guided by obtaining an easier interpretation of results.

However, in the large majority of situations and for large sample sizes, one weak answer to this question is the analogy with the parametric solution based on normality, or more generally on the exponential family, associated with the behaviour of the induced permutation CDF $F_T(t|\mathcal{X}_{/\mathbf{X}})$ in regular conditions. This analogy often leads to test statistics of the form $T^* = \sum_i X^*_{1i}$. In addition, as P is assumed to be unknown, we cannot know the analytic form of a best parametric test statistic for the given testing problem because this is essentially based on the behaviour of the likelihood ratio, which in turn implies knowledge of P. In any case, in mild regularity conditions the following proposition holds:

Proposition 6. *If T is a best statistic for a given parametric family of distributions and if its unconditional critical region does not depend on any specific alternative, then its permutation counterpart is asymptotically equivalent to it; hence, as they share the same limit power function, the permutation version of T is asymptotically best for the same family.*

A formal proof of this statement can be found in Hoeffding (1952). An explanation for the asymptotic permutation behaviour of T in H_0 is essentially based on the following argument. Let $F_T(t; n, P, H_0)$ denote the parametric CDF induced by T in H_0, and $F_T(t; P, H_0)$ its limit as n tends to infinity, where the symbols indicate dependence on the population distribution P. Both CDFs $F_T(t; n, P, H_0)$ and $F_T(t; P, H_0)$ may be regarded as obtained by an infinite Monte Carlo simulation from P. Let us adopt the same statistic T in a permutation framework and denote by $F^*_T(t|\mathcal{X}^{(n)}_{/\mathbf{X}})$ its permutation CDF related to the data set $\mathbf{X} = \mathbf{X}^{(n)} = \{X_i; i = 1, \ldots, n\}$. We know that there is a one-to-one relation between the EPM $\hat{P}_n(A) = \sum_i \mathbb{I}(X_i \in A)/n$, $A \in \mathcal{A}$, and the data set $\mathbf{X}^{(n)}$, except for a proper rearrangement of the data (see Proposition 2 and Remark 1, 2.1.3). Hence, they essentially contain the same relevant information with respect to the underlying distribution P, so that the whole EPM $\hat{P}_n$ is also sufficient. In very mild conditions, according to the well-known Glivenko–Cantelli theorem, as n tends to infinity, $\hat{P}_n$ converges to P almost surely and uniformly in $A \in \mathcal{A}$. Thus, by the theorems on continuous functionals of empirical processes (see Shorack and Wellner, 1986; Borovkov, 1987), $F^*_T(t|\mathcal{X}^{(n)}_{/\mathbf{X}})$ converges in probability to $F_T(t; P, H_0)$ as n tends to infinity, provided that T is a continuous function of $\mathbf{X}^{(n)}$ and F is continuous and monotonically increasing with respect to a suitable metric measuring the distance between any two members of the parametric family $\mathcal{P}$ to which P belongs by assumption. Moreover, if T is any test statistic, the power behaviour of its permutation counterpart being conditional on a set of sufficient statistics is at least improved (see Cox and Hinkley, 1974; Lehmann, 1986).

The study of T in H_1 is rather more complex. However, in general regular conditions, it is possible to argue that, at least in the vicinity of H_0, the parametric and permutation solutions are still asymptotically coincident.

From a practical point of view, as test statistics such as $T^* = \sum_i X^*_{1i}$ behave better than other competitors when underlying population distributions are at least approximately symmetric with finite second moments, a suggestion for choosing a suitable statistic is to use data transformations $\varphi(X)$ so that the resulting distribution is at least approximately symmetric and the response model is additive as in (M.i) in Sections 1.8.1 and 1.10.1. Moreover, in a large set of simulations with exchangeable, symmetrically distributed errors, including some heavy-tailed distributions, we found that the divergence of sample means is a proper test statistic for one- and two-sample problems, provided that second moments are finite; instead, divergence of sample medians is a proper test statistic for one- and two-sample problems regardless of the existence of finite moments of underlying distributions. For heavy-tailed distributions and/or asymmetric distributions, the Wilcoxon–Mann–Whitney T_{MW} also has a good power behaviour (see Lehmann, 2006).

In this way, for finite sample sizes, T may only be considered a *reasonable choice*. Of course, as P is unknown, in general it is not possible to provide T with any optimal property (see Section 3.6). However, if a single test statistic is felt not to provide an adequate summary of information, it may sometimes be convenient to apply $k > 1$ different test statistics $(T_1, \ldots, T_k)$, each sensitive to a specific aspect of the alternative distribution, and so able to summarize specific pieces of information, and then combine their related p-values, using a sort of two-stage choice. This multi-aspect testing strategy, discussed in Example 3, 4.6, is often used in subsequent chapters.

Remark 1. In the conditions required by the Glivenko–Cantelli theorem, asymptotic permutation (conditional) inferences become unconditional. Although we may regard permutation tests as being essentially robust with respect to underlying distributions, it would be interesting to investigate the notion of robustness for test statistics in the permutation framework (see Lambert, 1985; Welch and Gutierrez, 1988; Hettmansperger and McKean, 1998). This investigation is left to the reader.

(b) The answer to the second question concerning the rule that 'large values are significant' is twofold and seems heuristically similar to regular situations. On the one hand there is no loss of generality because all statistical tests can be equivalently transformed in such a way that in the alternative their distributions are stochastically larger than in the null hypothesis. Indeed, we are able to prove in very mild conditions that for one-sided alternatives the alternative permutation distribution of any test statistic T, as in the list in Section 2.3, is stochastically larger than in H_0 (see Section 3.1). This implies that the rule 'large values are significant' gives rise to tests whose conditional and unconditional power functions are monotonic increasingly related to treatment effects, independently of underlying distributions P and sample sizes. For two-sided alternatives we know that with not too small sample sizes, $|T|$ is almost always stochastically larger in H_1 than in H_0, a condition which also justifies the rule. In this way, for finite sample sizes, T may only be considered a *reasonable choice*.

On the other hand, we know that if P is provided with a monotonic likelihood ratio, then the UMP and UMPU critical regions for positively one-sided alternatives and two-sided alternatives have the form $T \geq T_\alpha$ and $|T| \geq |T_\alpha|$, respectively. Of course, in a nonparametric framework we cannot invoke such a property for P because it is assumed to be unknown. However, if n is not too small and if T is a symmetric function (i.e. invariant with respect to data entry) of a continuous transformation of data as in $T^* = \sum_i \varphi(X^*_{1i})$, then a PCLT can be applied for regular P, so that the permutation limit distribution of T tends to enjoy the normal property. It thus seems reasonable to choose a test statistic T according to asymptotic arguments.

2.6 Examples of One-Sample Problems

In this section, some examples of one-sample permutation testing problems are briefly discussed.

Example 1. *Testing for symmetry.*

Let us assume that we are given a sample $\mathbf{X} = \{X_1, \ldots, X_n\}$ of n i.i.d. observations from a continuous variable X with unknown distribution P on the real line. Suppose that we wish to test the null hypothesis H_0 that P is symmetric around the origin,

$$H_0 : \{P(z) = 1 - P(-z), z \in \mathcal{R}^1\},$$

against $H_1 : \{P$ is not symmetric around the origin$\}$.

To deal with this problem, we can regard the data as differences given by underlying paired observations. Thus we have observable variables $X_i = Y_{Ai} - Y_{Bi}, i = 1, \ldots, n$, where A and B are two occasions of measurement. From Section 1.8.2, we know that the distribution of differences X_i is symmetric around 0 if $Y_{Ai} \stackrel{d}{=} Y_{Bi}$, that is, when paired responses are exchangeable within each unit. We also know that the random attribution of the sign to the ith difference X_i is equivalent to considering a random permutation of paired measurements, conditional upon these paired observations (see Remark 3, 2.1.2). Thus, conditional testing of symmetry is equivalent to the conditional testing with paired exchangeable observations. Therefore, we obtain the permutation sample space $\mathcal{X}_{/\mathbf{X}} = \{\bigcup_{\mathbf{S}^*}[X_i S_i^*, i = 1, \ldots, n]\}$, where the $S_i^* = 1 - 2\mathcal{B}n(1, 1/2)$ are n i.i.d. random signs. A suitable test for H_0 against H_1 is $T = \sum_i X_i$, the permutation distribution of which is obtained by considering the permutation support $\mathcal{T}(\mathbf{X}) = \{T^* = T(\mathbf{X}^*) = \sum_i X_i \cdot S_i^*, \mathbf{X}^* \in \mathcal{X}_{/\mathbf{X}}\}$.

Permutationally equivalent to T is $K = \sum_i X_i / \left(\sum_i X_i^2\right)^{1/2}$, the permutation distribution of which is obtained by taking $K^* = \sum_i X_i \cdot S_i^* / \left(\sum_i X_i^2\right)^{1/2}$. Note that the denominator in K^* is invariant at all points of the permutation sample space $\mathcal{X}_{/\mathbf{X}}$. Since $\mathbb{E}_{\mathcal{X}_{/\mathbf{X}}}(K^*) = 0$ and $\mathbb{V}_{\mathcal{X}_{/\mathbf{X}}}(K^*) = 1$ (see Section 1.9.4), it follows that K^* is the standardized sum of n independent variables, so that if the ratio $(\sum_i X_i^4)/(\sum_i X_i^2)^2$ is small and the standard deviation σ_X is finite, when n is large the approximate distribution of K^*, by the PCLT, is standard normal: $K^* \sim \mathcal{N}(0, 1)$. Note that this limit distribution is obtained without involving the Student's t distribution because, in the permutation framework, with $\mathbf{X}$ given the role of a fixed finite population, the denominator $\sum_i X_i^2$ is the exact conditional variance of the numerator (see Section 3.8).

Remark 1. Testing for symmetry may also be used for testing location on one-sample problems. To be specific, let us suppose that $H_0 : \{\delta = \delta_0\}$ and that the observed data $\{Y_i, i = 1, \ldots, n\}$ are symmetrically distributed around δ, so that data transformations $X_i = Y_i - \delta_0$ are symmetrically distributed around 0 if and only if H_0 is true. With clear notation, in H_1 the observed value of T is such that $T^o(\delta) = T^o(0) + n\delta$ and so it is stochastically larger than in H_0. In Example 8, 4.6, these ideas are used for testing on the generalized Behrens–Fisher problem. When Y is not symmetrically distributed around δ_0, testing for $H_0 : \{\mathbb{E}(X) = 0\}$ by using the test statistic $T^* = \sum_i X_i \cdot S_i^*$ cannot in general lead to exact solutions (see also Example 9 in this section). In fact T^* can be either conservative or anticonservative, according to whether asymmetry is positive or negative. In order to get inferences that are robust, that is, asymptotically exact and almost exact for finite sample sizes, it is useful to adopt transformations to symmetry such as those suggested by Hinkley (1975). One of these transformations involves finding a power a such that, for an integer r, $\tilde{X}^a - X_{(r)}^a = X_{(n-r-1)}^a - \tilde{X}^a$, where $X_{(r)}$ is the rth order statistic of the X_i, $i = 1, \ldots, n$, and $\tilde{X} = \mathbb{M}d(X_1, \ldots, X_n)$ is the median. Often, for positive Y and positive asymmetry, transformations such as $\sqrt{Y}$, $\sqrt[3]{Y}$ or $\log Y$ are satisfactory for good approximations.

Example 2. *Testing independence of two variables.*

Let $\mathbf{X} = \{(X_1, Y_1), \ldots, (X_n, Y_n)\}$ be a sample of n i.i.d. observations from a bivariate response variable (X, Y) with unknown bivariate distribution P on the Euclidean space $\mathcal{R}^2$. We wish to test the null hypothesis H_0 that $P(x, y)$ has independent marginals $P_1(x)$ and $P_2(y)$

(see Bell and Doksum, 1967),

$$H_0 : \{P(x, y) = P_1(x) \cdot P_2(y), \forall (x, y) \in \mathcal{R}^2\}$$

against the alternative $H_1 : \{H_0$ is not true$\}$. Note that true H_0 implies that each X_i can be equally likely associated with any Y_j. Thus, the permutation sample space is $\mathcal{X}_{/\mathbf{X}} = \{\mathbf{X}^* = \{(X_1, Y_{u_1^*}), \ldots, (X_n, Y_{u_n^*})\}\}$, where $(u_1^*, \ldots, u_n^*)$ is any permutation of the basic labels $\{1, \ldots, n\}$, the cardinality of which is $M^{(n)} = n!$. Here, we have implicitly assumed that the jth marginal component takes values in sample space $\mathcal{X}_j$ and that $\mathcal{X}$ is the cartesian product $\mathcal{X} = \mathcal{X}_1 \times \mathcal{X}_2$.

To cope with this testing problem, we require a suitable test statistic. The most common one is based on the cross product $T = \sum_i X_i \cdot Y_i$, the permutation distribution of which is obtained by considering $T^* = \sum_i X_i \cdot Y_{u_i^*}$, and consists of examining the linear dependence between X and Y. A more appropriate test statistic assumes the form $T = \varphi(P, P_1 P_2)$, where φ is a metric on the nonparametric family of bivariate distributions $\mathcal{P}$ measuring distances of P from $P_1 \cdot P_2$. Following Romano (1989), a typical choice for T, in the spirit of the Kolmogorov–Smirnov test statistic, is

$$T^* = n^{1/2} \cdot \sup_{A \in \mathcal{B}} \left| \hat{P}^*(A) - \hat{P}_1^*(A) \cdot \hat{P}_2^*(A) \right|,$$

where $\mathcal{B}$ is a suitable non-empty collection of events, and $\hat{P}^*$, $\hat{P}_1^*$ and $\hat{P}_2^*$ are proper permutation sample estimates of the probability distributions involved.

Example 3. *Testing for linear regression.*

Let $\mathbf{X} = \{(X_1, Y_1), \ldots, (X_n, Y_n)\}$ be a sample of n i.i.d. observations from a bivariate response variable (X, Y) with unknown bivariate distribution P on $\mathcal{R}^2$ and where two variables are linked by a linear regression: $\mathbb{E}(Y|X = x) = a + \beta x$. We wish to test the null hypothesis $H_0 : \{\beta = 0\}$ against, for instance, $H_1 : \{\beta > 0\}$ under the assumption that in H_0 responses Y_i can be permuted with respect to covariate X, so that permutation sample space is $\mathcal{X}_{/(\mathbf{X},\mathbf{Y})} = \{\bigcup_{\mathbf{u}^*}[(X_i, Y_{u_i^*}), i = 1, \ldots, n]\}$, where $\mathbf{u}^*$ is a permutation of unit labels $(1, \ldots, n)$. Supposing that a suitable indicator for regression coefficient β is $\hat{\beta} = \left(\sum_i X_i Y_i - n\bar{X}\bar{Y}\right) / \left[\sum_i (X_i - \bar{X})^2 \cdot \sum_i (Y_i - \bar{Y})^2\right]$, then a test statistic is $T_\beta^* = \sum_i X_i Y_i^*$ which is permutationally equivalent to $\hat{\beta}^*$ (see Problem 15, 2.6.2).

It is worth noting here that, in order for permutations in $\mathcal{X}_{/(\mathbf{X},\mathbf{Y})}$ to be equally likely under H_0, we had to assume exchangeability of responses with respect to covariate. This assumption, which is common for instance to Spearman's and Kendall's rank tests, more than non-correlation, implies (quasi-)independence of two variables in H_0.

Example 4. *Testing for a change point.*

Let us assume that we have a sample $\mathbf{X} = \{X_1, \ldots, X_n\}$ of $n > 2$ observations from n independent variables $X_1, \ldots, X_n$, taking values in the same sample space $\mathcal{X}$ and respectively with unknown distributions $P_1, \ldots, P_n$. We wish to test the null hypothesis H_0 that the observations X_i have a common distribution P, against the alternative H_1 that, for some I, $1 \leq I \leq n$, $\{X_1, \ldots, X_I\}$ are i.i.d. from a distribution P_1 and that $\{X_{I+1}, \ldots, X_n\}$ are i.i.d. from a different distribution P_2. That is,

$$H_0 : \left\{\prod_i P_i(X_i) = \prod_i P(X_i)\right\}$$

against the alternative $H_1 : \left\{\prod_{i=1}^I P_1(X_i) \cdot \prod_{i=I+1}^n P_2(X_i)\right\}$.

An appropriate test statistic for this problem is

$$T^* = \max_{1 \leq i \leq n} \left[\gamma_i \cdot \sup_{A \in \mathcal{B}} \left| \hat{P}_1^*(A) - \hat{P}_2^*(A) \right|\right],$$

where $\mathcal{B}$ is a suitable non-empty collection of events, γ_i are normalizing constants, and $\hat{P}_1^*$ and $\hat{P}_2^*$ are proper permutation estimates. This test statistic (see Romano, 1989) may be simplified if we assume that variables X_i have finite means $\mu_i = \mathbb{E}(X_i), i = 1, \ldots, n$, and that their distributions are such that $P_i(z) = P(z - \mu_i)$, so that all variables have the same distribution except for the mean (a kind of *generalized homoscedasticity*). In this case the hypotheses are $H_0 : \{\mu_1 = \ldots = \mu_n\}$ and $H_1 : \left\{[\mu_1 = \ldots = \mu_I] \neq [\mu_{I+1} = \ldots = \mu_n]\right\}$. By defining $W_i = X_1 + \ldots + X_i$, a possible test statistic is

$$T^* = \max_{1 \le i \le n} \left[\left(\frac{i}{n} W_n - W_i^* \right)^2 \cdot \{i(n-i)\}^{-1} \right],$$

where $W_i^* = X_{u_1^*} + \ldots + X_{u_i^*}$, in which $(u_1^*, \ldots, u_n^*)$ is any permutation of the basic labels $(1, \ldots, n)$. Both test statistics have an intuitive appeal. In particular, the latter has some similarities with the Smirnov–Anderson–Darling goodness-of-fit test for discrete distributions (see James et al., 1987; see also Section 2.8.3 on goodness-of-fit testing for categorical variables). In Chapter 6 we will encounter different arguments for such a problem.

Example 5. *Testing exchangeability.*

Let $\mathbf{X} = \{(X_{11}, \ldots, X_{V1}), \ldots, (X_{1n}, \ldots, X_{Vn})\}$ be a sample of n i.i.d. observations from a quantitative V-dimensional variable $\mathbf{X}$, $V \ge 2$, with unknown V-dimensional distribution P. We wish to test the null hypothesis H_0 that the V components of $\mathbf{X}$ are exchangeable, that is, $H_0 : \{P(X_1, \ldots, X_V) = P(X_{v_1^*}, \ldots, X_{v_V^*})\}$, where $(v_1^*, \ldots, v_V^*)$ is any permutation of the labels $(1, \ldots, V)$, against the alternative H_1 that H_0 is not true.

For this problem the permutation sample space is $\mathcal{X}_{/\mathbf{X}} = \{\mathbf{X}^* = (X_{v_{11}^*1}, \ldots, X_{v_{q1}^*}1), \ldots, (X_{v_{1n}^*n}, \ldots, X_{v_{qn}^*n})\}$, where $\{(v_{1i}^*, \ldots, v_{Vi}^*), i = 1, \ldots, n\}$, are permutations of labels $(1, \ldots, V)$ relative to component variables $X_h, h = 1, \ldots, V$.

Let us consider a map $\chi : \mathcal{P} \to \mathcal{P}_0$, characterizing the null hypothesis H_0. That is, $\mathcal{P}_0$ is the set of probability distributions satisfying $\chi(P) = P$ (see Romano, 1989). Thus, to test H_0 against H_1, an appropriate test statistic is

$$T = \sqrt{n} \cdot \sup_{A \in \mathcal{A}} |P(A) - \chi(P(A))|,$$

where $\mathcal{A}$ is a suitable non-empty collection of events.

Example 6. *Some extensions of McNemar's test.*

Let us refer to the McNemar binomial solution to the problem with paired data discussed in Section 1.8.6. A slight extension of the binomial test useful for one and two-sided alternatives may be obtained in the following way. Let us assume, for instance, that observed differences are classified into the four categories: *very positive*, $C(++)$; *positive*, $C(+)$; *negative*, $C(-)$; and *very negative*, $C(--)$, in place of the binary categories *plus* and *minus*. Let us denote by $f(j)$ the observed frequency of category $C(j)$, that is, $f(j) = \sum_{i \le n} \mathbb{I}(X_i \in C_j)$, $j = [(++), \ldots, (--)]$. In $H_0 : X$ is symmetric around 0, the testing problem reduces to testing for symmetry of ordered categorical variables in which the differences are symmetrically distributed around zero. Let us denote by $\nu_{++} = f(++) + f(--)$, $\nu_+ = f(+) + f(-)$, and $\nu = \sum_j f(j) = \nu_{++} + \nu_+$, frequencies of symmetric categories and their sum (if the number of categories is odd there is one category, the central one, with null differences), respectively. Since permutations are within each individual pair of responses and independently with respect to individuals (see Remark 3, 2.1.2), the permutation frequency $f^*(c) = \sum_i \mathbb{I}[(X_i S_i^*) \in C_c]$, $c = [(++), (+)]$, where the i.i.d. $S_i^* = 1 - 2\mathcal{B}n(1, 1/2)$, is binomially distributed, $f^*(c) = \mathcal{B}n(\nu_c, 1/2)$. Thus, one suitable test statistic for one-sided alternatives is

$$T^* = \sum\nolimits_c \frac{[f^*(c) - \nu_c/2]}{\sqrt{\nu_c/4}},$$

corresponding to the sum of two independent standardized binomial distributions, whose null distribution is then approximately $\mathcal{N}(0, 2)$. Of course, based on the PCLT theorem, this approximation is better for large sample sizes. For instance, if in the IPAT example (Section 1.8) we decide that very positive differences correspond to $X > +4$, positive to $0 < X \leq +4$, negative to $-4 \leq X < 0$ and very negative to $X < -4$, we have $f(++) = 7$, $f(+) = 10$, $f(-) = 3$ and $f(--) = 0$, so that $\nu_{++} = 7$, $\nu_{+} = 13$, and $T^o = 4.986$, leading to rejection at $\alpha = 0.001$ (although in the present case approximation to the normal distribution is not satisfactory, because sample size n and ν_h frequencies are not sufficiently large; in turn, a more reliable result is obtained by the permutation solution, which with $B = 5000$ iterations gives $\hat{\lambda} = 0.0004$). It is, however, worth observing that the traditional McNemar standardized solution, since $U^* = \sum_c f^*(c) \sim \mathcal{B}n(\nu, 1/2)$, that is, $T_U^* = (U^* - \nu/2)/\sqrt{\nu/4}$, is approximately standard normally distributed, say $T_U^* \simeq \mathcal{N}(0, 1)$ – which in turn corresponds to the standardized sum of two independent binomials. With $U^o = 17$, as $T_U^o = 3.130$, the null hypothesis is rejected at $\alpha = 0.001$. Due to the more accurate management of information, T is expected to be somewhat more efficient than U when large variations (from $C(++)$ to $C(--)$) are considered more important than small variations (from $(+)$ to $(-)$), because in U there might be compensations of effects. Problems 8 and 9, 2.6.2, suggest the extension of this kind of solution to a $2k$, $k > 2$, symmetric partition of the set of differences.

As a second extension, suitable for two-sided alternatives, let us consider a test statistic with the chi-square form

$$X^{*2} = \sum_{j=(++)}^{(--)} \frac{[f^*(j) - \hat{f}_0(j)]^2}{\hat{f}_0(j)} = \sum_{c=(++)}^{(+)} [f^*(c) - \nu_c/2]^2/(\nu_c/4),$$

where $\hat{f}_0(j) = \nu_j/2$, $j = [(c), (-c)]$, in which $(-c)$ means $(--)$ if $c = (++)$, and so on.

In H_0, X^{*2} is approximately distributed as a central χ^2 with 2 d.f., because it is the sum of two independent standardized squared binomial distributions. For instance, with the IPAT data we have $\hat{f}_0(++) = \hat{f}_0(--) = 3.5$, $\hat{f}_0(+) = \hat{f}_0(-) = 6.5$. Thus, although the approximation to the χ^2 distribution is not satisfactory, because the sample size n and expected frequencies $\hat{f}_0(i)$ are not sufficiently large, we have $X^2 = 10.769$, which leads to the rejection of H_0 at $\alpha = 0.005$. Of course, for small sample sizes all these approximations may not be satisfactory so CMC or exact numeric calculations are needed.

Example 7. *Testing for symmetry in an ordered categorical variable.*

Solutions from Example 6 can be extended to testing for symmetry in an ordered categorical variable. Let us suppose, to this end, that the support of a categorical variable X is $(A_1 \prec A_2 \prec \ldots \prec A_k)$, where $\prec$ stands for 'inferior to', and suppose that H_0 is that the distribution of X is symmetric over its support, $\Pr\{A_h\} = \Pr\{A_{k-h+1}\}$, $h = 1, \ldots, k$. This problem can be thought of in terms of an underlying variable Y being observed on each individual before and after an administered treatment, so that 'differences' $X = \phi(Y_1, Y_2)$ are classified according to ordered categories $(A_1, \ldots, A_k)$. According to this interpretation, the null hypothesis can equivalently be transformed into $H_0 \equiv \{Y_1 \overset{d}{=} Y_2\}$, giving rise to a symmetric distribution over $(A_1, \ldots, A_k)$. The solution is then obtained by conditioning with respect to the observed data set $\mathbf{X} = \{X_i, i = 1, \ldots, n\}$, which is sufficient in H_0 for the underlying problem. The related sample space $\mathcal{X}_{/\mathbf{X}} = \{X_i S_i^*, i = 1, \ldots, n\}$, where the i.i.d. $S_i^* = \pm 1$ each with probability 1/2, is obtained similarly to Example 6. Since $f^*(c) = \sum_i \mathbb{I}(X_i S_i^* \in A_c) \sim \mathcal{B}n(\nu_c, 1/2)$, where the i.i.d. $S_i^* \sim 1 - 2\mathcal{B}n(1, 1/2)$, $\nu_c = f(c) + f(k - c + 1)$, and if all 'differences' $X_i = Y_{1i} - Y_{2i}$, $i = 1, \ldots, n$, are regarded as equally important although differently informative, then a general solution for one-sided alternatives is

$$T^* = \sum_{c=1}^{\lfloor k/2 \rfloor} \frac{[f^*(c) - \nu_c/2]}{\sqrt{\nu_c/4}},$$

where $\lfloor k/2 \rfloor$ is the integer part of $k/2$. Observing that T^* corresponds to the sum of $\lfloor k/2 \rfloor$ standardized binomials, its distribution is approximately $\mathcal{N}(0, \lfloor k/2 \rfloor)$. For two-sided alternatives a solution in the spirit of chi-squared is then $X^{*2} = \sum_{c=1}^{\lfloor k/2 \rfloor} [f^*(c) - \nu_c/2]^2/(\nu_c/4)$, the asymptotic distribution of which is χ^2 with $\lfloor k/2 \rfloor$ d.f.

If 'differences' have different degrees of importance, so that for instance a variation from A_1 to A_k is more important than a variation from A_3 to A_{k-2}, and so 'scores' ω_c are assigned to 'differences' from A_c to A_{k-c+1}, then a test statistic for one-sided alternatives is

$$T_\omega^* = \sum_{c=1}^{\lfloor k/2 \rfloor} \frac{\omega_c[f^*(c) - \nu_c/2]}{\sqrt{\nu_c/4}},$$

that is, the weighted sum of $\lfloor k/2 \rfloor$ independent standardized binomials, the null distribution of which is approximately $\mathcal{N}(0, \sum_c \omega_c^2)$. For two-sided alternatives, the solution in the spirit of chi-squared is $X_\omega^{*2} = \sum_{c=1}^{\lfloor k/2 \rfloor} \omega_c[f^*(c) - \nu_c/2]^2/(\nu_c/4)$, the distribution of which is then the weighted sum of $\lfloor k/2 \rfloor$ independent χ^2 with 1 d.f. This is not available in closed form.

Example 8. *Testing with paired data from an ordered categorical variable.*

This example considers an extension of solutions shown in Example 7. Suppose that the support of a variable Y is $(A_1 \prec A_2 \prec \ldots \prec A_k)$ and that Y is observed on n units before treatment, occasion 1, and after treatment, occasion 2. Thus the data set is $\mathbf{Y} = \{(Y_{1i}, Y_{2i}), i = 1, \ldots, n\}$. The null hypothesis is that treatment is ineffective, $H_0 : \{Y_1 \overset{d}{=} Y_2\}$. Thus, within each unit two observations are exchangeable. According to Remark 3, 2.1.2, the set of pairs $(Y_{1i}, Y_{2i}), i = 1, \ldots, n$, is the set of sufficient statistics for the problem in H_0. Thus, the related permutation sample space $\mathcal{X}_{/\mathbf{X}}$ is obtained by noting that for each unit the conditional probability of being in class A_j before treatment and in class A_h after treatment is equal to the probability of being in A_h before and in A_j after, that is, $P_i\{(A_j, A_h)|(Y_{1i}, Y_{2i})\} = P_i\{(A_h, A_j)|(Y_{1i}, Y_{2i})\} = 1/2$, and that units are independent. So $\mathcal{X}_{/\mathbf{X}}$ contains 2^n points. Thus, 'differences' $X_i = \phi(Y_{1i}, Y_{2i})$, $i = 1, \ldots, n$, lie in the support $(C_{11}, C_{12}, \ldots, C_{kk})$, where $X_i = C_{hj}$ means that ith subject has moved from category A_h at occasion 1 to A_j at occasion 2. Observe that, due to the exchangeability within individuals stated by H_0, we have that $\Pr\{C_{hj}\} = \Pr\{C_{jh}\}$, $j, h = 1, \ldots, k$; that is, H_0 implies that the distribution of 'differences' is symmetric with respect to the main diagonal. Thus, we have the permutation frequency $f^*(C_{hj}) = \sum_i \mathbb{I}(X_i S_i^* \in A_{hj}) \sim \mathcal{B}n(\nu_{hj}, 1/2)$, where the i.i.d. $S_i^* \sim 1 - 2\mathcal{B}n(1, 1/2)$ and $\nu_{hj} = f(C_{hj}) + f(C_{jh})$. In this context two permutation tests for one-sided alternatives are

$$T^* = \sum_{h>j}^{k} \frac{[f^*(C_{hj}) - \nu_{hj}/2]}{\sqrt{\nu_{hj}/4}},$$

corresponding to the sum of $k(k-1)/2$ standardized binomials, and

$$T_\omega^* = \sum_{h>j}^{k} \frac{\omega_{hj}[f^*(C_{hj}) - \nu_{hj}/2]}{\sqrt{\nu_{hj}/4}},$$

corresponding to the weighted sum of standardized binomials, where ω_{hj} are suitable assigned scores. MATLAB code is available from the book's website for weights $\omega_{hj} = |h - j|$.

And one solution for two-sided alternatives, in the spirit of chi-squared, is $X^{*2} = \sum_{h>j}^{k} [f^*(C_{hj}) - \nu_{hj}/2]^2 4/\nu_{hj}$.

Example 9. *Testing the median.*

Let us suppose that the response X and the related error deviates Z do not possess a mean value, that is, $\mathbb{E}(|Z|)$ is not finite, so that the sample mean $\bar{X}$ is not a proper indicator for treatment effect δ. In such cases it is wise to use test statistics based on the divergence of *robust* indicators such as sample medians or trimmed means (see Example 5, 4.5.3). To be precise, suppose that in a one-sample problem the data model is $\mathbf{X} = \{X_i = \delta + Z_{1i}, i = 1, \ldots, n\}$, where the meanings of symbols are clear. Suppose the hypotheses are $H_0 : \{\mathbb{M}d(X) = 0\} = \{\delta = 0\}$ against $H_1 : \{\delta > 0\}$. If Z were symmetrically distributed around 0, the problem would be solved with the test statistic $T^* = \mathbb{M}d[X_i S_i^*]$ in which the i.i.d. $S_i^* \sim 1 - 2\mathcal{B}n(1, 1/2)$. This would be the solution if paired data were observed, that is, if the data were $\{(Y_{1i} = \eta + Z_{1i}, Y_{2i} = \eta + \delta + Z_{2i}), i = 1, \ldots, n\}$ and individual differences $X_i = Y_{2i} - Y_{1i}$ were used. In such a case this test is exact because in H_0 data are exchangeable within each unit. If instead error deviates Z were not symmetric around 0, or we know a data transformation φ such that $\varphi(Z)$ becomes symmetric around 0, the solution would be $T^* = \mathbb{M}d[\varphi_i S_i^*]$, unless $\varphi(Z)$ has finite mean value in which case we might also use $T^* = \sum_i X_i S_i^*$. Alternatively, we can use the drastic sign transformation $\varphi_i = \{-1$ if $X_i < 0$, 0 if $X_i = 0$, and $+1$ if $X_i > 0\}$ and the McNemar test statistic $T^* = \sum_i \varphi_i S_i^* \sim \mathcal{B}n(\nu, 1/2)$, where $\nu = n - \sum_i \mathbb{I}(\varphi_i = 0)$. Since, in H_0, $\Pr\{\varphi_i = -1\} = \Pr\{\varphi_i = +1\}$, the latter test is exact. It is, however, worth noting that a test statistic like $T^* = \mathbb{M}d[X_i S_i^*]$ with non-symmetric deviates Z cannot be exact, because of lack of exchangeability (see Remark 1 above). Actually it can be either conservative or anticonservative according to the kind of asymmetry, although for large sample size it becomes almost exact.

2.6.1 A Problem with Repeated Observations

This testing problem arises when each experimental unit is observed on k occasions. To be more specific, let us assume that n units are each observed k times, with reference to k occasions of measurement which can be considered as playing the role of symbolic time treatment. Thus, realizations of a non-degenerate univariate response variable X are represented in a matrix layout such as $\mathbf{X} = \{X_{ji}, i = 1, \ldots, n, j = 1, \ldots, k\} = \{(X_{1i}, \ldots, X_{ki}), i = 1, \ldots, n\}$, where the $(X_{1i}, \ldots, X_{ki})$ are called *individual profiles*.

This layout, when units play the role of blocks and we are not interested in the block effect, may correspond to a two-way ANOVA with one observation per unit and treatment level, in which only the so-called treatment effects are of interest and no interaction is assumed between units and occasions of measurement. The block effect may be assumed either to be present or not, in the sense that we do not assume that response profiles are identically distributed with respect to units, so that we need not assume $P_i = P$, $i = 1, \ldots, n$. As usual, let us assume that these underlying unspecified distributions are non-degenerate and also that:

(i) The n profiles are independent.
(ii) Individual responses $(X_{1i}, \ldots, X_{ki})$ are homoscedastic and exchangeable within units in the null hypothesis (note that this interpretation is appropriate for most cases of repeated observations when there are no interaction effects between units and time).
(iii) The underlying model for fixed effects is additive. Thus, the model is assumed to behave formally in accordance with $X_{ji} = \mu + \eta_i + \delta_j + \sigma_i \cdot Z_{ji}$, where μ is a population constant, η_i is the block effect corresponding to the ith unit, δ_j is the treatment effect, error terms Z_{ji} are assumed to be exchangeable within units (i.e. equally distributed with respect to occasions of measurement and independent with respect to units, where the common distribution P_Z is unknown), and σ_i are unknown scale coefficients which may vary with respect to units (an extended response model is discussed in Pesarin, 2001, Chapters 7 and 11).

In particular, we also assume that error terms satisfy the condition $\mathbb{E}(Z_{ji}) = 0$, $\forall i, j$, so that we can adopt sample means as proper indicators for treatment effects. Note that this response model corresponds to a two-way ANOVA without interaction (see Chapter 11 for analysis of two-way

ANOVA). The null hypothesis is that responses within units are equal in distribution or, equivalently, that no treatment effect is present:

$$H_0 : \left\{ \bigcap_{i=1}^{n} (X_{1i} \stackrel{d}{=} \ldots \stackrel{d}{=} X_{ki}) \right\} = \{\delta_1 = \ldots = \delta_k = 0\},$$

against $H_1 : \{H_0 \text{ is not true}\}$.

Note that this formulation of H_0 assumes equality in distribution within each unit, jointly for all units. Thus, this problem appears both as a multi-sample problem and as an extension of that with paired observations.

Friedman's Rank Test

Under continuity of observed data, so that the ordinary rank transformation can be applied, our problem is solved by Friedman's nonparametric rank test, in which the so-called block or individual effect is ignored. Remember that Friedman's test (see Friedman, 1937) is based on the permutation rank statistic,

$$T_F = \sum_{j=1}^{k} \left[\bar{R}_j - \frac{k+1}{2} \right]^2 \frac{12 \cdot n}{k(k+1)},$$

where $\bar{R}_j = \sum_i R_{ji}/n$, $i = 1, \ldots, n$, $j = 1, \ldots, k$, is the jth mean rank and $R_{ji} = \sum_{h \leq k} \mathbb{I}(X_{hi} \leq X_{ji})$ is the rank of the jth observation within those related to the ith unit. For large sample sizes, in the null hypothesis, T_F is approximately distributed according to a central χ^2 with $k - 1$ degrees of freedom.

Of course, if responses were homoscedastic and normally distributed, this kind of problem would be solved by Snedecor's well-known F test.

A Permutation Solution

Note that assumptions (i)–(iii) imply that, in H_0, data within each unit are exchangeable with respect to occasions of measurement, and that they are independent with respect to units. Thus, the permutation sample space $\mathcal{X}_{/\mathbf{X}}$ contains $(k!)^n$ points. In fact, for each of the n independent units there are $k!$ permutations, which in H_0 are equally likely. Thus, a CMC solution implies taking into consideration a suitable test statistic, which may reasonably have the same form as in the parametric normal case. Hence, a suitable permutation test statistic is

$$T_R^* = \frac{\sum_{j=1}^{k} (\bar{X}_{j\cdot}^* - \bar{X}_{\cdot\cdot})^2}{\sum_{ji} (X_{ji}^* - \bar{X}_{\cdot i} - \bar{X}_{j\cdot}^* + \bar{X}_{\cdot\cdot})^2},$$

where $\bar{X}_{j\cdot}^* = \sum_i X_{ji}^*/n$, $j = 1, \ldots, k$, $\bar{X}_{\cdot i} = \bar{X}_{\cdot i}^* = \sum_j X_{ji}^*/k$, $i = 1, \ldots, n$, and $\bar{X}_{\cdot\cdot} = \bar{X}_{\cdot\cdot}^* = \sum_{ji} X_{ji}^*$ are the column, row and global means, respectively.

It is interesting to note that this test statistic, except for a constant coefficient, is nothing more than the permutation ratio of estimated variances of the treatment effect by that of errors.

The multivariate extension of the above problem is presented in Chapter 11, while a complete solution for fixed effects two-way ANOVA, balanced and unbalanced, is discussed in Pesarin (2001) and Basso et al. (2009a). Some multivariate problems for repeated observations and missing data are discussed in Chapter 7.

Remark 1. The CMC algorithm implies considering n independent permutations of the k data, one for each unit. This is done by the matrix of random permutations $\{u_{1i}^*, \ldots, u_{ki}^*, i = 1, \ldots, n\}$, where the ith row $u_{1i}^*, \ldots, u_{ki}^*$ represents a random permutation of integers $(1, \ldots, k)$ related to the ith unit and where permutations related to different units are independent.

Remark 2. As permutations are taken within units, this solution is not appropriate for the so-called block effect. In order to obtain a separate test for the block effect, a rather different strategy must be followed (see Pesarin, 2001, Chapter 8). However, when testing for the block effect is not important, the above solution is effective. From this point of view, this two-way ANOVA layout for repeated observations may be seen as specific to the absence of interaction effects.

Remark 3. When assumption (i) or (ii) or both are violated, the permutation principle is also violated. Thus, in these circumstances, some of the associated conclusions may become improper.

An Example

Let us consider an example from Landenna and Marasini (1990, p. 272) concerning observations of blood testosterone in 11 women observed five times during one day at 08.00, 08.30, 09.00, 10.00, and 15.00 hours. The data, expressed in milligrams per litre, are reported in Table 2.2.

The purpose of this experiment is to evaluate whether the level of testosterone in the blood is subject to change during the day. This example has the peculiarity that the observations are dependent since they are recorded on the same women at different times. Moreover, the problem of testing for possible differences among individuals, the so-called block effect, was regarded as unimportant, in the sense that its existence is well known and there is therefore no particular interest in testing for it. Therefore, under H_0 we can permute the data inside the rows of the data set independently. This problem can be viewed as a two-way ANOVA model without interaction, where the main factors are 'time' (factor B) and 'woman' (factor A: blocking factor, not of interest).

```
setwd("C:/path")
data<-read.csv("Testosterone.csv",header=TRUE)
Y = rep(seq(1,5),each=11) ; Time = colnames(data)
boxplot(unlist(data) Y,xlab="Time",ylab="Testosterone",names=Time)
lines(seq(1,5),apply(data,2,mean),lty="dotted")
```

Table 2.2 Repeated blood testosterone levels in 11 women

i	0800	0830	0900	1000	1500
1	320	278	236	222	232
2	478	513	415	359	292
3	921	701	645	526	458
4	213	230	261	253	199
5	273	338	323	332	222
6	392	302	289	305	172
7	469	443	292	235	233
8	422	389	359	331	185
9	613	649	626	588	636
10	395	318	298	269	328
11	462	400	360	247	284

The commands above assign the data set to the object `data` and represent it with a box-plot, the dotted line linking the sample means at each time. The total deviance SST can be decomposed as $SST = SSA + SSB + SSR$, where SSA and SSB are the deviances due to the main effects and SSR is the residual deviance. Note that SST is constant at each permutation, and so is SSA since we permute observations within the rows of `data`. On the other hand, SSB and SSR vary at each permutation. The test statistic for the time effect is $F_B = (df_{SSR}/df_{SSB}) \times SSB/SSR$. Leaving out the degrees of freedom df_{SSB} and df_{SSR}, which are permutationally invariant, the easiest way to obtain the residual deviance at each permutation is to write it as $SSR^* = SST - SSA - SSB^*$. Therefore, the F statistic can be written as $F^* = SSB^*/(SS - SSB^*)$, where the sum of squares SS is constant at each permutation. It is easy to see that F^* is a monotone function of $T^* = SSB^*$.

```
n=dim(data)[1] ; p = dim(data)[2]; B=1000;
m = mean(mean(data))
m.col = apply(data,2,mean)
SSB = n*sum((m.col-m)^2)

T<-array(0,dim=c((B+1),1))
T[1] = SSB

data.star = data
for(bb in 2:(B+1)){

U = matrix(runif(n*p),nrow=n) ## U is n x p
R = apply(U,1,rank)            ## R is p x n

for(i in 1:n){
data.star[i,] = data[i,R[,i]]
}

m.col = apply(data.star,2,mean)
SSB =  n*sum((m.col-m)^2)
T[bb] = SSB
}
t2p(T)[1]

[1] 0.0003
```

The p-value of this example is extremely significant, therefore we can conclude that testosterone levels varies according to time of day. With $B = 5000$ CMC iterations we obtain $\hat{\lambda} = 0.0003$, which leads to the rejection of H_0 at $\alpha = 0.001$. This result fits with that of Friedman's rank test: for the present data set $T_F = 19.709$, the permutation null distribution of which is well approximated by a central χ^2 with 4 d.f.

The corresponding MATLAB code is given below:

```
D=xlsimport('TestosteroneMATLAB.xls');
reminD(D)
[P T options] = NP_ReM('Y','Time','seq',1000,-1);
P2=NPC(P,'F');
```

The data set and the corresponding software codes are available from the `examples_chapters_1-4` folder on the book's website.

2.6.2 *Problems and Exercises*

1) Discuss a solution for a problem with repeated measurements in which the response model is $X_{ji} = \mu + \delta_j + Z_{ij}$, $i = 1, \ldots, n$, $j = 1, \ldots, k$ (note that in $H_0 : \{\delta_1 = \ldots = \delta_k = 0\}$, responses are exchangeable with respect to treatment levels *and also* with respect to individuals).

2) In the spirit of Remark 3, 2.6.1, prove that the test T_R^* remains valid even when $\mathbb{V}(Z_{ji}) = \sigma_i^2$, $i = 1, \ldots, n$, corresponding to 'within-unit homoscedasticity'. Discuss the distributional conditions for error component Z in which the permutation solution remains valid.

3) With reference to the problem of repeated measurements discussed in this section, extend the permutation solution to the case in which responses are binary (this corresponds to an extension of McNemar's test to $k > 2$ measurement occasions).

4) Prove that the test statistic T_R^* in Section 2.6.1 is permutationally equivalent to $\sum_j (\bar{X}_{j.}^*)^2$, which is easier for computation.

5) Assume that responses are repeated twice on the same units at two treatment levels and that units are partitioned into k groups, $\mathbf{X} = \{X_{tji}, t = 1, 2, j = 1, \ldots, k, i = 1, \ldots, n_j\}$. Assuming suitable homoscedasticity, discuss a solution to the problem in which $H_0 : \{X_{1j} \stackrel{d}{=} X_{2j}, j = 1, \ldots, k\}$ against a reasonable set of alternatives.

6) With reference to the response model of Problem 5 above, discuss a solution to the problem where $H_0 : \{(X_{11} - X_{12}) \stackrel{d}{=} \ldots \stackrel{d}{=} (X_{k1} - X_{k2})\}$ against a reasonable set of alternatives.

7) Prove that, with reference to the problem of repeated measurements discussed in this section, the orbit $\mathcal{X}_{/\mathbf{X}}$ associated with the data set $\mathbf{X}$ contains $(k!)^n$ elements and that the set of sufficient statistics in the presence of block effects is $(\mathbf{X}_1; \ldots; \mathbf{X}_n)$, that is, the set of individual profiles. In accordance with this result, note that permutations are allowed only within units, and that permutations related to different units are independent.

8) With reference to the solution of Example 7, 2.6, find a suitable solution when observed differences X are partitioned into $2k$ symmetric classes, with $k > 2$, where symmetry is with respect to the origin.

9) Discuss a test on paired observations with the assumption that data are ordered categorical and that paired observations permit us to establish whether responses are either better, equivalent or worse on one occasion of measurement than on the other. Show that this leads to a version of McNemar's test.

10) Extend the test solution for paired observations to the case where the model for responses takes the multiplicative form $X_{Bi} = \rho \cdot X_{Ai} + \varepsilon_i$, $i = 1, \ldots, n$, so that $H_0 : \{\rho = 1\}$ against $H_1 : \{\rho > 1\}$.

11) Discuss a permutation solution for paired observations in the case where $\sigma_A \neq \sigma_B$.

12) Show that when there are ties in Problem 9, that is, when there is at least one instance of no variations in the categorical responses, a solution not restricted to non-null differences implies auxiliary randomization (Lehmann, 1986).

13) With reference to Remark 3, 2.1.2, prove that if the ith unit has density distribution f_i, then the set of absolute differences $\{|X_i|, i = 1, \ldots, n\}$ is a set of jointly minimal sufficient statistics in H_0.

14) With reference to the problem of Example 6, 2.6.1, show that one suitable permutation solution when observed differences X are partitioned into $2k$ symmetric classes, with $k > 2$, where symmetry is with respect to the origin and when the alternatives are one-sided, as in $H_1 = \{Y_1 \stackrel{d}{>} Y_2\}$, is $T_>^* = \sum_{(k+1) \leq i \leq 2k} [f^*(i) - f_0(i)]/\sqrt{f_0(i)}$. Show that if $f_0(i)$ are not small, the permutation asymptotic distribution of $T_>^*$ is normal with mean 0 and variance k.

15) With reference to Example 3, 2.6, prove that two tests $T^* = \left(\sum_i X_i Y_i^* - n\bar{X}\bar{Y}^*\right) / \left[\sum_i (X_i - \bar{X})^2 \cdot \sum_i (Y_i^* - \bar{Y}^*)^2\right]^{1/2}$ and $T_\beta^* = \sum_i X_i Y_i^*$ are permutationally equivalent.

2.7 Examples of Multi-Sample Problems

In this section we present some typical examples of multi-sample permutation problems.

Example 1. *Testing for the equality of two distributions.*

Let us assume that observations from a response variable X on n units are partitioned into two groups corresponding to two levels of a symbolic treatment, of respectively n_1 and n_2 units. Let us also assume that the response variables in the two groups have unknown distributions P_1 and P_2, both defined on the probability space $(\mathcal{X}, \mathcal{B})$, where $\mathcal{X}$ is the sample space and $\mathcal{B}$ is an algebra of events. Hence, the sample data are

$$\mathbf{X} = \left\{X_{ji}, i = 1, \ldots, n_j, j = 1, 2\right\}.$$

It is generally of interest to test the null hypothesis $H_0 : \{P_1 = P_2\}$, that the two groups have the same underlying distribution, against the alternative $H_1 : \{P_1 < (\text{or} \neq, \text{or} >)\ P_2\}$, that events $A \in \mathcal{B}$ exist such that $P_1(A) < (\text{or} \neq, \text{or} >)\ P_2(A)$.

This problem may be dealt with in several ways, according to specific *side-assumptions* regarding the meaning of the concept of *inequality in distribution*. The next four examples cover some particular specifications of this concept. Two more specifications, the solution of which involves the NPC method, are discussed in Examples 3–6 and 8, 4.6.

Example 2. *Comparison of two locations.*

The first specification of the concept of inequality in distribution is concerned with the so-called *comparison of two means* (or, more generally, of two locations). This presumes that side-assumptions for the problem are such that the response data behave according to an *additive model* such as

$$X_{ji} = \mu + \delta_j + \sigma(\delta_j) \cdot Z_{ji}, \quad i = 1, \ldots, n_j, j = 1, 2,$$

where Z_{ji} are exchangeable random deviates with null mean values and unknown distribution P, μ is an unknown population constant, δ_j is the so-called fixed effect at the jth treatment, and $\sigma(\delta_j)$ is a scale coefficient which may be a monotonic function of δ_j or of $|\delta_j|$ and satisfies the condition $\sigma(0) = \sigma$, provided that the two underlying CDFs F_1 and F_2 do not cross each other (see Section 2.1.1). Some other specifications of this model are presented in Section 1.10.1.

Thus, the hypotheses become $H_0 : \{\delta_1 = \delta_2\}$ against $H_1 : \{\delta_1 < (>)\ \delta_2\}$, and a test statistic is $T = \sum_i X_{1i}/n_1 - \sum_i X_{2i}/n_2$, or any of its permutationally equivalent forms (we will see in Section 3.1 that this test is conditional and unconditional uniformly unbiased). Noting that H_0 implies exchangeability of the data between two groups, the set of sufficient statistics in H_0 is the pooled data set $\mathbf{X} = \mathbf{X}_1 \biguplus \mathbf{X}_2$ (see Section 2.1.2). It should be emphasized here that errors $\sigma(\delta_j) \cdot Z_{ji}$ are exchangeable between the two groups *only* in H_0.

Remark 1. In order to establish whether the test T is an exact permutation test (see Remark 1, 2.2.4 and Proposition 2, 3.1.1), let us consider its *permutation structure*, that is, the representation of a generic permutation of T in terms of treatment effects and errors.

To this end, let us imagine that ν^* data are randomly exchanged between two groups. Hence, after very simple calculations, we see that the permutation structure of T is

$$\begin{aligned} T^* &= \sum_i X_{1i}^*/n_1 - \sum_i X_{2i}^*/n_2 \\ &= \left(1 - \frac{2\nu^*}{n_1}\right)\delta_1 - \left(1 - \frac{2\nu^*}{n_2}\right)\delta_2 + \sum_i \sigma(\delta_1^*)Z_{1i}^*/n_1 - \sum_i \sigma(\delta_2^*)Z_{2i}^*/n_2. \end{aligned}$$

This structure shows that if and only if H_0 is true, T^* depends only on a permutation of exchangeable errors because all other quantities (μ, δ_1, and δ_2) simplify. Obviously, in H_1, the permutation structure depends also on treatment effects, but not on μ. Hence, this is an exact permutation test for testing $H_0 : \{\delta_1 = \delta_2\}$ against $H_1 : \{\delta_1 < (>)\ \delta_2\}$. Moreover, with random effects Δ such that $\Pr\{\Delta \geq 0\} = 1$ and $\Pr\{\Delta = 0\} > 0$, that is, when treatment is ineffective with some units, it is easy to prove that the test statistic $T^* = \sum_i X_{1i}^*$ is exact and unbiased (see Problems 13, 2.9 and 41, 3.9). Furthermore, as the non-null permutation structure is monotonically ordered with respect to $\delta_2 - \delta_1$ (see Theorem 2, 3.1.1), this test is at least unbiased.

Remark 2. Finding a proper solution to a problem such as the divergence of two medians or of two trimmed means is also straightforward. In order to obtain such an extension, we should assume that sample medians are proper indicators of treatment effects. Thus, one effective testing strategy is to find data transformations $Y = \varphi(X)$, so that an additive model such as that of the example is appropriate, at least approximately. Indeed, it is known that when an additive model for responses occurs, then test statistics based on sample means, such as T above, are often approximately 'good' tests provided that error components Z have finite mean. In Example 3, 4.6, a *multi-aspect* permutation solution is presented for jointly testing for a set of different transformations of interest within the NPC method.

Example 3. *Testing for equality of two distributions.*

The second specification of the concept of inequality in distribution is concerned with the so-called *comparison of two distributions* P_1 and P_2. This is a typical *goodness-of-fit* problem which gives rise to quite a large family of solutions, the best-known representatives of which are the Kolmogorov–Smirnov and Anderson–Darling test statistics. Both are based on the divergence of two EPMs, one from each data group. For more details, see, for example, D'Agostino and Stephens (1986).

Let us use $\hat{P}_j(A) = \sum_i \mathbb{I}(X_{ji} \in A)/n_j$, $j = 1, 2$, where A is any event of the algebra $\mathcal{A}$ (see Definition 2, 2.1.3) to denote the two EPMs. Hence, a proper permutation test for $H_0 : \{P_1 = P_2\}$ against $H_1 : \{P_1 \neq P_2\}$, in the spirit of the Kolmogorov–Smirnov statistic, is

$$T_{KS}^* = c_{n_1 n_2} \cdot \sup_{A \in \mathcal{B}} \left| \hat{P}_1^*(A) - \hat{P}_2^*(A) \right|,$$

where $c_{n_1 n_2}$ is a normalizing constant, $\mathcal{B}$ is a suitable non-empty collection of events and $\hat{P}_j^*(A) = \sum_i \mathbb{I}(X_{ji}^* \in A)/n_j$, $j = 1, 2$, are two permutation EPMs. Note that we usually only require $\mathcal{B}$ to be a proper subset of $\mathcal{A}$, although they may coincide. Observe that this solution does not assume continuity for X, as it may be applicable to any kind of variable. For numeric variables for which probability measures are equivalent to CDFs, it may be simplified into

$$T_{KS}^* = c_{n_1 n_2} \cdot \sup_{t \in \mathcal{R}^1} \left| \hat{F}_1^*(t) - \hat{F}_2^*(t) \right|,$$

where the permutation EDFs are $\hat{F}_j^*(t) = \sum_i \mathbb{I}(X_{ji}^* \leq t)/n_j$, $j = 1, 2$.

Of course, a solution in the spirit of the Anderson–Darling test is

$$T_{AD}^{*2} = n \cdot \int_{-\infty}^{\infty} \left(\hat{F}_1^*(t) - \hat{F}_2^*(t) \right)^2 \cdot \left(\hat{F}(t)[1 - \hat{F}(t)] \right)^{-1} d\hat{F}(t),$$

where $\hat{F}(t) = [n_1 \cdot \hat{F}_1(t) + n_2 \cdot \hat{F}_2(t)]/n$ is the pooled EDF, which in turn is a permutation invariant function (see Proposition 2, 2.1.3) the role of which is to standardize differences $\hat{F}_1^*(t) - \hat{F}_2^*(t)$ for every t.

In the case of stochastic dominance alternatives, when $H_1 : \{F_1(h) \geq F_2(h), h = 1\ldots, k\}$ where inequality becomes strict with probability greater than zero and the reverse inequality is assumed not possible, the Anderson–Darling type permutation test becomes

$$T^*_{AD} = \sum_{i=1}^{n} \left[\hat{F}^*_1(X_i) - \hat{F}(X_i)\right] \cdot \left(\hat{F}(X_i)[1 - \hat{F}(X_i)]\right)^{-1/2}.$$

Example 4. *Goodness-of-fit test for ordered categorical variables.*

The third specification of the concept of inequality in distribution is concerned with the so-called goodness-of-fit for ordered categorical variables (see also Section 2.8).

Let us assume that a given ordered categorical variable X is partitioned into $k \geq 2$ classes $\{A_h,\ h = 1, \ldots, k\}$, in the sense that relationships such as $A_h \prec A_j$ have a clear meaning for every pair $(1 \leq h < j \leq k)$. Let us also assume that the data are partitioned according to two levels of a symbolic treatment. In other words, given two independent random samples $\mathbf{X}_j = \{X_{ji},\ i = 1, \ldots, n_j\}$, $j = 1, 2$, we wish to test the hypotheses

$$H_0 : \left\{X_1 \stackrel{d}{=} X_2\right\} = \{F_1(h) = F_2(h),\ \forall h = 1, \ldots, k\},$$

that is, $H_0 : \left\{\bigcap_h [F_1(h) = F_2(h)]\right\}$, against the non-dominance alternative $H_1 : \{X_1 \stackrel{d}{\neq} X_2\} = \left\{\bigcup_h [F_1(h) \neq F_2(h)]\right\}$, where $F_j(h) = \Pr\{X_j \leq A_h\}$, $j = 1, 2$, represent an analogue of a CDF for categorical variable X_j in class A_h. Of course, H_1 defines the inequality in distribution of X_1 with respect to X_2. Observed data are usually organized in a $2 \times k$ contingency table. Cumulative distribution functions F_j are estimated by the corresponding EDFs: $\hat{F}_j(h) = \sum_i \mathbb{I}(X_{ji} \leq A_h)/n_j$, $h = 1, \ldots, k$, $j = 1, 2$. Observe that in this setting the whole data set $\mathbf{X} = \mathbf{X}_1 \biguplus \mathbf{X}_2$ and the set of marginal frequencies $\{n_1, n_2, f_{\cdot 1}, \ldots, f_{\cdot k}\}$, where $f_{\cdot h} = \sum_j \mathbb{I}(X_{ji} \in A_h)$, are both sets of sufficient statistics. Hence, a proper permutation test, in accordance with the Anderson–Darling approach, is

$$T^{*2}_{AD} = \sum_{h=1}^{k-1} \left[\hat{F}^*_1(h) - \hat{F}^*_2(h)\right]^2 \cdot \left(\hat{F}(h)[1 - \hat{F}(h)]\right)^{-1},$$

where $\hat{F}(h) = \sum_{ji} \mathbb{I}(X_{ji} \leq A_h)/n$ and $F^*_j(h) = \sum_i \mathbb{I}(X^*_{ji} \leq A_h)/n_j$, $j = 1, 2$. Note that all summands in T^{*2}_{AD} are standardized quantities, except for a permutationally invariant coefficient common to all h.

Of course, in the case of dominance alternatives, where $H_1 : \{F_1(x) \geq F_2(x), x \in \mathcal{R}^1\}$, this test becomes

$$T^*_{AD} = \sum_{h=1}^{k-1} \left[\hat{F}^*_1(h) - \hat{F}^*_2(h)\right] \cdot \left(\hat{F}(h)[1 - \hat{F}(h)]\right)^{-1/2}.$$

Example 5. *Equality of two nominal distributions.*

The fourth specification of the concept of equality in distribution is concerned with the goodness-of-fit for nominal (unordered) categorical variables, also referred to as equality of two nominal distributions. As in the previous example, let us assume that a given nominal variable X is partitioned into $k \geq 2$ non-overlapping classes $\{A_h,\ h = 1, \ldots, k\}$, and that the data are grouped according to two levels of a symbolic treatment. In other words, given two independent random samples $\mathbf{X}_j = \{X_{ji}, i = 1, \ldots, n_j\}$, $j = 1, 2$, we wish to test the hypotheses

$$H_0 : \left\{X_1 \stackrel{d}{=} X_2\right\} = \left\{\bigcap_h [p_1(h) = p_2(h)]\right\}$$

against

$$H_1 : \left\{X_1 \overset{d}{\neq} X_2\right\} = \left\{\bigcup\nolimits_h [p_1(h) \neq p_2(h)]\right\},$$

where $p_j(h) = P_j(A_h)$. In this setting, as in Example 4 above, the set of marginal frequencies $\{n_1, n_2, f_{\cdot 1}, \ldots, f_{\cdot k}\}$, where the marginal frequencies $f_{\cdot h} = \sum_{ji} \mathbb{I}(X_{ji} \in A_h)$ and the whole data set $\mathbf{X} = \mathbf{X}_1 \biguplus \mathbf{X}_2$ are still sets of sufficient statistics in H_0. A very popular permutation test is Pearson's well-known chi-square,

$$T^{*2} = \sum\nolimits_{jh} \frac{n_j \cdot [p_j^*(h) - \hat{p}(h)]^2}{\hat{p}(h)},$$

where $p_j^*(h) = \sum_i \mathbb{I}(X_{ji}^* \in A_h)/n_j = f_{jh}^*/n_j$ and $\hat{p}(h) = f_{\cdot h}/n$, $j = 1, 2$, $h = 1, \ldots, k$. Note that if for some classes $f_{\cdot h} = 0$, then these may be discarded from the analysis without loss of generality. In the null hypothesis, the permutation distribution of T^{*2}, when sample sizes are large and $|2p_j(h) - 1|$ are not too close to 1, is well approximated by that of a central chi-square with $k - 1$ degrees of freedom. However, it is worth noting that T^{*2} is an exact permutation test because its null distribution depends only on exchangeable errors (see Remark 1, 2.2.4 and Proposition 2, 3.1.1); thus, avoiding well-known difficulties related to asymptotic approximations, its null distribution may be evaluated by a CMC method or by exact calculation.

Essentially similar to T^{*2} and partially in the spirit of Anderson–Darling, with obvious notation, is the test statistic

$$T_{fAD}^{*2} = \sum\nolimits_{jh} \left(\frac{f_{jh}^*}{n_j} - \frac{f_{\cdot h}}{n}\right)^2 \bigg/ \left[f_{\cdot h} \cdot (n - f_{\cdot h}) \frac{n - n_j}{n_j}\right].$$

Remark 3. Note that T^{*2} and T_{fAD}^{*2} may also be used for the ordered categorical testing problem, whereas T_{AD}^{*2} in Example 4 can only be used for testing with ordered categorical variables.

Remark 4. There are many different permutation solutions for testing the equality in distribution of two nominal variables; see, for instance, Cressie and Read (1988), Agresti (2002) and Section 2.8; see Chapter 6 for multivariate situations.

Example 6. *Testing for equality of $C > 2$ distributions.*

This problem is simply an extension to $C > 2$ distributions of those shown in Examples 1–3. Let us assume that data from a real-valued response variable X and observed on n units are partitioned into $C > 2$ groups, in accordance with C levels of a symbolic treatment of respectively $n_1, \ldots, n_C$ units. Let us also assume that responses behave in accordance with unknown distributions $P_1, \ldots, P_C$, all defined on the same probability space $(\mathcal{X}, \mathcal{A})$. Hence, the sample data are $\mathbf{X} = \{X_{ji}, i = 1, \ldots, n_j, j = 1, \ldots, C\}$.

We present two different situations: one in which the null hypothesis is $H_0 : \{P_1 = \ldots = P_C\}$, that the C groups have the same distribution, against the alternative $H_1 : \{H_0 \text{ is not true}\}$, that at least one distribution is different from one of the others; and a second concerned with tests for the equality of $C > 2$ means, the so-called *one-way ANOVA* design.

The problem concerning the *equality of C continuous distributions* may be solved, for example, by an Anderson–Darling type test, which is based on the statistic

$$T_{AD}^{*2} = \sum_{j=1}^{C} \sum_{i=1}^{n_j} \left[\hat{F}_j^*(X_{ji}) - \hat{F}_\cdot(X_{ji})\right]^2 \cdot \left(\hat{F}_\cdot(X_{ji})[1 - \hat{F}_\cdot(X_{ji})]\right)^{-1},$$

where $\hat{F}.(x) = \sum_j n_j \cdot \hat{F}_j(x)/n$ is the pooled EDF, and all the other symbols have obvious meanings.

In order to solve the one-way ANOVA problem, let us presume that side-assumptions are such that the response data behave according to an additive model such as

$$X_{ji} = \mu + \delta_j + \sigma(\delta_j) \cdot Z_{ji}, i = 1, \ldots, n_j, j = 1, \ldots, C,$$

where the coefficients have the same meanings as in Example 2 above and treatment effects satisfy the constraint $\sum_j \delta_j = 0$. In particular, when $\sigma(\delta_j) = \sigma$, we have the so-called *homoscedastic* situation. Other specifications of this model are presented in Section 1.11.

Of course, without loss of generality, the hypotheses for this problem become $H_0 : \{\delta_1 = \ldots = \delta_C = 0\}$ against $H_1 : \{H_0 \text{ is not true}\}$, and a test statistic is $T = \sum_j n_j \bar{X}_j^2$, where $\bar{X}_j = \sum_i X_{ji}/n_j$. Note that this statistic is permutationally equivalent to $\sum_j n_j(\bar{X}_j - \bar{X}.)^2/\sum_{ji}(X_{ji} - \bar{X}_j)^2$, where $\bar{X}. = \sum_{ji} X_{ji}/n$ is the pooled mean. A set of sufficient statistics in H_0 is the pooled set of observed data $\mathbf{X} = \mathbf{X}_1 \uplus \ldots \uplus \mathbf{X}_C$. It should be observed that errors $\sigma(\delta_j) \cdot Z_{ji}$ are exchangeable with respect to groups only in H_0.

In order to prove that T is an exact permutation test (see Remark 1, 2.2.4 and Proposition 2, 3.1.1), let us consider its *permutation structure* (see Remark 1 above), and assume that ν_{hj}^* data are randomly moved from the hth to the jth group, where $\sum_h \nu_{hj}^* = n_j$, $j = 1, \ldots, C$, and where ν_{hh}^* represents the number of data which remain in the hth group. After very simple calculations, we see that $T^* = \sum_j \left[\sum_i \left(\delta_j^* + \sigma(\delta_j^*)Z_{ji}^*\right)\right]^2 /n_j$. This permutation structure shows that, if and only if H_0 is true, T^* depends only on a permutation of exchangeable errors, whereas in H_1 it depends essentially on treatment effects as well. Hence, as the permutation null distribution of T, given $\mathbf{X}$, depends only on exchangeable errors, T is an exact permutation test. In the next chapter, we shall discuss the properties of conditional and unconditional unbiasedness.

Remark 5. The literature deals with two concepts related to exactness for permutation tests. The most important one is related to the exchangeability of observed data in H_0 and to a statistic T, the permutation distribution of which depends *only* on exchangeable errors (see Remark 1, 2.2.4 and Proposition 2, 3.1.1). The other concept is related to the algorithms for evaluating the permutation distribution of a given test statistic, either approximate or exact, according to the previous concept. For instance, when a CMC procedure is used to evaluate the permutation distribution of any test, then an unbiased estimate of this distribution is obtained. Of course, as the number B of CMC iterations increases, this estimate becomes stochastically more accurate. In this case, we say that the exact permutation distribution is known except for a statistical estimation. Of course, when proper routines for exact calculations are available, the distribution of a test may be known exactly.

Example 7. *Two-way ANOVA without interaction.*

The two-way ANOVA design admits some simplifications, in accordance with assumptions related to the response model. One of these assumes that the interaction effect is not present and that there is only one unit per block, so that factor B is generally of no practical interest (see Example 9 below). In the notation of Example 6 above, the related response model becomes

$$\mathbf{X} = \{X_{ji} = \mu + a_j + b_i + Z_{ji}, j = 1, \ldots, J, i = 1, \ldots, n\}.$$

Here the hypotheses of interest are $H_0 : \{a_j = 0, j = 1, \ldots, J\}$ against $H_1 : \{\text{some } a_j \neq 0, j = 1, \ldots, J\}$. We emphasize that a set of sufficient statistics in H_0 is the set of individual response profiles, $\mathbf{X} = (\mathbf{X}_1; \ldots; \mathbf{X}_n)$, because individual (or block) effects b_i, playing the role of unknown nuisance entities, may assume different values. This implies that observations are exchangeable only within each unit and that permutations associated to different units are independent (see Remark 3, 2.1.1). Accordingly, an appropriate test statistic (see also Section 2.6.1) is

$$T_R^* = \frac{\sum_j n_j(\bar{X}_{j.}^* - \bar{X}_{..})^2}{\sum_{ji}(X_{ji}^* - \bar{X}_{j.}^* - \bar{X}_{.i} + \bar{X}_{..})^2},$$

where $\bar{X}_{j.}^* = \sum_i X_{ji}^*/n$, $j = 1, \ldots, J$, $\bar{X}_{.i} = \sum_j X_{ji}/J = \sum_j X_{ji}^*/J$, $i = 1, \ldots, n$, because exchanges are within each unit, and $\bar{X}_{..} = (nJ)^{-1}\sum_{ji} X_{ji}$ respectively are column, row and global means.

A solution based on within-unit rank transformation is the well-known Friedman's rank test in Section 2.6.1 (see Friedman, 1937).

Remark 6. It is interesting to note that here the permutation structure of the test $T_R^* = \sum_j(\bar{X}_{j.}^* - \bar{X}_{..})^2/\sum_{ji}(X_{ji}^* - \bar{X}_{j.}^* - \bar{X}_{.i} + \bar{X}_{..})^2$ does not depend on block effects b_i, so that it is exact for testing H_0 against H_1 independently of block effects (see Problem 12, 2.9).

Example 8. *General two-way ANOVA.*

In the balanced fixed effects homoscedastic two-way ANOVA design, in which two factors A and B are presumed to be tested at respectively J and I levels, responses are assumed to behave according to the model

$$\begin{aligned}\mathbf{X} &= \{\mathbf{X}_{ji}, j = 1, \ldots, J, i = 1, \ldots, I\}\\ &= \{X_{jir} = \mu + a_j + b_i + (ab)_{ji} + Z_{jir}, j = 1, \ldots, J, i = 1, \ldots, I, r = 1, \ldots, n\},\end{aligned}$$

where X_{jir} are the responses, μ is a population constant, a_j is the effect of factor A at the jth level, b_i is the effect of factor B at the ith level, $(ab)_{ji}$ is the jith interaction effect, n is the number of independent runs of jith treatment, Z_{jir} are exchangeable random errors with null mean value and unknown distribution P (note homoscedasticity and $\mathbb{E}(Z) = 0$), $\mathbf{X}_{ji}$ are the data of the jith group, and the effects satisfy the side-conditions

$$\sum_j a_j = \sum_i b_i = \sum_j (ab)_{ji} = \sum_i (ab)_{ji} = 0.$$

The null overall hypothesis is generally written as

$$H_0 : \left\{\left[\bigcap_j (a_j = 0)\right] \bigcap \left[\bigcap_i (b_i = 0)\right] \bigcap \left[\bigcap_{ji} ((ab)_{ji} = 0)\right]\right\},$$

and the overall alternative as $H_1 : \{H_0 \text{ is not true}\}$. Let us briefly discuss these hypotheses (see also Chapter 11). The experimenter's greatest interest is usually in testing separately for main effects and interactions. Hence, in the present case there are three separate null hypotheses of interest: (i) $H_{0A} : \{a_j = 0, j = 1, \ldots, J\}$ against $H_{1A} : \{H_{0A} \text{ is not true}\}$, irrespective of the truth of H_{0B} and/or H_{0AB}; (ii) $H_{0B} : \{b_i = 0, i = 1, \ldots, I\}$ against H_{1B}, irrespective of the truth of H_{0A} and/or H_{0AB}; and (iii) $H_{0AB} : \{(ab)_{ji} = 0, j = 1, \ldots, J, i = 1, \ldots, I\}$ against H_{1AB}, irrespective of the truth of H_{0A} and/or H_{0B}. Thus, the aim is to find three separate and possibly uncorrelated tests.

This problem is fully discussed in Pesarin (2001) and Basso et al. (2009a), for both balanced and unbalanced situations within the conditionality principle, when conditioning on a minimal set of jointly sufficient statistics under three partial null hypotheses $H_{0A} \bigcup H_{0B} \bigcup H_{0AB}$. A set of sufficient statistics for such a design is $\mathbf{X} = \{\mathbf{X}_{11}; \ldots; \mathbf{X}_{JI}\}$, that is, the data set is partitioned in accordance with treatment groups. This implies that no datum can be exchanged between two different blocks without compromising the separability of effects, so that within naive exchanges effects remain confounded. So we should look at a kind of *restricted* permutation strategy.

To find that $\{\mathbf{X}_{11};\ldots;\mathbf{X}_{JI}\}$ is a set of sufficient statistics and to characterize the related permutation sample space $\mathcal{X}_{/\mathbf{X}}$, let us denote by f the density corresponding to the underlying unknown distribution P. The global likelihood associated with the data set is $L(\mathbf{X};\mathbf{a},\mathbf{b},\mathbf{ab}) = \prod_{ijr} f[X_{ijr};a_i,b_j,(ab)_{ij}]$. Two points, $\mathbf{X}$ and $\mathbf{X}'$, lie in the same orbit of a minimal sufficient set of statistics (see Section 2.1.2) if and only if the likelihood ratio

$$\frac{L(\mathbf{X};\mathbf{a},\mathbf{b},\mathbf{ab})}{L(\mathbf{X}';\mathbf{a},\mathbf{b},\mathbf{ab})} = \frac{\prod_{ijr} f[X_{ijr};a_i,b_j,(ab)_{ij}]}{\prod_{ijr} f[X'_{ijr};a_i,b_j,(ab)_{ij}]} = \rho_f(\mathbf{X},\mathbf{X}')$$

is independent of all effects $(\mathbf{a},\mathbf{b},\mathbf{ab})$ and of the underlying likelihood model f. This occurs if and only if $\mathbf{X}' = (\mathbf{X}'_{11};\ldots;\mathbf{X}'_{JI})$, where $\mathbf{X}'_{ji}$ is a proper permutation of $\mathbf{X}_{ji}$. Thus, the associated permutation sample space is $\mathcal{X}_{/\mathbf{X}} = \mathcal{X}_{/\mathbf{X}_{11}} \times \ldots \times \mathcal{X}_{/\mathbf{X}_{JI}}$, which corresponds to the cartesian product of JI separate subspaces.

Remark 7. It is worth noting that the pooled data set $\mathbf{X}_{11} \biguplus \ldots \biguplus \mathbf{X}_{JI}$ is not sufficient for separate testing of three null hypotheses $(H_{0A}, H_{0B}, H_{0AB})$ in a replicated complete factorial design, whereas it is sufficient for $H'_0 : \{H_{0A} \bigcap H_{0B} \bigcap H_{0AB}\}$. Straightforward proofs of these statements are left to the reader.

Remark 8. It is well known that solutions based on the assumption of normality for errors are known to be positively correlated. For a simple proof of this fact, let us suppose that variance estimates of effects and errors are respectively $\hat{\sigma}^2_A$, $\hat{\sigma}^2_B$, $\hat{\sigma}^2_{AB}$ and $\hat{\sigma}^2_Z$. Then the relationship

$$\mathbb{E}\left[\frac{\hat{\sigma}^2_A}{\hat{\sigma}^2_Z}\cdot\frac{\hat{\sigma}^2_B}{\hat{\sigma}^2_Z}\right] = \mathbb{E}(\hat{\sigma}^2_A)\cdot\mathbb{E}(\hat{\sigma}^2_B)\cdot\mathbb{E}(1/\hat{\sigma}^4_Z) > \mathbb{E}(\hat{\sigma}^2_A)\cdot\mathbb{E}(\hat{\sigma}^2_B)\cdot\left[\mathbb{E}(1/\hat{\sigma}^2_Z)\right]^2$$

is always true provided that $0 < \mathbb{E}(1/\hat{\sigma}^2_Z) < \infty$. Thus, the two statistics $\hat{\sigma}^2_A/\hat{\sigma}^2_Z$ and $\hat{\sigma}^2_B/\hat{\sigma}^2_Z$ are positively correlated, as are all the others.

Hence, no uncorrelated testing for factors and interactions occurs within a parametric setting, unless σ^2_Z is known or unless an arbitrary random partition of $\hat{\sigma}^2_Z$ into three independent components is taken into consideration. Moreover, in a nonparametric rank based setting, only heuristic solutions have been proposed in the literature since Friedman (1937). Among these only a few are proper exact solutions in very specific situations. For instance, when interaction is null by assumption, the Friedman test, based on ranks, is exact (see Example 7 above).

Synchronized Permutations

To introduce the concept of synchronized permutations as a suitable kind of restricted permutation, let us refer to a replicated 2 × 2 complete factorial design. Table 2.3 lists the effects of combinations of factor levels where $\mathbf{X} = (\mathbf{X}_{11};\mathbf{X}_{12};\mathbf{X}_{21};\mathbf{X}_{22})$ is the set of jointly sufficient statistics in the separate set of null hypotheses $\{H_{0A}, H_{0B}, H_{0AB}\}$.

Table 2.3 Effects of treatment combinations in a 2^2 factorial

	A_1	A_2
B_1	a, b, ab	$-a, b, -ab$
B_2	$a, -b, -ab$	$-a, -b, ab$

Let us first consider the permutation structure of two intermediate statistics for comparing factor A separately at levels 1 and 2 of factor B : $T_{11/21} = {}^aT_{A|1} = \sum_r Y_{11r} - \sum_r Y_{21r}$ and $T_{12/22} = {}^aT_{A|2} = \sum_r Y_{12r} - \sum_r Y_{22r}$.

Let us imagine that ν_1^* data from block A_1B_1 are exchanged with ν_1^* data from block A_2B_1. In addition, let us suppose that ν_2^* data are exchanged from blocks A_1B_2 and A_2B_2. After elementary calculations, the permutation structures of the two intermediate statistics are respectively

$$ {}^aT_{A|1}^* = 2(n - 2\nu_1^*)(a + ab) + n(\overline{Z}_{11}^* - \overline{Z}_{21}^*) $$

and

$$ {}^aT_{A|2}^* = 2(n - 2\nu_2^*)(a - ab) + n(\overline{Z}_{12}^* - \overline{Z}_{22}^*), $$

where $\overline{Z}_{ij}^* = \sum_r Z_{ijr}^*/n$ are sample means of permutation errors relative to the ijth block.

Effects a and ab are confounded in both ${}^aT_{A|1}^*$ and ${}^aT_{A|2}^*$. However, if we synchronize the permutations of two intermediate statistics by imposing that $\nu_1^* = \nu_2^* = \nu^*$, then ${}^aT_A^* = {}^aT_{A|1}^* + {}^aT_{A|2}^*$ and ${}^aT_{AB}^* = {}^aT_{A|1}^* - {}^aT_{A|2}^*$ have respective permutation structures given by

$$ {}^aT_A^* = 4(n - 2\nu^*) \cdot a + n(\overline{Z}_{11}^* + \overline{Z}_{12}^* - \overline{Z}_{21}^* - \overline{Z}_{22}^*) $$

and

$$ {}^aT_{AB}^* = 4(n - 2\nu^*) \cdot ab + n(\overline{Z}_{11}^* - \overline{Z}_{12}^* - \overline{Z}_{21}^* + \overline{Z}_{22}^*). $$

Thus, ${}^aT_A^*$, being dependent only on effect a and on a linear combination of exchangeable errors, gives a separate exact permutation test for H_{0A}, independent of the truth of H_{0B} and/or H_{0AB}. Separately, ${}^aT_{AB}^*$, being dependent only on the interaction effect ab and on a linear combination of exchangeable errors, gives an exact permutation test for H_{0AB}, independent of the truth of H_{0A} and/or H_{0B}. Note that by observing that two error components are mutually orthogonal, the two separate tests, ${}^aT_A^*$ and ${}^aT_{AB}^*$, are uncorrelated.

In order to complete the analysis we must take into consideration: (i) the intermediate statistics for contrasting factor B separately for levels 1 and 2 of factor A, $T_{11/12} = {}^bT_{B|1} = \sum_r Y_{11r} - \sum_r Y_{12r}$ and $T_{21/22} = {}^bT_{B|2} = \sum_r Y_{21} - \sum_r Y_{22}$; and (ii) the intermediate statistics for cross-comparison of A_1B_1 with A_2B_2, and A_1B_2 with A_2B_1, $T_{11/22} = \sum_r Y_{11r} - \sum_r Y_{22r}$ and $T_{12/21} = \sum_r Y_{12r} - \sum_r Y_{21r}$. Of course, these new intermediate statistics are considered independently of ${}^aT_{A|1}$ and ${}^aT_{A|2}$ and obtained by independent synchronized permutations. Thus, we consider permutations between paired blocks (A_1B_1, A_1B_2), (A_2B_1, A_2B_2), and (A_1B_1, A_2B_2), (A_1B_2, A_2B_1) again by randomly exchanging ν^* data.

Therefore, the permutation structure of ${}^bT_B^* = {}^bT_{B|1}^* + {}^bT_{B|2}^*$ is given by $4(n - 2\nu^*) \cdot b + n(\hat{Z}_{11}^* - \hat{Z}_{12}^* + \hat{Z}_{21}^* - \hat{Z}_{22}^*)$ and that of ${}^bT_{AB}^* ={}^b T_{B|1}^* - {}^bT_{B|2}^*$ by $4(n - 2\nu^*) \cdot ab + n(\hat{Z}_{11}^* - \hat{Z}_{12}^* - \hat{Z}_{21}^* + \hat{Z}_{22}^*)$, where $\hat{Z}_{ij}^*$ are permutations of sample means of error components when ν^* elements are randomly exchanged between each pair of blocks. As ${}^bT_B^*$ depends only on effect b and on a linear combination of exchangeable errors, it gives a separate exact permutation test for H_{0B}, independent of the truth of H_{0A} and/or H_{0AB}. Moreover, ${}^bT_{AB}^*$ depends only on effect ab and on a linear combination of exchangeable errors, so that it gives a separate exact permutation test for H_{0AB}, independent of the truth of H_{0A} and/or H_{0B}.

In addition, the permutation structure of $\tilde{T}_A^* = T_{11/22}^* + T_{12/21}^*$ is $4(n - 2\nu^*) \cdot a + n(\tilde{Z}_{11}^* + \tilde{Z}_{12}^* - \tilde{Z}_{21}^* - \tilde{Z}_{22}^*)$ and that of $\tilde{T}_B^* = T_{11/22}^* - T_{12/21}^*$ is $4(n - 2\nu^*) \cdot b + n(\tilde{Z}_{11}^* - \tilde{Z}_{12}^* + \tilde{Z}_{21}^* - \tilde{Z}_{22}^*)$, where $\tilde{Z}_{ij}^*$ are permutations of sample means of error components in cross-comparisons. Thus, they provide separate exact permutation tests for a and b, respectively.

Remark 9. In this analysis, we have two partial tests for each effect: for separately testing $\{a = 0\}$, we have ${}^aT_A^*$ and $\tilde{T}_A^*$; for $\{b = 0\}$, ${}^bT_B^*$ and $\tilde{T}_B^*$; for $\{ab = 0\}$, ${}^aT_{AB}^*$ and ${}^bT_{AB}^*$. It should be observed that these pairs of partial tests are based on independent permutations; in addition, all six partial tests are uncorrelated (e.g. ${}^aT_A^*$ and ${}^aT_{AB}^*$ are uncorrelated), because their error components are based on orthogonal combinations of permuted sample means of errors. Thus, in order to complete the test procedure, we use one combination for each pair of tests on the same effect, $T_A''^* = \psi_A({}^aT_A^*, \tilde{T}_A^*)$, $T_B''^* = \psi_B({}^bT_B^*, \tilde{T}_B^*)$ and $T_{AB}''^* = \psi_{AB}({}^aT_{AB}^*, {}^bT_{AB}^*)$, where ψ_h, $h = A, B, AB$, are suitable combining functions (see Section 4.2.4).

Alternatively, and in order to save computation time, instead of using cross-comparisons, we may consider using only one partial test for each effect, such as ${}^aT_A^*$, ${}^bT_B^*$, and ${}^aT_{AB}^*$ or ${}^bT_{AB}^*$. As a result of this choice, we have a kind of *weakly randomized* solution given by $T_A^* = {}^aT_A^*$, $T_B^* = {}^bT_B^*$, and $T_{AB}^* = {}^aT_{AB}^*$ or $T_{AB}^* = {}^bT_{AB}^*$. This solution is compatible with the idea of using only one realignment for each main effect (see Chapter 11). However, we observe that, in general, $T_h''^*$ and T_h^*, $h = A, B, AB$, always give rise to almost coincident inferences, because in both cases all the observed data participate in the synchronized permutation procedure, although in T_h^* they are not permuted in all possible ways. However, both are invariant with respect to the experimenter's choice. In this sense, the impact of this choice on inferential conclusions is substantially irrelevant. Thus, henceforth, we mainly consider these weakly or almost non-randomized solutions.

Remark 10. Using the same arguments above, we observe that if interaction effects are assumed not to be present, so that $(ab)_{ji} = 0$, then the two separate sets of partially pooled data $\mathbf{X}_{11} \biguplus \mathbf{X}_{12}$ and $\mathbf{X}_{21} \biguplus \mathbf{X}_{22}$ are separately sufficient for H_{0A}, conditionally on levels B_1 and B_2, respectively. This allows us to obtain two independent partial tests for $H_{0A|B_1}$ and $H_{0A|B_2}$. Hence, for testing H_{0A}, any combination of two independent tests gives the solution.

Remark 11. The permutation distribution of $T_h''^*$, $h = A, B, AB$, in H_{0h} depends only on permutations of exchangeable errors, so that each test is permutationally exact. Moreover, the three separate tests are uncorrelated. In addition, $T_h''^*$ are proper tests for one-sided alternatives, whereas $(T_h''^*)^2$ or $|T_h''^*|$ are proper tests for two-sided alternatives.

Remark 12. Let us consider the data set $(\mathbf{X}_{11}; \mathbf{X}_{12}; \mathbf{X}_{21}; \mathbf{X}_{22})$ as a point in the sample space, in the sense that all its coordinates are geometrically determined. Thus (i) if the permutation procedure in the first pair of blocks exchanges elements in exactly the same corresponding coordinates as that applied to the other pair of blocks, so that exactly the same permutation is applied to each pair of blocks, giving rise to constrained synchronized permutations (CSP), then the cardinality of the permutation support of intermediate and separate tests for restricted alternatives is $\binom{2n}{n}$, where n is the number of replicates in each block. Otherwise (ii) if permutations applied to each pair of blocks exchange the same number of elements *after shuffling data in each block*, giving rise to unconstrained synchronized permutations (USP), then the cardinality of the permutation support of separate tests for restricted alternatives becomes $\sum_{\nu^*} \binom{n}{\nu^*}^4$. We observe that CSP allow us to control the minimum attainable α-size of the test, although in general USP give lower minimum attainable α-sizes. For their use, see Basso et al. (2009a) and Chapter 11.

Example 9. *A problem with repeated measurements (revisited).*

One more simplification assumes that n units are partitioned into $C \geq 2$ groups and that a variable X is observed. Groups are of size $n_j \geq 2$, $j = 1, \ldots, C$, with $n = \sum_j n_j$. Units belonging to the jth group are presumed to receive a treatment at the jth level. All units are observed on k fixed time occasions $\tau_1, \ldots, \tau_k$, where k is a finite integer. Hence, for each unit we observe the profile of a stochastic process, and profiles related to different units are assumed to be stochastically independent. A profile may be viewed either as the outcome of an underlying stochastic process or as a k-dimensional random variable.

Let us assume that the response model is $X_{ji}(t) = \mu + \eta_j(t) + \delta_{ji}(t) + \sigma(\eta_j, t) \cdot Z_{ji}(t)$, $i = 1, \ldots, n_j$, $j = 1, \ldots, C$, $t = 1, \ldots, k$, where $Z_{ji}(t)$ are error terms assumed to be i.i.d. with respect to units and treatment levels but not independent of time; μ is a population constant; coefficients $\eta_j(t)$ represent treatment effects and may depend on time; coefficients $\delta_{ji}(t)$ represent the so-called individual effects; and $\sigma(\eta_j, t)$ are time-varying scale coefficients which may depend, through monotonic functions, on treatment effects η_j or $|\eta_j|$, provided that the stochastic ordering of responses is satisfied. Hence, the whole set $\mathbf{X}$ of observed data is organized in a two-way layout. Alternatively, $\mathbf{X}$ may be organized as a one-way layout of profiles $\mathbf{X} = \{\mathbf{X}_{ji}, i = 1, \ldots, n_j, j = 1, \ldots, C\}$, where $\mathbf{X}_{ji} = \{X_{ji}(t),\ t = 1, \ldots, k\}$ indicates the jith observed profile. Let us assume that the null hypothesis of interest is that there are no differences in time effects due to treatment, so that $H_0 : \{\mathbf{X}_1 \stackrel{d}{=} \ldots \stackrel{d}{=} \mathbf{X}_C\}$ is equivalent to $H_{0\eta} : \{\eta_1(t) = \ldots = \eta_C(t), \forall t\}$ against H_1 : {at least one equality does not hold}. Distributional assumptions on responses allow us to suppose that the pooled set of individual profiles $\mathbf{X} = \mathbf{X}_1 \biguplus \ldots \biguplus \mathbf{X}_C$ is a set of sufficient statistics for the problem in H_0. Moreover, it should be emphasized that H_0 implies that the observed individual time profiles are exchangeable with respect to treatment levels (see Chapter 7 for a wider discussion).

Remark 13. The situation presented in this example is standard for most experimental designs when units are randomly assigned to treatment levels and are assumed to be homogeneous with respect to the most important experimental conditions, such as age, sex and health. Thus, the permutation testing principle applies to observed time profiles.

Remark 14. As the testing problems with repeated measurements are quite complex, especially when there are more measurements per individual than there are individuals, or when there are missing values, we shall postpone their discussion to Chapter 7, after the theory and methods of NPC have been developed.

Example 10. *Approximate testing for equality of scale coefficients.*

Suppose that we are interested in testing the equality of scale coefficients of two univariate distributions. To be more specific, let us assume that data are collected according to a response model $X_{ji} = \mu_j + \sigma_j \cdot Z_{ji}, i = 1, \ldots, n_j,\ j = 1, 2$, and that the hypotheses are written as

$$H_0 : \{\sigma_1 = \sigma_2\} = \left\{(X_1 - \mu_1) \stackrel{d}{=} (X_2 - \mu_2)\right\}$$

against $H_1 : \{\sigma_1 <$ (or $\neq$, or $>$) $\sigma_2\}$, where μ_1 and μ_2 are unknown nuisance location parameters and error components Z_{ji} are i.i.d. with null mean and unknown distribution P. Note that under these assumptions the Z_{ji} are exchangeable in both H_0 and H_1.

This problem has been widely studied within the nonparametric approach based on ranks; see, for instance, Ansari and Bradley (1960), Moses (1963) and Witting (1995). As it does not admit of any exact nonparametric non-randomized solution, unless μ_1 and μ_2 are known, here we would like to examine an approximate permutation solution.

For this problem, the pair of data groups $\mathbf{X} = (\mathbf{X}_1; \mathbf{X}_2)$ is a set of sufficient statistics (see Section 2.1.2), so that any permutation solution should be referred to it or, equivalently, to $(\bar{X}_1; \bar{X}_2; \mathbf{Y}_1; \mathbf{Y}_2)$, where $\mathbf{Y}_j = \{Y_{ji} = X_{ji} - \bar{X}_j, i = 1, \ldots, n_j\}$, $\bar{X}_j = \sum_i X_{ji}/n_j$, $j = 1, 2$, because there is a one-to-one relationship between the two sets. It should be noted that, as $Y_{ji} = \sigma_j(Z_{ji} - \bar{Z}_j)$, then sample deviates Y_{ji} are not exactly exchangeable in H_0 (see point (ii) in Remark 2, 2.1.2). However, if μ_1 and μ_2 were known, then the pooled set of true deviates $\mathbf{Y}^\dagger = \{X_{ji} - \mu_j, i = 1, \ldots, n_j, j = 1, 2\}$ would be sufficient for the problem and exchangeability in H_0 would be satisfied. Thus, as μ_1 and μ_2 are unknown, we can only proceed approximately by conditioning with respect to pooled sample deviates $\mathbf{Y}_1 \biguplus \mathbf{Y}_2$, which may be used to estimate $\mathbf{Y}^\dagger$.

Then, according to Good (2000), this problem may be approximately solved, for instance, by the test statistic

$$T_\sigma^* = \varphi\left(\sum_i Y_{1i}^{*2}/n_1 - \sum_i Y_{2i}^{*2}/n_2\right),$$

the permutations are obtained as in Section 2.2.5, and the function $\varphi(\cdot)$ corresponds to $+(\cdot)$ if the alternatives are '>', $-(\cdot)$ if '<', and the absolute value $|\cdot|$ if '$\neq$'. Note that this solution becomes exact asymptotically. The test statistic for comparison of mean absolute deviates, $T_{A\sigma}^* = \varphi\left(\sum_i |Y_{1i}^*|/n_1 - \sum_i |Y_{2i}^*|/n_2\right)$, may also be of interest.

Example 11. *Testing for equality of two means when a covariate is observed.*

Suppose that a covariate X is observed together with a response variable Y, so that in a two-sample design the data set is $(\mathbf{Y}, \mathbf{X}) = \{Y_{1i} = \mu + \delta + \beta(X_{1i}), i = 1, \ldots, n_1; Y_{2i} = \mu + \beta(X_{2i}), i = 1 \ldots, n_2\}$ where β is a regression function as in point (iv) of Remark 2, 2.1.2. Suppose also that the hypotheses are $H_0 : \{(Y_1, X_1) \overset{d}{=} (Y_2, X_2)\} \equiv \{\delta = 0\}$ against $H_1 : \{Y_1 \overset{d}{>} Y_2 | X_1 \overset{d}{=} X_2\} \equiv \{\delta > 0\}$. H_1 emphasizes that treatment effect operates only on the response Y. This means that the covariate X and regression function β are not affected by treatment. This problem can be solved, by ignoring the covariate, as in Section 1.10.3. However, as we will see in Remark 1, 3.2.1, the conditional and unconditional power of a permutation test essentially depends on the *signal to noise ratio* δ/σ, in the sense that the larger δ/σ the larger the power. Thus if, as is usual, by taking account of covariate X the standard deviation reduces we may improve testing power. To this end we should provide an estimate $\hat{\beta}$ of regression β based on the pooled data set $(\mathbf{Y}, \mathbf{X})$. With such an estimate we obtain the empirical deviates $\hat{Y}_{ji} = [Y_{ji} - \hat{\beta}(X_{ji})]$ which are exchangeable and so the resulting test statistic becomes $T_\beta^* = \sum_i \hat{Y}_{1i}^*$. It is worth noting that, when $\hat{\beta}$ is the least squares estimate, the residual variance $\sigma_\beta^2(Y|\mathbf{X})$ does not exceed $\sigma^2(Y)$. Thus there is a gain of inferential efficiency, at least asymptotically. In particular, if β is linear the residual variance is $\sigma_\beta^2(Y|\mathbf{X}) = (1 - \rho^2)\sigma^2(Y)$, where ρ is the correlation coefficient, in which case the signal to noise ratio increases of a factor of $1/\sqrt{1 - \rho^2}$.

2.8 Analysis of Ordered Categorical Variables

2.8.1 General Aspects

The statistical analysis of categorical variables (see also Examples 4 and 5, 2.7) is one of the oldest problems in the area of testing statistical hypotheses (see Cressie and Read, 1988; Agresti, 2002). Among the several solutions in the literature, it is worth mentioning the geometric solution of Berger et al. (1998), based on a convex hull test. This section investigates this problem and presents some solutions from the viewpoint of permutation testing, in particular by considering alternatives of the so-called stochastic dominance type for ordered categorical variables, also referred to as problems with restricted alternatives or alternatives under order restrictions (see Silvapulle and Sen, 2005). In Chapter 6, together with multivariate extensions of stochastic dominance problems, we shall discuss other solutions which are asymptotically good in the framework of the NPC of a set of dependent permutation tests. In Section 6.7, a problem of isotonic inference for categorical variables and an extension to multivariate responses are also discussed.

Without loss of generality, we describe the notation and define the goodness-of-fit testing problems for univariate ordered categorical variables by means of a two-sample design. We mainly take into consideration stochastic dominance alternatives because they are rather difficult to cope with using parametric approaches, especially within the framework of likelihood ratio tests (see Sampson and Whitaker, 1989; El Barmi and Dykstra, 1995; Wang, 1996; Cohen and Sackrowitz, 1998; Silvapulle and Sen, 2005), and because they are very frequently encountered in practical problems. A rather serious difficulty with the maximum likelihood ratio test is that its asymptotic null distribution depends upon the true unknown nuisance parameters. Thus, it is difficult to justify its use in practice.

For unrestricted or non-dominance alternatives, we only discuss a few solutions in the spirit of goodness-of-fit methods. The most common problems related to unrestricted alternatives and C-sample situations, $C > 2$, are mentioned as simple extensions of those in Examples 4 and 5, 2.7.

Let us assume that the support of a categorical variable X is partitioned into $k \geq 2$ ordered classes $\{A_i, i = 1, \ldots, k\}$, in the sense that relationships such as $A_i \prec A_j$ have a clear meaning for every pair of subscripts i, j such that $1 \leq i < j \leq k$. A typical situation occurs when, for instance, A_1 = 'highly opposed', A_2 = 'opposed', up to A_k = 'highly in favour' or the like and where $\prec$ stands for *inferior to*. In this setting, we assume the existence of a suitable underlying statistical model for responses $(\mathcal{X}, \mathcal{A}, P \in \mathcal{P})$ where, as usual, $\mathcal{X}$ is the sample space of X, $\mathcal{A}$ is an algebra of events, and $\mathcal{P}$ a nonparametric family of non-degenerate probability distributions on $(\mathcal{X}, \mathcal{A})$. Moreover, classes A_i, $i = 1, \ldots, k$, may represent either qualitative or quantitative categories, according to the nature of X. We also assume that data are classified according to two levels of a symbolic treatment, the expected effect of which is to decrease X_2 with respect to X_1 towards smaller categorical values. In other words, given two independent random samples $\mathbf{X}_j = \{X_{ju}, u = 1, \ldots, n_j\}$, $j = 1, 2$, we wish to test the hypothesis

$$H_0 : \left\{X_1 \stackrel{d}{=} X_2\right\} = \{F_1(A_i) = F_2(A_i), i = 1, \ldots, k\}$$

against

$$H_1 : \left\{X_1 \stackrel{d}{>} X_2\right\} = \{F_1(A_i) \leq F_2(A_i), i = 1, \ldots, k-1\},$$

where at least one inequality is strict and $F_j(A_k) = \Pr\{X_j \leq A_k\}$ plays the role of CDF for X_j, $j = 1, 2$. In this context, by assuming that no reverse inequality such as $F_1(A_i) > F_2(A_i)$, $i = 1, \ldots, k-1$, is possible, the alternative can also be written as $H_1 : \{\bigcup_{i=1}^{k-1}[F_1(A_i) < F_2(A_i)]\}$. Note that H_1 defines the stochastic dominance of X_1 with respect to X_2.

Observed data are usually organized in a $2 \times k$ contingency table as in Table 2.4, where A_i are the ordered classes, $f_{ji} = \sum_u \mathbb{I}(X_{ju} \in A_i)$ are the observed frequencies, $N_{ji} = \sum_{s \leq i} f_{js}$ and $N_{\cdot i} = N_{1i} + N_{2i}$ are the cumulative frequencies, $f_{\cdot i} = f_{1i} + f_{2i}$ are the marginal frequencies, $n_j = \sum_i f_{ji} = N_{jk}$ are the sample sizes, and $n = n_1 + n_2 = N_{\cdot k}$ is the total sample size.

Note that the CDFs F_j are respectively estimated by the corresponding EDFs: $\hat{F}_j(A_i) = N_{ji}/n_j$, $i = 1, \ldots, k$, $j = 1, 2$.

In order to simplify computational problems, we assume that the marginal frequencies $f_{\cdot i}$, $i = 1, \ldots, k$, are all positive, in the sense that we remove class i from the analysis if $f_{\cdot i} = 0$. Observe that, in this setting, the pooled data set $\mathbf{X} = \mathbf{X}_1 \biguplus \mathbf{X}_2$ and the set of marginal frequencies $\{n_1, n_2,$

Table 2.4 A typical $2 \times k$ contingency table

A_i	f_{1i}	f_{2i}	$f_{\cdot i}$	N_{1i}	N_{2i}	$N_{\cdot i}$
A_1	f_{11}	f_{21}	$f_{\cdot 1}$	N_{11}	N_{21}	$N_{\cdot 1}$
A_i	f_{1i}	f_{2i}	$f_{\cdot i}$	N_{1i}	N_{2i}	$N_{\cdot i}$
A_k	f_{1k}	f_{2k}	$f_{\cdot k}$	N_{1k}	N_{2k}	$N_{\cdot k}$
	n_1	n_2	n	—	—	—

$f_{\cdot 1}, \ldots, f_{\cdot k}\}$ are equivalent sets of sufficient statistics for P in H_0, being related by a one-to one relation except for an irrelevant rearrangement of the data (see Problem 1, 2.9).

Note that the underlying response model is similar to model (M.i) of Section 1.10.1 when extended to ordered categorical variables. The model for this stochastic dominance problem may then be formalized by a notation such as $X_1 = \varphi(Y_1) \stackrel{d}{=} \varphi(Y_2 + \Delta)$, where Y_j, $j = 1, 2$, represent underlying real-valued responses, φ is a function which transforms Y into ordered categorical data, and Δ represents a non-negative stochastic effect. This notation is suitable for simulation algorithms when underlying continuous models are supposed to be generated before data transformation into ordered classes. This analogy allows us to extend the use of terminology adopted for quantitative variables to the case of ordered categorical variables. Also note that the solution when $k = 2$ corresponds to Fisher's well-known *exact probability test*, which is UMPS conditionally on the set of sufficient statistics provided by the set of marginal frequencies.

In addition, it is worth noting that H_0 implies that the data of two groups are exchangeable, so that the permutation testing principle may be properly applied. This implies taking into consideration the permutation sample space $\mathcal{X}_{/\mathbf{X}}$ generated by all permutations of pooled data set $\mathbf{X}$, that is, the set of all possible tables in which the marginal frequencies are held fixed.

Remark 1. The CMC analysis for discrete distributions, especially in multivariate situations, becomes easier if, in place of the usual contingency tables, the unit-by-unit representation for sample data is used, that is, the same representation as for quantitative variables obtained by listing the n individual data $\mathbf{X} = \{X_{ju}, u = 1, \ldots, n_j, j = 1, 2\}$. For instance, in the example of Table 2.4, the unit-by-unit representation of the pooled data set is the vector $\mathbf{X} = \{X(u), u = 1, \ldots, n; n_1, n_2\}$, in which the first n_1 responses are the data belonging to group 1 and the other n_2 are those belonging to group 2. Of course, in this representation, individual responses are categorical: $X(u) \in \{A_1, \ldots, A_k\}$, $u = 1, \ldots, n$.

2.8.2 A Solution Based on Score Transformations

One way of solving the dominance testing problem is by attribution of real-valued scores ω to classes. This implies transforming A_i into ω_i, $i = 1, \ldots, k$, where scores must satisfy the conditions $\omega_i < \omega_j$ if $i < j$. Hence, one solution is by a permutation comparison of sample means of scores. That is, by using a permutation test statistic such as

$$T_\omega^* = \sum_{i=1}^{k} \omega_i \cdot (f_{1i}^*/n_1 - f_{2i}^*/n_2) = n_1\bar{\omega}_1^* - n_2\bar{\omega}_2^*,$$

where $f_{ji}^* = \sum_{r \le n} \mathbb{I}(X_{jr}^* \in A_i)$, $j = 1, 2$, $i = 1, \ldots, k$, are permutation frequencies related to class A_i, $\bar{\omega}_j^*$ are permutation sample means of scores, and $\mathbf{X}^* = \{X(v_u^*), u = 1, \ldots, n; n_1, n_2\}$ is a permutation obtained by the algorithm in Section 2.2.5. Note that, as ω_i and $f_{1i}^* + f_{2i}^* = f_{\cdot i}$, $i = 1, \ldots, k$, are fixed values, then T_ω^* is permutationally equivalent to $T_\omega^* = \sum_i \omega_i \cdot f_{1i}^* \approx n_1 \cdot \bar{\omega}_1^*$. Of course, the observed value of this statistic is $T_\omega^o = T_\omega(\mathbf{X}) = \sum_i \omega_i \cdot f_{1i}$.

General arguments for this kind of solution are discussed, for instance, in Chakraborti and Schaafsma (1996) and Gautman (1997). However, it should be noted that, as scores are arbitrarily established, this solution is often questionable.

Remark 1. If $f_{\cdot i} = 1$, $i = 1, \ldots, k$, which corresponds to the case in which all data have distinct (categorical) values, in accordance with a sort of quasi-continuity of categorical variables, and if $\omega_i = i = \mathbb{R}(A_i)$, that is, by assigning ordinary ranks to ordered classes, then this solution corresponds to a rank test of the Wilcoxon–Mann–Whitney type, whose null distribution can be well approximated by a PCLT.

On the other hand, in the general case where $f_{\cdot i} \geq 1, i = 1, \ldots, k$, frequencies may be regarded as ties for ranks and thus the null permutation distribution cannot be satisfactorily approximated by a CLT, unless sample sizes are very large. Hence, it must be evaluated by direct full calculations or estimated by a CMC. Note that $S^* = \sum_i i \cdot (f_{1i}^* - f_{2i}^*)$ is just the divergence of mean-rank indicators (see Problem 9, 2.9).

Remark 2. It is of interest to use scores $\omega_i = N_{\cdot i-1} + f_{\cdot i}/2$, where $N_{\cdot i-1} = 0$ if $i = 1$. These scores, which correspond to a sort of average rank common to all values in the ith class, are related to the so-called normalized EDF as introduced by Ruymgaart (1980) and frequently used in analysis of discrete data (e.g. Brunner et al., 1995; Munzel, 1999). However, the corresponding statistic $T_\omega^* = \sum_i \omega_i \cdot f_{1i}^*$ also leads to a permutation Wilcoxon–Mann–Whitney type test (see Remark 1, 2.3).

2.8.3 Typical Goodness-of-Fit Solutions

In order to avoid the arbitrary act of assigning scores to classes, in the spirit of goodness-of-fit methods we may use the same permutation test statistics as in Example 4, 2.7, that is,

$$T_{AD}^* = \sum_{i=1}^{k-1} N_{2i}^* \cdot [N_{\cdot i} \cdot (n - N_{\cdot i})]^{-1/2},$$

where $N_{\cdot i} = N_{1i} + N_{2i} = N_{1i}^* + N_{2i}^*$, in which $N_{ji}^* = \sum_{s \leq i} f_{js}^*$, $i = 1, \ldots, k-1$, $j = 1, 2$, are permutation cumulative frequencies. Note that T_{AD} corresponds to the discrete version of a statistic following the Anderson–Darling goodness-of-fit test for dominance alternatives, consisting of a comparison of two EDFs, standardized by the reciprocals of permutation standard deviations.

Of course, in the spirit of goodness-of-fit testing, many other test statistics may be used. Examples for restricted alternatives are: (a) $T_{KS}^* = \max(F_{2i}^* - F_{1i}^*)$, which is a discretized version for restricted alternatives of the Kolmogorov–Smirnov test; (b) $T_{CM}^* = \sum_i (F_{2i}^* - F_{1i}^*)$, which is a discretized version of the Cramér–von Mises test. In both these examples, $F_{ji}^* = N_{ji}^*/n_j$, $i = 1, \ldots, k$, $j = 1, 2$. It is straightforward to prove that T_{CM}^* is permutationally equivalent to S^* in Remark 1, 2.8.2. Also of interest, in a multinomial parametric context, is the likelihood ratio test for restricted alternatives as discussed in El Barmi and Dykstra (1995) and Wang (1996).

An Example

Let us consider the numerical example in Table 2.5, which gives fictitious data for two samples, each with 40 units, partitioned into five ordered categories. With this data set, using $B = 2000$ CMC iterations, we obtain $\hat{\lambda}_\omega = 0.0894$, where scores $\omega_i = i$, $i = 1, \ldots, k$, were used for T_ω.

Table 2.5 A 2 × 5 contingency table of fictitious data

A_i	f_{1i}	f_{2i}	N_{1i}	N_{2i}	$f_{\cdot i}$
A_1	8	17	8	17	25
A_2	9	6	17	23	15
A_3	6	6	23	29	12
A_4	8	3	31	32	11
A_5	9	8	40	40	17
	40	40			80

The data in Table 2.5 can be obtained by inserting the absolute frequencies of groups 1 and 2, obtaining their cumulative sum and binding together the four vectors:

```
f1 = c(8,9,6,8,9) ; f2 = c(17,6,6,3,8);
N1 = cumsum(f1) ; N2 = cumsum(f2);
N = f1+f2 ; n = sum(N);
cbind(f1,f2,N1,N2,N)
```

The test statistic T_D^* is a sum over $k-1=4$ categories of the quantities $D_i = N_{2i}/[N_{\cdot i} \cdot (n - N_{\cdot i})]^{1/2}$, $i = 1, \ldots, 4$. It is then easier to create a vector containing the D_i and then sum its element to obtain T_D^*. For the observed data:

```
B=10000
T = array(0,dim=c((B+1),1))
D = N2/sqrt(N*(n-N))
T[1] = sum(D[1:4])
```

A random permutation of data can be obtained by re-creating the original data that gives the observed frequencies `f1` and `f2`. Indeed, we have to obtain all possible configurations of frequencies $\{f_{i1}^*, f_{i2}^*\}$ that satisfy the row and column totals. To do that, first we create the vectors `X1` and `X2` containing the categories of each observation in each sample (for simplicity the categories are indicated by the numbers $1, \ldots, 5$). Later on, concatenate the vectors `X1` and `X2` in the vector `X` and let `X.star` be a random permutation of `X`. Create a vector of labels `Y` indicating the sample. In this example `X1` and `X2` are vectors of lengths 40, `X` and `Y` have lengths equal to 80. `X1` and `X2` are such that `table(X1) = f1` and `table(X1) = f2`.

Finally, the frequency table corresponding to a random permutation can be obtained by applying the function `table` to the elements of `X.star` belonging to the first and second sample, respectively. The permutation values of the test statistic are then obtained as above. Note that this way of proceeding guarantees that the marginal distributions of Table 2.5 are fixed, therefore we only need to obtain the frequency distribution in the second sample.

```
X1<-rep(seq(1,5),f1)
X2<-rep(seq(1,5),f2)
X<-c(X1,X2)
Y <-rep(c(1,2),c(sum(f1),sum(f2)))

for(bb in 2:(B+1)){
X.star=sample(X)
f2.star = table(X.star[Y==2])
N2.star = cumsum(f2.star)
D.star = N2.star/sqrt(N*(n-N))
T[bb] = sum(D.star[1:4])
}
t2p(T)[1]

[1] 0.0635
```

The corresponding MATLAB code is given below:

```
F=[8 17
9 6
6 6
```

```
8 3
9 8];

DATA=zeros(0,2);
for i=1:size(F,1)
    DATA=[DATA;[[repmat(1,F(i,1),1); repmat(2,F(i,2),1)]
    repmat(i,sum(F(i,:)),1) ]];
end

[P T] = NP_Cs_categ(DATA(:,2),DATA(:,1),1000,2,'A2');

opts.tail=-1
[P T] = NP_Cs_categ(DATA(:,2),DATA(:,1),1000,2,'AD',opts);

%it is different from A2. A2 does the weighted sum of
%squares (squared values)
%instead AD with tail=0 sum
%absolute values
opts.tail=0
[P T] = NP_Cs_categ(DATA(:,2),DATA(:,1),1000,2,'AD',opts);
```

The data set and the corresponding software codes are available from the `examples_chapters_1-4` folder on the book's website.

Incidentally, it is worth noting that the estimated p-value of T_D is a little smaller than the others, due to its generally better power behaviour with respect to other test statistics based on frequency distances.

2.8.4 Extension to Non-Dominance Alternatives and C Groups

Let us again assume that $C = 2$ and that the hypotheses are $H_0 : \{X_1 \stackrel{d}{=} X_2\}$ against $H_1 : \{X_1 \stackrel{d}{\neq} X_2\} = \{\bigcup_i [F_1(A_i) \neq F_2(A_i)]\}$, that is, a two-sample problem with two-sided or non-dominance alternatives. In this setting, according to, for instance, permutation one-way ANOVA testing on score transformations, the test statistic T_ω becomes $T_\omega^{*2} = (\bar{\omega}_1^*)^2$, and T_{AD} becomes $T_{AD}^{*2} = \sum_{i=1}^{k-1} \left(N_{2i}^* - N_{1i}^*\right)^2 [N_{\cdot i} \cdot (n - N_{\cdot i})]^{-1}$, which corresponds to a two-sample Anderson–Darling test statistic adjusted for discrete variables.

Note that for non-dominance alternatives and nominal categorical variables, the most popular test statistic is Pearson's chi-square for $2 \times k$ contingency tables. Also appropriate is the Cochran's likelihood ratio test (Cochran, 1952),

$$T_{LR}^{*2} = 2 \cdot \sum_{i=1}^{k} \sum_{j=1}^{2} f_{ji}^* \cdot \log(f_{ji}^* / \hat{f}_{ji}),$$

where $\hat{f}_{ji} = f_{\cdot i} \cdot n_j / n$ are the so-called expected frequencies in H_0. This test statistic, in H_0, is asymptotically distributed as a central χ^2 with $k - 1$ degrees of freedom. However, it must be stressed that both Pearson's χ^2 and Cochran's T_{LR}^2, although useful for nominal variables, are inadequate for ordered categorical variables because they do not take account of the ordering property of responses.

Of course many other test statistics may be adopted for nominal variables, such as the Freeman–Tukey square root divergence or the Cressie–Read power divergence families of tests (see Cressie and Read, 1988, for a full discussion). Here, within the framework of goodness-of-fit methods, we only present one permutation solution which is alternative to the chi-square and based on the statistic

$$T_f^{*2} = \sum_{i=1}^{k} \left(f_{2i}^* - \frac{n_2 \cdot f_{\cdot i}}{n} \right)^2 [f_{\cdot i} \cdot (n - f_{\cdot i})]^{-1}.$$

This test statistic, again following the Anderson–Darling approach, in H_0 is equivalent to the sum of standardized squared summands, except for a permutationally invariant coefficient. The rationale for this solution will be clarified in Chapter 4 because it is essentially an NPC of several dependent partial tests, by means of the so-called *direct combining function* (see (g) in Section 4.2.4). However, its behaviour is very close to that of the chi-square.

Extensions of hypotheses and tests to $C > 2$ groups are straightforward. Here, we only mention one extension of T_D^2 for ordered variables and one extension of T_f^2 for nominal variables and, of course, for unrestricted alternatives. In terms of our notation, such extensions are clearly

$$T_{AD}^{*2} = \sum_{j=1}^{C} \sum_{i=1}^{k-1} \left(F_{ji}^* - \bar{F}_i \right)^2 \left[\bar{F}_i \cdot (1 - \bar{F}_i) \cdot (n - n_j)/n_j \right]^{-1},$$

where $\bar{F}_i = N_{\cdot i}/n$ and $N_{\cdot i} = \sum_j N_{ji}$, and

$$T_{fAD}^{*2} = \sum_{j=1}^{C} \sum_{i=1}^{k} \left(\frac{f_{ji}^*}{n_j} - \frac{f_{\cdot i}}{n} \right)^2 \left[f_{\cdot i} \cdot (n - f_{\cdot i}) \cdot (n - n_j)/n_j \right]^{-1}.$$

In Chapter 6 two more solutions to this problem and an extension to multivariate situations within the context of NPC of several dependent permutation tests are presented. Moreover, in Section 6.9 an approximate permutation test for dominance in heterogeneity (concentration) in a two-sample design with nominal variables is discussed.

2.9 Problems and Exercises

1) Show that in standard contingency tables, related to univariate responses, marginal frequencies $\{f_{\cdot i},\ i = 1, \ldots, k,\ n_1, \ldots, n_C\}$ are equivalent to the pooled set of observations $\mathbf{X}$, that is, both are minimal sufficient in H_0 for the underlying multinomial distribution (establish a one-to-one relationship between unit-by-unit representation and the contingency table).

2) Show that when conditioning on marginal frequencies, the standard chi-square test for equality in distribution of two categorical variables in a two-sample problem is a permutation test.

3) Show that when conditioning on marginal frequencies, the standard chi-square test for equality in the distribution of C categorical variables in a C-sample problem is a permutation test.

4) Using unit-by-unit representation of sample data, prove that the standard chi-square test in a C-sample problem may be viewed as a one-way ANOVA problem for categorical variables.

5) Using unit-by-unit representation of sample categorical data, show that Cochran's likelihood ratio test is a permutation test and that it may be viewed as a solution to a one-way ANOVA problem.

6) Using unit-by-unit representation of sample categorical data, find the permutation distribution for a stochastic dominance problem in a 2×2 contingency table, that is, the distribution of Fisher's exact probability test.

7) Prove that the test statistic T_{fAD}^{*2} in Section 2.8.4 is permutationally equivalent to the chi-square statistic when $k = 2$.

8) With reference to Remark 1, 2.8.2, prove that if all marginal frequencies $f_{\cdot i} = 1, i = 1, \ldots, k = n$, then $\sum_i i \cdot f^*_{1i}$ is permutationally equivalent to the Wilcoxon–Mann–Whitney test (note that there is only one observation per category).

9) Prove that if the response variables in the two-sample problem are binary ($X_{ji} = 0$ if $X_{ji} \in \mathcal{X}_0$ and $X_{ji} = 1$ if $X_{ji} \in \mathcal{X}_1$, $i = 1, \ldots, n_j$, $j = 1, 2$, where $\mathcal{X}_0 \bigcup \mathcal{X}_1 = \mathcal{X}$ and $\mathcal{X}_0 \bigcap \mathcal{X}_1 = \emptyset$), then the permutation test for two-sided alternatives $T = |\bar{X}_1 - \bar{X}_2|$ is permutationally equivalent to Pearson's chi-square test for the resulting 2×2 contingency table.

10) Prove that if the response variables in the two-sample problem are binary, then the permutation test for restricted alternatives $T = \bar{X}_1 - \bar{X}_2$ is permutationally equivalent to Fisher's exact probability test for the resulting 2×2 contingency table.

11) With reference to the two-way design in Example 8, 2.7, show that the pooled data set $\mathbf{X}_{11} \biguplus \ldots \biguplus \mathbf{X}_{JI}$ is not sufficient for separate testing of three null hypotheses (H_{0A}, H_{0B}, H_{0AB}), whereas it is sufficient for $H'_0 : \{H_{0A} \bigcap H_{0B} \bigcap H_{0AB}\}$.

12) With reference to Remark 6, 2.7, show that the permutation structure (Remark 1, 2.7) of the test statistic $T^*_R = \sum_j (\bar{X}^*_{j\cdot} - \bar{X}_{\cdot\cdot})^2 / \sum_{ji} (X^*_{ji} - \bar{X}^*_{j\cdot} - \bar{X}_{\cdot i} + \bar{X}_{\cdot\cdot})^2$ in a two-way ANOVA design does not depend on block effects b_i, so that it is exact for testing H_0 against H_1 independently of block effects.

13) Show that in a standard two-sample design for one-sided alternatives, if random effects Δ are such that $\Pr\{\Delta \geq 0\} = 1$ and $\Pr\{\Delta = 0\} > 0$, that is, if treatment is ineffective with some units, then the test statistic $T^* = \sum_i X^*_{1i}$ is exact.

3

Further Properties of Permutation Tests

3.1 Unbiasedness of Two-sample Tests

3.1.1 *One-Sided Alternatives*

Introduction and Notation

For simplicity and without loss of generality, let us assume that the real one-dimensional variable X takes value on sample space $\mathcal{X}$, with probability distribution P defined on the measurable space $(\mathcal{X}, \mathcal{A})$, and that in H_1 the distribution of X_1 is shifted by a random quantity Δ with respect to that of X_2, so that the two respective CDFs are such that $F_1(x) \leq F_2(x)$, $x \in \mathcal{R}^1$, showing stochastic dominance. Most of the results on unbiasedness we obtain are also valid for ordered categorical variables and for paired observation designs. We assume that random effects Δ are non-negative, that is, $\Pr\{\Delta \geq 0\} = 1$ and $\Pr\{\Delta > 0\} \leq 1$, so that $H_0 : \{X_1 \stackrel{d}{=} X_2\}$ and $H_1 : \{X_1 \stackrel{d}{>} X_2\}$. Of course, if Δ takes negative values, such that $\Pr\{\Delta \leq 0\} = 1$, it will suffice to convert subscript 1 into 2 in response variables and in all consequent statistics, for instance by writing the alternative as $H_1 : \{X_2 \stackrel{d}{>} X_1\}$, and so on. Problems for which $0 < \Pr\{\Delta \geq 0\} < 1$, in which there are positive effects on some units and negative on others, are much more complex and are briefly discussed in Example 5, 4.6. To represent data sets in the alternative we use the notation $\mathbf{X}(\Delta) = \{X_{ji} = \mu + \Delta_{ji} + Z_{ji}, i = 1, \ldots, n_j, j = 1, 2\}$, where μ is a finite nuisance quantity common to all units, and the random deviates Z_{ji} are exchangeable and have unknown distribution $P \in \mathcal{P}$. If the existence of the mean value of Z is assumed, $\mathbb{E}(|Z|) < \infty$, we put $\mathbb{E}(Z) = 0$, so that errors are centred variables. We assume that random effects are such that $\Delta_{1i} = \delta_1 + Q_{1i} \stackrel{p}{\geq} 0$, $i = 1, \ldots, n_1$, with strict inequality for at least one $i \in (1, \ldots, n_1)$, in which δ_1 plays the role of average effect and deviates Q_{1i} may produce heteroscedasticity in the alternative and may depend on (μ, Z_{1i}); we also assume that $\Pr\{\Delta_{2i} = 0\} = 1$, $i = 1, \ldots, n_2$, i.e. there are no effects on X_2. In Section 4.5.3 we will see two ways of obtaining random effects dependent on the Z deviates.

In the following we use symbols Δ and δ when referring to random and fixed effects, respectively. Of course, fixed effects imply $\Pr\{Q_1 = 0\} = 1$. Since μ is a nuisance quantity common to all units and thus inessential for comparing X_1 to X_2 (see Problem 16, 2.4.2), we may model data sets as $\mathbf{X}(\Delta) = (\mathbf{Z}_1 + \Delta, \mathbf{Z}_2)$, where $\Delta = (\Delta_{11}, \ldots, \Delta_{1n_1})$. The latter notation emphasizes that effects Δ are assumed to be active only on units of the first sample, in accordance with the convention that units of the second sample receive the placebo. Connected with such a notation, we may equivalently express the hypotheses as $H_0 : \Pr\{\Delta = 0\} = 1$ and $H_1 : \Pr\{\Delta > 0\} > 0$.

The notion of unbiasedness for a test statistic is related to its comparative rejection behaviour in H_1 with respect to that in H_0. To this end, let us consider a non-degenerate test statistic $T : \mathcal{X}^n \rightarrow \mathcal{R}^1$, where typically we have $T(\mathbf{X}) = S_1(\mathbf{X}_1) - S_2(\mathbf{X}_2)$ corresponding to the comparison of two non-degenerate sample (measurable) statistics. Typically, S_j, $j = 1, 2$, corresponds to $\sum_i \varphi(X_{ji})/n_j$, or $\mathbb{M}d(X_{ji})$, etc. (a list of the most commonly used test statistics is given in Section 2.3). In order to be suitable for evaluating the related sampling diversity, we assume that functions S satisfy the following conditions:

(a) Statistics S_j are symmetric functions, that is, invariant with respect to rearrangements of data input, $S_j(\mathbf{X}_j) = S_j(\mathbf{X}_j^{\dagger})$, with $\mathbf{X}_j^{\dagger}$ any rearrangement (i.e. within-sample permutation) of $\mathbf{X}_j$, $j = 1, 2$, and provided with the same analytic form that may only differ for sample sizes so that both are indicators of the same quantity.
(b) S_j are monotonic non-decreasing, that is, $S_j(\mathbf{X} + \mathbf{Y}) \geq S_j(\mathbf{X})$, $j = 1, 2$, for any data set $\mathbf{X}$ and non-negative $\mathbf{Y} \overset{p}{\geq} 0$, so that large values of T are evidence against H_0.

With obvious notation, a test statistic is said to be unbiased if its rejection probability is such that $\Pr\{T(\mathbf{X}(0)) \geq T_\alpha | H_0\} \leq \alpha \leq \Pr\{T(\mathbf{X}(\Delta)) \geq T_\alpha | H_1\}$ for every type I error rate $\alpha \in (0, 1)$ and every specific alternative in H_1, where T_α is the critical value, assumed to be finite. The probability distribution involved is that which is generated by T when $\mathbf{X}$ takes values on sample space $\mathcal{X}^n$ according to the underlying distribution P^n. This notion corresponds to the traditional notion for unbiasedness, also called *unconditional* or *population unbiasedness*.

A well-known sufficient condition for unconditional unbiasedness of T is that its unconditional null distribution is at least weakly dominated by every specific distribution from the alternative: $T(\mathbf{X}(0)) \overset{d}{\leq} T(\mathbf{X}(\Delta))$.

In permutation testing we meet a somewhat more stringent notion of unbiasedness, which in turn is sufficient but not necessary for the unconditional form. We call it the *conditional* or *permutation unbiasedness*. A test statistic T is said to be conditionally or permutationally unbiased if the p-values $\lambda_T(\mathbf{X}) = \Pr\{T(\mathbf{X}^*) \geq T(\mathbf{X}) | \mathcal{X}_{/\mathbf{X}}\}$ are such that, for each $\mathbf{X} \in \mathcal{X}^n$ and for any $\Delta \in H_1$,

$$\Pr\{\lambda_T(\mathbf{X}(\Delta)) \leq \alpha_a | \mathcal{X}_{/\mathbf{X}(\Delta)}\} \geq \Pr\{\lambda_T(\mathbf{X}(0)) \leq \alpha_a | \mathcal{X}_{/\mathbf{X}(0)}\} = \alpha_a,$$

where: $\mathbf{X}(\Delta) = (\mathbf{Z}_1 + \Delta, \mathbf{Z}_2)$, $\mathbf{X}(0) = (\mathbf{Z}_1, \mathbf{Z}_2)$; $\mathbf{X}^*$ is interpreted as a random permutation of $\mathbf{X}$; and α_a is any attainable α-value. Similarly to Section 2.2.3, we suppress the subscript a and use the symbol α to denote attainable α-value. A sufficient condition for conditional unbiasedness is

$$\begin{aligned}\lambda_T(\mathbf{X}(\Delta)) &= \Pr\left\{T(\mathbf{X}^*(\Delta)) \geq T^o(\Delta) = T(\mathbf{X}(\Delta)) | \mathcal{X}_{/\mathbf{X}(\Delta)}\right\} \\ &\leq \Pr\left\{T(\mathbf{X}^*(0)) \geq T^o(0) = T(\mathbf{X}(0)) | \mathcal{X}_{/\mathbf{X}(0)}\right\} = \lambda_T(\mathbf{X}(0)).\end{aligned}$$

It is worth noting that the weak dominance relation $\leq$ on p-values can be uniform (i.e. satisfied by all data sets $\mathbf{X} \in \mathcal{X}^n$) or in distribution (i.e. satisfied by the majority of the $\mathbf{X}$).

Since the unconditional rejection probability of permutation test T is

$$\int_{\mathcal{X}^n} \Pr\{\lambda_T(\mathbf{X}(\Delta)) \leq \alpha \, | \mathcal{X}_{/\mathbf{X}(\Delta)}\} \, dP^n(\mathbf{X}(\Delta)) \geq \alpha,$$

because the integrand is greater than or equal to α and $dP^n(\mathbf{X}(\Delta)) \geq 0$, $\forall \mathbf{X} \in \mathcal{X}^n$, the following property is trivially true:

Proposition 1. *If a permutation test statistic T is conditionally unbiased for every $\mathbf{X} \in \mathcal{X}^n$, then it is also unconditionally unbiased for whatever underlying distribution P.*

Since the converse is not true, in that unconditional unbiasedness does not imply conditional unbiasedness, the notion of conditional (permutation) unbiasedness is somewhat more stringent than that of unconditional (see Problem 40, 3.9). Moreover, if a conditionally unbiased permutation test T is applied to subjects which were previously randomized to treatments, then the observed data set $\mathbf{X}$, instead of being i.i.d., can be selected from $\mathcal{X}^n$ by any selection-bias procedure (for a discussion, see Pesarin, 2002; see also Section 3.5) without compromising unconditional unbiasedness, provided that data are exchangeable in H_0. In this sense, permutation unbiasedness, also more stringent, is more frequently and usefully applicable to real problems than its unconditional counterpart.

Characterizing Conditional Unbiasedness

In order to look further into the conditional (permutation) unbiasedness of a test statistic T, let us consider its permutation structure, that is, the behaviour of values it takes in the observed data set $\mathbf{X}$ and in any of its permutations $\mathbf{X}^* \in \mathcal{X}^n_{/\mathbf{X}}$ under both H_0 and H_1 (see the informal definition in Remark 1, 2.7). Two observed values under H_0 and H_1 are respectively $T^o(0) = T(\mathbf{X}(0))$ and $T^o(\Delta) = T(\mathbf{X}(\Delta))$; two permutation values are $T^*(0) = T(\mathbf{X}^*(0))$ and $T^*(\Delta) = T(\mathbf{X}^*(\Delta))$, where $\mathbf{X}^*(\Delta) = \{[Z(u_i^*) + \Delta(u_i^*)], i = 1, \ldots, n; n_1, n_2\}$, $\mathbf{X}^*(0) = \{Z(u_i^*), i = 1, \ldots, n; n_1, n_2\}$, and $\mathbf{u}^* = (u_1^*, \ldots, u_n^*)$ is any permutation of $\mathbf{u} = (1, \ldots, n)$. Let us define the increments of T due to random effects Δ, that is, the difference of two values, in $\mathbf{X}(\Delta)$ and $\mathbf{X}(0)$ by $D_T(\mathbf{X}(\Delta)) = T(\mathbf{X}(\Delta)) - T(\mathbf{X}(0))$, and in $\mathbf{X}^*(\Delta)$ and $\mathbf{X}^*(0)$ by $D_T(\mathbf{X}^*(\Delta)) = T(\mathbf{X}^*(\Delta)) - T(\mathbf{X}^*(0))$.

The assumptions that Δ is non-negative, T is non-decreasing with respect to Δ in the first n_1 arguments and non-increasing in the second n_2 arguments, and that large values of T are evidence against H_0, imply that $T(\mathbf{X}(\Delta)) - T(\mathbf{X}(0)) = D_T(\mathbf{X}(\Delta)) \geq 0$. Thus, the related p-values in H_0 and H_1 are respectively

$$\lambda_T(\mathbf{X}(0)) = \Pr\left\{T(\mathbf{X}^*(0)) \geq T(\mathbf{X}(0)) | \mathcal{X}_{/\mathbf{X}(0)}\right\},$$

and

$$\begin{aligned}\lambda_T(\mathbf{X}(\Delta)) &= \Pr\left\{T(\mathbf{X}^*(\Delta)) \geq T(\mathbf{X}(\Delta)) | \mathcal{X}_{/\mathbf{X}(\Delta)}\right\} \\ &= \Pr\left\{T(\mathbf{X}^*(0)) + D_T(\mathbf{X}^*(\Delta)) - D_T(\mathbf{X}(\Delta)) \geq T(\mathbf{X}(0)) | \mathcal{X}_{/\mathbf{X}(0)}\right\}.\end{aligned}$$

In the latter expression it is worth noting that a one-to-one *pointwise* relationship between two conditional permutation spaces $\mathcal{X}_{/\mathbf{X}(\Delta)}$ and $\mathcal{X}_{/\mathbf{X}(0)}$ has been used. Indeed, focusing on a generic permutation $\mathbf{u}^*$ of $\mathbf{u}$, for each point $\mathbf{X}^*(0) = \{Z(u_i^*), i = 1, \ldots, n;\ n_1, n_2\} \in \mathcal{X}_{/\mathbf{X}(0)}$ we may obtain the corresponding point of $\mathcal{X}_{/\mathbf{X}(\Delta)}$ by simply writing $\mathbf{X}^*(\Delta) = \{[Z(u_i^*) + \Delta(u_i^*)], i = 1, \ldots, n;\ n_1, n_2\} = \mathbf{X}^*(0) + \boldsymbol{\Delta}^*$, where $\boldsymbol{\Delta}^* = \{\Delta(u_i^*), i = 1, \ldots, n;\ n_1, n_2\}$. Conversely, for any $\mathbf{X}^*(\Delta) \in \mathcal{X}_{/\mathbf{X}(\Delta)}$, we may write the corresponding point $\mathbf{X}^*(0) \in \mathcal{X}_{/\mathbf{X}(0)}$. Because of this, the notation $\Pr\left\{T(\mathbf{X}^*(\Delta)) \geq T(\mathbf{X}(\Delta)) | \mathcal{X}_{/\mathbf{X}(\Delta)}\right\}$ has the same meaning as $\Pr\left\{T(\mathbf{X}^*(\Delta)) \geq T(\mathbf{X}(\Delta)) | \mathcal{X}_{/\mathbf{X}}\right\}$. This one-to-one pointwise relationship is illustrated in Figure 3.1.

It is well known that:

- any test statistic T is built up to evaluate the sampling diversity of data distribution in H_1 with respect to that in H_0;
- the observed data $\mathbf{X}(\Delta)$ are such that effects Δ are active only on the first sample;
- the difference of test statistics in any data permutation $D_T(\mathbf{X}^*(\Delta))$, since some effects are exchanged between two groups, tends to be smaller than that of its observed value $D_T(\mathbf{X}(\Delta))$ which contains all non-null effects.

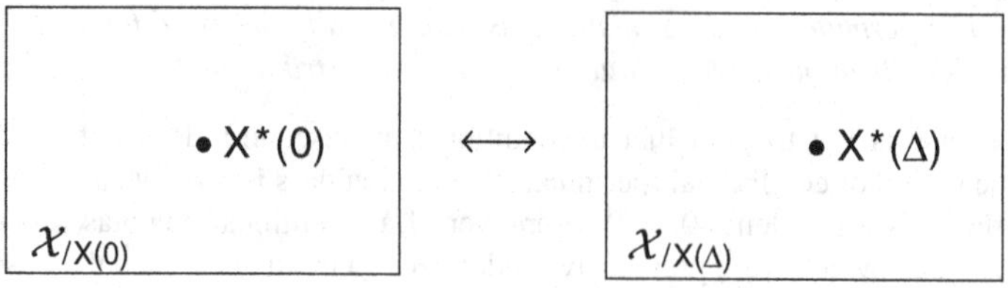

Figure 3.1 Pointwise relationship

Thus, we expect that $D_T(\mathbf{X}^*(\Delta)) - D_T(\mathbf{X}(\Delta))$ is likely to assume non-positive values. Therefore, depending on how this occurs, we can have different notions of permutation unbiasedness. More explicitly, therefore, we have the following:

(i) If for any $\Delta \overset{p}{\geq} 0$, $D_T(\mathbf{X}^*(\Delta)) - D_T(\mathbf{X}(\Delta)) \leq 0$ pointwise, that is, for all possible permutations $\mathbf{X}^* \in \mathcal{X}_{/\mathbf{X}}$, and uniformly for all data sets $\mathbf{X} \in \mathcal{X}^n$, then $\lambda_T(\mathbf{X}(0)) \geq \lambda_T(\mathbf{X}(\Delta))$. This uniform weak dominance leads to quite a stringent form of conditional unbiasedness, called *strictly uniform*. Associative statistics that satisfy condition (a) above, such as sample φ-means, $\bar{\varphi}_j = \sum_i \varphi(X_{ji})/n_j$, $j = 1, 2$, satisfy $D_T(\mathbf{X}^*(\Delta)) - D_T(\mathbf{X}(\Delta)) \leq 0$ pointwise. Pointwise weak dominance is also satisfied by non-associative S statistics for fixed effects: $\Delta \overset{p}{=} \delta$.

(ii) If uniformly for all data sets $\mathbf{X} \in \mathcal{X}^n$, $D_T(\mathbf{X}^*(\Delta)) - D_T(\mathbf{X}(\Delta)) \leq 0$ in permutation distribution, which implies $\lambda_T(\mathbf{X}(0)) \overset{d}{\geq} \lambda_T(\mathbf{X}(\Delta))$, then we have a slightly weaker form of conditional uniform unbiasedness. Non-associative statistics that satisfy condition (a) above, such as sample quantiles, satisfy $D_T(\mathbf{X}^*(\Delta)) - D_T(\mathbf{X}(\Delta)) \leq 0$ in permutation distribution and uniformly for all data sets $\mathbf{X} \in \mathcal{X}^n$ (see below).

(iii) If for some statistics with some data sets and/or some permutations $D_T(\mathbf{X}^*(\Delta)) - D_T(\mathbf{X}(\Delta)) > 0$, and if the permutation probability $\Pr\{D_T(\mathbf{X}^*(\Delta)) - D_T(\mathbf{X}(\Delta)) > 0|\mathcal{X}_{/\mathbf{X}(\Delta)}\}$ is not sufficiently smaller than $\Pr\{D_T(\mathbf{X}^*(\Delta)) - D_T(\mathbf{X}(\Delta)) < 0|\mathcal{X}_{/\mathbf{X}(\Delta)}\}$, then the conditional unbiasedness might not occur. Such a situation may sometimes arise when the dominance condition of the CDFs is violated, that is, when two sets $A_> = \{t : F_1(t) > F_2(t)\}$ and $A_< = \{t : F_1(t) < F_2(t)\}$ have positive probability with respect to both distributions. Moreover, associative and non-associative statistics for two-sided alternatives may fall within this framework, and so related tests might not be unbiased (see Section 3.1.2).

The strictly uniform conditional unbiasedness (i) is the most useful in practice. Here we will prove it for all tests based on sampling diversity of associative statistics such as

$$T^*_\varphi(\Delta) = \sum_i \varphi[X^*_{1i}(\Delta)] - \sum_i \varphi[X^*_{2i}(\Delta)],$$

which in turn, if φ is any non-degenerate measurable non-decreasing function, is permutationally equivalent to $\bar{\varphi}^*_1 - \bar{\varphi}^*_2$. For general statistics, including non-associative forms, the weaker form of conditional unbiasedness (ii) is slightly more difficult to prove and to apply, unless effects are fixed, that is, $\Delta \overset{p}{=} \delta$. Its proof, however, is also given below.

Uniform Conditional Unbiasedness

For simplicity, we prove conditional unbiasedness of tests based on divergence of associative statistics first. We then consider the permutation structures of T^* in H_0 and in H_1. Since $X^*_{ji}(\Delta) =$

$Z^*_{ji} + \Delta^*_{ji}$, with obvious notation, we note that $\sum_i[\varphi(Z^*_{ji} + \Delta^*_{ji}) - \varphi(Z^*_{ji})] = \sum_i d_T(Z^*_{ji}, \Delta^*_{ji}) = D_T(\mathbf{Z}^*_j, \Delta^*_j) \geq 0$, $j = 1, 2$, where $d_T(Z^*_{ji}, \Delta^*_{ji}) = \varphi(Z^*_{ji} + \Delta^*_{ji}) - \varphi(Z^*_{ji}) \geq 0$, because some Δ^*_{ji} are positive quantities and φ is non-degenerate and non-decreasing by assumption. Of course, if $\Delta^*_{ji} = 0$, $d_T(Z^*_{ji}, 0) = 0$, whereas $D_T(\mathbf{Z}^*_j, \Delta^*_j) = 0$ implies $\Delta^*_{ji} = 0$, $i = 1, \ldots, n_j$, $j = 1, 2$. The two observed values are then $T^o(0) = \sum_i \varphi(Z_{1i}) - \sum_i \varphi(Z_{2i})$ and $T^o(\Delta) = \sum_i \varphi(Z_{1i} + \Delta_{1i}) - \sum_i \varphi(Z_{2i}) = T^o(0) + D_T(\mathbf{Z}_1, \Delta_1)$, because $\Delta_{2i} = 0, i = 1, \ldots, n_2$. Moreover, the two permutation values are $T^*(0) = \sum_i \varphi(Z^*_{1i}) - \sum_i \varphi(Z^*_{2i})$ and $T^*(\Delta) = \sum_i \varphi(Z^*_{1i} + \Delta^*_{1i}) - \sum_i \varphi(Z^*_{2i} + \Delta^*_{2i}) = T^*(0) + D_T(\mathbf{Z}^*_1, \Delta^*_1) - D_T(\mathbf{Z}^*_2, \Delta^*_2)$. In addition, the following pointwise relations clearly occur: $D_T(\mathbf{Z}^*_1, \Delta^*_1) \leq D_T(\mathbf{Z}_1, \Delta_1)$, because for $u^*_i > n_1$ and $i \leq n_1$ the corresponding $\Delta^*_{1i} = \Delta(u^*_i) = 0$; and $D_T(\mathbf{Z}^*_2, \Delta^*_2) \geq 0$, because for $u^*_i \leq n_1$ and $i > n_1$ the corresponding $\Delta^*_{2i} = \Delta(u^*_i) \geq 0$.

Therefore, the related p-values are such that

$$\begin{aligned}\lambda_T(\mathbf{X}(\Delta)) &= \Pr\left\{T^*(\Delta) \geq T^o(\Delta) \,|\mathcal{X}_{/\mathbf{X}(\Delta)}\right\}\\ &= \Pr\left\{T^*(0) + D_T(\mathbf{Z}^*_1, \Delta^*_1) - D_T(\mathbf{Z}^*_2, \Delta^*_2) - D_T(\mathbf{Z}_1, \Delta_1) \geq T^o(0) \,|\mathcal{X}_{/\mathbf{X}(0)}\right\}\\ &\leq \Pr\left\{T^*(0) \geq T^o(0) \,|\mathcal{X}_{/\mathbf{X}(0)}\right\} = \lambda_T(\mathbf{X}(0)),\end{aligned}$$

from which we see that p-values in every alternative of H_1 are not larger than in H_0, and this holds for any underlying distribution P, for any associative test statistic T, and uniformly for all data sets $\mathbf{X} \in \mathcal{X}^n$, because $D_T(\mathbf{Z}^*_1, \Delta^*_1) - D_T(\mathbf{Z}^*_2, \Delta^*_2) - D_T(\mathbf{Z}_1, \Delta_1)$ is pointwise non-positive and because $\Pr\{T^* - W \geq t\} \leq \Pr\{T^* \leq t\}$ for any $W \geq 0$.

To establish conditional unbiasedness for any kind of statistic of the form $T(\mathbf{X}(\Delta)) = S_1(\mathbf{X}_1(\Delta)) - S_2(\mathbf{X}_2)$, with special attention to non-associative statistics and with obvious notation, let us observe that:

- $T^o(0) = S_1(\mathbf{Z}_1) - S_2(\mathbf{Z}_2)$.
- $T^o(\Delta) = S_1(\mathbf{Z}_1 + \Delta_1) - S_2(\mathbf{Z}_2) = S_1(\mathbf{Z}_1) + D_S(\mathbf{Z}_1, \Delta_1) - S_2(\mathbf{Z}_2) = T^o(0) + D_S(\mathbf{Z}_1, \Delta_1)$, where $D_S(\mathbf{Z}_1, \Delta_1) \geq 0$.
- $T^*(0) = S_1(\mathbf{Z}^*_1) - S_2(\mathbf{Z}^*_2)$.
- $T^*(\Delta) = S_1(\mathbf{Z}^*_1 + \Delta^*_1) - S_2(\mathbf{Z}^*_2 + \Delta^*_2) = T^*(0) + D_S(\mathbf{Z}^*_1, \Delta^*_1) - D_S(\mathbf{Z}^*_2, \Delta^*_2)$.
- $D_S(\mathbf{Z}^*_2, \Delta^*_2) \geq D_S(\mathbf{Z}^*_2, 0) = 0 = D_S(\mathbf{Z}_2, 0)$, because effects Δ^*_{2i} from the first group are non-negative.
- $D_S(\mathbf{Z}^*_1, \Delta^*_1) \leq D_S(\mathbf{Z}^*_1, \Delta_1)$ pointwise, because in $D_S(\mathbf{Z}^*_1, \Delta_1)$ there are non-negative effects assigned to units from the second group. For example, suppose that $n_1 = 3$, $n_2 = 3$, and $\mathbf{u}^* = (3, 5, 4, 1, 2, 6)$. Then $(\mathbf{Z}^*_1, \Delta^*_1) = [(Z_{13}, \Delta_{13}), (Z_{22}, 0), (Z_{21}, 0)]$, and so $(\mathbf{Z}^*_1, \Delta_1) = [(Z_{13}, \Delta_{13}), (Z_{22}, \Delta_{11}), (Z_{21}, \Delta_{12})]$ or $(\mathbf{Z}^*_1, \Delta_1) = [(Z_{13}, \Delta_{13}), (Z_{22}, \Delta_{12}), (Z_{21}, \Delta_{11})]$. It is to be emphasized that $X(u^*_i) = Z(u^*_i) + \Delta(u^*_i)$ if $u^*_i \leq n_1$, that is, units from the first group maintain their effects, whereas the rest of the effects are randomly assigned to units from the second group.
- $D_S(\mathbf{Z}^*_1, \Delta_1) \stackrel{d}{=} D_S(\mathbf{Z}_1, \Delta_1)$, because $\Pr\{\mathbf{Z}^*_1|\mathcal{X}_{/\mathbf{X}(0)}\} = \Pr\{\mathbf{Z}_1|\mathcal{X}_{/\mathbf{X}(0)}\}$ by Proposition 1, 2.1.2.

Thus, $D_S(\mathbf{Z}^*_1, \Delta^*_1) - D_S(\mathbf{Z}^*_2, \Delta^*_2) \leq D_S(\mathbf{Z}_1, \Delta_1)$ in permutation distribution and so

$$\begin{aligned}\lambda_T(\mathbf{X}(\Delta)) &= \Pr\left\{T(\mathbf{X}^*(\Delta)) \geq T(\mathbf{X}(\Delta))|\mathcal{X}_{/\mathbf{X}(\Delta)}\right\}\\ &= \Pr\left\{T^*(0) + D_S(\mathbf{Z}^*_1, \Delta^*_1) - D_S(\mathbf{Z}^*_2, \Delta^*_2) - D_S(\mathbf{Z}_1, \Delta_1) \geq T^o(0)|\mathcal{X}_{/\mathbf{X}(0)}\right\}\\ &\leq \Pr\left\{T^*(0) \geq T^o(0)|\mathcal{X}_{/\mathbf{X}(0)}\right\} = \lambda_T(\mathbf{X}(0)),\end{aligned}$$

where emphasis is on the dominance in permutation distribution of $\lambda_T(\mathbf{X}(\Delta))$ with respect to $\lambda_T(\mathbf{X}(0))$, uniformly for all data sets $\mathbf{X} \in \mathcal{X}^n$, for all underlying distributions P, and for all associative and non-associative statistics $T = S_1(\mathbf{X}_1) - S_2(\mathbf{X}_2)$ which satisfy conditions (a) and (b) above. If effects were fixed, $\Delta \stackrel{p}{=} \delta$ say, then $D_S(\mathbf{Z}_1^*, \delta_1^*) - D_S(\mathbf{Z}_2^*, \delta_2^*) \leq D_S(\mathbf{Z}_1, \delta_1)$ pointwise, instead of merely in permutation distribution. The simple proof of this is left to the reader as an exercise.

These results allow us to prove the following theorem:

Theorem 1. (Uniform conditional unbiasedness of T). *Permutation tests for random shift alternatives ($\Delta \stackrel{p}{>} 0$) based on divergence of associative or non-associative statistics of non-degenerate measurable non-decreasing transformations of the data, i.e.* $T^*(\Delta) = S_1(\mathbf{X}_1^*(\Delta)) - S_2(\mathbf{X}_2^*(\Delta))$, *are conditionally unbiased for every attainable* $\alpha \in \Lambda^{(n)}_{\mathbf{X}(0)}$, *every population distribution* P, *and uniformly for all data sets* $\mathbf{X} \in \mathcal{X}^n$. *In particular,*

$$\Pr\{\lambda(\mathbf{X}(\Delta)) \leq \alpha | \mathcal{X}_{/\mathbf{X}(\Delta)}\} \geq \Pr\{\lambda(\mathbf{X}(0)) \leq \alpha | \mathcal{X}_{/\mathbf{X}(0)}\} = \alpha.$$

One important consequence of the uniform conditional unbiasedness property of test statistics T is that we are allowed to extend the conditional inference (associated with the observed data set $\mathbf{X}$) unconditionally to the whole population from which data $\mathbf{X}$ are generated (see Section 3.5). This extension is useful, for instance, when in a randomized experiment on a drug compared to a placebo the conditional inferential conclusion is in favour of H_1, that is, by noting that the drug is effective on the present units. Indeed, the same conclusion can be extended to all the populations $P \in \mathcal{P}$, such that $dP(\mathbf{X})/d\xi > 0$, by concluding that the *drug is effective* and this irrespective of whether subjects are enrolled by random sampling or selection-bias sampling, provided that such subjects are randomized to treatments so that data exchangeability is satisfied in H_0. It is worth observing that if a selection-bias sampling is used in a parametric framework, in general no population (unconditional) inference can correctly be obtained (for a discussion see Pesarin, 2002), unless the selection mechanism is well defined, suitably modelled, and properly estimated.

In addition, by using the same reasoning to prove Theorem 1 above, it is also straightforward to prove the following theorem.

Theorem 2. (Uniform stochastic ordering of p-values). *If* $\Delta' \stackrel{p}{<} 0 \stackrel{p}{<} \Delta \stackrel{p}{<} \Delta''$ *are ordered random effects, where it is intended that* $\Delta \stackrel{p}{<} \Delta''$ *implies* $\Delta_{1i} \leq \Delta''_{1i}$, $i = 1, \ldots, n_1$, *then for every test* T *based on divergence of associative or non-associative statistics that satisfy conditions (a) and (b) above, the p-values are such that* $\lambda\left(\mathbf{X}(\Delta')\right) \stackrel{u}{\geq} \lambda\left(\mathbf{X}(0)\right) \stackrel{u}{\geq} \lambda\left(\mathbf{X}(\Delta)\right) \stackrel{u}{\geq} \lambda\left(\mathbf{X}(\Delta'')\right)$ *uniformly for all data sets* $\mathbf{X} \in \mathcal{X}^n$ *and every underlying distribution* P. *Hence, with respect to* Δ, *p-values are non-decreasingly uniformly ordered random variables.*

The proof is left as an exercise.

One consequence of Theorem 2 is that the conditional power of T,

$$W[(\Delta, \alpha, T)|\mathcal{X}_{/\mathbf{X}}] = \Pr\{\lambda_T(\mathbf{X}(\Delta)) \leq \alpha | \mathcal{X}_{/\mathbf{X}(\Delta)}\},$$

is such that for any attainable α-value,

$$W[(\Delta'', \alpha, T)|\mathcal{X}_{/\mathbf{X}}] \geq W[(\Delta, \alpha, T)|\mathcal{X}_{/\mathbf{X}}] \geq \alpha = W[(0, \alpha, T)|\mathcal{X}_{/\mathbf{X}}] \geq W[(\Delta', \alpha, T)|\mathcal{X}_{/\mathbf{X}}].$$

From the latter set of inequalities in particular we see that permutation test T is conditionally uniformly unbiased also for composite hypotheses with fixed effects such as $H_0 : \{\delta \leq 0\}$ versus $H_1 : \{\delta > 0\}$, and also for $H_0 : \{\delta \leq \delta_0 < 0\}$ versus $H_1 : \{\delta > \delta_1 > 0\}$. Of course, in these cases tests become conservative. Moreover, it is worth noting that for $\Delta \stackrel{p}{=} 0$, the conditional power essentially depends on random deviates $\mathbf{Z}$ only. In such a case, the test statistic T is said to provide an *exact test.* Due to its importance, the exactness property of permutation tests is stated by the following proposition:

Proposition 2. *A permutation test statistic T is said to be an exact test if its null distribution essentially depends on exchangeable deviates* $\mathbf{Z}$ *only.*

Although being permutationally exact is a known important property for a test statistic T, this does not imply that T is also a good test. Indeed, there are many permutationally exact tests for the same hypotheses with the same data $\mathbf{X}$, and some are better than others in terms of unbiasedness, power behaviour, etc. Furthermore, there are exact tests which are not unbiased, exact unbiased tests which are not consistent (see Section 4.3.2), and also consistent tests which are neither exact nor unbiased, such as permutation solutions of the Behrens–Fisher problem (see Example 8, 4.6).

One more consequence of Theorem 2 is that as $\lambda\,(\mathbf{X}(0))$ is uniformly distributed over its support, $\Lambda^{(n)}_{\mathbf{X}(0)} \in (0, 1)$ only when $\Delta \overset{p}{=} 0$, whereas if $\Delta \overset{p}{<} 0$ ($\overset{p}{>}0$), with respect to the underlying distribution P^n, $\lambda\,(\mathbf{X}(\Delta))$ is stochastically larger (smaller) than the uniform distribution. On the one hand this shows that the power function increases with Δ for fixed n; on the other the exchangeability condition, which is satisfied when $\Delta \overset{p}{=} 0$, is not strictly necessary for defining a proper permutation test. We have the following proposition:

Proposition 3. *What is really necessary for T to be a proper permutation test statistic is that, for instance, in the continuous case and for fixed effects, there exists a value δ', not belonging to H_1, such that $\lambda\left(\mathbf{X}(\delta')\right)$ is uniformly distributed over $\Lambda^{(n)}$ and that for every $\delta_0 \in H_0$ and $\delta_1 \in H_1$, $\lambda\,(\mathbf{X}(\delta_0)) > \lambda\,(\mathbf{X}(\delta_1))$, where this dominance is uniform for all $\mathbf{X} \in \mathcal{X}$. Hence, the value δ', for which data exchangeability is satisfied, must not be a member of H_1. Of course, when it is not a member of H_0, as when, for instance, $H_0 : \{\delta \le \delta_0\}$ versus $H_1 : \{\delta > \delta_1\}$, where we can have $\delta_0 < \delta' < \delta_1$, then the test becomes conservative.*

The results of Theorems 1 and 2 and of Propositions 1 and 3 can be easily extended to one-sample designs for one-sided alternatives. Such extensions, by observing that test statistics T are required to satisfy conditions (a) and (b) above similarly to statistics S, are left to the reader as exercises.

The results of Theorem 2 can also be used to obtain exact permutation confidence intervals for fixed effects δ and to deal with testing problems of non-inferiority (see Example 6, 4.6). Moreover, within the NPC methodology it also allows us to deal with testing for two-sided alternatives and with cases in which random effects can be positive on some subjects, null on others, and negative on the rest (see Example 5, 4.6).

Remark 1. The permutation test T allows for relaxation of the homoscedasticity condition in H_1, without compromising its exactness or its unbiasedness. It should be noted that the two-sample testing problem, when we may assume that treatment effects, together with locations, may also act on other aspects of interest, may be conveniently examined through $k > 1$ different statistics, each appropriate for one particular aspect. Problems of this kind and related solutions by *multi-aspect tests* are examined in Example 3, 4.6.

Remark 2. The assumption of non-negativity for stochastic effects Δ is equivalent to the dominance condition of two responses. Thus, two CDFs F_1 and F_2 are such that $\forall x \in \mathcal{R}^1$, $F_1(x) \le F_2(x)$ (see Problem 31, 3.9). This condition may sometimes be achieved by a model such as that of Example 2, 2.7, provided that the resulting CDFs do not cross. One general way of achieving this dominance condition in fixed effect models is through a model such as $\{X_{ji} = X(\mu, \delta_j, Z_{ji}),$ $i = 1, \ldots, n_j, j = 1, 2\}$, where responses are represented by a suitable monotonic function (increasing or decreasing) with respect to each argument. Observe that, in particular, one specification of this model is $\{X_{ji} = \mu + \delta_j + \sigma(\delta_j) \cdot Z_{ji}, i = 1, \ldots, n_j,\ j = 1, 2\}$, where scale coefficients may depend on treatment effects through a monotonic function of δ_j or of their absolute values $|\delta_j|$, provided that ordering conditions on CDFs are not violated. Note that the latter model is consistent with the notion of randomization (see Section 1.5), where units are randomly assigned to treatments, which in turn may also affect scale coefficients. It should be noted that, in general, this problem has no exact parametric solution.

If the dominance condition on CDFs, $F_1(x) \leq F_2(x)$, $\forall x \in \mathcal{R}^1$, is violated in the alternative, so that two CDFs cross each other, then the resulting permutation test may be biased, although it remains exact because the exchangeability condition is satisfied in H_0. For instance, there are situations connected with the Behrens–Fisher problem in which the acceptance probability in H_1 may be higher than in H_0.

Remark 3. With regard to Remark 2 above, we stress that the assumption of non-negativity for stochastic effects Δ (see Theorem 1 above) and the consequent stochastic dominance of the CDFs are only sufficient conditions for the unbiasedness of T. We guess that they may be partially relaxed. However, finding necessary conditions seems to be quite a difficult problem, which we leave to the reader.

3.1.2 Two-Sided Alternatives

General Aspects

Let us now turn to two-sided alternatives, that is test statistics for $H_0 : X_1 \stackrel{d}{=} X_2$ against $H_1 : X_1 \stackrel{d}{\neq} X_2$, where the direction of active effect is undefined. The notion of conditional and unconditional unbiasedness for two-sided alternatives is much more intriguing than for one-sided alternatives (see Lehmann, 1986, for a discussion). One of the reasons for this is that random effects $\Delta \stackrel{p}{\neq} 0$ might not imply the truth of either $\Delta \stackrel{p}{<} 0$ or $\Delta \stackrel{p}{>} 0$ and, except for quite special cases such as balanced designs, it is generally difficult to find suitable data transformations $\varphi : \mathcal{X} \rightarrow \mathcal{R}^n$ such that $\varphi(X(\Delta)) \stackrel{d}{>} \varphi(X(0))$, for any $\Delta \stackrel{p}{<} 0$ or $\Delta \stackrel{p}{>} 0$. Of course, if such a data transformation is known for the problem at hand, then the testing problem can be converted into an equivalent one-sided problem, the related permutation solutions of which have already been discussed. Two further reasons are that when unconditionally unbiased parametric tests exist, their acceptance regions, unless the population distribution is symmetric, are generally not symmetric or the rejection probabilities on both sides are not balanced, that is, not equal to $\alpha/2$. Furthermore, they are generally too impractical to use (see Cox and Hinkley, 1974, for a discussion), and essentially based on merely academic arguments. As a result, in practice, symmetric or balanced regions are commonly used. Of course, this may imply that we have to accept some slight bias in consequent inferences.

A practical and commonly used argument in population testing consists then in introducing a sort of extra condition requiring some *weak indifference principle* similar to the following: since we do not know the exact direction of the active alternative, we behave as if $|X_1 + \Delta - X_2| \stackrel{d}{>} 0$ would imply $H_1 : |\Delta| \stackrel{d}{>} 0$ with the possible exception of some small and irrelevant values of Δ. The sense of this weak indifference principle is that for some small values of Δ it is not particularly important to be able to distinguish between H_0 and H_1, especially if the bias is reasonably bounded and asymptotically vanishing, so that if sample sizes are sufficiently large, we can be confident the test remains unbiased.

In practice, the most commonly used two-sided tests are obtained by considering, for instance, statistics according to:

(a) the absolute mean divergence,

$$T_2(\Delta) = \left| \sum_i \varphi[X_{1i}(\Delta)]/n_1 - \sum_i \varphi(X_{2i})/n_2 \right|;$$

(b) the squared mean divergence,

$$T^2(\Delta) = \left[\sum_i \varphi[X_{1i}(\Delta)]/n_1 - \sum_i \varphi(X_{2i})/n_2\right]^2;$$

(c) the absolute divergence of general S statistics, either associative or non-associative,

$$T_S(\Delta) = |S_1[\mathbf{X}_1(\Delta)] - S_2(\mathbf{X}_2)|;$$

(d) the two-sided balanced confidence interval for fixed effects δ, $[\underline{\delta}_{\alpha/2}(\mathbf{X}), \overline{\delta}_{\alpha/2}(\mathbf{X})]$, where H_0 is accepted if 0 lies inside the confidence interval, i.e. if $\underline{\delta}_{\alpha/2}(\mathbf{X}) \leq 0 \leq \overline{\delta}_{\alpha/2}(\mathbf{X})$ (see Section 3.4);
(e) by considering symmetric acceptance regions $-t_{\alpha/2} < T(0) < t_{\alpha/2}$, where $t_{\alpha/2}$ is such that $\Pr\{T(0) \leq t_{\alpha/2}\} = 1 - \alpha/2$;
(f) if the null density distribution $f_T(\cdot)$ of T is unimodal, by the highest-density acceptance regions, $f_T(t_{1\alpha}) = f_T(t_{2\alpha})$ and $\Pr\{t_{1\alpha} \leq T(0) \leq t_{2\alpha}\} = 1 - \alpha$;
(g) the p-value, calculated as

$$\lambda_2(\mathbf{X}(\Delta)) = 1 - \Pr[-|T^o| < T < |T^o|].$$

An Example of an Unconditionally Biased Test

However, it is worth observing that if the unconditional sample distribution $F_T(t;\Delta)$ induced by a test statistic T in H_1 does not dominate the one induced in H_0, that is, if $F_T(t;\Delta) \not\leq F_T(t;0)$, $\forall t \in \mathcal{R}^1$, then in general there exist associated with T no unconditionally unbiased two-sided tests of the form (a)–(g). As a simple counterexample let us consider the following:

(1) the unconditional balanced acceptance region of T is $(T_{\alpha/2} \leq T \leq T_{1-\alpha/2})$, where T_π is the null π-quantile $F_T(T_\pi;0) = \pi$, and $F_T(t;0)$ is the null CDF of T;
(2) the unconditional sample distribution of T for fixed effects $0 < \delta < \infty$ is $F_T(t;\delta) = 1 - \exp(-t\,\delta)$, which is the distribution of a form of a chi-squared variable with two degrees of freedom and scale 2δ;
(3) the hypotheses are $H_0 : \delta - 1 = 0$ and $H_1 : \delta - 1 \neq 0$, and the two-sided null rejection probabilities are $\alpha/2 = 0.05$ on both sides.

The rejection probability for some values of δ is reported in Table 3.1. Exactly the same results are obtained if, instead of T, the rejection rule based on the balanced confidence interval (d) is used.

The maximum bias for $\alpha = 0.10$ is about 0.0156, which is obtained for $\delta = 1.40$ and is considerably smaller than $\alpha/2$. These results show a slight bias of T because its power is not monotonically non-decreasing with respect to $|\delta|$, and the region for which the test remains biased is $1 < \delta < 2$. Of course, if this bias is considered not to be particularly important, then according to the weak indifference principle we may regard the test as 'practically' unbiased.

Moreover, it is worth noting that as sample sizes increase, two-sided consistent tests become unconditionally unbiased. For instance, if the non-centrality of a chi-squared test with 2 d.f. is proportional to $n \cdot (\delta - 1)$, the set of δ values for which the test remains biased is $1 < \delta < 1 + 1/n$.

Table 3.1 Rejection probability of a two-sided test (balanced regions)

δ	0.90	1.00	1.10	1.40	1.80	2.00
$W(\delta)$	0.1126	0.1000	0.0919	0.0844	0.0928	0.1000

Hence, if sample sizes are not too small, so that the set of biased δ-values tends to vanish and the maximum bias is bounded, in general we can be confident of unconditional unbiasedness. It is, however, worth observing that as sample sizes increase, we are generally stricter and stricter with performance testing because for increasing sample sizes we usually wish to distinguish even small effects with increasing probability.

An Example of a Conditionally Biased Test

In conditional testing we find similar problems. Indeed, it is easy to find counterexamples against two-sided conditional unbiasedness for permutation tests. Moreover, since in nonparametric problems the population distribution is generally unknown, unbiased two-sided test statistics cannot be used, and so we are forced to confine ourselves to statistics of the form (a)–(f).

As a simple example, showing biasedness of permutation tests for testing $H_0 : \delta = 0$ against $H_1 : |\delta| > 0$, let us consider the following:

(1) the observed data sets are $\mathbf{Z}_1 = \{1, 3\}$ and $\mathbf{Z}_2 = \{2, 4\}$;
(2) the fixed effects are $\delta = \{-2; -1; 0; 1; 2; 3; 4\}$;
(3) the test statistic is $T_2^*(\delta) = \left|\sum_i X_{1i}^*(\delta) - \sum_i X_{2i}^*(\delta)\right|$.

Permutation test statistics and related p-values are reported in Table 3.2, from which biasedness of T_2 can easily be seen, since p-values are not non-increasing with respect to $|\delta|$. Indeed, as $\lambda_T(\delta = 1) = 6/6 > \lambda_T(0) = \lambda_T(2) = 4/6$, p-values seem non-increasing with respect to $|\delta - 1|$ but not to $|\delta - 0|$. However, if sample sizes increase, two-sided consistent tests become conditionally unbiased and so in general we can be confident of unbiasedness of permutation tests as well. Thus, the notion of permutation (conditional) consistency is of great importance.

In permutation contexts it is common to use the same kind of statistics as in population contexts, for example,

$$T_2(\Delta) = \left|\sum_i \varphi[X_{1i}^*(\Delta)]/n_1 - \sum_i \varphi[X_{2i}^*(\Delta)]/n_2\right|$$

for tests based on the absolute divergence;

$$T^{2*}(\Delta) = \left[\sum_i \varphi[X_{1i}^*(\Delta)]/n_1 - \sum_i \varphi[X_{2i}^*(\Delta)]/n_2\right]^2$$

Table 3.2 Permutation distribution of $T_2^*(\delta)$

$\mathbf{u}^* \setminus \delta$	−2	−1	0	1	2	3	4
1,2;3,4	|0−6|	|2−6|	|4−6|	|6−6|	|8−6|	|10−6|	|12−6|
1,3;2,4	|1−5|	|2−6|	|3−7|	|4−8|	|5−9|	|6−10|	|7−11|
1,4;2,3	|3−3|	|4−4|	|5−5|	|6−6|	|7−7|	|8−8|	|9−9|
2,3;1,4	|3−3|	|4−4|	|5−5|	|6−6|	|7−7|	|8−8|	|9−9|
2,4;1,3	|5−1|	|6−2|	|7−3|	|8−4|	|9−5|	|10−6|	|11−7|
3,4;1,2	|6−0|	|6−2|	|6−4|	|6−6|	|6−8|	|6−10|	|6−12|
$\lambda_T(\delta)$	2/6	4/6	4/6	6/6	4/6	4/6	2/6

for tests based on squared divergence;

$$T_S^*(\Delta) = \left|S_1[\mathbf{X}_1^*(\Delta)] - S_2[\mathbf{X}_2^*(\Delta)]\right|$$

for tests based on absolute divergence of S statistics; $[\underline{\delta}_{\alpha/2}(\mathbf{X}), \overline{\delta}_{\alpha/2}(\mathbf{X})]$, where H_0 is accepted if 0 lies inside the confidence interval (i.e. if $\underline{\delta}_{\alpha/2}(\mathbf{X}) \leq 0 \leq \overline{\delta}_{\alpha/2}(\mathbf{X})$), for a test based on the two-sided confidence interval for fixed effects; or by using the p-value of T^* calculated as

$$\lambda_2\,(\mathbf{X}(\Delta)) = 1 - \Pr[-|T^o| < T^* < |T^o| \,\big|\mathcal{X}_{/\mathbf{X}(\boldsymbol{\Delta})}]$$

or

$$\lambda_2\,(\mathbf{X}(\Delta)) = 2 \cdot \min\left\{\Pr[T^* \geq T^o|\mathcal{X}_{/\mathbf{X}(\boldsymbol{\Delta})}],\ \Pr[T^* \leq T^o|\mathcal{X}_{/\mathbf{X}(\boldsymbol{\Delta})}]\right\}.$$

It is, however, worth observing that for any given data set $\mathbf{X}$ and for fixed effects δ, it possible to find a value $\delta_{\mathbf{X}}$ such that for all $|\delta| > |\delta_{\mathbf{X}}|$ the two-sided test is uniformly unbiased. In contrast, for $|\delta| < |\delta_{\mathbf{X}}|$, the two-sided test might be biased, although its bias cannot be larger than $\alpha/2$. Moreover, this $\delta_{\mathbf{X}}$ equals zero when in the null hypothesis data $\mathbf{X}$ are symmetric with respect to a point μ, otherwise it is generally very close to zero, and goes to zero as sample sizes increase.

Within multi-aspect testing, in Example 4, 4.6, we will see a different way of obtaining two-sided alternatives by combining two tests, one for positive T^+ and one for negative alternatives T^-. This strategy may also yield the benefit of testing which arm is active while controlling the inferential errors.

Remark 1. From the arguments and examples above it is clear that for two-sided alternatives we have to tolerate some slight bias in consequent inferences. Indeed, unbiasedness in two-sided permutation tests, for either one-sample, two-sample or C-sample problems, is generally unattainable for finite sample sizes. It is in general achievable for large sample sizes or when the permutation null distribution is symmetric. For instance, in ANOVA designs the usual tests are not unbiased. However, when in two-sample designs, for every $t \in \mathcal{R}^1$, either $F_1(t) \geqslant F_2(t)$ or $F_1(t) \leqslant F_2(t)$, so that there is dominance in distribution between two variables and the test is consistent, then for sufficiently large sample sizes the two-sided test is also unbiased. This same property is also valid for ANOVA designs, provided that for any pair of distributions either $F_j(t) \geqslant F_h(t)$ or $F_j(t) \leqslant F_h(t)$, $j \neq h = 1, \ldots, C$.

3.2 Power Functions of Permutation Tests

3.2.1 *Definition and Algorithm for the Conditional Power*

As a guide, let us refer to a two-sample design for one-sided alternatives and fixed effects. Extensions to two-sided alternatives, one-sample, and C-sample designs are straightforward and are left to the reader as exercises. We consider a given test statistic T applied on the data set $\mathbf{X}(\delta) = (\mathbf{Z}_1 + \boldsymbol{\delta}, \mathbf{Z}_2)$, where fixed effects are $\boldsymbol{\delta} = (\delta_i = \delta > 0, i = 1, \ldots, n_1)$, and the deviates $\mathbf{Z}$, since we are arguing conditionally, are assumed to have the role of unobservable fixed quantities. With obvious notation, the *conditional power function* is defined as

$$\begin{aligned} W[(\delta, \alpha, T)|\mathcal{X}_{/\mathbf{X}}] &= \Pr\{\lambda_T(\mathbf{X}(\delta)) \leq \alpha|\mathcal{X}_{/\mathbf{X}(\delta)}\} \\ &= \mathbb{E}\{\mathbb{I}[\lambda(\mathbf{X}^\dagger(\delta)) \leq \alpha]|\mathcal{X}_{\mathbf{X}^\dagger(\delta)}\}, \end{aligned}$$

where its dependence on T, α, δ, n, and $\mathbf{Z}$ is self-evident, and the mean value is taken with respect to all possible $\mathbf{X}^\dagger(\delta)$. It should be emphasized that, due to Theorem 2, 3.1.1, $\delta < \delta'$ implies that

$W[(\delta, \alpha, T)|\mathcal{X}_{/\mathbf{X}}] \leq W[(\delta', \alpha, T)|\mathcal{X}_{/\mathbf{X}}]$ for every $\mathbf{X} \in \mathcal{X}$ and any $\alpha \in \Lambda_{\mathbf{X}}$. It is also worth noting that $\lambda(\mathbf{X}^{\dagger}(\delta))$ is the p-value calculated on the data set $\mathbf{X}^{\dagger}(\delta) = (\mathbf{Z}_1^{\dagger} + \boldsymbol{\delta}, \mathbf{Z}_2^{\dagger})$, where $\mathbf{Z}^{\dagger} \in \mathcal{X}_{/\mathbf{Z}}$ is a random permutation of unobservable deviates $\mathbf{Z}$. Indeed, the randomization principle essentially involves a random assingment of a subset $\mathbf{Z}_1^{\dagger}$ of deviates $\mathbf{Z}$ to treated units for which δ is active and the rest to the untreated, so that $\mathbf{Z}_1^{\dagger} + \boldsymbol{\delta}$ are the data $\mathbf{X}_1^{\dagger}$ of the first sample. From this point of view, the actual data set $\mathbf{X}(\delta)$ is just one of the possible sets $\mathbf{X}^{\dagger}(\delta)$ that can be obtained by a re-randomization of deviates to treatments. And so the notion of conditional power uses as many data sets $\mathbf{X}^{\dagger}$ as there are re-randomizations in $\mathcal{X}_{/\mathbf{Z}}$.

An algorithm for evaluating conditional power is based on the following steps:

1. Consider the pooled set of deviates $\mathbf{Z} = \mathbf{Z}_1 \biguplus \mathbf{Z}_2$ and the effects $\boldsymbol{\delta}$.
2. Take a re-randomization $\mathbf{Z}^{\dagger}$ of $\mathbf{Z}$ and the corresponding data set $\mathbf{X}_r^{\dagger}(\delta) = (\mathbf{Z}_{r1}^{\dagger} + \boldsymbol{\delta}, \mathbf{Z}_{r2}^{\dagger})$.
3. Using the algorithm in Section 2.2.5, based on B CMC iterations, calculate the p-value $\hat{\lambda}_T(\mathbf{X}_r^{\dagger}(\delta))$.
4. Independently repeat steps 2 and 3 R times.
5. The conditional power is then evaluated as $\hat{W}[(\delta, \alpha, T)|\mathcal{X}_{/\mathbf{X}}] = \sum_r \mathbb{I}[\hat{\lambda}_T(\mathbf{X}_r^{\dagger}(\delta)) \leq \alpha]/R$.

The extension to random effects Δ is straightforward if these, from step 2, are assumed independent of deviates $\mathbf{Z}$.

Remark 1. In particular, it should be noted that in a homoscedastic context, if conditions for a PCLT are satisfied (see Section 3.8) and sample sizes are large enough, the conditional power may be approximated by

$$W[(\delta, \alpha, T)|\mathcal{X}_{/\mathbf{X}}] \cong 1 - \Phi\left(z_{\alpha} - \frac{\delta}{\hat{\sigma}}\sqrt{\frac{n_1 \cdot n_2}{n}}\right),$$

where Φ is the standard normal CDF and $\hat{\sigma} = \left(\sum_{ji}(X_{ji} - \bar{X}_j)^2/n\right)^{1/2}$ and we note its essential dependence on the empirical *signal to noise ratio* $\delta/\hat{\sigma}$.

Remark 2. According to Theorem 2, 3.1.1, it is easy to prove that for increasing α the conditional power $W[(\delta, \alpha, T)|\mathcal{X}_{/\mathbf{X}}]$ cannot decrease. The simple proof of this is left to the reader as an exercise.

In order for this algorithm to be effectively carried out, in the given data set we should be able to separate the contributions of random deviates Z from those of effects δ. This is generally not possible in practice, because the $\mathbf{Z}$ and $\boldsymbol{\delta}$ components are not separately observable. Thus, conditional power is essentially a virtual notion in the sense that it is well defined but is not calculable. However, in place of $W[(\delta, \alpha, T)|\mathcal{X}_{/\mathbf{X}}]$, for a fixed effect model $\mathbf{X}(\delta) = (\mathbf{Z}_1 + \boldsymbol{\delta}, \mathbf{Z}_2)$, we may then attain the so-called *empirical post-hoc conditional power* $W[(\delta, \hat{\delta}, \alpha, T)|\mathcal{X}_{/\mathbf{X}}]$. This can be evaluated by the following algorithm.

1. Based on a suitable indicator T, consider the estimate $\hat{\delta}$ of δ from the pooled data set $\mathbf{X}(\delta)$ and the consequent empirical deviates $\hat{\mathbf{Z}} = (\mathbf{X}_1 - \hat{\delta}, \mathbf{X}_2)$. Note that, in accordance with point (ii) in Remark 2, 2.1.2, empirical deviates $\hat{\mathbf{Z}}$ are exchangeable.
2. Take a random re-randomization $\hat{\mathbf{Z}}^{\dagger} = \{\hat{Z}(u_i^{\dagger}), i = 1, \ldots, n\}$ of $\hat{\mathbf{Z}}$, where $(u_i^{\dagger}, i = 1, \ldots, n)$ is a permutation of $(1, \ldots, n)$, and for any chosen δ the corresponding data set $\hat{\mathbf{X}}_r^{\dagger}(\delta) = (\hat{\mathbf{Z}}_{r1}^{\dagger} + \boldsymbol{\delta}, \hat{\mathbf{Z}}_{r2}^{\dagger})$.
3. Using the algorithm in Section 2.2.5, based on B CMC iterations, calculate the p-value $\hat{\lambda}_T(\hat{\mathbf{X}}_r^{\dagger}(\delta))$.
4. Independently repeat steps 2 and 3 R times.

5. The empirical post-hoc conditional power for δ is then evaluated as $\hat{W}[(\delta, \hat{\delta}, \alpha, T)|\mathcal{X}_{/\mathbf{X}}] = \sum_r \mathbb{I}[\hat{\lambda}_T(\hat{\mathbf{X}}_r^{\dagger}(\delta)) \leq \alpha]/R$.
6. To obtain a function in δ and α, repeat steps 2–5 for different values of δ and α. With $\delta = \hat{\delta}$ we obtain the *actual* post-hoc conditional power $\hat{W}[(\hat{\delta}, \hat{\delta}, \alpha, T)|\mathcal{X}_{/\mathbf{X}}]$.

The actual post-hoc conditional power may be used to assess how reliable the testing inference associated with $(T, \mathbf{X})$ is, in the sense that if by chance the probability of obtaining the same inference with $(T, \mathbf{X}^{\dagger})$ as with $(T, \mathbf{X})$ is greater than (say) $1/2$, then the actual inferential conclusion, given the set of units underlying $\mathbf{X}$, is reproducible more often than not. For instance, using the data of Section 1.10 on job satisfaction in 20 subjects, we take $\alpha = 0.05$, $B = 1000$ CMC iterations and $R = 1000$ re-randomizations. The estimate of the effect is $\hat{\delta} = \bar{X}_1 - \bar{X}_2$, and $\hat{\mathbf{Z}} = [\mathbf{X}_1 - \hat{\delta}\mathbf{1}_{n_1}, \mathbf{X}_2]$, where $\mathbf{X}_1$ and $\mathbf{X}_2$ are the data vectors of the first and second sample, and $\mathbf{1}_{n_1}$ is a $+1$ vector of length n_1.

```
setwd("C:/path")
R=1000; B=1000
data<-read.csv("Job.csv",header=TRUE)
attach(data)

delta = mean(X[Y==1])-mean(X[Y==2])
Z = c(X[Y==1]-delta,X[Y==2])
```

Each resampling is then obtained by considering a random permutation of $\mathbf{Z}$ and by adding $\hat{\delta}$ to the first n_1 elements of $\mathbf{Z}^*$:

```
set.seed(100)
p.val = array(0,dim=c(R,1))

for(cc in 1:R){

Z.star=sample(Z)
Z.star[Y==1]=Z.star[Y==1] + delta

T<-array(0,dim=c((B+1),1))

T[1] = mean(Z.star[Y==1])-mean(Z.star[Y==2])

for(bb in 2:B){
Z.perm=sample(Z.star)
T[bb]= mean(Z.perm[Y==1])-mean(Z.perm[Y==2])
}## end bb

p.val[cc] = mean(T[-1]>=T[1])
print(cc)
}## end cc
```

As in the previous example, the vector `p.val` contains the p-values related to each resampling. The (estimated) conditional power is then:

```
pow = mean(p.val<=0.05); pow
[1] 0.998
```

Table 3.3 Empirical post-hoc conditional power: two-sample data

δ	0.00	9.00	13.00	17.292	19.00
$\hat{W}$	0.048	0.704	0.943	0.998	1.000

The data set and the corresponding software codes are available from the `examples_chapters_1-4` folder on the book's website. The empirical post-hoc conditional power for $\alpha = 0.05$ and various values of δ is shown in Table 3.3. From these results it is worth noting that the actual post-hoc conditional power at $\alpha = 0.05$ calculated on the estimated point $\hat{\delta} = 17.292$ is $\hat{W}[(\hat{\delta}, \hat{\delta}, \alpha, T)|\mathcal{X}_{/\mathbf{X}}] = 0.998$, that is, the test rejects H_0 on almost all re-randomized data sets. Thus, for this specific problem the rejection of equality of job satisfaction is highly reliable.

A similar conclusion can be obtained using the IPAT paired data of Section 1.8.

The conditional power is obtained by subtracting the estimate of the average difference $\hat{\delta} = \bar{Y}_1 - \bar{Y}_2$ from the vector `X`, i.e. by letting `Z = X - delta`, obtaining `Z.star`, a random resampling of `Z`, adding `delta` to `Z.star` and finally performing a paired one-sample permutation test on `Y.star` with `B` permutations ($B = 1000$).

```
setwd("C:/path")
data<-read.csv("IPAT.csv",header=TRUE)
n = dim(data)[1]; C=1000 ; B=1000

Z = data[,1]-data[,2]-sum(d)
```

We repeat this procedure for all possible resamplings of `Z` (in fact we use `R=1000` random resamplings) and each time store the p-value of the test in the vecrtor `p.val`:

```
p.val =array(0,dim=c(R,1))

for(cc in 1:R){ ## n of resamplings

Z.star=sample(Z)
Z.star=Z.star + sum(d)
print(cc)
T<-array(0,dim=c((B+1),1))

T[1] = sum(d)

for(bb in 2:(B+1)){
T[bb] = t(d)%*%(1-2*rbinom(n,1,.5))
}## end bb

p.val[cc]=t2p(T)[1]
}## end cc
```

The conditional power is then computed as the proportion of times the null hypothesis has been rejected at level α (here we let $\alpha = 0.005$):

```
alpha = 0.0005
pow = mean(p.val <= alpha)
pow
```

Table 3.4 Empirical post-hoc conditional power: paired data

δ	0.00	1.50	2.00	2.50	3.10
$\hat{W}$	0.0054	0.291	0.570	0.844	0.991

Indeed, by again using $B = 1000$, $R = 1000$, $\alpha = 0.005$, $\hat{\delta} = 3.10$ and various δ we obtain the results shown in Table 3.4.

Now the actual post hoc power with 0.005 is $\hat{W}[(\hat{\delta}, \hat{\delta}, \alpha, T)|\mathcal{X}_{/\mathbf{X}}] = 0.991$, and so the related rejection of H_0, with $\hat{\lambda} = 0.0003$, is highly reliable. Of course, the algorithm for paired data designs changes steps 1, 2 and 3 respectively into the following:

1′. Based on a suitable indicator T, consider the estimate of δ from the pooled data set $\mathbf{X}(\delta) = \mathbf{Z}_2 + \boldsymbol{\delta} - \mathbf{Z}_1$ and the consequent empirical deviates $\hat{\mathbf{X}} = \mathbf{X}(\delta) - \hat{\boldsymbol{\delta}}$, where $\hat{\delta} = \sum_i X_i/n$.
2′. Take a random rearrangement $\hat{\mathbf{X}}_r^\dagger = \{\hat{X}_i \cdot S_i^\dagger, i = 1, \ldots, n\}$, where $S_i^\dagger = 1$ or -1 each with probability 1/2, and for any chosen δ the corresponding data set $\hat{\mathbf{X}}_r^\dagger(\delta) = \hat{\mathbf{X}}_r^\dagger + \boldsymbol{\delta}$.
3′. Using the algorithm in Section 1.9.3, based on B CMC iterations, calculate the p-value $\hat{\lambda}_T(\hat{\mathbf{X}}_r^\dagger(\delta))$.

In Section 4.3.4 within the nonparametric combination procedure we will see extensions of these algorithms to multidimensional problems.

Remark 3. It is worth noting that, as was mentioned in Section 1.9.2, the null differences, $\mathbf{X}(\delta) = 0$, cannot be discarded from the permutation analysis because the algorithm takes into consideration the empirical deviates $\hat{\mathbf{X}} = \mathbf{X}(\delta) - \hat{\boldsymbol{\delta}}$ and the connected hypothesized deviates $\hat{\mathbf{X}} + \boldsymbol{\delta}'$.

3.2.2 *The Empirical Conditional ROC Curve*

If in step 6 of the algorithm in Section 3.2.1 above, we consider the pair $(\alpha, \hat{W}[(\hat{\delta}, \hat{\delta}, \alpha, T)|\mathcal{X}_{/\mathbf{X}}])$ for various α-values, we obtain the so-called empirical conditional ROC curve. This, calculated for various test statistics T, can be used to choose a 'practically best' test statistic with the present data set $\mathbf{X}$. A more in-depth look at the empirical conditional ROC curve is left to the reader.

3.2.3 *Definition and Algorithm for the Unconditional Power: Fixed Effects*

Of course, to define the unconditional version we must obtain the mean value of $W[(\delta, \alpha, T)|\mathcal{X}_{/\mathbf{X}}]$ with respect to the underlying population distribution P^n. That is:

$$\begin{aligned} W(\delta, \alpha, T, P, n) &= \mathbb{E}_{\mathcal{X}^n \backslash \mathcal{X}_{/\mathbf{X}}}\{\mathbb{E}[W((\delta, \alpha, T, n)|\mathcal{X}_{/\mathbf{X}})]\} \\ &= \mathbb{E}_{\mathcal{X}}\{W[(\delta, \alpha, T, n)|\mathcal{X}_{/\mathbf{X}}]\} \\ &= \int_{\mathcal{X}^n} \mathbb{I}\left[\lambda_T(\mathbf{X}(\delta)) \le \alpha \,|\mathcal{X}_{/\mathbf{X}}\right] dP^n(\mathbf{X}(\delta)). \end{aligned}$$

Note that in order to properly define the unconditional power $W(\delta, \alpha, T, P, n)$, the underlying population distribution P must be fully specified, that is, defined in its analytical form and all its parameters. Also note that averaging with respect the whole sample space $\mathcal{X}^n$ implies taking the mean with respect to each conditional distribution over $\mathcal{X}_{/\mathbf{X}}$ and then taking the mean of these with respect to the distribution over $\mathcal{X}^n \backslash \mathcal{X}_{/\mathbf{X}}$. A practical algorithm for evaluating the unconditional power can be based on a standard Monte Carlo simulation from P with MC iterations as follows:

1. Choose a value of δ.
2. From the given population distribution P draw one set of n deviates $\mathbf{Z}_r$, and then add δ to the first n_1 errors to define the data set $\mathbf{X}_r(\delta) = (\mathbf{Z}_{r1} + \boldsymbol{\delta}, \mathbf{Z}_{r2})$.
3. Using the algorithm in Section 2.2.5, based on B CMC iterations, calculate the p-value $\hat{\lambda}_T(\mathbf{X}_r(\delta))$.
4. Independently repeat steps 2 and 3 MC times.
5. Evaluate the estimated power as $\hat{W}(\delta, \alpha, T, P, n) = \sum_r \mathbb{I}[\hat{\lambda}_T(\mathbf{X}_r(\delta)) \leq \alpha]/MC$.
6. To obtain a function in δ, α, T and n, repeat steps 1–5 with different values of δ, α, T and n.

The latter result allows us to assess whether the α-size of any given test statistic T is α by means of a Monte Carlo simulation from any given fully specified distributions P, when $\delta = 0$. We sometimes use this property when wishing to examine the behaviour of a permutation test in H_0, especially when complex problems are involved. It is also used to assess whether a test statistic T is unbiased, that is, whether $\hat{W}(\delta, \alpha, T, P, n) \geq \alpha$. Extensions to multidimensional problems are straightforward within the NPC methodology (see Section 4.3.4).

Note that the unconditional power function $W(\delta, \alpha, T, P, n)$ depends on the underlying distribution P and, of course, on the test statistic T. Also note that for increasing sample sizes, it is not always possible to compare related power functions in discrete cases. Suppose, indeed, that we are given two data sets $\mathbf{X}$ and $\mathbf{X}'$ with respective sample sizes n and n', where $n < n'$. It may be that the intersection of related sets of attainable p-values $\Lambda_{\mathbf{X}}$ and $\Lambda_{\mathbf{X}'}$ is empty, since they may have no common point. Because of this, the power function may apparently decrease at some (desired) $\alpha = \alpha_d$ for increasing sample sizes especially when $\alpha_d > \alpha > \alpha'$ where α and α' are the largest values in $\Lambda_{\mathbf{X}}$ and in $\Lambda_{\mathbf{X}'}$ which are smaller than α_d. In such situations, either we consider the power behaviour for *large* sample sizes so that the two sets $\Lambda_{\mathbf{X}}$ and $\Lambda_{\mathbf{X}'}$ are not distinguishable in practice (in fact they are asymptotically coincident), or we consider randomized permutation tests ϕ_R. With regard to the latter, it is worth noting that ϕ_R satisfies $W(\delta, \alpha, T, P, n) \geq \alpha$ and $W(0, \alpha, T, P, n) = \alpha$ without restriction, for all $\mathbf{X} \in \mathcal{X}^n$, all P, all exact T and all n.

Remark 1. It is worth noting that, since the unconditional power is the expectation of the conditional power, $W(\delta, \alpha, T, P, n) = \mathbb{E}_{\mathcal{X}}\{W[(\delta, \alpha, T, n)|\mathcal{X}_{/\mathbf{X}}]\}$, the latter may be interpreted as a least squares estimate of the former, that is, with obvious notation, $\bar{W}[(\delta, \alpha, T, n)|\mathcal{X}_{/\mathbf{X}}] = W(\delta, \alpha, T, P, n)$. Moreover, it is straightforward to see that for any exact T, any P and any n, $W(\delta, \alpha, T, P, n)$ is non-decreasing in both δ and α.

3.2.4 Unconditional Power: Random Effects

If the distribution $Q(\Delta|\mathbf{X})$ of random effects $\Delta \in \Omega$, given the observable data $\mathbf{X} \in \mathcal{X}$ and the population distribution $P^n[\mathbf{X}(\Delta)]$, is known, the unconditional power function $W(\Delta, \alpha, T, P, Q, n)$ with random effects is defined as

$$W(\Delta, \alpha, T, P, Q, n) = \int_{\mathcal{X}^n \times \Omega} \mathbb{I}\left[\lambda_T\left(\mathbf{X}(\Delta)\right) \leq \alpha \left|\mathcal{X}_{/\mathbf{X}}\right.\right] dQ(\Delta|\mathbf{X})\, dP^n[\mathbf{X}(\Delta)].$$

The algorithm for its evaluation is straightforward. It is also straightforward to see that for any exact T, any P and any n, $W(\Delta, \alpha, T, P, Q, n)$ is is non-decreasing in both Δ and α.

3.2.5 Comments on Power Functions

Unfortunately it is generally difficult for permutation tests to express conditional and unconditional power functions in closed form for finite sample sizes, for use in actual calculations. This is often true for parametric tests as well when unknown nuisance parameters occur. For instance, when using the Student's t, in practice it is impossible to express its power function without knowledge of σ because the power depends essentially on the standardized non-centrality parameter, i.e. the signal to noise ratio δ/σ. Thus, power functions remain hypothetical entities measuring the probability of a test rejecting the null hypothesis when it is false. Power functions are hypothetical because they may be evaluated when the treatment effect and the population distribution are hypothesized to be δ and P, respectively.

On the one hand, we know that permutation conditional and unconditional power functions, for any given population distribution P, any given test statistic T and any given sample size n, are well-defined functions of δ, α, and $\mathbf{X}$. In particular, at least in principle, both may be evaluated by numerical calculation, Monte Carlo simulation or asymptotic approximations (see Remark 1, 3.2.1 and Section 3.7). On the other hand, the empirical conditional power, given a data set $\mathbf{X}(\delta)$, where the treatment effect is hypothesized to be δ, may be evaluated by appropriate algorithms; one is illustrated in Section 3.2.1. Such evaluation is useful when comparing various kinds of test statistics applied to the same data set $\mathbf{X}(\delta)$ or for all uses of conditional power functions. For instance, in problems of determining a permutation confidence interval or estimating a sample size which guarantees a specified testing performance, we do not know the so-called true value of δ or the underlying distribution P, so that the power function must be estimated in some way, typically the so-called post-hoc procedure.

The unconditional power function may sometimes be estimated by estimating P through EDF estimates, nonparametric density estimates, semiparametric approximations, saddlepoint approximations, Edgeworth expansions (Albers et al., 1976; Bickel and Van Zwet, 1978; Robinson, 1982; John and Robinson, 1983a; De Martini, 1998), and the *plug-in* principle by approaches such as the bootstrap, smoothed bootstrap, etc.

3.3 Consistency of Permutation Tests

When for any fixed α, T, and P, and any given $\delta > 0$, the limit as $n \to \infty$ of the unconditional power function is $\lim_{n\uparrow\infty} W(\delta, \alpha, T, P, n) = 1$, the test statistic T is said to be consistent in the traditional sense. Quite a different notion of consistency connected with multivariate variables is discussed in Section 4.5 when, in place of the divergent number n of units, it is the number of informative variables V that diverges.

Proof of traditional consistency may easily be obtained by observing that, in conditions for asymptotically finite critical values T_α, test statistics of the form $T[\mathbf{X}^{(n)*}(\delta)] = \sum_i X_i^*(\delta)/\sqrt{n} = \sum_i X_i^*/\sqrt{n} + \delta\sqrt{n}$ are such that, $\forall\alpha > 0$,

$$\lim_{n\to\infty} \Pr\{T(\mathbf{X}^{(n)*}(\delta)) \geq T_\alpha(\mathbf{X}^{(n)}(\delta))|\mathbf{X}^{(n)}(\delta)\} = 1.$$

The proof of this is straightforward and left to the reader (see Problem 1, 3.9). Note that for divergent sample sizes, there is no possibility of distinguishing between the concepts of *conditional* and *unconditional consistency*. The problem of finding general conditions for $\lim T_\alpha(\mathbf{X}^{(n)}) = T_\alpha$, that is, the convergence of permutation critical values to a constant, is postponed to Section 3.7 (see Hoeffding, 1952; Romano, 1990).

3.4 Permutation Confidence Interval for δ

In this section we assume fixed effects δ. Confidence intervals have a clear interpretation for fixed δ whereas for random effects Δ they are difficult to interpret and to obtain (for a hint, see Remark 2 below). The monotonic ordering property with respect to δ of conditional p-values and of permutation CDFs (related proof is given in Theorem 2, 3.1.1) allows for the construction of permutation confidence intervals of level $1-\alpha$ for δ by quite a natural procedure.

With reference to a two-sample design, let us assume that the pooled data set is $\mathbf{X}(\delta) = \mathbf{X}_1(\delta) \biguplus \mathbf{X}_2$, in which $\mathbf{X}_1(\delta) = (\mu + \delta + \sigma Z_{1i}, i = 1, \ldots, n_1)$ and $\mathbf{X}_2 = (\mu + \sigma Z_{2i}, i = 1, \ldots, n)$, where μ and σ are finite unknown nuisance quantities. The solution to this problem implies determining two functions of the data, $\underline{\delta}^*(\mathbf{X}(\delta))$ and $\overline{\delta}^*(\mathbf{X}(\delta))$, with the role of lower and upper limits respectively, in such a way that the permutation coverage probability

$$\Pr\{\underline{\delta}^*(\mathbf{X}(\delta)) \leq \delta \leq \overline{\delta}^*(\mathbf{X}(\delta)) | \mathcal{X}_{/\mathbf{X}(\delta)}\} = 1 - \alpha$$

holds for any chosen value of $0 < \alpha < \frac{1}{2}$, for any unknown δ, for whatever non-degenerate data set $\mathbf{X}(\delta)$, and of course independently of the underlying population distribution P. To this end, we recall the well-known rule that a confidence interval for δ *contains all those values* δ° for which, by using a given test statistic T, the null hypothesis $H_0(\delta^\circ) : \{(X_1(\delta) - \delta^\circ) \stackrel{d}{=} X_2\}$, against $H_1(\delta^\circ) : \{(X_1(\delta) - \delta^\circ) \stackrel{d}{\neq} X_2\}$, is accepted at level α. Of course, the one-sided permutation confidence interval for δ consists of all values δ^o which would be accepted at level α, if the null hypothesis $H_0 : \{\delta = \delta^o\}$ were tested against the alternative $H_1 : \{\delta <$ (or $>$) $\delta^o\}$ (see Noether, 1978; Gabriel and Hall, 1983; John and Robinson, 1983b; Robinson, 1987).

Thus, given the pair $(T, \mathbf{X})$ we have to determine the set of values of δ such that, up to a given numeric accuracy, the associated two-sided permutation p-value leads to the acceptance of the null hypothesis. Note that, in determining such an interval, the algorithm and related computations are easier if T is permutationally equivalent to a convenient sampling indicator $\hat{\delta}$ for δ. For instance, if the response model is linear, namely $X_1(\delta) = \mu + \delta + \sigma Z$ and $X_2 = \mu + \sigma Z$, where deviates Z are such that $\mathbb{E}_P(Z) = 0$, then a *natural* sampling indicator for δ is the test statistic $T = \sum_i X_{1i}/n_1 - \sum_i X_{2i}/n_2 = \hat{\delta}$. An extension of permutation confidence intervals to random effects Δ is suggested in Problem 6, 3.4.1.

An algorithm for evaluating the lower limit $\underline{\delta}^*(\mathbf{X}(\delta))$, solved within the CMC framework by using B iterations, with an estimated error of preset width $\varepsilon > 0$ on confidence level $1 - \alpha/2$, consists of the following steps:

(S.a) Choose a value for ε; of course ε must be reasonably related to B: the smaller ε is, the larger B is.

(S.b) Choose a negative number η and subtract $\hat{\delta} + \eta$ from every value of the first data group, obtaining the vector $\mathbf{X}_1(\eta) = \{X_{1i}(\delta) - (\hat{\delta} + \eta), i = 1, \ldots, n_1\}$ and the resulting pooled data set $\mathbf{X}(\eta) = \mathbf{X}_1(\eta) \biguplus \mathbf{X}_2$.

(S.c) Using a CMC procedure based on B iterations, compute $\hat{F}^*_B(T^o_\eta)$ on the statistic $T^*(\eta) = \bar{X}^*_1(\eta) - \bar{X}^*_2(\eta)$, where $T^o_\eta = T[\mathbf{X}(\eta)]$.

(S.d) Repeat steps (S.b) and (S.c) with different values for η until the condition $|1 - \hat{F}^*_B(T^o_\eta) - \frac{\alpha}{2}| < \frac{\varepsilon}{2}$ is satisfied, then assign $\underline{\delta}^*(\mathbf{X}(\delta)) = \hat{\delta} + \eta$.

In order to evaluate the upper confidence limit $\overline{\delta}^*(\mathbf{X}(\delta))$ change step (S.d) to:

(S.d′) Repeat steps (S.b) and (S.c) with different positive values for η, until the condition $|\hat{F}^*_B(T^o_\eta) - \frac{\alpha}{2}| < \frac{\varepsilon}{2}$ is satisfied, then assign $\overline{\delta}^*(\mathbf{X}(\delta)) = \hat{\delta} + \eta$.

It is worth observing that the resulting permutation confidence interval for δ is the one induced by the statistic T, given the data set $\mathbf{X}(\delta)$. Also observe that, since in the nonparametric setting no statistic $T : \mathcal{X} \rightarrow \mathcal{R}^1$ can be claimed to maintain sufficiency for the underlying population distribution P, then with the same data set different intervals are obtained if different non-equivalent statistics T are used. With the data from Section 1.10, related to job satisfaction, using $B = 10\,000$, $\alpha = 0.05$, and $\varepsilon = 0.001$, we obtain $\hat{\delta} = 17.29$ and $\underline{\delta}^*(\mathbf{X}(\delta)) = 8.542 \leq \delta \leq \overline{\delta}^*(\mathbf{X}(\delta)) = 25.947$, which is slightly asymmetric but very close to that based on the Student's t: $8.58 \leq \delta \leq 26.00$.

To speed up computations, the search can start with values resulting from the Student's t approach. It is worth observing that, conditionally on observed data $\mathbf{X}$, the Student's t approximates the permutation distribution. The procedure may easily be extended to any functional of the form $\delta_\varphi = \mathbb{E}[\varphi(X_1(\delta))] - \mathbb{E}[\varphi(X_2)]$, where φ is a monotonic transformation of the data which may be chosen in order that δ_φ has a suitable physical meaning. Also observe that these confidence intervals are balanced, leaving $\alpha/2$ of error probability on both sides. As permutation distributions $F(t;\delta|\mathcal{X}_{/\mathbf{X}})$ may be asymmetric with respect to δ, these permutation intervals might not be the shortest ones. However, if $F(t;\delta|\mathcal{X}_{/\mathbf{X}})$ is not far from symmetry, as is often the case, balanced intervals are approximately close to the shortest intervals.

Remark 1. This same procedure is also valid for paired data designs. Only here we must take into consideration that permutations are now in accordance with Remark 3, 2.1.2 – that is to say, step 2 of the algorithm in Section 2.2.5 is converted into step (S.b) of the algorithm in Section 1.9.3 (see Problem 2, 3.4.1). In this regard, with the IPAT data of Chapter 1, using $B = 10\,000$, $\alpha = 0.05$, and $\varepsilon = 0.001$, we obtain $\underline{\delta}^*(\mathbf{X}(\delta)) = 1.768$ and $\overline{\delta}^*(\mathbf{X}(\delta)) = 4.435$, which is very close to the values based on the Student's t: $1.760 \leq \delta \leq 4.440$. It is worth observing that null differences enter the process for determining confidence intervals as well as non-null differences, and so they cannot be discarded from analysis. Instead, for determining the (observed) p-value they can be discarded since their permutation values are null in any case (see Section 1.9.2). Also observe that the permutation confidence interval in the example is included in the one based on Student's t. This is not surprising because the underlying distribution, being related to the sum of a finite number of discrete items, cannot be normal and so the estimated sample deviation cannot be independent of the sample mean, making doubtful the applicability of Student's t distribution. Often with rounded normal data, the permutation confidence interval may be included in the Student's t counterpart because rounding data makes the estimate of σ smaller, and so gives a larger t and an anticonservative confidence interval.

Remark 2. In developing a complete theory for permutation confidence intervals with a given pair $(T, \mathbf{X})$, suppose that the permutation support $\mathcal{T}(\mathbf{X})$, CDF $F[(t;\delta)|\mathcal{X}_{/\mathbf{X}}]$ and related attainable α-values are defined in discrete sets; in particular, $F(t;\delta|\mathcal{X}_{/\mathbf{X}})$ has jumps on points of $\mathcal{T}(\mathbf{X})$. Hence, confidence limits $\underline{\delta}^*(\mathbf{X})$ and $\overline{\delta}^*(\mathbf{X})$ of size α may be determined up to intervals related to $T_{(h)} \leq t < T_{(h+1)}$ and with probability jumps $F[(T_{(h+1)};\delta)|\mathcal{X}_{/\mathbf{X}}] - F[(T_{(h)};\delta)|\mathcal{X}_{/\mathbf{X}}]$, where $T_{(h)}$ is the hth ordered element of $\mathcal{T}(\mathbf{X})$. Note that the lengths of $(T_{(h)}, T_{(h+1)}]$ and jumps of $F[(t;\delta)|\mathcal{X}_{/\mathbf{X}}]$ both depend on the sample size n, statistic T and data set $\mathbf{X}$. Moreover, in developing confidence intervals for random effects Δ, assuming that $\Delta = \delta + Q_\Delta$, where deviates Q_Δ are such that $\mathbb{E}(Q_\Delta) = 0$, we can obtain confidence intervals on the mean effect value δ by exactly the same algorithm for fixed effects. It should be noted, however, that data sets now have the structure of $\mathbf{X}_1(\Delta) = [\mu + \delta + (Q_\Delta + \sigma Z_{1i}), i = 1, \ldots, n_1]$ and $\mathbf{X}_2 = (\mu + \sigma Z_{2i}, i = 1, \ldots, n)$ where random deviates on data are $\sigma \cdot Z + Q_\Delta$ for $X_1(\Delta)$ and $\sigma \cdot Z$ for X_2, respectively (note non-homoscedasticity in the alternative). In Section 4.3.5 we will see an extension to the multidimensional case.

Remark 3. (*Permutation likelihood*). Benefiting from the statement in Theorem 2, 3.1.1, and the algorithm for confidence intervals, we may define the so-called permutation likelihood induced by T given $\mathbf{X}$ which, although derived differently from the one introduced by Owen (1988), may be seen as a form of empirical likelihood. To this end, let us assume a fixed effect model for responses. Therefore, given the pair $(T, \mathbf{X})$, and with obvious notation, p-values may also be expressed by

$$\lambda[\mathbf{X}(\delta)] = \int_{T^o(\delta)}^{\infty} dF_T[(t;\delta)|\mathcal{X}_{/\mathbf{X}}] = \sum_{t \geq T^o(\delta)} \Pr\{T^*(\delta) = t|\mathcal{X}_{/\mathbf{X}}\},$$

where, of course, the sum includes all points t of the conditional support $\mathcal{T}(\mathbf{X}(\delta))$ of T which are not smaller than $T^o(\delta)$, and the CDF $F_T[(t;\delta)|\mathcal{X}_{/\mathbf{X}}]$ for $t \in \mathcal{R}^1$ is obtained by considering the pooled data set $\mathbf{X}(\delta) = \mathbf{X}_1(\delta) \biguplus \mathbf{X}_2$. Note that the cardinality M of the conditional support $\mathcal{T}(\mathbf{X}(\delta))$ is related to that of the permutation sample space $\mathcal{X}_{/\mathbf{X}}$. When there are no ties, the two sets share the same cardinality with probability one. Thus, in general, $dF_T[(t;\delta)|\mathcal{X}_{/\mathbf{X}}] = 1/M$ if $t \in \mathcal{T}(\mathbf{X}(\delta))$ and zero elsewhere, where the differential is with respect to t.

However, depending on δ and $\mathbf{X}(\delta)$, and so also on n, points in $\mathcal{T}(\mathbf{X}(\delta))$ tend to concentrate, in the sense that for any $\varepsilon > 0$ the frequency of points in the interval $t \pm \varepsilon$ is $F_T[(t+\varepsilon;\delta)|\mathcal{X}_{/\mathbf{X}}] - F_T[(t-\varepsilon;\delta)|\mathcal{X}_{/\mathbf{X}}] = D_\varepsilon F_T[(t;\delta)|\mathcal{X}_{/\mathbf{X}}]$. From this point of view, according to a naive kernel estimate (see Wasserman, 2006), the quantity $D_\varepsilon F_T[(t;\delta)|\mathcal{X}_{/\mathbf{X}}]/2\varepsilon$ may be regarded as similar to a nonparametric density estimate of points around t given δ. Therefore, for fixed values of t, the behaviour of

$$f_T^{(\varepsilon)}[(t;\delta)|\mathcal{X}_{/\mathbf{X}}] = D_\varepsilon F_T[(t;\delta)|\mathcal{X}_{/\mathbf{X}}]/2\varepsilon$$

is formally and conceptually similar to that of a likelihood function for δ. We may call this quantity the *smoothed permutation likelihood* of T given $\mathbf{X}$. In this sense, the p-value of a test T, given $\mathbf{X}$, corresponds to a form of *integrated permutation likelihood induced by* T. Indeed, we may write

$$\lambda[\mathbf{X}(\delta)] \simeq \int_{T^o(\delta)}^{\infty} f_T^{(\varepsilon)}[(t;\delta)|\mathcal{X}_{/\mathbf{X}}] \cdot dt.$$

For practical purposes if any, in order to obtain good evaluations of $f_T^{(\varepsilon)}[t;\delta|\mathcal{X}_{/\mathbf{X}}]$ we can choose the window bandwidth for instance as $\varepsilon \simeq 1.06 s_T^*/M^{1/5}$, where $s_T^* = \min\{\sigma_T^*, (T_{Q3}^* - T_{Q1}^*)/1.34\}$ in which $\sigma_T^* = \sqrt{\mathbb{E}[T^* - \mathbb{E}(T^*)]^2}$ is the permutation standard deviation, and T_{Q3}^* and T_{Q1}^* are the third and first permutation quartiles of T. When sample sizes tend to the infinity, so that points in $\mathcal{T}(\mathbf{X}(\delta))$ become dense, the permutation likelihood $f_T^{(\varepsilon)}[(t;\delta)|\mathcal{X}_{/\mathbf{X}}]$ converges to a proper likelihood function. Of course, in place of this naive estimate it is possible to use any kernel estimate.

However, as the permutation likelihood is induced by a test statistic T, given $\mathbf{X}$, it should be emphasized that in general $f_T^{(\varepsilon)}[(t;\delta)|\mathcal{X}_{/\mathbf{X}}]$ cannot be directly used to find a *best* nonparametric estimator for δ because T, used as a statistical indicator, is pre-established with respect to any consequent inference. Moreover, it is worth noting that, similarly to most permutation entities, the permutation likelihood $f_T^{(\varepsilon)}[(t;\delta)|\mathcal{X}_{/\mathbf{X}}]$ cannot be expressed in closed form; so it has to be evaluated numerically through quite computer-intensive methods. The development of specific algorithms for it is outside the scope of this book. However, these algorithms may benefit from those established for determining the confidence interval.

Remark 4. (*Bayesian permutation inference*). In accordance with Remark 3 above, we may use the concept of the permutation likelihood induced by $(T, \mathbf{X})$ on a functional δ in order to define a kind of Bayesian permutation inference, in terms of tests of hypotheses, estimators and confidence intervals on δ, etc. Of course, similarly to parametric contexts, a Bayesian permutation approach implies referring to a prior distribution $\pi(\delta)$ for δ.

From this point of view, in general a functional such as δ is a function of all parameters defining the specific population distribution P within the nonparametric family $\mathcal{P}$. Of course associated with P there is the likelihood $f_T(t;\delta,P)$ induced by T. Therefore, the Bayesian permutation approach is slightly different from the traditional Bayesian parametric approach. In fact, the former is nonparametric and strictly conditional on the observed data set $\mathbf{X}$ (for a discussion on conditioning in parametric Bayesian inferences see Celant and Pesarin, 2000, 2001). Furthermore, although it is essentially different from the nonparametric Bayesian approach introduced by Ferguson (1973), it does seem to be of interest as a further way of making inference. For instance, if we use $\pi(\delta)$ to denote the prior density distribution of δ, which should be defined over $(\Omega, \mathcal{A}_\Omega)$, where Ω is the sample space for δ and $\mathcal{A}_\Omega$ is a σ-algebra of subsets of Ω, then the *posterior permutation distribution*, given $(T, t, \mathbf{X})$, is

$$\pi_T(\delta|(t, \mathbf{X}, \varepsilon)) = \frac{\pi(\delta) \cdot f_T^{(\varepsilon)}[(t; \delta)|\mathcal{X}_{/\mathbf{X}}]}{\int_\Omega f_T^{(\varepsilon)}[(t; \delta)|\mathcal{X}_{/\mathbf{X}}] \cdot \pi(\delta) \cdot d\delta},$$

where the roles played by test statistic T, sample point $t \in \mathcal{T}(\mathbf{X})$ and prior distribution π are emphasized. Of course, the most important sample point t is the observed value $T^o = T(\mathbf{X})$. For instance, the *best* Bayesian permutation estimate of δ under quadratic loss, given $(T, \mathbf{X})$, assuming that $\mathbb{E}[\delta^2|(T^o, \mathbf{X}, \varepsilon; \pi)]$ is finite, is the posterior mean

$$\hat{\delta}_{\mathbf{X},\pi} = \int_\Omega \delta \cdot \pi_T(\delta|(T^o, \mathbf{X}, \varepsilon)) \cdot d\delta,$$

and the *best* $(1 - \alpha)$ Bayesian permutation confidence interval, based on the notion of highest posterior density provided that this is sufficiently regular, is $\underline{\delta}_{\mathbf{X},\pi} \le \delta \le \bar{\delta}_{\mathbf{X},\pi}$ in such a way that

$$\pi_T(\underline{\delta}_{\mathbf{X},\pi}|(T^o, \mathbf{X}, \varepsilon)) = \pi_T(\bar{\delta}_{\mathbf{X},\pi}|(T^o, \mathbf{X}, \varepsilon)),$$

and

$$\int_{\underline{\delta}_{\mathbf{X},\pi}}^{\bar{\delta}_{\mathbf{X},\pi}} \pi_T(\delta|(T^o, \mathbf{X}, \varepsilon)) \cdot d\delta = 1 - \alpha.$$

Further developments in Bayesian permutation procedures are left to the reader (see Problems 15–17, 3.9).

It is worth noting that since Bayesian permutation inference is strictly conditional on $\mathcal{X}_{/\mathbf{X}}$, that is, on the data set $\mathbf{X}$ (more precisely, conditioning is with respect to the unobservable exchangeable deviates $\mathbf{Z}$), due to sufficiency of $\mathcal{X}_{/\mathbf{X}}$ (see Section 2.1.3) is completely unaffected by the underlying population distribution P. This implies a noticeable difference with respect to the parametric Bayesian inference based on the posterior distribution $\pi_T(\delta|\mathbf{X}, P)$, which also is conditional with respect to the actual data set $\mathbf{X}$ but through the population likelihood associated with P, namely $f_P(t)$. For instance, two researchers with the same prior $\pi(\delta)$, the same pair $(T, \mathbf{X})$, but with different likelihoods, f_1 and f_2 (say), arrive at exactly the same inference within the Bayesian permutation approach, since the posterior permutation distribution $\pi_T(\delta|(t, \mathbf{X}, \varepsilon))$ does not depend on f; whereas within the traditional Bayesian approach they may arrive at different inferences, since their posterior distributions are $\pi_T(\delta|\mathbf{X}, f_1)$ and $\pi_T(\delta|\mathbf{X}, f_2)$ respectively.

3.4.1 *Problems and Exercises*

1) Having proved the statement in Remark 2, 3.2.1, show that in the paired data design null differences cannot be discarded when considering permutation confidence intervals for effect δ.

2) Write an algorithm for a confidence interval for δ in the paired data design.

3) Write an algorithm to evaluate the shortest confidence interval of any functional of the form $\delta_\varphi = \mathbb{E}[\varphi(X_1)] - \mathbb{E}[\varphi(X_2)]$.

4) With reference to the two-sample design, discuss the confidence interval for a functional $\delta_\varphi = \mathbb{E}[\varphi(X_1(\delta))] - \mathbb{E}[\varphi(X_2)]$, where φ is a monotonic measurable transformation of the data.

5) Analogously to Problem 4 above, discuss the confidence interval for fuctional $\delta_\varphi = \mathbb{E}[\varphi(X(\delta))]$ in the paired data design.

6) With reference to the two-sample design, with random effect $\Delta = \delta + D$, where random deviates D are such that δ is common to all units, $\mathbb{E}(D) = 0$ and $\Pr\{\Delta \geq 0\} = 1$, show that the confidence interval on δ is the same as above, provided that $D + \sigma Z_1$ stochastically dominates σZ_2.

7) Extend the results of Problem 6 above to the paired data design.

8) Extend the notion of confidence interval to the functional $\delta^2 = \sum_j n_j \delta_j^2$ in a one-way ANOVA design and provide an algorithm for it (in which, by assumption, $\sum_j n_j \delta_j = 0$).

3.5 Extending Inference from Conditional to Unconditional

It was shown in Section 3.1.1 that the non-randomized permutation test ϕ based on a given test statistic T on divergence of symmetric functions S of the data, possesses both conditional unbiasedness and similarity properties, the former satisfied by *all population distributions P and all data sets* $\mathbf{X} \in \mathcal{X}^n$, the latter satisfied for continuous, non-degenerate variables and *almost all data sets*. These two properties are jointly sufficient for extending conditional inferential conclusions to unconditional ones, that is, for the extension of inferences related to the specific set of actually observed units (e.g. *the drug is effective on the observed units*) to conclusions related to the population from which units have been obtained (e.g. *the drug is unconditionally effective*). Such an extension is done with weak control of inferential errors. To this end, and with obvious notation, we observe the following:

(i) For each attainable $\alpha \in \Lambda$ and all sample sizes n, the similarity property implies that the power of the test in H_0 satisfies the relation

$$W(0, \alpha, T, P, n) = \int_{\mathcal{X}^n} \Pr\{\lambda_T(\mathbf{X}(0)) \leq \alpha | \mathcal{X}^n_{/\mathbf{X}}\} \cdot dP^n(\mathbf{X}) = \alpha,$$

because $\Pr\{\lambda(\mathbf{X}(0)) \leq \alpha | \mathcal{X}^n_{/\mathbf{X}}\} = \alpha$ for almost all samples $\mathbf{X} \in \mathcal{X}^n$ and all continuous, non-degenerate distributions P, independently of how data are selected.

(ii) The conditional unbiasedness for each $\alpha \in \Lambda_{\mathbf{X}}$ and all sample sizes n implies that the unconditional power function for each $\delta > 0$ satisfies

$$W(\delta, \alpha, T, P, n) = \int_{\mathcal{X}^n} \Pr\{\lambda_T(\mathbf{X}(\delta)) \leq \alpha | \mathcal{X}^n_{/\mathbf{X}}\} \cdot dP^n(\mathbf{X}) \geq \alpha,$$

for all distributions P, independently of how data are selected and provided that the generalized density is positive, i.e. $dP^n(\mathbf{X})/d\xi^n > 0$ (see Chapter 2), because in these conditions the integrand $\Pr\{\lambda(\mathbf{X}(\delta)) \leq \alpha | \mathcal{X}^n_{/\mathbf{X}(\delta)}\}$ is greater than or equal to α.

As a consequence, if for instance the inferential conclusion related to the actual data set $\mathbf{X}$ is in favour of H_1, so that we say that 'the data $\mathbf{X}$ are evidence of treatment effectiveness on observed units', due to (i) and (ii) we are allowed to say that this conclusion is also valid unconditionally for all populations P such that $dP^n(\mathbf{X})/d\xi^n > 0$. Thus, the extended inferential conclusion becomes that 'the treatment is likely to be effective'.

The condition that $dP^n(\mathbf{X})/d\xi^n > 0$ implies that inferential extensions must be carefully interpreted. In order to give a simple illustration of this aspect, let us consider an example of an experiment in which only males of a given population of animals are observed. Hence, based on the result actually obtained, the inferential extension from the observed units to the selected subpopulation $(X, \mathcal{X}_{/M}, P_M)$ is immediate, where $\mathcal{X}_{/M}$ is the reference set of the male sub-population and P_M its associated distribution. Indeed, on the one hand, rejecting the null hypothesis with the actual data set means that *the actual data are evidence for a non-null effect of treatment*, which in turn means that *treatment appears to be effective* irrespective of how data are collected, provided that they are exchangeable in the null hypothesis, because exchangeability, conditional unbiasedness and similarity guarantee that $W(\delta, \alpha, T, P, n) \geq \alpha$. On the other hand, if females of that population, due to the selection procedure, have a probability of zero of being observed, so that $dP_F^n(\mathbf{X})/d\xi^n = 0$, then regarding them in general we can say nothing reliable, because it may be impossible to guarantee that the test statistic which has been used for male data satisfies conditional unbiasedness and/or similarity properties for female data as well. As an example of this, suppose that on males (the actual data) the treatment effect is additive on deviates from the population mean, whereas on females it is multiplicative. Thus, $T = \sum X_{1i}$ possesses the similarity and conditional unbiasedness properties for male data, but there is no guarantee that it has such properties for female data, in which case T may be essentially inadequate. In general, the problem of establishing if that statistic possesses the required properties, when it is impossible to observe any data, remains essentially undefined. In these cases, therefore, the statistical properties and the inferential extensions may remain hypothetical and with no clear guarantees. In general, the extension (i.e. essentially the extrapolation) of any inferential conclusion to populations which cannot be observed can only be formally done with reference to assumptions that lie outside the control of experimenters, and so should be carefully considered. For instance, extensions to humans of inferential conclusions obtained from experiments on laboratory animals require specific hypothetical assumptions beyond those connected with the distributional properties of the actual data.

It is worth noting that properties (i) and (ii) are jointly sufficient (but not necessary) for inferential extension, because only the unbiasedness or only the similarity may provide insufficient guarantees. The following two examples illustrate this point: (a) a purely randomized test, for which the rejection is a function of only an auxiliary continuous random quantity independent of $\mathbf{X}$, is uniformly of size α for all possible data sets and thus it is strongly similar but not consistent, its type II error rate being fixed at α for all sample sizes and all non-centrality parameters; (b) a test that rejects H_0 with probability one for all $\mathbf{X}$ is conditionally and unconditionally unbiased and with maximum power in H_1, but in H_0 its type I error rate, being equal to 1, is strictly greater than α, and so it is strongly anticonservative.

We observe that for parametric tests, when there are nuisance entities to eliminate, the extension of inferential conclusions from conditional to unconditional can generally only be done if the data set is obtained through well-designed sampling procedures and applied to the entire target population. This is because if data are obtained by selection-bias procedures, then the associated inferences can be conditionally biased, or the estimates of nuisance parameters cannot be boundedly complete, or both (see Section 1.4). Thus, similarity or conditional unbiasedness, or both, are not guaranteed and so there is no guarantee of controlling required inferential properties. When selection-bias data $\mathbf{X}$ are observed and the selection mechanism is not well designed, due to the impossibility of writing a credible likelihood function, there no point in staying outside the conditioning with respect the associated orbit $\mathcal{X}_{/\mathbf{X}}$ and the related discrete distribution induced by the chosen statistic T. On the one hand, this implies adopting the permutation principle of testing; on the other, no parametric approach can be invoked for obtaining credible inferences.

In order to illustrate the latter point, among the infinite possibilities of carrying out selection-bias procedures, let us consider two situations which may be common in both experimental and observational studies.

Suppose that, in one-sample situations and normally distributed errors, the sample space $\mathcal{X}$ is partitioned into the subsets $\mathcal{X}_{/A}$ and $\mathcal{X}_{/B}$ of points respectively defined by *small* and *large* sample standard deviations, so that $\mathcal{X} = \mathcal{X}_{/A} \bigcup \mathcal{X}_{/B}$. Suppose also that the specific selection-bias procedure only considers sample points from $\mathcal{X}_{/A}$ and that the pair of estimates $(\bar{X}_A, \hat{\sigma}_A)$ is obtained. Figure 3.2 describes such a situation. Since $\hat{\sigma}_A$ is stochastically smaller than the corresponding estimate $\hat{\sigma}$ we would have by random sampling from the entire target population (note that in this context $\bar{X}_A$ has the same marginal distribution as $\bar{X}$), the Student's t statistic $t_A = (\bar{X}_A - \mu_0)\sqrt{n}/\hat{\sigma}_A$, being stochastically larger than the corresponding $t = (\bar{X} - \mu_0)\sqrt{n}/\hat{\sigma}$, clearly becomes anticonservative. Indeed, its type I error rate satisfies the relation $\alpha \leq \alpha_A < 1$. Thus, the parametric t test is strongly

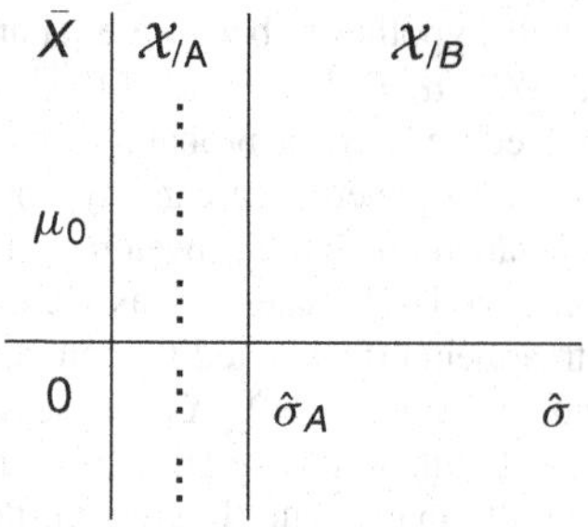

Figure 3.2 A case of selection-bias sampling from a normal population

anticonservative and so inferences based on selection-bias procedures from normal populations are generally conditionally biased, in the sense that we are unable to guarantee their unbiasedness. In these situations, therefore, Student's t test cannot be used since both invariance and boundedly complete properties are not satisfied. However, we may note that if Student's t is carried out with selection-bias data, results can correctly be extended to the actually selected (and usually unknown) population, but not to the target.

The second situation considers a very special case of the so-called weighted distributions (see Patil and Rao, 1977) in which it is supposed that

$$f_P(x) \cdot \varphi(x) = K_\varphi \cdot \exp\{-(x - \mu_\varphi)^2/2\sigma_\varphi^2\},$$

where $f_P(x)$ is the density corresponding to the target population P, $\varphi(x)$ is a non-negative weight function, K_φ is a normalizing constant, and μ_φ and σ_φ are the mean and standard deviation of the selected population P_φ (note normality). Thus, Student's t is a proper test for P_φ, but related inferential results cannot be extended to the target population P, unless the weight function φ is well specified and the corresponding parameters and coefficients well estimated.

Remark 1. Note that in the context of selection-bias sampling, it may be that sample means and variances are stochastically independent even if the target population is not normal, as in the second situation. Thus, in general we know that statistics based on selection-bias procedures may violate important parts of most of the theorems based on the notion of random sampling. Therefore, when sampling experiments are not well designed, to some extent parametric unconditional inferences become improper and so cannot be used correctly for inferential purposes.

Remark 2. It is, however, to be emphasized that with selection-bias data, we can extend testing conclusions but we cannot extend estimates of the size effect δ because no population functional can be correctly estimated on the target population unless the sample design has a known probabilistic structure. That is to say, we can extend test conclusions but not sample estimates.

3.6 Optimal Properties

It is well known that the unconditional power function (see Section 3.4) is particularly important in selecting an optimal test ϕ from a class of test statistics when testing H_0 against H_1.

Permutation tests differ from parametric tests mainly in that permutation critical values, $T_\alpha(\mathbf{X})$, vary as $\mathbf{X}$ varies in $\mathcal{X}$ (see Section 2.2.1). This fact makes it difficult to express in closed form the exact evaluation of the conditional and unconditional permutation power functions, as shown in Section 3.2. Moreover, the unconditional power function also depends on the underlying unknown distribution P generating points $\mathbf{X} \in \mathcal{X}$.

However, in order to appreciate the optimal properties of permutation tests, some useful results are stated in this section without proof. Proofs may be found in the references quoted. Let us begin with the following lemma.

Lemma 1. (Lehmann and Stein, 1949; Lehmann, 1986). *If ψ is any test of a hypothesis of invariance for a class of density functions with respect to a dominating measure ξ, and if the size of ψ is less than or equal to α, then there exists a permutation similar test ϕ such that $\int \phi dP \geq \int \psi dP$, for all probability distributions P.*

In practice, Lemma 1 states that permutation similar tests are at least as good as any other test of invariance and, from this point of view, they lie within a class of admissible tests. In this sense, optimal tests of invariance may be found within permutation similar tests. A test ϕ is said to be *admissible* if there is no other test which is at least as powerful as ϕ against some alternatives in H_1 and more powerful against all other alternatives (see Lehmann, 1986).

Section 3.7 presents some asymptotic approximations for the permutation power. Here we present a somewhat different way of defining an optimal permutation test in terms of parametric alternatives.

Theorem 3. (Lehmann and Stein, 1949; Lehmann, 1986). *Let H_0 be a hypothesis of invariance under a finite group of transformations* **G**, *and let f_P be the density function corresponding to a distribution P in H_1. For any $\mathbf{X} \in \mathcal{X}$, we denote by $\mathbf{X}^*_{(1)}, \ldots, \mathbf{X}^*_{(M)}$ the M points of the associated permutation sample space $\mathcal{X}_{/\mathbf{X}}$, arranged so that $f_P(\mathbf{X}^*_{(1)}) \geq \ldots \geq f_P(\mathbf{X}^*_{(M)})$. For testing H_0 against H_1 the most powerful randomization test of size α is given by*

$$\phi_R(\mathbf{X}) = \begin{cases} 1 & \text{if } f_P(\mathbf{X}) > f_\alpha, \\ \gamma & \text{if } f_P(\mathbf{X}) = f_\alpha, \\ 0 & \text{if } f_P(\mathbf{X}) < f_\alpha, \end{cases}$$

where, for any $\alpha \in (0, 1)$, $f_\alpha = f_P(\mathbf{X}^*_{(M_\alpha)})$ is the critical value, $M_\alpha = \lfloor \alpha \cdot M \rfloor$ is the number of points in the critical region, and $\gamma = [\alpha - \Pr\{f(\mathbf{X}) > f_\alpha|\mathbf{X}\}]/\Pr\{f(\mathbf{X}) = f_\alpha|\mathbf{X}\}$.

Theorem 3 shows that, in order to obtain an optimal permutation test, we must know the population distribution P in H_1. In the general nonparametric situation, this distribution is unknown, so that this theorem is difficult to apply. However, as we sometimes have approximate knowledge of P, we have a weak guideline enabling us to establish a reasonable choice for good tests.

Remark 1. We know that permutation tests ϕ work when H_0 implies invariance of the conditional distribution $P_{|\mathbf{X}}$ with respect to a finite group of transformations **G**. This basic structure may be relaxed in a few special cases. Permutation tests ϕ may sometimes become valid asymptotically. For more details on this subject, see the interesting results provided by Romano (1990), and some of the results in Chapters 4–12.

3.6.1 Problems and Exercises

1) Prove Proposition 2, 3.1.1, i.e. that if the permutation distribution of a test T may only be expressed in terms of exchangeable errors Z_i, then it is a *permutationally exact test*.

2) Extend Theorem 1, 3.1.1 to $H_1 : \{\Delta \overset{d}{<} 0\}$, in which random effects are non-positive.

3) Draw a block diagram for the test of symmetry of Example 1, 2.6.

4) Give a formal proof of the ordering property established in Theorem 2, 3.1.1.

5) Using the notation in Section 3.1, formally prove unbiasedness of the randomized test ϕ; give expressions for the conditional and unconditional power functions and show that they are consequences of the monotonic ordering property stated in Theorem 2, 3.1.1.

6) Prove that both conditional and unconditional power functions of the permutation test T for the problem with paired observations are monotonically non-decreasing with respect to δ.

7) With reference to Theorem 2, 3.1.1, prove that, in a two-sample design, the test for composite hypotheses, i.e. for $H_0 : \{\delta \leq \delta_0\}$ against $H_1 : \{\delta > \delta_0\}$, $T^* = \bar{X}_1^*$ is unbiased and conservative.

8) Using the notation in Section 3.1, prove that the permutation CDF $F(t;\, \delta|\mathcal{X}_{/\mathbf{X}}) = \Pr\{T^*(\delta) \leq t|\mathcal{X}_{/\mathbf{X}}\}$ is non-decreasing with respect to δ, for all $t \in \mathcal{R}^1$.

9) Using the notation of Section 3.1.1 in a paired data design (see Section 1.8), prove that the test $T^* = \sum_i X_i S_i^*$ is unbiased for testing $H_0 : \{\delta = 0\}$ against $H_1 : \{\delta > 0\}$, i.e. prove that

$$\Pr\{T^*(\delta) \geq T^o(\delta)|\mathcal{X}_{/\mathbf{X}}\} = \Pr\{T^*(0) + \delta(n - \textstyle\sum_i S_i^*) \geq T^o(0)|\mathcal{X}_{/\mathbf{X}}\}.$$

10) With reference to the Problem 9 above, prove unbiasedness of $T^* = \sum_i X_i S_i^*$ in paired data designs with (non-negative) random effects Δ.

11) Assume that the response model for paired observations is $\{Y_{1i} = \mu + \eta_i + \sigma_i \cdot Z_{1i}, Y_{2i} = \mu + \eta_i - \delta + \sigma_i(\delta) \cdot Z_{2i}, i = 1, \ldots, n\}$, where (Z_{1i}, Z_{2i}) are identically distributed within units and independent with respect to units, and scale coefficients $\sigma_i(\delta)$ are such that $\sigma_i(0) = \sigma_i > 0$. Show that the test $T^* = \sum_i X_i \cdot S_i^*$ is unbiased for testing $H_0 : \{\delta = 0\}$ against $H_1 : \{\delta > 0\}$ (note that location and scale functionals depend on units and in the null hypothesis scale coefficients are equal, hence exchangeability is satisfied).

12) Prove that, in the alternative, the probability distribution of points on the permutation sample space $\mathcal{X}_{/\mathbf{X}}$ depends on the treatment effect δ.

13) With reference to the paired data design, show that the permutation sample median $\tilde{X}^* = \mathbb{M}d(X_i S_i^*)$ is unbiased for testing $H_0 : \{\delta = 0\}$ against $H_1 : \{\delta > 0\}$.

14) With reference to the paired data design, show that the permutation sample trimmed mean of order b,

$$\bar{X}_b^* = \sum\nolimits_{b<i<n-b} X_{(i)}^*/(n - 2b),\;\; 0 \leq b \leq (n-1)/2,$$

where $X_{(i)}$ and $X_{(i)}^* = X_{(i)} S_i^*$ are the increasing order statistics associated with $\mathbf{X}$ and $\mathbf{X}^*$ respectively, is unbiased for testing $H_0 : \{\delta = 0\}$ against $H_1 : \{\delta > 0\}$.

15) When dealing with a paired data design the underlying response model is $Y_{1i} = \mu + \eta_i + \sigma_1 \cdot Z_{1i}, Y_{2i} = \mu + \eta_i - \delta + \sigma_2 \cdot Z_{2i}, i = 1, \ldots, n$, where the two unknown scale coefficients σ_1 and σ_2 are assumed not equal, due to the lack of exchangeability within units, prove that no general exact solution is possible.

16) Under the same conditions as in Problem 15, and in the very special case where error terms Z_{1i} and Z_{2i} are independent and symmetrically distributed around zero, prove that the permutation exact solutions are those based on testing for symmetry (see Example 1, 2.6). Also prove that one particular solution is provided by the sign or McNemar test (note that if two variables are symmetric around the same finite quantity, then their difference is symmetric around zero).

3.7 Some Asymptotic Properties

3.7.1 *Introduction*

This section presents some asymptotic properties of permutation tests. Proofs of theorems may be found in the literature. We refer here to the randomized version of the permutation test ϕ_R (see Section 2.2.2).

Permutation tests ϕ_R differ from parametric tests mainly because critical values T_α and p-values λ, if expressed in terms of the random data set $\mathbf{X}$, are essentially random quantities. In fact, for any finite sample size n, permutation functionals $F(t|\mathcal{X}_{/\mathbf{X}})$, T_α and λ, associated with $(T, \mathbf{X})$, vary as $\mathbf{X}$ randomly varies in the sample space $\mathcal{X}$. Specifically, the permutation conditional power function is also a random entity, which takes on different values when we start from different sample points $\mathbf{X}$ (see Section 2.2.1). This makes comparisons between permutation tests and other tests difficult.

Study of the asymptotic behaviour of permutation tests ϕ_R may be of some help in overcoming this difficulty. We shall see that, in quite general conditions, the permutation conditional distribution and parametric unconditional sample distribution of the test statistic T converge to the same limit, as sample size n goes to infinity. This gives us a theoretical basis on which to compare permutation tests with their parametric or nonparametric rank-based counterparts.

3.7.2 Two Basic Theorems

Let us examine the asymptotic behaviour of the permutation critical value $T_\alpha(\mathbf{X}^{(n)}) = T_{\alpha n}$, where its dependence on the data set $\mathbf{X}^{(n)}$ based on sample size $n \geq 2$ (see Sections 2.1 and 2.2) is emphasized. Assume that elements $\mathbf{X}^{(n)} \in \mathcal{X}^n$, $P^{(n)}$, $\mathcal{X}_{/\mathbf{X}^{(n)}}$, M^n, $\phi_R(\mathbf{X}^{(n)})$, $T(\mathbf{X}^{(n)})$, $T_{\alpha n}$, etc., are defined on an infinite sequence of positive integers n. Also assume that $M^n \to \infty$ and $\sum_{\mathcal{X}_{/\mathbf{X}^{(n)}}}(T^* < T_\alpha)/M^n \to 1-\alpha$, as $n \to \infty$. Following Hoeffding (1952), let us suppose that for a given sequence $P^{(n)}$ of distributions of $\mathbf{X}^{(n)}$ the following assumptions are satisfied:

(A.1) Test statistics T are expressed in such a way that there is a constant ξ_α, so that $T_{\alpha n} \xrightarrow{p} \xi_\alpha$, as $n \to \infty$.

(A.2) There is a function $K(y)$, continuous at $y = \xi_\alpha$, so that for every y, for which $K(y)$ is continuous, $\Pr\{T(\mathbf{X}^{(n)}) \leq y\} \to K(y)$.

From the definition of a permutation test ϕ_R (see Section 5.1.1), we have

$$\Pr\{T(\mathbf{X}^{(n)}) > T_{\alpha n}|\mathbf{X}^{(n)}\} \leq \mathbb{E}_{\mathcal{X}_{/\mathbf{X}^{(n)}}}\{\phi_R(\mathbf{X}^{(n)})\}$$
$$\leq \Pr\{T(\mathbf{X}^{(n)}) \geq T_{\alpha n}|\mathbf{X}^{(n)}\}.$$

Therefore, it follows that conditions (A.1) and (A.2) imply that

$$\mathbb{E}_{\mathcal{X}^{(n)}_{/\mathbf{X}}}\{\phi_R(\mathbf{X}^{(n)})\} \to 1 - K(\lambda).$$

To better identify situations when assumption (A.2) holds, we may introduce a stronger condition, valid for any sequence $\{P^{(n)}, n \geq 2\}$, as follows:

(A.3) $F(y|\mathcal{X}^{(n)}_{/\mathbf{X}}) \xrightarrow{p} F(y)$ for every y at which $F(y)$ is continuous, where $F(y)$ is a distribution function, the equation $F(y) = 1 - \alpha$ has only one solution $y_\alpha = \xi_\alpha$, and $F(y)$ is continuous at $y = \xi_\alpha$ (see also the discussion of question (a) in Section 2.5).

We report here, without proof, some results obtained by Hoeffding (1952), showing that (A.3) generally implies (A.2).

Theorem 4. (Hoeffding, 1952). *Under assumption* (A.3)*, there is a constant* ξ_α*, such that* $T_{\alpha n} \xrightarrow{p} \xi_\alpha$*, as* $n \to \infty$.

Let $\mathbf{X}^{(n)*}$ be any permutations of $\mathbf{X}^{(n)}$, randomly chosen in the permutation sample space $\mathcal{X}^{(n)}_{/\mathbf{X}}$. Moreover, let $\mathbf{X}^{(n)\prime}$ be one more random element from $\mathcal{X}^{(n)}_{/\mathbf{X}}$, having the same distribution as $\mathbf{X}^{(n)*}$, and let $\mathbf{X}^{(n)*}, \mathbf{X}^{(n)\prime}$ and $\mathbf{X}^{(n)}$ be mutually independent.

Theorem 5. (Hoeffding, 1952). *Assume that, for some sequences of distributions* $\{P^n, n \geq 2\}$, $T(\mathbf{X}^{(n)*})$ *and* $T(\mathbf{X}^{(n)\prime})$ *have the limiting joint distribution function* $F(y) \cdot F(y')$. *Thus, for every* y *at which* $F(y)$ *is continuous* $F(y|\mathcal{X}^{(n)}_{/\mathbf{X}}) \xrightarrow{p} F(y)$, *and if equation* $F(y) = 1 - \alpha$ *has only one solution* $y_\alpha = \xi_\alpha$, $T_{\alpha n} \xrightarrow{p} \xi_\alpha$, *as* $n \to \infty$.

Note that if $F(y|\mathcal{X}^{(n)}_{/\mathbf{X}}) \xrightarrow{p} F(y)$ for a sequence of distributions invariant under the action of a group of transformations $\mathbf{G}^n$, then $T(\mathbf{X}^{(n)})$ has asymptotic distribution $F(y)$.

Let ϕ' be a test defined as follows:

$$\phi'(\mathbf{X}^{(n)}) = \begin{cases} 1 & \text{if } T(\mathbf{X}^{(n)}) > t_{\alpha n}, \\ \gamma_n' & \text{if } T(\mathbf{X}^{(n)}) = t_{\alpha n}, \\ 0 & \text{if } T(\mathbf{X}^{(n)}) < t_{\alpha n}. \end{cases}$$

Assume that $t_{\alpha n}$ and γ_n' are such that ϕ' has size α for testing H_0. For instance, ϕ' may be a parametric test for H_0, satisfying some desirable property. If (A.3) is satisfied, then $t_{\alpha n} \to \xi_\alpha$. Moreover, if (A.2) holds, then

$$\mathbb{E}_{P^{(n)}}\{\phi'(\mathbf{X}^{(n)})\} \to 1 - K(\lambda).$$

If $\mathcal{P}_\xi$ denotes the class of all sequences $\{P^{(n)}; n \geq 2\}$ for which assumptions (A.3), with ξ_α fixed, and (A.2), with some $K'(y)$, are satisfied, and if $\mathcal{P}_\xi$ contains all sequences induced by H_0, then the powers of permutation tests ϕ_R and ϕ' tend to the same limit for every $\{P^{(n)}; n \geq 2\}$ in $\mathcal{P}_\xi$. The permutation test ϕ_R is thus asymptotically as powerful with respect to $\mathcal{P}_\xi$ as ϕ'. If ϕ' is the most powerful, or more generally 'optimum' in terms of power, the permutation test ϕ_R asymptotically captures the same properties.

Example 1. *Testing symmetry revisited* (see Example 1, 2.6).

It can be shown that assumption (A.3) is satisfied when approximating the permutation distribution of a test statistic by the standard normal distribution. Let us consider a situation in which $\mathbf{X}^{(n)}$ is a sample of n i.i.d. observations from a normal distribution. Thus, to test $H_0 : \{\delta = 0\}$ against $H_1 : \{\delta > 0\}$, the permutation test ϕ_R based on the test statistic $T = \sum_i X_i$ has the same asymptotic power as the standard one-sided Student's t test of size α.

Example 2. *One-way ANOVA revisited* (see Example 6, 2.7).

The permutation test ϕ_R for the ANOVA may be based on the test statistic $T = T(\mathbf{X}^{(n)})$ defined (in familiar notation) as

$$T = \frac{\sum_{j=1}^{C} n_j(\bar{X}_j - \bar{X}_.)^2}{\sum_{j=1}^{C}\sum_{i=1}^{n_j}(X_{ji} - \bar{X}_j)^2}.$$

Assume that $\mathbf{X}^{(n)}$ is generated from C independent normal distributions. Thus, the permutation test ϕ_R of size α, based on the test statistic T, has the same asymptotic power as the Snedecor F test.

Remark 1. In these examples, we tackle the problem of studying the asymptotic behaviour of permutation tests ϕ_R by checking assumption (A.3). This is perhaps the easiest way of achieving asymptotic results. Other ways include using versions of the so-called PCLT for approximating permutation CDFs $F(t|\mathcal{X}^{(n)}_{/\mathbf{X}})$ by the standard normal CDF $\Phi(t)$, $t \in \mathcal{R}^1$. This is done in the next section.

3.8 Permutation Central Limit Theorems

3.8.1 Basic Notions

Let us assume that a sample $\mathbf{X}^{(n)} = \{X_1, \ldots, X_n\}$ of n i.i.d. observations from X is given, that the point $\mathbf{X}^{(n)}$ takes values on a sample space $\mathcal{X}^n$, and that $P^{(n)}$ denotes probability distributions on $\mathcal{X}^n$ induced by the sampling experiment producing $\mathbf{X}^{(n)}$. Also assume that the null hypothesis H_0 to be tested implies that permutations of elements of $\mathbf{X}^{(n)}$ are equally likely (see Proposition 1, 2.1.3). We define a *linear permutation test statistic* $T = T(\mathbf{X}^{(n)})$ as

$$T = \sum_{i=1}^{n} A_i \cdot X_i,$$

where $\mathbf{A}_n = \{A_1, \ldots, A_n\}$ is a suitable vector of real numbers, not all equal to zero. We define $\bar{X} = n^{-1}\sum_{i=1}^{n} X_i$, $\bar{A} = n^{-1}\sum_{i=1}^{n} A_i$, $V_2(\mathbf{X}^{(n)}) = n^{-1}\sum_{i=1}^{n}(X_i - \bar{X})^2$ and $V_2(\mathbf{A}_n) = n^{-1}\sum_{i=1}^{n}(A_i - \bar{A})^2$. From the permutation distribution of T, it turns out that

$$\mathbb{E}_{\mathcal{X}^{(n)}_{/\mathbf{X}}}\{T(\mathbf{X}^{(n)*})|\mathcal{X}^{(n)}_{/\mathbf{X}}\} = n\bar{X}\bar{A},$$

because only $\{X_1, \ldots, X_n\}$ are permuted, whereas $\{A_1, \ldots, A_n\}$ remain fixed. Moreover,

$$\mathbb{E}_{\mathcal{X}^{(n)}_{/\mathbf{X}}}\{(T(\mathbf{X}^{(n)*}) - n\bar{X}\bar{A})^2|\mathcal{X}^{(n)}_{/\mathbf{X}}\} = (n-1)^{-1}\{n^2 V_2(\mathbf{X}^{(n)}) V_2(\mathbf{A}_n)\}^2.$$

Let us consider the standardized form,

$$Z = (n-1)^{1/2}(T - n\bar{X}\bar{A})\{n^2 V_2(\mathbf{X}^{(n)}) V_2(\mathbf{A}_n)\}^{-1}.$$

We know that the permutation CDF of the statistic Z, $F_Z(t|\mathcal{X}^{(n)}_{/\mathbf{X}})$, $t \in \mathcal{R}^1$, may be exactly determined in principle by enumerating all permutations of elements of $\mathbf{X}^{(n)}$. Assume that $\{\mathbf{X}^{(n)}; n \geq 2\}$ and $\{\mathbf{A}_n; n \geq 2\}$ are independently defined, each on an infinite sequence of positive integers n. In this situation, there are versions of the PCLT related to the question of how to find conditions on sequences $\{\mathbf{X}^{(n)}; n \geq 2\}$ and $\{\mathbf{A}_n; n \geq 2\}$, in which $F_Z(t|\mathcal{X}^{(n)}_{/\mathbf{X}})$ may be approximated by the standard normal CDF $\Phi(t)$, $t \in \mathcal{R}^1$.

3.8.2 Permutation Central Limit Theorems

In order to present some of the most important conditions on sequences $\{\mathbf{X}^{(n)}; n \geq 2\}$ and $\{\mathbf{A}_n; n \geq 2\}$ the limit distribution of which is found by a PCLT, it is convenient to consider a generic sequence $\{\mathbf{D}_n; n \geq 2\}$, so that we can refer conditions (C.1)–(C.4) below to both sequences $\{\mathbf{X}^{(n)}; n \geq 2\}$ and $\{\mathbf{A}_n; n \geq 2\}$. Following Puri and Sen (1971, Chapter 3), and Sen (1983, 1985), let us define $\bar{D} = n^{-1}\sum_{i=1}^{n} D_i$, $V_r(\mathbf{D}_n) = n^{-1}\sum_{i=1}^{n}(D_i - \bar{D})^r$, $r = 2, 3, 4, \ldots$, $W_r(\mathbf{D}_n) = n^{-1}\sum_{i=1}^{n}|D_i - \bar{D}|^r$, $r \geq 0$, and $R(\mathbf{D}_n) = \max_{1\leq i\leq n}(D_i) - \min_{1\leq i\leq n}(D_i)$.

We discuss the PCLT by taking into consideration conditions such as the following:

(C.1) $\prod_{i=1}^{I} P_{\kappa}(X_i) \cdot \prod_{i=I+1}^{n} P_{\kappa\kappa}(X_i) V_r(\mathbf{D}_n)\{V_2(\mathbf{D}_n)\}^{-r/2} = O(1)$ for all $r = 3, 4, \ldots$ (Wald and Wolfowitz, 1944).

(C.2) $V_r(\mathbf{D}_n)\{V_2(\mathbf{D}_n)\}^{-r/2} = o(n^{r/2-1})$, for all $r = 3, 4, \ldots$ (Noether, 1949).

It is easy to verify that condition (C.1) implies (C.2). Moreover, (C.2) is equivalent to either of the following two conditions (see Hoeffding, 1951b):

(C.3) For some $r > 2$, $W_r(\mathbf{D}_n)\{V_2(\mathbf{D}_n)\}^{-r/2} = o(n^{r/2-1})$.
(C.4) $R(\mathbf{D}_n)\{V_2(\mathbf{D}_n)\}^{-1/2} = o(n^{1/2})$.

Theorem 6. (PCLT). *If the sequence* $\{\mathbf{A}_n; n \geq 2\}$ *satisfies condition* (C.1), *and* $\{\mathbf{X}^{(n)}; n \geq 2\}$ *satisfies condition* (C.2), *and if*

$$\{nV_2(\mathbf{X}^{(n)})\}^{-1}\left\{\max_{1\leq i\leq n}|X_i - \bar{X}|\right\} \to 0,$$

as $n \to \infty$, *then, for every* $t \in \mathcal{R}^1$, $F_Z(t|\mathcal{X}^{(n)}_{/\mathbf{X}}) \xrightarrow{p} \Phi(t)$.

Remark 1. More general versions of Theorem 6 may be found in the literature. See, among others, Hoeffding (1951b), Motoo (1957), Hájek (1961), Jogdeo (1968) and Shapiro and Hubert (1979). Their results are especially useful when treating the so-called bilinear permutation statistics. For a review of multivariate generalizations of Theorem 6, see Sen (1983, 1985).

Remark 2. Some versions of the PCLT may be directly applied to the asymptotic theory of rank tests. The asymptotic counterparts of rank tests are generally easier to determine than those of permutation tests, because distributions induced by statistics on ranks are usually easily determined, essentially because ranks associated with data and critical values of related tests are fixed numbers. For more details, see Fraser (1957), Sen (1983, 1985), Hettmansperger (1984), Lehmann (1986), Hájek and Šidák (1967) and Hájek et al. (1999).

Remark 3. The bootstrap was introduced by Efron as a technique for assessing the statistical accuracy of estimates, mainly in nonparametric settings. It may also be used to test statistical hypotheses; see Efron and Tibshirani (1993) and Davison and Hinkley (1988) for a review and for related results. Specifically, at least in their simpler form, bootstrap inferences are based on a distribution obtained by giving the same probability mass to elements of a given sample $\mathbf{X}^{(n)}$, say $\mathbf{X}^{(n)\prime} = \{X_1', \ldots, X_n'\}$, which are randomly drawn *with replacement* from $\mathbf{X}^{(n)}$ (WRE). The distribution induced by a statistic T is called the *bootstrap distribution associated with* $\mathbf{X}^{(n)}$. In this sense, permutation tests have a parallel meaning, because random resampling can generally be interpreted to occur *without replacement* (WORE) from $\mathbf{X}^{(n)}$ (see consequences of Proposition 2, 2.1.3).

Let $J_n(t)$, $t \in \mathcal{R}^1$, be the bootstrap distribution induced by the test statistic T, given $\mathbf{X}^{(n)}$, which we wish to consider for testing the null hypothesis H_0. Assume that in H_0 permutations are equally likely. That is, we imagine a situation in which both a permutation test and a bootstrap test may be applied. Recall that for any test statistic T, $F(t|\mathcal{X}^{(n)}_{/\mathbf{X}})$ and $F(t)$, $t \in \mathcal{R}^1$, are respectively the permutation distribution and its limit, which is assumed to exist. Let us assume that the test statistic T is written as $T = n^{1/2}D(P^{(n)}, \tau P^{(n)})$, where $P^{(n)} \in \mathcal{P}^{(n)}$ is the underlying distribution which has generated $\mathbf{X}^{(n)}$, D is a metric measuring distances between members of $\mathcal{P}^{(n)}$, and τ is some mapping characterizing H_0 within $\mathcal{P}^{(n)}$.

In Romano (1988, 1989, 1990), it is shown that, as $n \to \infty$, the following two equations hold:

$$\sup_{-\infty<t<+\infty} |J_n(t) - F(t)| \xrightarrow{p} 0$$

and

$$\sup_{-\infty<t<+\infty} |J_n(t) - F(t|\mathcal{X}_{/\mathbf{X}}^{(n)})| \xrightarrow{p} 0.$$

As a consequence, the permutation and bootstrap tests, both based on $T = n^{1/2}D(P^{(n)}, \tau P^{(n)})$, as $n \to \infty$, tend to have the same inferential behaviour. They are asymptotically equivalent also in terms of power. Most of the test statistics described in the examples in Chapter 2 satisfy this structure. Similar results are given in Pallini (1992a), when the test statistic T is a suitable function of covariance matrices in a general multivariate setting.

Remark 4. Bootstrap tests (see Hinkley, 1989) are more flexible than permutation tests, because they may also be used when H_0 is not a hypothesis of invariance, and in particular in some cases when the exchangeability condition is not satisfied.

However, it should be emphasized that they are data-dependent without being strictly conditional procedures. In order to reach this conclusion, we may observe that if $\mathbf{X}$ and $\mathbf{X}'$ are two sample points of $\mathcal{X}$ with at least one (but not all) common element(s), then two associated bootstrap sample spaces are not separated, in the sense that their intersection is not empty. Thus, as there are sample points in $\mathcal{X}$ which are members of more than one bootstrap sample space, the set of all bootstrap sample spaces provides only a *covering* of $\mathcal{X}$ and not a *partitioning*. Conversely, as each sample point belongs to only one permutation sample space, the set of all $\mathcal{X}_{/\mathbf{X}}$ defines a partitioning of $\mathcal{X}$ (see Section 2.1.2). This means that, for finite sample sizes, inferential interpretations of bootstrap tests are not completely clear, because they are neither conditional nor unconditional procedures, although they seem closer to unconditional ones.

Hence, bootstrap tests may be effective for exploratory purposes, but are only vaguely useful for inferential objectives. Moreover, when both are applicable, permutation tests are preferable also because they are of exact size α and because, being conditional on a set of sufficient statistics, they enjoy desirable properties which make conditional and unconditional inference interpretations effective and essentially clear.

3.9 Problems and Exercises

1) With reference to a two-sample problem, discuss the consistency of T when comparing two means.

2) With reference to a two-sample problem, discuss the consistency of T when comparing two medians.

3) Extend the validity of Theorem 1, 3.1.1, when the alternative is $H_1\colon \{\Delta \overset{d}{<} 0\}$, i.e. show that $-T$ is significant for large values and is unbiased.

4) Discuss on the validity of Theorem 1, 3.1.1, for two-sided alternatives $H_1\colon \{\delta \neq 0\}$.

5) With reference to Example 2, 2.7, discuss the consistency of test statistics $T = \left(\bar{X}_1 - \bar{X}_2\right)^2$ and $W = |\bar{X}_1 - \bar{X}_2|$ for two-sided alternatives $H_1\colon \{\delta \neq 0\}$.

6) According to the ordering property of the permutation distribution established in Theorem 2, 3.1.1, and using the same notation, prove that if $\delta > \delta' > 0$, then $\Pr\{T^*(\delta) \geq T^o(\delta)|\mathbf{X}(\delta)\} \leq \Pr\{T^*(\delta') \geq T^o(\delta')|\mathbf{X}(\delta')\} \leq \Pr\{T^* \geq T^o|\mathbf{X}\}$, that is, the p-values are stochastically ordered (non-increasingly) with respect to δ.

7) Prove that both conditional and unconditional power functions of the permutation test T above are monotonically non-decreasing with respect to δ.

8) In accordance with Problems 6 and 7 above and Section 3.1.2, discuss conditions so that the permutation distributions of suitable tests for two-sided alternatives are ordered with respect to $|\delta|$. In particular, establish when $|\delta| > |\delta'| > 0$ implies $\Pr\{T^*(|\delta|) \geq T^o(|\delta|)|\mathbf{X}(|\delta|)\} \leq \Pr\{T^*(|\delta'|) \geq T^o(|\delta'|)|\mathbf{X}(|\delta'|)\} \leq \Pr\{T^* \geq T^o|\mathbf{X}\}$, i.e. permutation p-values are ordered non-increasing with respect to $|\delta|$.

9) Extend the validity of Theorem 1, 3.1.1, to the so-called individually varying fixed effects situation, in which $\delta_{1i} > 0$, $i = 1, \ldots, n_1$, and $\delta_{2i} = 0$, $i = 1, \ldots, n_2$, and show that, within a conditional inference framework, there is no real distinction between this model and the one with stochastic effects (see (M.1) in Section 1.10.1).

10) Prove that the non-randomized test ϕ, that is $\phi = 1$ when $T^o \geq T_a$ and 0 elsewhere, and defined according to the notation in Section 2.2.2, is unbiased.

11) Express the unconditional and conditional power functions of the test ϕ of Problem 10 above.

12) Discuss consistency of the test of symmetry in the standardized form K in Section 1.9.4 for testing $H_0 : \{\delta = 0\}$ against $H_1 : \{\delta > 0\}$.

13) With reference to the problem on paired data design (see Section 1.9), prove formally that if $\sigma^2(X) < \infty$, then the permutation distribution of $K^* = \sum_i X_i \cdot S_i^* / (\sum_i X_i^2)^{1/2}$, as n tends to infinity, converges to the distribution of a standard normal variable.

14) Show that the permutation test $T^* = \sum_j n_j(\bar{X}_j^*)$ in a one-way ANOVA design is consistent when all sample sizes n_j, $j = 1, \ldots, C$, tend to infinity.

15) Adapt arguments in Remarks 3 and 4, 3.4, on permutation likelihood and Bayesian inference to the paired data design.

16) With reference to Remark 4, 3.4, on Bayesian permutation inference, derive a Bayesian test procedure in a two-sample problem for simple hypotheses, $H_0 : \{\delta_1 = 0\}$ against $H_1 : \{\delta_1 = \delta_a\}$, where it is presumed that treatment is administered only to the first group, δ_a is a fixed value, and the prior distribution is given by $\pi(0) = p$ and $\pi(\delta_a) = 1 - p$.

17) With reference to Problem 16 above, derive a Bayesian test procedure in a two-sample problem for composite hypotheses, $H_0 : \{\delta_2 \leq 0\}$ against $H_1 : \{\delta_2 > 0\}$, where it is presumed that treatment is administered only to the second group and the prior density distribution is $\pi(\delta)$.

18) Prove that if, in the two-sample problem, the nonparametric family $\mathcal{P}$, to which P_1 and P_2 belong, contains only continuous distributions, then $\mathbf{X}$ is minimal sufficient in H_0.

19) According to the ordering property of permutation distribution established in Theorem 2, 3.1.1, and using the same notation, discuss conditions so that the permutation distributions of suitable tests for two-sided alternatives are ordered with respect to δ^2. In particular, establish when $\delta^2 > \delta'^2 > 0$ implies $\Pr\{T^*(\boldsymbol{\delta}) \geq T^o(\boldsymbol{\delta})|\mathbf{X}(\boldsymbol{\delta})\} \leq \Pr\{T^*(\boldsymbol{\delta}') \geq T^o(\boldsymbol{\delta}')|\mathbf{X}(\boldsymbol{\delta}')\} \leq \Pr\{T^* \geq T^o|\mathbf{X}\}$, that is, the permutation p-values are non-increasingly stochastically ordered with respect to δ^2.

20) Prove that both conditional and unconditional power functions of the permutation test T above are monotonically non-decreasing with respect to δ.

21) Prove the unbiasedness of the non-randomized test ϕ, defined according to the notation in Section 2.1.2.

22) Write the algorithm for the unconditional and conditional power functions of the non-randomized test ϕ of Problem 21 above.

23) Discuss on the unbiasedness of the test statistic T for the one-way ANOVA, defined in Section 1.11.2, when the response model is $\{X_{ji} = \mu + \Delta_{ji} + \sigma \cdot Z_{ji}, i = 1, \ldots, n_j, j = 1, \ldots, C\}$, where $\Delta_{ji} = \delta_j + \sigma_\Delta \cdot D_{ji}$ represent the stochastic effects with mean value δ_j, D_{ji} represent random errors of effects, the distributions of which are not dependent on treatment levels, and σ_Δ is a scale parameter not dependent on treatment levels. (Note that random errors D to some extent may depend on errors Z. Note also that, for the sake of simplicity, the unknown nuisance quantity μ

may be defined in such a way that $\sum_{ji} \Delta_{ji} = 0$; hence, the stochastic effects represent deviates from μ.)

24) Discuss on the unbiasedness of the test statistic T, defined above, for the generalized one-way ANOVA, so that $H_0 : \{X_1 \stackrel{d}{=} \ldots \stackrel{d}{=} X_C\}$ against H_1: $\{H_0$ not true$\}$, with the restriction that, for any pair $(i \neq j,\ i, j = 1, \ldots, C)$, the corresponding response variables are pairwise stochastically ordered according to either $X_i \stackrel{d}{>} X_j$ or $X_i \stackrel{d}{<} X_j$, that is, $\forall x \in \mathcal{R}^1$, so that the CDFs are related according to $F_i(x) \leq F_j(x)$ or $F_i(x) \geq F_j(x)$.

25) With reference to Example 3, 2.7, prove that

$$T_{AD}^* = \sum_{i=1}^{k-1} \frac{\left(\hat{F}_1^*(h) - \hat{F}_2^*(h)\right)}{\{\hat{F}(h)[1 - \hat{F}(h)]\}^{1/2}}$$

is unbiased for testing $H_0 : \{X_1 \stackrel{d}{=} X_2\}$ against $H_1 : \{X_1 \stackrel{d}{>} X_2\}$. It may be useful to represent the observable categorical variables X_j as monotonic, non-decreasing and non-continuous transformations of an underlying continuous variable Y, e.g. $X_j = \varphi(Y + \delta_j)$, with $\delta_1 = 0$ and $\delta_2 > 0$, and to proceed according to the proof of Theorem 1, 3.1.1.

26) Following the same idea as Problem 25 above, find conditions such that T_{AD}^{*2}, as defined in Example 4, 2.7, is unbiased for testing $H_0 : \{X_1 \stackrel{d}{=} X_2\}$ against $H_1 : \{X_1 \stackrel{d}{\neq} X_2\}$.

27) Prove that the two tests T_{AD}^* and T_{AD}^{*2} defined in Problems 24 and 25 above are consistent for proper alternatives.

28) Prove the unbiasedness of the test T_R for repeated measurements according to the problem in Section 2.6.1, also discussed in Example 9, 2.7.

29) Prove the consistency of T_R of Problem 28 above.

30) Write the algorithm for the conditional and unconditional power functions of T_R of Problem 28 above.

31) With reference to the two-sample design, prove that the condition of non-negativity with probability one of random effects Δ implies that two CDFs satisfy the dominance relation $F_1(x) \geq F_2(x)$, $\forall x \in \mathcal{R}^1$, and vice versa.

32) Give a formal proof of the statement in Remark 1, 3.2.1.

33) According to Remark 6, 2.7 and the related response model, prove that the permutation structure (see Remark 1, 2.7 for its definition) in the two-way ANOVA layout without interaction is not dependent on block effects b_i and so the resulting permutation test is exact.

34) Write an algorithm similar to that of Section 3.2.1 for evaluating the conditional power of any test T in the one-sample problem.

35) Write an algorithm similar to that of Section 3.2.1 for evaluating the conditional power of any test T in the one-way ANOVA problem.

36) With reference to the two-sample problem for comparison of locations, prove that, when testing $H_0 : \{\delta = 0\}$ against $H_1 : \{\delta > 0\}$, the test statistic $T^* = \tilde{X}_1^* - \tilde{X}_2^*$ based on sample medians is unbiased.

37) With reference to the two-sample problem for comparison of locations, prove that, when testing $H_0 : \{\delta = 0\}$ against $H_1 : \{\delta > 0\}$, the test statistic $T^* = \bar{X}_{1b}^* - \bar{X}_{2b}^*$, based on trimmed sample means of order b, where $\bar{X}_{jb}^* = \sum_{b<i<n_j-b} X_{j(i)}^*/(n_j - 2b)$, $0 \leq b \leq (n_j - 1)/2$, $j = 1, 2$, and where $X_{j(i)}^*$ are the increasing order statistics associated with the jth group $\mathbf{X}_j^*$, is unbiased.

38) Prove that the permutation solution for a two-way ANOVA without interactions maintains its validity even when the underlying response model is $X_{ji} = \mu_i + \delta_j + Z_{ji}, i = 1, \ldots, n, j = 1, \ldots, k$.

39) With reference to Theorem 1, 3.1.1, show that for fixed effects $\Delta \stackrel{p}{=} \delta$ the relationship $D_S(\mathbf{Z}_1^*, \delta_1^*) - D_S(\mathbf{Z}_2^*, \delta_2^*) \leq D_S(\mathbf{Z}_1, \delta_1)$ is pointwise satisfied, instead of merely in permutation distribution, also for non-associative statistics $T = S_1(\mathbf{X}_1) - S_2(\mathbf{X}_2)$ which satisfy conditions (a) and (b) of Section 3.1.1.

40) Give a formal proof of Proposition 1, 3.1.1, i.e. prove that a conditionally unbiased test for every data set $\mathbf{X} \in \mathcal{X}^n$ is also unconditionally unbiased. Also prove that conditional unbiasedness is sufficient but not necessary for unconditional unbiasedness.

41) Show that in a standard two-sample design for one-sided alternatives, if random effects Δ are such that $\Pr\{\Delta \geq 0\} = 1$ and $\Pr\{\Delta = 0\} > 0$, i.e. if treatment is ineffective with some units, the test statistic $T^* = \sum_i X_{1i}^*$ is unbiased.

42) With reference to a standard two-sample design, prove that permutation tests such as $T^* = \sum_i X_{1i}^*$ with random effects Δ and with individually varying fixed effects δ_i, provided that $\Delta_{1i} \stackrel{d}{=} \delta_{1i}$, $i = 1, \ldots, n_1$, give rise to the same conditional power function but not in general to the same unconditional power function.

4

The Nonparametric Combination Methodology

4.1 Introduction

4.1.1 General Aspects

In previous chapters we have discussed a number of useful solutions to some typical univariate permutation testing problems. Some of them were heuristically motivated, others more rationally justified. Moreover, some theory within the framework of conditionality and sufficiency principles has also been developed. In the present chapter we present a natural extension of permutation testing to a variety of rather complex multivariate problems. In particular, we introduce and discuss the method of nonparametric combination methodology (NPC) of a finite number of dependent permutation tests as a general tool for multivariate testing problems when a set of quite mild conditions holds. In Section 4.5 we will see an extension of the NPC up to countable number of dependent permutation tests. Of course when, as in many V-dimensional problems ($V \geq 2$) for continuous or categorical variables, one single appropriate overall test statistic $T : \mathcal{R}^V \rightarrow \mathcal{R}^1$ is available (e.g. of the chi-square or Hotelling's T^2 type), then in terms of computational complexity related permutation solutions become equivalent to simple univariate procedures; for the use of statistics of this kind in some standard multivariate problems (see Barton and David, 1961; Mantel and Valand, 1970; Good, 2000). A similar simplicity is also encountered when there are suitable data transformations $\varphi : \mathcal{R}^V \rightarrow \mathcal{R}^1$ of the V-dimensional into univariate derived data $Y = \varphi(X_1, \ldots, X_V)$ (examples are given by Reboussin and DeMets, 1996; Hoh and Ott, 2000; Mielke and Berry, 2007; see also (h) in Section 4.2.4; a typical example occurs, for instance, when in repeated measurements the so-called area under the curve (AUC) is considered). In this and the subsequent chapters we shall mostly be interested in more complex problems for which such kinds of single overall tests are not directly available, or not easy to find, or too difficult to justify.

Often in testing for complex hypotheses, when many response variables are involved or many different aspects are of interest (see Chapters 5–12 for several multivariate problems; Examples 3–8, 4.6, for jointly testing for many different aspects and for a form of monotonic testing in univariate problems), to some extent it is natural, convenient and often easier for the interpretation of results, to firstly process data using a finite set of $k > 1$ different *partial tests* (note that the number k of sub-problems is not necessarily equal to the dimensionality V of responses). Such partial tests, possibly after adjustment for multiplicity (see Westfall and Young, 1993; Basso et al., 2009a), may be useful for marginal or separate inferences. But if they are jointly considered,

Permutation Tests for Complex Data: Theory, Applications and Software Fortunato Pesarin, Luigi Salmaso

they provide information on a general overall (i.e. global) hypothesis, which typically constitutes the objective of most multivariate testing problems.

In order to motivate the necessity and usefulness of the NPC method, let us consider, for instance, a two-sample bivariate problem in which one variable is ordered categorical, the other quantitative, and two variables are dependent. Moreover, let us assume that a symbolic treatment may influence both variables, for instance by 'positive increments', so that the distribution is shifted towards higher values on both components (the so-called componentwise stochastic dominance). Let us also assume that the alternatives of interest are restricted to positive increments, that is they are both one-sided. Due to its complexity, such a problem is usually solved by two separate partial tests, one for the quantitative and one for the ordered categorical variable, and analysis tends to dwell separately on each sub-problem. However, for the general testing problem, both are jointly informative regarding the possible presence of non-null effects. Thus, the necessity of taking account of all available information through the combination of two tests in one *combined test* naturally arises.

When partial tests are stochastically independent, this combination is easily obtained (for a review of combination of one-sided independent tests, see Folks, 1984, and references therein). But in the great majority of situations it is impossible to invoke such a complete independence among partial tests both because they are functions of the same data set $\mathbf{X}$ and because component variables in $\mathbf{X}$ are generally not independent. Moreover, the underlying dependence relations among partial tests are rarely known, except perhaps for some quite simple situations such as the multivariate normal case where all dependences are pairwise linear, and even when they are known they are often too difficult to cope with. Therefore, this combination must be done nonparametrically, especially with regard to the underlying dependence relations (why this combination is nonparametric is discussed in Section 4.2.5).

In addition, when testing for restricted alternatives, only quite difficult solutions in a few specific situations are provided in a parametric setting, even in cases of multivariate normality (for solutions and references see Kudo, 1963; Nüesch, 1966; Shorack, 1967; Barlow et al. 1972; Chatterjee, 1984; Robertson et al. 1988; El Barmi and Dykstra, 1995; Perlman and Wu, 1998; Silvapulle and Sen, 2005). Difficulties generally increase when dealing with categorical responses (Wang, 1996; Basso et al., 2009a). In this context, on the one hand asymptotic null distributions of related maximum likelihood ratio tests depend on unknown multinomial parameters. On the other, Cohen and Sackrowitz (1998) have given an example where the power of a test for restricted alternatives based on the likelihood ratio is suspected not to be monotonically related to treatment effects. These two undesirable properties of likelihood ratio solutions are difficult to accept. There are also problems with repeated measurements in which underlying parametric models are not identifiable, for example when there are many more measurements within each unit than there are units in the study, so that most of the related testing problems are unsolvable in the parametric context (see Crowder and Hand, 1990; Higgins and Noble, 1993; Diggle et al., 2002). However, we shall see that in a set of mild, simple and easy-to-check conditions, a general and effective solution may be found via the NPC of k dependent permutation tests.

4.1.2 Bibliographic Notes

In the literature there are a relatively small number of references on the combination of dependent tests. Chung and Fraser (1958) suggest using the sum of k partial tests in testing for multivariate location when k is larger than the total sample size (providing a sort of direct NPC; see (h) in Section 4.2.4 for a brief discussion). Boyett and Shuster (1977), Mukerjee et al. (1986), Higgins and Noble (1993) and Blair et al. (1994) discuss the 'max t test' in some cases of nonparametric one-sided multivariate testing. Berk and Jones (1978) discuss how the Bahadur asymptotic relative efficiency relates to a Tippett combination procedure in the case of dependence of partial tests, but they do not provide any practical solutions. Wei and Johnson (1985) consider a locally optimal combination of dependent tests based on asymptotic arguments. Westberg (1986) considers some cases of robustness with respect to the dependence of an adaptive Tippett type combination procedure. Edgington (1995), in connection with some multivariate permutation problems, suggests using the sum of 'homogeneous tests'.

Conservative solutions may be found via the Bonferroni inequality (see Kounias and Sotirakoglou, 1989; Kounias, 1995; Galambos and Simonelli, 1996), by analogy with multiple comparison methods (Zanella, 1973; Miller, 1981; Shuster and Boyett, 1979; Petrondas and Gabriel, 1983; Hochberg and Tamhane, 1987; Westfall and Young, 1993; Hsu, 1996; Basso et al., 2009a). The conservativeness of solutions obtained via the Bonferroni inequality is often unacceptable, for both theoretical and practical purposes. Moreover, it is worth noting that multiple comparison procedures have their starting points in an overall test and look for significant tests on partial contrasts. Conversely, combination procedures start with *a set of partial tests*, each appropriate for a partial aspect, and look for joint analyses leading to global inferences.

Major results for the problem of NPC of dependent permutation tests have been obtained by Ballin and Pesarin (1990), Fattorini (1996), Giraldo and Pesarin (1992, 1993), Pallini (1990, 1991, 1992a, 1992b, 1992c), Pallini and Pesarin (1990, 1992a, 1994), Pesarin (1988, 1989, 1990a, 1990b, 1990c, 1991a, 1991b, 1992, 1993, 1994, 1995, 1996a, 1996b, 1996c, 1997a, 1997b, 1999a, 1999b, 2001), Pesarin and Salmaso (1998a, 1998b, 1999, 2000a, 2000b), and Celant et al. (2000a, 2000b). Additional references may be found in Arboretti et al. (2000a, 2005a, 2005b, 2007a, 2007b, 2007c, 2007d, 2007e, 2007f, 2007g, 2008a, 2008b, 2009b, 2009c, 2009d), Celant et al. (2009a, 2000b), Dalla Valle et al. (2000, 2003), Abbate et al. (2001, 2004), Mazzaro et al. (2001), Corain et al. (2002, 2009a, 2009b), Pesarin and Salmaso (2002, 2006, 2009), Arboretti and Salmaso (2003), Salmaso (2003, 2005), Corain and Salmaso (2003, 2004, 2007a, 2007b, 2009a, 2009b), Basso et al. (2004, 2007a, 2007b, 2007c, 2008, 2009a, 2009b), Filippini et al. (2004), Finos and Salmaso (2004, 2005, 2006, 2007), Bonnini et al. (2005, 2006a, 2006b, 2009), Fava et al. (2005), Salmaso and Solari (2005, 2006), Basso and Salmaso (2006, 2007, 2009a, 2009b), Berti et al. (2006), Marozzi and Salmaso (2006), Bassi et al. (2007), Finos et al. (2007, 2008, 2009), Guarda-Nardini et al. (2007, 2008), Arboretti and Bonnini (2008, 2009), Brombin and Salmaso (2008, 2009), Klingenberg et al. (2008), Manfredini et al. (2008), Solari et al. (2008), Bertoluzzo et al. (2009) and Brombin et al. (2009). In these papers, the main theory of NPC and many applications to rather complex testing problems are discussed. Among the many applications referred to, we may mention: the one-way MANOVA, where some of the variables are quantitative and others categorical; analysis of repeated measures with dependent random effects and dependent errors; analysis of multivariate restricted alternatives; analysis of testing problems where some data are missing and the underlying missing process is not ignorable; some goodness-of-fit problems with multivariate ordered categorical data; the multivariate Behrens–Fisher problem; some multivariate stochastic dominance problems; exact testing for interactions in the two-way ANOVA; a multivariate extension of McNemar's test; a multivariate extension of Fisher's exact probability test; problems of isotonic inference; and multi-aspect testing problems, where treatment may act on more than one aspect of interest.

These rather difficult testing problems, which are not adequately taken into consideration in the literature, in spite of the fact that they are very frequently encountered in a great variety of practical applications, emphasize the versatility and effectiveness of the NPC methodology. It should also be emphasized that, as permutation tests are conditional on a set of sufficient statistics, in very mild conditions the NPC methodology frees the researcher from the need to model the dependence relations among responses. This aspect is particularly relevant in many contexts, such as multivariate categorical responses, in which dependence relations are generally too difficult to define and to model (see Joe, 1997). Furthermore, several Monte Carlo experiments show that the unconditional power behaviour of combined tests is generally close to that of their best parametric counterparts, in the conditions for the latter. In this chapter, we shall review the theory of NPC of dependent tests, including some specific asymptotic aspects. Chapters 5–12 discuss a number of application problems, the solutions of which are obtained through NPC.

4.1.3 Main Assumptions and Notation

Let us introduce the notation and main assumptions regarding the data structure, set of partial tests, and hypotheses being tested in NPC contexts. For the sake of clarity and without loss of generality, let us refer to a one-way MANOVA design. To this end:

(i) With obvious notation, let us denote a V-dimensional data set by $\mathbf{X} = \{\mathbf{X}_j,\ j = 1, \ldots, C\} = \{\mathbf{X}_{ji},\ i = 1, \ldots, n_j,\ j = 1, \ldots, C\} = \{X_{hji},\ i = 1, \ldots, n_j,\ j = 1, \ldots, C,\ h = 1, \ldots, V\}$. As usual, we represent the data set and V-dimensional response using the same symbol $\mathbf{X}$, the meaning generally being clear from the context. The response $\mathbf{X}$ takes its values on the V-dimensional sample space $\mathcal{X}$, for which a σ-algebra $\mathcal{A}$ and a (possibly unspecified) nonparametric family $\mathcal{P}$ of non-degenerate distributions are assumed to exist. The data set $\mathbf{X}$ consists of $C \geq 2$ samples or groups of size $n_j \geq 2$, with $n = \sum_j n_j$; the groups are presumed to be related to C levels of a treatment and the data $\mathbf{X}_j$ are supposed i.i.d. with distributions $P_j \in \mathcal{P},\ j = 1, \ldots, C$ (in place of independence, exchangeability may generally suffice; see Sections 1.2 and 2.1.2). Of course, if covariates are available, it is straightforward to refer to a one-way MANCOVA design. Again, for easier computer handling, it is useful to express data sets with the unit-by-unit representation $\mathbf{X} = \{\mathbf{X}(i),\ i = 1, \ldots, n;\ n_1, \ldots, n_C\}$, in which it is assumed that the first n_1 data vectors belong to the first group, the next n_2 to the second, and so on.

(ii) The null hypothesis refers to equality of multivariate distributions of responses on C groups:

$$H_0 : \{P_1 = \ldots = P_C\} = \left\{\mathbf{X}_1 \stackrel{d}{=} \ldots \stackrel{d}{=} \mathbf{X}_C\right\}.$$

Let us suppose that, related to the specific problem at hand, a set of side-assumptions holds, so that H_0 may be properly and equivalently broken down into a finite set of sub-hypotheses $H_{0i},\ i = 1, \ldots, k$, each appropriate for a partial aspect of interest. Therefore, H_0 is true if all the H_{0i} are jointly true; and so it may be written as $\left\{\bigcap_{i=1}^{k} H_{0i}\right\}$. In this sense, H_0 is also called the *global* or *overall null hypothesis*. Note that the dimensionality V of responses is not necessarily related to that of the sub-hypotheses, although for most multivariate location problems we have $k = V$. Also note that H_0 implies that the V-dimensional data vectors in $\mathbf{X}$ are exchangeable with respect to C groups.

(iii) In the same set of side-assumptions as for point (ii), the alternative hypothesis states that at least one of the null sub-hypotheses H_{0i} is not true. Hence, the alternative may be represented by the union of k sub-alternatives,

$$H_1 : \left\{\bigcup_{i=1}^{k} H_{1i}\right\},$$

stating that H_1 is true when at least one sub-alternative is true. In this context, H_1 is called the *global* or *overall alternative*.

(iv) $\mathbf{T} = \mathbf{T}(\mathbf{X})$ represents a k-dimensional vector of test statistics, in which the ith component $T_i = T_i(\mathbf{X}),\ i = 1, \ldots, k$, represents the non-degenerate ith *partial test* which is assumed to be appropriate for testing sub-hypothesis H_{0i} against H_{1i}. Without loss of generality, in the NPC context all partial tests are assumed to be marginally unbiased, consistent and significant for large values (see Section 4.2.1 for the concepts of marginal unbiasedness and consistency).

Remark 1. It should be emphasized that partial or *componentwise testing* may also be useful in a marginal sense, which can be separately tested, possibly after p-value adjustment due to multiplicity (see Chapter 5). Thus, on the one hand, partial tests may provide marginal information for each specific sub-hypothesis; on the other, they jointly provide information on the global hypothesis

H_0. It should also be emphasized that, if we may presume that the set of proper side-assumptions holds, the breakdown into k sub-hypotheses allows us to express overall hypotheses in equivalent forms. Thus, this approach is essentially a procedural technique in which problems always remain multidimensional (see Section 4.2.3). Substantially, this approach corresponds to a method of analysis carried out in two successive phases, the first focusing on k partial aspects, and the second on their combination.

Remark 2. For the sake of simplicity, in this presentation we refer to a typical multivariate C-sample problem, such as the one-way MANOVA. Of course, in general and in a straightforward way, we may also refer to other multivariate data designs. Several relevant examples are discussed in Section 4.6 and from Chapter 5 onwards.

Remark 3. For component tests T_i, $i = 1, \dots, k$, we prefer to use the term *partial tests* in place of *marginal tests*, because the latter may lead to misunderstandings when multivariate permutations are being used. In multivariate testing, in order to preserve all underlying dependence relations among variables, permutations *must always* be carried out on individual data vectors, so that all component variables and partial tests *must be jointly analysed* (see Section 4.2.3). The notion of marginal tests may sometimes wrongly lead us to think that we are allowed to consider even independent and separate permutations for each component test, so that multivariate dependence relations may not be preserved. It is therefore important to distinguish between partial and marginal tests, although they are often coincident. In Remark 1, 4.3.2, we shall see partial tests which are not marginal tests, and marginal tests which are not partial tests.

Remark 4. The requirement for partial tests to be significant for large values is not restrictive (see point (iv) above), because in general we may refer to permutationally invariant transformations in order to satisfy this requirement. In Sections 6.7, in connection with some problems of isotonic inference, and in Examples 4 and 5, 4.6, we shall see an extension of this condition, in which significance for either large or small values is required.

4.1.4 Some Comments

The side-assumptions allowing us to break down the hypotheses into a finite set of equivalent sub-hypotheses are generally quite natural and easy to justify. The most common situation relates to multivariate testing on locations where each sub-hypothesis concerns one component variable, or a subset of them. Moreover, the set of sub-hypotheses and related partial tests are quite general and may occur in many complex ways. The variable $\mathbf{X}$ may be continuous, discrete or mixed. For this reason, the NPC method may be usefully applied to a great variety of situations.

Although it is generally possible to think of more direct and exact solutions, the NPC problem is tackled here through a CMC procedure, because the associated algorithm makes it easy to justify and properly interpret the related inferential results. However, it must be emphasized that the NPC of dependent tests is a tool which leads to exact solutions when all conditions in points (i)–(iv) in Section 4.1.3 are jointly satisfied. In particular, it is important to emphasize that:

(a) the null hypothesis *must* take the form $H_0 : \{P_1 = \ldots = P_C\} = \{\mathbf{X}_1 \stackrel{d}{=} \ldots \stackrel{d}{=} \mathbf{X}_C\}$, which implies the exchangeability of individual data vectors with respect to groups, so that the permutation multivariate testing principle is properly applicable;
(b) based on a set of side-assumptions, hypotheses H_0 and H_1 must be broken down respectively into $\{\bigcap_i H_{0i}\}$ and $\{\bigcup_i H_{1i}\}$;
(c) a set of appropriate partial tests significant for large values is available;
(d) all k partial tests are jointly analysed.

If anything in these conditions is not satisfied, something in the solution may be wrong. In particular, when some of the side-assumptions are violated, leading to an improper breakdown of the hypotheses, then we may produce solutions to testing problems which are far from what is desired. Failures in the permutation principle, due for example to lack of exchangeability in H_0, may cause conclusions without any control of inferential errors, or may introduce approximations which must be carefully analysed – Example 8, 4.6, and Chapters 7 and 9 present some cases of violations leading to approximate solutions; for permutation testing in the presence of nuisance entities, see also Tsui and Weerhandi (1989) and Commenges (1996). If some partial tests are not appropriate for specific sub-hypotheses, then the overall testing solution may become improper.

4.2 The Nonparametric Combination Methodology

4.2.1 *Assumptions on Partial Tests*

This section specifies the assumptions regarding the set of partial tests $\mathbf{T} = \{T_i,\ i = 1, \ldots, k\}$ which are needed for NPC:

(A.1) All permutation partial tests T_i are marginally unbiased and significant for large values, so that they are both conditionally and unconditionally stochastically larger in H_1 than in H_0.
(A.2) All permutation partial tests T_i are marginally consistent, i.e. as sample sizes tend to infinity $\Pr\{T_i \geq T_{i\alpha}|H_{1i}\} \rightarrow 1,\ \forall\alpha > 0$, where $T_{i\alpha}$, which is assumed to be finite, is the critical value of T_i at level α.

Remark 1. Assumption (A.1) formally implies that

$$\Pr\{T_i \geq T_{i\alpha}|\mathcal{X}_{/\mathbf{X}}, H_{1i}\} \geq \alpha,\ \forall\alpha > 0,\ i = 1, \ldots, k,$$

and, for all $z \in \mathcal{R}^1$,

$$\begin{aligned}\Pr\{T_i \leq z|\mathcal{X}_{/\mathbf{X}}, H_{0i}\} &= \Pr\left\{T_i \leq z|\mathcal{X}_{/\mathbf{X}}, H_{0i} \cap H_i^{\dagger}\right\} \\ &\geq \Pr\{T_i \leq z|\mathcal{X}_{/\mathbf{X}}, H_{1i}\} \\ &= \Pr\left\{T_i \leq z|\mathcal{X}_{/\mathbf{X}}, H_{1i} \cap H_i^{\dagger}\right\},\ i = 1, \ldots, k,\end{aligned}$$

where irrelevance with respect to the complementary set of hypotheses $H_i^{\dagger} : \left\{\bigcup_{j\neq i}(H_{0j} \cup H_{1j})\right\}$ means that it does not matter which among H_{0j} and $H_{1j},\ j \neq i$, is true when testing for the ith sub-hypothesis. Similarly, assumption (A.2) implies that if sub-alternative H_{1i} is true, then $\Pr\{T_i \geq T_{i\alpha}|\mathcal{X}_{/\mathbf{X}}, H_{1i} \cap H_i^{\dagger}\} \rightarrow 1,\ \forall\alpha > 0$, independently of $H_i^{\dagger}$.

Assumption (A.2) can sometimes be relaxed because what is really required is that at least one partial test for which the sub-alternative is true must be consistent. To be specific, suppose there are values $h \in (1, \ldots, k)$ such that the sub-alternative H_{1h} is true and define $H_1^{(true)} = \bigcup_h H_{1h}$ the set of true sub-alternatives for that problem, then it is required that at least one partial test T_h must be consistent, that is, as sample sizes diverge, $\Pr\{T_h^* \geq T_h^o|\mathcal{X}_{/\mathbf{X}}\}$ weakly converges to 0 for at least one h such that $H_{1h} \in H_1^{(true)}$. Suppose, for instance, that in a two-sample one-dimensional problem the alternative is *either* $H_{1<} : \{X_1 \overset{d}{<} X_2\}$ *or* $H_{1>} : \{X_1 \overset{d}{>} X_2\}$ and that partial tests are $T_<^* = \bar{X}_2^*$ and $T_>^* = \bar{X}_1^*$, respectively. Then in the alternative only one of two partial tests is consistent (see Example 4, 4.6). However, the combined test is consistent (see Theorem 2, 4.3.1).

Assumption (A.1) implies that the set of p-values $\lambda_1, \ldots, \lambda_k$, associated with the partial tests in $\mathbf{T}$, are *positively dependent in the alternative* (see Lehmann, 1986, p. 176; see also Dharmadhikari and Joag-Dev, 1988, for the concept, analysis and consequences of positive dependence on random variables), and this irrespective of dependence relations among component variables in **X**.

Sometimes marginal unbiasedness, which is a sufficient condition for NPC (see Remark 1, 4.3.2), is only approximately satisfied. One important example occurs when H_{01} and H_{02} are respectively related to locations and scale coefficients in a two-sample testing problem where symbolic treatment may influence both and two CDFs are not ordered so that they can cross (see Example 8, 4.6). One more important example occurs with two-sided alternatives with fixed effects. In this framework, as the effect δ is either positive or negative, of two partial tests, $T_{>}^{*} = \bar{X}_1^{*}$ for $\delta > 0$ and $T_{<}^{*} = \bar{X}_2^{*}$ for $\delta < 0$, only one is marginally unbiased and so their combination for two-sided alternatives cannot be claimed to be unbiased (see Example 4, 4.6). Therefore, for the marginal unbiasedness to be satisfied, on the one hand T_1 must be unbiased for H_{01} against H_{11}, irrespective of whether H_{02} is true or not; on the other T_2 must be unbiased for H_{02} against H_{12}, irrespective of whether H_{01} is true or not. But, with statistics comparing locations, as in the Behrens–Fisher problem, and the ratio of scale indicators for testing on scale coefficients (see Examples 3 and 8, 4.6, for a NPC solution when the treatment may act on the first two moments), the marginal unbiasedness of T_1 is only approximately satisfied, unless we know in advance that H_{02} is true. The same happens for T_2, which is marginally unbiased only if it is known that H_{01} is true. Hence, in these situations marginal unbiasedness is satisfied only approximately.

Remark 2. When partial tests are separately referred to component variables in **X** and these are related by monotonic regressions, assumption (A.1) is naturally satisfied. However, it should be noted that we generally do not need to assume that regression relationships on component variables in **X** are monotonic. We only require that permutation partial tests do satisfy (A.1) as, for instance, in multi-aspect testing (see Example 3, 4.6). Thus, the condition of monotonic regression relationships on p-values of partial tests, being only a sufficient condition for the validity of NPC, is weaker and more generally satisfied than monotonic regressions among the V components in **X** (see Remark 1, 4.3.2).

4.2.2 Desirable Properties of Combining Functions

For the sake of simplicity and uniformity of analysis, but without loss of generality, we only refer to combining functions applied to p-values associated with partial tests. Because of assumption (A.1), 4.2.1, and Theorem 1, 2.4, partial tests are permutationally equivalent to their p-values: $T_i \approx \Pr\{T_i^{*} \geq T_i^{o} | \mathcal{X}_{/\mathbf{X}}\} = \lambda_i$, $i = 1, \ldots, k$. Of course, this is a direct consequence of the monotonic non-increasing behaviour with respect to t of significance level functions $L_i(t) = \Pr\{T_i^{*} \geq t | \mathcal{X}_{/\mathbf{X}}\}$ or $L_i^{(2)}(t) = 1 - \left|2 \cdot \Pr\{T_i^{*} \geq t | \mathcal{X}_{/\mathbf{X}}\} - 1\right|$ for one-sided and two-sided alternatives, respectively.

Thus, the NPC in one second-order test

$$T'' = \psi(\lambda_1, \ldots, \lambda_k)$$

is achieved by a continuous, non-increasing, univariate, measurable and non-degenerate real function $\psi : (0, 1)^k \rightarrow \mathcal{R}^1$.

Note that the continuity of ψ is required because it has to be defined irrespective of the cardinality of $(\Lambda_1, \ldots, \Lambda_k)$. Moreover, the measurability property of ψ is required because it is used as a test statistic which then must induce a probability distribution on which inferential conclusions are necessarily based.

In order to be suitable for test combination (see Pesarin, 1992, 1999b, 2001; see also Goutis et al., 1996), all combining functions ψ must satisfy at least the following reasonable properties:

(P.1) A combining function ψ must be non-increasing in each argument, $\psi(\ldots, \lambda_i, \ldots) \geq \psi(\ldots, \lambda_i', \ldots)$ if $\lambda_i < \lambda_i'$, $i \in \{1, \ldots, k\}$. Also, it is generally desirable that ψ is symmetric i.e. invariant with respect to rearrangements of the input arguments: $\psi(\lambda_{u_1}, \ldots, \lambda_{u_k}) = \psi(\lambda_1, \ldots, \lambda_k)$ where $(u_1, \ldots, u_k)$ is any permutation of $(1, \ldots, k)$.

(P.2) Every combining function ψ must attain its supremum value $\bar{\psi}$, possibly not finite, even when only one argument attains zero: $\psi(\ldots, \lambda_i, \ldots) \to \bar{\psi}$ if $\lambda_i \to 0,\ i \in \{1, \ldots, k\}$.

(P.3) For all $\alpha > 0$, the critical value T''_α of every ψ is assumed to be finite and strictly smaller than $\bar{\psi} : T''_\alpha < \bar{\psi}$.

These properties are quite reasonable and intuitive, and are generally easy to justify. Property (P.1) agrees with the notion that large values are significant; it is also related to the unbiasedness of combined tests. It means that if, for instance, $\psi(\ldots, \lambda'_i, \ldots)$ is rejected, then $\psi(\ldots, \lambda_i, \ldots)$ must also be rejected because it better agrees with the alternative. Moreover, symmetry is required to obtain inferences independent of the order partial tests enter the analysis. Properties (P.2) and (P.3) are related to consistency.

It should be noted that properties (P.1)–(P.3) define a class $\mathcal{C}$ of combining functions, which contains the well-known functions of Fisher, Lancaster, Liptak, Tippett, etc. (for a review of the combination of independent tests, see Oosterhoff, 1969; see also Littell and Folks, 1971, 1973; Folks, 1984). $\mathcal{C}$ also contains the Mahalanobis quadratic form for invariant testing against alternatives lying at the same quadratic distance from H_0. In addition, according to Birnbaum (1954, 1955), $\mathcal{C}$ contains the class $\mathcal{C}_A$ of admissible combining functions of independent tests characterized by convex acceptance regions, when these are expressed in terms of p-values. In this respect, since the acceptance region does not depend on how partial tests are dependent, admissibility of the sub-class of $\mathcal{C}$ characterized by convex acceptance regions holds even for dependent partial tests. In Section 4.2.6 we will see a sufficient condition for admissibility of ψ. Admissibility of a test, although weak, is quite an important property as it says that no other test exists which is uniformly better than it in terms of unconditional power. And so, when choosing a way to combine, if we stay within $\mathcal{C}_A$ we are sure that no other choice is uniformly better. Class $\mathcal{C}$ in particular contains all combining functions which take account nonparametrically of the underlying dependence relations among p-values $\lambda_i,\ i = 1, \ldots, k$ (for some examples, see Section 4.2.4).

$\mathcal{C}$ is much larger than it appears at first sight: it contains all continuous and strictly increasing transformations of its members. For example, if $\eta : \mathcal{R}^1 \to \mathcal{R}^1$ is one of these transformations, then two combining functions ψ and $\eta \circ \psi$ are permutationally equivalent (see Problem 18, 4.2.7).

One problem naturally arises: how to construct, for any given testing problem, a good combining function in $\mathcal{C}$. Finding a best solution appears to be impossible in the case of finite sample sizes without any further restrictions because the admissible class $\mathcal{C}_A$ has more than one member. At the moment only 'asymptotic optimal combinations' can sometimes be obtained. An example of an asymptotic argument for establishing a locally optimal combination is given in Wei and Johnson (1985). Section 4.4 presents some arguments on this. Moreover, if $\eta_i = \eta_i(T_i),\ i = 1, \ldots, k$, where η_i are continuous and monotonically increasing transformations of partial tests, then $\forall \psi \in \mathcal{C},\ T''_\eta = \psi(\lambda_{\eta 1}, \ldots, \lambda_{\eta k})$ is permutationally equivalent to $\psi(\lambda_1, \ldots, \lambda_k) = T''$, because the p-values are invariant under continuous monotonic increasing transformations of statistics. Some practical guidelines for reasonable selection of a combining function are reported in Remarks 4 and 5, 4.2.4.

Remark 1. If permutation partial tests are all exact, then the combined test T''_ψ is exact for every combining function $\psi \in \mathcal{C}$ (see Problem 13, 6.2.6).

Remark 2. In accordance with Problem 10, 4.3.4 (see also Sections 2.7.1 and 4.2.6), in the permutation context it is possible (see Section 4.3.5) to define conditional multivariate confidence regions on functionals $\delta_i = \mathbb{E}(T_i),\ i = 1, \ldots, k$.

4.2.3 A Two-Phase Algorithm for Nonparametric Combination

This subsection deals with a two-phase algorithm used to obtain a conditional Monte Carlo (CMC) estimate of the permutation distribution of combined tests.

The first phase is concerned with the estimate of the k-variate distribution of $\mathbf{T}$, and the second finds the estimate of the permutation distribution of the combined test T''_ψ by using the same CMC results of the first phase. Note that when it is clear from the context which combining function ψ has been used, in place of T''_ψ we simply write T''.

In this multivariate and multi-aspect setting, simulations from the permutation sample space $\mathcal{X}_{/\mathbf{X}}$ by a CMC method are carried out in analogy with the algorithm discussed in Section 2.2.5 for univariate problems.

The first phase of the algorithm for estimating the k-variate distribution of $\mathbf{T}$ includes the following steps:

(S.a$_k$) Calculate the vector of the observed values of tests $\mathbf{T} : \mathbf{T}^o = \mathbf{T}(\mathbf{X})$.

(S.b$_k$) Consider a random permutation $\mathbf{X}^* \in \mathcal{X}_{/\mathbf{X}}$ of $\mathbf{X}$ and the values of vector statistics $\mathbf{T}^* = \mathbf{T}(\mathbf{X}^*)$. According to Remark 1, 2.1.2, it is worth noting that in multivariate situations the permutation $\mathbf{X}^*$ is obtained by first taking a random permutation $(u_1^*, \ldots, u_n^*)$ of $(1, \ldots, n)$ and then by assignment of related individual data vectors to the proper group; thus, by using the unit-by-unit representation, $\mathbf{X}^* = \{\mathbf{X}(u_i^*) = [X_1(u_i^*), \ldots, X_V(u_i^*)],\ i = 1, \ldots, n;\ n_1, \ldots, n_C\}$ (see Table 4.1).

(S.c$_k$) Carry out B independent repetitions of step (S.b$_k$). The set of CMC results $\{\mathbf{T}_b^*,\ b = 1, \ldots, B\}$ is thus a random sampling from the permutation k-variate distribution of vector test statistics $\mathbf{T}$.

(S.d$_k$) The k-variate EDF

$$\hat{F}(\mathbf{t}|\mathcal{X}_{/\mathbf{X}}) = \frac{\left[\frac{1}{2} + \sum_b \mathbb{I}(\mathbf{T}_b^* \le \mathbf{t})\right]}{B+1}, \quad \forall \mathbf{t} \in \mathcal{R}^k,$$

gives a consistent estimate of the corresponding k-dimensional permutation CDF $F(\mathbf{t}|\mathcal{X}_{/\mathbf{X}})$ of $\mathbf{T}$. Moreover, the ESFs

$$\hat{L}_i(t|\mathcal{X}_{/\mathbf{X}}) = \frac{\left[\frac{1}{2} + \sum_b \mathbb{I}(T_{ib}^* \ge t)\right]}{B+1}, \quad i = 1, \ldots, k,$$

give consistent estimates $\forall t \in \mathcal{R}^1$ of the k marginal permutation SLF $L_i(t|\mathcal{X}_{/\mathbf{X}}) = \Pr\{T_i^* \ge t|\mathcal{X}_{/\mathbf{X}}\}$. Thus $\hat{L}_i(T_i^o|\mathcal{X}_{/\mathbf{X}}) = \hat{\lambda}_i$ gives a consistent estimate of the marginal p-value $\lambda_i = \Pr\{T_i^* \ge T_i^o|\mathcal{X}_{/\mathbf{X}}\}$, relative to test T_i.

Table 4.1 summarizes the observed data set and one V-variate permutation in a two-sample design. Table 4.2 summarizes the CMC procedure. It should be emphasized that in multivariate problems, as the multivariate permutation sample space $\mathcal{X}_{/\mathbf{X}}$ is obtained by permuting individual data vectors (see Remark 1, 2.1.2), the CMC operates accordingly, so that $\mathbf{X}^* = \{\mathbf{X}(u_i^*),\ i = 1, \ldots, n;\ n_1, \ldots, n_C\}$, as is explicitly displayed in the second part of Table 4.1. Thus, all underlying dependence relations which are in the component variables are preserved. From this point of view, CMC is properly a multivariate procedure (see Remark 4, 4.1.3).

Note that with respect to traditional EDF estimators, 1/2 and 1 have been added respectively to the numerators and denominators of equations in step (S.d$_k$). This is done in order to obtain estimated values of the CDF $F(\mathbf{t}|\mathcal{X}_{/\mathbf{X}})$ and of SLF $L(\mathbf{t}|\mathcal{X}_{/\mathbf{X}})$ in the open interval (0,1), so that transformations by inverse CDF of continuous distributions, such as $-\log(\lambda)$ or $\Phi^{-1}(1-\lambda)$, where Φ is the standard normal CDF (see also Section 4.2.4), are continuous and so are always well defined. However, since B is generally large, this minor alteration is substantially irrelevant because it does not modify test

Table 4.1 Representation of a two-sample multivariate permutation

$X_1(1)$	$\cdots$	$X_1(n_1)$	$X_1(1+n_1)$	$\cdots$	$X_1(n)$		T_1^o
$\vdots$		$\vdots$	$\vdots$		$\vdots$	$\longrightarrow$	$\vdots$
$X_V(1)$	$\cdots$	$X_V(n_1)$	$X_V(1+n_1)$	$\cdots$	$X_V(n)$		T_k^o

$X_1(u_1^*)$	$\cdots$	$X_1(u_{n_1}^*)$	$X_1(u_{1+n_1}^*)$	$\cdots$	$X_1(u_n^*)$		T_1^*
$\vdots$		$\vdots$	$\vdots$		$\vdots$	$\longrightarrow$	$\vdots$
$X_V(u_1^*)$	$\cdots$	$X_V(u_{n_1}^*)$	$X_V(u_{1+n_1}^*)$	$\cdots$	$X_V(u_n^*)$		T_k^*

Table 4.2 Representation of the CMC method in multivariate tests

$\mathbf{X}$	$\mathbf{X}_1^*$	$\cdots$	$\mathbf{X}_b^*$	$\cdots$	$\mathbf{X}_B^*$
T_1^o	T_{11}^*	$\cdots$	T_{1b}^*	$\cdots$	T_{1B}^*
$\vdots$	$\vdots$		$\vdots$		$\vdots$
T_k^o	T_{k1}^*	$\cdots$	T_{kb}^*	$\cdots$	T_{kB}^*

behaviour and consequent inferences, both for finite sample sizes and asymptotically. In practice, in place of 1/2 and 1, we may add any positive quantity ε and 2ε, provided that ε is very small compared to B (see Problem 11, 4.2.7). In particular, Remark 1, 2.1.3, now implies the following theorem:

Theorem 1. *As B tends to infinity, $\forall \mathbf{t} \in \mathcal{R}^k$, $\hat{F}(\mathbf{t}|\mathcal{X}_{/\mathbf{X}})$ and $\hat{L}(\mathbf{t}|\mathcal{X}_{/\mathbf{X}})$ almost surely converge to the permutation CDF $F(\mathbf{t}|\mathcal{X}_{/\mathbf{X}})$ and the permutation SLF $L(\mathbf{t}|\mathcal{X}_{/\mathbf{X}})$, respectively.*

The Glivenko–Cantelli theorem, which may be directly applied because $1/\{2(B+1)\}$ vanishes as B tends to infinity, gives a straightforward proof of this statement (see: Shorack and Wellner, 1986; Borovkov, 1987).

The second phase of the algorithm for simulating a procedure for NPC should include the following steps:

(C.a) The k observed p-values are estimated on the data $\mathbf{X}$ by $\hat{\lambda}_i = \hat{L}_i(T_i^o|\mathcal{X}_{/\mathbf{X}})$, where $T_i^o = T_i(\mathbf{X})$, $i = 1, \ldots, k$, represent the observed values of partial tests and $\hat{L}_i$ is the ith marginal ESF, the latter being jointly estimated by the CMC method on the data set $\mathbf{X}$, in accordance with step 5 of Section 2.2.5 and step (S.d$_k$) above.

(C.b) The combined observed value of the second-order test is evaluated through the same CMC results as the first phase and is given by:

$$T''^o = \psi(\hat{\lambda}_1, \ldots, \hat{\lambda}_k).$$

(C.c) The bth combined value of vector statistics (step (S.d$_k$)) is then calculated by

$$T_b''^* = \psi(\hat{L}_{1b}^*, \ldots, \hat{L}_{kb}^*),$$

where $\hat{L}_{ib}^* = \hat{L}_i(T_{ib}^*|\mathcal{X}_{/\mathbf{X}})$, $i = 1, \ldots, k$, $b = 1, \ldots, B$.

Table 4.3 Representation of nonparametric combination

T_1^o	T_{11}^*	$\ldots$	T_{1b}^*	$\ldots$	T_{1B}^*
$\vdots$	$\vdots$		$\vdots$		$\vdots$
T_k^o	T_{k1}^*	$\ldots$	T_{kb}^*	$\ldots$	T_{kB}^*
			$\downarrow$		
$\hat{\lambda}_1$	$\hat{L}_{11}^*$	$\ldots$	$\hat{L}_{1b}^*$	$\ldots$	$\hat{L}_{1B}^*$
$\vdots$	$\vdots$		$\vdots$		$\vdots$
$\hat{\lambda}_k$	$\hat{L}_{k1}^*$	$\ldots$	$\hat{L}_{kb}^*$	$\ldots$	$\hat{L}_{kB}^*$
			$\downarrow$		
T''^o	$T_1''^*$	$\ldots$	$T_b''^*$	$\ldots$	$T_B''^*$

(C.d) Hence, the p-value of the combined test T'' is estimated as

$$\hat{\lambda}_\psi'' = \sum_b \mathbb{I}(T_b''^* \geq T''^o)/B.$$

(C.e) If $\hat{\lambda}_\psi'' \leq \alpha$, the global null hypothesis H_0 is rejected at significance level α.

Of course, if proper routines for exact calculations were available, then the multivariate distribution $F(\mathbf{t}|\mathcal{X}_{/\mathbf{X}})$, the partial p-values $(\lambda_1, \ldots, \lambda_k)$, the distribution of the combined test $F_\psi(t|\mathcal{X}_{/\mathbf{X}})$, and the combined p-value λ_ψ'' can be evaluated exactly. Table 4.3 displays the second phase of the NPC algorithm.

According to Theorem 1, as by assumption k is a fixed finite integer and ψ is continuous, when B tends to infinity, the combined EDF $\hat{F}_\psi(t|\mathcal{X}_{/\mathbf{X}}) = \sum_b \mathbb{I}(T_b''^* \leq t)/B$, for every real t, tends to the combined CDF $F_\psi(t|\mathcal{X}_{/\mathbf{X}}) = \Pr\{T''^* \leq t|\mathcal{X}_{/\mathbf{X}}\}$, and the combined p-value $\hat{\lambda}_\psi''$ tends to $\lambda_\psi'' = \Pr\{T''^* \geq T^o|\mathcal{X}_{/\mathbf{X}}\}$, where both convergences are with probability one. Thus, the CMC gives unbiased and consistent estimates of both the true permutation distribution $F_\psi(t|\mathcal{X}_{/\mathbf{X}})$ and the true p-value λ_ψ''. Straightforward details of these properties are left to the reader.

It may be instructive to analyse how the CMC random errors on estimates of partial p-values and on partial SLFs influence errors on combined tests. This analysis is also left to the reader.

Remark 1. It is important to note that in NPC methods, for any combining function ψ, all parameters or coefficients, as well as all other functionals which may appear in the permutation distribution $F_\psi(t|\mathcal{X}_{/\mathbf{X}})$, are generally not directly related to the similar quantities which define underlying population distributions P, because they are conditional entities which are evaluated only within the permutation sample space $\mathcal{X}_{/\mathbf{X}}$. Thus, this combination is a proper nonparametric method for multivariate testing problems (see Section 4.2.5). In fact, it takes into consideration *only* the entire joint k-variate permutation distribution $F(\mathbf{t}|\mathcal{X}_{/\mathbf{X}})$ of $\mathbf{T}$, estimated by the EDF $\hat{F}_B(\mathbf{t}|\mathcal{X}_{/\mathbf{X}})$. In particular, it is nonparametric with respect to the latent dependence relations in the population distribution P. Sometimes, in very regular situations and when proper evaluating functions are available, CMC allows us to estimate all dependence coefficients in $F(\mathbf{t}|\mathcal{X}_{/\mathbf{X}})$. This may yield the derivation of proper overall tests by standard techniques such as the quadratic form of combination (see Section 4.2.4).

Remark 2. Another feature of NPC methods is that T'' is a combination of significance levels. Thus, it is a combination of integrated permutation likelihoods (see Theorem 2, 3.1.1, and Remark 3, 3.4) whereas parametric multivariate tests are generally 'combinations of likelihood transformations'. In this sense, we may expect that, when there is a one-to-one relationship between significance levels and likelihoods, regarded as leading terms of significance level functions, most NPC tests are asymptotically equivalent with respect to corresponding likelihood-based counterparts (see Section 4.4 for a few examples).

Remark 3. The NPC methodology allows for straightforward extension to multiple testing and multiple comparisons; also straightforward is the *closed testing* approach (see Chapter 5; Marcus et al., 1976; Simes, 1986; Westfall et al., 1999). In particular, Tippett's combining function (see (c) in Section 4.2.4) can perform step-down procedures which enable computation to be speeded up. For more detailed discussion we refer to Arboretti et al. (1999, 2000b), Finos et al. (2000a, 2000b) and Basso et al. (2009b).

4.2.4 Some Useful Combining Functions

This section presents a concise review of some practical examples of combining functions (for more details on the combination of one-sided independent tests see Birnbaum, 1954; Oosterhoff, 1969; Folks, 1984). Most of these, in particular (a)–(h), are members of the admissible sub-class $\mathcal{C}_A$ (see Section 4.2.6).

(a) The Fisher *omnibus* combining function is based on the statistic

$$T_F'' = -2 \cdot \sum\nolimits_i \log(\lambda_i).$$

It is well known that if the k partial test statistics are independent and continuous, then in the null hypothesis T_F'' follows a central χ^2 distribution with $2k$ degrees of freedom. T_F'' is the most popular combining function and corresponds to the so-called *multiplicative rule*. In a permutation framework, due to Theorem 1, 2.4, the constant 2 may be omitted.

(b) The Liptak combining function is based on the statistic

$$T_L'' = \sum\nolimits_i \Phi^{-1}(1 - \lambda_i),$$

where Φ is the standard normal CDF. Of course, if the k partial tests were independent and continuous, then in the null hypothesis T_L'' would be normally distributed with mean 0 and variance k (see Liptak, 1958).
A version of the Liptak function is based on logistic transformation type of the λ_i : $T_P'' = \sum_i \log[(1 - \lambda_i)/\lambda_i]$.
More generally, if G is the CDF of a continuous variable, the generalized Liptak function is $T_G'' = \sum_i G^{-1}(1 - \lambda_i)$. Of course, within the independent case, the use of T_G'' is made easier if G is provided with the reproductive property with respect to the sum of summands.

(c) The Tippett combining function is given by

$$T_T'' = \max_{1 \leq i \leq k} (1 - \lambda_i),$$

significant for large values (the equivalent form $T_T'' = \min(\lambda_i)$ is significant for small values). Its null distribution, if the k tests are independent and continuous, behaves according to the largest (smallest) of k random values from the uniform distribution in the open interval (0,1). Tippett's T_T'' was the first combining function reported in the literature. For dependent partial tests it allows for bounds on the rejection probability according to the Bonferroni inequality. Special cases of Tippett's combining functions are the 'max t test' and the 'max chi-square'

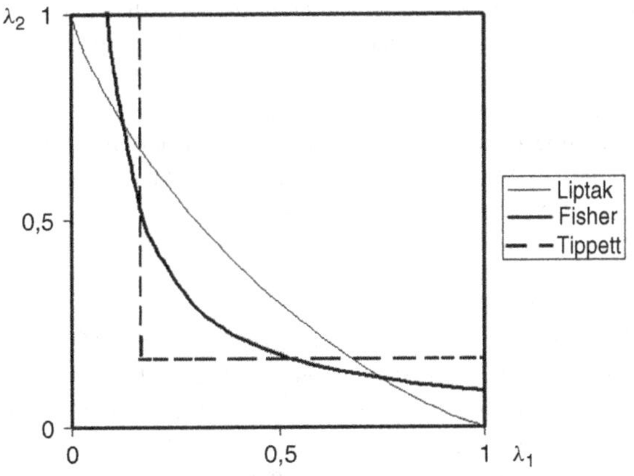

Figure 4.1 Critical regions of three combined tests

(see Chung and Fraser, 1958; Hirotsu, 1986, 1998a). Problem 7, 4.3.6, suggests using Tippett's combining function to test composite null hypotheses $H_0 : \left\{\bigcap_{1\leq i\leq k}(\delta_i \leq 0)\right\}$ against composite alternatives $H_1 : \left\{\bigcup_{1\leq i\leq k}(\delta_i > 0)\right\}$.

Figure 4.1 describes the critical regions of the Fisher, Liptak and Tippett combining functions in the very simple situation where $k = 2$ and two partial tests are independent. Note that three critical regions contain the entire lower border in the (λ_1, λ_2) representation (see Remark 1, 4.3.2). In particular, since $\lambda'' \leq \alpha$ implies that the global null hypothesis is rejected, the two points (0,1) and (1,0) of Liptak's solution lie in the rejection region, although the asymptotic probability of this event is always zero. Also note that, according to Birnbaum (1954), the three acceptance regions are convex (see Section 4.2.6).

A simple analysis of Figure 4.1 shows that Tippett's solution has a good power behaviour when one or a few, but not all, of the sub-alternatives are true; Liptak's is good when possibly all sub-alternatives are jointly true (see Remark 4, 4.4.2); Fisher's behaviour lies between the other two and so is to be preferred when no specific kind of sub-alternative is expected.

(d) The Lancaster combining solutions are based on statistics such as

$$T''_{\Gamma} = \sum\nolimits_i \Gamma^{-1}_{r,a}(1 - \lambda_i),$$

where $\Gamma_{r,a}$ represents the CDF of a central gamma distribution with known scale parameter a and r degrees of freedom. Of course, T''_{Γ} is a particular case of T''_G. If the k partial tests are independent, then the null distribution of T''_{Γ} is central gamma with scale parameter a and rk degrees of freedom. Of course, any inverse CDF transformation may be used in place of the inverse CDF of a gamma distribution. In particular, if $\Gamma^{-1}_{1,1/2}$ is the inverse CDF of a central χ^2 with 1 degree of freedom, then T''_{Γ} is distributed as a central χ^2 with k degrees of freedom. Another particular case of Lancaster's solution is connected with the following. Let us assume that all sub-alternatives H_{1i} are two-sided and related partial tests T_i are significant for either large or small values. In this setting, $\Phi^{-1}(1 - \lambda_i)$ is standard normally distributed in H_{0i}. Thus, a combining function is

$$T''_2 = \sum\nolimits_i \left[\Phi^{-1}(1 - \lambda_i)\right]^2,$$

which, if the partial tests are independent and continuous, is distributed as a central chi-square with k degrees of freedom in H_0. A special case of T_2'' is the 'cumulative chi-square' suggested by Hirotsu (1986).

(e) Again, if all sub-alternatives H_{1i} are unrestricted (two-sided), so that all partial tests T_i are significant for either large or small values, and if we are looking for power-invariant testing with respect to alternatives lying at the same quadratic distance from H_0, we may regard as a natural combining function the Mahalanobis quadratic form

$$T_Q'' = \mathbf{U}^\top \cdot (\mathbf{R}_\mathbf{U}^*)^{-1} \cdot \mathbf{U},$$

where $\mathbf{U}^\top = [\Phi^{-1}(1-\lambda_1), \ldots, \Phi^{-1}(1-\lambda_k)]^\top$ and $\mathbf{R}_\mathbf{U}^* = \{\mathbb{C}ov(U_j^*, U_i^*),\ j, i = 1, \ldots, k\}$ is the correlation matrix of the $\mathbf{U}$ transformations of permutation p-values λ^* (of course, the permutation correlation matrix $\mathbf{R}_\mathbf{U}^*$ is assumed to be positive definite; see also Theorem 6, 4.4.2). It should be emphasized that T_Q'' does not assume normality of responses or of permutation partial tests because only normality of $\mathbf{U}$ transformations is required. This property is based, at least asymptotically, on the uniform distribution of p-values λ in H_0 (see Proposition 1, 2.1.3). Therefore, the invariance testing property is related to alternatives lying at the same quadratic distance from H_0 measured on the space of inverse normal transformations of permutation p-values, that is, in terms of *normal probability distances*. It is also important to underline that $\mathbf{R}_\mathbf{U}^*$ corresponds to the *permutation covariance matrix* on $\mathbf{U}$ transformations, and that it is conditional on the data set $\mathbf{X}$. From this point of view, $\mathbb{C}ov(U_j^*, U_i^*)$ is estimated by $\sum_b U_{jb}^* \cdot U_{ib}^*/B$, because $\mathbb{C}ov(U_j^*, U_i^*) = \lim_{B\to\infty} \sum_{b\le B} U_{jb}^* \cdot U_{ib}^*/B$.

Remark 1. Most traditional multivariate tests may be viewed as particular cases of combination functions in suitable parametric frameworks, in which the dependence coefficients are either known or unconditionally estimated directly from the data set $\mathbf{X}$ within the assumed underlying family $\mathcal{P}$. In the NPC framework, conditional dependence coefficients are evaluated either from $\mathbf{X}$, when proper evaluating functions are available, or from simulation results $\{\mathbf{T}_b^*,\ b = 1, \ldots, B\}$. From this point of view we may think of NPC procedures as effective extensions of traditional multivariate testing procedures within the conditionality principle of inference. Lancaster's T_2'' in (d) above may be viewed as a particular case of quadratic combination in which dependence coefficients are omitted (see Example 2, 4.6, for a comparative discussion). Table 4.4 shows a pattern for the quadratic form of combination.

Table 4.4 Quadratic form of combination

$\hat{\lambda}_1$	$\hat{L}_{11}^*$	$\cdots$	$\hat{L}_{1b}^*$	$\cdots$	$\hat{L}_{1B}^*$
$\vdots$	$\vdots$		$\vdots$		$\vdots$
$\hat{\lambda}_k$	$\hat{L}_{k1}^*$	$\cdots$	$\hat{L}_{kb}^*$	$\cdots$	$\hat{L}_{kB}^*$

$\downarrow$

U_1^o	U_{11}^*	$\cdots$	U_{1b}^*	$\cdots$	U_{1B}^*
$\vdots$	$\vdots$		$\vdots$		$\vdots$
U_k^o	U_{k1}^*	$\cdots$	U_{kb}^*	$\cdots$	U_{kB}^*

$\downarrow$

$T_Q''^o$	$T_{Q1}''^*$	$\cdots$	$T_{Qb}''^*$	$\cdots$	$T_{QB}''^*$

In the NPC framework it appears that the number k of partial tests should not exceed the cardinality of $\mathcal{X}_{/\mathbf{X}}$ in order for the correlation matrix to be of full rank. However, in Section 4.5 we will see that this limitation does not apply since $k \in \mathbb{N}$ can diverge to infinity.

(f) Many other combining functions may be conveniently defined and used. Sometimes, confronted with specific problems, especially when the sub-hypotheses have different degrees of importance assigned to them, we need to use non-negative weights $w_i \geq 0,\ i = 1, \ldots, k$, where null weights imply discarding the corresponding partial tests from the combination process. Weighting partial tests may have important applications in quality control and clinical trials when component variables have different degrees of importance with respect to specific inferential purposes. For instance, Fisher's weighted combining function becomes

$$T'' = -\sum_i w_i \cdot \log(\lambda_i),$$

while a rather general weighted combining function is

$$T''_{\mathbf{w}} = \sum_i w_i \cdot \varphi(\lambda_i),$$

where φ is any positive, continuous, non-increasing and right-unbounded score function. It should be noted that the null distribution of the Fisher weighted combined test may be difficult to obtain outside a permutation framework, even in very simple situations. It should also be noted that, in parametric approaches, it is practically impossible to incorporate weights into testing contexts, especially if they are based on the likelihood ratio principle. Thus, from this point of view, the NPC method is much more flexible than ordinary parametric counterparts. A numerical example concerned with weighting partial tests is discussed in Section 7.11.

(g) When possible, one effective way to find combining functions in class $\mathcal{C}$ is by looking at their asymptotic behaviour under known circumstances. In particular, this approach can be adopted especially when, at least in principle, a parametric counterpart is available (see Theorems 5–, 4.4.2; see also the solution to the Behrens–Fisher problem in Example 8, 4.6). In fact, when the parametric unconditional estimation of underlying dependence coefficients is made asymptotically, for example, by consistent estimators, we may use the same parametric expression as a nonparametric combining function. In this case, for large sample sizes, the NPC is expected to behave very similarly to its parametric counterparts (see the discussion of question (a) in Section 2.5).

(h) An interesting subset of $\mathcal{C}$ is the set $\mathcal{C}_D$ of so-called *direct nonparametric combining functions*. When all partial test statistics are homogeneous, so that they share exactly the same asymptotic permutation distribution (e.g. they are all standard normal distributed, or of the chi-square type with the same degrees of freedom, and so on) and if their common asymptotic support is at least unbounded on the right (see Section 4.3.2), then we can take into consideration combining functions of the form $T''_D = \sum_i T_i$, $T''^{o}_D = \sum_i T^o_i$ and $T''^{*}_{Db} = \sum_i T^*_{ib}$, $b = 1, \ldots, B$ for the combined test, observed, and permutation values, respectively. Hence, according to (C.d) in Section 4.2.3, the combined p-value is given by $\hat{\lambda}'' = \sum_b \mathbb{I}(T''^{*}_{Db} \geq T''^{o}_D)/B$.
Observe that the permutation distributions of all partial tests can only be the same for quite large sample sizes. Thus, for finite sample sizes, this condition can be approximately satisfied because permutation distributions are essentially dependent on observed data. Therefore, quite large sample sizes are needed for effective use of the direct combination procedure, although expressing partial tests in standardized forms is often satisfactory, providing for good approximations.
In this book, this form of NPC is particularly used for combinations of statistics as in the complete two-way ANOVA, the complete analysis of 2^k replicated factorial designs, according to Hadamard's combination of statistics (see Pesarin 2001), the Behrens–Fisher problem, and

several other 'homogeneous' situations (e.g. most of the solutions in examples of Sections 2.6 and 2.7 related to the Anderson–Darling goodness-of-fit for ordered categorical variables fall within this procedure).

This kind of combination was first suggested by Chung and Fraser (1958) in connection with solving multivariate testing problems when the number of observed variables is larger than the number of units. Peto and Peto (1972) also use it without homoscedastic variables. Blair et al. (1994) study a direct combination solution and compare it with Hotelling's T^2. Edgington and Onghena (2007) also mention a sort of direct combination of partial tests for some multivariate testing problems. However, in these references, the discussion and analysis appear to be essentially heuristic and without the development of a proper theory.

Remark 2. The direct combination function allows us to avoid the quite intensive calculations involved in steps (S.b$_k$) and (C.a)–(C.c) of Section 4.2.3. In addition, although the direct combination function appears to operate as in univariate cases, it is essentially an NPC because: the testing problem is equivalently broken down into k sub-problems; one overall test statistic, $T : \mathcal{R}^k \rightarrow \mathcal{R}^1$, is not directly available; and the dependence relations among partial tests are implicitly 'captured' by the combining procedure. A typical example occurs in repeated measurements when for instance homoscedastic data are $\mathbf{X} = \{X_{hji}, i = 1, \ldots, n_j, j = 1, 2, h = 1, \ldots, k\}$, the hypotheses are $H_0 : \{X_{h1} \stackrel{d}{=} X_{h2}, h = 1, \ldots, k\}$ against for instance the stochastic dominance alternative $H_1 : \left\{\bigcup_h (X_{h1} \stackrel{d}{>} X_{h2})\right\}$, the k partial tests are $T_h^* = \sum_i X_{h1i}^*/n_1$, $h = 1, \ldots, k$, and the direct combination $T_D'' = \sum_h T_h^*$. In such a case, since $T_D'' = \sum_i \left[\sum_h X_{h1i}^*\right]/n_1$, the derived quantity $Y_{1i} = \sum_h X_{h1i}^*$ may be considered similar to the so-called area under the curve for the ith individual. It is exactly equivalent to the AUC when equal spacings are between time observations.

(i) As mentioned in Section 4.1, we sometimes encounter two kinds of simple multivariate problems. The first assumes that the dependence coefficients among the k partial tests are either known or unconditionally estimated from the data set $\mathbf{X}$. For instance, in Hotelling's two-sample statistic, as well as in many other relatively simple problems in which data are assumed normally distributed, we have the form $T_H^2 = \mathbf{T}^\top \cdot \mathbf{\Sigma}^{-1} \cdot \mathbf{T}$, where the partial tests are $T_i = (\bar{X}_{1i} - \bar{X}_{2i})$, $i = 1, \ldots, V$, and the positive definite covariance matrix $\mathbf{\Sigma}$ is directly estimated from $\mathbf{X}$. Therefore, for each permutation $\mathbf{X}^*$ we may use test statistics such as $T_H^* = \mathbf{T}^{*\top} \cdot \mathbf{\Sigma}^{-1} \cdot \mathbf{T}^*$ or $T_H^{**} = \mathbf{T}^{*\top} \cdot (\mathbf{\Sigma}^*)^{-1} \cdot \mathbf{T}^*$, where of course $\mathbf{\Sigma}^*$ is the covariance matrix evaluated on each permuted data set $\mathbf{X}^*$. Consequently, we may act in accordance with univariate situations. These simple multivariate situations may be referred to as *pseudo-parametric combination* problems. The second kind of relatively simple problem considers one derived univariate transformation of data for each unit, such as $Y_i = \varphi(X_{1i}, \ldots, X_{Vi})$, $i = 1, \ldots, n$. In general, this kind of transformation is used when it has a precise physical interpretation with respect to the specific inferential problem at hand. Thus the testing problem, which should be expressed by statements related to the derived variable Y, simply becomes univariate in its own right. Therefore, its solution becomes properly univariate. This kind of simple multivariate problem may be called *combination by derived variable*. Examples of combination by derived variable for quite special problems are given in Reboussin and DeMets (1996), Hoh and Ott (2000) and Mielke and Berry (2007). Of course, if several different aspects useful for physical interpretation are to be used, so that k different derived variables $\{Y_{hi} = \varphi_h(X_{1i}, \ldots, X_{Vi}),\ i = 1, \ldots, n,\ h = 1, \ldots, k\}$ are used for the analysis, then the problem remains multivariate and multi-aspect as in Example 3, 4.6.

Remark 3. When, in place of normality of responses, we assume that the permutation distribution of standardized partial tests $\mathbf{T}$, at least approximately, is k-variate normal with positive definite correlation matrix $\mathbf{R}^*$, then we may consider a test expression of the form

$$T_H'' = \mathbf{T}^\top \cdot (\mathbf{R}^*)^{-1} \cdot \mathbf{T},$$

Table 4.5 The iterated combination algorithm

ψ_1^o	ψ_{11}^*	$\dots$	ψ_{1b}^*	$\dots$	ψ_{1B}^*
$\vdots$	$\vdots$		$\vdots$		$\vdots$
ψ_s^o	ψ_{s1}^*	$\dots$	ψ_{sb}^*	$\dots$	ψ_{sB}^*

$\downarrow$

$\hat{\lambda}_1''$	$\hat{L}_{11}''^*$	$\dots$	$\hat{L}_{1b}''^*$	$\dots$	$\hat{L}_{1B}''^*$
$\vdots$	$\vdots$		$\vdots$		$\vdots$
$\hat{\lambda}_s''$	$\hat{L}_{s1}''^*$	$\dots$	$\hat{L}_{sb}''^*$	$\dots$	$\hat{L}_{sB}''^*$

$\downarrow$

$T_l'''^o$	$T_{l1}'''^*$	$\dots$	$T_{lb}'''^*$	$\dots$	$T_{lB}'''^*$

where the jith member of $\mathbf{R}^*$ is

$$\rho_{ji}^* = \mathbb{C}ov(T_j^*, T_i^*) = \sum_{b \le B} (T_{jb}^* - \tau_j^*)(T_{ib}^* - \tau_i^*)/B,$$

and where $\tau_j^* = \sum_{b \le B} T_{jb}^*/B$, $j, i = 1, \dots, k$, are the permutation means of partial tests. Note that, in H_1, τ_j^* are non-null functions of treatment effects, whereas under the multivariate additive model (extension of (M.i), 1.10.1) ρ_{ji}^* are not.

Remark 4. For any given data set $\mathbf{X}$, different combining functions due to different rejection regions may of course give slightly different overall p-values, although, due to their consistency (see Section 4.3.1), they are asymptotically equivalent in the alternative. However, in order to reduce this influence, we may *iterate* the combination procedure by applying more than one combining function $\psi_1, \dots, \psi_s$, $2 \le s$, to the same partial tests, and then combine the resulting second-order p-values $(\lambda_1'', \dots, \lambda_s'')$ into a third order of combination by means of one combining function, $\psi_l(\lambda_1'', \dots, \lambda_s'')$, say. From a series of Monte Carlo studies, provided that the second-order combination functions have different rejection regions, we obtained that the third-order p-values λ_l''' are almost invariant with respect to the choice of ψ_l within the class $\mathcal{C}$. Of course, this procedure may be iterated into a fourth order, and so on (examples are given in subsequent chapters). Table 4.5 illustrates the iterated combination algorithm.

Remark 5. The selection of one combining function ψ in $\mathcal{C}$ may appear to be a rather arbitrary act. However, the following practical guidelines may be useful:

(1) When our knowledge of sub-alternatives is such that we may argue asymptotically, we may use 'asymptotic optimal combinations' in the sense of Section 4.4, or at least in the sense of local optimality (Wei and Johnson, 1985).
(2) When we expect only one or a few, but not all, sub-alternatives to occur, we suggest using Tippett's combining function.
(3) When we expect all sub-alternatives to be jointly true, use of Liptak's or the direct combinations is generally justified; sometimes they are also asymptotically optimal (see Theorem 5, 4.4.2).
(4) When any such specific knowledge is available, we suggest using Fisher's combining function, because its behaviour is generally intermediate between those of Tippett and Liptak (see comments on Figure 4.1).

(5) When our preference is for a neutral form of combination, in accordance with Remark 4 above, we suggest iterating the combining procedure with different functions, until the final p-value becomes reasonably ψ-invariant.

4.2.5 *Why Combination is Nonparametric*

To justify the use of the adjective *nonparametric* for the combination procedures, especially with regard to underlying dependence coefficients, let us first consider the combination of two tests T_1 and T_2, with Liptak's rule. More specifically, let us suppose that sample sizes are sufficiently large, the underlying population distributions are sufficiently regular, and the whole permutation space $\mathcal{X}_{/\mathbf{X}}$ is examined so that in practice both supports $\mathcal{T}_1$ and $\mathcal{T}_2$ are almost continuous and the transformation $(U_1 = \Phi^{-1}(1 - \lambda_1),\ U_2 = \Phi^{-1}(1 - \lambda_2))$ in the alternative is well approximated by the bivariate normal distribution $(U_1, U_2) \sim \mathcal{N}_2[(\delta_1, \delta_2); \mathbf{R}]$, where δ_1 and δ_2 are the fixed shift effects on the U transformations, and $\mathbf{R} = \begin{bmatrix} 1 & \rho \\ \rho & 1 \end{bmatrix}$ is the correlation matrix. Note that since p-values (λ_1, λ_2) are positively dependent, the correlation coefficient $\rho = \rho(U_1, U_2)$ in $\mathbf{R}$ is non-negative. Suppose also that there are two statisticians: the first is supposed to know that the permutation distribution is bivariate normal $\mathcal{N}_2[(\delta_1, \delta_2); \mathbf{R}]$, the second is without this knowledge. In such conditions, the first decides to take his/her knowledge into account, including the dependence coefficient, and uses the combined statistic $T''_{\mathcal{N}} = (U_1 + U_2)/\sqrt{2(1+\rho)}$, whose reference distribution is $\mathcal{N}_1[(\delta_1 + \delta_2)/\sqrt{2(1+\rho)}, 1]$, which in the null hypothesis becomes $\mathcal{N}_1(0, 1)$. In this way he/she knows that the related p-value is $\lambda_{\mathcal{N}} = \Pr\{T''^{*}_{\mathcal{N}} \geq T''^{o}_{\mathcal{N}} | \mathcal{X}_{/\mathbf{X}}\} = 1 - \Phi(T''^{o}_{\mathcal{N}})$. The second decides to use Liptak's combination without explicitly taking the dependence coefficient into account, the value of which he/she ignores, and simply uses $T''_L = U_1 + U_2$. Of course, he/she is unaware that the distribution of T''_L is $\mathcal{N}_1[\delta_1 + \delta_2, 2(1+\rho)]$, which in H_0 becomes $\mathcal{N}_1[0, 2(1+\rho)]$, and so refers to the p-value $\lambda_L = \sum_{\mathcal{X}_{/\mathbf{X}}} \mathbb{I}[T''^{*}_L \geq T''^{o}_L]/M^{(n)}$ he/she knows numerically. However, it is worth noting that both come to exactly the same inferential conclusion, since $T''_{\mathcal{N}}$ and T''_L being one-to-one related, $T''_{\mathcal{N}} = T''_L/\sqrt{2(1+\rho)}$ say, are permutationally equivalent and so $\lambda_{\mathcal{N}} = \lambda_L = 1 - \Phi(T''^{o}_L/\sqrt{2(1+\rho)})$.

On the one hand, it is worth noting that in the NPC the correlation coefficient is taken into consideration implicitly by obtaining exactly the same distribution as we would have if the true correlation coefficient were used explicitly. On the other, the standardized non-centrality of T''_L, $(\delta_1 + \delta_2)/\sqrt{2(1+\rho)}$ say, shows that its power is maximal when $\rho = 0$, corresponding to independent partial tests, and is minimal when $\rho = 1$, that is when two partial tests are linearly related with probability one. Extending these notions to the k-dimensional case, we see that if the vector of transformed statistics $\mathbf{U} = (U_1, \ldots, U_k)$ is (approximately) k-variate normally distributed, $\mathbf{U} \sim \mathcal{N}_k(\boldsymbol{\beta}, \mathbf{R})$ say, then (approximately) $T''_L \sim \mathcal{N}_1(\eta, \sigma^2)$, where $\eta = \sum_i \eta_i$ and $\sigma^2 = \sum_{ji} \sigma_{ji}$. Thus, all dependence coefficients enter the distribution of T''_L without being explicitly evaluated and processed.

Accordingly, we may observe that when multivariate normality cannot be assumed, the combining procedure provides for the exact permutation distribution of T''_L, and in general of T''_{ψ}, $\psi \in \mathcal{C}$, which contains all the unknown dependence coefficients – coefficients which in general are specific to the actual data set $\mathbf{X}$ (so that they may vary as $\mathbf{X}$ varies in $\mathcal{X}^n$). Therefore, on the one hand, their number is indefinite, since they may be related to twofold, threefold, fourfold, ... partial tests; on the other, their associated regression forms are also indefinite (linear, quadratic, exponential, general monotonic, ...). Therefore, the nonparametric combining strategy is sufficiently general to cover almost all real situations of practical interest, without the necessity of properly defining a distribution model for the set of partial tests, especially with respect to their dependence structure. Thus, it is to be emphasized that the adjective *nonparametric* presumes that the number, and related values, of underlying dependence coefficients could be indefinite (see Section 1.2.1), while the reference null permutation distribution induced by combined statistic ψ is, in any case, the true one.

4.2.6 On Admissible Combining Functions

Birnbaum (1954, 1955) showed that in order for combined tests to be admissible, that is, if there is no other test which is at least as powerful as it is against some alternatives in H_1 and more powerful against all other alternatives, their rejection regions must be convex. Here we would like to provide a sufficient condition so that the combining function ψ, expressed in terms of p-values, satisfies such a requirement. Let us consider combining functions $T''_G = \sum_{i=1}^{k} G^{-1}(1-\lambda_i)$, where G^{-1} is the inverse CDF of a continuous random variable (see Problem 2, 4.2.7). This defines a family of combining functions which, among many others, contains Fisher's T''_F, Liptak's T''_L, and Lancaster's T''_Γ. Let us now find properties of G so that the related rejection region is convex for any α, and so T''_G is admissible, that is, $T''_G \in \mathcal{C}_A$ (see Section 4.2.2).

To this end, let us consider the case of two partial tests, that is, $T''_G = G^{-1}(x) + G^{-1}(y)$, where $x = 1 - \lambda_1$, and $y = 1 - \lambda_2$. Suppose also that, since we are arguing independently of any specific finite case, p-values λ_1 and λ_2 are at least approximately continuous and that $T''_G(x, y)$ is differentiable in x and y. Let us fix a value for $\alpha \in (0, 1)$ and call $T''_{G\alpha}$ the critical value of T''_G given $\mathbf{X}$. Consider the set of points $\{(x, y)_{G\alpha}\}$ such that $G^{-1}(x) + G^{-1}(y) = T''_{G\alpha}$, that is, the set of points which lie in the critical curve separating rejection from acceptance regions. This curve is implicitly defined as $y(x) = G[T''_{G\alpha} - G^{-1}(x)]$. A useful characterization for the convexity of $y(x)$ in the unit interval is that its derivative with respect to x is monotonically non-decreasing in that interval. Actually, we have $\frac{dy(x)}{dx} = -g[T''_{G\alpha} - G^{-1}(x)]/g[G^{-1}(x)]$, where g is the density associated with G with respect to the real line. Thus, all CDFs G such that $\frac{dy(x)}{dx} \le \frac{dy(x')}{dx'}$, with $x' > x$ and $y(x') = G[T''_{G\alpha} - G^{-1}(x')]$, give rise to convex rejection regions. This condition seems to exclude all G whose density g is not unimodal; however, normal (Liptak), exponential (Fisher), gamma (Lancaster), etc. do satisfy it. Convexity of the Tippett combining function can easily be proved directly (see Problem 20, 4.2.7). This proof and that for multivariate extension (see Problem 23, 4.2.7) are left to the reader as exercises. It is worth noting that the additive rule (see Section 4.3.3), although provided with a convex rejection region, is not a consistent test. This suggests that admissibility without unbiasedness and/or consistency is not a very useful property for a test.

4.2.7 Problems and Exercises

1) Find transformations $\varphi_1(\lambda)$ and $\varphi_2(\lambda)$ of p-values λ for restricted and unrestricted alternatives respectively, in such a way that, when partial tests are independent, for large sample sizes the two combined tests $T''_1 = \sum_i \varphi_1(\lambda_i)$ and $T''_2 = \sum_i \varphi_2(\lambda_i)$ are both distributed in H_0 as a central χ^2 with k degrees of freedom.

2) Prove that if G is the CDF associated with any given continuous univariate random variable, unbounded on the right, then $T''_G = \sum_i G^{-1}(1-\lambda_i)$ is a combined test belonging to class $\mathcal{C}$ and may be useful for NPC.

3) With reference to Problem 2 above, discuss what happens if G is the CDF of a discrete random variable.

4) Prove that the direct combination procedure, as described in (h) in Section 4.2.4, is in accordance with the general NPC, i.e. it is a member of $\mathcal{C}$.

5) With reference to (h) in Section 4.2.4, prove that $T''_\varphi = \sum_i \varphi(\lambda_i)$ is a direct combining procedure if the p-value transformations are such that $\varphi = L_T^{-1}$, where L_T is the common permutation SLF of all partial tests (note that, at least asymptotically, $T_i = L_T^{-1} \circ L_i(T_i)$, because L_T is one-to-one and L_i converges to L_T, by assumption).

6) Discuss the direct combination procedure when sample sizes are small. Prove that the resulting combination implies a sort of implicit weighted combination, where weights are data-dependent quantities.

7) Write an algorithm for the direct combination procedure.

8) Write an algorithm for the two kinds of simple combination: pseudo-parametric and derived variable, as outlined in (i) in Section 4.2.4.

9) Prove that the function $\tilde{T} = \sum_i \exp\{1 - \lambda_i\}$ is not a member of class $\mathcal{C}$; as a consequence, it is not a suitable combining function (see also Section 4.3.3).

10) Consider a standard two-sample bivariate testing problem, where the null hypothesis is $H_0 : \{(\mu_{11} = \mu_{12}) \bigcap (\mu_{21} = \mu_{22})\}$ against restricted alternatives of the form $H_1 : \{(\mu_{11} > \mu_{12}) \bigcup (\mu_{21} > \mu_{22})\}$, response variables are continuous and homoscedastic, and $\mu_{hj} = \mathbb{E}(X_{hj})$ is the finite expectation of the hth component variable in the jth group. Prove that:

(a) tests $T_h = \sum_i X_{h1i}$, $h = 1, 2$, are permutationally equivalent to $\bar{T}_h = (\bar{X}_{h1} - \bar{X}_{h2})/\{\sum_{ji} (X_{hji} - \bar{X}_{hj})^2\}^{1/2}$;

(b) $T'' = \bar{T}_1 + \bar{T}_2$ is a nonparametric combined test;

(c) $T' = T_1 + T_2$ is not in general a nonparametric combined test (give a counterexample).

11) Prove that, when B goes to infinity and if the SLF of (S.d$_k$) in Section 4.2.3 is defined by

$$\hat{L}_i^*(z, \varepsilon) = \frac{[\varepsilon + \sum_r \mathbb{I}(T_{ir}^* \geq z)]}{B + 2\varepsilon}, \quad i = 1, \ldots, k,\ 0 < \varepsilon < 1,$$

then all associated inferential conclusions are permutationally invariant with respect to ε, provided that the NPC function assumes the form $T'' = \sum_i \varphi[\hat{L}_i^*(T^o, \varepsilon)]$, where φ is a continuous function, unbounded on the right and strictly decreasing.

12) Show that if, for instance, the hypotheses are $H_0 : \left\{\bigcap_{1\leq i\leq k} (\mu_i \leq \mu_{0i})\right\}$ against $H_1 : \left\{\bigcup_{1\leq i\leq k} (\mu_i > \mu_{0i})\right\}$, i.e. if the multivariate hypotheses are broken down into a set of composite sub-hypotheses, the Tippett combining function gives correct solutions, provided that all partial tests are marginally unbiased and consistent.

13) Prove that if all partial tests are permutationally exact (see Remark 1, 2.2.4 and Proposition 2, 3.1.1), then their NPC is exact for whatever $\psi \in \mathcal{C}$.

14) Prove that if φ is a monotonically decreasing score function bounded on the right, so that $\lim \varphi(\lambda) = \bar{\varphi} < \infty$, as $\lambda \to 0$, then $T''_\varphi = \sum_i \varphi(\lambda_i)$ is not a member of $\mathcal{C}$ (see property (P.2), Section 4.2.2); thus T''_φ cannot be adopted as a combining function.

15) With reference to Problem 14 above, prove that if φ is a monotonically decreasing score function which is unbounded on the right, so that $\lim \varphi(\lambda) = \infty$, as $\lambda \to 0$, then $T''_\varphi = \sum_i \varphi(\lambda_i)$ is a member of $\mathcal{C}$ (see property (P.2), Section 4.2.2); thus T''_φ may be adopted as a combining function.

16) With reference to point (e) in Section 4.2.4, show that $T''_S = T_1^2 + T_2^2$ and T''_Q are not permutationally equivalent (although for any given data set $\mathbf{X}$, ρ^* is a permutationally invariant quantity, and there is no one-to-one relation between T''_S and T''_Q).

17) With reference to point (e) in Section 4.2.4, show that $T''_S = T_1^2 + T_2^2$ and T''_Q become asymptotically coincident when $\lim |\rho| = 1$.

18) Prove that $\mathcal{C}$ contains all continuous and strictly increasing transformations of its members. That is, if $\eta : \mathcal{R}^1 \to \mathcal{R}^1$ is one such transformation, then the combining functions ψ and $\eta \circ \psi$ are permutationally equivalent (see Section 2.4 for permutation equivalence of test statistics).

19) Find at least one member of the class $\mathcal{C}$ which is not a symmetric combining function (e.g. try with weights depending on observed p-values).

20) Prove that Tippett's combining function is admissible, i.e. its rejection region is convex.

21) Prove that Liptak's combining function is admissible, i.e. its rejection region is convex.

22) Prove that Fisher's combining function is admissible, i.e. its rejection region is convex.

23) With reference to Section 4.2.6 on admissible combining functions, extend the convexity property of rejection regions to more than two partial tests.

24) Discuss conditions such that Liptak's and Direct combining functions are asymptotically equivalent.

25) Prove that the direct combining functions are not members of class $\mathcal{C}$, unless the asymptotic common support of partial tests is unbounded on the right (see Problems 14 and 15 above).

4.3 Consistency, Unbiasedness and Power of Combined Tests

4.3.1 *Consistency*

This section examines aspects of the basic behaviour of NPC tests. We suppose here that assumptions (A.1) and (A.2) of Section 4.2.1 and properties (P.1)–(P.3) of Section 4.2.2 are satisfied. Theorems and other statements from Chapters 2 and 3 are also generally implicitly assumed.

In order to examine the unbiasedness and consistency of NPC tests T'', let us also assume that:

(a) when n goes to infinity, then so also do sample sizes of all groups, that is, $n \to \infty$ implies $\min_j(n_j) \to \infty$;
(b) the number B of CMC iterations goes to infinity;
(c) k and α are fixed.

Note that as the CMC method produces strongly consistent estimates, the role of $B \to \infty$ is identical to considering true exact permutation quantities in place of their estimates. Moreover, in dealing with consistency, we may assume that the cardinality $M^{(n)}$ of $\mathcal{X}_{/\mathbf{X}}$ and number B of CMC iterations are so large that all approximation errors arising from substitution of discrete distributions with their continuous images are sufficiently small as to be negligible.

Theorem 2. *If partial permutation tests T_i, $i = 1, \ldots, k$, are marginally unbiased and at least one is strongly consistent for respectively H_{0i} against H_{1i}, then $T'' = \psi(\lambda_1, \ldots, \lambda_k)$, $\forall \psi \in \mathcal{C}$, is a strongly consistent combined test for $H_0 : \left\{\bigcap_i H_{0i}\right\}$ against $H_1 : \left\{\bigcup_i H_{1i}\right\}$.*

Proof. To be strongly consistent, a combined test must reach its critical region with probability one if at least one sub-alternative H_{1i}, $i = 1, \ldots, k$, is true. Suppose that (see assumption (A.2) of Section 4.2.1) H_{1i} is true; then $\lambda_i \to 0$ with probability one as $n \to \infty$. Thus, by Theorem 1, 4.2.3, and properties (P.2) and (P.3), 4.2.2, $T'' \to \bar{\psi} > T''_\alpha$ with probability one, $\forall \alpha > 0$. Marginal unbiasedness is required to prevent from compensations as it occurs for instance in two-test problems with $\lambda_1 = 1 - \lambda_2$ (only one is unbiased, as in Example 4, 4.6) and $T''_G = \sum_i G^{-1}(1 - \lambda_i)$ with G the CDF of a symmetric distribution, where it is $T''_G \overset{d}{=} 0$.

For the weak consistency property of T'', see Problem 6, 4.3.6.

4.3.2 *Unbiasedness*

In order to achieve the unbiasedness of combined tests T''_ψ, $\forall \psi \in \mathcal{C}$, first let us consider the following lemma.

Lemma 1. *Let Y and W be two random variables defined on the same univariate probability space. If Y is stochastically larger than W, so that their CDFs satisfy $F_Y(t) \leq F_W(t)$, $\forall t \in \mathcal{R}^1$, and if φ is a non-decreasing measurable real function, then $\varphi(Y)$ is stochastically larger than $\varphi(W)$.*

The proof of this lemma is based on the following straightforward relationships:

$$\Pr\{\varphi(Y) \leq \varphi(t)\} = F_Y(t) \leq F_W(t) = \Pr\{\varphi(W) \leq \varphi(t)\}, \ \forall t \in \mathcal{R}^1.$$

Note that the measurability of φ is relevant in order for the above probability statements to be well defined.

Theorem 3. *If, given a data set* $\mathbf{X}$ *and any* $\alpha > 0$, *partial permutation tests* $\mathbf{T} = \{T_i,\ i = 1, \ldots, k\}$ *are all marginally unbiased for respectively* H_{0i} *against* $H_{1i},\ i = 1, \ldots, k$, *so that their associated p-values* $\lambda_i,\ i = 1, \ldots, k$, *are positively dependent, then* $T''_\psi = \psi(\lambda_1, \ldots, \lambda_k),\ \forall\psi \in C$, *is an unbiased combined test for* $H_0 : \left\{\bigcap_i H_{0i}\right\}$ *against* $H_1 : \left\{\bigcup_i H_{1i}\right\}$.

Proof. The marginal unbiasedness and positive dependence properties of partial tests T_i imply, with obvious notation, that $\Pr\{\lambda_i \leq z|\mathcal{X}_{/\mathbf{X}(0)}\} \leq \Pr\{\lambda_i \leq z|\mathcal{X}_{/\mathbf{X}(\Delta)}\},\ \forall z \in (0, 1),\ i = 1, \ldots, k$, because p-values $\lambda_i(\mathbf{X}(\Delta))$ are stochastically smaller than $\lambda_i(\mathbf{X}(0))$ (see Sections 3.1.1 and 3.2.5). Thus, by the non-decreasing property of combination function ψ (see (P.1) in Section 4.2.2) and Lemma 1 above, $\psi(\ldots, \lambda_i(\mathbf{X}(\Delta)), \ldots)$ is stochastically larger than $\psi(\ldots, \lambda_i(\mathbf{X}(0)), \ldots)$. Hence, by iterating from $i = 1$ to k, the unbiasedness of T''_ψ is achieved.

Remark 1. The marginal unbiasedness of partial tests, as stated in assumption (A.1) in Section 4.2.1, is only a sufficient condition for the unbiasedness of combined tests. In order to see this let us consider a counterexample. Suppose that, as is quite familiar in univariate one-way ANOVA, we decompose the overall hypotheses into the set of pairwise sub-hypotheses $\left\{\bigcap_{j>s}(X_j \overset{d}{=} X_s)\right\} = \{\bigcap_{j>s} H_{0js}\}$ against $\left\{\bigcup_{j>s}(X_j \overset{d}{\neq} X_s)\right\}$. To test these global hypotheses, let us consider the natural partial pairwise statistics $T^*_{js} = n_j(\bar{X}^*_j)^2 + n_s(\bar{X}^*_s)^2,\quad j > s = 1, \ldots, C-1$, followed by the direct combination $T''_D = \sum_{j>s} T^*_{js} = (C-1)\sum_{j\geq 1} n_j(\bar{X}^*_j)^2$. We note that T''_D is permutationally equivalent to the overall test discussed in Section 1.11 and Example 6, 2.7. However, as partial statistics T^*_{js} are jointly processed, so that overall permutations are considered, they are not marginal tests for separately testing H_{0js} against H_{1js}, because in $\mathcal{X}_{/\mathbf{X}}$ there are points $\mathbf{X}^*$ in which the coordinates $(\mathbf{X}^*_j \biguplus \mathbf{X}^*_s)$ related to the jth and sth groups have no data from $(\mathbf{X}_j \biguplus \mathbf{X}_s)$. Thus, T^*_{js} do not satisfy the requirements of marginal testing. In particular, they are not marginally unbiased (see Chapter 5 on multiple testing). Finding necessary and sufficient conditions seems to be rather more difficult. However, this search is beyond the scope of this book and is left to the reader.

Remark 2. An important consequence of Theorem 3 is that, if all partial tests are marginally unbiased for all $\alpha > 0$, so that they are ordered with respect to treatment effects (see Theorem 2, 3.1.1), then all nonparametric combined tests are also stochastically ordered with respect to treatment effects. To this end, without loss of generality, let us suppose that partial effects $0 \leq \Delta_i,\ i = 1, \ldots, k$, are non-negative and that $\{\Delta_i \overset{p}{\leq} \Delta'_i,\ i = 1, \ldots, k\}$ are two sets of stochastically ordered effects, where at least one inequality is strict. Thus, since all partial tests satisfy the relations $\lambda_i(0) \overset{d}{\geq} \lambda_i(\Delta_i) \overset{d}{\geq} \lambda_i(\Delta'_i),\ i = 1, \ldots, k$, due to the non-decreasing property of ψ, p-values of combined tests satisfy $\lambda''_\psi(0) \overset{d}{\geq} \lambda''_\psi(\Delta) \overset{d}{\geq} \lambda''_\psi(\Delta'),\ \forall\psi \in C$ (see Problem 15, 4.3.6).

Remark 3. The ordering property of nonparametric combined tests implies that the power function is monotonic non-decreasing with respect to increasing treatment effects (see Problem 16, 4.3.6).

Remark 4. To see another nice consequence of Theorem 3 and the remarks above, let us suppose the inclusion of a further informative partial test in the NPC analysis. With obvious notation, assuming its effect is $\Delta_{k+1} > 0$, a further partial test T_{k+1} is informative if the associated p-value is such that $\lambda''_{k+1}(\Delta_1, \ldots, \Delta_k; \Delta_{k+1}) \overset{d}{\leq} \lambda''_{k+1}(\Delta_1, \ldots, \Delta_k; 0)$. The problem is to find conditions so that the inclusion of T_{k+1} improves the power of the new combined test $\psi_{k+1} = \psi(\lambda_1, \ldots, \lambda_k; \lambda_{k+1})$ with respect to $\Psi_k = \psi(\lambda_1, \ldots, \lambda_k)$. We can easily obtain a solution to this within the Liptak combination and with large data sets. To this end, $T''_{L,k} = \sum_i^k \Phi^{-1}(1-\lambda_i) \sim N_1\left(\sum_i^k \Delta_i, \sum_{ji}^k \rho_{ji}\right)$, with $\rho_{ii} = 1$, and $T''_{L,k+1} = \sum_i^{k+1} \Phi^{-1}(1-\lambda_i) \sim N_1\left(\sum_i^{k+1} \Delta_i, \sum_{ji}^{k+1} \rho_{ji}\right)$. In this setting, the power of $T''_{L,k+1}$ is larger than that of $T''_{L,k}$ if $\sum_i^k \Delta_i/\left(\sum_{ji}^k \rho_{ji}\right)^{1/2} \overset{d}{\leq} \sum_i^{k+1} \Delta_i/\left(\sum_{ji}^{k+1} \rho_{ji}\right)^{1/2}$. For instance, with $k = 1$ and independent partial tests, this condition implies that if $\Delta_2 \overset{d}{\geq} \Delta_1(\sqrt{2}-1)$ then there is power improvement. Further details are left to the reader (see Problem 17, 4.3.6).

4.3.3 A Non-consistent Combining Function

It should be noted that all combining functions based on convex linear functions of bounded scores are not consistent. Indeed, they violate property (P.2) of Section 4.2.2, in the sense that, as sample size goes to infinity, the probability of attaining their critical region is smaller than unity when, for instance, only a subset of $k' < k$ of the sub-alternatives is true.

This means that combined tests such as $\sum_i(1-\lambda_i)$ or $\sum_i w_i \cdot \phi(\lambda_i)$, where weights w_i are finite positive quantities and $\phi(\lambda_i)$ are positive *bounded scores*, are not consistent (remember that $w_i = 0$ is equivalent to removing T_i from the combination process). Hence, they are not members of class $\mathcal{C}$ (see Problems 14 and 15, 4.2.7).

As a relevant example, let us consider the very simple situation related to the so-called *additive combining function*, where: (i) $k = 2$; (ii) tests T_1 and T_2 are independent, asymptotically continuous, and at least weakly consistent (independence is assumed here only in order to facilitate calculations); (iii) the combining function is the *additive rule*, $T'' = 2 - (\lambda_1 + \lambda_2)$; (iv) H_{01} and H_{12} are true; (v) B is very large, so that discrete distributions are replaced by their continuous images.

In this case, as n tends to infinity, $\lambda_2 \to 0$ at least in probability because H_{12} is true; whereas λ_1 remains uniformly distributed in the open interval $(0, 1)$ because H_{01} is true (see Proposition 5, 2.2.4). Noting that the critical region of T'' has the form represented by the shaded triangle in the (λ_1, λ_2) representation of Figure 4.2, we see that $\lim_{n\to\infty} \Pr\{T'' \geq T''_\alpha|(H_{01}, H_{12})\} = +\sqrt{2\alpha} < 1$, $0 < \alpha < 1/2$. Thus, the so-called *additive method of combining probability values* – the sum of partial p-values for the combination of independent partial tests – is not satisfactory. The example, which plays the role of counterexample, shows that this combining rule gives non-consistent tests (see also Goutis et al., 1996). Although the resulting test is exact, unbiased, with a monotonically non-decreasing power function (see Problem 12, 4.3.6), and with convex rejection region, its use is hard to justify.

Remark 1. In general, all combining functions, the critical regions of which in the $(\lambda_1, \ldots, \lambda_k)$ representation do not contain the entire lower border, are not consistent (see Problem 2, 4.3.6), so that they are not suitable for NPC. It is also worth noting that the additive combining function leads to a convex rejection region, but since it violates condition (P.2) it cannot be claimed to be a member of class $\mathcal{C}_A$.

4.3.4 Power of Combined Tests

Conditional Power Function

Let us define and give an algorithm for the empirical post-hoc conditional power function first. To this end let us argue for a two-sample design with two continuous variables and fixed effects; extensions to more complex cases are straightforward and left to readers as exercises.

With obvious notation, assume that the data set is $\mathbf{X}(\boldsymbol{\delta}) = \{(Z_{11i} + \delta_1, Z_{21i} + \delta_2), i = 1, \ldots, n_1; (Z_{12i}, Z_{22i}), i = 1, \ldots, n_2\} = (\mathbf{Z}_1 + \boldsymbol{\delta}, \mathbf{Z}_2)$, where errors $\mathbf{Z}$ are such that $\mathbb{E}(\mathbf{Z}) = \mathbf{0}$. Consider the testing for $H_0 : \{(\delta_1 = 0) \bigcap (\delta_2 = 0)\}$, against $H_1 : \{(\delta_1 > 0) \bigcap (\delta_2 > 0)\}$ by means of two partial tests $T_h^* = \bar{X}_{h1}^* - \bar{X}_{h2}^*$, $h = 1, 2$, and the combining function $\psi \in \mathcal{C}$. With reference to the univariate case in Section 3.2.1 and discussion therein, since the bivariate effect $\boldsymbol{\delta}$ cannot be separated from the corresponding errors $\mathbf{Z}$, the following algorithm evaluates the empirical post-hoc conditional power $W[(\boldsymbol{\delta}, \hat{\boldsymbol{\delta}}, \alpha, \mathbf{T}, \psi)|\mathcal{X}_{/\mathbf{X}}]$:

1. Based on a suitable bivariate indicator $\mathbf{T}$, consider the estimate $\hat{\boldsymbol{\delta}} = (T_1^o, T_2^o)$ of $\boldsymbol{\delta}$ from the pooled data set $\mathbf{X}(\boldsymbol{\delta})$ and the consequent empirical deviates $\hat{\mathbf{Z}} = (\mathbf{X}_1 - \hat{\boldsymbol{\delta}}, \mathbf{X}_2)$. In accordance with point (ii) in Remark 2, 2.1.2, empirical deviates $\hat{\mathbf{Z}}$ are exchangeable.

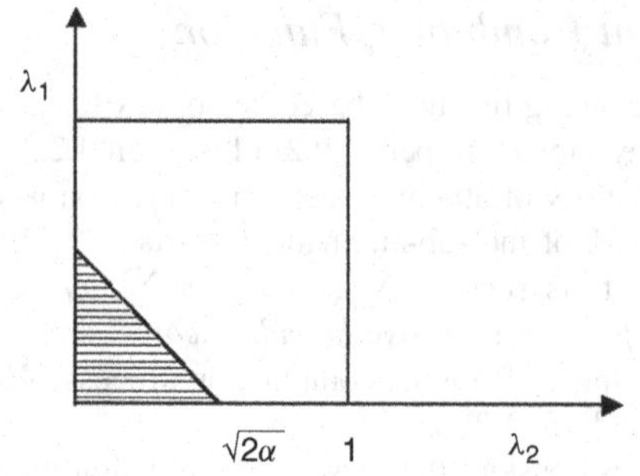

Figure 4.2 The additive combining function

2. Take a random re-randomization $\hat{\mathbf{Z}}_r^\dagger = \{[\hat{Z}_{r1}(u_i^\dagger), \hat{Z}_{r2}(u_i^\dagger)], i = 1, \ldots, n\}$ of $\hat{\mathbf{Z}}$, where $(u_i^\dagger, i = 1, \ldots, n)$ is a permutation of $(1, \ldots, n)$, and for any chosen $\boldsymbol{\delta} = (\delta_1, \delta_2)$ the corresponding data set $\hat{\mathbf{X}}_r^\dagger(\boldsymbol{\delta}) = (\hat{\mathbf{Z}}_{r1}^\dagger + \boldsymbol{\delta}, \hat{\mathbf{Z}}_{r2}^\dagger)$.
3. According to the NPC algorithm in Section 4.2.3, based on B CMC iterations, calculate the p-value $\hat{\lambda}_\psi''(\hat{\mathbf{X}}_r^\dagger(\boldsymbol{\delta}))$ of combined test T_ψ''.
4. Independently repeat steps 2 and 3 R times.
5. The empirical post-hoc conditional power for $\boldsymbol{\delta}$ is then evaluated as $\hat{W}[(\boldsymbol{\delta}, \hat{\boldsymbol{\delta}}, \alpha, \mathbf{T}, \psi)|\mathcal{X}_{/\mathbf{X}}] = \sum_r \mathbb{I}[\hat{\lambda}_\psi''(\hat{\mathbf{X}}_r^\dagger(\boldsymbol{\delta})) \leq \alpha]/R$.
6. To obtain a function in $\boldsymbol{\delta}$ and α, repeat steps 2 to 5 for different values of $\boldsymbol{\delta}$ and α. With $\boldsymbol{\delta} = \hat{\boldsymbol{\delta}}$ we obtain the *actual* post-hoc conditional power $\hat{W}[(\hat{\boldsymbol{\delta}}, \hat{\boldsymbol{\delta}}, \alpha, \mathbf{T}, \psi)|\mathcal{X}_{/\mathbf{X}}]$.

Similarly to the one dimensional case, the actual post-hoc conditional power may be used to assess how reliable the testing inference associated with $(\mathbf{T}, \psi, \mathbf{X})$ is, in the sense that if by chance the probability of obtaining the same inference with $(\mathbf{T}, \psi, \mathbf{X}^\dagger)$ as with $(\mathbf{T}, \psi, \mathbf{X})$ is greater than (say) $1/2$, then the actual inferential conclusion, given the set of units underlying $\mathbf{X}$, is reproducible more often than not. Of course, according to Section 3.2.2 it is also easy to define the NPC empirical conditional ROC curve by the pair $\left(\alpha, \hat{W}[(\hat{\boldsymbol{\delta}}, \hat{\boldsymbol{\delta}}, \alpha, \mathbf{T}, \psi)|\mathcal{X}_{/\mathbf{X}}]\right)$. Moreover, if $\boldsymbol{\delta}$ and $\mathbf{Z}$ could be separately considered it would be possible to evaluate the conditional power $W[(\boldsymbol{\delta}, \alpha, \mathbf{T}, \psi)|\mathcal{X}_{/\mathbf{X}}]$ in a straightforward way.

Unconditional Power Function

Consider now the unconditional power. Similarly to the univariate case, to define it we must consider the mean value of $W[(\boldsymbol{\delta}, \alpha, \mathbf{T}, \psi)|\mathcal{X}_{/\mathbf{X}}]$ with respect to the underlying population distribution P^n:

$$W(\boldsymbol{\delta}, \alpha, \mathbf{T}, \psi, P, n) = \mathbb{E}_{\mathcal{X}}\{W[(\boldsymbol{\delta}, \alpha, \mathbf{T}, \psi)|\mathcal{X}_{/\mathbf{X}}]\}$$
$$= \int_{\mathcal{X}^n} \mathbb{I}\left[\lambda_\psi''(\mathbf{X}(\boldsymbol{\delta})) \leq \alpha \,|\mathcal{X}_{/\mathbf{X}}\right] dP^n(\mathbf{X}(\boldsymbol{\delta})).$$

Note that to properly define the unconditional power $W(\boldsymbol{\delta}, \alpha, \mathbf{T}, \psi, P, n)$, the underlying population distribution P must be fully specified, that is, defined in its analytical form and all its parameters. Also note that averaging with respect the whole sample space $\mathcal{X}^n$ implies taking the mean over $\mathcal{X}_{/\mathbf{X}}$ and $\mathcal{X}^n \backslash \mathcal{X}_{/\mathbf{X}}$ as in the algorithm in Section 3.2.3. A practical algorithm for evaluating the unconditional power can be based on a standard Monte Carlo simulation from P with MC iterations as follows:

1. Choose a value of $\boldsymbol{\delta} = (\delta_1, \delta_2)$.
2. From the given population distribution P draw one set of n bivariate deviates $\mathbf{Z}_r$, and then add $\boldsymbol{\delta}$ to the first n_1 errors to define the data set $\mathbf{X}_r(\boldsymbol{\delta}) = (\mathbf{Z}_{r1} + \boldsymbol{\delta}, \mathbf{Z}_{r2})$.
3. Using the algorithm in Section 4.2.3, based on B CMC iterations, consider the p-value $\hat{\lambda}''_{\psi}(\mathbf{X}_r(\boldsymbol{\delta}))$ of combined test T''_{ψ}.
4. Independently repeat steps 2 and 3 MC times.
5. Evaluate the estimated power as $\hat{W}(\boldsymbol{\delta}, \alpha, \mathbf{T}, \psi, P, n) = \sum_r \mathbb{I}[\hat{\lambda}''_{\psi}(\mathbf{X}_r(\boldsymbol{\delta})) \leq \alpha]/MC$.
6. To obtain a function in $\boldsymbol{\delta}$, α, $\mathbf{T}$, ψ and n, repeat steps 1–5 with different values of $\boldsymbol{\delta}$, α, $\mathbf{T}$, ψ and n.

Remark 1. Theorem 2, 3.1.1, and property (P.1), 4.2.2, give sufficient conditions for both conditional post-hoc power $\hat{W}[(\boldsymbol{\delta}, \hat{\boldsymbol{\delta}}, \alpha, \mathbf{T}, \psi)|\mathcal{X}_{/\mathbf{X}}]$ and unconditional power $W(\boldsymbol{\delta}, \alpha, \mathbf{T}, \psi, P, n)$ to be non-decreasing in $\boldsymbol{\delta}$ and in attained α values.

4.3.5 Conditional Multivariate Confidence Region for $\boldsymbol{\delta}$

We may extend results of Section 3.4 related to confidence intervals for univariate cases to conditional multivariate confidence regions. As a guide, let us consider the two-sample bivariate design in which, with the same symbols as in Section 4.3.4, the data set is represented as $\mathbf{X}(\boldsymbol{\delta}) = (\mathbf{Z}_1 + \boldsymbol{\delta}, \mathbf{Z}_2)$.

The solution to this problem implies determining a bivariate region $\boldsymbol{\delta}^*_{\alpha}(\mathbf{X}(\boldsymbol{\delta}))$ in such a way that the permutation coverage probability $\Pr\{\boldsymbol{\delta} \in \boldsymbol{\delta}^*_{\alpha}(\mathbf{X}(\boldsymbol{\delta}))|\mathcal{X}_{/\mathbf{X}(\boldsymbol{\delta})}\} = 1 - \alpha$ holds for any given value of $\alpha \in (0, \frac{1}{2})$, for any unknown $\boldsymbol{\delta}$, for whatever non-degenerate data set $\mathbf{X}(\boldsymbol{\delta})$, and of course independently of the underlying population distribution P. To this end, once again we recall the rule that a confidence region for $\boldsymbol{\delta}$ contains all those values $\boldsymbol{\delta}^{\circ}$ for which, by using a given pair of test statistic $\mathbf{T}$ and a combining function $\psi \in C$ the null hypothesis $H_0(\boldsymbol{\delta}^{\circ}) : \{(\mathbf{X}_1(\boldsymbol{\delta}) - \boldsymbol{\delta}^{\circ}) \stackrel{d}{=} \mathbf{X}_2\}$, against $H_1(\boldsymbol{\delta}^{\circ}) : \{(\mathbf{X}_1(\boldsymbol{\delta}) - \boldsymbol{\delta}^{\circ}) \stackrel{d}{\neq} \mathbf{X}_2\}$, is accepted at level α.

The algorithm in Section 3.4 must be applied iteratively as follows:

1. Determine the sample estimate $\hat{\delta}_2 = \bar{X}_{21} - \bar{X}_{22}$ of effect δ_2.
2. Determine the $1 - \alpha$ confidence interval for δ_1 with $\delta_2 = \hat{\delta}_2$ at preset error width $\varepsilon > 0$, that is, $\underline{\delta}^*_1(\mathbf{X}(\delta_1, \hat{\delta}_2))$ and $\overline{\delta}^*_1(\mathbf{X}(\delta_1, \hat{\delta}_2))$, where the test statistic referred to is $T''_{\psi} = \psi(\lambda_1, \lambda_2)$. Note that the interval $[\underline{\delta}^*_1(\mathbf{X}(\delta_1, \hat{\delta}_2)), \overline{\delta}^*_1(\mathbf{X}(\delta_1, \hat{\delta}_2))]$ corresponds to the diameter on the δ_2 axis of confidence region $\boldsymbol{\delta}^*_{\alpha}(\mathbf{X}(\boldsymbol{\delta}))$.
3. For every $\delta_1 \in [\underline{\delta}^*_1(\mathbf{X}(\delta_1, \hat{\delta}_2)), \overline{\delta}^*_1(\mathbf{X}(\delta_1, \hat{\delta}_2))]$ determine the confidence interval for δ_2 at level $1 - \alpha$ at preset error width $\varepsilon > 0$, that is, $[\underline{\delta}^*_2(\mathbf{X}(\delta_1, \delta_2)), \overline{\delta}^*_2(\mathbf{X}(\delta_1, \delta_2))]$.

In this way, the confidence region becomes

$$\boldsymbol{\delta}^*_{\alpha}(\mathbf{X}(\boldsymbol{\delta})) = \{[\underline{\delta}^*_2(\mathbf{X}(\delta_1, \delta_2)), \overline{\delta}^*_2(\mathbf{X}(\delta_1, \delta_2))],\ \delta_1 \in [\underline{\delta}^*_1(\mathbf{X}(\delta_1, \hat{\delta}_2)), \overline{\delta}^*_1(\mathbf{X}(\delta_1, \hat{\delta}_2))]\}$$

since every pair (δ'_1, δ'_2) outside it is rejected with a probability greater than α.

4.3.6 Problems and Exercises

1) Provide an algorithm for conditional power on multivariate paired data designs.

2) Prove the statement in Remark 1, 4.3.3, that all combining functions, the critical regions of which, expressed in the $(\lambda_1, \ldots, \lambda_k)$ representation, do not contain the entire lower border, are not consistent.

3) Prove that for every combining function $\psi \in \mathcal{C}$, the combined p-value λ''_ψ is not an average value of partial p-values $\lambda_1, \ldots, \lambda_k$, i.e. it does not satisfy the condition $\min\{\lambda_1, \ldots, \lambda_k\} \leq \lambda''_\psi \leq \max\{\lambda_1, \ldots, \lambda_k\}$.

4) Prove that if a combining function φ leads to a p-value λ'' which is an average of partial p-values $\lambda_1, \ldots, \lambda_k$, then it is not a consistent combined test, and hence it is not a member of $\mathcal{C}$.

5) Prove that if $\mathbf{T} = \{T_i, i = 1, \ldots, k\}$ are weakly consistent partial permutation tests for H_{0i} against H_{1i} respectively, then $T''_\psi = \psi(\lambda_1, \ldots, \lambda_k)$, $\psi \in \mathcal{C}$, is a weakly consistent combined test for H_0 against H_1.

6) Prove that if at least one partial permutation test T_i, $i = 1, \ldots, k$, is weakly consistent for H_{0i} against H_{1i} respectively and all are marginally unbiased, then $T''_\psi = \psi(\lambda_1, \ldots, \lambda_k)$, $\psi \in \mathcal{C}$, is a weakly consistent combined test for H_0 against H_1.

7) With reference to Problem 12, 4.2.7, show that if a two-sample k-dimensional testing problem is such that $H_0 : \{\bigcap_{1 \leq i \leq k} (\mu_i \leq \mu_{0i})\}$ and $H_1 : \{\bigcup_{1 \leq i \leq k} (\mu_i > \mu_{0i})\}$, so that the multivariate hypotheses are broken down into a set of *composite* sub-hypotheses, the Tippett combining function, (c) in Section 4.2.4, provides for unbiased and consistent solutions.

8) Suppose that, in a two-sample k-dimensional testing problem, all partial p-values $(\lambda_1, \ldots, \lambda_k)$ in the alternative tend either to 0 or to 1, so that alternatives $H_1 : \{[\bigcup_i (\mu_i > \mu_{0i})] \text{ XOR } [\bigcup_i (\mu_i < \mu_{0i})]\}$ are either *dominating* or *dominated*, by an 'exclusive or relation', with respect to $H_0 : \{\bigcap_i (\mu_i = \mu_{0i})\}$. Prove that Liptak's combining function $T''_L = \sum_i \Phi^{-1}(1 - \lambda_1)$ is unbiased and consistent (see, in this framework, Example 4, 4.6 and Section 6.7).

9) Prove that if in weighted combining functions $T''_\mathbf{w} = \sum_i w_i \cdot \varphi(\lambda_i)$, as in (f) in Section 4.2.4, some but not all of the weights are zero and sub-alternatives to which tests are given zero weight may be true, then $T''_\mathbf{w}$ may not be consistent.

10) With reference to Remark 2, 4.3.2, give a formal proof of the ordering property of nonparametric combining functions ψ, for all $\psi \in \mathcal{C}$, with respect to the vector of partial treatment effects $\boldsymbol{\delta} = (\delta_i, i = 1, \ldots, k)$, provided that all partial tests are marginally unbiased for all $\alpha > 0$.

11) With reference to Remark 3, 4.3.2, give a formal proof of the monotonicity property, with respect to the vector of partial treatment effects $\boldsymbol{\delta} = (\delta_i, i = 1, \ldots, k)$, of the power function of nonparametric combining functions ψ, for all $\psi \in \mathcal{C}$, provided that all partial tests are marginally unbiased for all $\alpha > 0$.

12) With reference to Section 4.3.3, show that the non-consistent combined test $T'' = \sum_i (1 - \lambda_i)$, resulting from the additive rule, is provided with a monotonically non-decreasing power function which is not convergent to unity.

13) Draw a block diagram for the conditional confidence region $\boldsymbol{\delta}^*_\alpha(\mathbf{X}(\boldsymbol{\delta}))$ for the vector of functionals $\boldsymbol{\delta} = (\delta_1, \delta_2)$.

14) Extend algorithms of conditional and unconditional power in Section 4.3.4 to the k-dimensional case.

15) Provide an algorithm for confidence regions $\boldsymbol{\delta}^*_\alpha(\mathbf{X}(\boldsymbol{\delta}))$ for the three-dimensional case.

16) Provide an algorithm for unconditional power in a two-sample design for random effects Δ.

17) Provide details of statements in Remarks 2 and 4, 4.3.2. In particular, find that for fixed effects, independent partial tests and large data sets the limiting effect, with respect to k, for which there is power improvement on NPC is $\delta_{k+1} > 0$.

4.4 Some Further Asymptotic Properties

4.4.1 General Conditions

This section considers the asymptotic behaviour of some NPC tests and compares them to their parametric counterparts, in conditions allowing for the application of the latter. Of course, the consistency of NPC tests, either in strong or weak form, although of great interest, is only one of the asymptotic properties useful for characterizing them. In order to study some further properties, together with assumptions (a)–(c) of Section 4.3.1, let us assume the following:

(a) The hypotheses are expressed in terms of fixed treatment effects: $H_0 : \{\bigcap_i (\delta_i = 0)\}$ and $H_1 : \{\bigcup_i (\delta_i > 0)\}$ or $H_1 : \{\bigcup_i (\delta_i \neq 0)\}$.
(b) Parametric and permutation partial tests are calculated, except for population coefficients which may be known for parametric and unknown for permutation tests, by using the same standardized expressions $\mathbf{W} = \{W_1, \ldots, W_k\}$ and $\mathbf{T} = \{T_1, \ldots, T_k\}$, respectively. All partial tests are assumed to be significant for large values.
(c) Unless otherwise stated, the asymptotic k-variate distribution of parametric partial tests $\mathbf{W}$ is multivariate normal with (possibly unknown) finite mean vector $\{a_{(n)}\delta_i/\sigma_i,\ i = 1, \ldots, k\}$ and positive definite correlation matrix $\mathbf{R}$, where $a_{(n)}$ is a known function of sample sizes which diverges as n goes to infinity.
(d) Unless otherwise stated, the asymptotic k-variate distribution of permutation partial tests $\mathbf{T}$ is multivariate normal with (possibly unknown) finite mean vector $\{a_{(n)}\delta_i/\sigma_i^*,\ i = 1, \ldots, k\}$ and positive definite correlation matrix $\mathbf{R}^*$. It should be emphasized that permutation conditional quantities $\sigma_i^* = \sigma_i^*(\mathbf{X}),\ i = 1, \ldots, k$, and $\mathbf{R}^* = \mathbf{R}^*(\mathbf{X})$ are dependent on the data set and may be evaluated on the given data set or on data permutations.
(e) Conditions for the asymptotic optimality of partial permutation tests (see Sections 3.6–3.8) are assumed; in particular, as n goes to infinity, $\mathbf{R}^*$ and $\mathbf{R}_{\mathbf{U}}^* = \{\rho(U_i^*, U_j^*) = \mathbb{C}ov[\Phi^{-1}(1 - \lambda_i^*), \Phi^{-1}(1 - \lambda_j^*)],\ i, j = 1, \ldots, k\}$ both tend strongly to $\mathbf{R}$ and each partial permutation test is asymptotically coincident with its best asymptotic parametric counterpart.

4.4.2 Asymptotic Properties

Under conditions (a)–(e), the statements of the following theorems are presented.

Theorem 4. *If, in a given testing problem, permutation partial tests asymptotically follow a multivariate normal distribution, then the Liptak combined test* $T_L'' = \sum_i \Phi^{-1}(1 - \lambda_i)$ (see (b) in Section 4.2.4) *is almost surely asymptotically equivalent to* $T_{\mathcal{N}}'' = \sum_i T_i$.

Proof. For large values of n and B, Theorem 1, 4.2.3, implies that $\lambda_i = L(T_i) = 1 - F(T_i|\mathcal{X}_{/\mathbf{X}})$, $i = 1, \ldots, k$. Thus, by continuity of Φ and due to the given assumptions, we have $\Phi^{-1}(1 - \lambda_i) \simeq T_i$, because partial tests T_i are standardized. Therefore, as n tends to infinity, almost surely $\lim T_L'' = T_{\mathcal{N}}''$.

Remark 1. In this framework, $T_{\mathcal{N}}''$ is a direct combined test, as illustrated in (h) of Section 4.2.4. Also note that when permutation partial tests $T_i,\ i = 1, \ldots, k$, are based on statistics whose permutation means and variances are known functions of the data set $\mathbf{X}$, then $T_{\mathcal{N}}''$ may be preferred to T_L'' because it is less time-consuming. This is exactly the case when the T_i are linear functions of sample means. Also note that when the PCLT applies to all partial tests, the direct combination is asymptotically equivalent to Liptak's T_L''.

Remark 2. A similar result also holds in a straightforward way for the Lancaster combining function (see (d) in Section 4.2.4), which is suitable when all sub-alternatives H_{1i} are two-sided so that partial tests T_i are significant for either large or small values. In this case, we have

$$T_2'' = \sum_i \left[\Phi^{-1}(1-\lambda_i)\right]^2,$$

and now T_2'' converges almost surely to $T_S'' = \sum_i T_i^2$, which is also a direct combining function. Note that in this setting, the null asymptotic distribution of T_L'' is normal with mean value $\mathbb{E}(T_L'') = 0$ and variance $\mathbb{V}(T_L'') = \sum_{ij} \rho_{ij}^*$, where ρ_{ij}^* is the correlation coefficient between T_i and T_j, $i, j = 1, \ldots, k$. Calculation of the null asymptotic distribution of T_2'' presents other difficulties, since it is the sum of k *dependent* chi-squares with one degree of freedom each, so that in practice, we may refer to the CMC estimate.

The following is a consequence of arguments in Section 4.2.5 and Theorem 4.

Corollary 1. *If, in a given testing problem, parametric partial tests* $\mathbf{W}$ *are* k*-variate normally distributed and if the corresponding permutation partial tests are obtained by using the same computing expressions, then* T_L'' *is almost surely asymptotically equivalent to* $T_W = \sum_i W_i$.

According to the stated assumptions and supposing that each partial permutation test is asymptotically equivalent to its parametric counterpart and that the sum is a continuous function, the proof is straightforward.

Remark 3. The asymptotic equivalence of T_L'' and T_W is in both H_0 and H_1, so that they share the same asymptotic power behaviour.

Theorem 5. *If, for the same testing problem as in Theorem 4, the non-centrality parameters are such that* $\mathbf{R} \cdot \boldsymbol{\gamma} = \boldsymbol{\delta}$ *where* $\boldsymbol{\gamma} = (\gamma, \ldots, \gamma)$ *is a vector which has all components equal to the same number* $\gamma \geq 0$ *, then* $T_{\mathcal{N}}''$ *and* T_L'' *, as defined above, are asymptotically almost surely equivalent to* T_W *and UMP for testing* $H_0 : \{\gamma = 0\}$ *against* $H_1 : \{\gamma > 0\}$.

Proof. It is sufficient to note that, in the above-mentioned conditions, the likelihood ratio is a continuous monotonic function of T_W. In fact, for large values of n, we have

$$f_1/f_0 = \exp\left\{-\boldsymbol{\delta}^\top \cdot \mathbf{R}^{-1} \cdot \boldsymbol{\delta} \cdot \frac{a_{(n)}^2}{2} + \gamma \cdot T_W \cdot a_{(n)}\right\},$$

so that T_W is UMP. Moreover, according to Theorem 4 and Corollary 1, T_L'' and $T_{\mathcal{N}}''$ are almost surely asymptotically equivalent to T_W; thus, they are asymptotically UMP too.

Remark 4. Under the stated regularity conditions, leading to an asymptotically optimal parametric testing form (recall (g) in Section 4.2.4), Theorem 5 means that if all the k alternatives are jointly true with the same standardized non-centrality parameter (so that, when considered marginally, all partial tests are equally powerful), then T_L'' and $T_{\mathcal{N}}''$ are asymptotically equivalent to the best parametric test. Thus, it should be noted that, within the class C of all combining functions, there are members which are asymptotically best tests.

Theorem 6. *If permutation and parametric partial tests* $\mathbf{T}$ *and* $\mathbf{W}$ *are asymptotically multivariate normal, then the (parametric) quadratic form* $D^2 = \mathbf{W}^\top \cdot \mathbf{R}^{-1} \cdot \mathbf{W}$, *based on the Mahalanobis distance and useful for invariance testing of unrestricted* k*-dimensional alternatives* $H_1 : \left\{\bigcup_i (\delta_i \neq 0)\right\}$, *is almost surely asymptotically equivalent to* $T_{\mathcal{M}}'' = \mathbf{U}^\top \cdot (\mathbf{R}_{\mathbf{U}}^*)^{-1} \cdot \mathbf{U}$ *, where* $\mathbf{U}^\top = [\Phi^{-1}(1-\lambda_i),\ i = 1, \ldots, k]^\top$. *This corresponds to a Mahalanobis quadratic form on the* $\mathbf{U}$ *transformations of* p*-values.*

Proof. First, by Theorem 1, 4.2.3, the continuity of Φ, and assumption (e) of Section 4.4.1, the correlation matrix $\mathbf{R}^*_{\mathrm{U}}$ tends strongly to $\mathbf{R}$ as sample sizes diverge (see Pallini and Pesarin, 1992b). Secondly, when n and B are large, as observed in Theorem 4, we have $U_i \simeq T_i,\ i = 1, \ldots, k$. To complete the proof, note that only continuous transformations are involved.

Remark 5. Under the conditions for which D^2 is a best test, T''_M is also asymptotically almost surely best.

Corollary 2. *Under the same conditions as Theorem 6, when the permutation correlation matrix* $\mathbf{R}^*$ *of permutation partial tests* $\mathbf{T}$ *is known, the directly combined test* $T''_Q = \mathbf{T}^\top \cdot (\mathbf{R}^*)^{-1} \cdot \mathbf{T}$ *is asymptotically almost surely equivalent to* D^2.

The proof is straightforward by taking account of Theorem 4.

Remark 6. When permutation partial tests $T_i,\ i = 1, \ldots, k$, are based on statistics for which the permutation means and variances are known functions of the data set $\mathbf{X}$ (so that we are able to express these tests in standardized form, before considering B CMC iterations), then we may use the direct combination form T''_Q. We observe that T''_Q can be preferable to $T''_{\mathcal{M}}$ because it is much less demanding from a computational point of view. This is definitely the case when all permutation partial tests T_i are linear functions of sample means.

Remark 7. As a consequence of Theorem 6 and Corollary 2, in a multivariate two-sample problem under normality, Hotelling's T^2 and the permutation combined tests T''_Q and $T''_{\mathcal{M}}$ are all asymptotically almost surely equivalent with respect to the Mahalanobis D^2.

Theorem 7. *If* $\varphi : \mathcal{R}^k \to \mathcal{R}^1$ *is a continuous (measurable) function, then* $\varphi(T_1, \ldots, T_k)$ *is asymptotically almost surely equivalent to* $\varphi(U_1, \ldots, U_k)$.

The proof is straightforward, since assumptions (a)–(e) of Section 4.4.1 imply that, as B and n are large, T_i and U_i share the same distribution, together with the assumed continuity of φ.

Corollary 3. *If* $\varphi : \mathcal{R}^k \to \mathcal{R}^1$ *is a continuous (measurable) function, then* $\varphi[W_1, \ldots, W_k]$ *is asymptotically almost surely equivalent to* $\varphi(U_1, \ldots, U_k)$ *and to* $\varphi(T_1, \ldots, T_k)$.

The proof is straightforward.

Remark 8. Theorem 7 and Corollary 3 generalize Theorem 4 and Corollary 1 above, respectively.

Theorem 8. *If, for testing* $H_0 : \{\theta = \theta_0\}$ *against a given alternative* H_1, *a parametric test is a continuous measurable function of partial tests* $\mathbf{W}$, $\Lambda_W = \Lambda(W_1, \ldots, W_k)$, *where* $\mathbf{W}$ *are continuous multivariate but not necessarily normally distributed, and if the asymptotic marginal CDFs of corresponding permutation partial tests* $\{T_i,\ i = 1, \ldots, k\}$ *are* $\{F_i(t), i = 1, \ldots, k\}$, *which are continuous multivariate, then* $T''_\Lambda = \Lambda\{F_1^{-1}[\lambda_1(\mathbf{X})], \ldots, F_k^{-1}[\lambda_k(\mathbf{X})]\}$, λ_i *being the permutation partial p-values, is almost surely asymptotically equivalent to* Λ_W, *provided that the composite combining function* $\Lambda(F_1^{-1}, \ldots, F_k^{-1})$ *is a member of* $\mathcal{C}$.

The proof is again based on the almost sure convergence (by Theorem 1, 4.2.3) of $F_i^{-1}[\lambda_i(\mathbf{X})]$ to $W_i,\ i = 1, \ldots, k$, and on the assumed continuity of composite combining function $\Lambda(F_1^{-1}, \ldots, F_k^{-1})$.

As a consequence, the following two corollaries hold:

Corollary 4. *Under the same conditions as Theorem 8*, Λ_W *and* T''_Λ *are both almost surely asymptotically equivalent to* $T''_{\Lambda T} = \Lambda(T_1, \ldots, T_k)$.

Corollary 5. *Under the same conditions as Theorem 8, if* Λ_W *is any parametric test, which is in some sense a best test, then* T''_Λ *and* $T''_{\Lambda T}$ *are almost surely asymptotically best in the same sense.*

The proofs of these two corollaries are straightforward, as sample size goes to infinity, since $F_i^{-1}[\lambda_i(\mathbf{X})]$ converges almost surely to T_i, $i = 1, \ldots, k$.

Theorems 4–8 and their respective corollaries, especially the latter one, lead to best asymptotically NPCs, which may also be useful for finite sample sizes, although in this situation they cannot be regarded as best (recall the discussion in Section 2.5).

However, Tippett's combining function $T''_T = \max_i\{1 - \hat{\lambda}_i\}$, for instance, obeys the procedure of Berk and Jones (1978); thus, it is asymptotically relatively efficient according to the Bahadur definition under the conditions established by Berk and Jones (see also Littell and Folks, 1971, 1973).

Remark 9. In Pallini (1994) it is stated that, if all permutation partial tests are Bahadur optimal for their respective sub-hypotheses, then nonparametric combined tests, according to direct, Fisher's, Liptak's or Lancaster's combining functions, are also Bahadur optimal, irrespective of underlying monotonic dependence relations. In this sense, within the class C of all combining functions there are NPCs of optimal partial tests which are as fast as best parametric counterparts in rejecting H_1 when it is false.

Remark 10. The previous theorems and corollaries show that, within C, there are members which are asymptotically equivalent with respect to their best parametric counterparts when: (i) the conditions for the validity of parametric solutions hold; and (ii) permutation solutions are obtained by the same statistics of parametric counterparts.

4.5 Finite-Sample Consistency

4.5.1 Introduction

A quite important problem usually occurs in several multidimensional applications when sample sizes are fixed and the number of variables to be analysed is much larger than the sample sizes (Goggin, 1986). Typical examples are encountered in longitudinal analysis (Diggle et al., 2002), microarrays and genomics (Salmaso and Solari, 2005, 2006), brain imaging (Hossein-Zadeh et al., 2003; Friman and Westin, 2005), shape analysis (Dryden and Mardia, 1998; Bookstein, 1991), functional data (Bosq, 2000; Ramsay and Silverman, 1997, 2002; Feeraty and Vieu, 2006), finance data, etc. In Remarks 2, 3.2.1 and 1, 3.2.3 it is shown that, under very mild conditions, the power function of permutation tests monotonically increases as the related induced non-centrality functional increases. This is also true for multivariate situations. In particular, for any added variable the power does not decrease if this variable increases the induced non-centrality (see Remark 4, 4.3.2 and Problem 17, 4.3.6). We will investigate here the behaviour of the rejection rate for divergent number of variables. NPC method properties were obtained assuming that the number of partial tests is finite and possibly smaller than the cardinality of the permutation space. Hence the necessity to look further into the permutation methodology, especially in order to deal with such important problems. This analysis allows us to introduce the concept of *finite-sample consistency* (see Pesarin and Salmaso, 2009). Sufficient conditions are given in order that the rejection rate converges to one, at any attainable α-value for fixed sample sizes, when the number of variables diverges, provided that the non-centrality induced by test statistics also diverges.

Its application may be appropriate for problems related to discrete or discretized stochastic processes, as for instance when data are curves or images, for which at most a countable set of variables are observed or derived by Fourier or wavelet expansions or by functional principal component data transformations. Hence, the application range is rather broad and we see how the NPC applies with V variables (with $V \in \mathbb{N}$ a natural integer). Permutation tests for stochastic processes may be defined and applied in many cases. To be specific, the process may have stationary independent increments, or stationary symmetric increments, or be spherically exchangeable or exchangeable. See Bell et al. (1970) and Basawa and Prakasa-Rao (1980); see also Chapter 7 for some problems when data are discretized profiles of stochastic processes.

To discuss testing problems for stochastic dominance alternatives as are generated by symbolic treatments with non-negative V-dimensional random shift effects $\mathbf{\Delta}$, as usual we refer to one-sided two-sample designs as a guide. Extensions to non-positive, two-sided alternatives, and multi-sample designs are straightforward. Note that under H_0 two-sample data $\mathbf{X} = \{X_{hji},\ i = 1, \ldots, n_j,\ j = 1, 2,\ h = 1, \ldots, V\}$ are exchangeable, in accordance with the notion that units are randomized to treatments. Without loss of generality, we assume that effects in H_1 are such that $\mathbf{\Delta}_1 = \mathbf{\Delta} \overset{d}{>} \mathbf{0}$ and $\Pr\{\mathbf{\Delta}_2 = \mathbf{0}\} = 1$. Thus, the null hypothesis may also be written as $H_0 : \{\mathbf{\Delta} \overset{d}{=} \mathbf{0}\}$ and the alternative as $H_1 : \{\mathbf{\Delta} \overset{d}{>} \mathbf{0}\}$. We also assume that the n-sized V-dimensional data set is modelled as $\mathbf{X}(\mathbf{\Delta}) = \{\mathbf{Z}_1 + \mathbf{\Delta}, \mathbf{Z}_2\}$, where $\mathbf{\Delta} = (\Delta_{11}, \ldots, \Delta_{1n_1})$ and where $\mathbf{Z}_1, \mathbf{Z}_2$ have the role of random deviates the distribution $P_{\mathbf{Z}}$ of which is generally unknown. Of course, $\mathbf{X}(\mathbf{0}) = \{\mathbf{Z}_1, \mathbf{Z}_2\}$ represents data in H_0.

4.5.2 *Finite-Sample Consistency*

We investigate here the rejection behaviour of the permutation test $T(\mathbf{X}) = S_1(\mathbf{X}_1) - S_2(\mathbf{X}_2)$ (see Remarks 2, 3.2.1 and 1, 3.2.3) when for any reason, especially for divergent number V of variables, the random effect $\mathbf{\Delta}$ can diverge to infinity. Examples 1–5, 4.5.3, give some hints as to the application of this notion. Any test statistic is a mapping from the sample space to the real line, $T : \mathcal{X}^n \rightarrow \mathcal{R}^1$, and we investigate T by comparing its behaviour in H_0, $T(\mathbf{X}(\mathbf{0}))$, to its behaviour in H_1, $T(\mathbf{X}(\mathbf{\Delta}))$. It will be perfectly clear that such a comparison, together with the respective asymptotic behaviours, belongs within the permutation framework if we are able to write the related random variables in the form $T(\mathbf{X}(\mathbf{\Delta})) = T(\mathbf{X}(\mathbf{0})) + D_T(\mathbf{\Delta}, \mathbf{X}(\mathbf{0}))$, where the induced non-centrality $D_T(\mathbf{\Delta}, \mathbf{X}(\mathbf{0}))$ is a random function which may diverge in probability, that is, such that $\lim_{\Delta \uparrow \infty} \Pr\{D_T > t\} = 1$, for any real t. To this end, in Lemma 2 we investigate the behaviour of the conditional (permutation) rejection rate when sample sizes (n_1, n_2) *and* non-degenerate one-dimensional random deviates $\mathbf{Z} = (\mathbf{Z}_1, \mathbf{Z}_2)$ are held fixed as the fixed effect δ goes to infinity, according to some monotonic sequence $\{\delta_v, v \geq 1\}$. Then in Theorems 9 and 10 we investigate the unconditional (population) rejection rate when the set of i.i.d. random deviates $\mathbf{Z}$ vary in $\mathcal{X}^n$ according to the distribution $P_{\mathbf{Z}}$. The extension from fixed to random effects occurs in Theorem 11. Finally, using the equipower property of permutation tests, in Theorem 12 we extend the finite-sample consistency to the conventional notion, as is obtained for divergent sample sizes.

Since the main inferential conclusions associated with permutation tests are concerned with the observed data set $\mathbf{X}$ for the given set of n individuals, the notion of consistency that is most useful is the *weak* (or in probability) form which essentially states that for divergent values of non-centrality functional induced by the test statistic, the limit rejection probability of test T is one for any fixed $\alpha > 0$. This means that, for fixed sample sizes and large values of induced non-centrality, the rejection probability of T approaches one. We think that the *almost sure* version (strong or with probability one), although of great mathematical importance, is of limited relevance in the permutation context.

In Chapter 3 it is stated that neither conditional nor unconditional power functions of any associative or non-associative test statistic T for one-sided alternatives decrease as the effect increases.

That is, if $\delta < \delta'$, then for any attainable α-value, and with obvious notation,

$$\Pr\left\{\lambda(\mathbf{X}(\delta)) \leq \alpha | \mathcal{X}^n_{/\mathbf{X}(\delta)}\right\} \leq \Pr\left\{\lambda(\mathbf{X}(\delta')) \leq \alpha | \mathcal{X}^n_{/\mathbf{X}(\delta')}\right\}$$

and

$$\mathbb{E}_P\left[\Pr\left\{\lambda(\mathbf{X}(\delta)) \leq \alpha | \mathcal{X}^n_{/\mathbf{X}(\delta)}\right\}\right] \leq \mathbb{E}_P\left[\Pr\left\{\lambda(\mathbf{X}(\delta')) \leq \alpha | \mathcal{X}^n_{/\mathbf{X}(\delta')}\right\}\right]$$

respectively, where $\mathbb{E}_P(\cdot) = \int_{\mathcal{X}^n}(\cdot)dP^{(n)}$ is the mean value of $(\cdot)$ with respect to $P^{(n)}$. Similar relations also hold for random effects $\mathbf{\Delta}$.

Lemma 2. (Conditional finite-sample consistency of T). *Suppose that:*

(i) *T is any associative or non-associative test statistic for one-sided hypotheses;*
(ii) *the sample sizes (n_1, n_2) and the set of real deviates $\mathbf{Z} = \{\mathbf{Z}_1, \mathbf{Z}_2\} \in \mathcal{X}^n$ are fixed;*
(iii) *the data set is $\mathbf{X}(\delta) = (\mathbf{Z}_1 + \boldsymbol{\delta}, \mathbf{Z}_2)$, where $(\mathbf{Z}_1, \mathbf{Z}_2) \in \mathcal{X}^n$ are i.i.d. measurable real random deviates whose parent distribution is $P_Z(t) = \Pr\{Z \leq t\}$ and $\boldsymbol{\delta} = (\delta, \ldots, \delta)$ is the n_1-dimensional vector of non-negative fixed effects;*
(iv) *fixed effects δ diverge to infinity according to some monotonic sequence $\{\delta_v, v \geq 1\}$, the elements of which are such that $\delta_v \leq \delta_{v'}$ for any pair $v < v'$.*

If conditions (i)–(iv) *are satisfied, then the permutation (conditional) rejection rate of T converges to 1 for all α-values not smaller than the minimum attainable α_a; thus, T is conditional finite-sample consistent.*

Proof. For any chosen $\delta > 0$, let us consider the observed data set $\mathbf{X}(\delta) = (\mathbf{Z}_1 + \boldsymbol{\delta}, \mathbf{Z}_2)$. The permutation support induced by the test statistic T when applied to the data set $\mathbf{X}(\delta)$ is $\mathcal{T}_{\mathbf{X}(\delta)} = \{T^*(\delta) = T(\mathbf{X}^*(\delta)) : \mathbf{X}^*(\delta) \in \mathcal{X}^n_{/\mathbf{X}(\delta)}\}$. Depending on $\mathbf{Z}$, in the sequence $\{\delta_v, v \geq 1\}$ there is a value $\delta_\mathbf{Z}$ of δ such that the related observed value $T^o(\mathbf{X}(\delta_\mathbf{Z}))$ is right-extremal for the induced permutation support $\mathcal{T}_{\mathbf{X}(\delta_\mathbf{Z})}$, that is, $T^o(\mathbf{X}(\delta_\mathbf{Z})) = \max_{\mathcal{T}_{\mathbf{X}(\delta_\mathbf{Z})}}\{T^*(\delta_\mathbf{Z}) : \mathbf{X}^*(\delta_\mathbf{Z}) \in \mathcal{X}^n_{/\mathbf{X}(\delta_\mathbf{Z})}\}$. This $\delta_\mathbf{Z}$ can be determined by observing that a sufficient condition for right-extremal property of T^o is that $\min_{n_1}(Z_{1i} + \delta_\mathbf{Z}) > \max_{n_2}(Z_{2i})$, thus

$$\delta_\mathbf{Z} = \arg\min_{\delta_v}[\min_i(Z_{1i} + \delta_v) > \max_i(Z_{2i})].$$

Indeed, as the functions S are non-decreasing, we necessarily have that $S_1(\mathbf{Z}_1 + \boldsymbol{\delta}_\mathbf{Z}) > S_2(\mathbf{Z}_2)$ and so $T^o(\mathbf{X}(\delta_\mathbf{Z}))$ is right-extremal because for all permutations $\mathbf{X}^*(\delta_\mathbf{Z}) \neq \mathbf{X}(\delta_\mathbf{Z})$ we have $T^o(\mathbf{X}^*(\delta_\mathbf{Z})) < T^o(\mathbf{X}(\delta_\mathbf{Z}))$. Then the rejection rate relative to the minimum attainable α-value α_a, which for one-sided (two-sided) alternatives is $1/\binom{n}{n_1}$ $(2/\binom{n}{n_1})$, due to the monotonic behaviour with respect to δ, attains 1 for all $\delta > \delta_\mathbf{Z}$, hence, due to the monotonicity property with respect to α, it is also 1 for all α-values greater than α_a. Actually, the conditional power function of T, that is, $\Pr\{\lambda(\mathbf{X}(\delta)) \leq \alpha | \mathcal{X}^n_{/\mathbf{X}(\delta)}\}$, is 1 for all $\delta \geq \delta_\mathbf{Z}$ and $\alpha \geq \alpha_a$, and so it is also 1 in the limit.

On the one hand, the result of Lemma 2 may be seen as essentially trivial, as it says nothing unexpected. On the other hand, however, as this result depends on the fact that for any fixed set of real deviates $\mathbf{Z}$ there exists a value $\delta_\mathbf{Z}$ such that $T(\mathbf{X}(\delta_\mathbf{Z}))$ is right-extreme for the induced permutation support $\mathcal{T}_{\mathbf{X}(\delta_\mathbf{Z})}$, there is the difficulty of obtaining a result that is not only valid for almost all $\mathbf{Z} \in \mathcal{X}^n$, but also unconditionally valid. This implies considering the rejection behaviour of T irrespective of $\delta_\mathbf{Z}$, which in turn varies as $\mathbf{Z}$ varies in $\mathcal{X}^n$. The necessity of obtaining an unconditionally valid result arises from the fact that it is natural to require that the probability of rejecting H_0 when it is false and the non-centrality is large must be close to 1 independently of the observed units. We have the following theorem.

Theorem 9. (Weak unconditional finite-sample consistency of T). *Suppose that:*

(i) *T is any test statistic for one-sided hypotheses;*
(ii) *the sample sizes (n_1, n_2) are fixed and finite;*
(iii) *the data set is $\mathbf{X}(\delta) = (\mathbf{Z}_1 + \boldsymbol{\delta}, \mathbf{Z}_2)$, where $(\mathbf{Z}_1, \mathbf{Z}_2) \in \mathcal{X}^n$ are i.i.d. measurable real random deviates whose parent distribution is $P_Z(t) = \Pr\{Z \leq t\}$ and $\boldsymbol{\delta} = (\delta, \ldots, \delta)$ is the n_1-dimensional vector of non-negative fixed effects;*
(iv) *the fixed effects δ diverge to infinity according to the monotonic sequence $\{\delta_v, v \geq 1\}$ as in Lemma 2 above.*

If conditions (i)–(iv) *are satisfied, then the permutation unconditional rejection rate of test T converges to 1 for all α-values not smaller than the minimum attainable α_a; thus, T is weak unconditional finite-sample consistent.*

Proof. Observe that measurability of random deviates Z implies that $\lim_{t\downarrow-\infty} \Pr(Z \leq t) = 0$ and $\lim_{t\uparrow+\infty} \Pr(Z \leq t) = 1$. Also observe that, according to Lemma 2 above, a sufficient condition for the observed value $T^o(\mathbf{X}(\delta))$ to be right-extremal in the induced permutation support $\mathcal{T}_{\mathbf{X}(\delta)}$ is that $\min_{n_1}(Z_{1i} + \delta) > \max_{n_2}(Z_{2i})$. The probability of this event, as random deviates in $\mathbf{Z}$ are i.i.d., is

$$\Pr\left\{\min_{n_1}(Z_{1i} + \delta) > \max_{n_2}(Z_{2i})\right\} = \int_{\mathcal{X}} \left\{[1 - P_Z(t - \delta)]^{n_1}\right\} d\,[P_Z(t)]^{n_2},$$

the limit of which, as δ goes to infinity according to the given sequence $\{\delta_v, v \geq 1\}$, is 1 since (n_1, n_2) are fixed and finite and because, by the Lebesgue monotone convergence theorem (see Parthasarathy, 1977) according to which the limit of an integral is the integral of the limit, the associated sequence of probability measures $\{P_Z(t - \delta_v), v \geq 1\}$ converges to zero monotonically for any t.

An interpretation of this is that the probability of finding a set $\mathbf{Z} \in \mathcal{X}^n$ for which there does not exist a finite value of $\delta_{\mathbf{Z}} \in \{\delta_v, v \geq 1\}$ such that $\min_{n_1}(Z_{1i} + \delta_{\mathbf{Z}}) > \max_{n_2}(Z_{2i})$, converges to zero monotonically as δ diverges. Taking Lemma 2 into account, this implies that the unconditional rejection rate

$$W_\alpha(\delta) = \int_{\mathcal{X}} \Pr\{\lambda(\mathbf{X}(\delta)) \leq \alpha | \mathcal{X}^n_{/\mathbf{X}(\delta)}\}\, dP_{\mathbf{Z}}(\mathbf{z}),$$

where $P_{\mathbf{Z}}$ is the multivariate distribution of vector $\mathbf{Z}$, as δ tends to infinity, converges to 1 for all α-values not smaller than α_a.

It is to be emphasized that the notion of unconditional finite-sample consistency, defined for divergent fixed effects δ, is different from the conventional notion of (unconditional) consistency of a test, which in turn relates to the behaviour of the rejection rate for given δ when $\min(n_1, n_2)$ diverges. It is known that, in order to attain permutation unconditional consistency, random deviates Z must at least possess finite second moment (Sections 3.3 and 3.7; see also Hoeffding, 1952; Romano, 1990). Here we only require that T is measurable in H_0, so that in this respect it is to be emphasized that random deviates Z need not be provided with finite moments of any positive order. For instance, they can be distributed as Cauchy $\mathcal{C}y(0, \sigma)$ or Pareto $\mathcal{P}a(\theta, \sigma)$ with shape parameter $0 < \theta \leq 1$ and finite scale coefficients $\sigma > 0$. Note, however, that to investigate the conventional notion of consistency it is required that, with obvious notation, the sequence of test statistics $\{T_n; n \in \mathbb{N}\}$ can be written in the form $\{T(\mathbf{X}_n(\delta)) = T(\mathbf{X}_n(0)) + D_T(\delta, \mathbf{Z}_n); n \in \mathbb{N}\}$, where $D_T(\delta, \mathbf{Z}_n)$ diverges and $T(\mathbf{X}_n(0))$ is measurable in the limit. Thus, in one sense, the two notions may be seen as complementary to each other (see Theorem 12 for a parallel between the two notions).

Let us now consider the following theorem.

Theorem 10. *Suppose that random deviates* Z *and effects* δ *are such that:*

(i) *there exists a function* $\rho(\delta) > 0$ *of effects* δ *the limit of which is* 0 *as* δ *goes to infinity*;
(ii) T *is any test statistic, as above*;
(iii) *the data set is obtained by the transformation* $\mathbf{Y}(\delta) = \rho(\delta)\mathbf{X}(\delta)$;
(iv) $\lim_{\delta\uparrow\infty} \delta\rho(\delta) = \tilde{\delta} > 0$, *and* $\lim_{\delta\uparrow\infty} \Pr\{\rho(\delta)\cdot|Z| > \varepsilon\} = 0$, $\forall\varepsilon > 0$;
(v) *conditions* (iii) *and* (iv) *of Theorem 9 hold*.

If conditions (i)–(v) *hold then the unconditional rejection rate converges to 1 for all* α*-values not smaller than the minimum attainable* α_a; *thus,* T *is weak unconditional finite-sample consistent*.

Proof. Let us observe first that the data $\mathbf{Y}(\delta) = \rho(\delta)[\mathbf{Z}_1 + \boldsymbol{\delta}, \mathbf{Z}_2]$, as δ goes to infinity, collapse in distribution towards $[\tilde{\boldsymbol{\delta}}, 0]$. Also observe that, for any fixed set of random deviates $(\mathbf{Z}_1, \mathbf{Z}_2)$, $T(\mathbf{Y}(\delta))$ is right-extreme in the induced permutation support when $\min_{n_1}[(Z_{1i} + \delta)\rho(\delta)] > \max_{n_2}[Z_{2i}\rho(\delta)]$. Since $\rho(\delta)$ is positive, the event defined by this relation is equivalent to $\min_{n_1}[Z_{1i} + \delta] > \max_{n_2}[Z_{2i}]$, in the sense that the latter is true if and only if the former is true. Thus, by taking account of proof of Theorem 9, we have that

$$\Pr\left\{\min_{n_1}[(Z_{1i} + \delta)\rho(\delta)] > \max_{n_2}[Z_{2i}\rho(\delta)]\right\} = \Pr\left\{\min_{n_1}[Z_{1i} + \delta] > \max_{n_2}[Z_{2i}]\right\}$$
$$= \int_{\mathcal{X}} \{[1 - P_Z(t - \delta)]^{n_1}\}\, d\,[P_Z(t)]^{n_2},$$

the limit of which, as δ goes to infinity, is 1 because the associated sequence of probabilities $\{P_Z[t - \delta_v], v \geq 1\}$ monotonically converges to zero and (n_1, n_2) are fixed and finite. Thus, according to Theorem 9, the related rejection rate converges to 1 for all α-values not smaller than α_a; and so, T is weak unconditional finite-sample consistent.

Theorem 10 says that when, for divergent δ, the data distribution of $\{\rho(\delta_v)Z, v \geq 1\}$ collapses towards zero while $\{\delta_v\rho(\delta_v), v \geq 1\}$ is positive, then any permutation test statistic T applied to the transformed data set $\mathbf{Y}(\delta) = \rho(\delta)[\mathbf{Z}_1 + \boldsymbol{\delta}, \mathbf{Z}_2]$ is unconditionally finite-sample consistent (see Example 2, 4.5.3).

Theorem 11. (Weak unconditional finite-sample consistency for random effects). *The results of Lemma 2 and Theorems 9 and 10 can be extended to divergent random effects* Δ *according to some sequence* $\{\Delta_v, v \geq 1\}$ *whose elements are stochastically non-decreasing, that is*, $\Delta_v \overset{d}{\leq} \Delta_{v+1}, \forall v \geq 1$, *and provided that* $\lim_{v\uparrow\infty} \Pr\{\Delta_v > u\} \rightarrow 1$ *for every finite* u.

Proof. Actually, in order for the Lebesgue monotone convergence theorem to be applicable it suffices that $P_Z(t - \Delta'' \leq u)$ is stochastically dominated by $P_Z(t - \Delta' \leq u)$ for every u, whenever $\Delta' \overset{d}{\leq} \Delta''$, so that the associated sequence of probabilities $\{P_Z[t - \Delta_v], v \geq 1\}$ monotonically converges to zero.

Let us now examine a close parallel between finite-sample and conventional notions of consistency. To this end, let us assume that we have i.i.d. data from a one-dimensional variable and that sample sizes diverge.

Theorem 12. (Weak unconditional consistency of T). *Suppose that a typical two-sample problem, for one-sided alternatives with the data set* $\mathbf{X}(\delta) = (\boldsymbol{\delta} + \sigma\mathbf{Z}_1, \sigma\mathbf{Z}_2)$ *as above, is such that*:

(i) *the permutation test statistic* $T = \sum_{i\leq n_1} X_{1i}(\delta)/n_1$ *is assumed to be weak unconditional finite-sample consistent*;

(ii) *the conditions stated in Theorems 9–11 above are satisfied;*
(iii) *the one-dimensional random deviates Z have zero mean, that is,* $\mathbb{E}(Z) = 0$;
(iv) *the two-sample sizes* (n_1, n_2) *satisfy the relation* $(n_1 = vm_1, n_2 = vm_2)$, *so that they can diverge according to the sequence* $\{(vm_1, vm_2), v \geq 1\}$.

Then for any given $\delta > 0$ the unconditional rejection probability of T converges to 1, $\forall \alpha \geq \alpha_a$, as v goes to infinity; thus, T is weak unconditional consistent in accordance with the conventional notion of consistency.

Proof. Let us observe that the fixed effect δ is now an unknown constant and that sample sizes diverge, so that the conventional notion of consistency may be applied to T. For any integer $v \geq 1$, let us arrange the one-dimensional data sets $\mathbf{X}_1(\delta) = (\boldsymbol{\delta} + \sigma \mathbf{Z}_1) = \{\delta + \sigma Z_{1i}, i = 1, \ldots, n_1\}$ and $\mathbf{X}_2 = \sigma \mathbf{Z}_2 = \{\sigma Z_{2i}, i = 1, \ldots, n_2\}$ into respectively the v-dimensional sets $\mathbf{Y}_1(\delta) = \{Y_{11i} = X_{1i}, Y_{21i} = X_{1,m_1+i}, \ldots, Y_{v1i} = X_{1,m_1(v-1)+i}, i = 1, \ldots, m_1\}$ and $\mathbf{Y}_2 = \{Y_{12i} = X_{2i}, Y_{22i} = X_{2,m_2+i}, \ldots, Y_{v2i} = X_{2,m_2(v-1)+i}, i = 1, \ldots, m_2\}$, where $(n_1, n_2) = (vm_1, vm_2)$. Thus the data vector $\mathbf{X}(\delta)$, with one column and $n = n_1 + n_2$ rows, is organized into a matrix $\mathbf{Y}(\delta)$ with v columns and $m = m_1 + m_2$ rows. Of course, as v diverges, so does $\min(n_1, n_2)$. The test statistic T, when applied to the data set $\mathbf{Y}(\delta)$, as in Example 3, 4.5.3, is unconditionally finite-sample consistent, because the conditions of Theorem 10 are satisfied by assumption. Let us now observe that, for any $v \geq 1$, the observed value of T applied to $\mathbf{Y}(\delta)$ is $T(\mathbf{Y}(\delta)) = \sum_{i \leq m_1} \sum_{h \leq v} Y_{h1i}(\delta)/vm_1$ and applied to $\mathbf{X}(\delta)$ is $T(\mathbf{X}(\delta)) = \sum_{i \leq n_1} X_{1i}(\delta)/n_1$, and of course $T(\mathbf{Y}(\delta)) = T(\mathbf{X}(\delta))$. Moreover, we may also write $T(\mathbf{X}(\delta)) = T(\mathbf{X}(0)) + \delta/\sigma = T(\mathbf{Y}(\delta))$, stressing that two forms have the same null distribution and the same non-centrality functional which does not vary as v diverges, whereas the null component $T(\mathbf{X}(0))$ as v diverges collapses almost surely towards zero by the strong law of large numbers because, by assumption, the random deviates Z admits finite first moment and are i.i.d. Thus, by virtue of Theorem 10 the rejection probability for both ways converges to 1, for all $\delta > 0$. And so weak unconditional finite-sample consistency implies weak unconditional (conventional) consistency for all $\alpha \geq \alpha_a$.

Using the same arguments, it is straightforward to extend the results of Theorem 12 to any associative test statistic of Section 2.3. It is worth noting that the permutation sample space, when processing the n-row one-dimensional data set $\mathbf{X}(\delta)$, that is, $\mathcal{X}_{/\mathbf{X}(\delta)}$, has $\binom{n}{n_1}$ distinct elements, and when processing the data rearranged according to the m-row v-dimensional data set $\mathbf{Y}(\delta)$, that is, $\mathcal{X}_{/\mathbf{Y}(\delta)}$, it has $\binom{m}{m_1}$ elements. The two ways looking at permutation testing, having the same non-centrality and exactly the same likelihood, have the same unconditional power and so both are consistent for all α-values not smaller than the minimum attainable $\alpha_a = 1/\binom{m}{m_1}$. However, the two ways are not completely equivalent in inferential terms. In order to prove their complete equivalence, we have to prove that both are consistent for all $\alpha > 0$ and that convergence should be obtained for any kind of sequences such that $\min(n_1, n_2)$ diverges. What has been proved here is that an unconditional finite-sample consistent test for associative T is also unconditionally consistent in the conventional sense for all $\alpha \geq \alpha_a$ when the sequence of sample sizes is $\{(vm_1, vm_2), v \geq 1\}$. In practice, if we require consistency at least for α greater than a given value α°, and sample sizes are according to $\{(vm_1, vm_2), v \geq 1\}$, then we may find a pair of sample sizes (m_1, m_2) such that $\alpha^\circ > 1/\binom{m}{m_1}$ so that the two ways are equivalent at least for all $\alpha \geq \alpha^\circ$. Since for any arbitrarily chosen α° we may find a pair (m_1, m_2) such that $\alpha^\circ > \alpha_a$, then we may conclude that unconditional inferential conclusions associated with the two ways are always coincident, provided that sample sizes follow the sequence $\{(vm_1, vm_2), v \geq 1\}$. This can be seen as a proof that, if deviates Z have zero mean, then any unconditional finite-sample consistent associative test statistic is unconditionally consistent at any α-value at least when sample sizes diverge according to the sequence $\{(vm_1, vm_2), v \geq 1\}$. In this respect, however, we can at the moment only conjecture that the test statistic T is weak unconditional consistent for $\alpha > 0$ if and only if it is weak unconditional finite-sample consistent.

Theorems 9–12 can clearly be extended to a sequence of nominal and/or ordered categorical and/or discrete and/or real variables by using the NPC. They can also be extended to multidimensional variables.

4.5.3 *Some Applications of Finite-Sample Consistency*

We consider a (countably) large set of variables analysed by means of test statistics according to the so-called direct combination of several partial tests (see (h) in Section 4.2.4). The direct method of combination is chosen here mainly because proofs are easier to obtain. Extensions to other combining functions are essentially straightforward and left to the reader as exercises.

Example 1. By way of a typical situation, let us consider a two-sample design with $V \geq 1$ homoscedastic variables $\mathbf{X} = (X_1, \ldots, X_V)$, in which the observed data set is $X(\boldsymbol{\delta}) = \{\delta_h + Z_{h1i}, i = 1, \ldots, n_1; Z_{h2i}, i = 1, \ldots, n_2; h = 1, \ldots, V\}$, and the hypotheses are $H_0 : \{\mathbf{X}_1 \overset{d}{=} \mathbf{X}_2\} = \{\boldsymbol{\delta} = \mathbf{0}\}$ against $H_1 : \{\mathbf{X}_1 \overset{d}{>} \mathbf{X}_2\} = \{\boldsymbol{\delta} \geq \mathbf{0}\}$, where $\boldsymbol{\delta}$ is the vector of fixed effects, $\boldsymbol{\delta} = (\delta_1, \ldots, \delta_V)$, in which δ_h is the effect for the hth variable and $\mathbf{0}$ is the vector with V null components. Suppose that the permutation test statistic has the form

$$T^*(\boldsymbol{\delta}) = \psi(V) \sum\nolimits_{h \leq V} [\bar{X}^*_{h1}(\delta_h) - \bar{X}^*_{h2}(\delta_h)],$$

the observed value of which is $T^o(\boldsymbol{\delta}) = \psi(V) \sum_{h \leq V} [\bar{X}_{h1}(\delta_h) - \bar{X}_{h2}(0)]$, where $\psi(V)$ is such that the statistic $T(\mathbf{X}(\mathbf{0}))$ is measurable as V diverges, so that $\lim_{t \uparrow \infty} \Pr\{T(\mathbf{X}(\mathbf{0})) \leq t; P_{\mathbf{Z}}\} = 1$, and $\bar{X}^*_{hj}(\delta_h) = \sum_{i \leq n_j} X^*_{hji}(\delta_h)/n_j$, $j = 1, 2$, are permutation sample means of the hth variable. In other terms, the statistic T is a measurable sum of V partial tests. Suppose now that the non-centrality induced by the test statistic, that is, the global effect $\bar{\delta}_V = \psi(V) \sum_{h \leq V} \delta_h$, diverges as V diverges. To see the unconditional finite-sample consistency of T, let us consider the permutationally equivalent form of the test statistics

$$T^*(\boldsymbol{\delta}) = \psi(V) \sum_{h \leq V} \sum_{i \leq n_1} X^*_{h1i}(\delta_h) = \psi(V) \sum_{i \leq n_1} \sum_{h \leq V} X^*_{h1i}(\delta_h)$$

$$= \sum_{i \leq n_1} Y^*_{1i}(\boldsymbol{\delta}) = T^*(\mathbf{0}) + n_1 \bar{\delta}^*_V,$$

where the $Y_{1i}(\delta) = \psi(V) \sum_{h \leq V} X_{h1i}(\delta_h)$, $i = 1, \ldots, n_1$, are univariate data transformations which summarize the whole set of information on effects $\boldsymbol{\delta}$ collected by the V variables, $\bar{\delta}^*_V = \psi(V) \sum_{h \leq V} \delta^*_h$, $T^*(\mathbf{0})$ is the null permutation value of T which is a function only of random deviates $\mathbf{Z}^*_1 \in \mathbf{Z}$, and where, of course, the vectors $\mathbf{X}_{\cdot i} = (X_{1 \cdot i}, \ldots, X_{V \cdot i})$ relative to the n units are permuted. The right-hand side expression shows that a multivariate test statistic is reduced to a single one-dimensional quantity. Thus conditions of Theorem 9 are satisfied because, by assumption, $T^*(\mathbf{0})$ is measurable and $\bar{\delta}_V$ diverges. And so T is unconditionally finite-sample consistent.

A particular case occurs when all component variables $X_h(\delta_h)$, $h = 1, \ldots, V$, are provided with finite mean value, $\mathbb{E}[|X_h(\delta_h)|] < \infty$, $h = 1, \ldots, V$. Then we may set $\psi(V) = 1/V$. By the strong law of large numbers, $T(\mathbf{X}(\mathbf{0}))$ strongly converges to zero. If assumptions are such that $\bar{\delta}_V = \sum_{h \leq V} \delta_h / V$ is positive in the limit, we may apply Theorem 10 to achieve finite-sample consistency (see also Examples 3–5 below). We guess that in this case the unconditional finite-sample consistency is strong or with probability one.

Example 2. Let us suppose that the heteroscedastic data set is $\mathbf{X}(\boldsymbol{\delta}) = (\delta_h + \sigma_h Z_{h1i}, i = 1, \ldots, n_1, \sigma_h Z_{h2i}, i = 1, \ldots, n_2; h = 1, \ldots, V)$ for the hypotheses $H_0 : \{\mathbf{X}_1 \overset{d}{=} \mathbf{X}_2\} = \{\boldsymbol{\delta} = \mathbf{0}\}$ against

$H_1 : \{\mathbf{X}_1 \overset{d}{>} \mathbf{X}_2\} = \{\boldsymbol{\delta} \geq \mathbf{0}\}$, where δ_h and σ_h are the fixed effect and the scale coefficient of the hth variable. Suppose also that the test statistic has the form

$$T^*(\boldsymbol{\delta}) = \psi(V) \sum_{h \leq V} [\bar{X}^*_{h1i}(\delta_h) - \bar{X}^*_{h2}(\delta_h)]/S_h,$$

with observed value $T^o(\boldsymbol{\delta}) = \psi(V) \sum_{h \leq V} [\bar{X}_{h1}(\delta_h) - \bar{X}_{h2}(0)]/S_h$, where S_h is a permutation invariant statistic for the hth scale coefficient σ_h, that is, a function $S[X_{hji}(\delta_h), i = 1, \ldots, n_j, j = 1, 2]$ of pooled data, so that both conditional and unconditional distributions of $[\bar{X}_{h1}(\delta_h) - \bar{X}_{h2}(0)]/S_h$ are invariant with respect to scale σ_h, $h = 1, \ldots, V$, and $\psi(V)$ is such that the statistic $T^*(\mathbf{0})$ is measurable as V diverges. Therefore, the statistic T is a measurable sum of V scale-invariant partial tests. Since S_h is a function of random data, and thus is a random quantity, the scale-invariant non-centrality functional $\psi(V) \sum_{h \leq V} \delta_h / S_h$ becomes a random quantity which we may denote by $\bar{\Delta}_V$. Also, we may write the tests statistic as $T(\bar{\Delta}_V)$. Suppose now that the associated sequence of random effects $\{\bar{\Delta}_V, V \geq 1\}$, being the sum of V non-negative random quantities, diverges as V diverges.

To see the finite-sample consistency of $T(\bar{\Delta}_V)$, let us consider the permutationally equivalent form of the test statistics

$$T^*(\bar{\Delta}_V) = \psi(V) \sum_{h \leq V} \sum_{i \leq n_1} X^*_{h1i}(\delta_h)/S_h = \psi(V) \sum_{i \leq n_1} \sum_{h \leq V} X^*_{h1i}(\delta_h)/S_h$$

$$= \sum_{i \leq n_1} Y^*_{1i}(\bar{\delta}) = T^*(\mathbf{0}) + n_1 \bar{\Delta}^*_V,$$

where the $Y_{1i}(\bar{\delta})$, $i = 1, \ldots, n_1$, are univariate data transformations which summarize the whole set of information on effects $\boldsymbol{\delta}$ collected by the V variables and $\bar{\Delta}^*_V = \psi(V) \sum_{h \leq V} \delta^*_h / S_h$. The right-hand-side expression shows that a multivariate test statistic is reduced to one univariate.

In order for Theorem 11 to be applicable, it is worth noting that in the example we do not require that the δ_h are all positive; what is important is that $\bar{\Delta}_V$ diverges at least in probability as V diverges while $T^*(\mathbf{0})$ is measurable. Therefore, T is unconditional finite-sample consistent at least in the weak form. It is also emphasized that the V variables need not be independent – they can be dependent in any way – because their dependences are nonparametrically taken into consideration by the NPC procedure. What is important is that the distribution induced by $T(\mathbf{X}(\mathbf{0}))$ is measurable and that induced by $T(\mathbf{X}(\boldsymbol{\delta}))$ diverges at least in probability.

Remark 1. With reference to Example 2, it is worth observing that, since the statistics S_h are functions of the data, the resulting random effects Δ_V, being data dependent, are not independent of random deviates $\mathbf{Z}$ (see Section 2.1.1). We can see another way of dealing with random effects. Suppose the data are as in Example 2 and that the test statistic is $T = \sum_{hi} \varphi(X_{h1i})$, where φ is a suitable non-degenerate measurable non-decreasing function such that $\mathbb{E}[\varphi(X_h)]$ is finite for $1 \leq h \leq V$. Thus the data $\mathbf{X}(\delta)$ are transformed to $\mathbf{Y}(\bar{\delta}) = \varphi(\mathbf{X}(\bar{\delta})) = [\varphi(\delta_h + \sigma_h Z_{h1i}), i = 1, \ldots, n_1, \varphi(\sigma_h Z_{h2i}), i = 1, \ldots, n_2; h = 1, \ldots, V]$. In such a case, by denoting $\varphi(\sigma_h Z_{h2i}) = Y_{h2i}(0)$, we may write $\varphi(\delta_h + \sigma_h Z_{h1i}) = Y_{h1i}(0) + \Delta_\varphi(\delta_h, \sigma_h Z_{h1i})$, where $Y_{h1i}(0) = \varphi(\sigma_h Z_{h1i})$. Thus the resulting effects $\Delta_\varphi(\delta_h, \sigma_h Z_{h1i}) = \varphi(\delta_h + \sigma_h Z_{h1i}) - Y_{h1i}(0)$, containing random elements are random quantities which in turn are dependent on deviates Z.

Example 3. Let us consider a case where all V variables have null first moments and finite scale coefficients, $\{\mathbb{E}(Z_h) = 0, 0 < \sigma_h < \infty, h = 1, \ldots, V\}$, and have finite positive effects ($\delta_h > 0$, $h = 1, \ldots, V$). For the same hypotheses as Examples 1 and 2, consider the test statistic

$$T^*(\boldsymbol{\delta}) = (n_1)^{-1} \sum_{i \leq n_1} \sum_{h \leq V} X^*_{h1i}(\delta_h)/V = (n_1)^{-1} \sum_{i \leq n_1} Y^*_{1i}(\bar{\delta}_V)$$

where the $Y_{1i}(\bar{\delta}_V) = \sum_{h\leq V} \sigma_h Z_{h1i}/V + \sum_{h\leq V} \delta_h/V$, the mean value of which is $0 + \bar{\delta}_V$. As V diverges, provided that conditions for the weak law of large numbers for non-i.i.d. variables holds, we may state that $T^o(\boldsymbol{\delta}) = \sum_{i\leq n_1} Y_{1i}(\bar{\delta}_V)/n_1$ converges in probability to $\lim_{V\uparrow\infty} \bar{\delta}_V > 0$. Thus, since $\lim_{V\uparrow\infty} \sum_{h\leq V} \sigma_h Z_{h1i}/V \overset{P}{=} 0$, so that the whole null distribution collapses towards 0 in probability, we have that for every α-value not smaller than the minimum attainable α_a, the limit critical point of T is zero, and then in the limit $T^o(\boldsymbol{\delta})$ falls in the critical region in probability.

In order for Theorem 9 to be applicable, it is worth noting that we do not require the effects δ_h to be all positive; what is important is that $\bar{\delta}_V$ diverges as V diverges while $T^*(\mathbf{0})$ is measurable. Actually, on the one hand, we can ignore a finite number of variables and related effects without compromising the divergence of the sequence $\{\bar{\delta}_V, V \geq 1\}$; on the other hand, to fit its conditions we could extract from $\{\bar{\delta}_V, V \geq 1\}$ one monotonically divergent subsequence $\{\bar{\delta}_v, v \geq 1\}$ by discarding any finite number of its elements. Note also that we do not require the V variables to be independent. As a matter of fact they can be dependent in any way. What is important is that the distribution induced by $T(\mathbf{X}(\mathbf{0}))$ is measurable and that induced by $T(\mathbf{X}(\boldsymbol{\delta}))$ diverges at least in probability. In such a case unconditional finite-sample consistency is at least in accordance with the weak form. Of course, similar reasoning is applicable to Theorem 10.

Example 4. We now turn to a situation in which conditions of Theorem 10 do not occur. Consider a two-sample design where the data belong to a V-variate Cauchy distribution with independent homoscedastic components, with constant fixed effects and constant scale coefficients ($\delta_h = \delta, \sigma_h = \sigma, h = 1, \ldots, V$), and where the test statistic and hypotheses are as in Example 3 above. In such a case, since a permutationally equivalent test statistic has the form

$$T^*(\bar{\delta}_V) = (n_1)^{-1} \sum_{i\leq n_1} \sum_{h\leq V} X^*_{h1i}(\delta)/V = (n_1)^{-1} \sum_{i\leq n_1} Y^*_{1i}(\bar{\delta}_V),$$

which shows that, due to the well-known property of the Cauchy distribution (whereby the Y_{1i} being the arithmetic means of V i.i.d. components are Cauchy with location δ and scale σ), $T^*(0)$ is Cauchy located at 0 and the global effect $\bar{\delta}_V = \delta$ does not diverge as V diverges. Thus, in such a case this test, although unbiased, is not unconditionally finite-sample consistent for fixed sample sizes. In this case, since $\sum_{h\leq V} X_{h1i}(\delta)/V = \sum_{h\leq V}(\sigma Z_{h1i} + \delta)/V = \sum_{h\leq V} \sigma Z_{h1i}/V + V\delta/V$, we have that there does not exist a function $\rho(\delta)$ such that $Z\rho(\delta)$ collapses in probability towards zero whereas $\delta\rho(\delta)$ is positive. The same result applies to deviates Z with Pareto distribution $\mathcal{P}a(\theta, \sigma)$, with shape parameter $0 < \theta \leq 1$.

It is, however, to be emphasized that for fixed (n_1, n_2) and δ divergent, with random deviates distributed according to either Cauchy $\mathcal{C}y(0, \sigma)$ or Pareto $\mathcal{P}a(\theta, \sigma)$, the latter with shape parameter $0 < \theta \leq 1$, are conditional and unconditional finite-sample consistent. Whereas, when δ is fixed and V diverges both are not consistent, because in this case the law of large numbers does not apply.

Example 5. The results of Examples 1–4 can be generalized to variables which do not possess finite means, that is, the number of variables such that $\mathbb{E}(Z_h)$ does not exist as V diverges. Let us argue within a two-sample design and suppose that V homoscedastic deviates $\mathbf{Z} = (Z_1, \ldots, Z_V)$ have finite unique medians, that is, $\mathbb{M}_d(Z_h) = \eta_h$, $h = 1, \ldots, V$. Suppose also that the observed data set is $\mathbf{X}(\boldsymbol{\delta}) = \{\delta_h + \sigma_h Z_{h1i}, i = 1, \ldots, n_1, Z_{h2i}, i = 1, \ldots, n_2; h = 1, \ldots, V\}$, and the hypotheses are $H_0 : \{\boldsymbol{\delta} = \mathbf{0}\}$ against $H_1 : \{\boldsymbol{\delta} > \mathbf{0}\}$, where $\boldsymbol{\delta} = (\delta_1, \ldots, \delta_V)$. In this setting, consider the test statistic

$$T^*_{Md}(\tilde{\delta}_V) = \frac{1}{n_1} \sum_{i\leq n_1} \tilde{Y}^*_{1i}(\tilde{\delta}_V) - \frac{1}{n_2} \sum_{i\leq n_2} \tilde{Y}^*_{2i}(\tilde{\delta}_V),$$

where $\tilde{Y}^*_{ji}(\tilde{\delta}_V) = \mathbb{M}d[X^*_{hji}(\delta)/S_{Mh}, h = 1, \ldots V]$, $i = 1, \ldots, n_j$, $j = 1, 2$, is the median of V scale-free individual variables and $S_{Mh} = MAD_h = \mathbb{M}d[|X_{hji} - \tilde{X}_h|, i = 1, \ldots, n_j, j = 1, 2]$ is the median of absolute deviations from the median specific to the hth variable. As V diverges $\frac{1}{n_1}\sum^*_{i \le n_1} \tilde{Y}_{1i}(\tilde{\delta}_V)$ and $\frac{1}{n_2}\sum_{i \le n_2} \tilde{Y}_{2i}$, weakly converge to $\mathbb{M}d(Y_1(\tilde{\delta}_V)) > 0$ and 0, respectively. Thus, Theorem 10 applies and so $T^*_{Md}(\tilde{\delta}_V)$ is unconditional finite-sample consistent without requiring the existence of any positive moment for V variables.

The same problem can also be solved, for instance, by combining V Mann–Whitney's statistics (direct combination). To this end (see Section 2.4.1) let us consider the statistic

$$T^*_{MW}(\ddot{\delta}_V) = \frac{1}{Vn_1}\sum_{h \le V}\sum_{i \le n_1} \bar{F}_h[X^*_{h1i}(\delta)] = \frac{1}{n_1}\sum_{i \le n_1} \ddot{F}[X^*_{h1i}(\delta)],$$

where $\bar{F}_h(t) = \sum_{i=1}^n \mathbb{I}(X_{h \cdot i} \le t)/n$ is the pooled EDF for the hth variable and $\ddot{F}$ is the average EDF which is finite because all $\bar{F}_h$s are finite. It is thus possible to apply the law of large numbers for non-i.i.d. variables to show that $T^*_{MW}(0)$ converges to 0 and $T^*_{M,W}(\ddot{\delta}_V)$ converges to a positive number. Thus, as the conditions of Theorem 10 are satisfied, $T^*_{MW}(\ddot{\delta}_V)$ is finite-sample consistent.

Remark 2. It is straightforward to prove that T^*_{MW} is also appropriate for stochastic effects Δ and when some of the variables are ordered categorical and others numeric. If some variables are nominal and others ordered categorical or numeric, a proper strategy is as follows: (i) separately combine the nominal variable with T^*_1; (ii) combine the others with T^*_2; (iii) and then combine the former two with $T''' = \psi(\lambda_1, \lambda_2)$.

Remark 3. As an application of Theorem 12 above, let us consider the unconditional permutation power of a test statistic T for fixed sample sizes, with $V \ge 2$ i.i.d. variables and fixed effect δ, calculated in two ways: (i) by considering two V-dimensional samples sized m_1 and m_2 respectively; and (ii) by considering two unidimensional samples sized $n_1 = Vm_1$ and $n_2 = Vm_2$. Since the unconditional power essentially depends on the non-centrality induced by T, and two ways produce exactly the same non-centrality and the same underlying likelihood, we expect them to have the same power, at least approximately. Indeed, due to the discreteness of permutation distributions and consequent conservativeness of related tests, if sample sizes (m_1, m_2) are not too small, so that both permutation distributions are practically almost continuous and share about the same p-value support (i.e. $\Lambda^{(n)}_{\mathbf{X}(0)} = \Lambda^{(m)}_{\mathbf{X}(0)}$), then the two ways have approximately the same power. Thus, we can call this the *equipower property* of permutation tests. To give evidence of this, we report in Table 4.6 some Monte Carlo estimates of rejection rates for a two-sample design and one-sided alternatives, $\delta > 0$, of the test statistics $T^*_1(\delta) = \sum_{i \le n_1} X^*_{1i}(\delta)$ and $T^*_V = \sum_{i \le m_1} Y^*_{1i}(\delta)$, where $Y^*_{1i}(\delta) = \sum_{h \le V} X^*_{h1i}(\delta)$, with $n_1 = n_2 = 50$, at $\alpha = (0.01, 0.05, 0.10)$, with $MC = 4000$ Monte Carlo experiments, $B = 4000$ random inspections of permutation space, and data coming from population distributions Cauchy $\mathcal{C}y(0, 1)$ with no mean and infinite variance, Student's t with 2 degrees of freedom $\mathcal{S}t(2)$ with finite mean and infinite variance, Exponential $\mathcal{E}x(1)$ with finite mean and variance, and Normal $\mathcal{N}(0, 1)$.

The results of this simple Monte Carlo study confirm our thoughts about the approximate equipower property of permutation tests. The approximation seems to be better if the V-dimensional sample sizes (m_1, m_2) are not too small. For instance, the apparent power reduction of rejection rates for $V = 10$ with respect to those for $V = 1$ is substantially due to the fact that the true attained α-values are (0.00794, 0.0477, 0.0993) instead of the nominal values (0.01, 0.05, 0.10), respectively (in accordance with this, the rejection rates for the univariate $\mathcal{N}(0, 1)$ are 0.498, 0.781, and 0.875 instead of 0.553, 0.791, and 0.881).

There is, however, a difference in computing time between the two ways, because permuting B times $n = n_1 + n_2$ elements takes longer than permuting $m = m_1 + m_2 = n/V$ elements. Thus,

Table 4.6 Monte Carlo estimates of rejection rates for a two-sample design and one-sided alternatives

Distributions	α	V			
		1	2	5	10
	0.01	0.085	0.084	0.076	0.055
$\mathcal{C}y(0, 1)$, $\delta = 1$	0.05	0.187	0.188	0.186	0.175
	0.10	0.270	0.273	0.273	0.271
	0.01	0.263	0.269	0.218	0.159
$\mathcal{S}t(2)$, $\delta = 0.5$	0.05	0.484	0.488	0.452	0.424
	0.10	0.596	0.600	0.580	0.571
	0.01	0.573	0.559	0.537	0.336
$\mathcal{E}x(1)$, $\delta = 0.5$	0.05	0.801	0.795	0.789	0.733
	0.10	0.889	0.878	0.884	0.854
	0.01	0.553	0.538	0.511	0.332
$\mathcal{N}(0, 1)$, $\delta = 0.5$	0.05	0.791	0.790	0.782	0.720
	0.10	0.881	0.882	0.875	0.847

from the computing point of view, when possible it would be convenient to process V-dimensional data according to the design outlined in Theorem 12 above. When transforming a one-dimensional data set into a V-dimensional equivalent we need to randomly associate V observed data with each of m pseudo units – a process which implies a sort of auxiliary randomization in which respect there are known questionable problems connected with the possibility of introducing some non-objective elements: two statisticians may obtain different conclusions even though starting with the same data set (see Scheffé, 1944; Pesarin, 1984; Lehmann, 1986).

4.6 Some Examples of Nonparametric Combination

In this section we discuss some typical examples. The main aim of this discussion is to illustrate the potential, flexibility, utility and effectiveness of NPC methods. Example 1 shows that NPC methods really do take into account the underlying dependence relations among partial tests, making it unnecessary to specify them. Example 2 shows that the quadratic combination form converges to its parametric unconditional counterpart under the conditions for the latter. Examples 3–6 discuss multi-aspect testing problems, which are appropriate when specific side-assumptions allow us to hypothesize that treatments may influence more than one aspect of the underlying distributions. Example 7 deals with testing problems related to aspects of monotonic stochastic ordering; Example 8 refers to the so-called Behrens–Fisher problem. The section concludes with a real research example.

Example 1. *An artificial example.*
In order to appreciate that the NPC method does actually take into account the underlying dependence relations among partial tests, let us consider a somewhat artificial example. Table 4.7 gives data from Pollard (1977) on lengths of worms, also shown in Table 1.4; for the sake of simplicity only two groups, the first and the third, are analysed here. We wish to test whether the mean length of worms, variable X, is the same in both groups, against the alternative that group 3 is stochastically greater than group 1. Together with variable X, which is the only one actually observed,

Table 4.7 Lengths of worms in groups 1 and 3

Group 1			Group 3		
X_1	Y_1	W_1	X_3	Y_3	W_3
10.2	3.19	2.98	9.2	3.03	3.00
8.2	2.86	2.86	10.5	3.24	3.03
8.9	2.98	3.19	9.2	3.03	2.95
8.0	2.83	2.88	8.7	2.95	3.24
8.3	2.88	2.83	9.0	3.00	3.03
8.0	2.83	2.83			

let us consider two more artificial variables: $Y = +\sqrt{X}$ and W which, within each group, displays the same Y values except that they are assigned to different units, so that Y and W share the same sample means and variances as well as partial p-values $\lambda_Y = \lambda_W$, but are differently associated with X. Thus, the problem becomes artificially multivariate.

Observe that Y is simply a nonlinear monotonic transformation of X and therefore, in light of the concurrent multi-aspect testing problem, it may contain only a small amount of further relevant information on the distributional diversity of the two groups, especially if the assumption of additive effects is at least approximately satisfied (see also Example 3 below). Moreover, although within the two groups W has the same sample means as Y, it is differently associated with X, so it contains (essentially artificial) information on the distributional diversity between the two groups (it is well known that in multivariate problems information on distributional diversity is partially contained in the dependence relations).

Let us consider the following analyses:

(a) $H_{0a} : \{(X_1, Y_1) \stackrel{d}{=} (X_3, Y_3)\} = \{(X_1 \stackrel{d}{=} X_3) \bigcap (Y_1 \stackrel{d}{=} Y_3)\}$ against $H_{1a} : \{(X_1 \stackrel{d}{<} X_3) \bigcup (Y_1 \stackrel{d}{<} Y_3)\}$;

(b) $H_{0b} : \{(X_1, W_1) \stackrel{d}{=} (X_3, W_3)\} = \{(X_1 \stackrel{d}{=} X_3) \bigcap (W_1 \stackrel{d}{=} W_3)\}$ against $H_{1b} : \{(X_1 \stackrel{d}{<} X_3) \bigcup (W_1 \stackrel{d}{<} W_3)\}$.

It should be noted that all alternatives are restricted, so that parametric solutions are notoriously difficult. By using $B = 4000$ CMC iterations, we obtain estimated partial p-values $\hat{\lambda}_X = 0.096$, $\hat{\lambda}_Y = 0.084$ and $\hat{\lambda}_W = 0.082$.

Combination by Liptak's combining function gives the combined estimates $\hat{\lambda}''_a = 0.084$ and $\hat{\lambda}''_b = 0.043$, respectively. Of course, at $\alpha = 0.05$, the latter is significant while the former is not. We observe that in practice Y makes only a small contribution to the discrimination of the two groups. The small difference between the two p-values $\hat{\lambda}_X$ and $\hat{\lambda}_Y$ may be attributed to the possibly better discrimination effectiveness of Y with respect to X because the square root transformation seems to produce a sort of symmetrization on the resulting distribution, so that an additive effect model may fit Y better than X with the present data; the very small difference between $\hat{\lambda}_Y$ and $\hat{\lambda}_W$ is exclusively due to the CMC accuracy.

Although W gives exactly the same partial p-value as Y, due to being almost independent of X, it appears to contain artificial information useful for discrimination. Indeed, it plays the role of a complementary variable, and we know that if the monotonic dependence between X and W is 'small', the information increment furnished by W over that contained in X is 'higher' than if this dependence were almost sure, as occurs for Y. Actually, the minimal increment of information occurs when two variables are one-to-one, and the maximal when they are independent. However, it is important to stress that the NPC method does actually take into account the underlying dependence among all partial tests, without the necessity of explicitly incorporating any coefficient in the analysis.

Example 2. *Two quadratic combining functions.*
As a second example, let us consider the standard problem of the Mahalanobis D^2 in a *two-sample bivariate normal* situation. It is well known that the null distribution of

$$D^2 = \frac{W_1^2 + W_2^2 - 2\rho W_1 W_2}{1 - \rho^2}$$

is a central chi-square with 2 d.f.; here

$$W_j = \left(\bar{X}_{j1} - \bar{X}_{j2}\right) \frac{\sqrt{n_1 n_2/(n_1 + n_2)}}{\sigma_j}, \quad j = 1, 2,$$

are the standardized marginal normal tests, where the correlation coefficient ρ and σ_j, $j = 1, 2$, are population parameters,

As a competitor, let us consider the permutation solution provided by a quadratic combination (see (e) in Section 4.2.4) given by

$$T_H'' = \frac{T_1^{*2} + T_2^{*2} - 2\rho^* T_1^* T_2^*}{1 - \rho^{*2}},$$

where the permutation partial tests are

$$T_j^* = (\bar{X}_{j1}^* - \bar{X}_{j2}^*) \frac{\sqrt{n_1 n_2/(n_1 + n_2)}}{\sigma_j^*}, \quad j = 1, 2,$$

which have the same form as the corresponding marginal tests in D^2, except that σ_j^* are now calculated on the permutation data set $\mathbf{X}^*$ (see Remark 3, 4.2.4), and the correlation coefficient is $\rho^* = \mathbb{C}ov(T_1^*, T_2^*)$.

Note that both bivariate tests D^2 and T_H'' are based on quadratic forms and are appropriate for invariant testing with respect to alternatives lying at the same distance from H_0 and so they are comparable, at least asymptotically. In particular, it is worth noting that D^2 is a UMPU invariant test and that T_H'' is based *only* on permutation quantities. Recalling that as sample sizes go to infinity, the permutation correlation coefficient ρ^* almost surely converges to ρ and the observed values T_j^o almost surely converge to W_j, $j = 1, 2$, the permutation statistic T_H'' (see Theorem 6 and Remark 7, 4.4.2) almost surely converges to D^2, so that they are asymptotically equivalent. On the one hand, however, parametric solutions generally look for statistics presenting known reference null distributions, at least for large sample sizes. Thus, in their expressions they must incorporate all dependence coefficients present in P, where these coefficients must be either known or unconditionally estimated from the given data set $\mathbf{X}$, within the underlying nonparametric family $\mathcal{P}$. This is a task which may occur in quite simple situations. In practice, it can be attained when all underlying dependence relations are linear, and for invariance testing with respect to alternatives lying at the same quadratic distance from H_0, where this distance is a metric defined in the parameter space and expressed in terms of response values $\mathbf{X}$. On the other hand, NPC procedures look for simple expressions of test statistics and leave the task of providing the related reference distributions to the CMC procedure by taking account, in a nonparametric way, of all underlying dependences, including situations in which these are parametrically intractable.

Remark 1. In defining T_Q'' (see (e) in Section 4.2.4) we need not assume normality of responses or partial tests, so that the invariance testing property is assumed for alternatives lying at the same quadratic distance from H_0 and measured on the space of inverse normal transformations of permutation p-values, that is, in terms of *normal probability distances*.

Remark 2. In a permutation context, instead of T''_Q we may use the simpler direct combining solution $T''_D = T_1^{*2} + T_2^{*2}$ as a competitor of D^2. Of course, T''_Q and T''_D are not permutationally equivalent irrespective of ρ^* (see Problem 16, 4.2.7), although a set of Monte Carlo experiments shows that their unconditional power functions are very close to each other in all conditions (the maximum difference is for $\rho^* = 1/2$, whereas they coincide exactly when $\rho^* = 0$ and $\lim |\rho^*| = 1$; see Problem 17, 4.2.7). However, it should be noted that the amount of computation required by T''_D is much less than that of T''_Q, so that for practical purposes it may be regarded as an interesting competitor for unrestricted alternatives and for *quasi-invariant* testing.

Example 3. *Multi-aspect testing*.
As a third example, let us consider a two-sample dominance problem on *positive* univariate variables where, as usual, dominance means that in the alternative we have two CDFs related, for instance, by $F_1(x) \leq F_2(x)$, $x \in \mathcal{R}^1$. Let the side-assumptions for the problem be that the treatment may act on the first two moments of responses belonging to the first group. Moreover, and without loss of generality, let us assume that the data set and response model behave as $X_{1i} = \mu + \Delta_{1i} + Z_{1i}$, $X_{2i} = \mu + Z_{2i}$, $i = 1, \ldots, n_j$, $j = 1, 2$, where μ is a population nuisance constant, Z_{ji} are exchangeable random errors such that $\mu + Z_{ji} > 0$ in probability, and $\Delta_{1i} \geq 0$ are non-negative stochastic effects which may depend on $\mu + Z_{1i}$ and in addition satisfy the second-order condition $(\mu + \Delta_{1i} + Z_{1i})^2 \geq (\mu + Z_{1i})^2$, $i = 1, \ldots, n_1$.

Suppose that the hypotheses are $H_0 : \{X_1 \stackrel{d}{=} X_2\}$ against $H_1 : \{X_1 \stackrel{d}{>} X_2\} = \{X_1 - \Delta \stackrel{d}{=} X_2\}$, and that, focusing on the assumed side-conditions, we are essentially interested in the first two moments, so that the hypotheses become equivalent to $H_0 : \{(\mu_{11} = \mu_{12}) \bigcap (\mu_{21} = \mu_{22})\}$ and $H_1 : \{(\mu_{11} > \mu_{12}) \bigcup (\mu_{21} > \mu_{22})\}$, where $\mu_{rj} = \mathbb{E}(X_j^r)$ is the rth moment of the jth variable.

In order to deal with this typical *multi-aspect testing* problem (see also Fisher, 1935), we may first apply one partial permutation test to each concurrent aspect, $T_1^* = \sum_i X_{1i}^*$ and $T_2^* = \sum_i X_{1i}^{*2}$, followed by their NPC. By analysis of two permutation structures (see Remark 1, 2.7), it is easy to show that, in the null hypothesis, the joint distribution of two partial tests depends only on exchangeable errors, so that partial and combined permutation tests are all exact. Furthermore, the same analysis shows that two partial tests are marginally unbiased because both marginal distributions are ordered with respect to treatment effect. In order to see this, let us focus on one permutation in which ν^* elements are randomly exchanged between two groups, so that, in accordance with the pointwise representation and with obvious notation, we jointly have

$$T_1^*(\Delta) = \sum_i (\mu + \Delta_{1i}^* + Z_{1i}^*) \geq T_1^*(0) = \sum_i (\mu + Z_{1i}^*)$$

and

$$T_2^*(\Delta) = \sum_i (\mu + \Delta_{1i}^* + Z_{1i}^*)^2 \geq T_2^*(0) = \sum_i (\mu + Z_{1i}^*)^2,$$

because in both statistics there are ν^* elements where $\Delta_{1i}^* = 0$ and $n_1 - \nu^*$ where $\Delta_{1i}^* \geq 0$.

Thus, the NPC gives a proper solution. Note that if in Example 1 above we only consider two variables $Y = +\sqrt{X}$ and $Y^2 = X$, we may have one simple example of the present problem. Other examples will be presented in later chapters. One important application is related to the exact solutions of the univariate Behrens–Fisher problems in experimental situations, in which the null hypothesis assumes that data are exchangeable between groups (see Example 8). One more application is to approximate testing for joint equality of locations and scale coefficients (see Remarks 11 and 12).

Remark 3. Side-assumptions for multi-aspect testing in problems like those described above may reflect situations in which treatment is presumed to influence the first two moments in such a way that the two functionals μ_1 and μ_2 contain all the information on the effect. This condition may be appropriate when, in the alternative, all other aspects (moments or functionals) of the response distribution may be expressed in terms of the first two moments only, as when the rth moment $\mu_r(\Delta) = \mu_r[\mu_1(\Delta), \mu_2(\Delta)]$, for all integers such that $\mu_r(\Delta)$ is finite. This situation may occur when stochastic effects follow a model such as $0 \leq \Delta = \delta + \sigma_\Delta(\delta) \cdot W$, where random quantities W may depend on errors Z, $\delta > 0$ is the average effect, and $\sigma_\Delta(\delta)$ is a scale coefficient expressed as a function of δ, which satisfies the additional condition $\sigma_\Delta(0) = 0$. A very particular case occurs when responses are $X_{1i} = \mu + \delta + \sigma \cdot (1 + \delta\sigma_\Delta) \cdot Z_{1i}$, $X_{2i} = \mu + \sigma \cdot Z_{2i}$, $i = 1, \ldots, n_j$, $j = 1, 2$, where the scale coefficient on the first group is assumed to be linearly related to the location fixed effect.

Note that, in this context, all elements which characterize underlying distributions P and that are not functions of (μ_1, μ_2) are assumed to be unaffected by treatment. It is also worth noting that multi-aspect testing does not provide separate tests on effects δ and $\sigma_\Delta(\delta)$ since these are confounded in two partial tests.

Remark 4. The NPC solution discussed can easily be extended to C-sample problems if, together with the same assumptions for unbiasedness in one-way ANOVA (see Section 1.11), we may also assume that: (a) response variables are positive; (b) stochastic effects are non-negative and may act on the first two moments; (c) in the alternative, for every pair $j \neq h$, $j, h = 1, \ldots, C$, stochastic effects are pairwise ordered with probability one, so that either $\Delta_j \overset{d}{<} \Delta_h$ or $\Delta_j \overset{d}{>} \Delta_h$, and so, for every $x \in \mathcal{R}^1$ and pair $h \neq j = 1, \ldots, C$, corresponding CDFs obey either $F_j(x) \leq F_h(x)$ or $F_j(x) \geq F_h(x)$; (d) for every $i = 1, \ldots, n_j$ and $j = 1, \ldots, C$, responses satisfy the second-order condition $(\mu + \Delta_{ji} + Z_{ji})^2 \geq (\mu + Z_{ji})^2$. Under these conditions, the hypotheses $H_0 : \{X_1 \overset{d}{=} \ldots \overset{d}{=} X_C\}$ against $H_1 : \{H_0 \text{ is not true}\}$ become $H_0 : \left\{(\mu_{11} = \ldots = \mu_{1C}) \bigcap (\mu_{21} = \ldots = \mu_{2C})\right\}$ and $H_1 : \{\text{At least one equality is not true}\}$.

In order to deal with this problem, we may first consider the two permutation partial tests, $T_1^* = \sum_j n_j \cdot \left(\bar{X}_j^*\right)^2$ and $T_2^* = \sum_j n_j \cdot \left[\sum_i \left(X_{ji}^*\right)^2 / n_j\right]^2$, followed by their NPC (for marginal unbiasedness of two partial tests see Problem 20, 4.2.7, and for unbiasedness of the combined solution see Theorem 3, 4.3.2). Note that T_2^* corresponds to a one-way ANOVA test on squared data transformations, $Y_{ji} = (X_{ji})^2$.

Multivariate extensions to this problem are straightforward.

Let us extend the solution for two-sample dominance problems on positive univariate variables to k general aspects of interest by means of a *multi-aspect strategy*. To this end, suppose that the side-assumptions for the problem are that effects of symbolic treatment may act on $k \geq 2$ functionals $\eta_{rj} = \mathbb{E}[\eta_r(X_j)]$, $j = 1, 2$, $r = 1, \ldots, k$, where η_r are monotonic functions with finite expectation and none of them is a function of any other. Without loss of generality, let us again assume that the response model is $X_{1i} = \mu + \Delta_{1i} + Z_{1i}$, $X_{2i} = \mu + Z_{2i}$, $i = 1, \ldots, n_j$, $j = 1, 2$, where Z_{ji} are exchangeable random errors so that $\mu + Z_{ji} > 0$ in probability, and $\Delta_{1i} \geq 0$ are non-negative stochastic effects which satisfy the additional conditions $\eta_r(\mu + \Delta_{1i} + Z_{1i}) \geq \eta_r(\mu + Z_{1i})$, $i = 1, \ldots, n_1$, $r = 1, \ldots, k$.

Suppose that the hypotheses to test are $H_0 : \{X_1 \overset{d}{=} X_2\}$ against $H_1 : \{X_1 \overset{d}{>} X_2\} = \{X_1 - \Delta \overset{d}{=} X_2\}$. Let us now assume that the set of $2k$ functionals $\{\eta_{rj}, \ j = 1, 2, \ r = 1, \ldots, k\}$ contains all the information on the treatment effect; in particular, we note once again that, in this context, all elements which characterize underlying distributions P and that are not functions of $\{\eta_{rj}, \ j = 1, 2, \ r = 1, \ldots, k\}$ are assumed to be unaffected by treatment. Thus, with obvious notation, the hypotheses become $H_0 : \left\{\bigcap_r (\eta_{r1} = \eta_{r2})\right\}$ and $H_1 : \left\{\bigcup_r (\eta_{r1} > \eta_{r2})\right\}$, respectively. In order to deal with this problem, we may first apply k partial permutation tests $T_r^* = \sum_i \eta_r(X_{1i}^*)$, $r = 1, \ldots, k$, followed by their NPC. By analysis of k permutation structures (see Remark 1, 2.7), it is easy to show that, in the null hypothesis, as the joint distribution of k partial tests depends only on exchangeable errors, combined and partial tests are all exact; moreover, all partial tests are marginally unbiased, because related marginal distributions are ordered with respect to treatment effect. Thus, the NPC gives a proper unbiased solution. This solution may be useful in the case of multivariate derived variables (see point (i) in Section 4.2.4).

Remark 5. The multi-aspect approach may be applied to multivariate derived variable problems when, due to the necessity of physical interpretation, we consider two (or more) derived variables (or functionals) from the entire set of observed responses (see (i) in Section 4.2.4). What is needed is that partial tests are marginally unbiased. Extension to symmetry testing on one-sample problems, including paired data (see Example 1, 2.6), is straightforward. A multivariate extension of the latter is straightforward within the NPC methodology.

It is worth observing that multi-aspect testing and related solutions may be seen as essentially intermediate between the *comparison of two locations* and the *goodness-of-fit for equality of two distributions* (see Examples 2–4, 2.7). If treatment effects are assumed to act on a finite number k of functionals, we generally expect multi-aspect tests, solved through NPC, to be more appropriate and efficient than other nonparametric competitors, especially those based only on goodness-of-fit methods. In practice, we may see goodness-of-fit procedures as multi-aspect tests in which an infinite number of aspects are involved, the great majority of effects being of little importance.

In order to illustrate how the multi-aspect procedure works, we turn to the study discussed in Massaro and Blair (2003) in which the number of breeding yellow-eyed penguin (*Megadyptes antipodes*) pairs on Stewart Island (New Zealand), where cats are present, and on adjacent cat-free islands was compared. They found 79 pairs of yellow-eyed penguins breeding in $n = 19$ discrete locations on Stewart Island (4.2 average pairs per location), and 99 pairs breeding in $n = 10$ discrete locations on cat-free islands (9.9 average pairs per location). This study suggested that feral cats can pose a serious threat to penguin offspring on Stewart Island. Data are shown in Table 4.8 (178 pairs and 29 locations/colonies in total).

To compare the numbers of breeding yellow-eyed penguin pairs from the two groups, the authors performed a bootstrap test using the raw difference of sample means as a test statistic and obtained a significant result ($p = 0.009$). However, in this study, there was not only an empirical difference between the two means, but also between the standard deviations, since the variance was found to be much smaller on Stewart Island (Neuhäuser, 2007). Here we wish to analyse the same data using the NPC methodology (instead of the bootstrap test) and applying the multi-aspect procedure, in order to jointly evaluate location (T_{μ}^*) and scatter ($T_{\sigma^2}^*$) aspects, which are supposed to be responsible for the difference between the two groups. The null hypothesis $H_0 : Y_1 \stackrel{d}{=} Y_2$ implies the event $\{E[Y_1] = E[Y_2]\} \cap \{E[Y_1^2] = E[Y_2^2]\}$. In particular, we have examined the location aspect by means of a standard t-test on the raw data and the scatter aspect using transformed data (second moments of Y). Note that this solution is exact because $H_0 : Y_1 \stackrel{d}{=} Y_2$, stating the irrelevance of feral

Table 4.8 Observed colony sizes according to Massaro and Blair (2003, p. 110)

Group (X)	Colony sizes (Y)
Stewart Island	7 3 3 7 3 7 3 10 1 7 4 1 3 2 1 2 9 4 2
Cat-free islands	15 32 1 13 14 11 1 3 2 7

Table 4.9 Location, scatter and global p-values (Massaro and Blair, 2003, p. 110)

Aspect	p-values
T^*_μ	0.01240
$T^*_{\sigma^2}$	0.00345
Global, Tippett (G_T)	0.00623
Global, Liptak (G_L)	0.00071
Global, Fisher (G_F)	0.00095
Global G	0.00122

cats, and so the permutation testing principle applies. The main property of the present solution is that, under the alternative, two aspects are taken into consideration and this entails a gain in power. It is to be emphasized that this multi-aspect solution provides an exact solution to the well-known Behrens–Fisher problem when treatment effect, if any, is presumed to act not only on location but also on dispersion. The Tippett, Fisher and Liptak combining functions were selected in that order. For details on iterated combinations, see Salmaso and Solari (2006). Results are shown in Table 4.9. Both the location and scatter aspects are significant, along with all the global tests obtained after combining the two partial tests related to the aspects (G_T, G_L, G_F), and the global test G obtained after combining all the previously combined global tests.

The following MATLAB code was used to carry out the analysis:

```
B=10000;
[D,data,code]=xlsimport('massaroBLAIR');
reminD(D)

stats={'t',':Y.^2'};
[P, T] = NP_2s_MA('Y','group',100000,stats,'T',1,1);

pT=P(:,:,3);
pL=NPC(P(:,:,1:2),'L',1);
pF=NPC(P(:,:,1:2),'F',1);
pT=NPC(P(:,:,1:2),'T',1);
globalP=NPC([pT pL pF],'T',1);
```

Below we also provide R code for the same analysis. As stated, this is an example of a two-sample problem, where two aspects are jointly taken into account: location and scale parameters. We then use two test statistics related to the first and second moments of data Y:

$$T^*_\mu = \bar{Y}^*_1 - \bar{Y}^*_2, \quad T^*_{\sigma^2} = \frac{1}{n_1}\sum_{i=1}^{n_1} Y^{*2}_{i1} - \frac{1}{n_2}\sum_{i=1}^{n_2} Y^{*2}_{i2}.$$

We could also use $T^*_{\sigma^2} = s^{*2}_1/s^{*2}_2$, the ratio of permutation variances, as the test statistic, but in R the computation of the p-value will be done by applying the `t2p` function, intended for linear statistics only. Note that the null hypothesis might be rejected when at least one of the aspects considered is in conflict with the related null hypothesis (not necessarily that on locations), therefore this is not necessarily a Behrens–Fisher problem (see Example 8 below). The matrix `T` contains the null distribution of T_μ (first column) and T_{σ^2} (second column); `ID` is the vector of group labels. We use `B` $= 5000$ random permutations.

```
setwd("C:/path") ; source(t2p.r)
data<-read.csv("massaroBLAIR.csv",header=TRUE)

B=5000 ; ID = group
T=array(0,dim=c((B+1),2))
T[1,1] = mean(Y[ID==1])-mean(Y[ID==2])
T[1,2] = mean(Y[ID==1]^2)-mean(Y[ID==2]^2)

for(bb in 2:(B+1)){
y.star=sample(Y)
T[bb,1] = mean(y.star[ID==1])-mean(y.star[ID==2])
T[bb,2] = mean(y.star[ID==1]^2)-mean(y.star[ID==2]^2)
}
P = t2p(abs(T)) ; P[1,]

[1] 0.0136 0.0042
```

The partial p-values reported above allow us to evaluate the strength of the evidence against the null hypotheses concerned: here we see that both p-values are significant, but there is more evidence of discrepancy between the two groups on the second moments ($p = 0.0042$). In order to obtain a global p-value, we first apply Fisher's combining function and then call the `t2p` function again:

```
T1 = apply(P,1,function(x){-2*log(prod(x))})
t2p(T1)[1]

[1] 0.0078
```

The data set and the corresponding software codes are available from the `massaro_blair` folder on the book's website.

Example 4. *Testing two-sided alternatives separately.*
Within the multi-aspect context one can go somewhat further than traditional two-sided testing by using the NPC of two one-sided tests. That is, with obvious notation, by employing the following fourfold procedure:

(i) Let H_1 be a global alternative, i.e. $H_1 : \{H_1^+ \bigcup H_1^-\}$, where the two sub-alternatives are $H_1^+ : \{\Delta > 0\}$ and $H_1^- : \{\Delta < 0\}$. Of course, it is to be emphasized that in the traditional two-sided setting *one and only one* of H_1^+ and H_1^- is active.
(ii) The two related partial test statistics are $T_+^* = S_1(\mathbf{X}_1^*) - S_2(\mathbf{X}_2^*)$, and $T_-^* = S_2(\mathbf{X}_2^*) - S_1(\mathbf{X}_1^*)$, for the sub-alternatives H_1^+ and H_1^-, respectively, where the S_j, $j = 1, 2$, are proper symmetric statistics (see Section 2.3).
(iii) Let us use an NPC method on the associated p-values $\lambda^+ = \Pr\{T_+^* \geq T_+^o|\mathcal{X}_{/\mathbf{X}}\}$ and $\lambda^- = \Pr\{T_-^* \geq T_-^o|\mathcal{X}_{/\mathbf{X}}\}$, such as Tippett's or Fisher's.
(iv) Then, according to the theory of multiple testing and closed testing procedures (see Section 5.4), once H_0 is rejected, it is possible to make an inference on which sub-alternative is active, where it is to be emphasized that the associated error rates such as the FWE are exactly controlled.

Of course, in this framework, a third type of error might occur due to the false acceptance of one sub-alternative when the other is actually active. In any case, it is worth noting that the inferential conclusion becomes rather more rich than that offered by a simple two-sided testing as in Section 3.1.2. However, it is also worth noting that, since one of these partial tests is not unbiased, their NPC does not provide this procedure with two-sided unbiasedness.

Example 5. *Testing for multi-sided alternatives*.
Suppose that for some units in a two-sample design the random effect Δ is negative and for others it is positive, so that $\Pr\{\Delta < 0\} > 0$ and $\Pr\{\Delta > 0\} > 0$. On the one hand, such a situation is essentially different from that of Example 4, in that now two or three sub-hypotheses can be jointly true in the alternative. Actually, the hypotheses are $H_0 : \{\Pr[\Delta = 0] = 1\}$ against $H_1 : \{[(\Delta < 0) \bigcup (\Delta > 0)]$ and $\Pr[(\Delta \leq 0) \bigcap (\Delta \geq 0)] > 0\}$, where it is to be emphasized that the two sub-alternatives $H_1^- : \{\Delta < 0\}$ and $H_1^+ : \{\Delta > 0\}$ can be jointly active. On the other hand, this situation may occur, for instance, when a drug treatment can have genetic interaction, in that it is active with positive effects on some individuals, negative effects on others, and ineffective on the rest. Thus, starting for instance from an underlying unimodal distribution in H_0, the response distribution in the alternative may become bi- or trimodal. In order to deal with such an unusual situation, we may first apply two goodness-of-fit tests such as the Kolmogorov–Smirnov $T^*_{KS+} = \max_{i \leq n}[\hat{F}^*_2(X_i) - \hat{F}^*_1(X_i)]$ and $T^*_{KS-} = \max_{i \leq n}[\hat{F}^*_1(X_i) - \hat{F}^*_2(X_i)]$ and then proceed with their NPC.

In this framework, when working with paired data designs, we guess that it is also possible, by using a data-driven classification tool, to indicate which units had negative effects and which had positive. It is worth noting that in multi-sided testing more than two traditional errors may occur: (i) by rejecting H_0 when it is true; (ii) by accepting H_0 when is false; (iii) by rejecting $H_1^- : \{(\Delta < 0)\}$ when is true; (iv) by rejecting $H_1^+ : \{[(\Delta > 0)\}$ when is true. Error types (ii), (iii) and (iv) may occur jointly.

Example 6. *Testing for non-inferiority*.
By using the results in Theorem 2, 3.1.1, it is possible to deal with problems which are quite common in clinical trials, experimental pharmacology, industrial experimentation, and so on, such as $H_0 : \{X_1 \stackrel{d}{=} X_2\}$ against $H_1^+ : \{X_1 \stackrel{d}{>} X_2 + \delta^+\}$, that is, X_1 is superior to X_2 by a pre-established quantity $\delta^+ > 0$, and if H_1^+ is accepted against $H_1^{-\prime} : \{X_1 \stackrel{d}{>} X_2 - \delta^-\}$, for some pre-established quantity $\delta' > 0$, that is, X_1 is non-inferior to $X_2 - \delta^-$. The latter, which can easily be extended to vector variables, generalizes the well-known testing problem of non-inferiority (see Hung et al., 2003).

Example 7. *Testing for monotonic stochastic ordering*.
In this example we consider a C-sample univariate problem concerning an experiment where units are randomly assigned to C groups which are defined according to *increasing levels* of a treatment. Moreover, let us assume that responses are quantitative or ordered categorical, and the related model is $\{X_{ji} = \mu + \Delta_{ji} + Z_{ji},\ \ i = 1, \ldots, n_j,\ \ j = 1, \ldots, C\}$, where μ is a population constant, Z are exchangeable random errors with finite mean value, and Δ_j are the stochastic effects on the jth group. In addition, assume that effects satisfy the monotonic stochastic ordering condition $\Delta_1 \stackrel{d}{\leq} \ldots \stackrel{d}{\leq} \Delta_C$, so that the resulting CDFs satisfy $F_1(t) \geq \ldots \geq F_C(t),\ \ \forall t \in \mathcal{R}^1$. Testing $H_0 : \{X_1 \stackrel{d}{=} \ldots \stackrel{d}{=} X_C\} = \{\Delta_1 \stackrel{d}{=} \ldots \stackrel{d}{=} \Delta_C \stackrel{d}{=} 0\}$ against the alternative with monotonic order restriction $H_1 : \{X_1 \stackrel{d}{\leq} \ldots \stackrel{d}{\leq} X_C\} = \{\Delta_1 \stackrel{d}{\leq} \ldots \stackrel{d}{\leq} \Delta_C\}$, with at least one strict inequality, is a rather difficult problem. A parametric exact solution is difficult enough, especially when $C > 2$, and becomes very difficult, if not impossible, in multivariate situations. Note that these hypotheses define a problem of *isotonic inference* (see Hirotsu, 1998b). A nonparametric rank solution of this kind of problem is given by the Jonckheere–Terpstra test (see Randles and Wolfe, 1979; Hollander and Wolfe, 1999; see also

Shorack, 1967; Mansouri, 1990; Nelson, 1992). In the permutation context, this problem can be tackled in at least two ways.

(i) Let us suppose that responses are quantitative, errors Z have finite mean, $\mathbb{E}(|Z|) < \infty$, and that the design is balanced: $n_j = n$, $j = 1, \ldots, C$. Consider all pairwise comparisons, $T^*_{jh} = \bar{X}^*_j - \bar{X}^*_h$, $j > h = 1, \ldots, C-1$, all unbiased for testing the respective partial hypotheses H_{0jh} : $\{X_j \overset{d}{=} X_h\}$ against $H_{1jh} : \{X_j \overset{d}{>} X_h\}$; in fact, we may write $H_0 : \{\bigcap_{jh} H_{0jh}\}$ and $H_1 : \{\bigcup_{jh} H_{1jh}\}$. Application of the direct combining function gives $T^{*\prime\prime}_D = \sum_{jh} T^*_{jh} = \sum_j (2j - C - 1)\bar{X}^*_j$, which is nothing other than the covariance between the group ordering j and the related mean $\bar{X}^*_j$. Of course, it is assumed that the permutations are with respect to the pooled data set $\mathbf{X} = \biguplus_{j=1}^{C} \mathbf{X}_j$. Since all partial tests in H_0 are exact, unbiased and consistent, $T^{*\prime\prime}_D$ is exact, unbiased and consistent. This solution can easily be extended to unbalanced designs. In this context, within homoscedasticity of individual responses and by pairwise comparison of standardized partial tests, we get $T^{*\prime\prime}_D = \sum_j (2j - C - 1)\bar{X}^*_j \sqrt{n_j}$. Its proof and extension to ordered categorical multivariate responses, and to different combining functions, are left to the reader as exercises.

(ii) Let us imagine that for any $j \in \{1, \ldots, C-1\}$, the whole data set is split into two pooled pseudo-groups, where the first is obtained by pooling together data of the first j ordered groups and the second by pooling the rest. To be more specific, we define first pooled pseudo-group as $\mathbf{Y}_{1(j)} = \mathbf{X}_1 \uplus \ldots \uplus \mathbf{X}_j$ and the second as $\mathbf{Y}_{2(j)} = \mathbf{X}_{j+1} \uplus \ldots \uplus \mathbf{X}_C$, $j = 1, \ldots, C-1$, where $\uplus$ is the symbol for pooling data into one pseudo-group and $\mathbf{X}_j = \{X_{ji}, \ i = 1, \ldots, n_j\}$ is the data set in the jth group.

In the null hypothesis, data from every pair of pseudo-groups are exchangeable because related pooled variables satisfy the relationships $Y_{1(j)} \overset{d}{=} Y_{2(j)}$, $j = 1, \ldots, C-1$. In the alternative we see that $Y_{1(j)} \overset{d}{\leq} Y_{2(j)}$, which corresponds to the monotonic stochastic ordering (dominance) between any pair of pseudo-groups. This suggests that we express the hypotheses in the equivalent form $H_0 : \left\{\bigcap_j (Y_{1(j)} \overset{d}{=} Y_{2(j)})\right\}$ and $H_1 : \left\{\bigcup_j (Y_{1(j)} \overset{d}{\leq} Y_{2(j)})\right\}$, emphasizing a breakdown into a set of sub-hypotheses.

Consider the jth sub-hypotheses $H_{0j} : \{Y_{1(j)} \overset{d}{=} Y_{2(j)}\}$ and $H_{1j} : \{Y_{1(j)} \overset{d}{\leq} Y_{2(j)}\}$. We note that the related sub-problem corresponds to a two-sample comparison for restricted alternatives, a problem which has an exact and unbiased permutation solution (see Section 3.1). This solution is based on the test statistics $T^*_j = \sum_{1 \leq i \leq N_{2(j)}} Y^*_{2(j)i}$, where $N_{2(j)} = \sum_{r > j} n_r$ is the sample size of $\mathbf{Y}_{2(j)}$.

Thus, a set of suitable partial tests for the problem is $\{T^*_j, \ j = 1, \ldots, C-1\}$. Therefore, since these partial tests are all exact, marginally unbiased and consistent, their NPC provides for an exact overall solution. It is worth noting that partial tests T^*_j, $j = 1, \ldots, C-1$, generally do not play the role of marginal tests because permutations are on the whole pooled data set $\mathbf{X}$. Some application examples are discussed in Chapter 8.

Remark 6. If we may assume that stochastic effects act on the first two moments of the underlying distributions and our inferential interest is in both aspects, then we may also calculate partial tests on second moments, $T^*_{2j} = \sum_{1 \leq i \leq N_{2(j)}} [Y^*_{2(j)i}]^2$, $j = 1, \ldots, C-1$, and combine all $2(C-1)$ partial tests (see Example 3 above). This problem may be seen as one of *multi-aspect monotonic inference*.

Remark 7. The problem of monotonic inference may be extended to V-dimensional situations in a straightforward way. Let us suppose that the responses and related model are now $\{X_{hji} = \mu_h + \Delta_{hji} + Z_{hji}, \ i = 1, \ldots, n_j, \ j = 1, \ldots, C, \ h = 1, \ldots, V\}$, where Δ_{hj} are random effects on the hth variable of the jth group and $\mathbf{Z}$ are V-dimensional exchangeable errors.

Let us assume that random effects satisfy the V-dimensional ordering condition: $\mathbf{\Delta}_1 \overset{d}{\leq} \ldots \overset{d}{\leq} \mathbf{\Delta}_C$. This means that for any pair $1 \leq j < r \leq C$, $\mathbf{\Delta}_j \overset{d}{\leq} \mathbf{\Delta}_r$ implies $(\Delta_{hj} \overset{d}{\leq} \Delta_{hr}, \ h = 1, \ldots, V)$.

We may call this the *componentwise ordering* (see Sampson and Whitaker, 1989). Again, let us imagine that the whole data set is split into two pseudo-groups: $\mathbf{Y}_{1(j)} = \mathbf{X}_1 \biguplus \ldots \biguplus \mathbf{X}_j$ and $\mathbf{Y}_{2(j)} = \mathbf{X}_{j+1} \biguplus \ldots \biguplus \mathbf{X}_C$, $j = 1, \ldots, C-1$. Note that data vectors of these two pseudo-groups are exchangeable in the null hypothesis. This suggests breaking the hypotheses down into H_0 : $\{\bigcap_j [\bigcap_h Y_{h1(j)} \overset{d}{=} Y_{h2(j)}]\}$ and $H_1 : \{\bigcup_j [\bigcup_h Y_{h1(j)} \overset{d}{\leq} Y_{h1(j)}]\}$.

In order to deal with this problem, let us consider the $V(C-1)$ partial tests $T^*_{hj} = \sum_{1 \leq i \leq N_{2(j)}} Y^*_{h2(j)i}$, $j = 1, \ldots, C-1$, $h = 1, \ldots, V$. Since all these partial tests are exact and marginally unbiased and consistent, their NPC provides the required solution.

Note that the parametric solution of this *multivariate isotonic inference* is much more difficult than in the univariate case and no such solution is yet available in the literature.

A discussion of a similar problem, in a context of multivariate ordered categorical responses, is presented in Section 6.5. Two more applications are suggested in Example 8 in the context of an exact solution for the restricted multivariate Behrens–Fisher problem, and in Chapter 7 in the context of testing for stochastic ordering with repeated measurements.

Remark 8. An extension to so-called umbrella alternatives is discussed in Section 8.2.

Example 8. *Testing for the Behrens–Fisher problem.*
The well-known Behrens–Fisher problem is concerned with the comparison of locations of two distributions irrespective of scale coefficients, $H_0 : \{\mu_1 = \mu_2\}$ against $H_1 : \{\mu_1 < (\text{or} \neq \text{or} >)\mu_2\}$. Within the general family of Behrens–Fisher problems we may distinguish two main sub-families. The first contains problems for which we may assume that in the null hypothesis data are exchangeable with respect to two groups but are non-homoscedastic in the alternative. Most experimental designs in which units are randomly assigned to treatments and the effect may be assumed to be not only on locations but also on scale coefficients or on other functionals, fall within this sub-family (see Section 2.1.1). The second contains all other problems, especially those for which variances may differ in the null hypothesis. For instance, responses in some observational studies may fall within this sub-family. We may call these two sub-families of problems the *restricted* and the *generalized Behrens–Fisher*, respectively.

It is well known that when responses are normally distributed there exists no exact parametric solution for the general problem; in particular, no solutions based on the similarity property are possible (see Pfanzagl, 1974), except for that given by Scheffé (1943c) which uses an auxiliary randomization approach. However, for experimental designs producing stochastic dominance in the alternative, so that they belong to the restricted sub-family, we may consider that treatments have influence not only on locations but also on scale coefficients or on other functionals. Thus if multi-aspect testing is applied, we can obtain exact and effective permutation solutions using the NPC method (see Example 3 above). In this sense and from a pragmatic point of view the restricted sub-family covers the most important practical situations. The general sub-family, which admits only approximate solutions, appears somewhat academic and of interest only when the exchangeability of data with respect to groups cannot be assumed in H_0 or when responses in the alternative are not stochastically ordered and we are interested only in comparing locations. In the literature there are a large number of contributions to this very challenging problem; we discuss here only three univariate permutation solutions, one approximate solution using the Aspin–Welch statistic and two that use the multi-aspect procedure. One of the two multi-aspect solutions is exact for the restricted sub-family and one is approximate for the generalized one. A full discussion of the univariate and multivariate permutation solutions can be found in Pesarin (1995, 2001).

Let us consider an example from Johnson and Leone (1964, Vol. I, p. 226) concerning the percentage shrinkage of synthetic fibres measured at two temperatures: I, with 12 units at 120°C; and II, with 10 units at 140°C. The two-sample data $\mathbf{X} = \{X_{ji}, i = 1, \ldots, n_j, j = 1, 2\}$ are displayed in Table 4.10. We may assume that the two scale coefficients are possibly unequal in

Table 4.10 Percentage shrinkage of synthetic fibres (Johnson and Leone, 1964)

I			II		
3.45	3.62	3.60	3.72	4.01	3.54
3.49	3.64	3.56	3.67	4.03	3.40
3.52	3.53	3.57	3.96	3.60	3.76
3.44	3.56	3.43	3.91		

the alternative and that the underlying distributions may not be normal. The problem is to examine whether, at $\alpha = 0.01$, the distribution of percentage shrinkage at 140°C is stochastically larger than that at 120°C. A first approximate solution can be obtained by applying the well-known Aspin–Welch statistics in the permutation framework. With $B = 2000$ CMC iterations we obtain $\hat{\lambda}_{AW} = 0.0075$ for $T^*_{AW} = (\bar{X}^*_1 - \bar{X}^*_2)/\hat{\sigma}^*_E$, where the permutations are on data $\mathbf{X}$ and $\hat{\sigma}^*_E = [\sum_{ji}(X^*_{ji} - \bar{X}^*_j)^2/(n_j(n_j - 1))]^{1/2}$. This result is significant at $\alpha = 0.01$ and substantially in accordance with that of the parametric Aspin–Welch test ($t = -3.20$ with approximately 11 d.f.: $t_{0.005;11} = -3.106$).

To obtain an exact solution, let us observe that in this problem units are randomly assigned to treatments and so the two distributions and in particular the scale coefficients can be assumed to be equal in H_0, thus it looks like a typical experimental design. If we expect that, in the alternative, the treatments may act on the first two moments, provided that a stochastic dominance relation $X_1 \overset{d}{<} X_2$ is assumed, then we may apply a multi-aspect testing procedure. This interpretation is consistent with the idea that if temperature has no influence on fibre shrinkage, as H_0 asserts, then the two distributions are equal; conversely, in the alternative the distribution at 140°C is expected to dominate that at 120°C.

Again with $B = 2000$ CMC iterations on two partial tests $T^*_r = \sum_i (X^*_{2i})^r$, $r = 1, 2$, the first for comparing locations and the second for second moments, we obtain $\hat{\lambda}_1 = 0.0019$, $\hat{\lambda}_2 = 0.0016$ and $\hat{\lambda}'' = 0.0015$, respectively for the two partial tests and the combined test, the latter obtained by a direct combining function on standardized partial tests. It is worth noting that this solution gives a p-value lower than both the permutation and parametric approximate counterparts. Also note that, as both partial tests are exact and consistent having assumed that $\mathbb{E}(|X|)$ is finite, in these circumstances multi-aspect testing gives an exact and consistent solution.

To see a solution for the generalized problem, let us assume that each underlying distribution P_j is symmetric with respect to μ_j, $j = 1, 2$, with unknown scale coefficients σ_j. Thus, both distributions of $Y_j = (X_j - \mu)$, $j = 1, 2$, are symmetrically distributed around zero if and only if $H_0 : \{\mu_1 = \mu_2 = \mu\}$ is true. Moreover, the distributions of $Y_j = (X_j - \tilde{X})$, $j = 1, 2$, conditional on the pooled median $\tilde{X}$, defined as $\tilde{X} = (X_{(n/2)} + X_{(1+n/2)})/2$ if n is even and as $\tilde{X} = X_{((n+1)/2)}$ if n is odd, $X_{(i)}$ indicating the ith order statistic, are such that

$$\Pr\{Y_j < -z|\tilde{X} = -t\} = \Pr\{Y_j > z|\tilde{X} = t\}, \ \forall z, t \in \mathcal{R}^1.$$

Hence they are mutually symmetric. In practice, they are only slightly asymmetric around zero and their asymmetry vanishes as sample sizes increase. Accordingly, we may recall the main points of a proper permutation solution for univariate symmetric distributions as discussed in Example 1, 2.6. The reasons for a proper solution are graphically illustrated in Figure 4.3. Univariate hypotheses may be equivalently written as $H_0 : \{(Y_1$ is symmetric around $0) \bigcap (Y_2$ is symmetric around $0)\} = \{H_{01} \bigcap H_{02}\}$ and $H_1 : \{H_{01} \bigcup H_{02}\}$.

It is worth noting that in the generalized Behrens–Fisher problem, two symmetric distributions may be rather different. For instance, P_1 may be unimodal and P_2 multimodal.

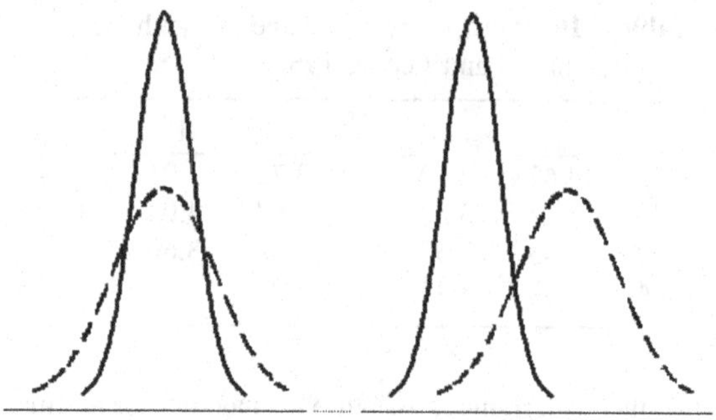

Figure 4.3 Comparisons of two symmetric distributions in H_0 and H_1

The distributions of the Y_j in H_0 are invariant with respect to the common location parameter μ, but they depend on scale coefficients σ_j and generally on underlying distributions P_j, $j = 1, 2$. Hence, in order to avoid this dependence, we may work within a permutation approach by conditioning with respect to a set of jointly sufficient statistics in H_0. This set is the pair of data groups $(\mathbf{X}_1; \mathbf{X}_2)$, in which data can only be permuted within groups, the permutation sample space being $\mathcal{X}_{/\mathbf{X}_1} \times \mathcal{X}_{/\mathbf{X}_2}$ (see Remark 4, 2.1.2 and Problem 15, 4.6.1). Equivalently, another jointly sufficient set of statistics is $(\tilde{X}; \mathbf{Y}_1; \mathbf{Y}_2)$, because $\tilde{X}$ is a permutation invariant quantity (see point (ii) in Remark 2, 2.1.2). The set $(\tilde{X}; |\mathbf{Y}_1|; |\mathbf{Y}_2|)$ may also be regarded as a set of sufficient statistics for the problem, but *only* in the univariate case (see Problems 16 and 17, 4.6.1).

Hence, in order to test H_0 against H_1 it is appropriate to first establish two separate partial tests of symmetry, one from each sample, followed by a suitable combination. A pair of conditional permutation tests for separately testing symmetry conditionally on $(\mathbf{X}_1; \mathbf{X}_2)$, in accordance with the theory of permutation testing for symmetry (see Example 1, 2.6), may be

$$T_j^* = \sum_{i=1}^{n_j} Y_{ji} \cdot S_{ji}^*/n_j, \; j = 1, 2,$$

where the random signs $\mathbf{S}^* = \{S_{ji}^*;\; i = 1, \ldots, n_j,\; j = 1, 2, \}$ can be regarded as a random sample of n i.i.d. observations from variable S, taking values -1 and $+1$, each with probability 1/2. It should be noted that, conditionally on $(\mathbf{X}_1; \mathbf{X}_2)$, the two partial tests are independent, because their joint permutation distribution is generated by independent signs and the Y play the role of fixed coefficients.

In H_0, being based on sample deviates $\mathbf{Y}$ (formally these are residuals), the permutation distributions of tests T_j^* are μ-invariant. Moreover, due to conditioning on sufficient statistics, they are (σ_1, σ_2)-invariant in H_0 and H_1, because these are constant coefficients within each data group. In addition, we note that conditioning on $(\tilde{X}; \mathbf{Y}_1; \mathbf{Y}_2)$ implies that the conditional distributions of $(\mathbf{Y}_j|\tilde{X})$ are not exactly symmetric around zero, so that test statistics T_j^* are only approximately exact for testing symmetry. This approximation is due to the fact that the Y are residuals, so that even within groups they are not completely independent with respect to units and exact joint exchangeability therefore fails. However, this approximation is generally quite good even for small sample sizes (see Pesarin, 1995, 2001).

At this stage, we may combine two partial tests by using the theory of combination of independent tests, or we can use NPC.

One such combination function, deduced by taking into consideration its asymptotic behaviour, is $\Psi^* = \varphi(T_2^* - T_1^*)$, where $\varphi(\cdot)$ corresponds to $+(\cdot)$, $-(\cdot)$ or the absolute value $|\cdot|$ respectively for alternatives corresponding to '$<$', '$>$' or '$\neq$'. Our preference for this direct form of combination is also due to the fact that the permutation mean value of Ψ^* is not dependent on $\mu - \tilde{X}$, so that the average distortion of Ψ^* induced by considering residuals with respect to $\tilde{X}$ vanishes. In practice, as both underlying distributions are symmetric by assumption, each partial test plays the role of a test for location (see Section 1.9). In fact, the response model being $X_{ji} = \mu_j + \sigma_j Z_{ji}$, $i = 1, \ldots, n_j$, $j = 1, 2$, the model for observed values of the jth partial statistic is $T_j = \delta_j + \sigma_j \bar{Z}_{j\cdot}$, where $\delta_j = (\mu_j - \mu)$ and $\bar{Z}_{j\cdot} = \sum_i Z_{ji}/n_j$. Thus, $T_2 - T_1 = (\mu_2 - \mu_1) + \sigma_2 \bar{Z}_{2\cdot} - \sigma_1 \bar{Z}_{1\cdot}$.

The permutation sample space $\mathcal{X}_{/(\mathbf{X}_1;\mathbf{X}_2)}$ and, if there are no ties, the permutation support of Ψ^* both contain 2^n points.

Let us denote the p-value of Ψ by

$$\lambda = \lambda(\mathbf{X}_1; \mathbf{X}_2) = \Pr\{\Psi^* \geq \Psi^o | \mathcal{X}_{/(\mathbf{X}_1;\mathbf{X}_2)}\},$$

where the observed value is $\Psi^o = \Psi(\mathbf{X}) = \varphi(\Sigma_i Y_{2i}/n_2 - \Sigma_i Y_{1i}/n_1)$. According to standard inferential rules, H_0 is rejected at significance level α if $\lambda \leq \alpha$.

For practical purposes, the evaluation of λ may be tackled using the algorithm in Section 1.9.3. On the data of the example and with $B = 2000$ CMC iterations, we obtain $\hat{\lambda}_\Psi = 0.0031$, a result which is slightly larger than the exact multi-aspect test $\hat{\lambda}''$ obtained under the assumption of stochastic dominance of P_2 with respect to P_1 in the alternative.

Although conditioning on $(\mathbf{X}_1; \mathbf{X}_2)$ leads to an approximate solution, the null distribution of Ψ_T^* is very close to being $(\sigma_1; \sigma_2)$-invariant for all $(n_1; n_2)$ so that it can be named *almost exact* (see Pesarin, 1995).

When underlying responses are asymmetrically distributed around their locations, we found in several Monte Carlo experiments that Ψ becomes slightly conservative. In order to overcome this slight drawback, we suggest the use of data transformations in order to symmetrize the responses before proceeding with the analysis (see Example 1, 2.6).

In order to show the asymptotic behaviour of Ψ^*, let us assume that: (i) response variables X_j have finite location and scale parameters (μ_j, σ_j), $j = 1, 2$; (ii) as sample sizes diverge, the ratio $\zeta = n_2/n_1$ converges to a positive number $\underset{\sim}{\zeta}$; (iii) as n diverges, the pooled median $\tilde{X}$ converges weakly or strongly, according to the properties of underlying distributions P_j, $j = 1, 2$, towards the pooled population location functional μ. Under these assumptions, Ψ_T^* is permutationally equivalent to

$$\Psi_Z^* = \frac{\varphi(\sum_i Y_{2i} S_{2i}^*/n_2 - \sum_i Y_{1i} S_{1i}^*/n_1)}{\left[\sum_i Y_{2i}^2/n_2^2 + \sum_i Y_{1i}^2/n_1^2 - (\mu_2 - \tilde{X})^2/n_2 - (\mu_1 - \tilde{X})^2/n_1\right]^{1/2}},$$

since the two statistics are related by an increasing one-to-one relationship with probability one, the denominator being a permutation constant (which is positive with probability one in the sample space $\mathcal{X}$).

Now, as n_1 and n_2 go independently to infinity, Slutsky's well-known theorem says that the two quantities

$$\frac{\sum_i (X_{ji} - \mu_j)^2 + n_j(\mu_j - \mu)^2}{\sum_i Y_{ji}^2}, \quad j = 1, 2,$$

both converge to 1 weakly or strongly, depending on how $\tilde{X}$ converges to μ. Let us now denote by Ψ_μ the test statistic for comparing two means when variances are known, the observed value of which is

$$\Psi_\mu(\mathbf{X}) = \frac{\varphi(\bar{X}_2 - \bar{X}_1)}{(\sigma_2^2/n_2 + \sigma_1^2/n_1)^{1/2}},$$

where the role of the whole data set $\mathbf{X}$ is emphasized. Hence, as n goes to infinity, the observed values of the two test statistics $\Psi_\mu(\mathbf{X})$ and

$$\Psi_Z(\mathbf{X}) = \frac{\varphi(\bar{Y}_2 - \bar{Y}_1)}{\left\{\sum_j \left[\sum_i Y_{ji}^2/n_j^2 - (\mu_j - \tilde{X})^2/n_j\right]\right\}^{1/2}},$$

where $\bar{Y}_j = T_j$, are such that, for every $\varepsilon > 0$,

$$\Pr\{|\Psi_Z(\mathbf{X}) - \Psi_\mu(\mathbf{X})| < \varepsilon\} \to 1,$$

weakly or strongly. Hence, the two tests are asymptotically equivalent for almost all $\mathbf{X} \in \mathcal{X}$, in H_0 and H_1.

This is the main argument in proving the following theorem:

Theorem 13. *If Ψ_μ is asymptotically 'optimal' in some sense (UMP or UMPU, depending on alternatives), then Ψ_Z, being convergent to Ψ_μ, is asymptotically optimal in the same sense.*

As a special case, we also have the following corollary:

Corollary 6. *If response variables are normally distributed, then Ψ_μ is an 'optimal' test, and Ψ_Z, being asymptotically equivalent to it, is also asymptotically optimal.*

The result of Corollary 6 shows that the asymptotic relative efficiency of Ψ or Ψ_Z with respect to Ψ_μ is 1 under the assumption of normality.

Extensions to $C > 2$ samples and to multivariate situations are straightforward within the NPC methodology (see Pesarin, 2001).

Remark 9. Monte Carlo evaluations of the stability of the permutation distribution of Ψ, as well as evaluation of its power behaviour with respect to Ψ_μ, under both normality and knowledge of σ_1 and σ_2, are reported in Pesarin (1995).

From these results, we may argue that its stability with respect to σ_1/σ_2 and n_1/n_2 is very good in all conditions, even when both are quite far from unity. Moreover, test convergence with respect to Ψ_μ is quite fast: in practice, for $n_1 \geq n_2 > 50$, the two tests are almost indistinguishable by means of Monte Carlo simulations.

Remark 10. By using the foregoing results together with that of Example 10, 2.7 on testing for equality of two scale coefficients, we may deal with the problem of approximate joint testing for location and scale coefficients. Indeed, again with reference to univariate response models such as $X_{ji} = \mu_j + \sigma_j Z_{ji}$, $i = 1, \ldots, n_j$, $j = 1, 2$, let us suppose that the hypotheses are $H_0 : \{(\mu_1 = \mu_2) \bigcap (\sigma_1 = \sigma_2)\} = H_{0\mu} \bigcap H_{0\sigma}$, against alternatives such as $H_1 : \{(\mu_1 <\neq> \mu_2) \bigcup (\sigma_1 <\neq> \sigma_2)\}$. This problem may be tackled by means of two separate tests, both consistent and approximately unbiased. The first involves testing the equality of locations irrespective of scales, as above for the Behrens–Fisher. The second tests the equality of scale coefficients irrespective of locations. Note that both tests are approximately marginally unbiased, irrespective of whether the other null hypothesis is true or not. Specifically, the two test statistics are (with obvious notation) $T_\mu^* = \varphi_\mu(T_1^* - T_2^*)$ and $T_\sigma^* = \varphi_\sigma\left(\sum_i Y_{1i}^{*2}/n_1 - \sum_i Y_{2i}^{*2}/n_2\right)$, where $\varphi_{\cdot}(\cdot)$ stands for $-(\cdot)$ if the alternative is $<$, for $+(\cdot)$ if $>$ and for $|\cdot|$ if $\neq$.

Note that the permutation signs $\mathbf{S}^*$, useful for T_μ^*, are completely independent among themselves and independent of the observed data; thus, they are also independent of permutations of sample deviates $\mathbf{Y}^*$, useful for obtaining T_σ^*. This implies that the two test statistics are at least approximately independent conditionally. Hence, their combination may be obtained by recourse to standard theory and methods for the combination of independent tests.

The multivariate extension of this problem is straightforward.

Remark 11. Let us assume that, in an experimental design (and with obvious notation), the model for positive responses is $X_{ji} = \mu + \delta_j + \sigma(\delta_j)Z_{ji}$, $i = 1, \ldots, n_j$, $j = 1, 2$, with $0 = \delta_1 \leq \delta_2$ and $\sigma(0) \leq \sigma(\delta_2)$, so that two population distributions satisfy the stochastic dominance relation $F_1(t) \leq F_2(t)$, $\forall t \in \mathcal{R}^1$. As these assumptions imply that, in the alternative, the two conditions $\mathbb{E}(X_1) \leq \mathbb{E}(X_2)$ and $\mathbb{E}(X_1^2) \leq \mathbb{E}(X_2^2)$ are jointly satisfied, then the problem of jointly testing for location and scale coefficients may be tackled within a multi-aspect testing method (see Example 3). Hence, two partial tests may be, for instance, $T_1^* = \sum_i X_{2i}^*$ and $T_2^* = \sum_i (X_{2i}^*)^2$, or permutational equivalents. The reader should prove that, under these conditions, this procedure gives an exact, unbiased and consistent solution for jointly testing for location and scale coefficients (see Problems 11–13, 4.6.1).

Example 9. *A real example for comparison of two species of flies*.
In this example we have two samples, one from each of two species of flies (*Leptoconops carteri* and *Leptoconops torrens*) and seven variables have been measured (Johnson and Wichern, 2007). We wish to test for the restricted alternative,

$$H_1 : \left\{ \left[\bigcup_{1 \leq h \leq 6} (\mu_{1h} < \mu_{2h}) \right] \bigcup (\mu_{17} > \mu_{27}) \right\},$$

where μ_{jh} is the mean of the hth variable in group j, $j = 1, 2$, $h = 1, \ldots, 7$. This can be done by carrying out one partial test for each variable (according to the related alternative), and then by combining the partial tests into a global test. A vector of contrasts will again be useful in obtaining the test statistic $T_h^* = \bar{x}_{2h}^* - \bar{x}_{1h}^*$. The first column of the data set contains an indicator variable of the species (1 = *L. carteri*, 2 = L. *torrens*). Since the `t2p` function follows the *large-is-significant* rule, we must change the signs of the permutation distribution of T_7^* before obtaining the partial p-value of X_7.

```
setwd("C:/path")
data = read.csv("Fly.csv",header=TRUE)
N = dim(data)[1] ; p = dim(data)[2]-1
n = table (data[,1]); B = 10000;
contr = as.vector(rep(c(-1/n[1],1/n[2]),each=n))
data = as.matrix(data[,-1])
T<-array(0,dim=c((B+1),p))
T[1,] = t(contr)%*%data

for(bb in 2:(B+1)){
data.star = data[sample(1:N),]
T[bb,] = t(contr)%*%data.star
}
T[,7] = -T[,7]
P = t2p(T)
partial.p = P[1,] ; names(partial.p) = colnames(data);
round(partial.p,digits=4)
```

```
      X1        X2        X3      X4      X5      X6      X7
  0.0248    0.2303    0.0000  0.3665  0.0000  0.3953  0.0619
```

The vector `partial.p` contains the partial p-values of each variable. The matrix `P` contains the null distribution of partial p-values. As a result X_1, X_3 and X_5 are strongly significant and X_7 is moderately non-significant. We combine the partial tests with the Fisher, Liptak and Tippett combining functions. Note that if Tippett's combining function is applied, the rule is *small is significant*, therefore we run the `t2p` function on the inverse of the vector containing Tippett's combination of the rows of `P`.

```
F = apply(P,1,function(x){-2*log(prod(x))})
L = apply(P,1,function(x){sum(qnorm(1-x))})
T = apply(P,1,min)

P.F = t2p(F); P.L = t2p(L) ; P.T = t2p(1/T);
globs = c(P.F[1],P.L[1],P.T[1])
names(globs) = c("Fisher","Liptak","Tippett")
globs
      Fisher       Liptak      Tippett
  0.00009999   0.00019998   0.00059994
```

The corresponding MATLAB code is given below:

```
D=textimport('Fly.csv',',',1);
remindD(D)
[P T opts] = NP_2s({'X1','X2','X3','X4','X5','X6','X7'},
'group',1000,[1 1 1 1 1 1 -1 ]);

P2=NPC(P,'F');
P2=NPC(P,'L');
P2=NPC(P,'T');
```

The data set and the corresponding software codes are available from the `examples_chapters_1-4` folder on the book's website.

4.6.1 Problems and Exercises

1) With reference to Example 8, 4.6, discuss a solution to the testing problem for two-sample designs when $\mathbb{V}(X_1) = \sigma^2(\delta)$ and $\sigma^2(0) = \sigma^2$. In particular, distinguish between two situations: (a) in the alternative the two CDFs do not cross; and (b) the two CDFs can cross.

2) With reference to point (iv) in Remark 2, 2.1.2 and Examples 11, 2.7 and 8, 4.6, discuss a testing solution when covariates $\mathbf{X}$ are assumed not exchangeable in H_0 between two groups and in particular when regression functions β_1 and β_2 cannot be assumed to be equal (show that this becomes a sort of Behrens–Fisher MANCOVA).

3) Extend the results of Examples 1 and 2, 4.6, to Fisher, Liptak and Tippett combining functions.

4) Extend the results of Examples 3 and 5, 4.6, to Fisher, Liptak and Tippett combining functions.

5) With reference to point (i) of Example 7, 4.6, on stochastic ordering, extend the given solution to Fisher, Liptak and Tippett combining functions.

6) Extend the approximate permutation solutions of Example 8, 4.6 to $C > 2$ groups.

7) Express the approximate permutation solutions of Example 8, 4.6 by using sample medians.

8) Find an approximate permutation procedure for evaluating a $1-\alpha$ confidence interval for the functional $\delta = \mathbb{E}(X_1 - X_2)$ within the univariate Behrens–Fisher framework.

9) With reference to Example 8, 4.6, give detailed extensions of approximate multivariate permutation solutions for the Behrens–Fisher problem.

10) Discuss how to find approximate confidence intervals for $\delta = \sigma_2/\sigma_1$, in accordance with the test for scale coefficients discussed in Example 10, 2.7.

11) Discuss the details of the multivariate extension of the joint test for locations and scale coefficients (Remark 11, 4.6).

12) Prove the statement of Remark 11, 4.6.

13) Extend the solution in Remark 11, 4.6, to V-dimensional problems.

14) With reference to the generalized Behrens–Fisher problem, prove that pooled data set $\mathbf{X}_1 \biguplus \mathbf{X}_2$ is not a set of sufficient statistics.

15) With reference to Problem 14 above, prove that the pair of data groups $(\mathbf{X}_1; \mathbf{X}_2)$, in which no data can be exchanged between two groups, is a set of sufficient statistics in the Behrens–Fisher context.

16) With reference to the generalized one-dimensional Behrens–Fisher problem, prove that if $\tilde{X}$ is the pooled median (see Example 8, 4.6) then $(\tilde{X}; \mathbf{Y}_1; \mathbf{Y}_2)$ and $(\tilde{X}; |\mathbf{Y}_1|; |\mathbf{Y}_2|)$ are sets of sufficient statistics both equivalent to $(\mathbf{X}_1; \mathbf{X}_2)$.

17) With reference to Problems 15 and 16 above, extend the results to multivariate situations and show that $(\tilde{\mathbf{X}}; |\mathbf{Y}_1|; |\mathbf{Y}_2|)$ is not a set of sufficient statistics.

18) With reference to the testing for the equality of scale coefficients in Example 10, 2.7 for univariate situations, show that the pair of data groups $(\mathbf{X}_1; \mathbf{X}_2)$ is a set of sufficient statistics. Also show that $(\bar{X}_1; \bar{X}_2; \mathbf{Y}_1; \mathbf{Y}_2)$, where $\mathbf{Y}_j = \{Y_{ji} = X_{ji} - \bar{X}_j, \ i = 1, \ldots, n_j\}$ and $\bar{X}_j = \sum_i X_{ji}/n_j, \ j = 1, 2$, is a set of sufficient statistics because there is a one-to-one relationship between two sets.

19) Give a formal proof that the solution of the restricted multivariate Behrens–Fisher problem is exact when multi-aspect testing (see Example 3, 4.6) is used.

20) Give a formal proof that the solution presented in Example 4, 4.6, related to the testing for two-sided alternatives separately, is consistent.

21) Give a formal proof that the solution presented in Example 5, 4.6, related to the testing for multi-sided alternatives, is consistent.

22) Give details of a permutation solution for non-inferiority testing presented in Example 6, 4.6.

4.7 Comments on the Nonparametric Combination

4.7.1 General Comments

The NPC of dependent permutation partial tests is a method for the combination of significance levels, that is, of rejection probabilities. In contrast, most parametric tests, based for instance on likelihood ratio behaviour, essentially combine discrepancy measures usually expressed by point distances in the sample space $\mathcal{X}$. In this sense, the NPC method appears to be a substantial extension of standard parametric approaches. From the next chapter onwards, we shall show that the NPC method is suitable and effective for many multivariate testing problems which, in a parametric framework, are very difficult or even impossible to solve.

As the NPC method is conditional on a set of sufficient statistics, it exhibits good general power behaviour (see Chapters 7–11 to appreciate its versatility and to evaluate unconditional power behaviour in some situations). Monte Carlo experiments, reported in Pesarin (1988, 1989, 1992, 1994, 1995, 1996a, 1996b, 2001), Ballin and Pesarin (1990) and Celant et al. (2000a, 2000b), show that the Fisher, Liptak or direct combining functions often have power functions which are quite close to the best parametric counterparts, even for moderate sample sizes. Thus, NPC tests are relatively efficient and much less demanding in terms of underlying assumptions with respect to parametric competitors.

In addition, the Fisher, Liptak, Lancaster, Tippett and direct combining functions for NPC are not at all affected by the functional analogue of multicollinearity among partial tests; indeed, the combination only results in a kind of implicit weighting of partial tests. In order to illustrate this, let us suppose that within a set of k partial tests, the first two are related with probability one, so that $T_1 \stackrel{p}{=} T_2$, and that Fisher, Liptak or direct combining functions are used. Thus, denoting $-\log(\hat{\lambda}_i)$, $\Phi^{-1}(1-\hat{\lambda}_i)$ or T_i by φ_i, for the Fisher, Liptak and direct combining functions respectively, the combined test becomes $T''_{\varphi} = 2\varphi_1 + \sum_{3 \leq i \leq k} \varphi_i$, which is a special case of an unbounded, convex, weighted linear combination function (see (f) in Section 4.2.4). Thus these solutions belong to $\mathcal{C}$ because they satisfy all the conditions of Section 4.2.

Of course, when a quadratic combining function is used, the matrix $\mathbf{R}^*$ must be positive definite (see Remark 3, 4.2.4). Thus, in order to obtain a computable function, we must avoid linear relationships with probability one among partial tests in this situation. However, except for the quadratic form, possible functional analogues of multicollinearity do not give rise to computational problems in NPC methods. In this sense, problems in which the number V of component variables is larger than the number n of subjects are generally easy to solve, provided that the conditions of Sections 4.1.3, 4.2.1 and 4.2.2 are satisfied.

4.7.2 Final Remarks

Except for the direct combining function, NPC procedures require intensive computation in order to find sufficiently accurate Monte Carlo estimates of the k-dimensional permutation distribution of partial tests and combined p-value. The availability of fast and relatively inexpensive computers, and of efficient software, makes the procedure effective and practical.

Although intensive computation is seldom avoidable, when the multivariate permutation CDF $F(\mathbf{t}|\mathcal{X}_{/\mathbf{X}})$ is available, by exact or approximate calculations (see Ives, 1976; Gail and Mantel, 1977; Mehta and Patel, 1980, 1983, 1999; Mielke and Iyer, 1982; Berry, 1982; Pagano and Tritchler, 1983; Berry and Mielke, 1983, 1984, 1985; Mehta et al., 1985, 1988a, 1988b; Zimmerman, 1985a, b; Berry et al., 1986; Lock, 1986; Spino and Pagano 1991; Mielke and Berry, 2007), Edgeworth expansions, saddlepoint approximations (Barndorff-Nielsen and Cox, 1979; Robinson, 1980, 1982; Davison and Hinkley, 1988; Boos and Brownie, 1989), CLTs, empirical likelihood methods, sequential approximations (Lock, 1991), or inversion of characteristic functions or Fourier transformations (Barabesi, 1998, 2000, 2001), etc., then more direct solutions may be used.

One major feature of the NPC of dependent tests, provided that the permutation principle applies, is that we must pay attention to a set of partial tests, each appropriate for the related sub-hypotheses, because the underlying dependence relations are nonparametrically and implicitly captured by the combining procedure. In particular, the researcher is not explicitly required to specify the dependence structure on response variables. This aspect is of great importance especially for non-normal and categorical variables in which dependence relations are generally too difficult to define and, even when well defined, are too hard to cope with (see Joe, 1997). The researcher is only required to make sure that all partial tests are marginally unbiased (see Remark 1, 4.2.1), a sufficient condition which is generally easy to check.

In Section 4.1 it was emphasized that the NPC procedure may be effective when one overall test is not directly available. In such a situation it is usually convenient to analyse data by firstly examining a set of k partial aspects each interesting in a marginal sense, and then combining all captured information, provided that side-assumptions allow for proper breakdown of hypotheses and the k partial tests are marginally unbiased. In principle it is possible to apply a proper single overall permutation procedure directly, if it were known, and then avoid the combination step. But in most complex situations such a single test is not directly available, or is not easy to justify, or it is useful to obtain separate tests within multiplicity control of inferential errors and in doing so enrich the study result.

In a way the NPC procedure for dependent tests may be viewed as a two-phase testing procedure. The first phase involves a simulation from the permutation sample space $\mathcal{X}_{/\mathbf{X}}$ by means of a CMC method based on B iterations, to estimate $L(\mathbf{t}|\mathcal{X}_{/\mathbf{X}})$. The second combines the estimated p-values of partial tests into an estimate of the overall p-value λ'' by using the same CMC results in the first phase. Of course, the two phases are jointly processed, so that the procedure always remains multivariate in its own right. Furthermore, in the presence of a stratification variable, through a multi-phase procedure, the NPC allows for quite flexible solutions. For instance, we may first combine partial tests with respect to variables within each stratum, and then combine the combined tests with respect to strata. Alternatively, we may first combine partial tests related to each variable with respect to strata, and then combine the combined tests with respect to variables. In this respect, the nonparametric componentwise analysis (POSET method) – as suggested for instance, by Rosenbaum (2002) – can only permit the overall solution and nothing can be said as regards the stratified partial analyses.

As a final remark, from a general point of view and under very mild conditions, the NPC method may be regarded as a way of reducing the degree of complexity of most testing problems.

5

Multiplicity Control and Closed Testing

In many modern investigations (econometrics, biostatistics, machine learning, etc.) several thousand hypotheses may be of interest and, as a consequence, may be tested. In such cases, the problem of multiplicity control arises whenever the number of hypotheses to be tested is greater than one. With reference to this problem, two major issues may be identified: multiple comparisons and multiple testing problems. We start by defining raw and adjusted p-values, and then present a brief overview of multiple comparison procedures (MCPs) along with some definitions of the global type I error. Our main focus is on closed testing procedures for multiple comparisons and multiple testing. Some hints are also given with reference to weighted methods for controlling FWE and FDR, adjustment of stepwise p-values, and optimal subset procedures. While a detailed treatment of MCP issues is beyond the scope of this chapter, our goal is to give some hints, to discuss the basics, thus making the reader comfortable with the notion of multiplicity adjustment via closed testing performed when carrying out the analyses from real data sets by means of permutation combination-based tests. Recent developments in the field of MCPs are discussed, for example, in the papers by Bretz et al. (2009), Calian et al. (2008), Sonnemann (2008) and Westfall and Troendle (2008). We also refer the reader to those papers and references therein for a specialized discussion on multiplicity procedures within the permutation approach.

5.1 Defining Raw and Adjusted p-Values

Along with a raw (unadjusted) p-value, on many occasions it is recommended to calculate an *adjusted* p-value $\tilde{\lambda}_j,\ j = 1, \ldots, k$, for each test of H_{0j} against H_{1j}, so the decision to reject H_{0j} at $FWE = \alpha$ is obtained merely by noting whether $\tilde{\lambda}_j \leq \alpha$. Analogously to the definition of an ordinary unadjusted p-value, an adjusted p-value is defined as

$$\tilde{\lambda}_j = \inf\{\alpha | H_{0i} \text{ is rejected at } FWE = \alpha\}.$$

Hence, $\tilde{\lambda}_j$ is the smallest significance level for which we still reject H_{0j}, given a particular simultaneous procedure. Multiplicity-adjusted p-values of various forms have been considered by Shafer and Olkin (1983), Farrar and Crump (1988), Heyse and Rom (1988), Westfall and Young (1989) and Dunnett and Tamhane (1992). Alternatively, tabled critical values to accept or reject the null hypothesis (Westfall and Young, 1993) are commonly used to perform multiple testing.

Permutation Tests for Complex Data: Theory, Applications and Software Fortunato Pesarin, Luigi Salmaso

The concepts of raw and adjusted p-values are mathematically well defined in Dudoit and van der Laan (2008). Here we wish to provide some hints, reporting some useful definitions for raw and adjusted p-values, while recommending that readers consult the original book by Dudoit and van der Laan (2008). Assume that it is of interest to test k null hypotheses H_{0j}, $j = 1, \ldots, k$, individually at level α, based on test statistics $T = \{T_j; j = 1, \ldots, k\}$, with unknown probability distribution P and assumed null distribution P_0 (Dudoit and van der Laan, 2008). The null hypothesis H_{0j} is rejected at the single-test nominal type I error level α if T_j belongs to the rejection region (RR), i.e. $T_j \in RR_{j;\alpha}$. The rejection regions $RR_{j;\alpha} = RR(Q_{0j;\alpha})$, based on the marginal null distributions $Q_{0j;\alpha}$, are such that the probability of making a type I error is at most α for each test,

$$\Pr_{Q_{0j}}(T_j \in RR_{j;\alpha}) \leq \alpha,$$

and the *nestedness assumption*, stating that

$$RR(j; T, Q_0, \alpha_1) \subseteq RR(j; T, Q_0, \alpha_2), \quad \forall \alpha_1 \leq \alpha_2,$$

is satisfied. The *raw* (unadjusted) p-value $\lambda_{0j} = P(T_j, Q_{0j})$, for the single test of null hypothesis H_{0j} is then defined as

$$\begin{aligned}\lambda_{0j} &= \inf\{\alpha \in [0, 1] : \text{reject } H_{0j} \text{ at single test nominal level } \alpha\} \\ &= \inf\{\alpha \in [0, 1] : T_j \in RR_{j;\alpha}\}, \quad j = 1, \ldots, k,\end{aligned}$$

that is, the unadjusted p-value λ_{0j}, for null hypothesis H_{0j}, is the smallest nominal type I error level of the single hypothesis testing procedure at which we would reject H_{0j}, given T_j. The smaller the unadjusted p-value λ_{0j}, the stronger the evidence against the corresponding null hypothesis H_{0j}.

This concept may be easily extended to multiple testing problems (MTPs). Let us consider any multiple testing procedure, with rejection regions $RR_{j;\alpha} = RR(j; T, Q_0, \alpha)$. Hence, the k-vector of *adjusted* p-values may be defined as

$$\begin{aligned}\tilde{\lambda}_{0j} &= \inf\{\alpha \in [0, 1] : \text{reject } H_{0j} \text{ at nominal MTP level } \alpha\} \\ &= \inf\{\alpha \in [0, 1] : T(j) \in RR_{j;\alpha}\}, \quad j = 1, \ldots, k,\end{aligned}$$

that is, the adjusted p-value $\tilde{\lambda}_{0j}$, for the null hypothesis H_{0j}, is the smallest nominal type I error level of the multiple hypothesis testing procedure at which one would reject H_{0j} given T. As in single hypothesis tests, the smaller the adjusted p-value $\tilde{\lambda}_{0j}$, the stronger the evidence against the corresponding null hypothesis H_{0j}. Thus, we reject H_{0j} for small adjusted p-values $\tilde{\lambda}_{0j}$. Note that the unadjusted p-value λ_{0j}, for a single test of a null hypothesis H_{0j}, corresponds to the special case $k = 1$ (Dudoit and van der Laan, 2008).

5.2 Controlling for Multiplicity

5.2.1 Multiple Comparison and Multiple Testing

The terms *multiple comparisons* and *multiple tests* are synonymous and are often used interchangeably. However, we refer to multiple comparisons when we are interested in comparing mean values of different groups (e.g. group 1 against group 2, group 1 against group 3, group 2 against group 3, etc.). By multiple tests we mean multiple tests carried out when handling multivariate problems. For example, when evaluating the effectiveness of a treatment, it might be considered effective if

Table 5.1 Main characteristics of multiple comparisons and multiple tests (from Westfall et al., 1999)

Multiple comparisons	Multiple tests
The main goal is to compare means obtained in AN(C)OVA type problems	Refer to more general inferential problems, concerning multivariate data
Inference based on *confidence intervals*	Inference based on *tests of hypotheses*
Single-step methods	Stepwise methods

it reduces the disease symptoms, or speeds up the recovery time, or reduces the occurrences of side-effects. Hence the principal distinction between multiple comparisons and multiple tests is that with multiple comparisons it is possible to compare three or more means, in pairs or combinations, of the same measurements (Westfall et al., 1999). In contrast, multiple testing procedures consider multiple measurements. See Table 5.1, which summarizes some of the common characteristics of multiple comparisons and multiple tests.

5.2.2 *Some Definitions of the Global Type I Error*

Type I error is defined as the probability of rejecting the null hypothesis H_0 when it is true. Hence, intuitively, the global type I error increases together with the number of tested true null hypotheses. In fact, if the tests are each conducted at α level, the type I error may take the minimum value α if only one of the hypotheses is true. Otherwise, when increasing the number of true null hypotheses, it may assume greater values. MCPs have been developed to control the multiplicity issue which will be discussed in Section 5.3.

We use the term 'family' to refer to the whole set of hypotheses that should be tested to achieve the aims of a study; for example, the set of all possible pairwise comparisons is a family.

Global type I error may be defined in different ways. Let us consider the following quantities (see Table 5.2):

- U = number of true null hypotheses correctly not rejected;
- V = number of true null hypotheses wrongly rejected;
- W = number of false null hypotheses wrongly not rejected;

Table 5.2 Defining different types of global errors

		Decision taken		
		Not rejected	Rejected	
State	True null hypotheses	U	V	m_0
of nature	False null hypotheses	W	S	m_1
		$m - R$	R	m

- S = number of false null hypotheses correctly rejected.

In what follows, the terms listed below will often be used:

- familywise error rate, $FWER = \Pr\{V > 0\}$;
- false discovery proportion, $FDP = V/R$, i.e. the number of false rejections divided by the total number of rejections (defined as 0 if there are no rejections);
- false discovery rate, $FDR = \mathbb{E}[FPD|R > 0]\Pr\{R > 0\}$, where $FDP = V/R$, i.e. the expected proportion of false rejections within the class of all rejected null hypotheses.

A property that is generally required is strong control of the familywise error rate (FWER), that is, the probability of making one or more errors on the entirety of the hypotheses considered (Marcus et al., 1976). On the other hand, weak control of the FWER means simply controlling α for the global test (i.e. the test where all hypotheses are null). Although the latter is a more lenient control, it does not allow the selection of active variables because it simply produces a global p-value that does not allow interesting hypotheses to be selected, so the former is usually preferred because it makes inference on each (univariate) hypothesis (Finos and Salmaso, 2005). A precise definition of strong and weak FWE may be found in Blair et al. (1996). FWE is defined as the probability that we reject one or more of the true null hypotheses in a set of comparisons. If a multiple testing procedure produces an FWE rate less than or equal to a specified level α when all null hypotheses in the set of comparisons are true, then we say that the procedure maintains control of FWE in a weak sense. By contrast, with FWE maintained at a level less than or equal to α for any subset of true null hypotheses, we then say that the procedure maintains control in a strong sense. The choice of error type to control has to be made very carefully and should be based on considerations regarding the particular study at hand. An alternative approach to multiplicity control is given by the false discovery rate (FDR). This is the maximum proportion of type I errors in the set of elementary hypotheses. The FWE guarantees stricter control than the FDR, which in fact only controls the FWE in the case of global null hypotheses, that is, when all the hypotheses involved are under H_0 (Benjamini and Hochberg, 1995). In confirmatory studies, for example, it is usually better to strongly control the FWE, thus ensuring an adequate inference when we wish to avoid making even one error. It is a stricter requirement than the FDR and guarantees 'stronger' inferences.

In contrast, when it is of interest to highlight a pattern of potentially involved variables, especially when dealing with a large number of variables, the FDR would appear to be a more reasonable approach. In this way it is accepted that part (no greater than the α proportion) of the rejected hypotheses are in fact under the null (Finos and Salmaso, 2005).

In this chapter we deal with the FWE-controlling procedures for two reasons: it can be shown that $\mathbb{E}[V/R] \leq \Pr\{V > 1\}$, hence FDR procedures are less conservative than the FWE ones (Horn and Dunnet, 2004) and they do not control the FWE (Westfall, 1997), while the contrary holds (Westfall et al., 1999). Therefore the FWE is the natural extension of the definition of type I error in the case of single comparisons, and it is closer to the logic underlying the traditional inference.

5.3 Multiple Testing

When dealing with multiple testing problems, false positives must be controlled over all tests. The standard measure is the probability of any type I errors (i.e. the FWE). The problem of multiplicity control arises in all cases where the number of sub-hypotheses to be tested is greater than one. Such partial tests, one for each sub-hypothesis, possibly after adjustment for multiplicity (Westfall and Young, 1993), may be useful for marginal or separate inferences. However, when jointly considered, they provide information on the global hypothesis. Very often this represents the main goal in multivariate testing problems. In order to produce a valid test for the combination of a large number of p-values, we must guarantee that such a test is unbiased and produces, therefore, p-values below the significance level with a probability less or equal to α itself. This combination can be very troublesome unless we are working in a permutation framework. A Bonferroni correction is valid, but the conservativeness of this solution is often unacceptable for both theoretical and practical purposes. Moreover, this combination loses power when there is dependence between p-values. In contrast, using appropriate permutation methods, dependencies may be controlled. Multiple testing procedures have their starting point in an overall test and look for significant partial tests. Conversely, combination procedures start with a set of partial tests, each appropriate for a partial aspect, and look for joint analyses leading to global inferences (see the algorithms for NPC discussed in Chapter 4). The global p-value obtained through the NPC procedure of p-values associated with sub-hypotheses is an exact test, thus providing weak control of the multiplicity. The inference in this case must be limited to the global evaluation of the phenomenon. Due to the use of NPC methods, a more detailed analysis may be carried out. Actually what is important is to select potentially active hypotheses (i.e. under the alternative). A correction of each single p-value is therefore necessary in this case. A possible solution within a nonparametric permutation framework is represented by closed testing procedures, which are quite powerful multiple inference methods (Westfall and Wolfinger, 2000). Such methods include the so-called single-step and stepwise procedures, allowing us to significantly reduce the computational burden.

5.4 The Closed Testing Approach

It would be no exaggeration to say that most of the recent developments in stepwise multiple comparison procedures (MCPs) are strictly connected with the closure principle, formulated by Marcus et al. (1976). In particular, the methods based on this concept are called *closed testing procedures* since they refer to families of hypotheses that are closed under intersection.

Definition 1. *A closed family is a family for which any subset intersection hypothesis involving members of the family of tests is also a member of the family (Westfall et al.,* 1999*).*

Closed testing methods are special cases of stepwise methods. The goal of multiple testing procedures is to control the 'maximum overall type I error rate', that is, the maximum probability that one or more null hypotheses are rejected incorrectly. For details, we refer the reader to Westfall and Wolfinger (2000) and Westfall and Young (1993).

Closed testing methods have recently attracted more and more interest, since they are flexible, potentially adaptable to any situation and are able produce coherent decisions.

When dealing with closed testing procedures, coherence is a necessary property, while consonance is only desirable (see Westfall et al., 1999).

Marcus et al. (1976) showed that the closed testing procedure controls the FWE in the strong sense. The real strength of the closed testing approach lies in its generality. In fact the performance of the procedure may depend on the choice of the tests used for both the minimal and the composite hypotheses of the closure tree. The whole set of hypotheses may be properly assessed using both parametric and nonparametric tests. This fact has a clear advantage: nonparametric methods, and in particular NPC tests, are found to be very powerful when the structure (and nature) of the data do not meet the requirements for the correct application of traditional parametric tests. Moreover, they implicitly capture the whole dependence structure of the data. Hence, combining the closed testing approach with nonparametric procedures allows us to improve the performance of the method. However, when the assumptions for valid application of parametric tests are satisfied and the dependence structure of the data is known, using closed testing procedures in a parametric framework produces the best results, even if such a situation rarely occurs in practice. Testing the composite hypotheses of the closure tree using methods capable of handling the dependence structure of the C minimal hypotheses is of primary importance, especially in the case of all pairwise comparisons. In the following sections, we show how to reduce the closure tree, thus providing significant improvements in the procedure's power behaviour.

5.4.1 Closed Testing for Multiple Testing

There are two important classes of multiple testing procedures: *single-step* and *stepwise* methods. The latter type of method allows a substantial reduction in the computational burden. An example of the single-step method is the simple Bonferroni procedure. Assuming there are k hypotheses of interest, the Bonferroni procedure rejects any null hypothesis H_{0i}, $i = 1, \ldots, k$, if the corresponding p-value is less than or equal to α/k, where α is the desired FWE level. In contrast, stepwise methods make use of significance levels larger than α/k, allowing us to detect more significant differences and offering greater power.

The basic idea underlying closed testing procedures is quite simple, but some notation is required. Suppose that we wish to test the hypotheses H_1, H_2, H_3, allowing the comparison of three treatment groups with a common control group or a single treatment against a single control on the basis of three different variables (see Figure 5.1). The closed testing method works as follows:

1. Test each minimal hypothesis H_1, H_2 and H_3 using an appropriate α-level test.
2. Create the closure of the set, which is the set of all possible intersections among H_1, H_2 and H_3 (in this case the hypotheses H_{12}, H_{13}, H_{23} and H_{123}).
3. Test each intersection using an appropriate α-level test.
4. Any hypothesis H_i, with control of the FWE, may be rejected when the following conditions both hold:
   - the test of H_i itself yields a statistically significant result; and
   - the test of every intersection hypothesis that includes H_i is statistically significant.

When using a closed testing procedure, the adjusted p-value for a given hypothesis H_i is the maximum of all p-values for tests that include H_i as a special case (including the p-value for the

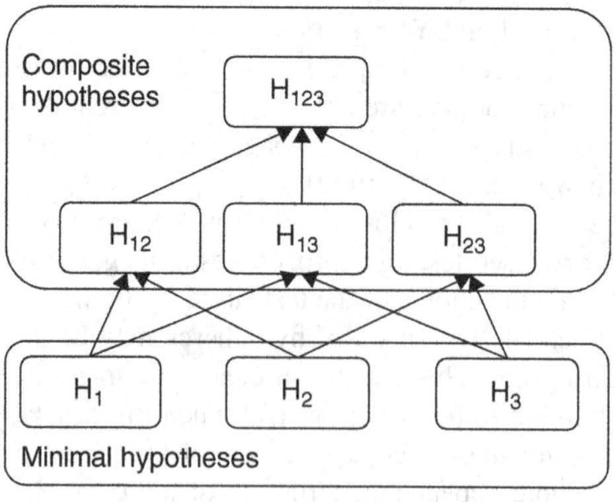

Figure 5.1 The closure tree (Westfall and Wolfinger, 2000)

H_i test itself), hence $\tilde{\lambda}_i = \max(\lambda_i, \lambda_{ik}, \lambda_{ikl}, \ldots)$, the maximum of all p-values for tests including H_i as component.

Hence, in order to test the null hypotheses, all the related closed families, that is, all the families for which any subset intersection hypothesis involving members of the family of tests is also a member of the family, must be considered. The rule for rejecting the original null hypothesis is defined as follows:

- test every member of the closed family using a (suitable) level test;
- a hypothesis can be rejected provided that (a) its corresponding test was significant at level α and (b) every other hypothesis in the family that implies it, can also be rejected by its corresponding α-level test.

The choice of method to calculate the appropriate level of significance is different depending on the situation; we will discuss this later.

It is in fact possible to obtain step-down procedures by means of the closure principle, and by applying the related single-step methods to the closed family hypotheses. In *step-down* procedures, the hypotheses corresponding to the most significant test statistics are considered in turn, with successive tests depending on the outcome of previous ones. As soon as one fails to reject a null hypothesis, no further hypotheses are rejected. These procedures are referred to as *shortcut* versions of the closed testing procedure. For example, the Bonferroni–Holm method is the shortcut version of the closed testing procedure using Bonferroni's procedure.

The closure principle has recently been applied to nonparametric tests, leading to improvements in terms of power. Westfall and Young (1993) show that the Min$P$ (the minimum significant test p-value) which, in the classic version of the closed testing method using the Bonferroni inequality, has to be compared with the α/k level, where k is the number of minimal hypotheses considered, in this framework may be compared with the α-quantile of the Min$P$ distribution under H_0. This distribution is unknown, but may easily be obtained within the permutation framework (Finos et al., 2003). In the following, we briefly present the shortcut version of the closed testing procedures for the step-down Min$P$ Bonferroni procedure and the step-down Tippett procedure (Finos et al., 2003).

Remark 1. It is well known (Dmitrienko et al., 2003) that Bonferroni-based tests are conservative for large correlation, and may all be non-significant even when all unadjusted p-values are significant. The method is very general as it is applicable to any situation; however, its severe conservativeness, especially when the endpoints are correlated, constitutes a barrier to its usefulness (Huque and Alosh, 2008). Resampling-based procedures (Westfall and Young, 1993) may solve the former problem, while the use of the Simes test (Simes, 1986) may mitigate the latter problem.

5.4.2 Closed Testing Using the MinP Bonferroni–Holm Procedure

Let us deal with a multivariate two-sample problem, supposing it is of interest to test for differences between the means of the two groups over three variables.

Definition 2. (Westfall and Young, 2000). *When considering a closed testing procedure, the adjusted p-value $\tilde{\lambda}_i$ for a given hypothesis H_i is the maximum of all p-values for tests including H_i as a special case (including the p-value for the H_i test itself).*

In order to apply Bonferroni's Min$P$ test, the minimum p-value of the individual component tests should be compared with α/k, where α is the desired FWE level and k is the number of minimal hypotheses, and reject the composite hypothesis when Min$P \leq \alpha/k$.

Holm (1979) proposed a modification of the simple Bonferroni procedure. Indeed, the Min$P$ Bonferroni–Holm procedure is the most classical version of closed testing (multi-level) procedures and is called the sequentially rejective Bonferroni procedure. Holm demonstrated that it is not necessary to calculate the p-values for the entire tree. It is sufficient to calculate the p-values for the nodes of the tree corresponding to the ordered p-values. The Bonferroni–Holm closed Min$P$-based method may be obtained as follows:

1. Let H_i, $i = 1, \ldots, k$, be the set of the minimal hypotheses, and let $\lambda_{(1)}, \ldots, \lambda_{(k)}$ be the vector containing the (increasing) ordered p-values corresponding to the set of minimal hypotheses.
2. $\tilde{\lambda}_{(1)} = k \cdot \lambda_{(1)}$; if $\tilde{\lambda}_{(1)} \leq \alpha$ the corresponding hypothesis $H_{(1)}$ is rejected and the procedure is continued; otherwise the hypotheses $H_{(1)}, \ldots, H_{(k)}$ are accepted and the procedure stops.
3. $\tilde{\lambda}_{(j)} = \max((k - j + 1) \cdot \lambda_{(j)}, \tilde{\lambda}_{(j-1)})$; if $\tilde{\lambda}_{(j)} \leq \alpha$ the corresponding hypothesis $H_{(j)}$ is rejected and the procedure is continued; otherwise the hypotheses $H_{(j)}, \ldots, H_{(k)}$ are accepted and the procedure stops, for $j = 2, \ldots, k$.

Let us now briefly introduce the data set used by Westfall and Wolfinger (2000) to test for differences between the means of two groups, $G = 0$ and $G = 1$ (two-sample test), for each of three variables Y_1, Y_2, and Y_3 (see the data in Table 5.3). The p-values are shown in Figures 5.2 and 5.3. The data are analysed using MATLAB and R in Section 5.5.

With reference to the example in Figure 5.2, the hypotheses H_{12} and H_{13} do not need to be tested. In fact, the ordered vector of the minimal hypotheses is $\lambda_3 < \lambda_2 < \lambda_1$ and $H_{123} : \lambda_{123} = 3 \cdot \lambda_3 = 3 \cdot 0.0067 = 0.0201$. Let us assume that $\alpha = 0.05$, in which case the hypothesis H_{123} should be rejected and as a consequence automatically all the hypotheses containing H_3 will be rejected (i.e. H_3, H_{13} and H_{23}). Hence, an adjusted p-value equal to $\tilde{\lambda}_3 = 0.0201$ will be associated with the hypothesis H_3. Then in the next step the composite hypothesis which has still not been rejected, H_{12}, will be evaluated and its significance will be obtained as $H_{12} : \lambda_{12} = 2 \cdot \lambda_2 = 2 \cdot 0.0262 = 0.0524$. With $\alpha = 0.05$, the hypotheses H_1 and H_2 will be accepted.

Table 5.3 `mult` data (Westfall and Wolfinger, 2000)

Group	Y_1	Y_2	Y_3
0	14.4	7.00	4.30
0	14.6	7.09	3.88
0	13.8	7.06	5.34
0	10.1	4.26	4.26
0	11.1	5.49	4.52
0	12.4	6.13	5.69
0	12.7	6.69	4.45
1	11.8	5.44	3.94
1	18.3	1.28	0.67
1	18.0	1.50	0.67
1	20.8	1.51	0.72
1	18.3	1.14	0.67
1	14.8	2.74	0.67
1	13.8	7.08	3.43
1	11.5	6.37	5.64
1	10.9	6.26	3.47

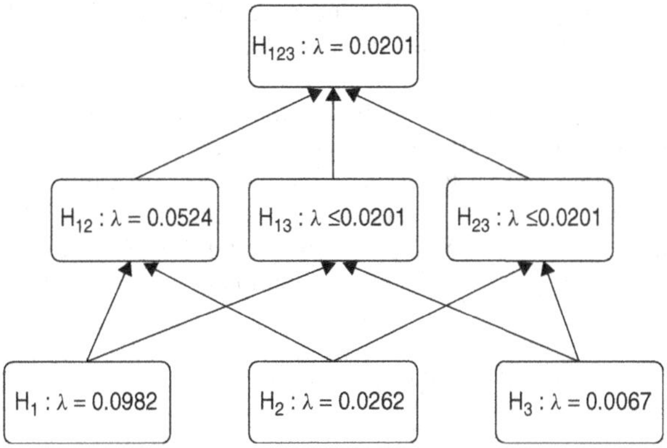

Figure 5.2 Step-down Min$P$ Bonferroni–Holm procedure (Finos et al., 2003)

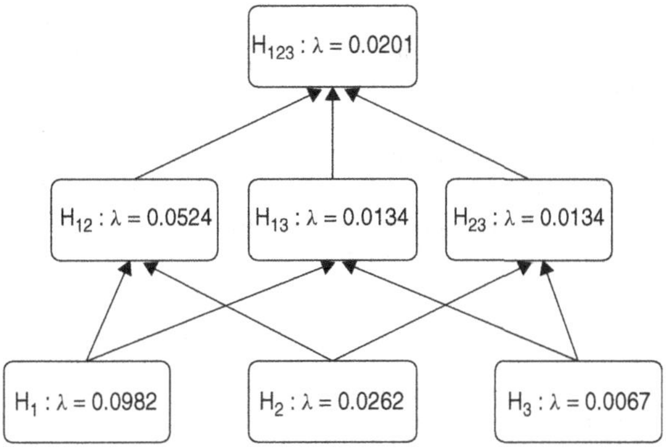

Figure 5.3 Bonferroni *single-step* procedure (Finos et al., 2003)

By applying Bonferroni's inequality (see Figure 5.3) to a single-step procedure ($\lambda_{H_i} : \max(\lambda_{H_j} : H_j \supseteq H_i)$), we would have obtained the same results, but the hypotheses would be tested instead of calculating the significance for only two hypotheses.

On the one hand, by means of single-step methods, the critical value for all tests is found in only one step, since each test is calculated without reference to the significance or non-significance of remaining inferences. On the other hand, sequentially rejective or stepwise methods differ from single-step methods in that the result of a given test depends on the results of the other test. However, by means of these methods it is possible to substantially increase power while retaining FWE control (Westfall et al., 1999). $\lambda_{123} = 3 \cdot \min(\lambda_1, \lambda_2, \lambda_3)$ and $\lambda_{12} = 2 \cdot \min(\lambda_1, \lambda_2)$ and the minimal hypotheses have been tested using a standard Student's t test.

It is well known that the Bonferroni–Holm Min$P$ test is very conservative, especially when the correlation structure among variables is strong. Westfall and Young (1993) proposed to compare the observed Min$P$ (denoted by Minp^o) for a given composite hypothesis with the α-quantile of the Min$P$ distribution under the null hypothesis, instead of comparing it with α/k. This corresponds to calculating the p-value $\lambda = \Pr(\text{Min}P \leq \text{Min}p^o)$, where Min$P$ is the random value of the minimum p-value for the given composite hypothesis, and Minp^o denotes the minimum p-value observed for the hypothesis under study. Usually the distribution of Min$P$ is unknown but can be evaluated by means of permutation methods as shown in Finos et al. (2003).

Westfall and Young (1993) showed that, like the Bonferroni–Holm method, this method may be used to generate a stepwise procedure. Hence, it is no longer necessary to compute all the p-values, but it is enough to use an algorithm similar to that proposed by Holm. In conclusion, the estimate of the α-quantile of the Min$P$ null distribution obtained by means of resampling methods without replacement is equivalent to Tippett's nonparametric combining function. The significance is calculated as $\lambda = \Pr\{\min_{1\le i\le k}(\lambda_i)^* \le \min_{1\le i\le k}(\lambda_i)\}$, where $\min_{1\le i\le k}(\lambda_i)^*$ indicates the permutation distribution of Min$P$ and $\min_{1\le i\le k}(\lambda_i)$ denotes the minimum p-value of the composite hypothesis (i.e. Minp^o):

$$\begin{aligned}\lambda &= \Pr(\min_{1\le i\le k}(\lambda_i)^* \le \min_{1\le i\le k}(\lambda_i))\\ &= \Pr(1-\min_{1\le i\le k}(\lambda_i)^* \ge 1-\min_{1\le i\le k}(\lambda_i))\\ &= \Pr(\max_{1\le i\le k}(1-\lambda_i)^* \ge \max_{1\le i\le k}(1-\lambda_i))\\ &= T_T'',\end{aligned}$$

where T_T'' denotes Tippett's combination function (i.e. $T_T'' = \max_{1\le i\le k}(1-\lambda_i)$).

The shortcut is obtained as follows.

1. Let $\lambda_{(1)},\ldots,\lambda_{(k)}$ be the vector containing the (increasing) ordered p-values corresponding to the set of minimal hypotheses.
2. $\tilde{\lambda}_{(1)} = \lambda''_{(1),\ldots,(k)\,\text{Tippett}}$; if $\tilde{\lambda}_{(1)} \le \alpha$, reject the corresponding hypothesis $H_{(1)}$ and go on; otherwise retain the hypotheses $H_{(1)},\ldots,H_{(k)}$ and stop.
3. $\tilde{\lambda}_{(i)} = \max(\lambda''_{(1),\ldots,(k)\,\text{Tippett}}, \tilde{\lambda}_{(i-1)})$; if $\tilde{\lambda}_{(i)} \le \alpha$, reject (also) $H_{(i)}$ and go on; otherwise retain the hypotheses $H_{(i)},\ldots,H_{(k)}$ and stop, for $i = 2,\ldots,k$.

Remark 1. Within the context of multiple testing methods, we mention a recent paper by Goeman and Mansman (2008), where a focus level procedure (i.e. a sequentially rejective multiple testing method), is presented. The procedure actually combines Holm's (1979) procedure with the closed testing procedure of Marcus et al. (1976). The proposed procedure strongly controls the FWE without any additional assumptions on the joint distribution of the test statistics used, while explicitly making use of the directed acyclic graph structure of gene ontology. We refer the reader to the original paper by Goeman and Mansman (2008) for further details.

Remark 2. If present, a stratification variable (also called a block variable) may be included in the analyses. The resulting p-values are combined over levels of the stratification variable and multiplicity adjustments are performed independently among strata levels. As an example, see the analysis performed using the washing test data set, discussed in Section 5.6.

5.5 Mult Data Example

5.5.1 Analysis Using MATLAB

We now analyse the Mult data set, previously discussed in Westfall and Wolfinger (2000), to test for equality of the multivariate distribution of three variables Y_1, Y_2, and Y_3 in the two groups labelled by the binary variable X (two independent samples test). The aim of this example is to show how the closed testing procedure performs when different combining functions are applied.

Table 5.4 Example from Westfall and Wolfinger (2000)

	λ (raw p-value)	$\tilde{\lambda}$ (Fisher)	$\tilde{\lambda}$ (Tippett)	$\tilde{\lambda}$ (Liptak)
Y_1	0.1004	0.1004	0.1004	0.1004
Y_2	0.0313	0.0445	0.0561	0.0426
Y_3	0.0097	0.0234	0.0199	0.0270
p-GLOB	·	0.0220	0.0199	0.0221

We show the results for the closed testing procedure, using Tippett, Fisher and Liptak combining functions (see Table 5.4). It should be noted that no uniformly most powerful (UMP) combined test exists when $k > 2$. The MATLAB code is as follows:

```
MATLAB code:

B=10000
[data,code,DATA]=xlsimport('mult');
Y=data(:,2:4);
X=data(:,1);
p=NP_2s(Y,X,B,0);
NPC_FWE(p,'F');
NPC_FWE(p,'T');
NPC_FWE(p,'L');
```

5.5.2 Analysis Using R

The aim of this example is to show how the closed testing procedure performs when different combining functions are applied. The R code that follows is basically the body of the `FWE.minP` function, except for Fisher's and Liptak's combining functions which have not been implemented (if needed, one may just add an argument indicating the desired combining function and make some adjustments in the body with `if` conditions, and add the lines that are signed by the symbol `#`).

The procedure is as follows. Let $p_{(1)} \leq p_{(2)} \leq p_{(3)}$ be the (increasingly) ordered raw p-values; choose a combining function (whose arguments are raw p-values), and obtain a global test involving all three variables. The adjusted p-value related to $p_{(1)}$ is the p-value of the global test on all the three variables. In the second step, obtain a global test involving the variables corresponding to $p_{(2)} \leq p_{(3)}$ only; the adjusted p-value related to $p_{(2)}$ is the p-value of the global test. In the third step, obtain a global test involving the variable generating $p_{(3)}$ only; that is, there is no need to perform the procedure on the variable with largest raw p-value. A monotonicity condition must be satisfied, that the adjusted p-value at each step is not smaller than the adjusted p-value at the previous step.

First of all, we carry out the testing procedure on the three variables and obtain the raw p-values. We consider the difference in means as the test statistic, which can be obtained by multiplying the data matrix by a contrast vector. We consider a two-sided alternative and $B = 5000$ random permutations.

```
setwd("C:/path") ; source("t2p.r")
data<-read.csv("mult", header = TRUE)
```

```
attach(data)

n=table(X) ; p = dim(data)[2]-1
contr<-rep(c(1/n[1],-1/n[2]),n)

B=5000 ; n = dim(data)[1]
Y = as.matrix(data[,-1])
T = array(0,dim=c((B+1),p))

T[1,] = t(Y)%*%contr

for(bb in 2:(B+1)){

Y.star = Y[sample(1:n),]
T[bb,] = t(Y.star)%*%contr
}

P = t2p(abs(T)) ; p.raw = P[1,]
p.raw

[1] 0.0948 0.0296 0.0088
```

The vector `p.raw` contains the raw p-values related to each variable. Now we order the raw p-values increasingly, and store their original position in the vector `o` (this will be useful for printing the adjusted p-values in their original positions). The matrix `P.ord` contains the null distribution of the raw p-values, and its columns are ordered with respect to the elements of the first row (i.e. the observed raw p-values).

```
p.ord<-sort(p.raw,decreasing=FALSE)
o<-order(p.raw,decreasing=FALSE)

B=dim(P)[1]-1
p=dim(P)[2]

p.ris<-array(0,dim=c(p,1))
P.ord<-P[,o]
```

Now we obtain the adjusted p-values as the global p-values obtained by combining the last $3 - j$ columns of `P.ord`, $j = 0, 1, 2$. We can do this by choosing one of the three combining functions proposed (Tippett's, Fisher's and Liptak's). The following code computes the adjusted p-values by applying Tippett's combining function; in order to change combining function, uncomment (i.e. remove the symbol '`#`') the rows with the desired combining function and comment the others. Note that the computation of the p-values depends on the chosen combining function (the rule *small is significant* applies when Tippett's combining function is applied).

```
T=apply(P.ord,1,min)                                 #Tippett's
#T=apply(P.ord,1,function(x){-2*log(prod(x))}) #Fisher's
#T=apply(P.ord,1,function(x){sum(qnorm(1-x))}) #Liptak's

p.ris[1] = mean(T[-1]<=T[1]) #Tippett's
#p.ris[1] = mean(T[-1]>=T[1]) #Fisher's & Liptak's
```

```
if(p>2){
for(j in 2:(p-1)){

T=apply(P.ord[,j:p],1,min)                                  #Tippett's
#T=apply(P.ord[,j:p],1,function(x){-2*log(prod(x))}) #Fisher's
#T=apply(P.ord[,j:p],1,function(x){sum(qnorm(1-x))}) #Liptak's

p.ris[j] = max(mean(T[-1]<=T[1]),p.ris[(j-1)])              #Tippett's
#p.ris[j] = max(mean(T[-1]>=T[1]),p.ris[(j-1)])             #Fisher's & Liptak's
}
}
p.ris[p] = max(p.ord[p],p.ris[p-1])
p.ris[o]=p.ris

rownames(p.ris)=colnames(data)[-1]
p.ris
     [,1]
Y1 0.0992
Y2 0.0520
Y3 0.0212
```

	Raw p	Fisher's	Tippett's	Liptak's
Y1	0.0992	0.0992	0.0992	0.0992
Y2	0.0288	0.0468	0.0520	0.0466
Y3	0.0098	0.0226	0.0212	0.0232

The data set and the corresponding software codes are available from the `mult` folder on the book's website.

5.6 Washing Test Data

5.6.1 Analysis Using MATLAB

Let us now introduce a real case study concerned with the development of a new detergent, where an instrumental performance study modelled by a C independent sample design has been carried out. The R&D division of a chemical company wishes to assess the level of performance of a set of eight products ($C = 8$) on 25 types of stain, which may be classified into three main domains (or categories) – general detergency, bleachable and enzymatic. A suitable experiment has been designed. For each of the eight products, four washing machines (therefore, four replicates/observations in each group) are used to wash a piece of fabric soiled with one of the 25 stains. The experimental response variable is the reflectance, that is, the percentage of stain removed for the 25 types of stain. Hence, it is of interest to compare the eight products on the basis of their reflectance (the variable Y), using, as stratification variable Z, the type of stain. Here we wish to show that it is possible to include a stratification variable while controlling the FWE (see Remark 1, 5.4.2). Indeed, once the various strata of experimental units have been defined, the observation vector may be resampled independently within strata to simulate the complete null hypothesis of no treatment effect for any stratum. The test statistics are then recomputed by combining over

Table 5.5 Controlling for multiplicity over types of stain and domains

Domain	Strata (type of stain)	No.	p-value	p-FWE	p-value	p-FWE
Bleach	Blueberry juice	6	0.0080	0.1489		
Bleach	Coffee	10	0.0030	0.0679		
Bleach	Grass	12	0.3007	0.9501		
Bleach	Tea	20	0.0330	0.3736		
Bleach	Tomato	21	0.7912	0.9910		
Bleach	Wine	25	0.0150	0.2268	0.0180	0.0260
Detergency	Artificial sebum on PES/cot	1	0.0060	0.1209		
Detergency	Artificial sebum on cotton	2	0.0190	0.2518		
Detergency	Bacon grease	3	0.1009	0.6883		
Detergency	Butter	7	0.0509	0.4745		
Detergency	Frying fat	11	0.0150	0.2268		
Detergency	Lanolin	13	0.3047	0.9501		
Detergency	Lipstick1	14	0.3636	0.9501		
Detergency	Lipstick2	15	0.6813	0.9910		
Detergency	Make-up1	16	0.6294	0.9910		
Detergency	Make-up2	17	0.2068	0.8721		
Detergency	Olive oil	18	0.0140	0.2268		
Detergency	Used Motor oil2	22	0.7642	0.9910		
Detergency	Used Motor oil3	23	0.8332	0.9910		
Detergency	Vegetable oil	24	0.0010	0.0240	0.0140	0.0260
Enzymatic	Blood	4	0.1359	0.7582		
Enzymatic	Blood1	5	0.0330	0.3736		
Enzymatic	Chocolate	8	0.0010	0.0240		
Enzymatic	Cocoa	9	0.0110	0.1888		
Enzymatic	Rice starch	19	0.0030	0.0679	0.0050	0.0050
p-Global				0.0240		0.0050

strata, thus obtaining a test statistic for each stratum. Then multiplicity adjustment for multiple tests that have been combined over strata may easily be done following the standard procedure (see Section 5.4.1). Raw and adjusted p-values, along with the global p-value, are displayed in Table 5.5. MATLAB code is given below.

```
B=1000
[data]=xlsimport('WashingTest');
strata=data(2);
Y=data(4);
X=data(1);
[P]=by_strata(strata, 'NP_Cs', Y,X, B, 1);
P2=NPC_FWE(P,'T',1);
d1=[6 10 12 20 21 25];
d2=[1 2 3 7 11 13 14 15 16 17 18 22 23 24];
d3=[4 5 8 9 19];
bleach=NPC(P(:,d1),'T',1);
```

```
detergency=NPC(P(:,d2),'T',1);
enzymatic=NPC(P(:,d3),'T',1);
NPC_FWE([bleach detergency enzymatic],'T',1);
```

5.6.2 *Analysis Using R*

The R code for analysing the washing test data is as follows:

```
setwd("C:/path")
data<-read.csv("WashingTest.csv",header=TRUE)
attach(data)

C = length(unique(Product))
p = length(unique(Stain))
r = dim(data)[1]/(C*p)

B=1000
```

It is easy to show that the F statistic for one-way ANOVA is permutationally equivalent to the statistic $T_j^* = \sum_{i=1}^{C} n_i (\bar{X}_{ji}^*)^2$, $j = 1, \ldots, p$. We first obtain, for each type of stain, the vector of sample means $\mathbf{M}_j = [\bar{X}_{j1}, \ldots, \bar{X}_{jC}]$, $j = 1, \ldots, p$. We do this by multiplying the vector of responses of the jth stratum for the diagonal block-matrix of contrasts:

$$\texttt{contr} = \begin{bmatrix} 1/r\mathbf{I}_r & \vdots & \ldots & \vdots \\ \vdots & 1/r\mathbf{I}_r & \ldots & \vdots \\ \vdots & \vdots & \ddots & \vdots \\ \vdots & \vdots & \ldots & 1/r\mathbf{I}_r \end{bmatrix}$$

The matrix `contr` has dimension $(r \times C) \times C$, and $\mathbf{I}_r$ is the identity matrix of rank r. The observed values of the partial test statistics T_j^o are then obtained by adding the squared elements of $\mathbf{M}_j$.

```
contr = array(0,dim=c(r*C,C))
for(cc in 1:C){
contr[((cc-1)*r+1):(cc*r),cc] = rep(1/r,r)
}

T = array(0,dim=c((B+1),p))
M = array(0,dim=c((B+1),p,C))

for(j in 1:p){
M[1,j,] = Reflectance[((j-1)*r*C+1):(j*C*r)]%*%contr
T[1,j] = sum(M[1,j,]^2)
}
```

This is a stratified analysis, where strata are defined by the type of stain (25 categories), therefore we consider independent within-strata permutations. The vector `R.star` is a random permutation of the response elements of the jth stratum.

```
for(bb in 2:(B+1)){
print(bb)
for(j in 1:p){
R = Reflectance[((j-1)*r*C+1):(j*C*r)]
n = length(R)
R.star = R[sample(1:n)]
M[bb,j,] = R.star%*%contr
T[bb,j] = sum(M[bb,j,]^2)
}
}
```

The matrix `M` stores the sample means of the C products of each variable on each permutation. The matrix `T` contains the null distribution of T_j^* for each variable. Finally, we obtain the partial p-values of each stratum and adjust these p-values for multiplicity by running the `FWE.minP` function.

```
source("t2p.r")
P = t2p(T)
colnames(P) = unique(Stain)

source("FWEminP.r")
p.FWE = FWE.minP(P)
res=data.frame(p = P[1,],p.fWE = p.FWE)

res
                            Category      p     p.fWE
Tea                           Bleach  0.0430   0.4334
Coffee                        Bleach  0.0020   0.0406
Blueberry juice               Bleach  0.0120   0.2086
Wine                          Bleach  0.0120   0.2086
Grass                         Bleach  0.2890   0.9302
Tomato                        Bleach  0.8006   0.9946
Used Motor oil3          Detergency  0.8332   0.9946
Olive oil                Detergency  0.0116   0.2086
Bacon grease             Detergency  0.1126   0.7214
Frying fat               Detergency  0.0196   0.2712
Vegetable oil            Detergency  0.0004   0.0096
Butter                   Detergency  0.0464   0.4334
Make-up2                 Detergency  0.2228   0.8910
Lipstick2                Detergency  0.6740   0.9946
Artificial sebum on cotton  Detergency  0.0238   0.3036
Make-up1                 Detergency  0.6376   0.9946
Artificial sebum on PES/cot Detergency  0.0042   0.0814
Lipstick1                Detergency  0.3538   0.9366
Used Motor oil2          Detergency  0.7612   0.9946
Lanolin                  Detergency  0.3246   0.9366
Chocolate                 Enzymatic  0.0000   0.0000
Blood                     Enzymatic  0.1296   0.7476
Rice starch               Enzymatic  0.0024   0.0514
Blood1                    Enzymatic  0.0390   0.4258
Cocoa                     Enzymatic  0.0132   0.2086
```

The type of stain considered can be combined according to the three domains (bleach, detergency, enzymatic). Applying Fisher's combining function, we obtain that there are significant differences within all domains.

```
T.dom = array(0,dim=c((B+1),3))

dom=c(rep(1,6),rep(2,14),rep(3,5))

for(dd in 1:3){
T.dom[,dd] = apply(P[,dom==dd],1,function(x){-2*log(prod(x))})
}

P.dom = t2p(T.dom)
p.dom = P.dom[1,]
p.d.fwe = FWE.minP(P.dom)
res.dom = data.frame(p.dom,p.d.fwe)
colnames(res.dom) = c('Bleach', 'Detergency', 'Enzymatic')
res.dom

    p.dom    p.d.fwe
1       0          0
2       0          0
3       0          0
```

The data set and the corresponding software codes are available from the `Washing_test` folder on the book's website.

5.7 Weighted Methods for Controlling FWE and FDR

When dealing with a very large number of variables (say, hundreds of thousands), the standard type I error rate criterion is no longer recommended (Finos and Salmaso, 2007). On the other hand, Bonferroni's method used to control the FWE may be excessively conservative. In fact when k tests are available, a test will be considered significant if its p-value is less than or equal to α/k. As previously seen, the Bonferroni–Holm (Holm, 1979; Ludbrook, 1998) step-down approach represents another valid solution but it does not provide great power when k is large. The method rejects the hypothesis corresponding to the most significant test (endowed with the smallest p-value) if its p-value is less than α/k; if this hypothesis is rejected, then the second smallest p-value is compared to $\alpha/(k-1)$, and so on.

An alternative is the FDR-controlling method of Benjamini and Hochberg (1995). Once again, FDR may become exceedingly conservative for large k. Furthermore, since FDR-controlling procedures do not generally control the FWE, they allow a fraction of detected significances to be in error.

In Finos and Salmaso (2007) weighted methods controlling FWE and FDR are presented. Weighted methods represent an important feature of multiplicity control methods. The weights must usually be chosen *a priori* on the basis of experimental hypotheses. Under some conditions, however, they can be chosen by making use of information from the data, therefore *a posteriori*, while maintaining multiplicity control.

Weighted methods are in fact useful when some H_i hypotheses are considered more important than others on the basis of *a priori* knowledge, for example in clinical trials, where patients' endpoints may be ranked *a priori*, thus planning the testing procedure in order to emphasize the most important hypotheses. The simplest weighted multiple testing procedure, discussed in

Rosenthal and Rubin (1983), consists in the rejection of H_i whenever $\lambda_i \leq w_i\alpha$, where the weights w_i lie in the simplex $w_i \geq 0$, $\sum w_i = 1$, and λ_i is the p-value of the ith test, $i = 1, \ldots, k$. The choice of w_i may be based purely on the *a priori* importance of the hypotheses or, to optimize power, on prior information (Spjøtvoll, 1972; Westfall and Krishen, 2001).

When weights are properly chosen from the concurrent data set, an increase in terms of power may be obtained without invalidating significance levels. Finos and Salmaso (2007) provide a review of weighted methods for FWE (both parametric and nonparametric) and FDR control and a review of data-driven weighted methods for FWE control. Moreover, the authors propose a class of weighted FWE-and FDR-controlling procedures which take advantage of information from the available data (data-driven weights), thus improving, under some conditions, power performance. They also suggest new data-driven weighted procedures controlling FDR in the case of independence between variables and under positive regression dependency on the subset of variables corresponding to null hypotheses. We refer the reader to the paper by Finos and Salmaso (2007) for an in-depth examination of these topics.

5.8 Adjusting Stepwise p-Values

Stepwise methods for variable selection are frequently used to determine the predictors of an outcome in generalized linear models; see Miller (1984) and Hocking (1976) for a comprehensive overview of model selection methods. Despite their widespread use, it is well known that the tests on the explained deviance of the selected model are biased. Several solutions have been proposed to overcome this problem. In particular, Copas and Long (1991) propose a correction for forward selection in multiple linear models for orthogonal regressors; Grechanovsky and Pinsker (1995) generalize it to general forward selection for linear models; Harshman and Lundy (2006) describe a computer-intensive method that accurately estimates stepwise p-values by using a modified randomization permutation test procedure that empirically determines the appropriate null distributions.

The Harshman–Lundy method also corrects for bias due to the advantage of getting the best current option and the disadvantage of not getting the even better alternatives chosen at prior steps. Actually, all these solutions control the FWER strongly but are restricted to linear models under forward selection. For a complete literature review, see Finos et al. (2010). Hence, the p-values of stepwise regression can be highly biased, and care must be taken with the evaluation of results from stepwise generalized linear models (GLMs). This biasedness arises from the fact that the traditional test statistics upon which these methods are based were intended for testing pre-specified hypotheses; instead the tested model is selected through a data-driven procedure. In particular, the problem of equipping them with statistically valid stopping rules remains unresolved. At each forward selection step, the covariate that explains more of the residual response than any other remaining regressor is entered into the regression, if this explained residual response is large enough; otherwise the forward selection stops. What constitutes large enough is decided by a stopping rule, that is, by a significance test performed at each step of this multi-step decision-making process. Typically, computer packages provide a p-value based on the F statistic calculated based on the observed data and computed from tables of the F-distribution. Otherwise, the deviance, scaled deviance and χ^2 distribution represent the corresponding goodness-of-fit criteria and approximate distribution in the GLM. As pointed out by many authors, this distribution is correct only if all previously entered regressors have not been data-driven. However, since the forward selection searches for the best regressor and may discontinue this process at any step, the F-distribution is not valid (Grechanovsky and Pinsker, 1995). A multiplicity problem therefore arises. In Finos et al. (2010) a nonparametric procedure to adjust the p-value of the selected model of any stepwise selection method is discussed. The proposed method is unbiased and consistent. Moreover, it is proven that the procedure controls the FWE in a weak sense only (Westfall and Young, 1993). This means that it does not correct the

p-values of each selected variable but only the p-value of the model. Therefore, it is only able to assess whether at least one variable among those selected is associated with the dependent variable. Although control is weak, it has the clear advantage of being applicable to any GLM (not only to linear models) and to any stepwise methods. The procedure does not therefore ask practitioners to change the way they model phenomena and select variables, but simply gives a minimum validity inferential criterion to their findings.

5.8.1 Showing Biasedness of Standard p-Values for Stepwise Regression

Here we present an explanatory simulation study. We have generated standard normal distributed independent covariates that are unrelated and independent of the outcome. We have also chosen the backward mode of stepwise searching. In the first simulation study the number of covariates was set at $V = 10$, the number of cases at $n = 20$ and the number of Monte Carlo replications at 1000. In the second simulation study we set the number of covariates at $V = 20$, the number of cases at $n = 30$ and the number of Monte Carlo replications at 1000. Table 5.6 shows the estimated probability of finding a non-real significant model after stepwise regression under the null hypothesis H_0. For $\alpha = 0.05$, under H_0 the probability of finding a significant model is more than 50% in the first simulation study and more than 80% in the second simulation study.

5.8.2 Algorithm Description

As mentioned in the previous section, when dealing with model selection, the problem of multiplicity control arises because of the multitude of models explored by the stepwise method. With V covariates, $M = 2^V - 1$ potential models could be obtained.

Let Ω define the set of all these models (i.e. with cardinality $\mathcal{C}(\Omega) = M$), where ω are its elements and λ_ω is the p-value associated with the model ω. Furthermore, let $\Omega_0 \subset \Omega$ denote the subset with all models ω under the null hypothesis of no association with $\mathbf{Y}$. It is well known that $P_n(\lambda_\omega \leq \alpha | \omega \in \Omega_0) = P_n(\min_{\omega \in \Omega_0} \lambda_\omega \leq \alpha) \leq \alpha$ does not hold in all non-trivial situations. Therefore, the probability that at least one model is wrongly rejected is out of control. This fact gives rise to the practical problem of p-values tending to be (very) *small* also when response $\mathbf{Y}$ is *not* associated with any of the covariates $\mathbf{X}$.

The familiar Bonferroni inequality guarantees an upper bound to this probability, $P_n(\min_{\omega \in \Omega_0} \lambda \leq \alpha) \leq \alpha\mathcal{C}(\Omega_0)$. In practical applications, however, $\mathcal{C}(\Omega_0)$ is unknown and $\mathcal{C}(\Omega)$ is used instead. The Bonferroni correction is very conservative (i.e. has low power) whenever $\mathcal{C}(\Omega)$ is high and/or the p-values are dependent. Both conditions hold in our case: $\mathcal{C}(\Omega) = 2^V - 1$ increases exponentially with V and the dependence among p-values of two different models can be very high since dependence exists whenever two models share at least one variable.

Moreover, the exhaustive search of all elements in Ω is not always feasible. All stepwise methods aim to avoid the cost of this exhaustive research by exploring a subset of elements in Ω in a (more or less) 'efficient' way. However, the price paid is that the selected model is sub-optimal and the p-value is not always the minimum (actually different methods lead to different selected models).

Table 5.6 Type I error out of control

nominal α level	0.01	0.05	0.10	0.20	0.30	0.50
1st simulation	0.187	0.521	0.733	0.892	0.935	0.939
2nd simulation	0.530	0.840	0.938	0.983	0.996	0.998

Thus, a Bonferroni correction would appear to be a valid but not a useful correction in order to control the FWER in a weak sense (i.e. the probability of at least one wrong rejection when $\Omega_0 = \Omega$).

The method proposed in Finos et al. (2009), under the global null hypothesis that all covariates are unrelated to the outcome variable, corrects the p-value of the selected model in such a way that it controls the type I error at level α and ensures unbiasedness and consistency of the p-values of the selected model.

A possible correction for the p-value of the selected model can be based on the following algorithm:

(1) Perform a standard stepwise regression (backward or forward) in a GLM (e.g. linear, logistic, Poisson, Cox models) for the model $\mathbf{Y} = h(\mathbf{X})$, where $\mathbf{Y}$ indicates the response variable and $\mathbf{X}$ are the covariates.
(2) Extract the p-value associated with the F test on residual deviance for the GLM. This p-value is called the observed (raw) p-value.
(3) Carry out B independent permutations of the response variable $\mathbf{Y}$ and repeat steps (1) and (2).
(4) The adjusted p-value $\tilde{\lambda}$ is exactly the fraction of permutation p-values that are less or equal to the observed one.

In conclusion, the p-values obtained after performing a stepwise regression may be highly biased. In particular, the evaluation of stepwise GLM must be done with care, mainly when regressors have been data-driven. However, it is possible to correct p-values in a very simple manner; for example, Finos et al. (2009) suggest a nonparametric permutation solution that is exact, flexible and potentially adaptable to most different applications of model selection. The correction becomes more severe when many variables are processed by the stepwise machinery.

5.8.3 Optimal Subset Procedures

Optimal subset procedures are another approach to multiplicity control. Here we merely provide some hints and refer the reader to the original paper by Finos and Salmaso (2005) for details. This approach is in fact based on the selection of the best subset of partial (univariate) hypotheses such that, once conveniently combined, they provide the minimal p-value.

Optimal subset procedures constitute a less stringent multiplicity control than FWE and FDR. Indeed, the control of the FWE is weak, that is, it is correct only in the case of a global null configuration (all hypotheses are null).

Finos and Salmaso (2005) also show how stepwise regression may be seen as a special case of the optimal subset procedures and that it is possible to adjust for multiplicity the p-value corresponding to the selected model. As is known, in the model selection context, the multiplicity issue arises because of the multitude of models explored by the stepwise method. As mentioned before, instead of controlling the multiplicity of the univariate tests, these procedures select the multivariate hypotheses which produce the most significant combined tests. The p-values of these multivariate hypotheses are then adjusted for multiplicity.

In this way, therefore, the optimal subset procedures supply a *global* response on the model and not a specific response on the single partial tests. By way of compensation, however, they show sensitivity in identifying the hypotheses under the alternative.

Optimal subset procedures are found to be more sensitive when it comes to identifying the hypotheses under the alternative than other procedures, such as the Bonferroni–Holm (Holm, 1979) and FDR procedures (Benjamini and Hochberg, 1997). Due to their main features (i.e. weaker multiplicity control and greater power), these procedures are recommended in all studies involving a large number of variables, especially when a global assessment of the phenomenon, rather than the strong inference on single univariate hypotheses, is of interest.

6

Analysis of Multivariate Categorical Variables

6.1 Introduction

This chapter deals with a permutation approach to some multivariate problems of hypothesis testing with regard to categorical data in a nonparametric framework. The problems considered concern the definition of the underlying distribution of data and the possibility of applying the likelihood ratio test or other parametric solutions. Practical examples from real application problems are discussed and some scripts, useful for carrying out the required analyses in MATLAB, R and SAS, are also illustrated. Section 6.2 deals with a typical multivariate paired data problem, leading to the solution of a natural multivariate extension of McNemar's test. Other testing problems of interest are the multivariate goodness-of-fit test for ordered variables and the MANOVA test with nominal categorical data, described in Sections 6.3 and 6.4, respectively. In many pharmacological studies, where the effects of different dose levels are compared and where responses are represented by ordered categorical variables, it is generally of interest to test certain order relations on the distributions of the response variables and to consider more than one response variable simultaneously. Section 6.5 is devoted to this type of stochastic ordering problem.

A very common testing problem consists in evaluating the independence in distribution between two categorical variables. The classical solution is the χ^2 test for contingency tables in the presence of sparse data. This may prove to be anticonservative. A nonparametric solution which decomposes the problem and considers each category as a dummy variable giving a solution based on the NPC of dependent tests is described in Section 6.6.

When the goal of a genetic study is the identification of genes causing a given pathology, discovering the major susceptibility locus can be the starting point for advances in the understanding of the causes of a disease. Section 6.7 describes a permutation approach to testing allelic association and genotype-specific effects in the genetic study of a disease. The term used to identify this testing problem is 'isotonic inference'.

In several sciences complex univariate and multivariate testing problems may arise when categorical data are present. In Section 6.8, the problem of testing whether the response related to one treatment is stochastically larger than that of another is considered for univariate and multivariate ordinal categorical data. The solution based on the parametric test on moments implies the transformation of categorical response variables into numeric variables and the breaking down of the original hypotheses into partial sub-hypotheses related to the moments of the transformed variables.

Permutation Tests for Complex Data: Theory, Applications and Software Fortunato Pesarin, Luigi Salmaso

One application problem which is very frequently encountered, yet almost ignored in the literature, is that of establishing whether the distribution of a categorical variable is more heterogeneous (less homogeneous or less concentrated) in one population than in another. The nonparametric solution illustrated in Section 6.9 is based on a permutation test, in some aspects similar to the permutation solution for stochastic dominance. The peculiarity of the methodology illustrated is that exchangeability under the null hypothesis is not exact (but well approximated in practice) because the solution is based on sample estimates of the true ordering of unknown probabilities.

Section 6.10 describes an application problem regarding a comparative performance evaluation of PhD programs. The first part concerns an extension of the NPC of dependent rankings for the computation of a composite indicator measuring satisfaction with some aspects of PhD programs. The second part looks at a problem of hypothesis testing, where a permutation ANOVA is applied to compare the performances of different PhD programs. The chapter concludes with a brief description of an SAS macro for the practical application of methodologies discussed in this chapter.

6.2 The Multivariate McNemar Test

In this section we illustrate one of the specifications of a multivariate paired data problem, leading to the solution of a natural multivariate extension of McNemar's test (see also Section 1.8.6 and Examples 6–8, 2.6). In the univariate situation, one variable Y is observed in two experimental conditions on each of n units, leading to a set of independent pairs $\mathbf{Y} = \{(Y_{1i}, Y_{2i}), i = 1, \ldots, n\}$ and the associated set of differences $\mathbf{X} = \{X_i = Y_{1i} - Y_{2i}, i = 1, \ldots, n\}$. The response variable is ordered categorical with two classes (or binary, or even continuous transformed into binary) in such a way that it is possible to evaluate within each unit whether there is a positive variation $\phi(Y_{1i}, Y_{2i}) = +1$ when $Y_{1i} < Y_{2i}$, a negative variation $\phi(Y_{1i}, Y_{2i}) = -1$ when $Y_{1i} > Y_{2i}$, or a tie $\phi(Y_{1i}, Y_{2i}) = 0$ when $Y_{1i} = Y_{2i}$, $i = 1, \ldots, n$. The test aims to establish whether the data agree or disagree with the hypothesis of a null treatment effect, so that $H_0 : \{Y_1 \overset{d}{=} Y_2\} = \{\Pr(+1) = \Pr(-1)\}$. It is known that this univariate problem, assuming that $\Pr(+1)$ and $\Pr(-1)$ are unknown nuisance quantities, can be solved within permutation arguments using a binomial test.

Let us now assume that the response variables are V-dimensional binary, $\mathbf{Y} = (Y_1, \ldots, Y_V)$, so that the data set is

$$\begin{aligned}\mathbf{Y} &= \{(Y_{h1i}, Y_{h2i}),\ i = 1, \ldots, n,\ h = 1, \ldots, V\}\\ &= \{(\mathbf{Y}_{1i}, \mathbf{Y}_{2i}),\ i = 1, \ldots, n\},\end{aligned}$$

and the related hypotheses become

$$H_0 : \left\{\mathbf{Y}_1 \overset{d}{=} \mathbf{Y}_2\right\} = \left\{\bigcap_{h=1}^{V} [\Pr(+1)_h = \Pr(-1)_h]\right\}$$

against alternatives of the form

$$H_1 : \left\{\mathbf{Y}_1 < \overset{d}{\neq} > \mathbf{Y}_2\right\} = \left\{\bigcup_{h=1}^{V} [\Pr(+1)_h < \neq > \Pr(-1)_h]\right\},$$

where it is emphasized that some of the sub-alternatives are one-sided and others two-sided. This way of illustrating the problem shows that it is a particular case of the more general problem concerning multivariate paired observations. Hence, its solution takes into consideration the set of differences

$$\mathbf{X} = \{X_{hi} = \phi(Y_{h1i}, Y_{h2i}),\ i = 1, \ldots, n,\ h = 1, \ldots, V\}$$

and the set of partial tests

$$T_h^* = \varphi_h\left(\sum_{i=1}^{n} X_{hi} S_i^*\right), \quad h = 1, \ldots, V,$$

where to preserve dependence relations the $S_i^* = 1 - 2\mathcal{B}n(1, 1/2)$ are h-invariant random signs, and φ_h corresponds respectively to the sign '+', the 'absolute value', or '−' according to whether the specific hth sub-alternative is '<', '≠', or '>'. Thus, as all partial tests are marginally unbiased, because each of them is separately related to one component variable, the NPC method provides for a proper overall solution. For instance, the direct combination leads to

$$T_D^{*\prime\prime} = \sum_h \varphi_h\left[\frac{\sum_i X_{hi} S_i^*}{(\sum_i X_{hi}^2)^{1/2}}\right]$$

because $\sum_i X_{hi}^2$ is the conditional (permutation) variance of $\sum_{i=1}^n X_{hi} S_i^*$ (see Section 1.9.4).

It is worth noting that in this context we cannot use asymptotic approximations unless we know the dependence relations among component binomials (in Pesarin, 2001, some hints in this direction are suggested), and so a CMC approach based on B iterations seems to be an appropriate way forward. Brown and Hettmansperger (1989) considered the particular case in which data are bivariate and alternatives are two-sided, i.e. unrestricted (see also Klingenberg and Agresti, 2006).

Remark 1. It is worth observing that all partial tests T_h are permutationally equivalent to $U_h = \#(X_{hi} = +1)$, $h = 1, \ldots, V$, the permutation marginal distributions of which are binomial $\mathcal{B}n(\nu_h, \vartheta_h)$, where $\nu_h = \#(X_{hi} \neq 0)$, $\vartheta_h = 1/2$ in H_{0h}, and $\vartheta_h <\neq> 1/2$ in H_{1h}. Of course, the V-variate permutation distribution of $\mathbf{U} = (U_1, \ldots, U_V)$ is not multinomial because $\sum_h \vartheta_h$ is generally different from 1; however, it is such that all its univariate marginals are binomials. From this point of view, a multinomial variable may be seen as either multivariate or *univariate multiparametric*, whereas $\mathbf{U} = (U_1, \ldots, U_V)$ is typically multivariate.

From the foregoing, we may argue that it is rather difficult to find general solutions for these problems outside the NPC method. As an example of an application to a real problem, let us consider a data set related to an epidemiological study on the quality of care in β-thalassaemia, conducted on 446 subjects in 15 Italian centres specializing in the management and care of this disease (Pesarin, 2001).

The study was carried out shortly before the introduction of a modified formulation of a specific drug D currently used for treatment of this disease (occasion 1), and approximately one year later, when the modified formulation Dm was available to patients (occasion 2). Each patient was examined on both occasions with respect to two different binary variables, so that observations are paired bivariate.

The modified formulation Dm aims to reduce the side-effects which may occur during subcutaneous infusion of the drug. These side-effects may appear as adverse reactions at the injection site, called *local adverse events* (variable Y_1), such as skin irritation, swelling or itching. Occasionally infusion of this kind of drug may also cause so-called *systemic symptoms* (variable Y_2), such as myalgia, fever or headaches.

The occurrence of adverse events may be related to several factors such as the mechanical effects of the infusion process, or a subjective reaction to either the active ingredient or to by-products generated during the preparation of the drug. Hence, evaluation of the tolerability profile and of the comparative incidence of both local and systemic adverse events between the two formulations D and Dm of the drug was included as part of a broader epidemiological study on the quality of care in β-thalassaemia (Arboretti et al., 1997).

Table 6.1 Adverse events in β-thalassaemia patients

Y_{11}	Y_{12}	Y_{21}	Y_{22}	f
0	0	0	0	340
0	1	0	0	15
0	1	0	1	13
1	0	0	0	22
1	0	1	0	23
1	1	0	0	6
1	1	0	1	7
1	1	1	0	3
1	1	1	1	17

Table 6.2 Two marginal tables derived from Table 6.1

$Y_{11}\backslash Y_{12}$	0	1		$Y_{21}\backslash Y_{22}$	0	1
0	340	28		0	383	20
1	45	33		1	26	17

Paired bivariate data from this study are reported in Table 6.1, where only the presence ($\equiv 1$) or absence ($\equiv 0$) of adverse events is considered. For instance, for the response point ($Y_{11} = 0,\ Y_{12} = 1,\ Y_{21} = 0,\ Y_{22} = 0$), the second row contains 15 individuals. Note that all omitted combinations of responses have no individuals (e.g. the response point ($Y_{11} = 0, Y_{12} = 0,\ Y_{21} = 0,\ Y_{22} = 1$) has no individuals). Two marginal tables are displayed in Table 6.2.

Formally, we test $H_0 : \{(Y_{11} \overset{d}{=} Y_{12}) \bigcap (Y_{21} \overset{d}{=} Y_{22})\}$ against $H_1 : \{(Y_{11} \overset{d}{>} Y_{12}) \bigcup (Y_{21} \overset{d}{>} Y_{22})\}$ because there is a specific interest in whether the new formulation Dm causes an overall stochastic reduction in adverse events. The test is thus a bivariate McNemar case. With $B = 4000$ CMC iterations, the two partial p-values are $\hat{\lambda}_{Y_1} = 0.025,\ \hat{\lambda}_{Y_2} = 0.264$, and the combined p-value, obtained by using a direct combination function on standardized partial tests, is $\hat{\lambda}'' = 0.051$, which is not significant at $\alpha = 0.05$. Standardized partial tests have the form $T_h^* = \sum_i X_{hi} S_i^* / (\sum_i X_{hi}^2)^{1/2},\ j = 1, 2$, where $X_1 = Y_{12} - Y_{11}$ and $X_2 = Y_{22} - Y_{21}$.

The combined p-value, being close to significance, may suggest further examination of the properties of the modified drug formulation, possibly by considering responses with more than two ordered categories, according to the theory in next subsection, or by including other informative variables. Note that the marginal p-values obtained by normal approximation of related binomials are $\lambda_1 = 0.0233$ and $\lambda_2 = 0.288$ respectively.

6.2.1 An Extension of the Multivariate McNemar Test

Let us consider the extensions of univariate McNemar tests, discussed above, where k ordered categories for 'differences' $(A_1, \ldots, A_k)$ were considered, instead of three $(-, 0, +)$ see Example 6, 2.6. As a multivariate extension, let us consider the multivariate 'difference' response

$[X_h \in (A_{h1}, \ldots, A_{hk_h}), h = 1, \ldots, V]$, where classes on the hth component variable satisfy $k_h > 2$, where the hypotheses are:

$$H_0 : \left\{\mathbf{Y}_1 \stackrel{d}{=} \mathbf{Y}_2\right\} = \left\{\bigcap_{h=1}^{V} \bigcap_{c=1}^{\lfloor k_h/2 \rfloor} [\Pr(A_{hc}) = \Pr(A_{hk_h-c+1})]\right\}$$

against alternatives of the form

$$H_1 : \left\{\mathbf{Y}_1 \stackrel{d}{<\neq>} \mathbf{Y}_2\right\} = \left\{\bigcup_h \bigcup_c [\Pr(A_{hc}) <\neq> \Pr(A_{hk_h-c+1})]\right\}.$$

As an extension of the test $T_\omega^* = \sum_{c=1}^{\lfloor k/2 \rfloor}[f^*(c) - \nu_c/2]/\sqrt{\nu_c/4}$, let us consider the direct combination of standardized partial tests, that is,

$$T''^* = \sum_{h=1}^{V} \varphi_h \left\{ \sum_{c=1}^{\lfloor k_h/2 \rfloor} [f_h^*(c) - \nu_{hc}/2] \left(\frac{\nu_{hc}}{4}\right)^{-1/2} \right\},$$

where again φ_h corresponds respectively to '+', the absolute value or '−', according to whether the specific hth sub-alternative is '<', '≠', or '>'. Of course, instead of direct combination, we may also consider any other combining function. Extensions when different degrees of importance for 'differences' related to each component variable, ω_{hc} (say), are to be considered, and those in relation to chi-square are left to the reader as exercises.

6.3 Multivariate Goodness-of-Fit Testing for Ordered Variables

Let us first observe that all goodness-of-fit test statistics presented in Section 2.8 are nothing more than NPCs by direct combining functions of a set of partial tests. For instance, tests for the response variable $X \in (A_1 \prec \ldots \prec A_k)$ in relation to the Anderson–Darling $T_D^* = \sum_{j=1}^{k-1} N_{2j}^* \cdot \left[N_{\cdot j} \cdot (n - N_{\cdot j})\right]^{-1/2}$, and used for testing the null hypothesis

$$H_0 : \left\{X_1 \stackrel{d}{=} X_2\right\} = \left\{\bigcap_{j=1}^{k} F_1(A_j) = F_2(A_j)\right\}$$

against the alternative

$$H_1 : \left\{X_1 \stackrel{d}{>} X_2\right\} = \left\{\bigcup_j [F_1(A_j) < F_2(A_j)]\right\},$$

can be seen as the direct combination of standardized partial tests $T_j^* = N_{2j}^* \cdot \left[N_{\cdot j} \cdot (n - N_{\cdot j})\right]^{-1/2}$, each suitable for testing the partial hypotheses $H_{0j} : F_1(A_j) = F_2(A_j)$ against $H_{1j} : F_1(A_j) < F_2(A_j),\ j = 1, \ldots, k-1$.

As a multivariate extension, let us consider one representation of the two-sample multivariate problem. This situation to some extent corresponds to a problem of componentwise *multivariate monotonic inference*. To this end, let us assume that the response variable is V-dimensional ordered categorical, $\mathbf{X} = (X_1, \ldots, X_V)$, with respective numbers of ordered classes $\mathbf{k} = (k_1, \ldots, k_V)$. Units

n_1 and n_2 are independently observed from $\mathbf{X}_1$ and $\mathbf{X}_2$ respectively, each related to a treatment level. The hypotheses to be tested are

$$H_0 : \left\{\mathbf{X}_1 \overset{d}{=} \mathbf{X}_2\right\} = \left\{\bigcap_{h=1}^{V} \bigcap_{j=1}^{k_h-1} (F_{h1j} = F_{h2j})\right\}$$

against $H_1 : \{\mathbf{X}_1 \overset{d}{>} \mathbf{X}_2\} = \{\bigcup_h \bigcup_j (F_{h1i} < F_{h2i})\}$, where F_{h1j} and F_{h2j} play the role of CDFs for the jth category of the hth variable in the first and second groups, respectively.

This problem (see Wang, 1996; Silvapulle and Sen, 2005) is considered to be extremely difficult when approached using the likelihood ratio method and, until now, no satisfactory solutions have been proposed in the literature (see Silvapulle and Sen, 2005).

Within the NPC approach the solution to this extended problem is now straightforward. We may consider the same kind of partial tests previously discussed, for instance, $T_{hj}^* = N_{h2j}^*$, or $T_{hj}^* = N_{h2j}^* \cdot \left[N_{\cdot j} \cdot (n - N_{\cdot j})\right]^{-1/2}$ $j = 1, \ldots, k_h - 1$, $h = 1, \ldots, V$; hence, we may proceed with Fisher's NPC, $T_F^{*\prime\prime} = -\sum_h \sum_j \log(\hat{L}_{hj}^*)$, or with the direct combination $T_D^{*\prime\prime} = \sum_{hj} N_{h2j}^* \cdot \left[N_{\cdot j} \cdot (n - N_{\cdot j})\right]^{-1/2}$.

In a two-sample problem of non-dominance, where the alternative is $H_1 : \{\mathbf{X}_1 \overset{d}{\neq} \mathbf{X}_2\} = \left\{\bigcup_{h=1}^{V} \bigcup_{j=1}^{k_h-1} (F_{h1j} \neq F_{h2j})\right\}$, the solution becomes

$$T_D^{*2} = \sum_{h=1}^{V} \sum_{j=1}^{k_h-1} \left(N_{2hj}^* - N_{1hj}^*\right)^2 \left[N_{h\cdot j}(n - N_{h\cdot j})\right]^{-1}.$$

It is worth noting that T_D^{*2} is not the square of $T_D^{*\prime\prime}$; however, it does correspond to one of the possible analogues of Hotelling's T^2 test for ordered categorical variables.

The extension to the multivariate C-sample case, $C > 2$, considers the hypotheses

$$H_0 : \left\{\mathbf{X}_1 \overset{d}{=} \ldots \overset{d}{=} \mathbf{X}_C\right\} = \left\{\bigcap\nolimits_{hi} (F_{h1i} = \ldots = F_{hCi})\right\}$$

against $H_1 = \{H_0 \text{ is not true}\}$. This extension, being related to a one-way MANOVA design for ordered categorical variables, is in some senses straightforward. In fact, we may consider partial test statistics such as

$$T_{hj}^* = \sum_{c=1}^{C} n_c (\hat{F}_{hcj}^*)^2, \quad j = 1, \ldots, k_h - 1, \; h = 1, \ldots, V,$$

where $\hat{F}_{hcj}^* = N_{hcj}^*/n_c$, and one NPC is Fisher's $T_F^{*\prime\prime} = -\sum_{hj} \log(\hat{L}_{hj}^*)$, where symbols and structures have obvious meanings.

Alternatively, we may make use of the direct combination function applied to partial Anderson–Darling type tests T_{ADh}^2, one for each component variable. Consequently,

$$T_{MD}^{*2} = \sum_{h=1}^{V} \sum_{c=1}^{C} \sum_{j=1}^{k_h-1} \left(\hat{F}_{hcj}^* - \bar{F}_{h\cdot j}\right)^2 \left[\bar{F}_{h\cdot j} \cdot (1 - \bar{F}_{h\cdot j}) \cdot (n - n_c)/n_c\right]^{-1},$$

where $\bar{F}_{h\cdot j} = N_{h\cdot j}/n$ and $N_{h\cdot j} = \sum_c N_{hcj}$.

6.3.1 Multivariate Extension of Fisher's Exact Probability Test

Let us now suppose that the responses of the goodness-of-fit multidimensional problem are binary, so that $X_{hci} = 0$ or 1, $i = 1, \ldots, n_j$, $c = 1, 2$, $h = 1, \ldots, V$. Again, the hypotheses under test are

$$H_0 : \left\{\bigcap_{h=1}^{V}[X_{h1} \overset{d}{=} X_{h2}]\right\} = \left\{\bigcap_{h=1}^{V} H_{0h}\right\}$$

against, say, the dominance alternative $H_1 : \{\bigcup_h [X_{h1} \overset{d}{>} X_{h2}]\} = \{\bigcup_h H_{1h}\}$. For this problem, appropriate permutation partial tests are

$$T_h^* = \sum_i X_{h1i}^*, \; h = 1, \ldots, V.$$

Note that if $V = 1$, this problem corresponds to Fisher's well-known exact probability test, in the sense that the sub-hypotheses are tested by partial Fisher exact probability tests. If $V > 1$, we have a natural multivariate extension of Fisher's exact test, which may be solved effectively using the NPC method.

6.4 MANOVA with Nominal Categorical Data

This section considers an extension of a multivariate permutation one-way ANOVA procedure to cases in which responses are nominal categorical . This kind of problem is quite common, especially in such areas as clinical trials, psychology and social sciences, in which many nominal categorical variables are observed.

For each unit $i = 1, \ldots, n$, let us assume that the observed responses have the form of V categorical variables $\mathbf{X} = (X_1, \ldots, X_V)$. Also assume that the data are partitioned into C groups or samples of size n_c, $c = 1, \ldots, C$, according to C levels of a symbolic treatment, and the V categorical variables may respectively present $k_1, \ldots, k_V$ classes. Thus, the hth variable takes values in the support $(A_{h1}, \ldots, A_{hk_h})$.

The data set $\mathbf{X} = \{X_{hci}, i = 1, \ldots, n_c, c = 1, \ldots, C, h = 1, \ldots, V\}$ may also be written in accordance with the unit-by-unit representation of individual records, such as $\mathbf{X} = \{X_h(i), i = 1, \ldots, n, h = 1, \ldots, V; n_1, \ldots, n_C\}$.

The following assumptions are made: (i) the model for treatment effects produces distributional differences on some (or all) of the V categorical variables without restrictions; (ii) data vectors are i.i.d. within groups, and groups are independent; (iii) all marginal distributions of the variables $\mathbf{X}$ are non-degenerate; (iv) component variables $(X_1, \ldots, X_V)$ are non-independent (see Joe, 1997, for the notion of multivariate dependence on categorical variables). The hypotheses under test are

$$H_0 : \{P_1 = \ldots = P_C\} = \left\{\mathbf{X}_1 \overset{d}{=} \ldots \overset{d}{=} \mathbf{X}_C\right\}$$

against $H_1 : \{H_0 \text{ is not true}\}$, where P_c is the underlying distribution of $\mathbf{X}$ in the cth group.

The hypotheses and assumptions are such that the permutation testing principle can be properly applied. In any case, the complexity of underlying dependence structures among the V variables is such that one overall test statistic is very difficult to find directly, especially in a full parametric setting. Thus, the problem can be solved by considering a set of V partial tests, one for each component variable, followed by their NPC, provided that all k partial tests are marginally unbiased. This procedure is appropriate because, by assumption, the hypotheses may be equivalently broken down into

$$H_0 : \left\{ \bigcap_{h=1}^{V} (X_{h1} \stackrel{d}{=} \ldots \stackrel{d}{=} X_{hC}) \right\} = \left\{ \bigcap_{h=1}^{V} H_{0h} \right\},$$

against $H_1 : \left\{ \bigcup H_{1h} \right\}$, which is much easier to process. In particular, $H_{0h} : \{X_{h1} \stackrel{d}{=} \ldots \stackrel{d}{=} X_{hC}\}$, $h = 1, \ldots, V$, indicates the equality in distribution among the C levels of the hth variable X_h.

We observe that for each of the V sub-hypotheses H_{0h}, a permutation test statistic such as Pearson's chi-square is almost always appropriate; that is,

$$X_h^{*2} = \sum_{c=1}^{C} \sum_{j=1}^{k_h} \left[\frac{\left(f_{hcj}^{*} - \hat{f}_{hcj} \right)^2}{\hat{f}_{hcj}} \right],$$

where, as usual, $\hat{f}_{hcj} = n \cdot f_{h \cdot j} / n_c$, $f_{h \cdot j} = \sum_c f_{hcj}$, $f_{hcj}^{*} = \sum_{i=N_{c-1}}^{N_c} \mathbb{I}[X_h^{*}(i) \in A_{hj}]$ and $N_c = \sum_{r \leq c} n_r$. Thus, the nonparametric direct combination $X_D^{*2} = \bigcup_h X_h^{*2}$ provides for a global solution. Alternatively, it is possible to apply any combining function provided that, according to theory, all partial tests are transformed into their permutation p-values.

The extension to mixed variables, some nominal, some ordered categorical and some quantitative, is now straightforward. This can be done by separately combining the nominal, ordered categorical and quantitative variables, and then by combining their respective p-values into one overall test. Details are left to the reader.

6.5 Stochastic Ordering

6.5.1 Formal Description

In many dose-response studies (see Example 7, 4.6), in the presence of ordered categorical responses, it is of interest to test certain order relations on the distributions of the response variables and to consider multivariate response variables. Let us suppose that, in a dose-response experiment, C different doses of a treatment are administered to C different groups of patients. Let $\mathbf{X}_{ci} = (X_{1ci}, \ldots, X_{Vci})$ be the response on V variables for the ith subject randomly assigned to treatment dose c, $c = 1, \ldots, C$, $i = 1, \ldots, n_c$, and the total number of observations $n = \sum_{c=1}^{C} n_c$.

Let us assume that the support of the hth categorical variable X_h $(h = 1, \ldots, V)$ is partitioned into $k_h \geq 2$ ordered categories $A_1 \prec \ldots \prec A_{k_h}$. Moreover, $X_{c1}, \ldots, X_{cn_c}$ are n_c i.i.d. V-dimensional variables with marginal COF$_s$ $F_{hc}(A_j) = \Pr\left\{X_{hc} \leq A_j\right\}$, $j = 1, \ldots, k_h$, $h = 1, \ldots, V, c = 1, \ldots, C$.

Stochastically ordered random vectors (see Marshall and Olkin, 1979; Silvapulle and Sen, 2005) have received much less attention than inference based on stochastically ordered univariate random variables. The C multivariate distributions are said to be stochastically ordered when $X_1 \stackrel{d}{\leq} X_2 \stackrel{d}{\leq} \ldots \stackrel{d}{\leq} X_C$, if and only if $\mathbb{E}[\varphi(X_1)] \leq \mathbb{E}[\varphi(X_2)] \leq \ldots \leq \mathbb{E}[\varphi(X_C)]$ holds for all increasing real functions $\varphi(\cdot)$ such that the expectation exists, where at least one inequality is strict.

We wish to test the null hypothesis $H_0 : \{X_1 \stackrel{d}{=} \ldots \stackrel{d}{=} X_C\}$, against the alternative $H_1 : \{X_1 \stackrel{d}{\leq} \ldots \stackrel{d}{\leq} X_C$ and $X_1, \ldots, X_C$ are not equal in distribution$\}$, that is, at least one inequality is strict.

Bacelli and Makowski (1989) prove that H_0 holds if and only if $X_1, \ldots, X_C$ have the same marginal distributions, namely

$$F_{h1}(x) = \ldots = F_{hC}(x), \quad \forall x, \; h = 1, \ldots, V.$$

As a consequence, H_1 holds if and only if $X_{h1} \overset{d}{\leq} \ldots \overset{d}{\leq} X_{hC}$, $h = 1, \ldots, V$, and $\mathbf{X}_1, \ldots, \mathbf{X}_C$ are not equal in distribution, expressed as

$$F_{h1}(x) \geq \ldots \geq F_{hC}(x), \quad \forall x, \ h = 1, \ldots, V,$$

with at least one strict inequality holding at some real x and for at least one h.

In complex problems it is preferable to regard H_0 as the intersection of a number of partial hypotheses and H_1 as the union of the same number of corresponding partial alternatives, in symbols

$$H_0 : \left\{\bigcap_{h=1}^{V} H_{0h}\right\} = \left\{\bigcap_{h=1}^{V} F_{h1}(x) = \ldots = F_{hC}(x), \ \forall x\right\}$$

and

$$H_1 : \left\{\bigcup_{h=1}^{V} H_{1h}\right\} = \left\{\bigcup_{h=1}^{V} F_{h1}(x) \geq \ldots \geq F_{hC}(x), \ \forall x \text{ and not } H_{0h}\right\}.$$

6.5.2 *Further Breaking Down the Hypotheses*

Let us consider the hth sub-problem of testing H_{0h} against H_{1h}. A nonparametric rank solution for this problem is proposed by the Jonckheere–Tepstra test (Jonckheere, 1954). El Barmi and Mukerjee (2005) proposed an asymptotic test by using the sequential testing procedure described in Hogg (1962). This procedure consists in sequentially testing $H_{0hj} : \{F_{h1} = \ldots = F_{h(j-1)} = F_{hj}\}$ against $H_{1hj} : \{F_{h1} = \ldots = F_{h(j-1)} \geq F_{hj}\}$, $j = 2, \ldots, C$, with strict inequality holding for some x, as in two-sample tests by pooling the first $j - 1$ samples for testing H_{0jh}. Thus H_{0h} and H_{1h} can be expressed as

$$H_{0h} : \bigcap_{j=2}^{C} \{H_{0jh}\} = \bigcap_{j=2}^{C} \{F_{h1} = \ldots = F_{h(j-1)} = F_{hj}\}$$

and

$$H_{1h} : \left\{\bigcup_{j=2}^{C} H_{1hj}\right\} = \left\{\bigcup_{j=2}^{C} [F_{h1} = \ldots = F_{h(j-1)} \geq F_{hj}]\right\},$$

respectively. However, H_{0h} and H_{1h} can also be expressed as

$$H_{h0} : \left\{\bigcap_{j=2}^{C} H_{0hj}\right\} = \left\{\bigcap_{j=2}^{C} [F_{h1} = \ldots = F_{h(j-1)} = F_{hj} = \ldots = F_{hC}]\right\},$$

$$H_{h1} : \left\{\bigcup_{j=2}^{C} H_{1hj}\right\} = \left\{\bigcup_{j=2}^{C} [F_{h1} = \ldots = F_{h(j-1)} \geq F_{hj} = \ldots = F_{hC}]\right\},$$

where for testing H_{0hj} against H_{1hj} we refer to a two-sample problem by pooling the first $j - 1$ and the last $C - j + 1$ samples, $j = 2, \ldots, C$.

6.5.3 *Permutation Test*

A parametric solution to this problem is very complex, especially when $C > 2$ (Wang, 1996). In order to tackle this problem in a permutation context (see also Example 7, 4.6), let us consider the solution proposed in Finos et al. (2007). For any $j \in \{2, \ldots, C\}$, the whole data set is split into two pooled pseudo-groups, where the first is obtained by pooling together data from the first $j-1$ (ordered) groups and the second by pooling the remaining $C-j+1$ groups. In other words, we define the first pooled pseudo-group as $\mathbf{X}_{1(j)} = \mathbf{X}_1 \biguplus \ldots \biguplus \mathbf{X}_{j-1}$ and the second as $\mathbf{X}_{2(j)} = \mathbf{X}_j \biguplus \ldots \biguplus \mathbf{X}_C$, $j = 2, \ldots, C$, where $\mathbf{X}_j = \{\mathbf{X}_{ji}, \ i = 1, \ldots, n_j\}$ is the V-dimensional data set in the jth group.

Under the null hypothesis, for each pair of pseudo-groups data are exchangeable among pseudo-groups because related pooled variables satisfy the relationships $\mathbf{X}_{1(j)} \stackrel{d}{=} \mathbf{X}_{2(j)}$, $j = 2, \ldots, C$. Under the alternative, the relation $\mathbf{X}_{1(j)} \stackrel{d}{<} \mathbf{X}_{2(j)}$ corresponds to the monotonic stochastic dominance between any pair of pseudo-groups. This suggests that the hypotheses can be written in the equivalent form

$$H_0 : \left\{ \bigcap_{h=1}^{V} \bigcap_{j=2}^{C} H_{0hj} \right\} = \left\{ \bigcap_{h=1}^{V} \bigcap_{j=2}^{C} [X_{h1(j)} \stackrel{d}{=} X_{h2(j)}] \right\},$$

against

$$H_1 : \left\{ \bigcup_{h=1}^{V} \bigcup_{j=2}^{C} H_{1hj} \right\} = \left\{ \bigcup_{h=1}^{V} \bigcup_{j=2}^{C} [X_{h1(j)} \stackrel{d}{\leq} X_{h2(j)}] \right\},$$

where $X_{h1(j)}$ and $X_{h2(j)}$ are the hth component variable of $\mathbf{X}_{1(j)}$ and $\mathbf{X}_{2(j)}$, respectively.

Let us look at the jth sub-hypothesis for the hth component variable $H_{0hj} : \{X_{h1(j)} \stackrel{d}{=} X_{h2(j)}\}$ against $H_{1hj} : \{X_{h1(j)} \stackrel{d}{\leq} X_{h2(j)}\}$. We note that the related sub-problem corresponds to a two-sample comparison for restricted alternatives, a problem which has an exact and unbiased permutation solution (see Pesarin, 2001; see also Example 7, 4.6).

Observed data correspond to a $2 \times k_h$ contingency table as in Table 6.3, where $f_{h1(j)r} = \#(X_{h1(j)} \in A_r)$ and $f_{h2(j)r} = \#(X_{h2(j)} \in A_r)$ are the observed frequencies, $N_{h1(j)r} = \sum_{s \leq r} f_{h1(j)s}$ and $N_{h2(j)r} = \sum_{s \leq r} f_{h2(j)s}$ are the cumulative frequencies, $N_{h\cdot(j)r} = N_{h1(j)r} + N_{h2(j)r}$ and $f_{h\cdot(j)r} = f_{h1(j)r} + f_{h2(j)r}$ are the marginal frequencies, $n_{1(j)} = \sum_r f_{h1(j)r}$ and $n_{2(j)} = \sum_r f_{h2(j)r}$ are the sample sizes, and $n = n_{1(j)} + n_{2(j)} = \sum_j n_j$ is the total sample size.

In order to avoid computational problems, let us assume that marginal frequencies $f_{h\cdot(j)r}$, $r = 1, \ldots, k_h$, are all positive, in the sense that we remove class r from analysis if $f_{h\cdot(j)r} = 0$.

Table 6.3 A typical $2 \times k_h$ contingency table

Categories	Absolute frequencies			Cumulative frequencies		
A_1	$f_{h1(j)1}$	$f_{h2(j)1}$	$f_{h\cdot(j)1}$	$N_{h1(j)1}$	$N_{h2(j)1}$	$N_{h\cdot(j)1}$
...	...	...	...	...	...	...
A_r	$f_{h1(j)r}$	$f_{h2(j)r}$	$f_{h\cdot(j)r}$	$N_{h1(j)r}$	$N_{h2(j)r}$	$N_{h\cdot(j)r}$
...	...	...	...	...	...	...
A_{k_h}	$f_{h1(j)k_h}$	$f_{h2(j)k_h}$	$f_{h\cdot(j)k_h}$	$N_{h1(j)k_h}$	$N_{h2(j)k_h}$	$N_{h\cdot(j)k_h}$
	$n_{1(j)}$	$n_{2(j)}$	n	–	–	–

Exchangeability under H_{0hj} implies that the permutation testing principle may be properly applied. This implies taking into consideration the permutation sample space generated by all permutations of pooled data set **X**, that is, the set of all possible tables in which the marginal frequencies are fixed.

To test H_{0hj} against H_{1hj}, we may consider the following as a permutation test statistic:

$$_DT^*_{hj} = \sum_{r=1}^{k_h-1} \left(N^*_{h1(j)r} - N^*_{h2(j)r}\right)\left[4\frac{N_{h\cdot(j)r}}{n} \cdot \frac{n - N_{h\cdot(j)r}}{n} \cdot \frac{n_{1(j)}n_{2(j)}}{n-1}\right]^{-1/2},$$

which is permutationally equivalent to

$$_DT^*_{hj} = \sum_{r=1}^{k_h-1} N^*_{h1(j)r} \cdot \left[N_{h\cdot(j)r} \cdot (n - N_{h\cdot(j)r})\right]^{-1/2},$$

with $N_{h\cdot(j)r} = N_{h1(j)r} + N_{h2(j)r} = N^*_{h1(j)r} + N^*_{h2(j)r}$, in which $N^*_{h1(j)r} = \sum_{s\le r} f^*_{h1(j)s}$ and $N^*_{h2(j)r} = \sum_{s\le r} f^*_{h2(j)s}$, $r = 1, \ldots, k_h - 1$, are permutation cumulative frequencies. Note that $_DT$ corresponds to the discrete version of a statistic following the Anderson–Darling goodness-of-fit test for dominance alternatives, which consists of a comparison of two EDFs, weighted by the reciprocals of permutation standard deviations (D'Agostino and Stephens, 1986).

Of course, many other test statistics may be useful. For example, we may consider: (a) $_{KM}T^*_{hj} = \sup_r(N^*_{h1(j)r} - N^*_{h2(j)r})$, which is a discretized version for restricted alternatives of the Kolmogorov–Smirnov test; (b) $_{CM}T^*_{hj} = \sum_r(N^*_{h1(j)r} - N^*_{h2(j)r})$, which is a discretized version of the Cramér–von Mises test.

Within the NPC approach the solution to this multivariate problem is now straightforward. To this end, we may consider the same types of partial tests introduced to solve the single partial problems, for instance $_DT^*_{hj}$, $h = 1, \ldots, V$, $j = 2, \ldots, C$. Hence, we may proceed with an NPC, for instance using Fisher's combining function $T''_F = -\sum_h \sum_j \log(\hat{\lambda}_{hj})$ where the $\hat{\lambda}_{hj}$ are the p-values of the partial tests according to the permutation distribution. This type of combination is rather computer-intensive, and due to its good power behaviour when sample sizes are sufficiently large, we suggest using a direct combination function on the standardized partial tests such as $_DT^*_{hj}$. Thus, one solution for restricted alternatives is $T''^*_{MD} = \sum_h \sum_j {_DT^*_{hj}}$. Since the partial tests are all exact and marginally unbiased, their NPC provides for an exact overall solution.

6.6 Multifocus Analysis

6.6.1 General Aspects

This section presents a nonparametric procedure which is useful for testing independence in distribution for categorical variables between two or more populations (see Finos and Salmaso, 2004). Let us consider a categorical variable and two independent samples selected from two different populations (groups). The aim is to test the independence in distribution of the two populations.

Data can be represented by means of a two-way contingency table of observed frequencies as in Table 6.4, where f_{ir} is the absolute frequency for the rth category in sample i, $f_{\cdot r} = \sum_i f_{ir}$ is the marginal frequency for the rth category, and n_i is the size of sample i ($i = 1, 2$; $r = 1, \ldots, k$). The classical solution is the χ^2 test which is generally valid but not very powerful when dealing with sparse tables (Agresti, 2002).

The following testing procedure, proposed by Finos and Salmaso (2004), called *multifocus analysis*, involves a nonparametric test which considers the categorical variable as a whole (global testing or weak FWE control), and each category as a single dummy variable (strong control; Hochberg

Table 6.4 Two-way contingency table for two independent samples

	Categories				
Sample	A_1	A_2	...	A_k	
1	f_{11}	f_{12}	...	f_{1k}	n_1
2	f_{21}	f_{22}	...	f_{2k}	n_2
	$f_{\cdot 1}$	$f_{\cdot 2}$	...	$f_{\cdot k}$	n

and Tamhane, 1987; and Sections 5.1–5.4). This property is essential when the inference is aimed not only at a global evaluation but also at the identification of the categories which most distinguish between two groups (e.g. cases from controls).

6.6.2 *The Multifocus Solution*

What distinguishes multifocus analysis is the decomposition of the contingency table into k two-by-two sub-tables with the aim of verifying the equality in distribution of each category in the two samples. The p-values referring to the respective sub-tables are nonparametrically combined according to the NPC of dependent permutation tests as discussed in Chapter 4.

Let us use X_i to denote the categorical response variable corresponding to ith population, taking values in $(A_1, A_2, \ldots, A_k)$, and π_{ir} to denote the probability that X_i falls in the rth category, $\pi_{ir} = \Pr\{X_i \in A_r\}$, $i = 1, 2$, $r = 1, \ldots, k$. The probability distributions of X_1 and X_2 are shown in Table 6.5.

The relative frequency f_{ir}/n_i (conditionally on the ith row of the contingency table) is a sample estimate of probability π_{ir}. The hypothesis of the testing problem can be written as

$$H_0 : \left\{\bigcap_{r=1}^{k} H_{0r}\right\} = \left\{\bigcap_{r=1}^{k} \pi_{1r} = \pi_{2r}\right\}$$

versus $H_1 : \left\{\bigcup_r H_{1r}\right\} = \left\{\bigcup_r \pi_{1r} \neq \pi_{2r}\right\}$. Hence, the (global) null hypothesis is true if each of the null sub-hypotheses is true, while the (global) alternative hypothesis is true if at least one of the alternative sub-hypotheses is true or equivalently if at least one null sub-hypothesis is false.

The generic table of frequencies (Table 6.4) can be decomposed into k sub-tables made up from the comparison of each single category against the remaining ones gathered into a single cell as shown in Table 6.6 and thus considering k separate tests.

In Table 6.6, $\bar{A}_r$ represents the set of all categories excluding A_r, that is, $\bar{A}_r = \{A_1, A_2, \ldots, A_k\} \setminus \{A_r\}$, and obviously $n_i - f_{ir} = \sum_{s \neq r} f_{is}$, $i = 1, 2$, $r = 1, \ldots, k$. A possible choice for

Table 6.5 Probability distributions of the two populations

	Categories				
Population	A_1	A_2	...	A_k	
X_1	π_{11}	π_{12}	...	π_{1k}	1
X_2	π_{21}	π_{22}	...	π_{2k}	1

Table 6.6 Contingency table for category A_r

Sample	Categories		
	A_r	$\overline{A_r}$	
1	f_{1r}	$n_1 - f_{1r}$	n_1
2	f_{2r}	$n_2 - f_{2r}$	n_2
	$f_{\cdot r}$	$f_{\cdot 2}$	n

the test statistic is given by the χ^2 statistic whose formulation is

$$
{}_\chi T_r = \frac{\left(f_{1r} - \hat{f}_{1r}\right)^2}{\hat{f}_{1r}} + \frac{\left[(n_1 - f_{1r}) - \left(n_1 - \hat{f}_{1r}\right)\right]^2}{\left(n_1 - \hat{f}_{1r}\right)} + \frac{\left(f_{2r} - \hat{f}_{2r}\right)^2}{\hat{f}_{2r}} + \frac{\left[(n_2 - f_{2r}) - \left(n_2 - \hat{f}_{2r}\right)\right]^2}{\left(n_2 - \hat{f}_{2r}\right)},
$$

where $\hat{f}_{ir} = n_i \cdot f_{\cdot r}/n,\ i = 1, 2,\ r = 1, \ldots, k$. Observing that ${}_\chi T_r$ is a monotonic function of $\left(f_{1r} - \hat{f}_{1r}\right)^2 / \hat{f}_{1r}$ and $\hat{f}_{1r}$ is constant with respect to each permutation (therefore to each point on the permutation space), the following test statistic is permutationally equivalent to ${}_\chi T_r$ because any monotonic transformation does not change the ordered statistic and therefore leaves the p-value calculation unchanged (Pesarin, 2001 and Section 2.4):

$$
T_r^* = \left(f_{1r}^* - \hat{f}_{1r}\right)^2.
$$

In order to show how to construct k separate tests for the k sub-hypothesis, let us now pass from the $2 \times k$ table representation shown in Table 6.4 to a matrix representation. In this case there will be $n_1 + n_2$ rows (n_1 and n_2 being the size of the two samples) and two columns: the first is the vector of group 1 or 2 labels (treatment factor), the second represents the variable categories for each individual observation.

The definition of k (dichotomous) dummy variables from the categorical variable (second column) represents a solution to our requirements. Testing the association of the rth dichotomous variable with the vector indicating the group of origin verifies the corresponding rth sub-hypothesis. Two possible choices are provided by Fisher's exact test or by the test proposed here. By adopting this test for the single sub-hypotheses it is also possible to evaluate directional alternative hypotheses (in this case permutationally equivalent to Fisher's exact test).

In order to find a test on the whole categorical variable of the two groups it is necessary to consider a test which combines the k values of p calculated on the sub-tables. In this case the problem of weak control of multiplicity arises (Hochberg and Tamane, 1987 and Sections 5.1–5.4). To produce a valid test for the combination of a multiplicity of p-values, we have to guarantee that such a global test is unbiased and thus produces p-values below the significance level α with a probability less than or equal to α itself.

Except for the field of permutation methods, such a combination seems problematic. The dependence between the k tests makes it impossible to use the classic Fisher combination ($T_F'' = -\sum_r \log(\lambda_r)$) with χ^2_{2k} asymptotic distributions (Folks 1984), where λ_r is the p-value of the

rth partial test. The alternative is provided by Bonferroni's combination, $\lambda''_B = k\ \min_r(\lambda_c)$. However, this combination is known for its loss of power when there is dependence between p-values (Hochberg and Tamhane, 1987; Westfall and Young, 1993) and the bond which arises due to the fixing of the marginal quantity leads to a dependence between the categories which cannot be omitted from the problem.

The adoption of an adequate permutation method makes it possible to immediately control the dependences. The global p-value obtained from the nonparametric combination of the p-values relating to the sub-hypotheses is an exact test and thus guarantees a weak control of the multiplicity. Such a p-value refers to the intersection of the sub-hypotheses previously shown. The inference in this case must be limited to the global evaluation of the phenomenon. Given the approach used, it is possible to give greater detail to the analysis using the tests on the single sub-hypotheses. This allows us to decide for which of the k categories of a variable the differences between the two samples are significant. Note that the tests relating to the sub-hypotheses are separate tests for the independence hypothesis for each individual category $r = 1, \ldots, k$, and the dependences between the tests are maintained by the permutation strategy, without having to model it.

A p-value adjustment for multiplicity is again necessary in this case. Here, however, the correction must concern each individual p-value since we are interested in making a decision about each one. A solution is provided by a closed testing procedure (Finos et al., 2003).

6.6.3 An Application

Let us consider an application concerning a case–control study in which the dependent variable is the haplotypical expression in a chromosome 17 marker. The aim of the study is to estimate the association between the group variable (cases/controls) and the marker (the categorical variable). Data are shown in Table 6.7.

The association test was performed using χ^2 (asymptotic and exact) and T_{SM} tests (Sham and Curtis, 1995). Moreover, the analysis followed the multifocus approach. Ten thousand CMC iterations were carried out for the four tests based on a permutation strategy.

All test p-values are less than 0.05, hence the tests are significant when $\alpha = 0.05$. In particular, ${}_{as}\lambda_{\chi^2} = 0.043$, ${}_{ex}\lambda_{\chi^2} = 0.031$ and $\lambda_{SM} = 0.018$. The tests based on multifocus analysis, with the Fisher and Tippett combining function, give the results $\lambda_{MuF} = 0.021$ and $\lambda_{MuT} = 0.009$, respectively.

Table 6.8 shows the raw p-values and those adjusted for multiplicity using Tippett's combination for each category. The alleles A and C are individually significant at a level of $\alpha = 0.05$ (0.0376 and 0.0015, respectively). After adjustment for multiplicity (Tippett's step-down procedure; Finos et al., 2003) only allele C remains significant (0.0097).

The analysis was performed in both R and MATLAB. Thus two scripts were created. From the `chrom17` folder on the book's website it is possible to download the file `chrom17m.txt`, where the R script is enclosed, and the file `chrom17m.m`, which allows the execution of the analysis in

Table 6.7 Frequencies for the two treatment groups in a chromosome 17 marker

	Haplotypes										
Group	A	B	C	D	E	F	G	H	I	J	
Cases	19	1	51	3	4	6	29	25	6	0	144
Controls	35	1	29	6	5	3	29	38	5	2	153
	54	2	80	9	9	9	58	63	11	2	297

Table 6.8 Association of the locus with the illness: single haplotypes using Tippett's combining function

Haplotypes	Raw p-values	Adjusted p-values
A	0.038	0.214
B	1.000	1.000
C	0.001	0.009
D	0.507	0.954
E	1.000	1.000
F	0.324	0.873
G	0.882	0.993
H	0.121	0.449
I	0.768	0.993
J	0.492	0.935

MATLAB. The data sets for the analysis in R and MATLAB are respectively stored in the files `chrom17m.csv` and `chrom17m.xls`.

6.7 Isotonic Inference

6.7.1 Introduction

Case–control studies are important in genetic epidemiology for establishing an association between genes and certain pathologies. A typical 3×2 contingency table for such studies is shown in Table 6.9. The n_1 cases (subjects with the disease) and the n_2 controls (healthy subjects) are classified according to the genotypes: we distinguish between subjects with zero (negative), one (heterozygous) and two (homozygous) copies of the rare allele A.

The odds ratios

$$\theta_{Aa} = \frac{[\Pr\{disease|Aa\} / \Pr\{no\ disease|Aa\}]}{[\Pr\{disease|aa\} / \Pr\{no\ disease|aa\}]}$$

and

$$\theta_{AA} = \frac{[\Pr\{disease|AA\} / \Pr\{no\ disease|AA\}]}{[\Pr\{disease|Aa\} / \Pr\{no\ disease|Aa\}]}$$

Table 6.9 Contingency table of a case–control study of association between genes and a given pathology

Genotypes	Cases	Controls	
aa	$f_{aa,1}$	$f_{aa,2}$	n_{aa}
Aa	$f_{Aa,1}$	$f_{Aa,2}$	n_{Aa}
AA	$f_{AA,1}$	$f_{AA,2}$	n_{AA}
	n_1	n_2	n

can be consistently estimated by $OR_{Aa} = f_{Aa,1} f_{aa,2} / f_{Aa,2} f_{aa,1}$ and $OR_{AA} = f_{AA,1} f_{Aa,2} / f_{AA,2} f_{Aa,1}$, respectively. The significance of the deviation of these ratios from one can be tested by the usual chi-square test or, for small samples, by Fisher's exact test.

In genetic epidemiology, association studies are useful for investigating candidate disease genes. Association studies are case–control population-based studies where unrelated affected and unaffected individuals are compared. Association between an allele A at a gene of interest and the disease under study means that the given allele occurs at a significantly higher frequency among affected individuals than among control individuals. For a biallelic locus with common allele a and rare allele A, individuals may carry none (subjects with genotype aa), one (subjects with genotype Aa) or double (subjects with genotype AA) copies of the A allele. Traditionally a test for allelic association involves testing for the distribution of case/control genotypes using the likelihood ratio chi-square statistic (asymptotically distributed as χ^2 with 2 d.f.) or the Fisher exact test.

The effect of an allele can be described as recessive, codominant or dominant. In the *recessive* case, there is an effect only in the presence of two copies of allele A (genotype AA). In the *codominant* case, allele A has an additive effect: genotype Aa carries a larger (or smaller) risk of illness than genotype aa, and AA carries a larger (or smaller) risk of illness than genotype Aa. Obviously, AA carries a larger (or smaller) risk of illness than genotype aa. In the *dominant* case, the effect of the A allele is the same in the AA and Aa genotype. In this situation, there is no relative risk (or protection) between AA and Aa, but only between AA (or Aa) and aa. For these reasons, differences in the risks should be tested for while maximizing over the restricted parameter space that corresponds to plausible biological models: risk of $AA \geq (\leq)$ risk of $Aa \geq (\leq)$ risk of aa.

In case–control studies it is easy to obtain genotype-specific relative risk from odds ratios, $\theta_{AA} = \rho_{AA}/\rho_{Aa}$ and $\theta_{Aa} = \rho_{Aa}/\rho_{aa}$, where ρ_h denotes the relative risk of h ($h = aa,\ Aa,\ AA$). The null and alternative hypotheses are

$$H_0 : \{\theta_{AA} = \theta_{Aa} = 1\}$$

and

$$H_1 : \{[(\theta_{AA} \geq 1) \cap (\theta_{Aa} \geq 1)] \text{ XOR } [(\theta_{AA} \leq 1) \cap (\theta_{Aa} \leq 1)]\},$$

where at least one inequality is strong. These particular hypotheses were defined by Chiano and Clayton (1998). From a statistical point of view, the alternative hypothesis is isotonic, i.e. the variables are ordered in one sense. However, there is the further complication due to the exclusive 'or' (XOR) relation. This approach allows us to study genetic diseases for which the relative effect of the putative allele (dominant, recessive or codominant) is not known, or in studies on related genetic polymorphism that may be protective or deleterious with respect to the disease.

6.7.2 Allelic Association Analysis in Genetics

In the genetic configuration introduced by Chiano and Clayton (1998), the statistical problem we are discussing can be formalized in the following way. Let us assume responses are bivariate and that observed subjects are partitioned into two groups (as per a typical case–control study), so that data may be represented as $\mathbf{X} = \{(X_{1ji}, X_{2ji}),\ i = 1, \ldots, nj,\ j = 1, 2\}$, where responses are binary ordered categorical. Of course, in a more general setting we may also consider real-valued responses, or any kind of ordered variables, with more than two dimensions and with more than two groups. The hypotheses we are interested in are

$$H_0 : \left\{\left(X_{11}, X_{21}\right) \stackrel{d}{=} \left(X_{12}, X_{22}\right)\right\} = \left\{\left(X_{11} \stackrel{d}{=} X_{12}\right) \bigcap \left(X_{21} \stackrel{d}{=} X_{22}\right)\right\},$$

against the special isotonic set of alternatives,

$$H_1 : \left\{ \begin{array}{c} \left(X_{11} \overset{d}{\geq} X_{12}\right) \bigcap \left(X_{21} \overset{d}{\geq} X_{22}\right) \text{ XOR} \\ \left(X_{11} \overset{d}{\leq} X_{12}\right) \bigcap \left(X_{21} \overset{d}{\leq} X_{22}\right) \end{array} \right\},$$

where at least one inequality is strong, under H_1 one and only one of the two bivariate stochastic dominance relations is true, and the variable X_{hj} represents the hth response for the jth group $(h,\ j = 1, 2)$.

To facilitate interpretation, it is often useful to introduce a response model such as $X_{hji} = \delta_{hj}(\mu_h + Z_{hji})$, where δ_{hj} is the effect on the hth variable in the jth group, all other symbols having obvious meanings. In accordance with this model, the hypotheses may be written as

$$H_0 : \{(\delta_{11} = \delta_{12} = 1) \bigcap (\delta_{21} = \delta_{22} = 1)\}$$

against

$$H_1 : \left\{ \left[(\delta_{11} \geq \delta_{12} = 1) \bigcap (\delta_{21} \geq \delta_{22} = 1)\right] \text{XOR} \left[(\delta_{11} \leq \delta_{12} = 1) \bigcap (\delta_{21} \leq \delta_{22} = 1)\right] \right\},$$

where at least one inequality in each 'sub-alternative' is strong. In the genetic context the alternative hypothesis means that a gene is associated with a given disease so that, on affected individuals (cases), at least one of the genotype frequencies with the putative allele increases XOR decreases with respect to non-affected individuals (controls).

6.7.3 Parametric Solutions

Let us consider two possible solutions to the testing problem based on a parametric approach. The first proposal, by Chiano and Clayton (1998), uses a reparametrization and Wilks's likelihood ratio chi-squared statistic. The second method, developed by El Barmi and Dykstra (1995) and Dykstra et al. (1995), is based on the maximum likelihood estimator calculated from a multinomial distribution.

Chiano and Clayton's Method

Chiano and Clayton (1998) consider the formalization of the problem based on the odds ratios but use the log transformations:

$$\gamma_{AA} = \log \theta_{AA} \text{ and } \gamma_{Aa} = \log \theta_{Aa}.$$

Thus the hypotheses become

$$H_0 : \{\gamma_{AA} = \gamma_{Aa} = 0\}$$

against

$$H_1 : \left\{(\gamma_{AA} \geq 0) \bigcap (\gamma_{Aa} \geq 0)\right\} \text{XOR} \left\{(\gamma_{AA} \leq 0) \bigcap (\gamma_{Aa} \leq 0)\right\}.$$

If regularity conditions hold, inference would be made with reference to Wilks's likelihood ratio chi-squared statistic:

$$\Lambda = 2 \sum nL\left(\widehat{\gamma}_{AA}, \widehat{\gamma}_{Aa}\right) \log \left[\frac{L\left(\widehat{\gamma}_{AA}, \widehat{\gamma}_{Aa}\right)}{L(0, 0)}\right]$$

distributed as a χ^2 with 2 d.f. Unfortunately, under such order restriction, Wilks's regularity assumptions are not met since the null point (origin) is on the boundary of the parametric space. The likelihood is therefore maximized subject to order constraints as follows. First, we obtain the unrestricted maximum likelihood estimate $(\widehat{\gamma}_{AA}, \widehat{\gamma}_{Aa})$ of $(\gamma_{AA}, \gamma_{Aa})$:

(1) if $(\widehat{\gamma}_{AA}, \widehat{\gamma}_{Aa}) \in \left\{(\gamma_{AA} \leq 0) \bigcap (\gamma_{Aa} \geq 0)\right\} \bigcup \left\{(\gamma_{AA} \geq 0) \bigcap (\gamma_{Aa} \leq 0)\right\}$ the maximum likelihood estimate is recalculated under the constraint $\widehat{\gamma}_{AA} = 0$ or $\widehat{\gamma}_{Aa} = 0$;
(2) otherwise the maximum likelihood estimate corresponds to unrestricted solution $(\widehat{\gamma}_{AA}, \widehat{\gamma}_{Aa})$.

Hence, it turns out that the distribution of the likelihood ratio chi-square statistic can be represented as a mixture of two chi-square distributions:

$$\Lambda \sim \lambda\left(\gamma_{AA}, \gamma_{Aa}\right) \chi_2^2 + \left[1 - \lambda\left(\gamma_{AA}, \gamma_{Aa}\right)\right] \chi_1^2,$$

where $\lambda\left(\gamma_{AA}, \gamma_{Aa}\right)$ is the probability that $(\widehat{\gamma}_{AA}, \widehat{\gamma}_{Aa})$ falls in the region indicated in (1), and it can be approximated to

$$\lambda\left(\gamma_{AA}, \gamma_{Aa}\right) = \cos^{-1}\left(\frac{\nu_{12}}{\sqrt{\nu_{11}\nu_{22}}}\right)/\pi,$$

where ν_{ij} is the element in the ith and jth column of the 2×2 matrix $\mathbb{V}(\widehat{\gamma}_{AA}, \widehat{\gamma}_{Aa}|H_0)$.

Maximum Likelihood Approach

The case–control contingency table in Table 6.9 can be interpreted as two independent vectors of data. The random sample of n_1 cases is taken from a multinomial distribution with probability vector $\boldsymbol{\pi}_1 = (\pi_{aa1}, \pi_{Aa1}, \pi_{AA1})$ while the n_2 controls are taken from a multinomial distribution (independent of the other) with parameter $\boldsymbol{\pi}_2 = (\pi_{aa2}, \pi_{Aa2}, \pi_{AA2})$. It is possible to derive the nonparametric maximum likelihood estimators (MLEs) $\widehat{\boldsymbol{\pi}}_1$ and $\widehat{\boldsymbol{\pi}}_2$ under the hypotheses

$$H_0 : \left\{\boldsymbol{\pi}_1 \overset{LR}{=} \boldsymbol{\pi}_2\right\}$$

against

$$H_1 : \left\{(\boldsymbol{\pi}_1 \overset{LR}{\leq} \boldsymbol{\pi}_2) \text{ or } (\boldsymbol{\pi}_1 \overset{LR}{\geq} \boldsymbol{\pi}_2)\right\},$$

with at least one strict inequality, where $\hat{\boldsymbol{\pi}}_j = (f_{aaj}/n_j, f_{Aaj}/n_j, f_{AAj}/n_j)$, $j = 1, 2$. The symbol $\overset{LR}{>}$ ($\overset{LR}{<}$) means that a likelihood ratio ordering exists between the distributions of two random variables. For example, $X \overset{LR}{>} Y$ means that for all a, b such that $a < b$, the conditional distribution of X given $X \in (a, b)$ is stochastically greater than the distribution of Y given $Y \in (a, b)$, or equivalently $f_X(t)/f_Y(t)$ is a non-decreasing function of t, where $f_X(t)$ and $f_Y(t)$ are the density functions of X and Y. Starting from the likelihood function $L(\boldsymbol{\pi}_1, \boldsymbol{\pi}_2) \propto \prod_s \pi_{s1}^{f_{s1}} \pi_{s2}^{f_{s2}}$, $s \in \{aa, Aa, AA\}$, through reparameterizations and following an algorithm proposed by Dykstra et al. (1995) and El Barmi and Dykstra (1995), it is possible to obtain the MLEs and perform a likelihood ratio test.

6.7.4 *Permutation Approach*

Considering the above testing problem, the pooled data set $\mathbf{X}$ is a sufficient statistic for the problem, therefore a partial test statistic could be

$$T_h^* = \sum_i \left|X_{h2i}^* - X_{h1i}^*\right|, \quad h = 1, 2,$$

and the null hypothesis should be rejected for high values of T_h. A combination of partial p-values λ_h allows us to calculate a global p-value which is useful for solving the global testing problem by the NPC methodology.

An allele A at a gene of interest is said to be associated with the disease if it occurs at significantly higher or lower frequencies among affected individuals than among control individuals. Given that the effect of an allele can be recessive, codominant or dominant, the problem's null hypothesis, defined in terms of sample odds ratios (OR) and according to the permutation approach, can also be expressed as

$$H_0 : \left\{\left(f_{AA,1} \cdot f_{Aa,2} \overset{d}{=} f_{AA,2} \cdot f_{Aa,1}\right) \cap \left(f_{Aa,1} \cdot f_{aa,2} \overset{d}{=} f_{Aa,2} \cdot f_{aa,1}\right)\right\}$$

which allows a computationally easy solution. The two partial tests can be written as

$$H_{0AA} : \left\{f_{AA,1} \cdot f_{Aa,2} \overset{d}{=} f_{AA,2} \cdot f_{Aa,1}\right\}$$

versus

$$H_{1AA} : \left\{\begin{bmatrix} f_{AA,1} \cdot f_{Aa,2} \overset{d}{\geq} f_{AA,2} \cdot f_{Aa,1} \end{bmatrix} \text{XOR} \atop \begin{bmatrix} f_{AA,1} \cdot f_{Aa,2} \overset{d}{\leq} f_{AA,2} \cdot f_{Aa,1} \end{bmatrix} \right\}$$

and

$$H_{0Aa} : \left\{f_{Aa,1} \cdot f_{aa,2} \overset{d}{=} f_{Aa,2} \cdot f_{aa,1}\right\}$$

versus

$$H_{1Aa} : \left\{\begin{bmatrix} f_{Aa,1} \cdot f_{aa,2} \overset{d}{\geq} f_{Aa,2} \cdot f_{aa,1} \end{bmatrix} \text{XOR} \atop \begin{bmatrix} f_{Aa,1} \cdot f_{aa,2} \overset{d}{\leq} f_{Aa,2} \cdot f_{aa,1} \end{bmatrix} \right\}.$$

The permutation solution can be based on the partial statistics $T_{AA} = OR_{AA}$ and $T_{Aa} = OR_{Aa}$. Considering all the permutations of the rows of data set $\mathbf{X}$, or (for computational convenience) a random sample B of such permutations, and recalculating the frequencies of Table 6.9 and the values T^*_{AA} and T^*_{Aa}, corresponding to each permutation (preserving the marginal values of the contingency table n_1, n_2, n_{aa}, n_{Aa} and n_{AA}), it is possible to calculate estimations of partial p-values $\widehat{\lambda}_{AA} = \#(T^*_{AA} \geq T^o_{AA})/B$ and $\widehat{\lambda}_{Aa} = \#(T^*_{Aa} \geq T^o_{Aa})/B$, where T^o_h is the observed value of the test statistic T_h, that is, the value calculated on the observed data set $\mathbf{X}$, $T^o_h = T_h(\mathbf{X})$, $h = 1, 2$. The application of a combining function $\psi(\cdot)$ allows us to obtain a p-value for the global test $\widehat{\lambda}_\psi = \psi\left(\widehat{\lambda}_{AA}, \widehat{\lambda}_{Aa}\right)$, to be compared with the significance level α according to the classical decisional rule of testing problems.

6.8 Test on Moments for Ordered Variables

6.8.1 General Aspects

An interesting and common testing problem is that of comparing two independent samples when the response variables are ordered categorical. In particular, one-sided tests are very useful in practical applications but very difficult to deal with. Traditional likelihood ratio tests need to know the distribution under the null hypothesis but it is often not possible to find a plausible distribution of data or to perform a test for it, thus asymptotic and bootstrap approaches (Wang, 1996) or nonparametric solutions (Pesarin, 2001) are required.

Let us denote by X_j $(j = 1, 2)$ the random variable representing the response for the jth treatment group. The hypothesis of interest is that response in treatment 1 is stochastically larger than in treatment 2. Formally, we are interested in testing

$$H_0 : \{X_1 \overset{d}{=} X_2\} \text{ against } H_1 : \{X_1 \overset{d}{>} X_2\}.$$

The parametric solution for the multivariate case is even more difficult, especially when a large number of component variables are present because it is difficult or impossible to describe the data distribution function and asymptotic solutions are often not suitable.

Let us assume that the support of X_j $(j = 1, 2)$ is partitioned into $k \geq 2$ ordered categories $\{A_r, r = 1, \ldots, k\}$ and that $\mathbf{X}_j = \{\mathbf{X}_{ji},\ i = 1, \ldots, n_j\},\ j = 1, 2$, are two samples of i.i.d. observations. The null hypothesis can be written in terms of cumulative distribution functions $F_j(A_r) = \Pr\{X_j \leq A_r\}$, $j = 1, 2$,

$$H_0 : \{F_1(A_r) = F_2(A_r), r = 1, \ldots, k - 1\},$$

and similarly the alternative hypothesis can be written as

$$H_1 : \{F_1(A_r) \leq F_2(A_r),\ r = 1, \ldots, k - 1,\ \text{AND } \exists r : F_1(A_r) < F_2(A_r)\}.$$

6.8.2 *Score Transformations and Univariate Tests*

The testing problem in Section 6.8.1 can be decomposed into $k - 1$ partial problems, or equivalently the hypotheses can be broken down into $k - 1$ sub-hypotheses, as

$$H_0 : \left\{\bigcap\nolimits_{r=1}^{k-1} [F_1(A_r) = F_2(A_r)]\right\},$$

and similarly the alternative can be written as

$$H_1 : \left\{\bigcup\nolimits_r [F_1(A_r) \leq F_2(A_r)],\ \text{AND } \exists r : F_1(A_r) < F_2(A_r)\right\}.$$

We observe that under the null hypothesis, data are exchangeable between groups, thus it is possible to calculate p-values according to the permutation distribution of a suitable test statistic and apply the classical decision rule for the testing problem. Several contributions on stochastic dominance problems can be found in the literature (see Silvapulle and Sen, 2005; Wilcoxon, 1945; Mann and Whitney, 1947; Brunner and Munzel, 2000).

The following solution is based on the joint analysis' of tests on sample moments. It requires an initial transformation of ordered categories $A_1, \ldots, A_k$ into numeric scores $w_1, \ldots, w_k$, that is, the application of an increasing function $g(\cdot)$ such that $g(A_k) = w_k$. Thus $\mathbb{E}\left[g(X)^s\right] = \sum_{r=1}^{k} \Pr\{X = A_r\}\, w_r^s$ is the sth moment of $g(X)$. It is well known that two distributions of discrete variables, defined on the same support with k distinct real values, coincide if and only if their first $k - 1$ moments are equal. As a matter of fact a characteristic function, like a probability generating function, depends only on the first $k - 1$ moments (Jacod and Protter,

2000). Thus, the global null hypothesis can be written as

$$H_0 : \left\{\bigcap_{s=1}^{k-1} \mathbb{E}\left[g\left(X_1\right)^s\right] = \mathbb{E}\left[g\left(X_2\right)^s\right]\right\}$$
$$= \left\{\bigcap_{s=1}^{k-1} \sum_{r=1}^{k} \left[\Pr\left\{X_1 = A_r\right\} - \Pr\left\{X_2 = A_r\right\}\right] w_r^s = 0\right\}.$$

Since the inequality $\bigcup_s \left\{\mathbb{E}\left[g\left(X_1\right)^s\right] > \mathbb{E}\left[g\left(X_2\right)^s\right]\right\}$ is not equivalent to the stochastic dominance of X_1 over X_2, the application of the test on moments requires us to assume $X_1 \overset{d}{\geq} X_2$, or equivalently to exclude $X_1 \overset{d}{<} X_2$. If this condition holds, then stochastic dominance is equivalent to moment inequalities and the directional alternative hypothesis can be written as

$$H_1 : \left\{\bigcap_{s=1}^{k-1} \mathbb{E}\left[g\left(X_1\right)^s\right] > \mathbb{E}\left[g\left(X_2\right)^s\right]\right\}$$

where the opposite inequality is never true by assumption.

A permutation partial test for each of the $k - 1$ sub-hypotheses can be carried out using the differences between the sample moments or permutationally equivalent statistics, such as the sample moments of the first sample, as test statistics. Hence, the statistic for the sth partial test could be

$$T_s^* = \frac{\sum_{r=1}^{k} w_r^s \cdot f_{1r}^*}{n_1}, \quad s = 1, \ldots, k - 1,$$

where f_{1r}^* and n_1 respectively denote the absolute permutation frequency of A_r and the sample size in the first sample. NPC of partial tests allows us to calculate a p-value for the global test on moments .

6.8.3 *Multivariate Extension*

The problem of two-sample tests on moments for restricted alternatives can be extended to the multivariate case. The alternative hypothesis in this case is also called the componentwise stochastic dominance hypothesis.

Let us use $X_j = \left(X_{1j,} \ldots, X_{Vj}\right)$, $j = 1, 2$, to denote the V-variate response for the jth population and $A_{h1}, \ldots, A_{hk_h}$ to denote the k_h ordered classes of the support of the hth component variable. The V-dimensional random vector X_1 is stochastically larger then X_2, or equivalently, X_1 dominates X_2 (in symbols $X_1 \overset{d}{>} X_2$) if and only if $\mathbb{E}[g(X_1)] \leq \mathbb{E}[g(X_2)]$ holds for all increasing real functions $g(\cdot)$ such that the expectation exists. Hence the multivariate extension of the testing problem can be represented by

$$H_0 : \left\{X_1 \overset{d}{=} X_2\right\} = \left\{\bigcap_{h=1}^{V} \left[X_{h1} \overset{d}{=} X_{h2}\right]\right\},$$
$$H_1 : \left\{X_1 \overset{d}{>} X_2\right\} = \left\{\bigcup_h \left[X_{h1} \overset{d}{>} X_{h2}\right]\right\}.$$

It is quite complex and often impossible to find a solution based on the maximum likelihood ratio test (Wang, 1996). According to the NPC methodology, the testing problem can be decomposed into V sub-problems: for each component variable a partial test can be carried out and the NPC makes it possible to solve the global testing problem taking into account the dependence structure of variables and consequently of the V dependent partial tests, even if the multivariate and univariate distributions are unknown. A solution for each partial testing problem can be given by the score transformation and the application of the test on moments. Thus, the global null hypothesis can be broken down into $k - V$ sub-hypotheses, where $k = \sum_h k_h$, and can be written as

$$H_0 : \left\{ \bigcap_{h=1}^{V} \bigcap_{s=1}^{k_h-1} \mathbb{E}\left[g\left(X_{h1}\right)^s\right] = \mathbb{E}\left[g\left(X_{h2}\right)^s\right] \right\}.$$

Assuming that $X_{h1} \overset{d}{\geq} X_{h2}$ for all $h \in \{1, \ldots, V\}$, the alternative hypothesis can be written as

$$H_1 : \left\{ \bigcup_h \bigcup_s \mathbb{E}\left[g\left(X_{h1}\right)^s\right] > \mathbb{E}\left[g\left(X_{h2}\right)^s\right] \right\},$$

where the opposite inequality is never true.

Thus, the solution is found in two steps:

1. the partial tests on moments for each variable are performed and combined to obtain V global tests on moments using the NPC;
2. the V partial tests obtained in the first step are combined to obtain the overall multivariate test.

The procedure described involve a multiple testing procedure, therefore it is necessary to control the maximum familywise error rate (FWE), that is, the maximum probability that one or more null hypotheses are erroneously rejected. Closed testing procedures, based on 'families' of multivariate tests and compatible with the NPC methodology, are an ideal solution for multiplicity control (Chapter 5; and Arboretti and Bonnini, 2009).

6.9 Heterogeneity Comparisons

6.9.1 *Introduction*

A typical problem of heterogeneity comparison is discussed in Corrain et al. (1977) describe an anthropological study on a number of Kenyan populations. Among these, the 'Ol Molo' is a nomadic population which therefore is expected to have more genetic exchanges with other ethnic groups than, for instance, the 'Kamba' population, which is non-nomadic and exhibits rather rigid endogamous behaviour. Hence the Ol Molo are likely to be characterized by a higher genetic *heterogeneity* than the Kamba. Corrain et al. (1977), whose interest lies in comparing genetic heterogeneity, take four genetic factors into consideration ($Gm(1)$, $Gm(2)$, $Gm(4)$ and $Gm(12)$), and all their phenotypic combinations, corresponding to $2^4 = 16$ nominal categories. The response variable is therefore nominal categorical and each category is a sequence of four signs ("+" or "−") indicating the presence or absence of these factors.

In descriptive statistics a variable X is said to be *minimally heterogeneous* when its distribution is degenerate, and is said to be *maximally heterogeneous* when its distribution is uniform over its support, that is, the set of categories. Thus, heterogeneity is related to the concentration of probabilities and its degree depends on these probabilities. Let us suppose that the response variable X takes values in $(A_1, \ldots, A_k)$, with probability distribution $\Pr\{X = A_r\} = \pi_r$, $r = 1, \ldots, k$. The following properties must be satisfied by an index for measuring the degree of heterogeneity:

1. It reaches its minimum when there is an integer $r \in (1, \ldots, k)$ such that $\pi_r = 1$ and $\pi_s = 0$, for all $s \neq r$.

2. It assumes increasingly large values as it moves away from the degenerate towards the uniform distribution.
3. It reaches its maximum when $\pi_r = 1/k$, for all $r \in (1, \ldots, k)$.

Various indicators satisfy the three properties and can be used to measure the degree of heterogeneity. Among the most commonly used are:

- Gini's index (Gini, 1912), $G = 1 - \sum_{r=1}^{k} \pi_r^2$.
- Shannon's entropy index (Shannon, 1948), $S = -\sum_{r=1}^{k} \pi_r \log(\pi_r)$, where $\log(\cdot)$ is the natural logarithm and by convention $0 \cdot \log(0) = 0$.
- A family of indexes proposed by Rényi (1996), called entropy of order δ and defined as $R_\delta = \frac{1}{1-\delta} \log \sum_{r=1}^{k} \pi_r^\delta$, with special cases

$$R_1 = \lim_{\delta \to 1} R_\delta = -\sum_{r=1}^{k} \pi_r \log(\pi_r) = S;$$

$$R_2 = -\log\left(\sum_{r=1}^{k} \pi_r^2\right) = -\log[1 - G];$$

$$R_\infty = \lim_{\delta \to \infty} \left[\left(\frac{1}{1-\delta} \log \sum_{r=1}^{k} \pi_r^\delta\right)\right] = -\log\left[\max_{1 \le r \le k} (\pi_r)\right].$$

A critical comparative discussion on heterogeneity indexes from a descriptive point of view can be found in Piccolo (2000).

6.9.2 *Tests for Comparing Heterogeneities*

Let us consider two independent samples with i.i.d. observations $\mathbf{X}_j = \{X_{ji},\ i = 1, \ldots, n_j > 2\},\ j = 1, 2$. As usual, the symbol $\mathbf{X}$ denotes the pooled data set $\mathbf{X} = \mathbf{X}_1 \biguplus \mathbf{X}_2$. Observed data can be displayed in a $2 \times k$ contingency table where the absolute frequencies are $\{f_{jr} = \sum_{i \le n_j} \mathbb{I}(X_{ji} = A_r),\ r = 1, \ldots, k,\ j = 1, 2\}$. The marginal frequency of the rth column (category) is denoted by $f_{\cdot r} = f_{1r} + f_{2r},\ r = 1, \ldots, k$.

Alternatively, data can be represented unit-by-unit by listing the $n = n_1 + n_2$ individual observations. The observed data set is then denoted by $\mathbf{X} = \{X(i),\ i = 1, \ldots, n;\ n_1, n_2\}$. Each permutation of the records of the data set is associated with a contingency table (the permuted table) with marginal frequencies equal to those of the observed contingency table. In other words, marginal frequencies are permutation invariant, that is, $f_{\cdot r} = f_{1r} + f_{2r} = f_{1r}^* + f_{2r}^* = f_{\cdot r}^*,\ r = 1, \ldots, k$, where $f_{jr}^*,\ j = 1, 2$, are the frequencies of the permuted table.

It is worth observing that if the exchangeability condition holds, in univariate two-sample designs the set of marginal frequencies $(n_1, n_2, f_{\cdot 1}, \ldots, f_{\cdot k})$, the pooled data set $\mathbf{X}$, and any of its permutations $\mathbf{X}^*$ are equivalent sets of sufficient statistics (see Section 2.8). A well-known relevant consequence of exchangeability is that any conditional inference is exact. If exchangeability is satisfied only asymptotically (i.e. it is approximately satisfied), only approximate solutions are possible. This is the case with the permutation tests for heterogeneity.

Given two populations P_1 and P_2, if we use $Het(P_j)$ to denote the heterogeneity of $P_j,\ j = 1, 2$, the testing problem can be expressed as $H_0 : Het(P_1) = Het(P_2)$ against some alternatives. For the example regarding the heterogeneity dominance of the Ol Molo with respect to the Kamba, the alternative can be written as $H_1 : Het(P_1) > Het(P_2)$.

If probabilities related to the populations $\{\pi_{jr}, r = 1, \ldots, k, \ j = 1, 2\}$ were known, they could be arranged in non-increasing order, $\pi_{j(1)} \geq \ldots \geq \pi_{j(k)}$, and an equivalent expression for the null hypothesis given is:

$$H_0 : \{Het(P_1) = Het(P_2)\} = \{\pi_{1(r)} = \pi_{2(r)}, \ r = 1, \ldots, k\}.$$

Under the null hypothesis, exchangeability holds and the permutation testing principle is applicable exactly. Therefore the one-sided alternative can be written as

$$\begin{aligned} H_1 \ &: \ \{Het(P_1) > Het(P_2)\} \\ &= \left\{ \sum_{s=1}^{r} \pi_{1(s)} \leq \sum_{s=1}^{r} \pi_{2(s)}, r = 1, \ldots, k \right\} \end{aligned}$$

where at least one strict inequality holds.

This problem is in some ways similar to that of stochastic dominance of cumulative probabilities as in Section 6.5. The difference between the two problems lies in the ordering criterion which is based on the probabilities in the test on heterogeneities and on the ordered categories in the traditional stochastic dominance test.

Unfortunately, parameters $\pi_{jr}, \ r = 1, \ldots, k, \ j = 1, 2$, are unknown and we can only estimate them using the observed relative frequencies, $\hat{\pi}_{jr} = f_{jr}/n_j$. Obviously the ordered parameters $\pi_{j(r)}, \ r = 1, \ldots, k, \ j = 1, 2$, are also unknown, thus the ordering within each population, estimated through relative frequencies within each sample, is a *data-driven ordering* and so may differ from the true one. The main implication is that exchangeability under H_0 is not exact. We notice that only asymptotically data-driven and true ordering are equal with probability one. Indeed, by virtue of the well-known Glivenko–Cantelli theorem (Shorack and Wellner, 1986), the exchangeability of data with respect to samples is asymptotically attained in H_0.

To solve the testing problem, a reasonable test statistic should be a function of the difference of the sample indices of heterogeneity, $T_I = I_1 - I_2$, where $I_j, \ j = 1, 2$, are the sample indices (in the example I_j stands for $G_j, \ S_j$, or $R_{\delta j}$, etc.). Large values of T_I lead to the rejection of H_0. It is worth noting that in the $2 \times k$ contingency table rearranged according to ordered frequencies (ordered table) $\{f_{j(r)}, \ r = 1, \ldots, k, \ j = 1, 2\}$, where $f_{j(1)} \geq \ldots \geq f_{j(k)}$, the rth column (ordered category) in group 1 corresponds to an original category which may differ from the rth column in group 2. Furthermore, where there are ties in frequencies, we can arbitrarily choose their order since this has no influence on the permutation analysis.

Once the data-driven ordering is obtained, it is possible to proceed in a similar fashion with the testing for stochastic dominance on ordered categorical variables. Operating with the ordered table, the observed value of the test statistics T_I^o, permutation values T_I^* and p-value λ_I are calculated. Using simulation studies Arboretti et al. (2009a) prove that under H_1 the power behaviours of T_G and T_S are very similar and better than that of $T_{R\infty}$, and that under H_0 all the tests are well approximated.

6.9.3 A Case Study in Population Genetics

Let us again consider the anthropological study of the Ol Molo and Kamba populations described above. Table 6.10 shows the sample frequencies of the 16 phenotypic combinations in the samples selected from the two populations.

From Table 6.10 it is possible to order the absolute frequencies separately for each sample and to calculate the ordered relative frequencies. The ordered table of relative frequencies provides an estimate of the ordered probabilities distributions (Table 6.11).

Table 6.10 Observed frequencies of *Gm* phenotypic combinations in Ol Molo and Kamba

Categories	Combinations				Frequencies	
	$Gm(1)$	$Gm(2)$	$Gm(4)$	$Gm(12)$	Ol Molo	Kamba
A_1	+	+	+	+	12	0
A_2	−	+	+	+	1	0
A_3	+	−	+	+	8	6
A_4	+	+	−	+	2	1
A_5	+	+	+	−	0	0
A_6	−	−	+	+	1	0
A_7	+	+	−	−	1	0
A_8	−	+	−	+	0	0
A_9	−	+	+	−	0	0
A_{10}	+	−	+	−	2	0
A_{11}	+	−	−	+	8	15
A_{12}	+	−	−	−	6	0
A_{13}	−	+	−	−	0	0
A_{14}	−	−	+	−	0	0
A_{15}	−	−	−	+	3	1
A_{16}	−	−	−	−	1	0
					45	23

Table 6.11 Observed relative frequencies of *Gm* phenotypic combinations in Ol Molo and Kamba

Population	(1)	(2)	(3)	(4)	(5)	(6)	(7)	(8)	(9)	(10)	(11)	···	(16)
Ol Molo	0.27	0.18	0.18	0.13	0.07	0.04	0.04	0.02	0.02	0.02	0.02	···	0
Kamba	0.66	0.26	0.04	0.04	0	0	0	0	0	0	0	···	0

As shown in Table 6.11, from a descriptive point of view, the heterogeneity of the Ol Molo appears higher than that of the Kamba. Using a CMC simulation with $B = 50\,000$ random permutations, we obtain the p-values $\widehat{\lambda}_S = 0.00003$ using the test statistic T_S based on Shannon's index, $\widehat{\lambda}_G = 0.00012$ using the test statistic T_G based on Gini's index, and $\widehat{\lambda}_{R\infty} = 0.00183$ for Rényi's $T_{R\infty}$. It is clear that the three p-values are lower than the significance level $\alpha = 0.01$, hence the null hypothesis of equal heterogeneity has to be rejected in favour of the alternative that the genetic heterogeneity of the Ol Molo is greater than that of the Kamba.

In order to carry out the above analysis it is possible to use MATLAB or R. The MATLAB code can be found in the file `Kenya.m` (data set in `Kenya.xls`). The R script is given in the file `Kenya.txt` (data set in `Kenya.csv`). All the files can be downloaded from the `Kenya` folder on the book's website.

6.10 Application to PhD Programme Evaluation Using SAS

6.10.1 Description of the Problem

The quality evaluation of a PhD programme requires complex analysis because it will be based on several aspects. Performance evaluation is a multidimensional phenomenon and, from the statistical

Table 6.12 Relative frequencies of satisfaction judgements for the three PhD macro-areas

Judgement	MB	ST	EL
Education–employment relationship			
very satisfied	0.30	0.61	0.91
quite satisfied	0.35	0.22	0.09
quite unsatisfied/not very satisfied	0.20	0.10	0.00
very unsatisfied/not at all satisfied	0.15	0.07	0.00
Education–employment coherence			
very satisfied	0.10	0.39	0.65
quite satisfied	0.45	0.17	0.30
quite unsatisfied/not very satisfied	0.25	0.26	0.05
very unsatisfied/not at all satisfied	0.20	0.18	0.00
Use of education at work			
very satisfied	0.25	0.26	0.57
quite satisfied	0.30	0.30	0.35
quite unsatisfied/not very satisfied	0.24	0.22	0.08
very unsatisfied/not at all satisfied	0.21	0.22	0.00

point of view, the main methodological difficulties relate to the synthesis of information in the construction of a composite indicator, and the need to take the dependence structure among the univariate component variables in the multivariate distribution and the categorical nature of data into account.

Some Italian universities have carried out surveys on postdoctoral students' opinions on the education provision and researcher activities during their PhD courses. In 2004 the University of Ferrara carried out such a survey. This survey involved a random sample of 120 persons: four cohorts of 30 postdocs who graduated between 2001 and 2004, grouped into three macro-areas, namely economic-legal (EL), medical-biological (MB) and scientific-technological (ST).

To express their satisfaction with the various aspects of their work, education received and organization of the PhD course, postdocs gave an integer score from 1 to 4 corresponding to four ordered categories: 'not at all satisfied', 'not very satisfied', 'quite satisfied', 'very satisfied'. To evaluate external effectiveness, one of the composite aspects considered in the survey was 'education–employment relationship', a three-dimensional notion involving: (1) coherence between education and employment; (2) use in employment of acquired abilities; (3) adequacy of the PhD training for work. Relative frequencies are shown in Table 6.12.

This section presents two macros which are useful for computing a composite nonparametric performance index and performing a permutation ANOVA test, to obtain a comparative performance evaluation of the three PhD programmes from the 2004 University of Ferrara survey (Arboretti et al., 2009b). Although we are dealing with an observational study, the assumption of exchangeability in the null hypothesis is plausible in this context.

6.10.2 Global Satisfaction Index

The following procedure allows for the computation of a combined indicator of a set of k ordered categorical variables representing satisfaction judgements given by a set of experts or evaluators, following the method proposed by Arboretti et al. (2007g). Let us suppose that each marginal variable has a given (non-negative) degree of importance and can assume m ordered scores. Let us also suppose that large values correspond to higher satisfaction levels.

The NPC of dependent rankings (Lago and Pesarin, 2000) can be used to obtain a single combined ranking of the statistical units under evaluation which summarizes many partial (univariate) rankings. Starting with the component variables $X_1, \ldots, X_p$, each providing information about a partial aspect, and their weights $\omega_1, \ldots, \omega_p$, the global index is obtained by means of a real function $\phi : \mathcal{R}^{2p} \rightarrow \mathcal{R}^1$, which combines the available information. A few easy-to-check minimal properties have to be satisfied by ϕ: (i) ϕ must be continuous in all its $2k$ arguments, so that small changes in a subset of arguments imply small changes in ϕ; (ii) ϕ must be a non-decreasing function of all the partial ranks; (iii) ϕ must be symmetric with respect to rearrangements of the p pairs of arguments $(\omega_1, X_1), \ldots, (\omega_p, X_p)$. These properties are satisfied by several combining functions, such as $T_F = -\sum_h \omega_h \log(1 - \lambda_h)$, $T_L = \sum_h \omega_h \Phi^{-1}(\lambda_h)$, $T_T = \sum_h \omega_h \log\left(\frac{\lambda_h}{1-\lambda_h}\right)$ and others, where λ_h is the relative rank (or normalized score) for a given unit according to the hth variable (ranking) and $\Phi(\cdot)$ is the standard normal CDF.

To compare the absolute evaluations with target values, the concept of *extreme satisfaction profiles* is introduced. These involve hypothetical frequency distributions of variables corresponding to minimum or maximum satisfaction of the set of evaluators. In order to include the extreme satisfaction profiles in the analysis, original values are transformed according to the following rules:

1. Scores 1 and 2, corresponding to expressions of dissatisfaction, are transformed by adding $(1 - p_{hr}) \cdot 0.5$, $r = 1, 2$, $h = 1, \ldots, p$, where p_{hr} is the relative frequency of evaluators who chose the rth judgement or category for the hth aspect or variable.
2. Scores 3 and 4, corresponding to expressions of satisfaction, are transformed by adding $p_{hr} \cdot 0.5$, $r = 3, 4$, $h = 1, \ldots, p$.
3. Using z_{hi} to denote the transformed scores related to the hth variable and ith unit ($h = 1, \ldots, p$; $i = EL$, MB or ST), the normalized scores λ_{hi} are computed: $\lambda_{hi} = (z_{hi} - z_{h,\min} + 0.5)/(z_{h,\max} - z_{h,\min} + 1)$, where $z_{h,\min}$ and $z_{h,\max}$ are the transformed scores corresponding to minimum and maximum satisfaction respectively, according to the extreme satisfaction profiles.
4. For each unit, a combined value is calculated using a combining function (e.g. Fisher's) according to NPC theory: $T_i = \phi\left(\lambda_{1i}, \ldots, \lambda_{pi}; \omega_1, \ldots, \omega_p\right) = -\sum_h \omega_h \log\left(1 - \lambda_{hi}\right)$, $i = EL, MB, ST$.

This method is particularly useful in overcoming two methodological problems related to the computation of composite performance indicators: (1) the synthesis of information to reduce the data dimensionality; (2) the criticism of Bird et al. (2005), who stress that rank is a relative datum and extreme ranking positions do not immediately equate to genuinely worst or best performances.

To facilitate the reading of the final results a further transformation can be applied to the T_i values to obtain normalized scores: $S_i = (T_i - T_{\min})/(T_{\max} - T_{\min})$, where $T_{\min}$ and $T_{\max}$ are the theoretical minimum and maximum values for T_i. Performing the analysis on data illustrated in Table 6.12, it is evident that postdocs in the economic-legal field are the most satisfied and those in the medical-biological field are the least satisfied: $S_{EL} = 0.66 > S_{ST} = 0.47 > S_{MB} = 0.34$.

The analysis can be performed with the SAS macro `npc_ranking` (dataset, cod, w, k, m, t, list). The input parameters are:

- `dataset`: SAS data set name;
- `cod`: name of variable identifying statistical units;
- `w`: list of weights for the variables (weights must sum to the number variables, `k`);
- `k`: number of ordered variables;
- `m`: number of ordered categories, representing ordered discrete scores, with value 1 corresponding to lower satisfaction and m to higher satisfaction;

- `t`: number of least values corresponding to satisfaction judgements;
- `u`: percentage frequency of subjects with value m in the extreme satisfactory profile;
- `l`: percentage frequency of subjects with value 1 in the extreme satisfactory profile;
- `list`: list of names of ordered variables.

The macro generates the SAS temporary data set called `Fisher` containing the variable '`y_def`' representing the combined index varying from 0 to 1. The file `NPCRanking.sas`, downloadable from the `PhDdata` folder on the book's website, contains the script to perform the macro.

6.10.3 Multivariate Performance Comparisons

To perform the ANOVA permutation test, in accordance with the NPC test theory (Sections 1.11 and 4.2), in the SAS macro NPC_2samples the input `npc(dati, var_byn, var_cat, var_con, dom_byn, dom_con, weights, clas, nsample, strato, paired, unit, missing)` can be used.

The null hypothesis of the testing problem can be written as

$$H_0 : \left\{ \bigcap_{h=1}^{3} X_{h,EL} \stackrel{d}{=} X_{h,MB} \stackrel{d}{=} X_{h,ST} \right\} = \left\{ \bigcap_{h=1}^{3} H_{0h} \right\},$$

where $X_{h,i}$ represents the performance of the ith group (macro-area) considering the hth variable (partial aspect). The alternative hypothesis can be written as $H_1 : \left\{ \bigcup_h H_{1h} \right\}$.

According to the NPC of dependent permutation tests, each of the three sub-problems can be solved by performing a univariate permutation test, and the combination of partial p-values using Fisher's function gives a global p-value which is useful for solving the global testing problem. In this application case the final p-value for the global testing problem, with $B = 10\,000$, takes the value $0.000 < 0.01 = \alpha$, thus the null hypothesis has to be rejected in favour of the alternative hypothesis.

The input parameters of the procedure are:

- `dati`: name of the data set;
- `var_byn`: list of binary variables;
- `var_cat`: list of categorical, non-binary variables;
- `var_con`: list of continuous variables;
- `dom_byn`: list of directional marginal sub-hypotheses for binary variables:
  - if XAu< XBu specify LESS
  - if XAu> XBu specify GREAT
  - if XAu$\neq$ XBu specify NOTEQ;
- `dom_con`: list of directional marginal sub-hypotheses for continuous variables, see above;
- `weights`: list of weights for the variables, first specify weights for binary variables and then weights for categorical variables and continuous variables;
- `clas`: variable defining the two groups (character variable);
- `nsample`: number of conditional resamplings;
- `strato`: variable defining strata (character variable);
- `paired`: paired data (yes/no);
- `unit`: variable identifying paired observations;
- `missing`: presence of missing values (yes/no).

SAS macros can be downloaded from the book's website.

7

Permutation Testing for Repeated Measurements

7.1 Introduction

Repeated measures designs are used in observational and experimental situations in which each subject is observed on a finite or at most a countable number of occasions, usually in time or space, so that successive responses are dependent. In practice, responses of one unit may be viewed as obtained by a discrete or discretized stochastic process. With reference to each specific subject, repeated observations are also called the *response profiles*, and may be viewed as a multivariate variable.

Without loss of generality, we discuss general problems which can be referred to in terms of a one-way MANOVA layout for response profiles. Hence, we refer to testing problems for treatment effects when: (a) measurements are typically repeated a number of times on the same units; (b) units are partitioned into C groups or samples, there being C levels of a treatment; (c) the hypotheses being tested aim to investigate whether the observed profiles do or do not depend on treatment levels; (d) it is presumed that responses may depend on time or space and that related effects are not of primary interest. For simplicity, we henceforth refer to time occasions of observation, where 'time' is taken to mean any sequentially ordered entity, including space, lexicographic ordering, etc.

In the context of this chapter, repeated measurements, panel data, longitudinal data, response trajectories and profiles are considered as synonymous. As the proposed solutions essentially employ the method of NPC of dependent permutation tests, each obtained by a partial analysis on data observed on the same ordered occasion (this is called *time-to-time analysis* in Pesarin, 1997a, 1997b, 1999b, 2001; Celant et al., 2000a, 2000b), we assume that the permutation testing principle holds. In particular, in the null hypothesis, in which treatment does not induce differences with respect to treatment levels, we assume that the individual response profiles are exchangeable with respect to groups.

To be more specific, let us refer to a problem in which n units are partitioned into C groups and a univariate variable X is observed. Groups are of size $n_j \geq 2$, $j = 1, \ldots, C$, with $n = \sum_j n_j$. Units belonging to the jth group are presumed to receive a treatment at the jth level. All units are observed at k fixed ordered occasions $\tau_1, \ldots, \tau_k$, where k is an integer. For simplicity, we refer to time occasions by using t to mean τ_t, $t = 1, \ldots, k$. Hence, for each unit, we observe the discrete or discretized profile of a stochastic process, and profiles related to different units are assumed to be stochastically independent. Thus, within the hypothesis that treatment levels have no effect on response distributions, profiles are exchangeable with respect to groups.

Permutation Tests for Complex Data: Theory, Applications and Software Fortunato Pesarin, Luigi Salmaso

7.2 Carry-Over Effects in Repeated Measures Designs

When planning and carrying out repeated measures designs, the experimenter should take two sources of bias into account: practice effects and differential carry-over effects. *Practice effects* occur when subjects, exposed to all the conditions, gradually become familiar with the procedures and the experimental context, thus getting practice. If any effect is produced, it is due to a real learning process. As a result, potential variations (in the subsequent measurements) may be ascribed to the acquired practice rather than to the administered treatment. Using a sufficiently large number of testing orders, thus ensuring the equal occurrence of each experimental condition at each stage of practice during the trial, minimizes this drawback. In practice, this could be achieved by means of counterbalancing, and the resulting experimental design becomes a type of Latin square design. In contrast, *differential carry-over* or *residual effects* cannot be controlled by counterbalancing. These are distinguishing effects in that the treatment administered first (or previously) influences a patient's subsequent (or current) responses in one way in one testing order, and in a different way in another testing order. For example, when comparing the active treatment (A) with placebo (B), patients assigned to the AB sequence may experience a substantial carry-over effect in the second administration, whereas patients in the BA sequence might have a low B carry-over effect in the second period. By allowing wash-out periods, these effects can be reduced and practice effects may have the same influence on all treatment conditions. In the general definition, *carry-over effects* represent the continuation or the protraction of previous treatment administration on subsequent measurements. As a result, the first treatment may improve or decrease the effectiveness of the second, thus reducing the potential capability of repeated measures experiments to show treatment effects. If subsequent administrations of treatment A and B are close together, there is a sort of mixture, which could mask actual differences between the two treatments. In order to minimize carry-over effects, successive treatment times for each subject should be far enough apart by setting appropriate wash-out periods. Although the experimenter should check on carry-over effects in order to run a valid repeated measures experiment, carry-over effects do not invalidate a repeated measures randomization test. In any case, two distinct kinds of carry-over effects can be identified: carry-over effects that are the same for all treatments, and differential carry-over effects that vary from one treatment to another. The latter in particular may invalidate a design with repeated measures. Only 'true' treatment effects should emerge, hence it is important to investigate whether or not treatments A or B are different in the absence of carry-over effects. Repeated measures designs are appealing in that the results obtained from such experiments allow us to infer what would happen in independent experimental groups exposed to the same treatments (Edgington and Onghena, 2007). The subsequent administration of treatments to a subject should barely be influenced by the previous ones. In conclusion, minimizing the presence of carry-over effects common to the various treatments may reduce the presence of differential carry-over effects.

We observe that in the literature such a problem has not been dealt with within the permutation approach. Later on in this chapter we will discuss a permutation solution for the standard cross-over design without taking into account carry-over effects.

7.3 Modelling Repeated Measurements

7.3.1 A General Additive Model

Let us assume that there are no missing values (see Section 7.13 for analysis of problems with missing values). Moreover, let us refer to a univariate stochastic time model with additive effects, covering a number of practical situations. Extensions of the proposed solution to multivariate response profiles are generally straightforward, by analogy with those given for the one-way MANOVA layout.

The symbol $\mathbf{X} = \{X_{ji}(t),\ i = 1, \ldots, n_j,\ j = 1, \ldots, C,\ t = 1, \ldots, k\}$ indicates that the whole set of observed data is organized as a two-way layout of univariate observations. Alternatively, especially when effects due to time are not of primary interest, $\mathbf{X}$ may be organized as a one-way layout of profiles, $\mathbf{X} = \{\mathbf{X}_{ji},\ i = 1, \ldots, n_j,\ j = 1, \ldots, C\}$, where $\mathbf{X}_{ji} = \{X_{ji}(t),\ t = 1, \ldots, k\}$ denotes the jith observed profile.

The general additive response model referred to in this section is

$$X_{ji}(t) = \mu + \eta_j(t) + \Delta_{ji}(t) + \sigma(\eta_j(t)) \cdot Z_{ji}(t),$$

$i = 1, \ldots, n_j,\ j = 1, \ldots, C,\ t = 1, \ldots, k$. In this model the $Z_{ji}(t)$ are generally non-Gaussian error terms distributed as a stationary stochastic process with null mean and unknown distribution P_Z (i.e. a generic white noise process); these error terms are assumed to be exchangeable with respect to units and treatment levels but, of course, not independent of time. Moreover, μ is a population constant; coefficients $\eta_j(t)$ represent the *main treatment effects* and may depend on time through any kind of function, but are independent of units; quantities $\Delta_{ji}(t)$ represent the so-called *individual effects*; and $\sigma(\eta_j(t))$ are time-varying scale coefficients which may depend, through monotonic functions, on main treatment effects η_j, provided that the resulting CDFs are pairwise ordered so that they do not cross each other, as in $X_j(t) \overset{d}{<}$ (or $\overset{d}{>}$) $X_r(t)$, $t = 1, \ldots, k$, and $j \neq r = 1, \ldots, C$. When the $\Delta_{ji}(t)$ are stochastic, we assume that they have null mean values and distributions which may depend on main effects, units and treatment levels. Hence, random effects $\Delta_{ji}(t)$ are determinations of an unobservable stochastic process or, equivalently, of a k-dimensional variable $\boldsymbol{\Delta} = \{\Delta(t),\ t = 1, \ldots, k\}$. In this context, we assume that $\boldsymbol{\Delta}_j \sim \mathcal{D}_k\{\mathbf{0}, \boldsymbol{\beta}(\boldsymbol{\eta}_j)\}$, where $\mathcal{D}_k$ is any unspecified distribution with null mean vector and unknown dispersion matrix $\boldsymbol{\beta}$, indicating how unit effects vary with respect to main effects $\boldsymbol{\eta}_j = \{\eta_j(t),\ t = 1, \ldots, k\}$. Regarding the dispersion matrix $\boldsymbol{\beta}$, we assume that the resulting treatment effects are pairwise stochastically ordered, as in $\Delta_j(t) \overset{d}{<}$ (or $\overset{d}{>}$) $\Delta_r(t)$, $t = 1, \ldots, k$, and $j \neq r = 1, \ldots, C$. Moreover, we assume that the underlying bivariate stochastic processes $\{\Delta_{ji}(t),\ \sigma(\eta_j(t)) \cdot Z_{ji}(t), t = 1, \ldots, k\}$ of individual stochastic effects and error terms, in the null hypothesis, are exchangeable with respect to groups. This property is easily justified when subjects are randomized to treatments.

This setting is consistent with a general form of dependent random effects fitting a very large number of processes that are useful in most practical situations. In particular, it may interpret a number of the so-called *growth processes*. Of course, when $\boldsymbol{\beta} = \mathbf{0}$ with probability one for all t, the resulting model has fixed effects.

For evaluating the inherent difficulties in statistical analysis of real problems when repeated observations are involved, see, for example, Laird and Ware (1982), Azzalini (1984), Ware (1985), Crowder and Hand (1990), Diggle et al. (2002), Lindsay (1994) and Davidian and Giltinan (1995). In particular, when dispersion matrices $\boldsymbol{\Sigma}$ and $\boldsymbol{\beta}$ have no known simple structure, the underlying model may not be identifiable and thus no parametric inference is possible. Also, when $k \geq n$, the problem cannot admit any parametric solution; see Chung and Fraser (1958) and Blair et al. (1994) in which heuristic solutions are suggested under normality of errors $\mathbf{Z}$ and for fixed effects.

One among the many possible specifications of models for individual effects assumes that terms $\Delta_{ji}(t)$ behave according to an AR(1) process:

$$\Delta_{ji}(0) \overset{d}{=} 0; \quad \Delta_{ji}(t) = \gamma(t) \cdot \Delta_{ji}(t-1) + \beta(\eta_j(t)) \cdot W_{ji}(t),$$

$i = 1, \ldots, n_j,\ j = 1, \ldots, C,\ t = 1, \ldots, k$, where $W_{ji}(t)$ represent random contributions interpreting deviates of individual behaviour; $\gamma(t)$ are autoregressive parameters which are assumed to be independent of treatment levels and units, but not time; and $\beta(\eta_j(t))$, $t = 1, \ldots, k$, are time-varying scale coefficients of autoregressive parameters, which may depend on the main effects. By assumption, the terms $W_{ji}(t)$ have null mean value, unspecified distributions, and are possibly time-dependent, so that they may behave as a stationary stochastic process.

A simplification of the previous model considers a regression-type form such as

$$\Delta_{ji}(t) = \gamma_j(t) + \beta(t) \cdot W_{ji}(t),\ i = 1, \ldots, n_j, j = 1, \ldots, C, t = 1, \ldots, k.$$

Of course, many other models of dependence errors might be considered, including situations where matrices $\mathbf{\Sigma}$ and $\boldsymbol{\beta}$ are both full.

Remark 1. This problem corresponds to a special case of a two-way layout, in which effects due to time are supposed to be of no interest, in the sense that such effects are generally taken for granted. Also, it should be noted that this layout is unbalanced, because we do not assume that $n_j = n/C$.

Remark 2. As individual profiles are assumed to be exchangeable in the null hypothesis, a set of sufficient statistics for this problem is the pooled vector of observed profiles

$$\mathbf{X} = \{\mathbf{X}_{ji},\ i = 1, \ldots, n_j, j = 1, \ldots, C\} = \left\{\mathbf{X}_1 \biguplus \ldots \biguplus \mathbf{X}_C\right\}.$$

Note that exchangeability of observations with respect to time cannot generally be assumed.

7.3.2 Hypotheses of Interest

The hypotheses we wish to test are

$$H_0 : \left\{\mathbf{X}_1 \stackrel{d}{=} \ldots \stackrel{d}{=} \mathbf{X}_C\right\} = \left\{X_1(t) \stackrel{d}{=} \ldots \stackrel{d}{=} X_C(t),\ t = 1, \ldots, k\right\}$$

$$= \left\{\bigcap_{t=1}^{k}\left[X_1(t) \stackrel{d}{=} \ldots \stackrel{d}{=} X_C(t)\right]\right\} = \left\{\bigcap_{t=1}^{k} H_{0t}\right\}$$

against $H_1 : \{\bigcup_t [H_{0t} \text{ is not true}]\} = \{\bigcup_t H_{1t}\}$, in which a decomposition of the global hypotheses into k sub-hypotheses according to time is highlighted. This decomposition corresponds to the so-called *time-to-time* analysis.

Distributional assumptions imply that $\mathbf{X} = \mathbf{X}_1 \biguplus \ldots \biguplus \mathbf{X}_C$ is a set of sufficient statistics for the problem in H_0. Moreover, H_0: $\{\mathbf{X}_1 \stackrel{d}{=} \ldots \stackrel{d}{=} \mathbf{X}_C\}$ implies that the observed profiles are exchangeable with respect to treatment levels. Thus, the permutation testing principle applies to observed time profiles.

Note that by decomposition into k partial sub-hypotheses, each sub-problem is reduced to a one-way ANOVA. Also note that, from this point of view, the associated two-way ANOVA, in which effects due to time are not of interest, becomes equivalent to a one-way MANOVA. Hence, for partially testing sub-hypotheses H_{0t} against H_{1t} we may use any suitable permutation test T_t^* as in Sections 1.11, 2.7, 2.8.4 and 6.4. For real-valued univariate data, such tests may have the standard C-sample form. In addition, each partial test is marginally unbiased and consistent, and large values are significant. Hence, NPC may be used.

7.4 Testing Solutions

7.4.1 Solutions Using the NPC Approach

In the given conditions, k partial permutation tests $T_t^* = \sum_j n_j \cdot (\bar{X}_j^*)^2$, where $\bar{X}_j^* = \sum_i X_{ji}^*(t)/n_j$, $t = 1, \ldots, k$, are appropriate for time-to-time sub-hypotheses H_{0t} against H_{1t}. Thus, in order to achieve a global complete solution for H_0 against H_1, we must combine all these partial tests. Of course, due to the complexity of the problem and to the unknown k-dimensional distribution of $(T_1, \ldots, T_k)$ (see Crowder and Hand, 1990; Diggle et al., 2002), we are generally unable to evaluate all dependence relations among partial tests directly from $\mathbf{X}$. Therefore, this combination should be nonparametric and may be obtained through any combining function $\psi \in \mathcal{C}$. Of course, when the underlying model is not identifiable, and so some or all of the coefficients cannot be estimated, this NPC becomes unavoidable. Moreover, when all observations come from only one type of variable (continuous, discrete, nominal, ordered categorical) and thus partial tests are homogeneous, a direct combination of standardized partial tests, such as $T_t^* = \sum_j n_j \cdot [\bar{X}_j^*(t) - \bar{X}_\cdot(t)]^2 / \sum_{ji} [X_{ji}^*(t) - \bar{X}_j^*(t)]^2$, may be appropriate especially when k is large. This may not be the case when observations are on variables of different types (e.g. some continuous and others categorical).

Remark 1. When baseline observations $X_{ji}(0)$ are present and are assumed to influence subsequent responses, so that they are not negligible, then all partial tests should be conditional on them; that is, the $X_{ji}(0)$ must be considered as playing the role of covariates. In practice, in this case, the testing process may be carried out by considering k ANCOVA layouts. Sometimes, according to proper models describing how covariates influence responses, it may suffice to consider data transformations as, for example: time increments, $X_{ji}(t) - X_{ji}(t-1)$; increments with respect to the baseline, $X_{ji}(t) - X_{ji}(0)$; relative time increments, $[X_{ji}(t) - X_{ji}(t-1)]/X_{ji}(t-1)$, etc. Of course, in these cases, the k partial tests are modified accordingly.

Remark 2. When, in accordance with multi-aspect testing and, for instance, we assume that effects may act on the first two time-to-time moments, so that the null hypothesis H_0 becomes

$$H_0 : \left\{ \left[\bigcap_{t-1}^{k} (\mathbb{E}(X_1(t)) = \ldots = \mathbb{E}(X_C(t))) \right] \bigcap \left[\bigcap_{t-1}^{k} (\mathbb{E}(X_1^2(t)) = \ldots = \mathbb{E}(X_C^2(t))) \right] \right\},$$

then we may apply two partial tests for each time, $\{(T_{1t}, T_{2t}), t = 1, \ldots, k\}$, followed by their NPC.

Remark 3. In the case of testing for monotonic stochastic ordering, in which the alternative is expressed by

$$H_1 : \left\{ \mathbf{X}_1 \overset{d}{\leq} \ldots \overset{d}{\leq} \mathbf{X}_C \right\} = \left\{ X_1(t) \overset{d}{\leq} \ldots \overset{d}{\leq} X_C(t),\ t = 1, \ldots, k \right\},$$

the inequality being strict for at least one index j and/or t, the testing problem may be solved as follows. For each $j \in (1, \ldots, C-1)$, according to Example 7, 4.6 and Chapter 8, we may split the whole data set into two pooled pseudo-groups, where the first is obtained by pooling together data of the first j (ordered) groups and the second by pooling the rest. Thus, the $C-1$ pairs of pooled pseudo-groups to consider are $\mathbf{Y}_{1(j)} = \mathbf{X}_1 \biguplus \ldots \biguplus \mathbf{X}_j$ and $\mathbf{Y}_{2(j)} = \mathbf{X}_{j+1} \biguplus \ldots \biguplus \mathbf{X}_C$, $j = 1, \ldots, C-1$, where $\mathbf{X}_j = \{X_{ji}(t),\ i = 1, \ldots, n_j,\ t = 1, \ldots, k\}$ is the data set of the jth group.

In the overall null hypothesis, time profiles of every pair of pseudo-groups are exchangeable, because $\mathbf{Y}_{1(j)} \overset{d}{=} \mathbf{Y}_{2(j)}$, whereas in the alternative there is monotonic stochastic dominance over time $\mathbf{Y}_{1(j)} \overset{d}{\leq} \mathbf{Y}_{2(j)} = \bigcup_t [Y_{1(j)}(t) \overset{d}{\leq} Y_{2(j)}(t)]$, where symbols have obvious meanings. This suggests to express the global hypotheses in the equivalent form

$$H_0 : \left\{ \bigcap\nolimits_t \bigcap\nolimits_j \left[Y_{1(j)}(t) \overset{d}{=} Y_{2(j)}(t) \right] \right\}$$

and $H_1 : \{\bigcup_t \bigcup_j [Y_{1(j)}(t) \overset{d}{\leq} Y_{2(j)}(t)]\}$, where a (time-to-time) decomposition into an appropriate set of sub-hypotheses is emphasized.

Therefore, for each $j = 1, \ldots, C-1$ and each $t = 1, \ldots, k$, a proper time-to-time partial test is $T_j^*(t) = \sum_{1 \leq i \leq N_{2(j)}} Y_{2(j)i}^*(t)$, where $N_{2(j)} = \sum_{r>j} n_r$ is the sample size of $\mathbf{Y}_{2(j)}$. Since under the conditions of the problem all these $k(C-1)$ partial tests are exact and marginally unbiased, their NPC provides for an exact overall solution.

7.4.2 *Analysis of Two-Sample Dominance Problems*

Let us assume that responses behave in accordance with the additive model in Section 7.3.1. A rather special and interesting problem arises when $C = 2$ and when, for instance, we are interested in testing whether the first process is stochastically dominated by the second: $\{X_1(t) \overset{d}{<} X_2(t), t = 1, \ldots, k\}$. In such a case and referring to models with stochastic coefficients, the hypotheses are

$$H_0 : \left\{\bigcap_{t=1}^{k}\left[X_1(t) \overset{d}{=} X_2(t)\right]\right\} = \left\{\bigcap_{t=1}^{k}[\eta_1(t) = \eta_2(t)]\right\} = \left\{\bigcap_{t=1}^{k} H_{0t}\right\}$$

against $H_1 : \{\bigcup_t [X_1(t) \overset{d}{<} X_2(t)]\} = \{\bigcup_t [\eta_1(t) < \eta_2(t)]\} = \{\bigcup_t H_{1t}\}$, in which a stochastic dominance problem and a suitable decomposition of hypotheses are highlighted. Observe that the alternative is now broken down into k one-sided (restricted) sub-alternatives. Hence, for each sub-hypothesis, a one-tailed partial test for comparison of two locations should be considered.

The overall solution for this is now straightforward because: (i) as the exchangeability of individual profiles with respect to treatment levels is assumed in H_0, the permutation principle applies; (ii) the hypotheses are broken down into a set of sub-hypotheses; (iii) by assumption, for each sub-hypothesis there exists a test which is exact, marginally unbiased, consistent, and significant for large values. Hence, the sufficient conditions for the NPC method are satisfied and consequently overall permutation solutions are exact, unbiased and consistent. In the present case, a set of permutation partial test statistics might be $\{T_t^* = \bar{X}_2^*(t), t = 1, \ldots, k\}$. These partial tests are marginally unbiased, exact, significant for large values and consistent.

Extensions of the solution to V-dimensional repeated observations ($V > 1$) and to multi-aspect testing procedures are left as exercises. Extensions to missing value situations are discussed in Section 7.13.

7.4.3 *Analysis of the Cross-Over (AB-BA) Design*

The main feature of a cross-over design (see also Remark 5, 2.1.2) is that each patient receives more than one of the treatments under study. In its complete version, patients in each group receive each treatment. In order to evaluate patient response and treatment effectiveness, each patient is randomized to receive one of the possible sequence of treatments. All possible sequences are used. In this way, it is possible to obtain a measure of experimental control, thus providing a valid statistical analysis of the results. In its simplest version, only two treatments (A and B) are available: the so-called "two-period, two-treatment" cross-over or AB-BA design. A period is each time interval in which one of the treatments under study is administered. In this design, there are two treatment periods: in the first period patients randomized to the sequence AB receive treatment A and those randomized to the sequence BA receive treatment B; in the second period, the AB group receives B and the BA group receives A. Obviously it is of interest to compare treatments A and B and, since this comparison is made within subjects, a reliable evaluation of treatment differences can be obtained, rather than considering two independent groups. Once the first treatment cycle is completed, to avoid carry-over effects it is advisable to have a washout period, allowing the first treatment administered to be eliminated (reduced until negligible) from the body. Since two treatment sequences (AB and BA) are employed, the possibility of confusing treatment effect with time period effect is minimized. If patients are randomized to receive one or the other sequence, it is unlikely that all responsive subjects will receive the same sequence of treatments, although it is possible to specify, when planning the design of the experiment, how many of the n subjects receive AB. However, when randomization is not possible, data collected during the experiment allow us to assess whether observed treatment differences are accidental or causative. Although extensively used in clinical trials, since treatment sequences are administered consecutively, these designs have often been criticized on ethical grounds.

7.4.4 Analysis of a Cross-Over Design with Paired Data

A cross-over design with paired observations may be seen as a particular case of an unbalanced two-way ANOVA for twice repeated measurements. More specifically, let us suppose for example that n experimental units are partitioned into two groups: those belonging to the first receive *placebo before treatment*, and the others receive *treatment before placebo*. In this kind of cross-over experiment, two rather different testing problems are generally of interest: (i) whether or not treatment is effective; (ii) whether or not the two ways treatment is administered are equivalent.

Let us assume that response data behave according to

$$Y_{jti} = \mu + \alpha_{ji} + \beta_{jt} + Z_{jti}, \ i = 1, \ldots, n_j, \ t = 1, 2, \ j = 1, 2,$$

where μ is an unknown population constant; α_{ji} are the so-called individual effects, which are generally considered nuisance entities and so are not of interest for analysis; β_{jt} are (possibly random) effects due to treatment t on the jth group; and Z_{jti} are random errors assumed to be independent with respect to but not within units. Within units, as observations are paired, differences behave according to

$$X_{jti} = Y_{j1i} - Y_{j2i} = \beta_{j1} - \beta_{j2} + Z_{j1i} - Z_{j2i} = \delta_j + \sigma(\delta_j) \cdot Z_{ji},$$

where individual nuisance effects are compensated, the δ_j represent incremental treatment effects, Z_{ji} are the error components symmetrically distributed around 0, and $\sigma(\delta_j)$ are scale coefficients, which may depend on main effects but are invariant with respect to units.

The two separate sets of hypotheses of interest are $H_0' : \{\delta_1 = \delta_2 = 0\}$ against $H_1' : \{(\delta_1 \neq 0) \bigcup (\delta_2 \neq 0)\}$ and $H_0'' : \{X_1 \stackrel{d}{=} X_2\} = \{\delta_1 = \delta_2\}$ against $H_1'' : \{\delta_1 \neq \delta_2\}$, where the second set becomes of major interest especially when H_0' is false. But, of course, even if H_0' is accepted, we do not know whether it is true or false. Hence, we must act independently and jointly for both sub-problems (see Remark 5, 2.1.2).

A set of joint sufficient statistics for testing both sub-problems is $\mathbf{X} = \{\mathbf{X}_1, \mathbf{X}_2\}$, which corresponds to the whole set of differences partitioned into two groups. Note that to test H_0' against H_1', data are exchangeable within each unit because this sub-problem is equivalent to testing for symmetry of differences with respect to 0, so that there are 2^n possible permutations in the related permutation sample space (see Remark 3, 2.1.2). To test H_0'' against H_1'', data are exchangeable between two groups because this sub-problem is equivalent to testing for equality of locations (of differences) in a two-sample problem and there are $n!/(n_1!n_2!)$ possible permutations.

As a consequence, two separate permutation tests are: $T_1^* = \sum_i X_{1i} \cdot S_{1i}^* - \sum_i X_{2i} \cdot S_{2i}^*$, where S_{ji}^* are random signs, and $T_2^* = \sum_i X_{1i}^*/n_1 - \sum_i X_{2i}^*/n_2$; moreover, due to the way permutations are obtained, two tests T_1^* and T_2^* are independent in H_0.

Remark 1. The permutation structures of two test statistics (see Remark 1, 2.7) are $T_1^* = \sum_i(\delta_1 + \sigma(\delta_1)Z_{1i}) \cdot S_{1i}^* - \sum_i(\delta_2 + \sigma(\delta_2)Z_{2i}) \cdot S_{2i}^*$ and $T_2^* = \sum_i(\delta_1^* + \sigma(\delta_1^*)Z_{1i}^*)/n_1 - \sum_i(\delta_2^* + \sigma(\delta_2^*)Z_{2i}^*)/n_2$, respectively. Both structures, in the respective null sub-hypotheses, depend only on permutations of exchangeable errors. Hence, both are exact permutation tests for H_0' and H_0'' against, respectively, H_1' and H_1''. Moreover, in their respective alternatives, both permutation structures are ordered with respect to treatment effects, so that they are unbiased and, if the errors possess finite first moment, consistent (see Theorem 12, 4.5.2).

Remark 2. The multivariate extension of the present cross-over problem is straightforward. It can be done by considering solutions to the testing problem with paired multivariate responses and by the multivariate two-sample problem.

7.5 Testing for Repeated Measurements with Missing Data

The rest of this chapter provides solutions to problems with repeated measures when some of the data are missing. Experimental units are partitioned into $e \geq 2$ groups and the hypotheses to be tested are whether the observed profiles of repeated measures over time are dependent on treatment levels. The NPC solution is compared to two different parametric approaches to the problem of missing values: Hotelling's T^2 with deletion of units with at least one missing datum, and Hotelling's T^2 with data imputation by the EM algorithm (Dempster et al., 1977; Little and Rubin, 1987). We start by presenting characteristic features of the nonparametric permutation solution for multivariate testing problems in the presence of missing data missing completely at random (MCAR) and missing not at random (MNAR).

7.6 General Aspects of Permutation Testing with Missing Data

Here we introduce the permutation analysis of missing values for problems of multivariate testing of hypotheses when, in the null hypothesis, both observed and missing data are assumed to be exchangeable with respect to groups associated with treatment levels. Such multivariate testing problems are solvable by the NPC of dependent permutation tests. In particular, the hypotheses are presumed to be broken down into a set of sub-hypotheses, and related partial tests are assumed to satisfy the conditions discussed in Chapter 4: specifically, they are assumed to be marginally unbiased, significant for large values and consistent. Some of the ideas for the permutation analysis of missing values were first introduced in Pesarin (1990b, 1991a, 2001), Giraldo and Pesarin (1992, 1993), Giraldo et al. (1994) and Bertacche and Pesarin (1997).

Although some solutions presented in this chapter are exact, the most important ones are approximate because the permutation distributions of the test statistics concerned are not exactly invariant with respect to permutations of missing data, as we shall see. However, the approximations are quite accurate in all situations, provided that the number of effective data in all data permutations is not too small. To this end, without relevant loss of information we may remove from the permutation sample space, associated with the whole data set, all data permutations in which the actual sample sizes of really observed data are not sufficient for approximations. We must establish a kind of restriction on the permutation space, provided that this restriction does not imply biased effects on inferential conclusions.

7.6.1 Bibliographic Notes

In statistical analyses the problem of missing data is very common and has various solutions through a variety of approaches; see, for example, Rubin (1976), Dempster et al. (1977), Little (1982), Schemper (1984), Wei and Lachin (1984), Wei and Johnson (1985), Servy and Sen (1987),

Little and Rubin (1987), Laird (1988), Barton and Cramer (1989) and Maritz (1995). The large majority of contributions to this problem concern the estimation of parameters or functionals, such as regression coefficients and covariance matrices. Entsuah (1990) used a resampling procedure in an examination of treatment-related withdrawals from a study. An interesting resampling contribution was provided by Efron (1992), where imputation bootstrap methods were suggested, mainly for interval estimation.

Greenlees et al. (1982), in the framework of income surveys, developed the so-called selection model method for imputing missing values when the probability of non-response depends on the variables being imputed. The missing data problem is viewed as parallel to that of parametric estimation in a regression model with stochastic censoring of one dependent variable. Having solved this estimation problem by numerical maximization of log-likelihood, they use a prediction approach to impute logarithms of missing income values. Glynn et al. (1986) discuss the performance of two alternative approaches: the selection model approach of Greenlees et al. (1982) and the mixture model approach for obtaining estimates of means and regression coefficients when non-responses depend on the outcome variables.

The case of categorical variables with outcomes subject to non-ignorable non-responses is examined in Fuchs (1982), Baker and Laird (1988), Fay (1986), Nordheim (1984), Oshungade (1989), Mandel (1993), Phillips (1993), Park and Brown (1994), Conaway (1994) and, in a permutation framework, Bertacche and Pesarin (1997).

The main, fundamental assumption in almost all solutions from the literature is that missing data are *missing completely at random*. The definition of this is given in Rubin (1976). Roughly speaking, it means that the probability that an observation is missing does not depend on the value it would have assumed. When this rather strong assumption fails, it is necessary to specify the process producing the missing data, and the consequent analysis in general becomes much more difficult, especially within a parametric framework, unless the missing mechanism is well defined and related nuisance parameters well estimated from the data.

Section 7.7 illustrates when the process producing the missing data can be ignored and when the necessity to specify it occurs. Sections 7.8–7.10 discuss nonparametric solutions within a permutation framework for some typical multivariate testing problems with data missing completely at random and some quite general situations of data missing not at random. Section 7.11 presents an example of an application in which the probability of missing data depends on treatment levels, so that, in the alternative, data are not missing completely at random. Lastly, Section 7.12 presents a problem in which missing data are assumed to be missing completely at random.

7.7 On Missing Data Processes

In all kinds of problems, missing data are usually assumed to originate from an underlying random process, which may or may not be related to the observation process. Thus, within a parametric approach, in order to make valid inferences in the presence of missing data, this process must in general be properly specified. But when we assume that the probability of a datum being missing does not depend on its unobserved value, so that the missing data are missing at random, then we may ignore this process and so need not specify it. Rubin (1976) formalizes the definition of data missing at random and gives conditions in which it is appropriate to ignore the process producing missing data when making inferences about the distribution of observed data.

7.7.1 Data Missing Completely at Random

Let us use θ to denote the parameter regulating the distribution of the observable variable and ϕ to denote that of the missing data process; thus, the vector (θ, ϕ) identifies the whole probability distribution of observed data within a family $\mathcal{P}$ of non-degenerate distributions. The ignorability of the missing data process depends on the method of inference and on three conditions which the data generating process must satisfy.

According to Rubin (1976):

> The missing data are *missing at random* (MAR) if for each possible value of the parameter ϕ, the conditional probability of the observed pattern of missing data given the missing data and the value of the observed data, is the same for all possible values of the missing data. The observed data are *observed at random* (OAR) if for each possible value of the missing data and the parameter ϕ, the conditional probability of the observed pattern of missing data given the missing data and the observed data, is the same for all possible values of the observed data. The parameter ϕ is distinct from θ if there are no *a priori* ties, via parametric space restrictions or prior distributions, between ϕ and θ.

If the missing data are MAR and the observed data are OAR, the missing data are *missing completely at random* (MCAR). In this case, missingness does not depend on observed or unobserved values, and observed values may be considered as a random subsample of the complete data set. In these situations, therefore, it is appropriate to ignore the process that causes missing data when making inferences on θ.

7.7.2 *Data Missing Not at Random*

If the missing data are missing not at random (MNAR), then in order to make valid parametric inferences the missing data process must be properly specified. Typically in experimental situations this occurs when the treatment acts on the missingness mechanism through the probability of a datum being missing or the probability of its being observed. In most observational situations, missing data come from sample surveys where the circumstances behind non-responses are varied and complex. In general, it is very unlikely that a single model may correctly reflect all the implications of non-responses in all instances. Thus, the analysis of MNAR data is much more complicated than that of MCAR data because inferences must be made by taking into consideration the data set as a whole and by specifying a proper model for each specific situation. In any case, the specification of a model which correctly represents the missing data process seems the only way to eliminate the inferential bias caused by non-responses in a parametric framework.

In the literature, various models have been proposed, most of which concern cases in which non-responses are confined to a single variable.

7.8 The Permutation Approach

Without loss of generality, we discuss the permutation solution by referring to a one-way MANOVA layout, which is to test the equality of $C \geq 2$ V-dimensional distributions, where some of the data are supposed to be missing.

Suppose that we have C groups of V-dimensional responses $\mathbf{X}_j = \{\mathbf{X}_{ji} = (X_{hji},\ h = 1, \ldots, V),\ i = 1, \ldots, n_j\}$, $j = 1, \ldots, C$, respectively with distribution function P_j, $\mathbf{X}_{ji} \in \mathcal{R}^V$, where $n = \sum_j n_j$ is the total sample size. As usual, we assume that the null hypothesis is $H_0 : \{P_1 = \ldots = P_C = P\} = \{\mathbf{X}_1 \stackrel{d}{=} \ldots \stackrel{d}{=} \mathbf{X}_C\}$ and the alternative is $H_1 : \{H_0 \text{ is not true}\}$.

As regards the data, we assume that they are exchangeable with respect to C groups in the null hypothesis. Note that this requirement concerns both observed and missing data. In addition, we also assume that the model for treatment effects is such that resulting CDFs satisfy the pairwise dominance condition, so that locations of suitable transformations φ_h, $h = 1, \ldots, V$, of the data are useful for discrimination, where φ_h may be specific to the hth variable. This assumption leads us to consider sample means of transformed data as proper indicators for treatment effects. The reason for this kind of statistical indicator, and consequently for this kind of assumption, is that in this situation we are able to derive an effective solution, as we shall see in Section 7.10. Therefore, we assume that the analysis is based on the transformed data

$$\mathbf{Y} = \{Y_{hji} = \varphi_h(X_{hji}),\ i = 1, \ldots, n_j,\ j = 1, \ldots, C,\ h = 1, \ldots, V\}.$$

Hence, consequent permutation partial tests should be based on proper functions of sample totals $S^*_{hj} = \sum_{i \le n_j} Y^*_{hji}$, $j = 1, \ldots, C$, $h = 1, \ldots, V$.

In order to take into account that, for whatever reason, some of the data are missing, we must also consider the associated *inclusion indicator*, that is, the complementary set of binary values

$$\mathbf{O} = \{O_{hji},\ i = 1, \ldots, n_j,\ j = 1, \ldots, C,\ h = 1, \ldots, V\},$$

where $O_{hji} = 1$ if X_{hji} has been observed and collected, otherwise $O_{hji} = 0$. Thus, the inclusion indicator $\mathbf{O}$ represents the *observed configuration* in the data set. Hence, the *whole set of observed data* is summarized by the pair of associated matrices $(\mathbf{Y}, \mathbf{O})$. Moreover, we define by $\nu_{hj} = \sum_i O_{hji}$, $j = 1, \ldots, C$, $h = 1, \ldots, V$, the *actual sample size* of the really observed data in the jth group relative to the hth variable, and by $\nu_{h\cdot} = \sum_j \nu_{hj}$, $h = 1, \ldots, V$, the *total actual sample size* of the really observed data relative to the hth variable.

Thus, assuming that data are jointly exchangeable in the null hypothesis with respect to groups on both the $\mathbf{Y}$ and $\mathbf{O}$ components, we may express the hypotheses of interest as

$$H_0 : \left\{(\mathbf{Y}_1, \mathbf{O}_1) \stackrel{d}{=} \ldots \stackrel{d}{=} (\mathbf{Y}_C, \mathbf{O}_C)\right\}$$

against the alternative $H_1 : \{H_0 \text{ is not true}\}$.

The hypotheses and assumptions are such that the permutation testing principle applies. Note that in the present context P represents the joint multivariate distribution in the null hypothesis of $(\mathbf{Y}_j, \mathbf{O}_j)$, $j = 1, \ldots, C$. Also note that, with obvious notation, we may write $P = P_{\mathbf{O}} \cdot P_{\mathbf{Y}|\mathbf{O}}$, where the roles of the inclusion indicator $\mathbf{O}$ and of observed data $\mathbf{Y}$, conditionally on $\mathbf{O}$, are emphasized. It should be noted that the whole set of observed data $(\mathbf{Y}, \mathbf{O})$ in the null hypothesis is a set of jointly sufficient statistics for the underlying observed and missing data processes.

7.8.1 Deletion, Imputation and Intention to Treat Strategies

Most testing solutions within a parametric framework are based either on the so-called *deletion principle* or on *imputation methods*, both useful in MCAR situations. The former considers the part of the data set resulting from removing all incomplete vectors; the latter involves replacing missing data by means of suitable functions of actually observed data, in accordance with an imputation criterion.

Among deletion methods, we recall *listwise deletion* and *pairwise deletion* of the missing values. According to the listwise deletion approach (also called *complete case analysis*) individuals with at least one missing value are removed from the analysis. As a result, there is a considerable loss of data and of information. In contrast, pairwise deletion removes only specific missing values from the analysis and not the entire case, as the listwise method does. For example, in a multivariate problem, if correlations are of interest, then bivariate correlations will be calculated using only cases with valid data for those two variables. As a result, correlations will be calculated using different sample sizes.

Many imputation techniques have been proposed in literature. Some of the most commonly used, among them *hot-deck imputation* and *regression imputation*, will be discussed later. In a way, all these methods aim to satisfy the idea that analyses should be done on complete data sets because, in general, all parametric approaches require such a condition.

Furthermore, in the parametric framework, both adherence and missing data issues are often addressed using an approach based on the *intention to treat* (ITT) principle. We recall that in clinical research adherence and data missingness are distinct problems. Even patients who have dropped out may re-enter the study for final measurements, while adherent patients may be missed during the course of the study. In longitudinal trials, patients, randomized to receive a specific treatment, are repeatedly measured throughout the duration of the study. As unfortunately frequently happens, many patients may be missed – for example, they may die, miss a visit, stop taking their medication or just leave the trial at any time. Similar events must be considered when comparing treatment performances, especially if they are suspected of influencing the final outcome. Both explanatory and pragmatic analyses may be carried out when such off-treatment measures arise. While explanatory analyses aim to detect the biological effects of treatment, excluding non-compliant patients from analysis if needed, pragmatic analyses focus on the utility and effectiveness of the therapy. ITT analysis is an example of the latter. Hence, following the ITT principle, missing data should be replaced with suitable values. This problem may be dealt with using the *last observation carried forward* (LOCF) approach, where a patient's last measurement is used as the final one. Another strategy is to extrapolate missing values from the available data (e.g. regression substitution). Among imputation methods, we recall the hot-hand or hot-deck approach, whereby patients with missing data are assigned data values from other patients who share the same characteristics, such as age, sex, height or weight. A traditional approach is the so-called *mean substitution*, replacing the missing values with the mean of the observed values. Moreover, *propensity score* methods for missing data may also be used. All these approaches are simply different imputation methods, making use of the available observations to predict values for the missing ones.

In summary, according to the ITT principle, treatment groups are compared in terms of the treatment to which they were randomly assigned at the beginning of the study, regardless of deviations from clinical trial protocol or patient non-compliance. The widespread use of this approach is mostly due to the fact that ITT analysis preserves the randomized trial population (e.g. including non-compliant patients, irrespective of protocol deviations), thus providing an unbiased estimate of treatment effect and giving a clear overview of the real clinical situation. In this way, it also avoids potential overestimates of treatment effectiveness deriving from the removal of non-compliant patients – that is, ITT is a conservative data analysis approach. In conclusion, a 'quasi' ITT approach, excluding ineligible patients from the study, is recommended, especially when the proportion of ineligible participants is very high (Heritier et al., 2003).

However, the permutation approach considers data *as they are*, without deletion or imputation. When deleting incomplete vectors we also remove the possible partial information they contain, which may be valuable and useful for analysis. Furthermore, imputation methods may introduce information bias which may negatively influence the analysis.

7.8.2 Breaking Down the Hypotheses

The complexity of this testing problem is such that it is very difficult to find a single overall test statistic. In any case, as previously shown, the problem may be tackled by means of the NPC of a set of dependent permutation tests.

Let us suppose that in the testing problem we are interested in V different aspects (note that this corresponds to one aspect for each component variable; for an extension, see the example in Section 7.11). Hence, we consider a set of V partial tests which, due to our assumptions, may be simply reduced to V one-way bivariate ANOVAs, one for each marginal component in $(\mathbf{Y}, \mathbf{O})$, followed by their NPC. To this end, we observe that the null hypothesis may be equivalently written in the form

$$H_0 : \left\{ \bigcap_{h=1}^{V} \left[(Y_{h1}, O_{h1}) \overset{d}{=} \ldots \overset{d}{=} (Y_{hC}, O_{hC}) \right] \right\} = \left\{ \bigcap_{h=1}^{V} H_{0h} \right\},$$

where, as usual, a suitable and meaningful breakdown of H_0 is emphasized. Hence, the hypothesis H_0 against H_1 is broken down into V sub-hypotheses H_{0h} against H_{1h}, $h = 1, \ldots, V$, in such a way that H_0 is true if all the H_{0h} are jointly true and H_1 is true if at least one among the H_{1h} is true, so that H_1: $\{ \bigcup_h H_{1h} \}$.

Thus, to test H_0 against H_1, we consider a V-dimensional vector of real-valued test statistics $\mathbf{T} = \{T_1, \ldots, T_V\}$, the hth component of which is the univariate partial test for the hth sub-hypothesis H_{0h} against H_{1h}. Without loss of generality, we assume that partial tests are non-degenerate, marginally unbiased, consistent, and significant for large values. Hence, the combined test is a function of V dependent partial tests. Of course, the combination must be nonparametric, particularly with regard to the underlying dependence relation structure, because in this setting only very rarely may the dependence structure among partial tests be effectively analysed.

7.9 The Structure of Testing Problems

7.9.1 *Hypotheses for MNAR Models*

Let us first consider an MNAR model for missing data. If the missing data are MNAR, then H_0 must take into consideration the homogeneity in distribution with respect to the C samples of the actually observed and collected data $\mathbf{Y}$ jointly with that associated with the missing data process $\mathbf{O}$ because in this setting it is assumed that, in the alternative, the symbolic treatment may also influence missingness. In fact, the treatment may affect the distributions of both variables $\mathbf{Y}$ and of the inclusion indicator $\mathbf{O}$. Hence, in the case of MNAR missing data, the null hypothesis requires the joint distributional equality of the missing data process in the C groups, giving rise to $\mathbf{O}$, and of response variables $\mathbf{Y}$ conditional on $\mathbf{O}$, that is,

$$H_0 : \left\{ \left[\mathbf{O}_1 \overset{d}{=} \ldots \overset{d}{=} \mathbf{O}_C \right] \bigcap \left[\left(\mathbf{Y}_1 \overset{d}{=} \ldots \overset{d}{=} \mathbf{Y}_C \right) | \mathbf{O} \right] \right\}.$$

The assumed exchangeability in the null hypothesis of the n individual data vectors in $(\mathbf{Y}, \mathbf{O})$ with respect to the C groups implies that the treatment effects are null on *all* observed and unobserved variables. In other words, we assume that there is no difference in distribution for the multivariate inclusion indicator variables $\mathbf{O}_j$, $j = 1, \ldots, C$, and, conditionally on $\mathbf{O}$, for actually observed variables $\mathbf{Y}$. As a consequence, it is not necessary to specify both the missing data process and the data distribution, provided that marginally unbiased permutation tests are available. In particular, it is not necessary to specify the dependence relation structure in $(\mathbf{Y}, \mathbf{O})$ because it is nonparametrically processed. In this framework, the hypotheses may be broken down into the $2V$ sub-hypotheses

$$H_0 : \left\{ \left[\bigcap_h \left(O_{h1} \overset{d}{=} \ldots \overset{d}{=} O_{hC} \right) \right] \bigcap \left[\bigcap_h \left(Y_{h1} \overset{d}{=} \ldots \overset{d}{=} Y_{hC} \right) | \mathbf{O} \right] \right\}$$

$$= \left\{ H_0^{\mathbf{O}} \bigcap H_0^{\mathbf{Y}|\mathbf{O}} \right\} = \left\{ \left(\bigcap_h H_{0h}^{\mathbf{O}} \right) \bigcap \left(\bigcap_h H_{0h}^{\mathbf{Y}|\mathbf{O}} \right) \right\}$$

against

$$H_1 : \left\{ \left(\bigcup_h H_{1h}^{\mathbf{O}} \right) \bigcup \left(\bigcup_h H_{1h}^{\mathbf{Y}|\mathbf{O}} \right) \right\},$$

where $H_{0h}^{\mathbf{O}}$ indicates the equality in distribution among the C levels of the hth marginal component of the inclusion (missing) indicator process, and $H_{0h}^{\mathbf{Y}|\mathbf{O}}$ indicates the equality in distribution of the hth component of $\mathbf{Y}$, conditional on $\mathbf{O}$.

For each of the V sub-hypotheses $H_{0h}^{\mathbf{O}}$ a permutation test statistic such as Pearson's X^2, or other suitable tests for proper testing with binary categorical data, are generally appropriate (for testing with categorical variables see Cressie and Read, 1988; Agresti, 2002; and Chapter 6 above). For each of the k sub-hypotheses $H_{0h}^{\mathbf{Y}|\mathbf{O}}$, $\mathbf{O}$ is fixed at its observed value, so that we may proceed conditionally. Proper partial tests are discussed in Section 7.10.

Remark 1. The NPC of $2V$ partial tests may be done in at least three different ways: (a) using one single combining function on all $2V$ partial tests such as $T_a'' = \psi(\lambda_1^{\mathbf{O}}, \ldots, \lambda_V^{\mathbf{O}}; \lambda_1^{\mathbf{Y}|\mathbf{O}}, \ldots, \lambda_V^{\mathbf{Y}|\mathbf{O}})$; (b) using V second-order combinations, one for each component variable, $T_{bh}'' = \psi_h(\lambda_h^{\mathbf{O}}; \lambda_h^{\mathbf{Y}|\mathbf{O}})$, $h = 1, \ldots, V$, followed by a third order combination, $T_b''' = \psi(\lambda_{b1}'', \ldots, \lambda_{bV}'')$; (c) using two second-order combinations, $T_{c\mathbf{O}}'' = \psi_{\mathbf{O}}(\lambda_1^{\mathbf{O}}, \ldots, \lambda_V^{\mathbf{O}})$ and $T_{c\mathbf{Y}|\mathbf{O}}'' = \psi_{\mathbf{Y}}(\lambda_1^{\mathbf{Y}|\mathbf{O}}, \ldots, \lambda_V^{\mathbf{Y}|\mathbf{O}})$, respectively on the inclusion indicator $\mathbf{O}$ and on the actually observed data $(\mathbf{Y}|\mathbf{O})$, followed by a third order combination, $T_c''' = \psi(\lambda_{c\mathbf{O}}''; \lambda_{c\mathbf{Y}|\mathbf{O}}'')$. Note that if in all phases and in each of the three methods of combination the same combining function ψ is used, then T_a'', T_b''' and T_c''' are almost permutationally equivalent, except for approximations due to finite simulations and nonlinearity of combining functions.

In addition, due to assumptions on partial tests, the second-level partial test $T_{c\mathbf{Y}|\mathbf{O}}''$ is marginally unbiased for

$$H_0^{\mathbf{Y}|\mathbf{O}} : \left\{ \left[\left(\mathbf{Y}_1 \overset{d}{=} \ldots \overset{d}{=} \mathbf{Y}_C \right) |\mathbf{O} \right] \right\}$$

and so allows for a form of separate testing on actually observed data, conditional on $\mathbf{O}$, even though missing data are non-MAR. This is useful in many circumstances, especially when interest centres on actually observed data. Note that this solution coincides formally with that of MCAR problems (see the next subsection).

7.9.2 *Hypotheses for MCAR Models*

If the missing data are MCAR, we may, according to Rubin, proceed conditionally with respect to the observed inclusion indicator $\mathbf{O}$ and ignore $H_0^{\mathbf{O}}$ because in this context we have assumed that $\mathbf{O}$ does not provide any discriminative information about treatment effects. Thus, as sub-hypotheses on $\mathbf{O}$ are true by assumption, $H_0^{\mathbf{O}} : \{\mathbf{O}_1 \overset{d}{=} \ldots \overset{d}{=} \mathbf{O}_C\}$ may be ignored. Hence, we may write the null hypothesis in the relatively simpler form

$$H_0 = H_0^{\mathbf{Y}|\mathbf{O}} : \left\{ \bigcap_h \left[\left(Y_{h1} \overset{d}{=} \ldots \overset{d}{=} Y_{hC} \right) |\mathbf{O} \right] \right\} = \left\{ \bigcap_h H_{0h}^{\mathbf{Y}|\mathbf{O}} \right\}$$

against

$$H_1 : \left\{ \bigcup_h H_{1h}^{\mathbf{Y}|\mathbf{O}} \right\},$$

where the notation is the same as in Section 7.9.1. Of course, this problem is solved by the NPC

$$T'' = T_{c\mathbf{Y}|\mathbf{O}}'' = \psi_{\mathbf{Y}} \left(\lambda_1^{\mathbf{Y}|\mathbf{O}}, \ldots, \lambda_V^{\mathbf{Y}|\mathbf{O}} \right).$$

7.9.3 Permutation Structure with Missing Values

In order to deal with this problem using a permutation strategy, it is necessary to consider the role of permuted inclusion indicators $\mathbf{O}^* = \{O^*_{hji}, i = 1, \ldots, n_j, j = 1, \ldots, C, h = 1, \ldots, V\}$, especially with respect to numbers of missing data, in all points of the permutation sample space $(\mathcal{Y}, \mathcal{O})_{/(\mathbf{Y},\mathbf{O})}$ associated with the pair $(\mathbf{Y}, \mathbf{O})$.

The permutation actual sample sizes of really valid data for each component variable within each group, $\nu^*_{hj} = \sum_i O^*_{hji}$, $j = 1, \ldots, C$, $h = 1, \ldots, V$, vary according to the random attribution of unit vectors, and of relative missing data, to the C groups. This, of course, is because units with missing data participate in the permutation mechanism as well as all other units.

Table 7.1 shows an example of how the permutation strategy applies to a bivariate variable (W_1, W_2) when a generic permutation of data is considered and where some of the data are missing. In the example there are two groups with $n_1 = n_2 = 8$, the missing data are denoted by '?', and the permutation considered is $\mathbf{u}^* = (9, 15, 5, 6, 11, 13, 3, 14, 1, 8, 16, 10, 7, 2, 4, 12)$.

As the actual sample sizes of valid data vary according to the permutation of units with missing data, the $V \times C$ matrix $\{\nu^*_{hj}, j = 1, \ldots, C, h = 1, \ldots, V\}$ is not invariant in the permutation sample space. In other words, and focusing on the hth variable, if for some j we have $\nu_{hj} < n_{hj}$, so that the related number of missing data is positive, then the permutation probability $\Pr\{\bigcap_j(\nu^*_{hj} = \nu_{hj})\}$ of finding the C permutation of actual sample sizes of valid data all equal to those in $(\mathbf{Y}, \mathbf{O})$ is strictly less than one.

For instance, in the example, the actual sample size of valid data of variable V in the first group changes from three to four: $\nu_{11} = 3$, while $\nu^*_{11} = 4$, etc. In practice, for the data set of the same example, the permutation actual sample size of valid data related to the first group satisfies the inequalities $4 \leq \nu^*_{11} \leq 8$ and $3 \leq \nu^*_{21} \leq 8$.

Thus, the key to a suitable solution is to use partial test statistics, the permutation distributions of which are at least approximately invariant with respect to the permutation of actual sample sizes of valid data. This is done in the next section. However, these tests are also presented in Pesarin (2001).

Table 7.1 Example of a permutation with missing values

	i	W_1	W_2	u^*_i	W^*_1	W^*_2
	1	W_1	W_2	9	?	W_2
	2	W_1	?	15	W_1	W_2
	3	?	W_2	5	?	W_2
I	4	W_1	W_2	6	?	W_2
	5	?	W_2	11	W_1	?
	6	?	W_2	13	W_1	?
	7	W_1	W_2	3	?	W_2
	8	W_1	W_2	14	W_1	W_2
	9	?	W_2	1	W_1	W_2
	10	W_1	W_2	8	W_1	W_2
	11	W_1	?	16	W_1	?
II	12	W_1	?	10	W_1	W_2
	13	W_1	?	7	W_1	W_2
	14	W_1	W_2	2	W_1	?
	15	W_1	W_2	4	W_1	W_2
	16	W_1	?	12	W_1	?

Remark 1. When the missing data process is MCAR by assumption, so that treatment has no influence on missingness or observation probabilities and data are unidimensional, then we may ignore missing data, simply by discarding them from the data set, when testing $H_0 : \{(Y_1 \stackrel{d}{=} \ldots \stackrel{d}{=} Y_C)|\mathbf{O}\}$ against $H_1 : \{H_0$ is not true$\}$. Thus, actual sample sizes are reduced to $\nu_j \leq n_j$ units. In such a situation, when working with reduced data sets, we may obtain exact permutation solutions. Also note that, in this context, a univariate MCAR model corresponds to a *random censoring model*.

7.10 Permutation Analysis of Missing Values

7.10.1 Partitioning the Permutation Sample Space

Here we summarize the main arguments for an approximate permutation solution. Let us first consider an MCAR model and a vector of partial test statistics $\mathbf{T}$, based on functions of sample totals of valid data, and denote its multivariate permutation distribution by $F[\mathbf{t}|(\mathbf{Y}, \mathbf{O})]$, $\mathbf{t} \in \mathcal{R}^V$. We observe that the set of permutations $\mathbf{O}^*$ of $\mathbf{O}$, that is, the set of possible permuted inclusion indicators according to the random attribution of data to the C groups, induces a partition into sub-orbits on the whole permutation sample space $(\mathcal{Y}, \mathcal{O})_{/(\mathbf{Y},\mathbf{O})}$. These sub-orbits are characterized by points which exhibit the same matrix of permutation actual sample sizes of valid data $\{\nu_{hj}^*, j = 1, \ldots, C, h = 1, \ldots, V\}$.

This partition is displayed in Figure 7.1, which shows that the two points $(\mathbf{Y}_1^*, \mathbf{O}_1^*)$ and $(\mathbf{Y}_2^*, \mathbf{O}_2^*)$ lie on the same sub-orbit if the respective permutation actual sample sizes of valid data $\nu_{1hj}^* = \sum_i O_{1hji}^*$ and $\nu_{2hj}^* = \sum_i O_{2hji}^*$ are equal for every h and j, $h = 1, \ldots, V$, $j = 1, \ldots, C$.

Of course, if the permutation sub-distributions of the whole matrix of sample totals $\{S_{hj}^* = \sum_i Y_{hji}^* \cdot O_{hji}^*, j = 1, \ldots, C, h = 1, \ldots, V\}$, where it is assumed that $O_{hji}^* = 0$ implies $Y_{hji}^* \cdot O_{hji}^* = 0$, are invariant with respect to the sub-orbits induced by $\mathbf{O}^*$, then we may evaluate $F[\mathbf{t}|(\mathbf{Y}, \mathbf{O})]$ for instance by a simple CMC procedure, that is, by ignoring the partition into induced sub-orbits.

The distributional invariance with respect to permuted inclusion indicators $\mathbf{O}^*$ of sample totals $\mathbf{S}^*$ implies that the equality

$$F[\mathbf{t}|(\mathbf{Y}, \mathbf{O})] = F[\mathbf{t}|(\mathbf{Y}, \mathbf{O}^*)]$$

is satisfied for every $\mathbf{t} \in \mathcal{R}^V$, for every specific permutation $\mathbf{O}^*$ of $\mathbf{O}$, and for all data sets $\mathbf{Y}$. Except for some special situations (for one of these see Section 7.12) or for very large sample sizes, this distributional invariance is extremely difficult to satisfy exactly when $V > 1$ and for all finite sample sizes. Of course, when $V = 1$, so that the problem is one-dimensional, this distributional invariance

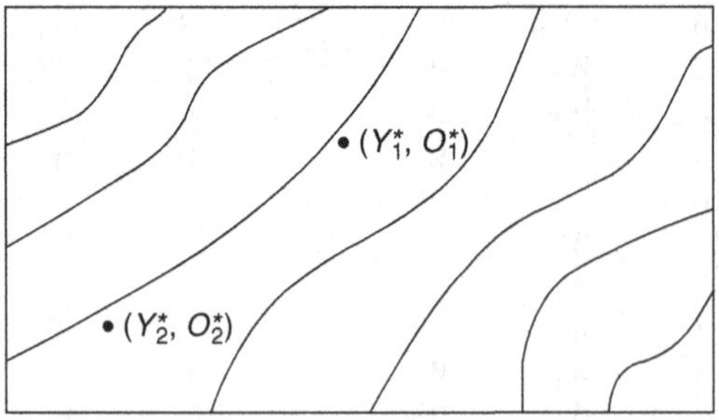

Figure 7.1 Partition of $(\mathcal{Y}, \mathcal{O})_{/(\mathbf{Y},O)}$ into sub-orbits induced by $\mathbf{O}^*$

may become exact in MCAR models because, conditionally, we are allowed to ignore missingness by removing all unobserved units from the data set.

Moreover, when problems involve multivariate paired data, so that numbers of missing differences are permutationally invariant quantities, then related tests become exact (see Section 7.12). Therefore, in general, we must look for approximate solutions.

7.10.2 Solution for Two-Sample MCAR Problems

Let us denote by $\boldsymbol{\nu} = \{\nu_{hj}, j = 1, \ldots, C, h = 1, \ldots, V\}$ the $V \times C$ matrix of actual sample sizes of valid data in the observed inclusion indicator $\mathbf{O}$. We have assumed that our test statistics are based on permutation sample totals of valid data $\{S^*_{hj} = \sum_i Y^*_{hji} \cdot O^*_{hji}, j = 1, \ldots, C, h = 1, \ldots, V\}$. Thus, supposing that the permutation distribution of the sample total S^*_{hj}, conditional on the whole data set $(\mathbf{Y}, \mathbf{O})$ considered as a finite population, depends essentially on the number ν^*_{hj} of summands, the previous distributional equality becomes equivalent to

$$F[\mathbf{t}|(\mathbf{Y}, \boldsymbol{\nu})] = F[\mathbf{t}|(\mathbf{Y}, \boldsymbol{\nu}^*)],$$

where $\boldsymbol{\nu}^* = \{\nu^*_{hj},\ j = 1, \ldots, C,\ h = 1, \ldots, V\}$ represents the $V \times C$ matrix of permutation of actual sample sizes of valid data associated with $\mathbf{O}^*$. Hence, we have to find test statistics the permutation null sub-distributions of which are invariant with respect to $\boldsymbol{\nu}^*$ and for all $\mathbf{Y}$.

Again, only very rarely can this condition be satisfied exactly (see Section 7.12 for one situation in which this condition is achieved). Therefore, in general, we must consider an approximate solution based on the distributional invariance of means and variances of partial tests $\mathbf{T}$ with respect to $\boldsymbol{\nu}^*$. In other words, we must look for statistics the means and variances of which are invariant with respect to the sub-orbits induced by $\mathbf{O}^*$ on permutation sample space $(\mathcal{Y}, \mathcal{O})_{/(\mathbf{Y},\mathbf{O})}$.

For simplicity, and without loss of generality, let us first consider the case of a univariate variable Y, so that we have only one test statistic T, and assume that the permutation tests to be considered are based on univariate sample totals of valid data, $S^*_j = \sum_i Y^*_{ji} \cdot O^*_{ji}$, $j = 1, \ldots, C$.

Of course, the overall total $S = \sum_j S_j$, which is assumed to be a non-null quantity, is permutationally invariant because in $(\mathcal{Y}, \mathcal{O})_{/(\mathbf{Y},\mathbf{O})}$, the equation

$$S = \sum\nolimits_{ji} Y_{ji} \cdot O_{ji} = \sum\nolimits_j S^*_j$$

is always satisfied.

Let us now consider the two-sample case ($C = 2$) and assume that the test statistic for $H_0^{\mathbf{Y}|\mathbf{O}}$ against $H_1^{\mathbf{Y}|\mathbf{O}}$ is a linear combination of S^*_1 and S^*_2. Thus, the test is expressed in the form

$$T^*(a^*, b^*|\boldsymbol{\nu}^*) = a^* \cdot S^*_1 - b^* \cdot S^*_2,$$

where a^* and b^* are two coefficients which are independent of the actually observed data $\mathbf{Y}$ but which may be permutationally non-invariant. These coefficients must be determined assuming that, in the null hypothesis, the variance $\mathbb{V}[T^*(a^*, b^*)|\boldsymbol{\nu}^*] = \zeta^2$ is constant, in the sense that it is independent of the permutation of actual sample sizes ν^*_j, $j = 1, 2$, and that the mean values should identically satisfy the condition $\mathbb{E}[T^*(a^*, b^*)|\boldsymbol{\nu}^*] = 0$.

In accordance with the technique of without-replacement random sampling from $(\mathbf{Y}, \mathbf{O})$ which, due to conditioning, is assumed to play the role of a finite population (see Section 2.1.3), we can write the following set of equations:

$$\nu^*_1 + \nu^*_2 = \nu,$$

$$S^*_1 + S^*_2 = S,$$

$$\mathbb{E}(S_j^*) = S \cdot \nu_j^*/\nu, \quad j = 1, 2,$$
$$\mathbb{V}(S_j^*) = \sigma^2 \cdot \nu_j^*(\nu - \nu_j^*)/(\nu - 1) = \mathbb{V}(\nu^*), \quad j = 1, 2,$$

where V is a positive function, and $\nu = \nu_1 + \nu_2$, $S = S_1 + S_2$ and $\sigma^2 = \sum_{ji}(Y_{ji} - S/\nu)^2 \cdot O_{ji}/\nu$ are permutationally invariant non-null quantities. Thus, for any given pair of positive permutation actual sample sizes (ν_1^*, ν_2^*), the two permutation sampling totals S_1^* and S_2^* have the same variance and their correlation coefficient is $\rho(S_1^*, S_2^*) = -1$, because their sum S is a permutation invariant quantity. Hence, we may write

$$\mathbb{E}[T^*(a^*, b^*)] = a^* \cdot S \cdot \nu_1^* - b^* \cdot S \cdot \nu_2^* = 0,$$
$$\mathbb{V}[T^*(a^*, b^*)] = a^{*2} V(\boldsymbol{\nu}^*) + 2a^* b^* V(\boldsymbol{\nu}^*) + b^{*2} V(\boldsymbol{\nu}^*) = (a^* + b^*)^2 V(\boldsymbol{\nu}^*).$$

The solutions to these equations are $a^* = (\nu_2^*/\nu_1^*)^{1/2}$ and $b^* = (\nu_1^*/\nu_2^*)^{1/2}$, ignoring an inessential positive coefficient.

Hence, for $C = 2$ and $V = 1$, the test statistic, the sub-distributions of which are approximately invariant with respect to permutation of actual sample sizes of valid data because they are permutationally invariant in mean value and variance, takes the form

$$T^* = S_1^* \cdot (\nu_2^*/\nu_1^*)^{1/2} - S_2^* \cdot (\nu_1^*/\nu_2^*)^{1/2}.$$

It is worth noting that, when there are no missing values, so that $\nu_j^* = n_j$, $j = 1, 2$, the latter test is permutationally equivalent to the standard two-sample permutation test for comparison of locations $T^* \approx \sum_i Y_{1i}^*$.

Remark 1. Since the test T^* is approximately exact, it is also approximately unbiased. In order to prove consistency, we recall that the size of missing data ν^* may diverge, provided that, as n tends to infinity, $\lim \nu^*/n \in (0, 1)$. Details are left to the reader.

Remark 2. In order for the given solution to be well defined, we must assume that ν_1^* and ν_2^* are jointly positive. This implies that, in general, we must consider a sort of restricted permutation strategy which consists of discarding from the analysis all points of the permutation sample space $(\mathcal{Y}, \mathcal{O})_{/(\mathbf{Y},\mathbf{O})}$ in which even a single component of the permutation matrix $\boldsymbol{\nu}^*$, of actual sample sizes of valid data, is zero. Of course, this kind of restriction has no effect on inferential conclusions.

7.10.3 Extensions to Multivariate C-Sample Problems

In the case of $C > 2$ and again with $V = 1$, one approximate solution is

$$T_C^* = \sum_{j=1}^{C} \left\{ S_j^* \cdot \left(\frac{\nu - \nu_j^*}{\nu_j^*} \right)^{1/2} - (S - S_j^*) \cdot \left(\frac{\nu_j^*}{\nu - \nu_j^*} \right)^{1/2} \right\}^2.$$

This test statistic may be seen as a direct combination of C partial dependent tests, each obtained by a permutation comparison of the jth group with respect to all other $C - 1$ groups pooled together. Also, in the case of complete data, when there are no missing values, this test is equivalent to the permutation test for a standard one-way ANOVA layout, provided that sample sizes are balanced, $n_j = m$, $j = 1, \ldots, C$, whereas in the unbalanced cases the two solutions, although not coincident, are very close to each other.

One more solution may be obtained by the direct NPC of all pairwise comparisons:

$$T_{2C}^{*} = \sum_{r<j} \left(T_{rj}^{*}\right)^{2},$$

where $T_{rj}^{*} = S_r^{*} \cdot (\nu_j^{*}/\nu_r^{*})^{1/2} - S_j^{*} \cdot (\nu_r^{*}/\nu_j^{*})^{1/2}$, $1 \leq r < j \leq C$.

Of course, if $V > 1$, a nonparametric combination will result. Hence, to test $H_0 : \{\bigcap_h H_{0h}^{\mathbf{Y}|\mathbf{O}}\}$ against $H_1 : \{\bigcup_h H_{1h}^{\mathbf{Y}|\mathbf{O}}\}$, the solution becomes $T'' = \psi(\lambda_1, \ldots, \lambda_V)$, where ψ is any member of the class $\mathcal{C}$, and λ_h is the partial p-value of either

$$T_{Ch}^{*} = \sum_{j=1}^{C} \left\{ S_{hj}^{*} \cdot \left(\frac{\nu_h - \nu_{hj}^{*}}{\nu_{hj}^{*}} \right)^{1/2} - (S_h - S_{hj}^{*}) \cdot \left(\frac{\nu_{hj}^{*}}{\nu_h - \nu_{hj}^{*}} \right)^{1/2} \right\}^{2}$$

or

$$T_{2Ch}^{*} = \sum_{r<j} \left(S_{hr}^{*} \cdot \left(\frac{\nu_{hj}^{*}}{\nu_{hr}^{*}} \right)^{1/2} - S_{hj}^{*} \cdot \left(\frac{\nu_{hr}^{*}}{\nu_{hj}^{*}} \right)^{1/2} \right)^{2},$$

each relative to the hth component variable, $h = 1, \ldots, V$.

7.10.4 Extension to MNAR Models

For MNAR models, again in a nonparametric way, we must also combine the V test statistics on the components of the inclusion indicator $\mathbf{O}$, provided that all partial tests are marginally unbiased (see Section 4.2.1). More specifically, to test $H_0 : \{[\bigcap_h H_{0h}^{\mathbf{O}}] \bigcap [\bigcap_h H_{0h}^{\mathbf{Y}|\mathbf{O}}]\}$ against $H_1 : \{[\bigcup_h H_{1h}^{\mathbf{O}}] \bigcup [\bigcup_h H_{1h}^{\mathbf{Y}|\mathbf{O}}]\}$ we must now combine V tests $T_h^{*\mathbf{O}}$ and V tests $T_h^{*\mathbf{Y}|\mathbf{O}}$, $h = 1, \ldots, V$. Hence (with obvious notation),

$$T'' = \psi(\lambda_1^{\mathbf{O}}, \ldots, \lambda_V^{\mathbf{O}}; \lambda_1^{\mathbf{Y}|\mathbf{O}}, \ldots, \lambda_V^{\mathbf{Y}|\mathbf{O}}).$$

For each of the V sub-hypotheses $H_{0h}^{\mathbf{O}}$ against $H_{1h}^{\mathbf{O}}$, a permutation statistic such as Pearson's chi-square or any other suitable test statistic for proper testing of categorical data may be used (for instance, when $C = 2$ and restricted alternatives are of interest, Fisher's exact probability test may be appropriate). Instead, for V sub-hypotheses $H_{0h}^{\mathbf{Y}|\mathbf{O}}$ against $H_{1h}^{\mathbf{Y}|\mathbf{O}}$, proper tests are discussed in Section 7.10.3.

This combined permutation test has good general asymptotic properties. In particular, under very mild conditions, if best univariate partial tests are used, then the combined test is asymptotically best in the same sense.

Remark 1. Partial tests $T_h^{*\mathbf{O}}$ on the components of $\mathbf{O}$ are exact, unbiased and consistent, whereas $T_h^{*\mathbf{Y}|\mathbf{O}}$ on the components of $\mathbf{Y}$ are consistent but approximately exact and unbiased, so that the combined test T'' is consistent and approximately exact and unbiased for all $\psi \in \mathcal{C}$ (see Remark 1, 7.10.2).

Remark 2. It is easy to extend the solution to problems with multivariate monotonic stochastic ordering defined by

$$H_0 : \left\{ (\mathbf{X}_1, \mathbf{O}_1) \stackrel{d}{=} \ldots \stackrel{d}{=} (\mathbf{X}_C, \mathbf{O}_C) \right\}$$

against for instance

$$H_1 : \left\{(\mathbf{X}_1, \mathbf{O}_1) \overset{d}{\leq} \ldots \overset{d}{\leq} (\mathbf{X}_C, \mathbf{O}_C)\right\},$$

where at least one inequality is strict. Details are left to the reader.

7.11 Germina Data: An Example of an MNAR Model

The illustrative example in this section, with a few changes, comes from Pesarin (1991b). In it $n_1 = 20$ seeds of a given plant are sown in a standard (untreated) plot of soil and $n_2 = 20$ more seeds are sown in soil treated with a fertilizer. By assumption, seeds are randomized to treatments. The expected treatment effects of interest are: (a) improved probability of germination, denoted by the variable O; (b) improved index of production, indicated by the bivariate quantitative response (X, Y), where X is an index measured by the weight of adult plants and Y is an index of the total surface area of their leaves.

Note that O is typically binary categorical, taking value 1 if the seeds germinate and 0 otherwise. X and Y are positive quantities when $O = 1$. The underlying non-degenerate distribution P is presumed to be unknown. O also plays the role of inclusion indicator of missing values with respect to quantitative responses (X, Y); the pair (X_{ji}, Y_{ji}) is actually observed on the jith unit if $O = 1$, whereas if $O = 0$, that is, the jith unit did not germinate, (X_{ji}, Y_{ji}) cannot be observed and so appears to be missing or censored, $i = 1, \ldots, n$, $j = 1, 2$. In a way, the data here appears to be jointly censored by a random mechanism, depending on treatment.

In such a context, we may write the test hypotheses as

$$\begin{aligned} H_0 &: \{P_1 = P_2\} = \{(O_1, X_1, Y_1) \overset{d}{=} (O_2, X_2, Y_2)\} \\ &= \left\{\left[\{O_1 \overset{d}{=} O_2\}\right] \bigcap \left[(X_1, Y_1) \overset{d}{=} (X_2, Y_2)|\mathbf{O}\right]\right\} \end{aligned}$$

against

$$H_1 : \left\{\left[O_1 \overset{d}{<} O_2\right] \bigcup \left[(X_1 \overset{d}{<} X_2)|\mathbf{O}\right] \bigcup \left[(Y_1 \overset{d}{<} Y_2)|\mathbf{O}\right]\right\},$$

because, conditionally on germination, the expected treatment effects are assumed to produce higher values in both component variables. Moreover, let us assume that partial expected effects on the first component variable $[X|\mathbf{O}]$ are marginally homoscedastic and additive on location, whereas effects on $[Y|\mathbf{O}]$ are presumed to act on both location and second moment (as in multi-aspect testing problems, see Example 3, 4.6).

These side-assumptions are, on the one hand, consistent with the impression that observed second-variable data show increments in both mean and variance; on the other hand, they are consistent with the fact that most of the weight in this particular kind of plant comes from its leaves. In addition, these side-assumptions contribute towards showing the versatility of NPC methods. Hence, the set of sub-alternatives may be written as: $H_{11}^{\mathbf{O}} : \{\mathbb{E}(O_1) < \mathbb{E}(O_2)\}$, $H_{12}^{(\mathbf{X},\mathbf{Y})|\mathbf{O}} : \{\mathbb{E}(X_1) < \mathbb{E}(X_2)\}$, $H_{13}^{(\mathbf{X},\mathbf{Y})|\mathbf{O}} : \{\mathbb{E}(Y_1) < \mathbb{E}(Y_2)\}$ and $H_{14}^{(\mathbf{X},\mathbf{Y})|\mathbf{O}} : \{\mathbb{E}(Y_1^2) < \mathbb{E}(Y_2^2)\}$.

Remark 1. In order to facilitate computation related to partial tests for quantitative responses of actually observed units, which concern the variables X, Y and Y^2, the data set should be in a unit-by-unit representation, $(\mathbf{O}, \mathbf{X}, \mathbf{Y}) = \{(O_{ji}, X_{ji}, Y_{ji}, Y_{ji}^2), i = 1, \ldots, n_j, j = 1, 2\}$, such that the record of the jith unit is the four-dimensional vector $(O_{ji}, X_{ji}, Y_{ji}, Y_{ji}^2)$.

Although, in this representation, the component Y_{ji}^2 is apparently uninformative, being related to Y_{ji} with probability one, it is useful to note that the related partial test possibly contains information on second-order effects (see the notion of multi-aspect testing).

Remark 2. If treatment effects are also expected to act on cross-functionals, as in $\mathbb{E}[\eta(X_1, Y_1)] < \mathbb{E}[\eta(X_2, Y_2)]$, then it is straightforward to include the specific partial tests in the analysis. Related computations may be facilitated if unit records are written as $[O_{ji}, X_{ji}, Y_{ji}, Y_{ji}^2, \eta(X_{ji}, Y_{ji})]$, $i = 1, \ldots, n_j$, $j = 1, 2$.

7.11.1 Problem Description

This example is a typical MNAR situation. We note that its solution within a parametric framework, due to its complexity, if not impossible, is extremely difficult. On the other hand, the permutation solution is relatively easy to obtain within the NPC methodology and the analysis of missing data.

The problem could be viewed as emblematic of a collection of rather difficult problems:

- analysis of four-dimensional restricted alternatives;
- analysis of data of mixed type (one categorical and two quantitative variables);
- analysis of MNAR missing values where, in H_1, the missing mechanism is assumed to depend on treatment;
- multi-aspect analysis on quantitative responses.

In a way, it was through dealing with this problem that we began to systematically study, develop a coherent theory, and suggest non-standard proposals within the area of NPC methods of dependent permutation tests.

In any case, it is worth noting that the key idea for a proper solution is that, in H_0, both the exchangeability and permutation principles are satisfied.

The data from two groups are reported in Table 7.2, where $n = 20$, $\nu_1 = 12$, and $\nu_2 = 17$ ($\nu_j = \sum_i O_{ji}$, $j = 1, 2$, is the number of germinated seeds in the jth group and represents the actual sample size for quantitative responses).

7.11.2 The Permutation Solution

The hypotheses are decomposed into four sub-hypotheses in which the alternatives are all one-sided. Thus, according to the previous theory, partial tests may be based on the following set of statistics:

$$T_1^* = \nu_2^*,$$

$$T_2^* = \sum\nolimits_i X_{2i}^* \cdot \gamma_2^* - \sum\nolimits_i X_{1i}^* \cdot \gamma_1^*,$$

$$T_3^* = \sum\nolimits_i Y_{2i}^* \cdot \gamma_2^* - \sum\nolimits_i Y_{1i}^* \cdot \gamma_1^*,$$

$$T_4^* = \sum\nolimits_i Y_{2i}^{*2} \cdot \gamma_2^* - \sum\nolimits_i Y_{1i}^{*2} \cdot \gamma_1^*,$$

where $\gamma_j^* = (\nu_k^*/\nu_j^*)^{1/2}$, $k \neq j = 1, 2$.

Note that all partial tests are significant for large values and that all are, at least approximately, marginally unbiased because the expected treatment effect operates on all variables towards stochastically larger values. Also note that T_1^* is permutationally equivalent to Fisher's exact probability test on the 2×2 table of germination frequencies, as reported in Table 7.3.

Table 7.2 Two groups of seeds, (I) without and (II) with fertilizer

	I			II		
i	O_1	X_1	Y_1	O_2	X_2	Y_2
1	0	?	?	0	?	?
2	0	?	?	0	?	?
3	0	?	?	0	?	?
4	0	?	?	1	3.31	18.49
5	0	?	?	1	6.56	19.20
6	0	?	?	1	3.16	9.85
7	0	?	?	1	4.07	15.83
8	0	?	?	1	2.09	6.16
9	1	6.03	12.54	1	6.72	17.58
10	1	4.20	14.81	1	3.93	19.29
11	1	4.49	16.71	1	2.56	10.77
12	1	2.00	7.53	1	8.30	18.31
13	1	2.84	7.02	1	4.21	10.56
14	1	3.88	8.09	1	1.86	9.48
15	1	2.04	5.76	1	3.09	12.54
16	1	5.48	18.01	1	5.09	18.35
17	1	2.31	8.81	1	4.08	11.84
18	1	1.90	8.17	1	3.63	11.44
19	1	1.75	6.62	1	2.61	7.66
20	1	3.02	7.69	1	5.21	12.00

Table 7.3 Germination data

	$20 - \nu$	ν
I	8	12
II	3	17

The complexity implied by the restricted alternatives, the particular treatment effects, which need multi-aspect testing, and the unusual structure of the data are such that ordinary parametric solutions are nearly impossible to apply.

Firstly, we wish to determine whether the fertilizer improves the probability of seed germination, or adds weight to adult plants, or produces an increase in the total surface of their leaves (first and second moments). Moreover, using the NPC methodology, we may be interested in evaluating whether the use of fertilizer results in a general improvement of all the variables (see global p-values).

As is shown in Table 7.4, we may conclude that fertilizer produces an increase in the total surface of the leaves (first and second moments) and a significant global improvement may be found using Fisher's combining function $\hat{\lambda}'' = 0.0201$.

The combination problem may be dealt with in at least two different ways (see Figure 7.2).

As already seen, we may perform an NPC of four partial tests, as in $T'' = \psi_1(T_1, T_2, T_3, T_4)$, giving each partial test the same weight, implicitly weighting the Y-component twice and the others once. Of course, if this is compatible with the inferential objectives of the experimenter, then this is the solution he or she is looking for. If the inferential objective considers the three

Table 7.4 Testing equality in distribution for two independent samples for continuous (or dichotomous) variables (with or without missing values)

Variables	Test	p-value
Germinate	T_1	0.0810
Weight	T_2	0.0965
Surface	T_3	0.0264
Surface^2	T_4	0.0363
p-Global (Fisher)	T''	0.0201

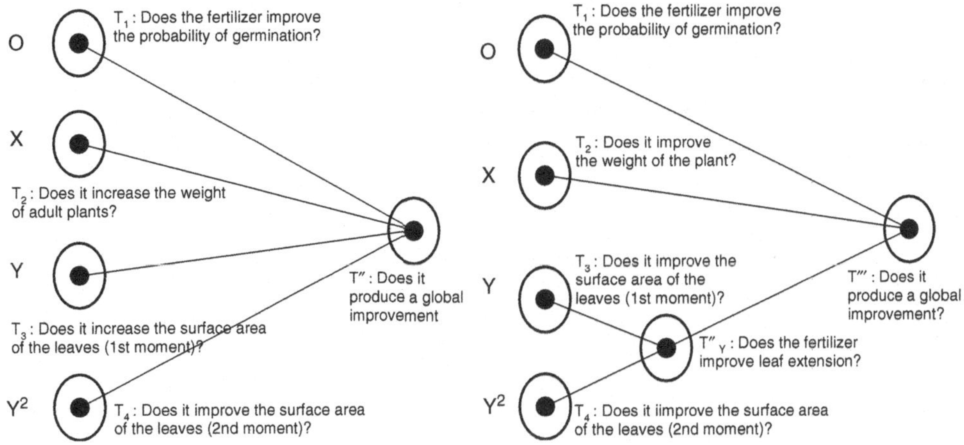

Figure 7.2 Different analyses for 'germina' data

variables (O, X, Y) as having the same importance, as seems reasonable in the present problem, then the overall solution becomes $T''' = \psi_3\left[T_1, T_2, T''_Y\right]$, where $T''_Y = \psi_2(T_3, T_4)$. Of course, in this analysis ψ_1, ψ_2, ψ_3 are suitable combining functions belonging to the class $\mathcal{C}$, which may differ from each other in accordance with specific combining properties. Table 7.5 reports the results of this different analysis.

Table 7.5 A solution to the combination problem by the multi-aspect approach

Variables	Test	p-value
Germinate	T_1	0.0810
Weight	T_2	0.0965
Surface+Surface^2	T''_Y	0.0305
p-Global (Fisher)	T'''	0.0205

7.11.3 Analysis Using MATLAB

In order to carry out the analysis for the `germina` example, we give the corresponding MATLAB code (the data set is in the Excel file `germina.xls`, available from the book's website).

```
B=5000;
[D, germina]=xlsimport('germina');
[P T options]=np_2s(D(2:end),D(1),B,1);
reminD(D)
[P T options]=np_2s({'Germinated','Weight','Surface','Surface2'},
'Fertilizer',B,1);
%p-values:
%VARIABLES Y
%  Y1            Y2            Y3            Y4
%0.080992     0.09649      0.026397     0.036296

NPC(P,'F',options);
%VARIABLE    Y1      Y2      Y3      Y4
%p-value 0.0810  0.0965  0.0264  0.0363
%Comb Funct Fisher
%p-GLOBAL 0.0201

options.labels.dims{2}={'Surface' 'Surface^2'};
options.labels.dims{3}=[D(1).name];
p_surf=NPC(P(:,[end-1:end]),'F',options);
%VARIABLE    Y1      Y2
%p-value 0.0264  0.0363
%Comb Funct Fisher
%p-GLOBAL 0.0305

options.labels.dims{2}=[D(2:3).name]
options.labels.dims{2}{3}='Surface MA';
options.labels.dims{3}=[D(1).name];
NPC_FWE([P(:,[1 2]) p_surf],'F',options);
%VARIABLE    Y1      Y2      Y3
%p-value          0.0810  0.0965  0.0305
%p-FWE            0.0810  0.0965  0.0444
%Comb Funct       Fisher
%p-GLOBAL         0.0205
```

In the above code, `xlsimport` imports an Excel file of data; `np_2s` is used to test the equality in distribution of two independent samples, whenever the data set contains continuous or dichotomous variables and even in the presence of missing values; and `NPC` allows the NPC methodology to be used and thus produces multivariate tests.

7.11.4 Analysis Using R

We now give the R code (data are contained in the`germina.csv` file available from the book's website). As already stated, the data set contains data of a study on 40 seeds trees. Twenty seeds were randomly chosen and treated with a placebo, while the remaining 20 seeds were treated with a fertilizer. The following variables were recorded: germination of the seed (0 = no, 1 = yes), average weight and leaf surface given that the seed has germinated. A further variable is `Surface2`, the square of `Surface`, which allows us to consider the second moment of `Surface` as well. Clearly, the last two columns of data have the functional relationship `Surface2 = Surface`^{2}, and the data of `Weight` and `Surface` are not available if `Germinated` is equal to zero. Missing data are indicated with the symbol `Na` (not available).

```
setwd("C:/path")
source("t2p.r")
data<-read.csv("germina.csv",header=TRUE,na.strings=" ")
```

We wish to establish whether the fertilizer has a positive impact on the tree growth. In particular, we wish to evaluate whether the fertilizer has a significant effect not only in terms of tree growth, but even on the germination of the seeds. Thus, if we denote by O_i the variable `Germinated`, by X_i the weight and by Y the surface area of the leaves, we wish to assess the global null hypothesis,

$$H_0 : \{(O_1 \stackrel{d}{=} O_0) \bigcap (X_1 \stackrel{d}{=} X_0) \bigcap (Y_1 \stackrel{d}{=} Y_0) \bigcap (Y_1^2 \stackrel{d}{=} Y_0^2) | O_1 \stackrel{d}{=} O_0\},$$

against the alternative,

$$H_1 : \{(O_1 \stackrel{d}{>} O_0) \bigcup (X_1 \stackrel{d}{>} X_0) \bigcup (Y_1 \stackrel{d}{>} Y_0) \bigcup (Y_1^2 \stackrel{d}{>} Y_0^2) | O_1 \stackrel{d}{=} O_0\}.$$

```
B=5000
n=dim(data)[1]
p=dim(data)[2]-1
T<-array(0,dim=c((B+1),p))
colnames(T)=c("O","X","Y","Y2")
```

As far as the quantitative variables are concerned, we take as our test statistic the weighted difference $T_j^* = \sqrt{\nu_0^{*2}/\nu_1^{*2}}\bar{Z}_{1j}^* - \sqrt{\nu_1^{*2}/\nu_0^{*2}}\bar{Z}_{0j}^*$, $j = 1, 2, 3$, where Z is the matrix with the continuous variable of the data set `germina`, and ν_0^* and ν_1^* are the number of germinated seeds in the samples treated with placebo and fertilizer, respectively, after a random assignment of the sample labels to the rows of data. Since we are performing a random assignment of the labels with respect to germinated seeds, when writing the R code, we have to consider the data matrix without the first column (i.e. `data[,-1]`). The number of non-missing observations on the observed data are $\nu_0 = 12$ and $\nu_1 = 17$. As far as the variable O is concerned a directional statistic that can be considered is the odds ratio (OR). It is easy to see that the OR is permutationally equivalent to the product of the main diagonal of the matrix `O` (a 2×2 table with the distribution of the variable `O` in the two samples). Of course, the test statistics T_j^* are computed on the non-missing observations only.

```
O = table(data[,1],data[,2]);
Z = data[,-c(1,2)]

T[1,1] = prod(diag(O))
nu = O[,2] ;

contr = rep(c(-sqrt(nu[2]/nu[1]),sqrt(nu[1]/nu[2])),nu)
T[1,-1] = t(Z[data[,2]>0,])%*%contr
```

Since the variables are recorded on the same trees, they are dependent. We can obtain the permutation distribution by considering $B = 5000$ random permutations of the rows of data (i.e.

`data[,-1]`). That is, we fix the variable denoting the sample and permute the vectors whose elements are in the remaining columns. Note that the number of non-missing data changes at each permutation, and must be computed each time.

```
set.seed(10)
for(bb in 2:(B+1)){
data.star = cbind(data[,1], data[sample(1:n),-1])
O.star = table(data.star[,1],data.star[,2]);
Z.star = data.star[,-c(1,2)]

T[bb,1] = prod(diag(O.star))
nu = O.star[,2] ;

contr.star = rep(c(-sqrt(nu[2]/nu[1]),sqrt(nu[1]/nu[2])),nu)
T[bb,-1] = t(Z.star[data.star[,2]>0,])%*%contr
}
```

At this point, the first column of `T` has the null distribution of the test statistic T_O^* =`O.star[1,1]*O.star[2,2]` and the remaining columns contain the null distribution of T_j^*. Each test statistic considered is significant for large values. We apply the `t2p` function in order to obtain the partial p-value:

```
P = t2p(T); cat("Partial p-values:","\n"); P[1,]

     O      X      Y     Y2
0.0758 0.0934 0.0256 0.0342
```

We can now combine the two partial tests related to the variable `Surface` and obtain a unique test related to this aspect: we compute Fisher's combining function on the elements of `T[,3:4]`.

```
T.Y = apply(P[,3:4],1,function(x){-2*log(prod(x))})
P.Y = t2p(T.Y)
P=cbind(P[,1:2],P.Y) ; colnames(P)[3] = "Y"
cat("Partial p-values:","\n"); P[1,]

Partial p-values:
     O      X      Y
0.0758 0.0934 0.0300
```

The p-values above are the partial results related to the three aspects considered: `Germination`, `Weight` and `Surface`. We can run the function `FWE.minP` in order to control for the multiplicity. The `FWE.minP` function performs a permutation Bonferroni–Holm step-down procedure and combines the partial p-values with Tippett's combining function.

```
source("FWEminP.r")
cat("Adjusted Partial p-values:","\n"); FWE.minP(P)

Adjusted Partial p-values:
    [,1]
O 0.1474
X 0.1474
Y 0.0624
```

To carry out a global test, we can combine the three partial tests with Fisher's combining function:

```
T.G = apply(P,1,function(x){-2*log(prod(x))})
p.G = t2p(T.G)[1]; p.G
```

```
[1] 0.0176
```

From these results we can conclude that there is a significant effect of the fertilizer on tree growth. The greatest effect seems to be related to the area of the leaves, rather than their weight. The fertilizer does not seem to be significantly associated with the germination of the seeds. The data set and the corresponding software codes are available from the `germina` folder on the book's website.

Remark 1. The data in the example appear randomly censored, where the censoring process may depend on treatment, because effective observations are obtained on germinated seeds, when $O = 1$, and no observation is possible on non-germinated seeds, when $O = 0$. The given solution is coherent with the point of view of a 'scientist' who is interested in showing whether fertilizer is effective by increasing the probability of germination and/or by increasing distributions of weight and/or leaf surfaces. Also, he may be interested in showing which of three aspects is most influenced by adjusting partial p-values according to solutions in Chapter 5. However, if the censoring process is considered of no concern or if he may assume that there is no influence of treatment on the probability of germination, so that the censoring processes may be ignored, then the present problem admits two quite different exact permutation solutions. The first of these may be found by considering the point of view of the 'fertilizer manufacturer' (which is similar to that of a 'farmer') whose objective is to see whether fertilizer improves the production per sown seed. Accordingly, he put production of 0 on both variables for non-germinated seeds, thus obtaining observed mixed variables with some data concentration at zero. By applying the NPC with $B = 10\,000$, we obtain $\lambda_X = 0.0148$, $\lambda_Y = 0.0045$, $\lambda_{Y2} = 0.0051$, $\lambda''_{23} = 0.0041$, and $\lambda_{global} = 0.0069$. The latter, agrees with the hypothesis that the fertilizer is effective, as do all partial results. The second exact solution supposes that if the inclusion indicator $\mathbf{O}$ can be ignored, then we may delete all non-germinated units from the data set (see Remark 1, 7.9.3). Thus, actually observed data may be viewed as if actual sample sizes were reduced from n_j to ν_j, $j = 1, 2$. Of course, this exact conditional analysis is only slightly different from that performed with T'''_{XY}, in the sense that the latter, being valid even for non-ignorable missing processes, corresponds to the point of view of a 'user' whose main objective is concerned with the so-called 'visible productivity' of the fertilizer. By still applying the NPC after deleting missing data with $B = 10\,000$, we obtain $\lambda_X = 0.0981$, $\lambda_Y = 0.0250$, $\lambda_{Y2} = 0.0322$, $\lambda''_{23} = 0.0285$, and $\lambda_{global} = 0.0446$. Of course, since they correspond to different objectives, the three solutions give different results.

7.12 Multivariate Paired Observations

Suppose that a V-dimensional non-degenerate real variable $\mathbf{X}$ is observed on k different time occasions on the same n units in two experimental situations, corresponding to two levels of a symbolic treatment. The whole data set is again denoted by

$$\begin{aligned}\mathbf{X} &= \{X_{hjit},\ t = 1, \ldots, k,\ i = 1, \ldots, n,\ j = 1, 2,\ h = 1, \ldots, V\} \\ &= \{\mathbf{X}_{jit},\ t = 1, \ldots, k,\ i = 1, \ldots, n,\ j = 1, 2\} \\ &= \{\mathbf{X}_{ji},\ i = 1, \ldots, n, j = 1, 2\},\end{aligned}$$

where $\mathbf{X}$ represents the V-variate response $(X_1, \ldots, X_V)$, the V-variate time profile $(\mathbf{X}_1, \ldots, \mathbf{X}_k)$ and the whole data set. Let us also suppose that some of the data are missing, giving rise to the inclusion indicator set

$$\mathbf{O} = \{O_{hjit},\ t = 1, \ldots, k,\ i = 1, \ldots, n,\ j = 1, 2,\ h = 1, \ldots, V\},$$

where O_{hjit} equals 1 if X_{hjit} is observed and 0 if it is missing.

Let us further assume that we are interested in testing for V effects δ_h irrespective of time, underlying dependences, and unknown distributions. Thus, if the missing data follow an MCAR model, then the hypotheses under consideration can be formally expressed (with obvious notation) as

$$H_0 : \left\{\bigcap_{t=1}^{k}\left(\mathbf{X}_{1t} \stackrel{d}{=} \mathbf{X}_{2t}\right)|\mathbf{O}\right\} = \left\{\bigcap_{t=1}^{k}\bigcap_{h=1}^{V} H_{0ht}\right\},$$

against alternatives of the form

$$H_1 : \left\{\bigcup_{t=1}^{k}\left(\mathbf{X}_{1t} \stackrel{d}{<\neq>} \mathbf{X}_{2t}\right)|\mathbf{O}\right\} = \left\{\bigcup_{t=1}^{k}\bigcup_{h=1}^{V} H_{1ht}\right\}.$$

In this context, partial permutation tests have the form

$$T_{ht}^{*} = \varphi_{ht}\left[\frac{\sum_i Y_{hit} \cdot O_{hit} \cdot S_i^{*}}{\left(\sum_i Y_{hit}^2 \cdot O_{hit}^2\right)^{1/2}}\right],\ t = 1, \ldots, k,\ h = 1, \ldots, V,$$

where $Y_{hit} = X_{h1it} - X_{h2it}$ are unit-by-unit, time-to-time and variable-by-variable observed differences, $O_{hit} = O_{h1it} \cdot O_{h2it}$ equals 1 if and only if X_{h1it} and X_{h2it} are both observed, S_i^* are permutation signs, and functions φ_{ht} correspond to the absolute value or to signs plus or minus, according to whether the htth sub-alternative H_{1ht} is two-sided ('$\neq$') or one-sided ('$>$' or '$<$') respectively.

Remark 1. Observe that all partial tests T_{ht}^*, $t = 1, \ldots, k$, $h = 1, \ldots, V$, are permutationally exact, provided that actual sample sizes $\nu_{ht} = \sum_i O_{hit}$ are positive, because ν_{ht} are permutationally invariant quantities and thus null distributions depend only on exchangeable errors. Moreover, they have null permutation means and unit variances and are marginally unbiased. Hence, their NPC is appropriate for multivariate testing.

Remark 2. Since all partial tests are expressed in standardized form, a direct combination is possible, although other combining functions may work better. However, in the framework of direct combination, for instance, we may have $T_h''^* = \sum_t T_{ht}^*$ for testing within each variable irrespective of time, and $T''' = \sum_h T_h''^*$ for overall testing.

Remark 3. The extension to MNAR models, provided that partial tests for the components of the inclusion indicator $\mathbf{O}$ are marginally unbiased, is straightforward. Observe that if partial tests for $H_0^{\mathbf{O}} : \{\bigcap_h H_{0h}^{\mathbf{O}}\}$ against $H_1^{\mathbf{O}} : \{\bigcup_h H_{1h}^{\mathbf{O}}\}$ are exact, then the NPC for $H_0 : \{[\bigcap_h H_{0h}^{\mathbf{O}}]\bigcap[\bigcap_h H_{0h}^{\mathbf{Y}|\mathbf{O}}]\}$ against $H_1 : \{[\bigcup_h H_{1h}^{\mathbf{O}}]\bigcup[\bigcup_h H_{1h}^{\mathbf{Y}|\mathbf{O}}]\}$ is also exact.

7.13 Repeated Measures and Missing Data

We consider a two-sample test on n_1 and n_2 units. Measurements on a response variable are repeated on the same units on fixed time occasions and the two samples have two levels of a treatment. The response model is that of Section 7.3.1. The hypotheses that we wish to test are

$$H_0 : \left\{\mathbf{X}_1 \stackrel{d}{=} \mathbf{X}_2\right\} = \left\{\bigcap_{1\le t\le k} [X_1(t) \stackrel{d}{=} X_2(t)]\right\} = \left\{\bigcap_{1\le t\le k} H_{0t}\right\}$$

against: $H_1 : \{H_0 \text{ is not true}\}$.

As some of the data are missing (see Section 7.8), we must also consider the associated inclusion indicator (data configuration)

$$\mathbf{O} = \{O_{ji}(t), i = 1, \ldots, n_j, j = 1, 2, t = 1, \ldots, k\},$$

where $O_{ji}(t) = 1$ if $X_{ji}(t)$ is actually observed, and $O_{ji}(t) = 0$ if it is missing. Thus, $\mathbf{O}$ represents the time profile of inclusion indicator. Hence, the complete set of observed data is summarized by the pair of associated matrices $(\mathbf{X}, \mathbf{O})$. Moreover, we denote by $\nu_j(t) = \sum_i O_{ji}(t)$, $j = 1, 2$, $t = 1, \ldots, k$, the actual sample size of valid data in the jth group relative to the tth time, and by $\nu_.(t) = \sum_j \nu_j(t)$, $t = 1, \ldots, k$, the total sample size of valid data relative to the tth time.

Thus, assuming that in the null hypothesis there is no difference in distribution due to treatment, jointly on all $(\mathbf{X}, \mathbf{O})$ components, we may express the null hypothesis as

$$H_0 : \{(\mathbf{X}_1, \mathbf{O}_1) \stackrel{d}{=} (\mathbf{X}_2, \mathbf{O}_2)\}$$

against the alternative $H_1 : \{H_0 \text{ is not true}\}$.

The hypotheses and assumptions are such that the permutation testing principle applies. In fact, in H_0, data are i.i.d., so that they are exchangeable with respect to the two groups. The null hypothesis may also be written as

$$H_0 : \left\{\left(\mathbf{O}_1 \stackrel{d}{=} \mathbf{O}_2\right) \bigcap \left[\left(\mathbf{X}_1 \stackrel{d}{=} \mathbf{X}_2\right) \mid \mathbf{O}\right]\right\}$$

$$= \left\{\left[\bigcap_{t=1}^{k}\left(O_1(t) \stackrel{d}{=} O_2(t)\right)\right] \bigcap \left[\bigcap_{t=1}^{k}\left(X_1(t) \stackrel{d}{=} X_2(t) \mid \mathbf{O}\right)\right]\right\}.$$

If missing data are assumed MCAR, then $\left\{\mathbf{O}_1 \stackrel{d}{=} \mathbf{O}_2\right\}$ is true by assumption. Hence, this part of the analysis is omitted. If missing data are MNAR, then this part should be considered because in this context it is assumed that the $\mathbf{O}$-components may contain useful information for discriminating between the null and alternative hypotheses.

In the case of MCAR data, each of the k sub-hypotheses is reduced to a two-sample test conditional on $\mathbf{O}$, where partial tests are related to $H_{0t}^{\mathbf{X}} : \{X_1(t) \stackrel{d}{=} X_2(t)|\mathbf{O}\}$, $t = 1, \ldots, k$. Hence, for partial testing of H_{0t} against H_{1t}, we may use any marginally unbiased test $T(t)$ and, in order to achieve a complete solution for H_0 against H_1, we must combine all partial tests through any nonparametric combining function. When missing data are MNAR, we must jointly take into consideration the k partial tests for $H_{0t}^{\mathbf{O}} : \{O_1(t) \stackrel{d}{=} O_2(t)\}$, $t = 1, \ldots, k$, as well. For each t, the latter may be Fisher's exact probability test. Therefore, in order to solve the whole problem, we must nonparametrically combine all $2k$ partial tests.

7.13.1 An Example

This example is concerned with a problem which is quite common, for instance, in clinical trials, in which data are obtained by repeated measurements on a set of individuals at fixed time intervals and individuals are partitioned into two or more levels of a symbolic treatment.

Table 7.6 Two-sample repeated observations, with MCAR missingness

X_1					X_2				
$t = 0$	1	2	3	4	0	1	2	3	4
89.15	78.66	72.25	?	59.21	91.12	75.18	72.77	?	60.71
91.54	77.67	77.74	70.00	61.89	77.34	70.01	66.44	60.16	56.35
99.55	?	81.37	72.15	66.36	66.98	65.08	58.48	54.16	53.78
82.85	82.47	71.97	68.01	?	70.46	65.65	64.26	58.68	58.42
96.77	85.01	77.14	67.99	60.72	82.05	73.99	66.54	63.05	60.09
79.03	71.03	?	59.74	63.25	91.10	76.69	?	59.81	59.34
67.08	64.78	72.85	66.17	62.93	88.24	75.92	71.67	63.01	59.80
78.27	73.45	68.11	67.34	62.91	76.73	?	62.65	65.03	63.28
72.45	68.92	64.19	58.86	62.11	93.24	77.82	76.99	67.89	65.65

Imagine that n individuals, homogeneous with respect to a given blood variable whose values are higher than normal, are randomized into two groups, the first of which (n_1 units) receives treatment A_1 and the other A_2. The aim of both treatments is to stabilize the variable by stochastically reducing it towards normal values (reducing means and, possibly, variances). Data are presumed to be observed at $k+1$ fixed time intervals and are interpreted as profiles of a stochastic process $\{X_{ji}(t),\ i = 1, \ldots, n_j,\ j = 1, 2,\ t = 0, \ldots, k\}$, where data at time 0 are taken as baseline values. In addition, independently of time and treatment level, we assume an MCAR model for missing values.

Suppose that the hypotheses of interest are

$$H_0 : \left\{ \bigcap_{t=1}^{k} \left[(X_1(t) - X_1(0)) \stackrel{d}{=} (X_2(t) - X_2(0)) \,|\mathbf{O} \right] \right\}$$

against

$$H_1 : \left\{ \bigcup_{t=1}^{k} \left[(X_1(t) - X_1(0)) \stackrel{d}{<} (X_2(t) - X_2(0)) \,|\mathbf{O} \right] \right\},$$

indicating the more rapid effectiveness of treatment A_2 than that of A_1.

The problem may be solved by using k two-sample tests followed by their NPC. Partial tests are

$$T_t^* = S_{2t}^* \cdot (\nu_{1t}^*/\nu_{2t}^*)^{1/2} - S_{1t}^* \cdot (\nu_{2t}^*/\nu_{1t}^*)^{1/2},\ t = 1, \ldots, k,$$

where $S_{jt}^* = \sum_i \left[X_{ji}^*(t) - X_{ji}^*(0) \right]$ and $\nu_{jt}^* = \sum_i O_{ji}^*(t)$, $j = 1, 2$.

The data, which are assumed to behave in accordance with the model in Section 7.3.1, are shown in Table 7.6, where missing data are denoted with ?.

With $B = 2000$ CMC iterations, the partial p-values are $\hat{\lambda}_1 = 0.185$, $\hat{\lambda}_2 = 0.039$, $\hat{\lambda}_3 = 0.027$ and $\hat{\lambda}_4 = 0.035$; using Fisher's combining function gives a combined p-value of $\hat{\lambda}'' = 0.029$. Hence, H_0 is rejected at $\alpha = 0.05$.

7.14 Botulinum Data

We consider a real case study related to a preliminary double-blind, placebo-controlled, randomized clinical trial with a six-month follow-up period, the aim of which is to assess the effectiveness of type A botulinum toxin in treating myofascial pain symptoms and to reduce muscle hyperactivity in bruxers. Twenty patients (10 males and 10 females, aged between 25 and 45) with clinical diagnosis of bruxism and with myofascial pain in the masticatory muscles were enrolled in the trial, with a treatment group (10 subjects) treated with type A botulinum toxin injections and a control group (10 subjects) treated with saline placebo injections. Several clinical variables (in the medical literature, the same as end-points) were assessed at baseline, one-week, one-month and six-month follow-up appointments, along with electromyography (EMG) recordings of muscle activity in different conditions. Clinical end-points are as follows:

- *pain at rest* (DR), *when phoning* (DF) and *when chewing* (DM), assessed by means of a Visual Analogue Scale (VAS) from 0 to 10, with the extremes being 'no pain' and 'pain as bad as the patient has ever experienced', respectively;
- *mastication efficiency* (CM), assessed by a VAS from 0 to 10, the extremes of which were 'eating only semi-liquid' and 'eating solid hard food';
- *maximum non-assisted* (Mas) and *assisted* (Maf) mouth opening (in millimetres), *protrusive* (Mp) and *laterotrusive left* (Mll) and *right* (Mlr) movements (in millimetres);
- *functional limitation* (LF) during usual jaw movements (0, absent; 1, slight; 2, moderate; 3, intense; 4, severe);

For further details on the case study, we refer to Arboretti (2008) and Guarda-Nardini et al. (2008). At the same time as the clinical evaluations, all patients underwent EMG recordings of left and right anterior and posterior temporalis muscles at rest (LTA, RTA, LTP, RTP, respectively) and left and right masseter muscles at rest (LMM, RMM); left and right anterior temporalis muscles during maximum voluntary clenching (LTA11, RTA11) and during clenching on cotton rolls (LTA11c, RTA11c); masseter muscles during maximum voluntary clenching (LMM11, RMM11) and during clenching on cotton rolls (LMM11c, RMM11c).

Hence, a V-dimensional non-degenerate variable is observed on k different time occasions on n units in two experimental situations, corresponding to two levels of a symbolic treatment. In our study, $V = 24$, $k = 4$, $n_1 = n_2 = 10$, $n = 20$. It is worth noting that in this longitudinal study there is a major obstacle to applying classic parametric tests: the number of observed variables at different time points is much higher than the number of subjects ($V \cdot k \gg n$). Furthermore, since all variables may be informative for differentiating the two groups, the NPC approach properly applies when analysing these data. Classic parametric tests or even rank tests in such situations may fail to take into account the dependence structure across variables and time points.

The whole data set is denoted by:

$$\begin{aligned}\mathbf{X} &= \{X_{hji}(t),\ t = 1, \ldots, k,\ i = 1, \ldots, n_j,\ j = 1, 2,\ h = 1, \ldots, V\} \\ &= \{\mathbf{X}_{hji},\ i = 1, \ldots, n_j,\ j = 1, 2,\ h = 1, \ldots, V\},\end{aligned}$$

where $\mathbf{X}_{hji} = \{X_{hji}(t),\ t = 1, \ldots, k\}$.

In order to take account of different baseline observations, assumed to have the role of covariates, the $k - 1$ V-dimensional differences $D_{hji}(t) = X_{hji}(1) - X_{hji}(t)$, $t = 2, \ldots, k$, $i = 1, \ldots, n_j$, $j = 1, 2$, $h = 1, \ldots, V$, are considered in the analysis. Hence, the hypothesis testing problem related to the hth variable may be formalized as

$$H_{0h} : \left\{\bigcap_{t=2}^{k}\left[D_{h1}(t) \stackrel{d}{=} D_{h2}(t)\right]\right\} = \left\{\bigcap_{t=2}^{k} H_{0ht}\right\},\quad h = 1, \ldots, V,$$

against the alternative

$$H_{1h} : \left\{\bigcup_{t=2}^{k} H_{1ht}\right\},\quad h = 1, \ldots, V,$$

where $H_{1ht}: \left\{\left[D_{h1}(t) \overset{d}{>} D_{h2}(t)\right] \text{ or } \left[D_{h1}(t) \overset{d}{<} D_{h2}(t)\right]\right\}$ according to which kind of stochastic dominance is of interest for the hth variable. The alternative hypothesis is that patients treated with the botulinum toxin had lower values than those treated with the placebo (i.e. differences between baseline values and follow-up values tend to increase, for which the $\overset{d}{>}$ dominance is appropriate), except for variables ME, Mas, Maf, Mp, Mll, Mlr, E, and T, where the placebo group is expected to have lower values than the toxin group, for which the $\overset{d}{<}$ is then appropriate.

Permutation tests were used to test the partial null hypotheses, and Tippett's combining function was chosen to perform global tests. Data missing from the data set are assumed to be missing completely at random (MCAR), but the presence of MNAR data could easily be checked. All the analyses were performed both in MATLAB and R. In the following, we display both MATLAB and R codes.

7.14.1 Analysis Using MATLAB

Table 7.7 shows the results of the analysis. Before adjusting p-values for multiplicity, variables `LTA11`, `RTA11`, `LMM11`, `LTA11c`, `RTA11c`, `LMM11c`, `RMM`11c, DM, DR and Maf are found to be

Table 7.7 Global p-values

	Comparison				
Variables	baseline−1week	baseline−1month	baseline−6months	p-value	p-FWE
LTA	0.5654	0.1319	0.6314	0.3856	0.6563
RTA	0.6344	0.0090	0.8302	0.1069	0.5574
LTP	0.4755	0.1129	0.1668	0.1878	0.6214
RTP	0.1818	0.2577	0.0589	0.1059	0.5574
LMM	0.2717	0.1958	0.4635	0.2767	0.6514
RMM	0.5495	0.1728	0.5005	0.3716	0.6563
LTA11	0.0470	0.0070	0.0190	**0.0110**	0.1638
RTA11	0.0190	0.0040	0.2438	**0.0150**	0.1978
LMM11	0.0120	0.0110	0.4306	**0.0190**	0.2218
RMM11	0.1449	0.0160	0.3786	0.0629	0.4955
LTA11c	0.0030	0.0060	0.0699	**0.0030**	0.0569
RTA11c	0.0010	0.0020	0.2917	**0.0030**	0.0569
LMM11c	0.0030	0.0010	0.2797	**0.0010**	**0.0220**
RMM11c	0.0010	0.0040	0.4006	**0.0010**	**0.0220**
CM	0.3397	0.0120	0.6743	0.0749	0.4955
DM	0.1459	0.0290	0.0020	**0.0060**	0.1029
DF	0.2847	0.0549	0.0450	0.0639	0.4955
DR	0.7902	0.0100	0.0539	**0.0240**	0.2677
LF	0.3487	0.0410	0.1169	0.0609	0.4955
Mas	0.1638	0.4196	0.0170	0.0669	0.4955
Maf	0.1199	0.0709	0.0040	**0.0150**	0.1978
Mp	0.4036	0.1668	0.0789	0.1349	0.5574
Mld	0.1948	0.3337	0.3027	0.2468	0.6514
Mls	0.2188	0.1179	0.1239	0.1209	0.5574
p-Global					0.0220

significant. After adjustment (p-FWE), only two variables, `LMM11c` and `RMM11c`, are significant at $\alpha = .05$. The original data are contained in the file `botulinum.xls` and are available from the `botulinum` folder on the book's website.

```
B=1000;
[code,data]=xlsimport('botulinum');

options.labels.dims{2}=[code(5:end).name code(5:end).name
code(5:end).name];
Y=data(:,5:end);
baseline=Y(1:20,:);
FU1=Y(21:40,:);
FU2=Y(41:60,:);
FU3=Y(61:80,:);

Y2=zeros(20, 24*3);

Y2(:,1:24)=FU1-baseline;
Y2(:,25:48)=FU2-baseline;
Y2(:,49:72)=FU3-baseline;

group=data(1:20,4);
alt=[1*ones(1,19) -1*ones(1,5) 1*ones(1,19) -1*ones(1,5)
1*ones(1,19) -1*ones(1,5)];
[P, T, options] = NP_2s(Y2,group,B,alt,options);


P=reshape(P,B+1,24,3);

for i=1:24
    options.labels.dims{2}{i}=options.labels.dims{2}{i}
(find((options.labels.dims{2}{i}==options.labels.dims{2}{24+i}).*
(options.labels.dims{2}{i}==options.labels.dims{2}{48+i})));
end
options.labels.dims{2}=options.labels.dims{2}(1:24);

options.labels.dims{3}={'baseline-1week',
'baseline-1month','baseline-6months'};

%pMat_Show(P,.05,options.labels,1);

%combining times
options.Combdims=3;
P2=NPC(P,'F',options);

%multiplicity control over variables
options.Combdims=2;
p2_FWE=NPC_FWE(P2,'T',options);
```

7.14.2 Analysis Using R

This is a multivariate case–control study with missing data and repeated measures. The aim of the study is to assess the effect of botulinum, that is expected to have opposite effects on the set of variables considered (i.e. we consider one-sided alternatives on each variable, following the ruke that either large is significant or small is significant, as appropriate). We consider the difference D between observations at time t_k, $k = 1, 2, 3$, and the observations at time t_0. We may organize data in an $n \times p \times 3$ matrix, so the third dimension contains the differences D_{ijk} related to the ith subject on the jth variable at time k. This will be useful when a repeated measures analysis is considered.

To begin with we read the data, collect information about samples (`paz`) and time (`Time`), and then remove the relevant columns.

```
setwd("C:/path")
data<-read.csv("botulinum.csv",header=TRUE,na.strings=' ')

n = length(unique(data[,1]))
p = dim(data)[2]-4
paz = ifelse(data[,4]=='botox',1,2)[1:20]
Time = data$Time+1

data=data[,-c(1:4)]
```

Now we put the record of each subject in an $n \times p$ matrix, so the third dimension contains the repeated measures. Then we obtain the differences D_{ijk} with respect to time zero observations and remove the vector of observations at t_0 (`D[,,1]`).

```
D=array(0,dim=c(n,p,4))

for(t in 1:4){ D[,,t]=as.matrix(data[Time==t,]) }

for(t in 2:4){ ### differences Tj-T0
D[,,t] = D[,,t]-D[,,1]
}

D = D[,,-1]
```

We first assess the effect of botulinum by performing a two-sample test with repeated measures. This requires us to modify the test statistic according to the number of non-missing observations ν. The vector `alternative` contains the directions of the alternative hypotheses. Since the distribution of non-missing observations varies from variable to variable, we may consider one variable at time, and do not record this information at any step. If a test for MAR observations is desired, this information should be collected as well.

The vector `O` is the indicator function (missing/not missing) applied to each variable, the vector `nu` contains the number of non-missing observations in the two samples, and `y` is the vector of observations with missing data replaced by zeros. The previous vectors are related to each variable at each time and are not saved. As usual, we store the observed values of the test statistic in the first row of the matrix `T`. The test statistic for the jth variable is $T_j^* = \sum_{k=1}^{3}[\gamma_{1j}^* D_{1jk}^* - \gamma_{2j}^* D_{2jk}^*]$, where $\gamma_{1j}^* = \sqrt{\upsilon_2^*/\upsilon_1^*} = (\gamma_{2j}^*)^{-1}$. Note that we have already combined, for the sake of simplicity, the repeated measures information with the direct combination (sum).

```
B=2000
T<-array(0,dim=c((B+1),p))
alternative = c(rep(-1,19),rep(1,5))

for(j in 1:p){
for(t in 1:3){

O = ifelse(is.na(D[,j,t])==TRUE,0,1)
y = ifelse(is.na(D[,j,t])==TRUE,0,D[,j,t])
nu = table(O,paz)
if(dim(nu)[1]>1){nu=nu[2,]}
D1 = sum(y[paz==1]) ;  D2 = sum(y[paz==2])

T[1,j] = T[1,j] + D1*sqrt(nu[2]/nu[1])-D2*sqrt(nu[2]/nu[1])
}
}
```

In order to obtain the null distribution, we permute observations in the usual way, by randomly exchanging the rows of `data` (i.e. the first dimension of `D`). This ensures that the inner dependencies among variables and due to the repeated mesures are maintained.

```
for(bb in 2:(B+1)){

D.star = D[sample(1:n),,]
for(j in 1:p){
for(t in 1:3){

O = ifelse(is.na(D.star[,j,t])==TRUE,0,1)
y = ifelse(is.na(D.star[,j,t])==TRUE,0,D.star[,j,t])
nu = table(O,paz)
if(dim(nu)[1]>1){nu=nu[2,]}

D1 = sum(y[paz==1]) ; D2 = sum(y[paz==2]);

T[bb,j] = T[bb,j] + D1*sqrt(nu[2]/nu[1])-D2*sqrt(nu[2]/nu[1])

}
}

print(bb)
}## end bb
```

We obtain the raw p-values by changing the signs of the columns of `T` according to the `alternative` vector and then apply the function `t2p`. We only show the results that are significant at the 5% nominal level.

```
for(j in 1:p){  T[,j]=T[,j]*alternative[j]  }

source("t2p.r") ; P=t2p(T)

res =  data.frame(colnames(data),P[1,]);
```

```
colnames(res) = c("Variable","Raw p-values")
res[res[,2]<.05,]

   Variable Raw p-values
7     LTA11       0.0070
8     RTA11       0.0055
9     LMM11       0.0210
10    RMM11       0.0345
11   LTA11c       0.0000
12   RTA11c       0.0000
13   LMM11c       0.0010
14   RMM11c       0.0000
16       DM       0.0065
18       DR       0.0450
21      Maf       0.0145
```

If we wish to account for multiplicity, we use the function `FWE.minP`, which returns adjusted *p*-values.

```
source("FWEminP.r")

p.adj = FWE.minP(P)
p.adj = data.frame(res[,1],p.adj)
colnames(p.adj) = colnames(res)
p.adj[p.adj[,2]<=0.05,]

   Variable Raw p-values
11   LTA11c        0.000
12   RTA11c        0.000
13   LMM11c        0.019
14   RMM11c        0.000
```

7.15 Waterfalls Data

7.15.1 Analysis Using MATLAB

The R&D division of a home-care company is studying five possible new fragrances (labelled s, t, v, w, x) of a given detergent to compare with their own presently marketed product (labelled r). The experiment is designed as follows: after testing one given product (using sense of smell), the panellist assigns three different scores to it, describing the three most important aspects of the product: strength, pleasantness (on a scale from 1 to 7), and appropriateness (yes or no, coded 1 or 2, respectively). In the Excel file, strength has been shortened to Stren, pleasantness to Pleas, and appropriateness to Appro. There are a total of seven panellists involved in the experiment. The same experiment is replicated under different assessment conditions (Bloom, Dry, Long, Neat and Wet), representing the situations in which customers will eventually make use of the product. In the Excel file Bloom is shortened to Blo, Long to Lon, Neat to Nea. Thus we have:

- Panelist, a quantitative variable (taking an integer value between 1 and 7) indicating the panellist doing the experiment;
- Fragr, a categorical variable denoting the fragrances, coded with the labels r, s, t, v, w, x;

- Blo_Stren, a quantitative variable (between 1 and 7) denoting the strength aspect for the bloom condition;
- Blo_Pleas, a quantitative variable (between 1 and 7) denoting the pleasantness aspect for the bloom condition;
- Blo_Appro, a binary variable (taking value 1 or 2) denoting the appropriateness aspect for the bloom condition;
- Dry_Stren, a quantitative variable denoting the strength aspect for the dry condition;
- Dry_Pleas, a quantitative variable denoting the pleasantness aspect for the dry condition;
- Dry_Appro, a binary variable denoting the appropriateness aspect for the dry condition;
- Lon_Stren, a quantitative variable indicating the strength aspect for the long condition;
- Lon_Pleas, a quantitative variable indicating the pleasantness aspect for the long condition;
- Lon_Appro, a binary variable indicating the appropriateness aspect for the long condition;
- Nea_Stren, a quantitative variable expressing the strength aspect for the neat condition;
- Nea_Pleas, a quantitative variable expressing the pleasantness aspect for the neat condition;
- Nea_Appro, a binary variable expressing the appropriateness aspect for the neat condition;
- Wet_Stren, a quantitative variable indicating the pleasantness aspect for the wet condition;
- Wet_Pleas, a quantitative variable indicating the pleasantness aspect for the wet condition;
- Wet_Appro, a binary variable indicating the appropriateness aspect for the wet condition.

Data are contained in the `Waterfalls.xls` file (17 variables and 42 observations in total). In the following, we present results along with MATLAB and R codes.

The goal of this analysis is to compare the new fragrances with the baseline/standard one (r). We expect new fragrances to be good competitors which may replace the standard product. This is a particular case of repeated measures analysis, since the repetitions are due to the fact that the same panellists evaluate each one of the five possible contrasts: r against each of the five possible new fragrances in turn. To take into account the intrinsic dependence among judgements (i.e. the evaluation of a product is obviously influenced by the previous product), we propose a stratified repeated measures analysis, taking as our stratification variable the panellist and comparing the standard fragrance to the new ones.

Let X_{hjz} be the score given by panellist z, $z = 1, \ldots, 7$, on the variable $h = 1, \ldots, 15$ for the product $j = r, s, t, v, w, x$. Hence, the following hypotheses are of interest:

$$H_0 : \left\{ \bigcap_{h=1}^{15} \left[\bigcap_{z=1}^{7} \left(X_{h(j=r)z} \stackrel{d}{=} X_{h(j \neq r)z} \right) \right], \ j = r, s, t, v, w, x \right\},$$

meaning that new fragrances and standard product are equal in distribution, against the alternative

$$H_1 : \left\{ \bigcup_{h=1}^{15} \left[\bigcup_{z=1}^{7} \left(X_{h(j=r)z} \stackrel{d}{=} X_{h(j \neq r)z} \right) \right], \ j = r, s, t, v, w, x \right\},$$

stating that at least one new fragrance may replace (is better than) the standard one.

Let us define the index c, containing the five possible contrasts. We calculate the matrix of statistics $\mathbf{T}$, containing all the differences between the standard product r and the new products (s, t, v, w, x), variable by variable, stratified by panellist:

$$T^*_{chz} = (X_{h(j=r)z} - X_{h(j \neq r)z}) \cdot S^*_{hz},$$

$z = 1, \ldots, 7$, $h = 1, \ldots, 15$, $j = r, s, t, v, w, x$, $c = 1, \ldots, 5$, where S^*_{hz} is a random sign satisfying $\Pr\{S^*_{hz} = +1\} = \Pr\{S^*_{hz} = -1\} = 1/2$ (as in a classical multivariate paired data problem).

To do this in MATLAB, we use the function `by_strata` along with the function `NP_ReM`, using the `Bal` option for the design matrix `DES`. Then, in order get the total score for a specific contrast/comparison, for each variable, we sum the scores given by the seven panellists. This will be used as the test statistic for subsequent analyses, thus obtaining the matrix **TT**:

$$TT^*_{ch} = \sum_{z=1}^{7} T^*_{chz}, \quad c = 1, \ldots, 5, h = 1, \ldots, 15.$$

As already mentioned, three aspects have been evaluated for each condition. Hence, it is also of interest to evaluate the following hypotheses. Let us denote the aspects by the index $a = 1, 2, 3$ and the conditions by the index $d = 1, \ldots, 5$. The same three aspects have been evaluated for each condition. Thus $h = 1, \ldots, d \cdot a_d$, where a_d denotes the aspect a for the condition d.

$$H_{0cd} = \left\{ \bigcap_{a=1}^{3} H_{0cd_a} \right\}, \quad c = 1, \ldots, 5, d = 1, \ldots, 5,$$

where H_{0cd_a} denotes the specific (partial) hypothesis for aspect a in condition d, against the alternative

$$H_{1cd} = \left\{ \bigcup_{a=1}^{3} H_{0cd_a)} \right\}, \quad c = 1, \ldots, 5, d = 1, \ldots, 5.$$

Tables 7.8 and 7.9 display the results obtained.

Alternatively, if we had ignored the fact that the same panellist evaluates the six products, hence if we had ignored the embedded dependence in the judgement, we would have compared the six products by performing a MANOVA analysis, thus considering the fragrances as independent samples.

Table 7.8 Assessing statistical significance of variables within each contrast

	Contrast				
	r v. s	r v. t	r v. v	r v. w	r v. x
Blo_Stren	0.9650	0.8601	0.5724	0.6014	0.9081
Blo_Pleas	1.0000	1.0000	0.0230	0.9940	1.0000
Blo_Appro	0.5275	0.1269	0.3656	0.5245	0.1269
Dry_Stren	0.8292	1.0000	0.1239	0.9540	1.0000
Dry_Pleas	0.8462	0.9311	0.0100	1.0000	0.5015
Dry_Appro	0.8711	0.9311	0.4815	1.0000	0.6663
Lon_Stren	0.9790	0.8332	0.7053	0.8492	1.0000
Lon_Pleas	0.3447	0.8382	0.2318	0.4216	0.9870
Lon_Appro	0.8332	0.5175	0.8182	0.7612	0.8881
Nea_Stren	0.2807	0.1788	0.3137	0.2607	0.1169
Nea_Pleas	0.8671	0.4645	0.6414	0.5215	0.4825
Nea_Appro	0.8841	0.9321	0.8871	0.8871	0.9740
Wet_Stren	0.0230	0.0609	0.0480	0.0160	0.0699
Wet_Pleas	0.6593	0.8312	0.4935	0.8701	0.2218
Wet_Appro	0.3297	0.5265	0.1409	0.1329	0.5175

Table 7.9 Combining conditions within each aspect. Raw and adjusted p-values, as well as the global p-value are given

		Contrast					
		r v. s	r v. t	r v. v	r v. w	r v. x	p-Global
Bloom	p-value	0.9161	0.4476	0.0649	0.7463	0.4695	
	p-FWE	0.9161	0.7223	0.4176	0.8761	0.7223	0.4156
Dry	p-value	0.9640	1.0000	0.0100	1.0000	0.7423	
	p-FWE	1.0000	1.0000	0.3886	1.0000	0.9770	0.3886
Long	p-value	0.8282	0.8871	0.5395	0.7073	1.0000	
	p-FWE	0.9660	0.9810	0.9441	0.9660	1.0000	0.9231
Neat	p-value	0.7253	0.3836	0.5465	0.4406	0.3147	
	p-FWE	0.7253	0.6254	0.7063	0.6963	0.6154	0.5914
Wet	p-value	0.0160	0.1748	0.0180	0.0100	0.0579	
		0.0440	0.1748	0.0539	0.0330	0.1029	0.0160

```
[data,code]=xlsimport('Waterfalls');
B=1000;
Y=data(:,3:end);
X=data(:,2);
panel=data(:,1);
options.labels.dims{2}=[data(3:end).name];
[P, T, options]= by_strata(panel,'NP_ReM',Y,X,'bal',B,-1,0);

TT=zeros([B+1 5 15]);
for b=1:(B+1)
for j=1:5
for i=1:15
    TT(b,j,i)=sum(-T(b,i,j,:));
end
end
end

[PP]=t2p(TT);
PP(B+1,:,:)
size(PP)

P=reshape(PP,[B+1 5 3 5]);
options.Combdims=3;
options.OUT=1;
P2=NPC(P,'F',options);
size(P2)
options.Combdims=2;
P3=NPC_FWE(P2,'F',options);
```

7.15.2 Analysis Using R

This is an all-to-one comparison problem: there are six fragrances tested on five experimental conditions, judged by $l = 7$ panellists. The aim of the study is to compare some new fragrances with a baseline one. The judgement is based on three aspects (strength, pleasantness and appropriateness), and therefore we have $p = 15$ response variables. Since the same panellist was asked to judge each fragrance under each condition and aspect, dependencies among each panellist's ratings are assumed, whereas panellists' judgements are assumed to be independent to each other.

We first read and organize the `data` into a three-dimensional matrix `D`, where the first dimension refers to the C fragrances, the second refers to the p variables, and the third to the judgements of the lth panellist.

Since our main interest is in comparing the new fragrances with a reference one (the first, named `r`), we first consider the differences of scores of the mth fragrance and the first fragrance within the same panellist's judgements, $m = 2, \ldots, 6$. Hence, after defining the auxiliary matrix `DD` (see R code below) we end up with a new matrix `D`, whose dimension is $(C - 1) \times p \times l$, where the element D_{ijk} refers to the ith comparison on the jth variable made by the kth panellist.

```
setwd("C:path/")
data <- read.csv("Waterfalls.csv",header=TRUE)

B = 1000
attach(data)
data = data[,-c(1,2)]

L = unique(Panelist) ; l = length(L)
n = dim(data)[1] ; p = dim(data)[2] ; C = n/l

D = array(0,dim=c(C,p,l))

 for(j in 1:l){
 D[,,j]=as.matrix(data[Panelist==j,])
 }

 DD = D  # auxiliary

 for(i in 1:l){
 for(j in 1:C){

 DD[j,,i]=D[j,,i]-D[1,,i]
 }
 }

 D = DD[-1,,]
```

The observed value of the test statistic of the ith comparison on the jth variable is $T_{ij\cdot} = \sum_{k=1}^{l} D_{ijk}$. Large (positive) values of D_{ijk} indicate that the kth panellist gave a score for the $(i + 1)$th fragrance that is better that the score for the reference one. Large values of $T_{ij\cdot}$ are significant in favour of the alternative hypothesis $X_{mj} \overset{d}{>} X_{1j}$, where X_{mj} is the score of the mth fragrance on the jth variable.

Since each panellist makes a judgement on all C fragrances, we have to consider permutations of a restricted kind to account for the dependence among the judgements of each panellist. Thus, we consider $C-1$ comparisons and treat them as $C-1$ paired observation problems. We consider $C-1$ independent permutations of paired observations. Thus, for each panellist we independently generate a vector of $C-1$ random signs S^*_{ij}, where $\Pr\{S^*_{ij}=1\}=\Pr\{S^*_{ij}=-1\}=1/2$, and obtain the permutation value of the test statistic as $T^*_{ij\cdot}=\sum_{k=1}^{l} D_{ijk}S^*_{ij}$.

```
 T = array(0,dim=c((B+1),p,(C-1)))

 for(j in 1:(C-1)){
 T[1,,] = t(apply(D,c(1,2),sum))
 }

for(bb in 2:(B+1)){
D.star=D

for(j in 1:l){
S.star =  1-2*rbinom((C-1),1,.5)
D.star[,,j] = D[,,j]*S.star
}

 for(j in 1:(C-1)){
 T[bb,,] = t(apply(D.star,c(1,2),sum))
 }

print(bb)
}## end bb

source("t2p.r")
P=t2p(T)

partial.p = P[1,,]
rownames(partial.p) = colnames(data)
colnames(partial.p) = c('s-r', 't-r','v-r','w-r','x-r')
)
```

The matrix `partial.p` contains the raw p-value of each comparison and variable. Wishing to combine the partial tests within the same experimental conditions:

```
dom = rep(c(1:5),each=3)
T.dom=array(0,dim=c((B+1),5,5))

for(d in 1:5){
T.dom[,d,] = apply(P[,dom==d,],c(1,3),function(x){-2*log(prod(x))})
}

P.dom = t2p(T.dom)
res = P.dom[1,,]
rownames(res) = c('Bloom','Dry','Long','Neat','Wet')
colnames(res) = c('r-s','r-t','r-v','r-w','r-x')
res

          r-s    r-t    r-v    r-w    r-x
```

```
Bloom 0.8836 0.5248 0.0660 0.7212 0.4786
Dry   0.9568 1.0000 0.0000 1.0000 0.7732
Long  0.8786 0.8710 0.5556 0.7228 0.9938
Neat  0.6198 0.3648 0.5602 0.4208 0.2894
Wet   0.0130 0.1650 0.0000 0.0000 0.0672
```

The data set and the corresponding software codes can be found in the `waterfalls` folder on the book's website.

8

Some Stochastic Ordering Problems

In many physical and biological phenomena covering a wide variety of scientific research fields, it is not unusual to encounter stochastic ordering problems where C groups (typically ordered with respect to time) are observed along with a (continuous, numeric or ordinal categorical) response. In multi-sample and multivariate problems, a parametric approach is usually very difficult to cope with. In this chapter we present some permutation approaches dealing with hypothesis testing for multivariate monotonic stochastic ordering with continuous and/or categorical variables and umbrella testing. Two applications are also discussed: one concerning the comparison of cancer growth patterns in laboratory animals and the other referring to a functional observational battery study designed to measure the neurotoxicity of perchloroethylene, a solvent used in dry cleaning.

8.1 Multivariate Ordered Alternatives

Let us assume we are interested in evaluating the dose–effect relationship in an experiment where C doses of a drug are administered to independent groups of patients. We denote by $\mathbf{X}_{ji} = (X_{1ji}, \ldots, X_{Vji}) \in \mathcal{R}^V$ the vector of responses on V variables for the ith patient randomly assigned to drug dose j, $j = 1, \ldots, C$, $i = 1, \ldots, n_j$, and let the total number of observations be $n = \sum_{j=1}^{C} n_j$. We assume that $\mathbf{X}_{j1}, \ldots, \mathbf{X}_{jn_j}$ are n_j i.i.d. random vectors with a continuous distribution function $\mathbf{F}_j$ defined on $\mathcal{R}^V$ and finite mean $\mathbb{E}(\mathbf{X}_j) = \boldsymbol{\mu}_j$, $j = 1, \ldots, C$, and we use F_{hj} to denote the hth marginal distribution for $\mathbf{F}_j$, $h = 1, \ldots, V$. When comparing increasing doses of a treatment, the C distributions are usually assumed to be stochastically ordered. While inference based on stochastically ordered univariate random variables has been studied extensively in the literature (for additional information, see Ross, 1983; Stoyan, 1983; see also Example 7, 4.6 and Section 6.5), less attention has been paid to the study of stochastically ordered random vectors (Marshall and Olkin, 1979). In fact, the C multivariate distributions are comparable for the stochastic ordering $\overset{d}{\leq}$, that is,

$$\mathbf{X}_1 \overset{d}{\leq} \ldots \overset{d}{\leq} \mathbf{X}_C$$

if and only if $\mathbb{E}[g(\mathbf{X}_1)] \leq \ldots \leq \mathbb{E}[g(\mathbf{X}_C)]$ holds for all increasing functions $g : \mathcal{R}^V \rightarrow \mathcal{R}$ such that the expectation exists. It is of interest to test the null hypothesis

$$H_0 : \left\{\mathbf{X}_1 \overset{d}{=} \ldots \overset{d}{=} \mathbf{X}_C\right\},$$

stating the equality in distribution ($\stackrel{d}{=}$) of the $\mathbf{X}_j$, $j = 1, \ldots, C$, against the alternative H_1: $\{\mathbf{X}_1 \stackrel{d}{\leq} \ldots \stackrel{d}{\leq} \mathbf{X}_C\}$, where at least one inequality is strict.

We point out that when applying a test of hypotheses, either H_0 or H_1 is assumed to be true, hence we must make the prior assumption that $\mathbf{X}_1 \stackrel{d}{\leq} \ldots \stackrel{d}{\leq} \mathbf{X}_C$ holds. From Corollary 3 in Bacelli and Makowski (1989), it follows that under the above assumption, H_0 holds if and only if $\mathbf{X}_1, \ldots, \mathbf{X}_C$ have the same marginal distributions, namely,

$$F_{h1}(t) = \ldots = F_{hC}(t), \quad \forall t \in \mathcal{R}, \quad h = 1, \ldots, V.$$

As a consequence, H_1 holds if and only if $X_{1h} \stackrel{d}{\leq} \ldots \stackrel{d}{\leq} X_{hC}$, $h = 1, \ldots, V$, and $\mathbf{X}_1, \ldots, \mathbf{X}_C$ are not equal in distribution, expressed as

$$F_{h1}(t) \geq \ldots \geq F_{hC}(t), \quad \forall t \in \mathcal{R}, \quad h = 1, \ldots, V,$$

with at least one strict inequality holding at some $t \in \mathcal{R}$ for at least one h. When dealing with more complex problems, the previous hypotheses may be written in the form

$$H_0 : \left\{\bigcap_{h=1}^{V} (F_{h1}(t) = \ldots = F_{hC}(t), \ \forall t \in \mathcal{R})\right\} = \left\{\bigcap_{h=1}^{V} H_{0h}\right\}$$

and

$$H_1 : \left\{\bigcup_{h=1}^{V} (F_{h1}(t) \geq \ldots \geq F_{hC}(t), \ \forall t \in \mathcal{R} \text{ and not } H_{0h})\right\} = \left\{\bigcup_{h=1}^{V} H_{1h}\right\},$$

where a suitable and meaningful breakdown of H_0 and H_1 into a set of partial hypotheses is emphasized. In the nonparametric framework, let us consider, for the sake of simplicity, the class $\mathcal{P}_C$ of all continuous distribution functions on $\mathcal{R}^V$. The simultaneous stochastically ordered random variable model is

$$\mathcal{P}_C^{\otimes C} = \{\mathbf{F}_1, \ldots, \mathbf{F}_C \in \mathcal{P}_C : F_{h1}(t) \geq \ldots \geq F_{hC}(t), \ \forall t \in \mathcal{R}, h = 1, \ldots, V\},$$

where the symbol $\otimes$ denotes the Kronecker product. If we assume that the distributions $\mathbf{F}_1, \ldots, \mathbf{F}_C$ differ only in their locations, that is, $\mathbf{F}_j(\mathbf{t}) = \mathbf{F}(\mathbf{t} + \boldsymbol{\delta}_j)$, $\forall \mathbf{F} \in \mathcal{P}_C$, $\forall \mathbf{t}, \boldsymbol{\delta}_j \in \mathcal{R}^V$, $j = 1, \ldots, C$, then the previous model comes down to the homoscedastic location model

$$\mathcal{P}_\delta = \left\{\mathbf{F} \in \mathcal{P}_C : \mathbf{F}_1(\mathbf{t} - \boldsymbol{\delta}_1) = \ldots = \mathbf{F}_C(\mathbf{t} - \boldsymbol{\delta}_C), \ \forall \mathbf{t}, \boldsymbol{\delta}_1 \leq \ldots \leq \boldsymbol{\delta}_C \in \mathcal{R}^V\right\},$$

where $\mathbf{a} \leq \mathbf{b}$ means $a_h \leq b_h$ for $h = 1, \ldots, V$. Once more, for the sake of simplicity, effects δ_j, $j = 1, \ldots, C$, are assumed to be fixed given that in the parametric framework it is easier to work with fixed effects than random effects. On the other hand, in the nonparametric framework, this problem does not occur (see the next section for the extension to stochastic effects). Let $\mathcal{P}_{\mathcal{N}_V}$ be the class of all V-variate normal distributions; the normal homoscedastic location model corresponds to

$$\mathcal{P}_{\mathcal{N}_V} = \left\{\mathbf{F} \in \mathcal{P}_{\mathcal{N}_V} : \mathbf{F}_1(\mathbf{t} - \boldsymbol{\delta}_1) = \ldots = \mathbf{F}_C(\mathbf{t} - \boldsymbol{\delta}_C), \ \forall \mathbf{x}, \boldsymbol{\delta}_1 \leq \ldots \leq \boldsymbol{\delta}_C \in \mathcal{R}^V\right\}.$$

Obviously, the inclusion relation $\mathcal{P}_{\mathcal{N}V} \subseteq \mathcal{P}_\delta \subseteq \mathcal{P}_C^{\otimes C}$ holds true. Under the homoscedastic location model, the testing problem becomes

$$H_0^\dagger : \left\{\bigcap_{h=1}^{V} (\mu_{h1} = \ldots = \mu_{hC})\right\} = \left\{\bigcap_{h=1}^{V} H_{0h}^\dagger\right\}$$

where $\boldsymbol{\mu}_{hj}, h = 1, \ldots, V, j = 1, \ldots, C$, is assumed finite, against

$$H_1^{\dagger} : \left\{ \bigcup_{h=1}^{V} \left(\mu_{h1} \leq \ldots \leq \mu_{hC}, \text{ and not } H_{0h}^{\dagger} \right) \right\} = \left\{ \bigcup_{h=1}^{V} H_{1h}^{\dagger} \right\}.$$

In the literature a variety of statistics have been proposed for testing the equality of vector means against a multivariate ordered alternative under the homoscedastic location model, that is, for testing $H_0^{\star}$ against $H_1^{\star}$. A brief outline of the literature is given in Finos et al. (2007). In particular, the authors propose a further decomposition of the global problem into sub-hypotheses, such that each component problem can be tested using simple two-sample test statistics for stochastic ordering alternatives. As test statistics they use the permutation counterpart of the traditional one-sided Student's t test, denoted by $T_{hk}(\mathbf{X})$. They then combine all the partial p-values by means of NPC methodology. As is well known, Tippett (ψ_T), Fisher (ψ_F) and Liptak (ψ_L) are other possible combining functions. In order to emphasize the chosen combining functions, the notation $T''^{o}_{\psi_s \psi_f}$ and $\lambda''^{o}_{\psi_s \psi_f}$, where $f, s \in \{T, F, L\}$, is used.

Of course, the p-value $\lambda''^{o}_{\psi_s \psi_f}$ of combined test $T''^{o}_{\psi_s \psi_f}$ depends in general on which combining functions ψ_f and ψ_s are used because different admissible combining functions have different convex acceptance regions. In particular they iterate the combining procedure (Salmaso and Solari, 2006) by applying the three combining functions T, F, L to the same partial tests, and then combining the resulting p-values using one combining function. Fisher's iterated combining function is denoted by

$$\psi_I = \psi_F \left(\lambda_{\psi_T}, \lambda_{\psi_F}, \lambda_{\psi_L} \right).$$

For details, see Finos et al. (2007).

8.2 Testing for Umbrella Alternatives

In a one-way ANOVA design, the response may stochastically increase up to a point as the treatment level increases, then decrease despite further increases in the treatment level. This up-then-down behaviour is known in the literature as *umbrella ordering* (see Mack and Wolfe, 1981; Basso and Salmaso, 2009a). In these circumstances it may be of interest to find the change-point group, that is, the group in which an inversion in the trend of the variable under study is observed. A change point is not merely a maximum (or minimum) of the regression function, since a further requirement is that the trend is monotonically increasing prior to that point, and monotonically decreasing afterwards. Despite the numerous solutions proposed in the parametric framework, there is a lack of literature concerning nonparametric permutation proposals for umbrella alternatives, apart from some hints in Manly (1997) and the recent paper by Neuhäuser et al. (2003). Basso and Salmaso (2009a) provide a solution within a conditional approach, by considering some NPCs of dependent tests for simple stochastic ordering problems. The proposed procedure is in fact found to be very flexible and potentially adaptable to trend and/or repeated measures problems and works even with very small sample sizes (says two replicates for each treatment) and/or in unbalanced cases. Hence, its use is recommended when small sample sizes are present or when data cannot be assumed to follow a known distribution. The context is that of a one-way ANOVA design, where the experimental factor levels (time, increasing doses of a drug, etc.) determine the treatments which identify the C groups.

Let us consider a fixed effects additive response model,

$$\begin{aligned}\mathbf{X} &= \left\{ X_{ji} = \mu + \delta_j + Z_{ji}, i = 1, \ldots, n_j, j = 1, \ldots, C \right\} \\ &= \left\{ \mathbf{X}_j, j = 1, \ldots, C \right\},\end{aligned}$$

where μ is the population mean, δ_j are fixed treatment effects which satisfy the contrast condition $\sum_j \delta_j = 0$, Z_{ji} are exchangeable errors with zero mean and finite variance σ^2 (typically i.i.d. random variables, independent of the δ_j), and n_j are fixed sample sizes. Note that, in this model, responses are assumed to be homoscedastic. For the sake of simplicity, fixed treatment effects have been considered; however, the solution remains valid when assuming generalized models with random effects Δ_{ji}. Let $F_j(x)$ be the CDF of the response variable in group j. Then we wish to assess the null hypothesis of no treatment effect,

$$H_0: \{F_1(t) = F_2(t) = \cdots = F_C(t), \quad \forall t \in \mathcal{R}\},$$

against the umbrella alternative hypothesis,

$$H_1: \{F_1(t) \geq \cdots \geq F_{j-1}(t) \geq F_j(t)$$
$$\leq F_{j+1}(t) \leq \cdots \leq F_C(t)\}$$

for some $j \in \{1, \ldots, C\}$, and with at least one strict inequality. That is to say, the aim of the study is to find the change-point group j (if it exists). In such problems, therefore, it is of interest to determine whether there is an umbrella behaviour due to the experimental factor, and which is the change-point group. A parametric solution to this problem is very difficult, especially when $C > 2$. Mack and Wolfe (1981) suggested a nonparametric rank solution. In particular, they suggested a test statistic for umbrella alternatives obtained as a weighted linear combination of standardized Mann–Whitney statistics, along with their null distributions, in a wide variety of situations. Basso and Salmaso (2009a) describe a permutation approach which is conditional on the observed data considered as a set of sufficient statistics in H_0 for the problem at hand. In fact, if the peak group is known, say the $\hat{j}$th group, the problem of umbrella alternatives can be simplified in the intersection of two simple stochastic ordering alternatives (one increasing and one decreasing) and detected by combining two partial tests for simple stochastic ordering alternatives, that is,

$$H_1 = H_{1\hat{j}}^{\nearrow} \bigcap H_{1\hat{j}}^{\searrow}$$

where: $H_{1\hat{j}}^{\nearrow}: \{F_1(t) \geq \cdots \geq F_{\hat{j}-1}(t) \geq F_{\hat{j}}(t)\}$ and $H_{1\hat{j}}^{\searrow}: \{F_{\hat{j}}(t) \leq F_{\hat{j}+1}(t) \leq \cdots \leq F_C(t)\}$.

8.2.1 Hypotheses and Tests in Simple Stochastic Ordering

Under the assumptions of the additive model with fixed effects previously discussed, let us consider the simple stochastic ordering problem for the first $\hat{j}$ samples to assess the null hypothesis $F_1(t) = \ldots = F_{\hat{j}}(t)$ against the alternative hypothesis $F_1(t) \geq \ldots \geq F_{\hat{j}}(t)$. Note that under the null hypothesis, all elements of the response $\mathbf{X}_{\hat{j}}$ are exchangeable. This will enable us to provide the null distribution of a proper test statistic. If $\hat{j} = 2$, then the stochastic ordering problem reduces to a two-sample problem with restricted alternatives. If $\hat{j} > 2$, then let us suppose that the whole data set is split into two pooled pseudo-groups, where the first is obtained by pooling together data from the first c groups (ordered with respect to the treatment levels), and the second by pooling together the remaining observations.

Let $\mathbf{Y}_{(1)c} = \mathbf{X}_1 \uplus \cdots \uplus \mathbf{X}_c$ be the first (ordered) pseudo-group and let $\mathbf{Y}_{(2)c} = \mathbf{X}_{c+1} \uplus \cdots \uplus \mathbf{X}_{\hat{j}}$ be the second (ordered) pseudo-group, $c = 1, \ldots, \hat{j} - 1$. Let $Y_{1(c)}$ and $Y_{2(c)}$ be the random variables describing the generic observation of the pooled vectors $\mathbf{Y}_{(1)c}$ and $\mathbf{Y}_{(2)c}$, respectively. In the null hypothesis, data of every pair of pseudo-groups are exchangeable because the related variables satisfy the relationships $Y_{1(c)} \stackrel{d}{=} Y_{2(c)}$, $c = 1, \ldots, \hat{j} - 1$. Under the alternative, let us assume $Y_{1(c)} \stackrel{d}{\leq}$

$Y_{2(c)}$, that is, the monotonic stochastic ordering (dominance) between any pair of pseudo-groups (i.e. for $c = 1, \ldots, \hat{j} - 1$). Hence, it is possible to formulate the hypotheses as

$$H_0 : \left\{ \bigcap_{c=1}^{\hat{j}-1} (Y_{1(c)} \overset{d}{=} Y_{2(c)}) \right\}$$

against

$$H_{1\hat{j}}^{\nearrow} : \left\{ \bigcup_{c=1}^{\hat{j}-1} (Y_{1(c)} \overset{d}{\leq} Y_{2(c)}) \right\},$$

where a breakdown into a set of sub-hypotheses is emphasized.

Let us turn our attention to the cth sub-hypothesis $H_{0c} : \{Y_{1(c)} \overset{d}{=} Y_{2(c)}\}$ against $H_{1c} : \{Y_{1(c)} \overset{d}{\leq} Y_{2(c)}\}$. Note that the related sub-problem corresponds to a two-sample comparison for restricted alternatives, a problem which has an exact and unbiased permutation solution. This solution may be based (among others) on the test statistics:

$$T_{c\nearrow}^* = \frac{\bar{Y}_{2(c)}^* - \bar{Y}_{1(c)}^*}{\sqrt{\hat{\sigma}_c^2 \left(\frac{1}{n_{(1)c}} + \frac{1}{n_{(2)c}} \right)}}, \quad c = 1, \ldots, \hat{j} - 1,$$

where $\bar{Y}_{2(c)}^*$ and $\bar{Y}_{1(c)}^*$ are permutation sample means of the second and first pseudo-groups respectively, $\hat{\sigma}_c^2$ is the pooled estimate of the error variance, and $n_{(1)c}$ and $n_{(2)c}$ are the lengths of $\mathbf{Y}_{(1)c}$ and $\mathbf{Y}_{(2)c}$ respectively. Note that standardized statistics are suggested here in order to apply the direct combining function. Large values of the test statistics $T_{c\nearrow}$ are significant against $H_{0(c)} : \{Y_{1(c)} \overset{d}{=} Y_{2(c)}\}$ in favour of the alternatives $H_{1(c)}^{\nearrow} : \{Y_{1(c)} \overset{d}{\leq} Y_{2(c)}\}$. The algorithm for obtaining the permutation test for $H_{0(c)}$ against $H_{1(c)}^{\nearrow}$ may easily be derived from the general algorithm for NPC.

8.2.2 *Permutation Tests for Umbrella Alternatives*

In practice, the peak group is generally unknown. Basso and Salmaso (2009a) therefore suggest detecting it by repeating the procedure for the known peak *as if* every group were the known peak group. In this context, we report only the algorithm and final results. However, we refer the reader to the original paper for further details and theoretical matters.

For each $j \in 1, \ldots, C$, let

$$\psi_{j\nearrow}^* = \sum_{c=1}^{j-1} T_{c\nearrow}^* \quad \text{and} \quad \psi_{j\searrow}^* = \sum_{c=j}^{C-1} T_{c\searrow}^*$$

be two partial tests to assess H_{0k}: $\{F_1(t) = \ldots = F_k(t)\}$ against respectively $H_{1k}^{\nearrow}$ and $H_{1k}^{\searrow}$ by applying the direct NPC of the partial tests $T_{j\nearrow}^*$ and $T_{j\searrow}^*$. Note that when $k = 1$ we actually test for decreasing ordering only, whereas when $k = C$ we only test for increasing ordering. The next steps are as follows:

- Obtain the partial p-values to assess H_{0j} against $H_{1j}^{\nearrow}$ and $H_{1j}^{\searrow}$, respectively. Let $(p''_{j\nearrow}, p''_{j\searrow})$ be the pair of p-values from the observed data.
- Obtain the null distribution of the pair of p-values to assess H_{0j} against $H_{1j}^{\nearrow}$ and $H_{1j}^{\searrow}$, respectively. This will be denoted by the pair $({}^b p''^*_{j\nearrow}, {}^b p''^*_{j\searrow})$, $b = 1, \ldots, B$. That is, $({}^b p''^*_{j\nearrow}, {}^b p''^*_{j\searrow})$ is obtained by applying the previous algorithm for simple stochastic ordering alternatives and by replacing $\mathbf{X}$ with $\mathbf{X}_b^*$.

- Obtain the observed value of the test statistic with Fisher's NPC function,

$$\Psi_j = -2\log(p''_{j\nearrow} \cdot p''_{j\searrow}).$$

- Obtain the null distribution of Ψ_j by computing

$$^b\Psi_j^* = -2\log(^b p''^*_{j\nearrow} \cdot {}^b p^{*''}_{j\searrow}), \quad b = 1, \ldots, B.$$

- Obtain the p-value for the umbrella alternative on group j as:

$$\lambda_j = \frac{\#[^b\Psi_j^* \geq \Psi_j]}{B}.$$

Note that if λ_j is significant, then there is evidence in the data of an umbrella alternative with peak group j. In order to evaluate whether there is a significant presence of any umbrella alternative, the p-values for the umbrella alternative of each group should be combined by means of NPC methodology. Hence, the following steps should be followed:

- Obtain the null distribution of the p-value for the umbrella alternative on group j as

$$^b\lambda_j^* = \frac{\#[\boldsymbol{\Psi}^* \geq {}^b\Psi_j^*]}{B}, \qquad b = 1, \ldots, B,$$

  where $\boldsymbol{\Psi}^*$ is the vector with the permutation null distribution of λ_j.
- Apply Tippett's combining function to the λ_j, providing the observed value of the global test statistic for the umbrella alternative in any group as $\boldsymbol{\lambda} = \min(\lambda_1, \ldots, \lambda_C)$. Note that small values of $\boldsymbol{\lambda}$ are significant against the null hypothesis.
- Calculate the null distribution of $\boldsymbol{\lambda}''$ given by $^b\boldsymbol{\lambda}^* = \min(^b\lambda_1^*, \ldots, {}^b\lambda_C^*), b = 1, \ldots, B.$
- Obtain the global p-value,

$$\boldsymbol{\lambda}'' = \frac{\#[^b\boldsymbol{\lambda}^* \leq \boldsymbol{\lambda}]}{B}.$$

Note that the combining functions are applied simultaneously to each random permutation, providing the null distributions of partial and global tests as well. The NPC methodology applies three times:

1. when obtaining simple stochastic ordering tests to assess $H_{1j}^{\nearrow}$ and $H_{1j}^{\searrow}$ for the jth group (by using the *direct* combining function), according to the multi-aspect procedure (see Example 3, 4.6);
2. when combining two partial tests for simple stochastic ordering alternatives, providing a test for umbrella for each group as if it were the known peak group (here by applying Fisher's combining function);
3. when combining the partial test for umbrella alternatives on each group (by using Tippett's combining function).

A significant global p-value indicates that there is evidence in favour of an umbrella alternative. The peak group is then identified by looking at the partial p-values for umbrella alternatives $\{\lambda_1, \ldots, \lambda_k, \ldots, \lambda_C\}$. The peak group (if any) is then the one with the minimum p-value.

The proposed algorithm may still apply with different combining functions in the first two steps, but not in the third. This is because Tippett's combining function is powerful even when only one of its arguments is in the alternative. As regards our choices in the first two steps of the algorithm, the direct combining function in step 1 was chosen for computational reasons, whereas Fisher's combining function in step 2 was applied because use of Fisher's combining function is recommended when no specific knowledge of the sub-alternatives is available.

In comparison with the Mack–Wolfe test, the authors find that the rejection rates of the proposed global test (Basso and Salmaso, 2009a), under the null hypothesis H_0, are close to the nominal rates and that, under the alternative hypothesis H_1, the power of the Mack–Wolfe test is very close to that of the proposed permutation test.

When simulating an anti-umbrella alternative, that is, data are not under the null hypothesis, but the alternative hypothesis is not of the umbrella type, the authors show that the rejection rates of the global tests are always lower than the related nominal levels.

Indeed, the simulations show that the probability of having ties (i.e. more than one estimated peak group) is higher when the Mack–Wolfe test is applied (see the complete simulation study in Basso and Salmaso, 2009a).

Umbrella alternative and trend problems are often associated with repeated measures experiments, where n units are subjected to the same treatment (e.g. increasing doses of drugs) and the response is measured at several time points. The permutation test for umbrella alternatives, proposed in Basso and Salmaso (2009a), can easily be modified to account for repeated measures (by changing the permutation strategy) and/or trend analysis (by looking at the first/last partial p-value). With reference to this topic, the authors also provide a comparison with Page's test (1963) for trend analysis with correlated data. Under H_0, they find that the permutation test controls the type I error, both when data are uncorrelated and when they are correlated, while Page's test seems to be conservative when data are uncorrelated. Under the alternative H_1, Page's test displays a gain in power when considering correlated data (see the complete simulation study in Basso and Salmaso, 2009a).

8.3 Analysis of Experimental Tumour Growth Curves

The problem described in this section concerns an application of multidimensional permutation methods to the comparison of cancer growth patterns in laboratory animals. The data come from an experiment which aims to reveal the anti-cancer activity of a taxol-like synthetic drug (Barzi et al., 1999a, 1999b).

The main characteristic of malignant tissue is rapid growth due to uncontrollable cellular division. The drug under trial attaches itself to intracellular structures and kills the tumour cells at the moment of their division, making cellular proliferation more difficult, if not impossible. This explains the regression and delay phase of growth in treated tumours observed during the initial period. After a certain time interval, tumour cells may show resistance to the taxol synthetic products and normal proliferation may start again (regrowth phase). We are interested in testing whether the growth inhibition effect of a treatment is statistically significant and, if so, evaluating how long the effect remains significant. It is also of interest to assess differences among groups in time intervals between drug administration and regrowth phase.

When comparing treated growths with untreated growths, several multivariate stochastic ordering problems for repeated measurement designs arise. A parametric approach to hypothesis testing is not possible because the number of experimental units is less than the number of repeated measures (Crowder and Hand, 1990; Diggle et al., 2002). Furthermore, we cannot really model the growth patterns as being Gompertz or exponential (Bassukas and Maurer-Schulze, 1988; Bassukas, 1994) because of the initial regression and subsequent regrowth of tumour volumes in treated groups. Hence, we propose a nonparametric permutation-based approach to analyse this kind of growth pattern.

The animals used in the experiment were nude mice. Tumour specimens about 2–3 mm in diameter were injected into both flanks of each animal. When tumour masses reached a diameter of 4–8 mm, the mice were randomized into four groups of six units each. The first (so-called control) group was treated with a placebo; three other groups were treated with increasing doses of a taxol synthetic product. Treatments were administered only three times at intervals of 4 days in order to reduce possible toxic side-effects of the drugs.

The tumour under study is a particular ovarian type which is resistant to the anti-tumour activity of non-synthetic taxol. Tumour mass volumes were determined every 4 days by calibrated measurements of two perpendicular diameters. Since not all the inoculated tumour specimens grew, we obtained the following sample sizes for the four groups under study: $n_1 = 11$, $n_2 = 9$, $n_3 = 7$, $n_4 = 9$. Measurements were repeated on all units on 18 time occasions. The response variable considered was the 'relative volume', $X_{jti} = V_{jti}/V_{j0i}$, $i = 1, \ldots, n_j$, $t = 1, \ldots, 17$, $j = 1, \ldots, 4$, where V_{j0i} is the baseline volume and V_{jti} is the tumour volume for the ith individual at time t in the jth group.

In order to test whether treatment significantly inhibits tumour growth or whether one treatment is significantly more efficient than another, we can model the problem according to the hypotheses

$$H_0 : \left\{ \bigcap_{t=1}^{17} \left(X_{1t} \overset{d}{=} \ldots \overset{d}{=} X_{4t} \right) \right\}$$

against

$$H_1 : \left\{ \bigcup_{t=1}^{17} \left(X_{1t} \overset{d}{>} X_{2t} \overset{d}{>} X_{3t} \overset{d}{>} X_{4t} \right) \right\},$$

where the latter emphasizes the monotonic stochastic ordering of time profiles with respect to increasing treatment levels. The related 17×3 partial tests are $T^*_{tj} = \sum_{1 \le i \le N_{1(j)}} Y^*_{1t(j)i}$, $j = 1, \ldots, 3$, $t = 1, \ldots, 17$, where $\mathbf{Y}_{1t(j)} = \mathbf{X}_{1t} \biguplus \ldots \biguplus \mathbf{X}_{jt}$ and $\mathbf{Y}_{2t(j)} = \mathbf{X}_{j+1t} \biguplus \ldots \biguplus \mathbf{X}_{4t}$, $j = 1, \ldots, 3$, are respectively the pseudo-groups obtained by an orderly pooling of the first j groups into $\mathbf{Y}_{1t(j)}$ and the others into $\mathbf{Y}_{2t(j)}$, $\mathbf{X}_{jt} = \{X_{jit}, i = 1, \ldots, n_j\}$ is the data set in the jth group at time t, and $N_{1(j)} = \sum_{r \le j} n_r$ is the sample size of $\mathbf{Y}_{1t(j)}$. As all these partial tests are marginally unbiased under the stated conditions, NPC provides a proper solution.

We obtained the results shown in Table 8.1 using MATLAB with different combination functions both for partial tests and for the overall test; the number of CMC iterations is $B = 10\,000$. The third-order combined p-value, using Fisher's combination function, for the overall hypothesis gives $\hat{\lambda}'''_F = 0.0030$; using Tippett's function it gives $\hat{\lambda}'''_T = 0.0010$; and using Liptak's it gives $\hat{\lambda}'''_L = 0.0040$. This leads to the rejection of the null hypothesis of absence of a stochastic ordering at a significance level of $\alpha = 0.001$. From the above analysis we may conclude that the inhibitory effect of the dose administered to the fourth group is significantly the most relevant, by controlling the FWE.

```
[D,data,code]=xlsimport('rats','Foglio1');
reminD(D)
[no Ts]=v(3:size(D,2));
[P,T] = NP_StOrd(Ts,'GROUP',1000,-1,'F',1);
pf=NPC(P,'F');
pfFWE=NPC_FWE(P,'T');
[P,T] = NP_StOrd(Ts,'GROUP',1000,-1,'L',1);
pl=NPC(P,'L');
plFWE=NPC_FWE(P,'T');
[P,T] = NP_StOrd(Ts,'GROUP',1000,-1,'T',1);
pt=NPC(P,'T');
ptFWE=NPC_FWE(P,'T');
```

We also show how to perform the same analysis using R. As already stated, the aim of the analysis is to test whether there exists a stochastic ordering with respect to groups, time by time. Here groups correspond to increasing doses of a new taxol synthetic product (group 1 was treated

Table 8.1 Partial and adjusted p-values from the tumour growth data analysis

	Fisher		Liptak		Tippett	
	p-value	p-FWE	p-value	p-FWE	p-value	p-FWE
X_1	0.7622	0.7622	0.8621	0.8621	0.6084	0.6084
X_2	0.0010	0.0030	0.0080	0.0160	0.0030	0.0010
X_3	0.0030	0.0060	0.0050	0.0140	0.0030	0.0010
X_4	0.0010	0.0030	0.0010	0.0040	0.0010	0.0010
X_5	0.0010	0.0030	0.0010	0.0040	0.0010	0.0010
X_6	0.0010	0.0030	0.0010	0.0040	0.0010	0.0010
X_7	0.0010	0.0030	0.0010	0.0040	0.0010	0.0010
X_8	0.0010	0.0030	0.0010	0.0040	0.0010	0.0010
X_9	0.0010	0.0030	0.0010	0.0040	0.0020	0.0020
X_{10}	0.0010	0.0030	0.0010	0.0040	0.0010	0.0010
X_{11}	0.0010	0.0030	0.0010	0.0040	0.0010	0.0010
X_{12}	0.0010	0.0030	0.0010	0.0040	0.0010	0.0010
X_{13}	0.0010	0.0030	0.0010	0.0040	0.0010	0.0010
X_{14}	0.0010	0.0030	0.0010	0.0040	0.0010	0.0010
X_{15}	0.0010	0.0030	0.0010	0.0040	0.0020	0.0020
X_{16}	0.0010	0.0030	0.0010	0.0040	0.0030	0.0010
X_{17}	0.0010	0.0030	0.0010	0.0040	0.0010	0.0010
Global (T)		0.0030		0.0040		0.0010

with a placebo). The alternative hypothesis is the existence of a significant decrease of the tumour mass as the dose of the taxol synthetic product increases. The analysis is run with the aid of the `stoch.ord` function, whose arguments are a vector of continuous data (here the columns corresponding to each time), and a categorical variable indicating time or grouping (here the vector `g`). Possible choices of the alternatives are `alt = -1` (decreasing trend with respect to grouping) and `alt = 1` (increasing trend). The testing algorithm is described at the beginning of this section, and we only considered Fisher's combining function to collect the information of the pseudo-groups. The function can process one variable at time; in order to maintain the dependence among the response variables the function is always run with a common `seed`, which will be the input to the `set.seed` function (thus providing exactly the same permutations each time).

```
setwd("C:/path")
source("t2p.r") ; source("stoch_ord.r") ; source("FWEminP.r")
data<-read.csv("rats.csv",header=TRUE)
g = data[,1]
B=10000
data=data[,-c(1,2)]
p=dim(data)[2]
P=array(0,dim=c((B+1),p))

for(j in 1:p){
P[,j] = stoch.ord(data[,j],g,alt=-1,B=B)
```

```
cat(j,"variable/",p,"\n")
}

colnames(P)=colnames(data)
P.FWE = FWE.minP(P)
res = cbind(P[1,],P.FWE)
colnames(res)=c("Raw-p","Adj-p")
res

      Raw-p  Adj-p
T_1   0.7840 0.7840
T_2   0.0009 0.0016
T_3   0.0000 0.0000
T_4   0.0000 0.0000
T_5   0.0000 0.0000
T_6   0.0000 0.0000
T_7   0.0000 0.0000
T_8   0.0000 0.0000
T_9   0.0000 0.0000
T_10  0.0000 0.0000
T_11  0.0000 0.0000
T_12  0.0000 0.0000
T_13  0.0000 0.0000
T_14  0.0000 0.0000
T_15  0.0000 0.0000
T_16  0.0000 0.0000
T_17  0.0000 0.0000
```

The above results are raw and adjusted p-values assessing the alternative hypothesis of a decreasing tumour mass as the dose of taxol synthetic product increases. It is evident that there is always a significant reduction of the tumoral mass except at time 1. The data set and the corresponding software codes can be found in the `rats` folder on the book's website.

8.4 Analysis of PERC Data

8.4.1 Introduction

As is known, a more realistic assessment of the benefits or risks of a treatment and its association with dose may be obtained by jointly considering the several collected variables, rather than ignoring some or analysing them separately (Klingenberg et al., 2008). Moreover, due to the fact that a dose effect may manifest itself in a subject in a wide variety of ways, it is crucial that these endpoints are observed jointly so as not to miss out on any effects or interactions.

The functional observational battery (FOB; Moser, 1989, 1992) is a biological screening assay comprising roughly 25 endpoints designed to evaluate neurophysiological effects in animals after exposure to a toxin. In such studies, it is assumed that an adverse effect of a substance is manifested as a distinct change in an animal's physiological or behavioural response (Han et al., 2004).

The FOB tries to group endpoints by a common domain, each domain describing a possibly distinct neurological function. It is often the case that some of the recorded endpoints are strongly connected to each other, hence they may express similar effects and thus be redundant. For this reason, Moser (1992) suggested grouping the individual endpoints into six domains, containing both discrete and continuous endpoints, referring to precise areas of neurological dysfunction.

Table 8.2 Variables and domains

Domain name	Code	Description
Autonomic	lacrimation	Lacrimation
	palp_closure	Palpebral closure
	salivation	Salivation
	urination	Urination
	defecation	Defecation
	pupil	Pupil response
Sensorimotor	approach	Approach response
	touch	Touch response
	click	Click response
	tail_pinch	Tail pinch response
Excitability	removal	Ease of removal
	handling	Handling reactivity
	clonic_mvts	Clonic movements
	tonic_mvts	Tonic movements
	vocalizations	Vocalizations
	arousal	Arousal
Activity	hc_posture	Home cage posture
	rears	Rearing
	motor_activity	Motor activity (counts)
Neuromuscular	gait	Gait score
	mobility	Mobility
	righting	Righting reflex
	foot_splay	Landing foot splay
	forelimb_grip	Forelimb grip strength
	hindlimb_grip	Hindlimb grip strength
Physiological	piloerection	Piloerection
	weight	Body weight
	temperature	Body temperature

These domains are assumed to be mutually exclusive, that is, each domain measures a specific component of neurotoxicity (Han et al., 2004). However, particular care is required in grouping these variables: if the endpoints are not properly grouped, this strategy may fail to identify true effects, thus increasing dependence across domains.

The FOB discussed here comprises home cage, handling, open-field and manipulative behavioural measures, as well as physiological measures (McDaniel and Moser, 1993). Table 8.2 summarizes the variables involved in an FOB study designed to measure the neurotoxicity of perchloroethylene (PERC), a solvent used in dry cleaning, which has many chlorinated hydrocarbons and is a central nervous system depressant (also suspected to cause leukaemia).

A total of 36 animals were randomly assigned to either placebo or one of four PERC exposure levels of (five dose groups in total: 0, 150, 500, 1500 and 5000g), and the FOB was administered at several time intervals (0, 4 and 24 hours). Each animal was evaluated at 28 endpoints (22 ordinal and 6 continuous). Hence, this data set comprises mixed variables. Moreover, the variables can be classified into six domains (autonomic, sensorimotor, excitability, activity, neuromuscular and physiological). Since each mouse has been observed three times, we have 108 observations in total. The statistical analysis of these data structures is challenging, since

- the endpoints are mixed in nature (continuous and discrete variables);
- the underlying contingency tables for the multivariate categorical responses are very sparse and unbalanced;
- the endpoints are closely related to each other. As a consequence, determining the extent of correlation between endpoints, within and across domains, is crucial to properly measure the effects.

Hence, the presence of a large number of endpoints, repeated measures, and the mixed nature of the variables evaluated in FOB studies complicate standard statistical analyses, making it necessary to model complex correlation patterns.

Han et al. (2004) use data from an FOB analysis to illustrate methods for testing a dose–response relationship with multiple correlated binary responses. They transformed the original ordinal responses into binary ones (absence/presence of endpoint), thereby losing information on severity. In fact, they propose an exact test for multiple binary outcomes under the assumption that the correlation among these items is equal and that the endpoints are independent across different domains (Han et al., 2004). This test is based both upon an exponential model described by Molenberghs and Ryan (1999) and an exact test for exchangeable correlated binary data for groups of correlated observations developed by Corcoran et al. (2001). Here we present a nonparametric permutation analysis of this data set, treating data as multivariate responses (both ordinal and continuous) with a general correlation structure that we do not need to model explicitly, since the dependencies among variables are implicitly captured by the permutation process itself.

8.4.2 A Permutation Solution

The whole data set may be represented as

$$\begin{aligned}\mathbf{X} &= \{X_{hji}(t), t = 1, \ldots, k, i = 1, \ldots, n_j, j = 1, \ldots, C, h = 1, \ldots, V\} \\ &= \{\mathbf{X}_{hji}, i = 1, \ldots, n_j, j = 1, \ldots, C, h = 1, \ldots, V\},\end{aligned}$$

where $\mathbf{X}_{hji} = \{X_{hji}(t), t = 1, \ldots, k\}$ denotes the hth response of the ith subject, at exposure level j and time t. Hence, the global hypothesis testing problem may be expressed as

$$H_0 : \left\{\bigcap_{t=1}^{k}\left[\bigcap_{h=1}^{V}\left(X_{h1}(t) \stackrel{d}{=} \ldots \stackrel{d}{=} X_{hC}(t)\right)\right]\right\},$$

against

$$H_1 : \left\{\bigcup_{t=1}^{k}\left[\bigcup_{h=1}^{V}\left(X_{h1}(t) \stackrel{d}{\geq} \ldots \stackrel{d}{\geq} X_{hC}(t)\right)\right]\right\}.$$

We are in fact in the context of testing multivariate ordered alternatives. Hence, by applying the NPC methodology, we may further decompose the global null hypothesis, including, for example, the information on domains. As mentioned before, Moser (1989) introduced the domain approach, thus classifying the endpoints into six mutually exclusive sets, which may be seen as latent variables, representing distinct areas of neurological functioning. Readers are reminded that in this case $C = 5$, because of the presence of five dose groups (0, 150, 500, 1500 and 5000g). Since three measurements are available for each variable (at 0, 4 and 24 hours), the testing hypothesis problem may be rewritten as

$$H_{0l} : \left\{ \bigcap_{t=1}^{3} \left[\bigcap_{h=1}^{V_l} \left(X_{h1}^{l}(t) \stackrel{d}{=} \ldots \stackrel{d}{=} X_{h5}^{l}(t) \right) \right] \right\}, l = 1, \ldots, 6,$$

against

$$H_{1l} : \left\{ \bigcup_{t=1}^{3} \left[\bigcup_{h=1}^{V_l} \left(X_{h1}^{l}(t) \stackrel{d}{>} \ldots \stackrel{d}{>} X_{h5}^{l}(t) \right) \right] \right\}, l = 1, \ldots, 6,$$

where $\mathbf{X}_j^l(t) = \left\{ \mathbf{X}_{1j}^l(t) \ldots, \mathbf{X}_{jV_l}^l(t) \right\}$ represents the lth domain response for the jth exposure level at time t.

In summary, for each of the three times, we test the monotonic stochastic ordering of responses with respect to increasing exposure levels. Then, by means of the NPC methodology, for each time, we evaluate this effect within each domain (six domains for each time) by combining the variables belonging to the same domain. We then combine with respect to time and adjust the p-values for multiplicity, thus obtaining the information for each domain. As usual, it is of interest to know specifically what domains are statistically significant at level α after proper adjustment for multiplicity. We apply Tippett's Min$P$ stepdown procedure. Finally, we may get global p-values by combining the domains and testing the hypothesis

$$H_0 : \left\{ \bigcap_{l=1}^{6} H_{0l} \right\}$$

against

$$H_1 : \left\{ \bigcup_{l=1}^{6} H_{1l} \right\}.$$

In the following sections we illustrate the results obtained for this analysis, and give the corresponding MATLAB and R codes.

8.4.3 Analysis Using MATLAB

Results of interest are shown in Table 8.6, where it can be seen that, after adjustment for multiplicity, domains 1 and 5 (respectively autonomic and neuromuscular) are significant, as well as the global p-value. Tables 8.3–8.5 and Table 8.7 show further or intermediate analyses, time by time.

Part of the MATLAB code for carrying out the analysis is given below. We refer the reader to the `PERCH` folder on the book's website for the full version. We wish to emphasize that due to the actual nature of the variables considered, to perform the stochastic ordering we need to separately process continuous and categorical ordinal variables. In the former case, we apply the `NP_StOrd`

Table 8.3 Results from stochastic ordering and combination within domains, controlling FWE (time t_0)

Time t_0	Variable		Domain	
	p-value	p-FWE	p-value	p-FWE
lacrimation	0.4625	1.0000		
palp_closure	1.0000	1.0000		
salivation	1.0000	1.0000		
urination	0.8502	1.0000		
defecation	0.2358	0.9770		
pupil	1.0000	1.0000	0.4955	0.9331
approach	0.5664	1.0000		
touch	0.9461	1.0000		
click	0.8921	1.0000		
tail_pinch	0.9720	1.0000	0.9660	0.9980
removal	0.7543	1.0000		
handling	0.1648	0.9021		
clonic_mvts	0.7203	1.0000		
tonic_mvts	1.0000	1.0000		
vocalizations	1.0000	1.0000		
arousal	0.0519	0.5355	0.2128	0.7353
hc_posture	0.6633	1.0000		
rears	0.1449	0.8861		
motor_activity	1.0000	1.0000	0.3237	0.8502
gait	1.0000	1.0000		
mobility	1.0000	1.0000		
righting	1.0000	1.0000		
foot_splay	0.7063	1.0000		
forelimb_grip	0.7832	1.0000		
hindlimb_grip	0.6723	1.0000	0.9541	0.9980
piloerection	1.0000	1.0000		
weight	0.7113	1.0000		
temperature	0.8322	1.0000	0.9131	0.9980
p-Global		0.5355		0.7353

Table 8.4 Results from stochastic ordering and combination within domains, controlling FWE (time t_4)

Time t_4	Variable		Domain	
	p-value	p-FWE	p-value	p-FWE
lacrimation	0.9910	1.0000		
palp_closure	1.0000	1.0000		
salivation	1.0000	1.0000		
urination	0.7602	0.9980		
defecation	0.2138	0.9081		
pupil	0.9850	1.0000	0.9111	0.9111
approach	0.0010	0.0170		
touch	0.0110	0.1249		
click	0.0010	0.0170		
tail_pinch	0.0020	0.0300	0.0010	0.0060
removal	0.0739	0.5634		
handling	0.0060	0.0779		
clonic_mvts	0.3047	0.9351		
tonic_mvts	1.0000	1.0000		
vocalizations	1.0000	1.0000		
arousal	0.0250	0.2607	0.0070	0.0210
hc_posture	0.5095	0.9870		
rears	0.2218	0.9081		
motor_activity	0.0010	0.0170	0.0030	0.0120
gait	1.0000	1.0000		
mobility	1.0000	1.0000		
righting	1.0000	1.0000		
foot_splay	0.9970	1.0000		
forelimb_grip	0.0010	0.0170		
hindlimb_grip	0.0020	0.0300	0.0210	0.0410
piloerection	1.0000	1.0000		
weight	0.5824	0.9920		
temperature	0.0010	0.0170	0.0020	0.0100
p-Global		0.0170		0.0060

Table 8.5 Results from stochastic ordering and combination within domains, controlling FWE (time t_{24})

Time t_{24}	Variable		Domain	
	p-value	p-FWE	p-value	p-FWE
lacrimation	0.9990	1.0000		
palp_closure	1.0000	1.0000		
salivation	1.0000	1.0000		
urination	0.1029	0.6903		
defecation	0.1918	0.7932		
pupil	1.0000	1.0000	0.7053	0.7053
approach	0.7872	0.9960		
touch	0.0020	0.0310		
click	0.0040	0.0519		
tail_pinch	0.0010	0.0180	0.0010	0.0050
removal	0.6464	0.9960		
handling	0.0110	0.1279		
clonic_mvts	0.0030	0.0430		
tonic_mvts	1.0000	1.0000		
vocalizations	1.0000	1.0000		
arousal	0.0010	0.0180	0.0010	0.0050
hc_posture	0.8452	0.9960		
rears	0.1578	0.7812		
motor_activity	0.0080	0.0949	0.0230	0.0460
gait	0.9990	1.0000		
mobility	1.0000	1.0000		
righting	0.9920	1.0000		
foot_splay	0.1738	0.7932		
forelimb_grip	0.0010	0.0180		
hindlimb_grip	0.0010	0.0180	0.0030	0.0110
piloerection	0.4476	0.9740		
weight	0.1419	0.7812		
temperature	0.0030	0.0430	0.0040	0.0120
p-Global		0.0180		0.0050

Table 8.6 Combining domains among times, controlling FWE (time t_{24}). ψ indicates the combining function; in this case Tippett's combining function has been used

	t_0	t_4	t_{24}	$\psi(\lambda_{t_0}, \lambda_{t_4}, \lambda_{t_{24}})$	p-FWE
Domain 1 *Autonomic*	0.4955	0.9111	0.7053	0.8332	0.8332
Domain 2 *Sensorimotor*	0.9660	0.0010	0.0010	0.0020	0.0020
Domain 3 *Excitability*	0.2128	0.0070	0.0010	0.0030	0.0150
Domain 4 *Neuromuscular*	0.3237	0.0030	0.0230	0.0090	0.0220
Domain 5 *Activity*	0.9541	0.0210	0.0030	0.0070	0.0220
Domain 6 *Physiologic*	0.9131	0.0020	0.0040	0.0060	0.0220
p-Global					0.0020

Table 8.7 Combining variables among times, controlling FWE (time t_{24}). ψ indicates the combining function; in this case Tippett's combining function has been used

	t_0	t_4	t_{24}	$\psi(\lambda_{t_0}, \lambda_{t_4}, \lambda_{t_{24}})$	p-FWE
lacrimation	0.4625	0.9910	0.9990	0.7802	0.9850
palp_closure	1.0000	1.0000	1.0000	1.0000	1.0000
salivation	1.0000	1.0000	1.0000	1.0000	1.0000
urination	0.8502	0.7602	0.1029	0.2657	0.9031
defecation	0.2358	0.2138	0.1918	0.4845	0.9650
pupil	1.0000	0.9850	1.0000	1.0000	1.0000
approach	0.5664	0.0010	0.7872	0.0030	0.0440
touch	0.9461	0.0110	0.0020	0.0060	0.0739
click	0.8921	0.0010	0.0040	0.0020	0.0170
tail_pinch	0.9720	0.0020	0.0010	0.0020	0.0170
removal	0.7543	0.0739	0.6464	0.2038	0.8881
handling	0.1648	0.0060	0.0110	0.0160	0.1598
clonic_mvts	0.7203	0.3047	0.0030	0.0080	0.0829
tonic_mvts	1.0000	1.0000	1.0000	1.0000	1.0000
vocalizations	1.0000	1.0000	1.0000	1.0000	1.0000
arousal	0.0519	0.0250	0.0010	0.0030	0.0440
hc_posture	0.6633	0.5095	0.8452	0.8581	0.9910
rears	0.1449	0.2218	0.1578	0.3656	0.9570
motor_activity	1.0000	0.0010	0.0080	0.0020	0.0170

(continued overleaf)

Table 8.7 (*continued*)

	t_0	t_4	t_{24}	$\psi(\lambda_{t_0}, \lambda_{t_4}, \lambda_{t_{24}})$	p-FWE
gait	1.0000	1.0000	0.9990	1.0000	1.0000
mobility	1.0000	1.0000	1.0000	1.0000	1.0000
righting	1.0000	1.0000	0.9920	0.9970	1.0000
foot_splay	0.7063	0.9970	0.1738	0.4276	0.9650
forelimb_grip	0.7832	0.0010	0.0010	0.0020	0.0170
hindlimb_grip	0.6723	0.0020	0.0010	0.0030	0.0440
piloerection	1.0000	1.0000	0.4476	0.6803	0.9780
weight	0.7113	0.5824	0.1419	0.2108	0.8881
temperature	0.8322	0.0010	0.0030	0.0030	0.0440
					0.0170

function, in the latter we need to apply both the `NP_Cs_Categ` and `expand_categ` functions. By means of the `expand_categ` function, doses are transformed into dummy variables.

```
[D]=xlsimport('perch');
remindD(D)

varYordinal=[5:22, 24:26, 30]
varYquant=[23, 27:29, 31:32]

B=1000

D_=D;
for i=1:length(D)
    D(i).vals=D(i).vals(D_(4).vals==4);
end
remindD(D)

B=Space_perm(length(v('dose')),B);

[Pquant, T,  options_q,p_sub, T_sub] = NP_StOrd(D(varYquant),
'dose',B,-1,'F');

[doseDUMMY, orig,w,labelsDUMMY] = expand_categ('dose',2);

options_o.labels.dims3=labelsDUMMY;
options_o.tail=-1;
[Pc, Tc, options_o, P_sub, T_sub] = NP_Cs_Categ(D(varYordinal),
doseDUMMY,B,2,'F',options_o);

Pordinal=NPC(Pc,'F',options_o);

options=options_o;
```

```
options.Combdims=2;

options.labels.dims{2}=[D([varYordinal(:,1:18) varYquant(:,1)
varYordinal(:,19:21) varYquant(:,2:4) varYordinal(:,22)
varYquant(:,5:6)]).name];
P=[Pordinal(:,1:18) Pquant(:,1) Pordinal(:,19:21) Pquant(:,2:4)
Pordinal(:,22) Pquant(:,5:6)];

fprintf('\n*************Results: time 4\n')
P2=NPC_FWE(P,'T',options);

domains=[1 1 1 1 1 1 2 2 2 2 3 3 3 3 3 3 4 4 4 5 5 5 5 5 5 6 6 6];

%Domains

for i=1:6 %i domini
    optionsD.labels.dims{2}=options.labels.dims{2}(domains==i)
    PD(:,i)=NPC(P(:,domains==i),'F',optionsD);
end

NPC_fwe(PD,'T');
P_4=P; %storing results for time=4
PD_4=PD;
```

Repeat the same analysis for times 0 and 24 and then combine domains over times. Once again, we refer the reader to the book's website for the complete code.

```
%combining times over domains (analysis of interest)

PD_times(:,:,3)=PD_24;
PD_times(:,:,2)=PD_4;
PD_times(:,:,1)=PD_0;

fprintf('\n*************Results: times combined\n')
options.Combdims=3;
Pdomain=NPC(PD_times,'T',options);

options.Combdims=2;
P2fwe=NPC_fwe(Pdomain,'T',options);

%combining times over variables (an additional analysis)

P_times(:,:,3)=P_24;
P_times(:,:,2)=P_4;
P_times(:,:,1)=P_0;

fprintf('\n*************Results: times combined\n')
options.Combdims=3;
P=NPC(P_times,'T',options);
options.Combdims=2;
P2fwe=NPC_fwe(P,'T',options);
```

8.4.4 Analysis Using R

As already stated, the aim of the analysis is to test for the presence of a (decreasing) stochastic ordering with respect to the increasing doses of the solvent.

We require the `stoch.ord2` function, a modification of the `stoch.ord` function that accounts for the presence of ordinal data. This function requires as entries one response variable `y` and one ordinal variable `x` (typically *time*, `dose` in this example). What the function basically does is carry out $k-1$ two-sample tests (grouping the k categories of `x` into all two possible ordered pseudo-groups), and test for the presence of at least one significant test (by applying Tippett's combining function). The test statistics we have considered for the two-sample tests are the t statistic when the response variable is continuous and the statistic $T^* = \sum_{i=1}^{C} N_{2i}^*[N_{\cdot i}^*(N - N_{\cdot i}^*)]^{-1/2}$ when it is ordinal (see also Chapter 6); here C is the number of categories of the ordinal response variable. The whole null distribution of the global test statistic Min$P$ is returned by the function; its first element is the global p-value assessing the presence of a decreasing/increasing stochastic ordering (`alt` being set equal to -1 or 1) with respect to the ordered increasing categories of `x`.

We will carry out a stochastic ordering test to assess the effect of `dose` on the jth response at the kth time, $j = 1, \ldots, p$, $k = 1, \ldots, t$. Thus, we will first combine the partial tests with respect to the domains, and finally combine with respect to time.

The function can process one variable at time; in order to maintain the dependence among the response variables the function is always run with a common `seed`, which will be the input to the `set.seed` function (thus providing exactly the same permutations each time). We run the function each time and store the vector of p-values returned by the `stoch.ord2` function in the three-dimensional array `P`. The option `cat` specifies whether the response variable considered is ordinal (1) or continuous (0). The partial test is not carried out whenever the response is constant (thus we initialize the elements of `P` equal to 1).

```
setwd("C:/path")
source("t2p.r") ; source("stoch_ord2.r")
class = c(rep("integer",2),rep("factor",23),rep("numeric",6))
data<-read.csv("perch.csv",header=TRUE,colClasses=class)

dose = data$dose
time = data$testtime
mouse = data$rat

data = data[,-c(1:3)]

B=1000
p=dim(data)[2]
t=length(unique(time))

P<-array(1,dim=c((B+1),p,t))
seed=101

for(k in 1:t){
tt = unique(time)[k] ; cat("Time = ",tt,"\n\n")

 for(j in 1:p){
cat('Processing variable:',j,"\n")
cat = ifelse(is.factor(data[,j]),1,0)
```

```
tab = table(data[time==tt,j])
if(length(tab[tab>0])>1){
dose.t = dose[time==tt]
P[,j,k] = stoch.ord2(data[time==tt,j],dose.t,alt=-1,B=B,
cat=cat,seed=seed)
}

}## end j
}## end k
```

The $p \times t$ matrix `P[1,,]` contains the observed p-values of each partial test. The raw p-values are contained in the matrix `res`.

```
res=P[1,,]
rownames(res) = colnames(data)
colnames(res) = c('T0','T4','T24')
res
                  T0    T4   T24
lacrimation     0.439 0.998 1.000
palp_closure    1.000 1.000 1.000
salivation      1.000 1.000 1.000
urination       0.917 0.581 0.070
defecation      0.133 0.176 0.145
pupil           1.000 0.984 0.999
approach        0.653 0.000 0.809
touch           0.968 0.000 0.000
click           0.901 0.000 0.004
tail_pinch      1.000 0.000 0.000
removal         0.997 0.555 0.449
handling        0.186 0.000 0.000
clonic_mvts     0.705 0.308 0.002
tonic_mvts      1.000 1.000 1.000
arousal         0.062 0.000 0.000
vocalizations   1.000 1.000 1.000
hc_posture      0.282 0.735 0.344
rears           0.301 0.250 0.122
motor_activity  1.000 0.216 0.043
gait            1.000 1.000 0.998
mobility        1.000 1.000 1.000
righting        1.000 1.000 0.993
foot_splay      0.695 0.999 0.218
forelimb_grip   0.802 0.000 0.000
hindlimb_grip   0.666 0.000 0.000
piloerection    1.000 1.000 0.445
weight          0.687 0.554 0.159
temperature     0.762 0.000 0.000
```

We now combine, within each time, the response variables according to specified domains (see Table 8.2). We apply Fisher's combining function. The raw p-values of each domain/time are in the matrix `res.dom`.

```
dom = c(rep(1,6),rep(2,4),rep(3,6),rep(4,3),rep(5,6),rep(6,3))
```

```
T.dom=array(0,dim=c((B+1),6,t))
for(d in 1:6){
T.dom[,d,] = apply(P[,dom==d,],c(1,3),function(x){-2*log(prod(x))})
}

P.dom=t2p(T.dom)
res.dom = P.dom[1,,]
rownames(res.dom)=c('Autonomic','Sensorimotor','Excitability',
'Neuromuscular','Activity','Physiologic')
colnames(res.dom)=c('T0','T4','T24')
res.dom

                 T0    T4    T24
Autonomic     0.337 0.846 0.618
Sensorimotor  0.977 0.000 0.000
Excitability  0.276 0.000 0.000
Neuromuscular 0.274 0.353 0.040
Activity      0.953 0.000 0.000
Physiologic   0.880 0.000 0.000
```

Finally, we combine with respect to time with Fisher's combining function and adjust the raw p-values for multiplicity by running the `FWE.minP` function.

```
T.time = array(0,dim=c(B+1,6))


for(d in 1:6){
T.time[,d] = apply(P.dom[,d,],1,function(x){-2*log(prod(x))})
}

P.time = t2p(T.time)
source("FWEminP.r")
res.glob.adj = FWE.minP(P.time)

res.glob = cbind(P.time[1,],res.glob.adj)
rownames(res.glob)=rownames(res.dom)
colnames(res.glob)=c('p','p.adj')
res.glob

                  p p.adj
Autonomic     0.692 0.692
Sensorimotor  0.000 0.000
Excitability  0.000 0.000
Neuromuscular 0.077 0.146
Activity      0.000 0.000
Physiologic   0.000 0.000
```

9

NPC Tests for Survival Analysis

9.1 Introduction and Main Notation

This chapter deals with permutation methods for problems of hypothesis testing in the framework of survival analysis. The field of survival analysis involves methods for the analysis of data on an event observed over time and the study of factors associated with the occurrence rates of this event. Survival analysis is a branch of statistics which deals with *death* in biological organisms and *failure* in mechanical systems. The same topic is called *reliability theory* or *reliability analysis* in engineering and *duration analysis* or *duration modelling* in economics and sociology. In this context, death or failure is considered an *event*. Therefore, we consider failure time data that arise when items are placed at risk under different experimental conditions. Major areas of application are biomedical studies of degenerative diseases and industrial life testing.

9.1.1 Failure Time Distributions

Let us assume that there is a given sequence of fixed probability spaces $(\Omega^{(n)}, \mathcal{A}^{(n)}, P^{(n)})$, where $n \in \mathbb{N}$, along with a family of right-continuous, non-decreasing complete sub-σ-algebras $\left\{\mathcal{A}_t^{(n)} : 0 \leq t < \infty\right\}$ which correspond to the history of the survival study up to and including time t, where n is the sample size of the study. In this chapter we assume that an individual can experience an event at most once. Thus, suppose we have two independent samples of size n_1 and n_2, respectively. Let $n = n_1 + n_2$ be the pooled sample size.

In the type I censoring model, T_{mj} $(m = 1, \ldots, n_j,\ j = 1, 2)$ usually represents the true survival time being tested (the length of time to the event) with CDF F_j (where the survival function $S_j(t) = \Pr\left(T_{mj} > t\right) = 1 - F_j(t)$) and C_{mj} usually designates the censoring variable with distribution function K_j and censoring survival function $G_j(t) = \Pr\left(C_{mj} > t\right) = 1 - K_j(t)$ for the longest time subject m can be observed. We assume that the censoring distributions are independent of the failure time distributions.

A typical right-censored survival data set consists of n independent realizations of the random pair (X, Δ); thus corresponding to subject m in sample j is the random vector (X_{mj}, Δ_{mj}) for $j = 1, 2$ and $m = 1, \ldots, n_j$. The variable X_{mj} represents the time for which subject m in group j is under study. If at the end of the observation period the event of interest has occurred, X_{mj} is said to be *uncensored*; otherwise it is *censored*. Now, let $B_j(t) = \Pr(X_{mj} \leq t)$. The variable Δ_{mj} is the censoring indicator, taking the value 1 if the observation on subject m is uncensored, and 0 otherwise. Briefly,

Permutation Tests for Complex Data: Theory, Applications and Software Fortunato Pesarin, Luigi Salmaso

$$X_{mj} = \min(T_{mj}, C_{mj}), \quad \Delta_m = \mathbb{I}(T_{mj} \le C_{mj}),$$

where $\mathbb{I}(A)$ denotes the indicator function of the event A.

The CDF $F(t) = \Pr(T \le t)$ of T is of primary interest, while that of C, $G(t) = \Pr(C \le t)$, is considered to be an unknown nuisance entity.

Furthermore, let $t_1 \le \ldots \le t_n$ denote the ordered pooled times (i.e. here the t_m consist of both event and censoring times) and $t_1 < \ldots < t_D$ denote the distinct event times in the pooled sample (i.e. here we consider only event times t_i). We shall take τ to be the largest of the observed event times. Finally, in the random censoring model, we suppose that at time t_i there are

$$d_{ij}(t_i) = d_{ij} = \sum_{m=1}^{n} [\mathbb{I}(X_{mj} = t_i, \Delta_{mj} = 1)]$$

events (sometimes simply referred to as deaths) and

$$c_{ij}(t_i) = c_{ij} = \sum_{m=1}^{n} [\mathbb{I}(X_{mj} = t_i, \Delta_{mj} = 0)]$$

censored observations, $j = 1, 2, i = 1, \ldots, D$. Let

$$Y_{ij}(t_i) = Y_{ij} = \sum_{m=1}^{n} [\mathbb{I}(X_{mj} \ge t_i)]$$

be the number of individuals who are at risk at time t_i. Furthermore, $d_i = d_{i1} + d_{i2}$ and $Y_i = Y_{i1} + Y_{i2}$ are respectively the total number of events and the number of subjects at risk in the pooled sample at time t_i. The quantity $\widehat{h}_j(t_i) = d_{ij}/Y_{ij}$, the so-called *hazard function*, provides an estimate of the conditional probability that an individual who survives just prior to time t_i experiences the event at time t_i. This is the basic quantity from which we will construct estimators of the survival function $S_j(t)$, the censoring survival function $G_j(t)$, and the cumulative hazard rate $H_j(t)$.

9.1.2 Data Structure

The whole set of observed data is summarized by the pair of associated matrices $(\mathbf{X}, \mathbf{\Delta})$:

$$(\mathbf{X}, \mathbf{\Delta}) = \{(\mathbf{X}_j, \mathbf{\Delta}_j), j = 1, 2\} = \left\{(X_{mj}, \Delta_{mj}), m = 1, \ldots, n_j, j = 1, 2\right\}.$$

Thus let us assume that observations from a random vector $(\mathbf{X}, \mathbf{\Delta}) = \left\{(\mathbf{X}_1, \mathbf{\Delta}_1) \biguplus (\mathbf{X}_2, \mathbf{\Delta}_2)\right\}$ on n units are partitioned into two groups of n_1 and n_2 individuals respectively, corresponding to two levels of a treatment. Let us also assume that the response variables in the two groups have unknown distributions $P_1 = P_{1\Delta} \cdot P_{1X|\Delta}$ and $P_2 = P_{2\Delta} \cdot P_{2X|\Delta}$ ($P_j \in \mathcal{P}$, where $\mathcal{P}$ is a possibly unspecified nonparametric family of non-degenerate distributions), both defined on the same probability space $(\Omega, \mathcal{A})$, where $\Omega = (\mathcal{X}, \mathcal{O})$ is the sample space and $\mathcal{A}$ is a σ-algebra of events. Hence, let $\Omega = (\mathcal{X}, \mathcal{O})$ be the support of the random vector $(\mathbf{X}, \mathbf{\Delta})$ and $\Omega_{/(\mathbf{X},\mathbf{\Delta})}$ the permutation sample space given $(\mathbf{X}, \mathbf{\Delta})$. So $(\mathcal{X}, \mathcal{O})_{/(\mathbf{X},\mathbf{\Delta})}$ is the orbit associated with the data set $(\mathbf{X}, \mathbf{\Delta})$ as the set containing all permutations $(\mathbf{X}^*, \mathbf{\Delta}^*)$.

In the permutation setting, let $(\mathbf{X}_j, \mathbf{\Delta}_j)$ be the observed data set of n_j elements related to the jth sample where $j = 1, 2$. Let us also use $\left\{(\mathbf{X}_j^{*b}, \mathbf{\Delta}_j^{*b}), j = 1, 2, b = 1, \ldots, B\right\}$ to denote a sample from the permutation sample space.

9.2 Comparison of Survival Curves

In survival analysis, it is often of interest to test whether or not two survival time distributions are equal. We assume that observations are available on the failure time of n individuals usually taken to be independent. Focusing on the case of two independent samples, the aim is often to compare two therapies, two products, two processes, two treatments, etc. The two samples are denoted by $(X_{11}, \ldots, X_{1n_1})$ and $(X_{21}, \ldots, X_{2n_2})$, respectively. The X_{1i} constitute a random sample on the random variable X_1 with CDF F_1, and the X_{2i} a random sample on the random variable X_2 with CDF F_2. One usually wishes to test the null hypothesis $H_0 : \{F_1(t) = F_2(t), \forall t \in \mathcal{R}^+\}$.

More precisely, the system of hypotheses of interest is

$$\begin{aligned} H_0 &: \{P_1(t) = P_2(t) = P(t), \forall t \leq \tau\} \\ &= \{[S_1(t) = S_2(t)] \bigcap [K_1(t) = K_2(t)], \forall t \leq \tau\}, \end{aligned}$$

against, in the case of treatment-independent censoring,

$$\begin{aligned} H_1 &: \{P_1(t) \leq\neq\geq P_2(t), \text{ some } t \leq \tau\} \\ &= \{S_1(t) \leq\neq\geq S_2(t), \text{ some } t \leq \tau\} \end{aligned}$$

or against, in the case of treatment-dependent censoring, as the censoring distributions K_j (with $j = 1, 2$) may vary among treatments,

$$H_1 : \left\{\left[S_1(t) \leq\neq\geq S_2(t)\right] \bigcup \left[K_1(t) \neq K_2(t)\right], \text{ some } t \leq \tau\right\}.$$

Remark 1. As frequently occurs in survival studies, time to event data are characterized by incompleteness due to censoring; in particular, right-censored data occur when the unobserved and unknown time to the event of interest is longer than the recorded time for which an individual was under observation. For instance, subjects in a survival study can be lost to follow-up due to transfer to a non-participating institution, or the study can finish before all the subjects have observed the event.

The focus here is the presence of complicated censoring patterns and, in particular, the type of censoring. In the right-censored survival data framework, censored data are usually assumed to originate from an underlying random process, which may or may not be related to treatment levels or to event processes. When we assume that the probability of a datum being censored does not depend on its unobserved value, then we may ignore this process and so need not specify it.

Nearly all statistical procedures for right-censored survival data are based on the assumption that censoring effects are, in a very specific sense, non-informative with respect to the distribution of survival time, that is, unaffected by treatment levels. If the censoring distributions are equal, the censoring process does not depend on group, and observed values may be considered as a random subsample of the complete data set. Thus, in these situations, it is appropriate to ignore the process that causes censored data when making inferences on X. Therefore, in the case of a treatment-independent censoring distribution, the process that causes censored data is called *ignorable* and analysis may be carried out conditionally on the actually observed data.

In contrast, when the censoring patterns are treatment-dependent, the observation pairs from the first sample do not have the same distribution as those from the second sample, even when the null hypothesis on pure survival times is true. In the case of treatment-dependent censoring distributions, in order to make valid inferences the censored data process must be properly specified. Thus, the analysis of treatment-dependent censoring data is much more complicated than that of treatment-independent censoring data because inferences must be made by taking into consideration the data set as a whole and by specifying a proper model for the censoring pattern. In any case, the specification of a model which correctly represents the censored data process seems the only way to eliminate the inferential bias caused by censoring.

If we assume that, in the null hypothesis, both event and censoring times are jointly exchangeable with respect to groups, then such multivariate testing problems are solvable by the NPC of dependent permutation tests. In particular, the hypotheses can be broken down into a set of sub-hypotheses, and related partial tests are assumed to be marginally unbiased, significant for large values and consistent.

Several solutions for permutation analysis of survival data analysis have been introduced in Callegaro et al. (2003) and Bonnini et al. (2005).

Although some solutions presented in this chapter are exact, the most important of them are approximate because the permutation distributions of the test statistics are not exactly invariant with respect to permutations of censored data, as we shall see. However, the approximations are quite accurate in all situations, provided that the number of observed data is not too small. To this end, we may remove from the permutation sample space associated with the whole data set, all data permutations where the permutation sample sizes of actually observed data are not large enough for approximations. In a way, similarly to the permutation missing data analysis (see Section 7.9), we must establish a kind of restriction on the permutation space, provided that this restriction does not imply relevant bias our inferential conclusions.

9.3 An Overview of the Literature

In this section we briefly present the most widely used statistics for testing the equality of two survival distributions based on independent randomly censored samples. The different nonparametric approaches can be classified as asymptotic or permutation procedures. In the framework of asymptotic nonparametric methods, several classes of tests have been described in the literature.

A first class of methods is based on integrated weighted comparisons of the estimated cumulative hazard functions in the two-sample design under the null and alternative hypotheses, based on the Nelson–Aalen estimator (Nelson, 1972; Aalen, 1978) within weighted log-rank statistics (or weighted Cox–Mantel statistics). In particular, these methods are based on the weighted differences between the observed and expected hazard rates. The test is based on weighted comparisons of the estimated hazard rates of the jth population under the null and alternative hypotheses. An important consideration in applying this class of tests is the choice of weight function to be used. Weights are used to highlight certain pieces of the survival curves. A variety of weight functions have been suggested in the literature, and, depending on the choice of weight function, the related tests are more or less sensitive to early or late departures from the hypothesized relationship between the two hazard functions (as specified in the null hypothesis). All these statistics are censored data generalizations of linear rank statistics. For instance, the most commonly used rank-based test statistic is the log-rank test proposed by Mantel (1966), Peto and Peto (1972) and Cox (1972), and it is a generalization for censored data of the exponential ordered scores test of Savage (1956). This statistic has good power when it comes to detecting differences in the hazard rates, when the ratio of hazard functions in the populations being compared is approximately constant. Gilbert (1962), Gehan (1965) and Breslow (1970) suggested a censored data generalization of the two-sample Wilcoxon–Mann–Whitney rank test. Peto and Peto (1972) and Kalbfleisch and Prentice (1980) described other generalizations of the Wilcoxon–Mann–Whitney test. They used a different estimate of the survival function based on the combined sample. Tarone and Ware (1977) proposed a class of multi-sample statistics for right-censored survival data that includes the log-rank test and the censored data generalized Wilcoxon–Mann–Whitney procedures.

Harrington and Fleming (1982) introduced a very general class of tests which includes, as special cases, the log-rank test and another version of the Wilcoxon–Mann–Whitney test. They use the Kaplan–Meier estimate of the survival function based on the combined sample at the previous event time. More recently, Gaugler et al. (2007) proposed a modified Fleming–Harrington test very similar to the original version, which includes, as a special case, the Peto–Peto and Kalbfleisch–Prentice test. Here the Peto–Peto and Kalbfleisch–Prentice estimate of the survival function is based on the pooled sample at the current event time. Jones and Crowley (1989) introduced a more general class of single-covariate nonparametric tests for right-censored survival data that includes the Tarone–Ware two-sample class, the Cox (1972) score test, the Tarone (1975) and Jonckheere (Gehan, 1965) C-sample trend statistics, the Brown et al. (1974) modification of the Kendall rank statistic, Prentice's (1978) linear rank statistics, O'Brien's (1978) logit rank statistic and several new procedures. This class can be generalized to include the Tarone–Ware C-sample class.

The statistical properties of the aforementioned test statistics have been studied by many authors. Here we mention, among others, works by Gill (1980), Fleming and Harrington (1981), Breslow et al. (1984), Fleming et al. (1987), Lee (1996), Korosok and Lin (1999), Shen and Cai (2001), and Wu and Gilbert (2002).

A variety of weight functions $W(t_i)$ have been proposed in the literature (see Table 9.1, where symbols have their obvious usual meanings).

These tests are sensitive to alternatives of ordered hazard functions (Fleming and Harrington, 1991). When they are applied in samples from populations where the hazard rates cross, they have little power because early positive differences in favour of one group are compensated by late differences in the rates with opposite signs, in favour of the other treatment.

A second class of procedures is based on the maximum of the sequential evaluation of the weighted log-rank tests at each event time; these are referred to as Rényi-type statistics. Such tests are presumed to have 'good' power to detect crossing hazards. These supremum versions of the weighted log-rank tests were proposed by Gill (1980) and are generalizations of the Kolmogorov–Smirnov statistic for comparing two censored data samples.

A third class of procedures is the weighted Kaplan–Meier (WKM) statistics based directly on integrated weighted comparisons of survival functions (rather than based on ranks) in the two

Table 9.1 Different types of weight function

$Z_j(\tau) = \sum_{i=1}^{D} W(t_i)\left\{d_{ij} - \hat{E}\left[d_{ij}\right]\right\}$	Weighted log-rank statistic
$Q = \frac{Z_1(\tau)}{\sqrt{\hat{\sigma}_{11}^2}}$	Weighted log-rank test
Weight function	Test name
$W(t_i) = 1, \quad \forall t_i$	Log-rank test
$W(t_i) = Y_i$	Gehan–Breslow test
$W(t_i) = \sqrt{Y_i}$	Tarone–Ware test
$W(t_i) = \tilde{S}(t_i)$	Peto–Peto and Kalbfleisch–Prentice test
$W(t_i) = \frac{\tilde{S}(t_i)Y_i}{Y_i+1}$	Modified Peto–Peto and Kalbfleisch–Prentice test (Andersen et al., 1982)
$W_{p,q}(t_i) = (\hat{S}(t_{i-1})^p - (1-\hat{S}(t_{i-1}))^q, p \geq 0; q \geq 0$	Fleming–Harrington test
$W_{p,q}(t_i) = (\tilde{S}(t_i)^p - (1-\tilde{S}(t_i))^q, p \geq 0; q \geq 0$	Modified Fleming–Harrington test (Gaugler et al., 2007)

samples under the null and alternative hypotheses, based on the Kaplan–Meier estimator. This class of tests was initially proposed by Pepe and Fleming (1989), whereas more recently Lee et al. (2008) studied an integrated version of the WKM. Note that two other versions of these two statistics can be obtained by replacing the Kaplan–Meier estimators with the Peto–Peto and Kalbfleisch–Prentice estimators. WKM statistics provide censored data generalizations of the two-sample z- or t-test statistics. Asymptotic distribution properties of the WKM statistics can be found in Pepe and Fleming (1989, 1991) and Lee et al. (2008).

A fourth class of procedures is a censored data version of the Cramér–von Mises statistics, based on the integrated squared difference between the two estimated cumulative hazard rates, based in turn on the Nelson–Aalen estimator. This is done to obtain a limiting distribution which does not depend on the relationship between the event and censoring times and because such tests arise naturally from counting process theory. Small-sample and asymptotic distribution properties of three versions of this rank test statistic can be found in Koziol (1978) and Shumacher (1984).

A fifth class of test statistics is a generalization of the two-sample median statistics for censored data, proposed by Brookmeyer and Crowley (1982), which is useful when we are interested in comparing the median survival times of the two samples rather than the difference in the hazard rate or the survival functions over time. Asymptotic distribution properties of the WKM statistics can be found in Brookmeyer and Crowley (1982).

Other two-sample procedures have been suggested in the literature. The most recent works include: a midrank modification of rank tests for exact, tied and censored data proposed by Hudgens and Satten (2002); a nonparametric procedure for use when the distribution of time to censoring depends on treatment group and survival time, proposed by DiRienzo (2003); an asymptotically valid C-sample test statistic proposed by Heller and Venkatraman (2004); and the randomization-based log-rank test proposed by Zhang and Rosenberger (2005).

9.3.1 *Permutation Tests in Survival Analysis*

When asymptotic tests are used to compare survival functions, it is possible to find situations in which the number of failures of interest is so small that it is reasonable to question the validity of an asymptotic test. In such situations the standard asymptotic log-rank test, for example, is frequently replaced by its corresponding log-rank permutation test. This provides an exact small-sample test when the censoring patterns in the two compared populations are equal. Indeed, when the censoring patterns are treatment-dependent, the observation pairs from the first sample do not have the same distribution as those from the second sample. The failure of the asymptotic log-rank test is not only due to an inappropriate asymptotic approximation, which in turn can be replaced by an exact evaluation, but it is also due to ignoring the interdependency of the observed risk sets $Y_{ij}(t_i)$, that is, the risk sets in the 2×2 contingency tables associated with the D event times (Heinze et al., 2003). Furthermore, since the two group sizes may be unbalanced, the asymptotic test may not be appropriate. Therefore, also their asymptotic distributions under the null hypothesis may not be appropriate. Some authors, among them Kellerer and Chmelevsky (1983), Chen and Gaylor (1986), Ali (1990) and Soper and Tonkonoh (1993), have suggested so-called exact procedures. These provide exact small-sample tests when the censoring patterns in the samples being compared are treatment-independent.

In situations like these, permutation tests may be helpful. In this chapter we give a brief description of methods that require extensive computing. In the survival analysis framework, we present a review of the most common two-sample permutation tests which have been suggested in the literature. The only assumptions that are necessary are the model assumptions (independence of the pairs (X_{mj}, Δ_{mj}) and independence of X_{mj} and C_{mj}).

There are three different unidimensional exact conditional procedures analogous to the asymptotic log-rank test and suitable for situations of treatment-dependent censoring. The first method was proposed by Heimann and Neuhaus (1998), while the other two were proposed by Heinze et al. (2003). Callegaro et al. (2003) proposed an exact Rényi-type test based on Rényi-type statistics and suitable in the case of treatment-dependent censoring where it is assumed that, in the alternative hypothesis, treatment may also influence the censoring process. An interesting permutation contribution is given by Galimberti and Valsecchi (2002) who introduced a multidimensional permutation test to compare survival curves for matched data when the number of strata increases, the stratum sizes are small, and the proportional hazard model is not satisfied.

9.4 Two NPC Tests

In the framework of permutation methods, it is possible to consider an analysis approach incorporating two successive stages, the first focusing on the D observed distinct event times in the pooled sample, and the second focusing on the combination of these D tests into an overall test for comparing the survival curves.

Therefore, the NPC procedure for dependent tests may be viewed as a two-phase (or multi-phase) testing procedure. In the first phase, let us suppose that $\Gamma_i : (\mathcal{X}, \mathcal{O}) \longrightarrow \mathcal{R}^1$ $(i = 1, \ldots, D)$ is an appropriate univariate partial test statistic for the ith sub-hypothesis H_{0i} against H_{1i}, for which (without loss of generality) we assume that Γ_i is non-degenerate, marginally unbiased, consistent and that large values are significant, so that they are stochastically larger in H_{1i} than in H_{0i} in both conditional and unconditional senses. In the second phase, we construct the global test statistic $\Gamma'' = \psi(\widehat{\lambda}_1, \ldots, \widehat{\lambda}_i, \ldots, \widehat{\lambda}_D)$ by combining the permutation p-values $\widehat{\lambda}_i = \widehat{\lambda}_{\Gamma_i}$ associated with the D partial tests through a suitable combining function $\psi \in \mathcal{C}$. Hence, the combined test is a function of D dependent partial tests.

Remark 1. When there is a more complex data configuration (where the more interesting cases are given by testing in the presence of stratification variables, closed testing, multi-aspect testing and repeated measures), however, the NPC may be like a multi-phase procedure characterized by several intermediate combinations, where we may, for instance, first combine partial tests with respect to variables within each stratum s (with $s = 1, \ldots, S$), and then combine the second-order tests with respect to strata into a single third-order combined test.

Indeed, in the analysis of survival right-censored data, we can view censored data as a missing data problem. In this sense, we can carry out a multidimensional permutation test based on the theory of permutation testing with missing data. In fact, if we assume that once we have fixed an observed time t_i, $i = 1, \ldots, D$, the data already censored may be considered similarly to missing data, then it is possible to extend the theory of permutation analysis of missing values set out in Chapter 7 to the right-censored survival analysis.

9.4.1 Breaking Down the Hypotheses

It is generally of interest to test for the global (or overall) null hypothesis that the two groups have the same underlying distribution,

$$H_0^G : \left\{(X_1, \Delta_1) \stackrel{d}{=} (X_2, \Delta_2)\right\},$$

against a one-sided (stochastic dominance) or a two-sided (inequality in distribution) global alternative hypothesis,

$$H_1^G : \left\{(X_1, \Delta_1) \stackrel{d}{\leq\neq\geq} (X_2, \Delta_2)\right\}.$$

Let us assume that under the null hypothesis the data (**X**, **Δ**) are jointly exchangeable with respect to the two groups on both **X** and **Δ** components. It is important to note that the pooled set of observed data (**X**, **Δ**) in the null hypothesis is a set of jointly *sufficient statistics* for the underlying observed and censored data process. Moreover, H_0^G implies the exchangeability of individual data vectors with respect to groups, so that the permutation multivariate testing principle is properly applicable.

The complexity of this testing problem is such that it is very difficult to find a single overall test statistic. However, the problem may be dealt with by means of the NPC of a set of dependent permutation tests.

Hence, we consider a set of D partial tests followed by their NPC. To this end, we observe that the null hypothesis may be equivalently written in the form

$$H_0^G : \left\{\bigcap_{i=1}^{D}\left[(X_{i1}, O_{i1}) \stackrel{d}{=} (X_{i2}, O_{i2})\right]\right\} = \left\{\bigcap_{i=1}^{D} H_{0i}\right\},$$

equivalent to

$$H_0^G : \left\{\left[\bigcap_{i=1}^{D}\left(O_{1i} \stackrel{d}{=} O_{2i}\right)\right]\bigcap\left[\bigcap_{i=1}^{D}\left(X_{1i} \stackrel{d}{=} X_{2i}\right)|\mathbf{O}\right]\right\} = H_0^{\mathbf{O}}\bigcap H_0^{\mathbf{X}|\mathbf{O}},$$

where a suitable and meaningful breakdown of H_0^G is emphasized. The overall alternative hypothesis may be written as

$$\begin{aligned} H_1^G &= \left\{\bigcup_{i=1}^{D}\left[(X_{1i}, O_{1i}) \le \stackrel{d}{\neq} \ge (X_{2i}, O_{2i})\right]\right\} = \left\{\bigcup_{i=1}^{D} H_{1i}\right\} \\ &= \left\{\left[\bigcup_{i=1}^{D}\left(O_{1i} \le \stackrel{d}{\neq} \ge O_{2i}\right)\right]\bigcup\left[\bigcup_{i=1}^{D}\left(X_{1i} \le \stackrel{d}{\neq} \ge X_{2i}\right)|\mathbf{O}\right]\right\} \\ &= H_1^{\mathbf{O}}\bigcup H_1^{\mathbf{X}|\mathbf{O}}. \end{aligned}$$

The hypothesis H_0^G against H_1^G is thus broken down into D sub-hypotheses H_{0i} against H_{1i}, $i = 1, \ldots, D$, in such a way that H_0^G is true if all the H_{0i} are jointly true and H_1^G implies that the inequality of the two distributions entails the falsity of at least one partial null hypothesis. Finally, note that the hypotheses and assumptions are such that the *permutation testing principle* applies.

Thus, to test H_0^G against H_1^G, we consider a D-dimensional vector of real-valued test statistics $\mathbf{\Gamma} = \{\mathbf{\Gamma}_1, \ldots, \mathbf{\Gamma}_D\}$, the ith component of which is the univariate partial test for the ith sub-hypothesis H_{0i} against H_{1i}. Without loss of generality, we assume that partial tests are non-degenerate, marginally unbiased, consistent, and significant for large values. Hence, the combined test is a function of D dependent partial tests and, of course, the combination must be nonparametric, particularly with regard to the underlying dependence relation structure.

9.4.2 *The Test Structure*

Let us consider $t_{(1)} < \ldots < t_{(D)}$, $i = 1, \ldots, D$, the ordered and distinct observed times of the event of interest. For each subject m within the jth group ($m = 1, \ldots, n_j$, $j = 1, 2$) and each t_i we calculate V_{mji} as

$$V_{mji} = \begin{cases} 1 & \text{if} \quad X_{mj} > t_i \\ 0 & \text{if} \quad X_{mj} \le t_i \text{ and } X_{mj} = T_{mj} \\ C & \text{if} \quad X_{mj} \le t_i \text{ and } X_{mj} = C_{mj}; \end{cases}$$

and O_{mji} as

$$O_{mji} = \begin{cases} 0 & \text{if } V_{mji} = C \\ 1 & \text{otherwise.} \end{cases}$$

Let us then define $\nu_{ji} = \sum_{m=1}^{n_j} O_{mji}$ as the number of observations that have not already been censored at time t_i in the jth group, and $\nu_i = \sum_{j=1}^{2} \nu_{ji}$ as the number of observations that have not already been censored at time t_i in the pooled sample.

9.4.3 NPC Test for Treatment-Independent Censoring

In this section we consider a multidimensional permutation test in the case of treatment-independent censoring (TIC-NPC). This test, proposed by Callegaro et al. (2003), is based on the assumption that censoring effects are non-informative with respect to the distribution of survival time. In the present context, we are interested in testing the global null hypothesis,

$$\begin{aligned} H_0^G &: \left\{\left[S_1(t_i) = S_2(t_i) \;\; \forall t_i,\; i = 1, \ldots, D\right] \bigcap \left[\Delta_1 \stackrel{d}{=} \Delta_2\right]\right\} \\ &= \left\{\left[X_1 \stackrel{d}{=} X_2\right] \bigcap \left[\Delta_1 \stackrel{d}{=} \Delta_2\right]\right\}, \end{aligned}$$

against the overall alternative,

$$\begin{aligned} H_1^G &: \{S_1(t_i) \leq\neq\geq S_2(t_i) \; \forall t_i,\; \exists t_i : S_1(t_i) <\neq> S_2(t_i)\} \\ &= \left\{X_1 \stackrel{d}{\leq\neq\geq} X_2\right\}. \end{aligned}$$

If the censored data are treatment-independent, we may proceed conditionally on the observed censoring indicator $\mathbf{O}$ and ignore $H_0^{\mathbf{O}}$ because in this context we have assumed that $\mathbf{O}$ does not provide any information about treatment effects. The global null and alternative hypotheses are broken down into D sub-hypotheses, one for each D time $t_{(1)} < \ldots < t_{(D)}$. Hence, we may equivalently write the null hypothesis in the relatively simpler form

$$H_0^G = H_0^{\mathbf{X}|\mathbf{O}} : \left\{\bigcap_{i=1}^{D}\left[\left(X_{i1} \stackrel{d}{=} X_{i2}\right) | \mathbf{O}\right]\right\} = \left\{\bigcap\nolimits_i H_{0i}^{\mathbf{X}|\mathbf{O}}\right\}$$

against

$$H_1^G : \left\{\bigcup_{i=1}^{D} H_{1i}^{\mathbf{X}|\mathbf{O}}\right\},$$

The partial permutation test statistics for testing the sub-hypothesis $H_{0i}^{\mathbf{X}|\mathbf{O}}$ against the sub-alternative $H_{1i}^{\mathbf{X}|\mathbf{O}}$ then take the form

$$\Gamma_i^{\mathbf{X}|\mathbf{O}*} = \bar{S}_2^*(t_i)\sqrt{\frac{\nu_{1i}^*}{\nu_{2i}^*}} - S_1^*(t_i)\sqrt{\frac{\nu_{2i}^*}{\nu_{1i}^*}},$$

where $S_j^*(t_i) = \sum_{m=1}^{n_j} V_{mji}^* O_{mji}^*$ is a proper function of the univariate sample totals of valid data (see Section 7.10.2).

Note that each test statistic $\Gamma_i^{\mathbf{X}|\mathbf{O}*}$ is permutationally invariant, in mean value and variance, with respect to the sample size $\nu_{ji}^* = \sum_{m=1}^{n_j} O_{mji}^*$, that varies according to the random attribution of units to the two groups, because units with censoring data participate in the permutation mechanism as well as all other units. Also note that when there are no censoring values, so that $\nu_{ji}^* = n_j$, $j = 1, 2$, the tests are permutationally equivalent to the traditional two-sample permutation test for comparison of locations.

Remark 1. Since the tests Γ_i^* are approximately exact, they are also approximately unbiased. In order to prove consistency, we recall that the size of missing data ν^* may diverge, provided that as n tends to infinity, $\lim \nu_i^* \in [0, 1]$.

Remark 2. In order for the given solution to be well defined, we must assume that ν_{1i}^* and ν_{2i}^* are jointly positive. This implies that, in general, we must consider a sort of restricted permutation strategy which consists of discarding from the analysis all points of the permutation sample space $(X, \mathcal{O})_{/(\mathbf{X},\mathbf{O})}$ in which even a single component of the permutation matrix $\boldsymbol{\nu}^*$, of actual sample sizes of valid data, is zero. Of course, this kind of restriction has no effect on inferential conclusions.

Therefore, the survival analysis may be solved by NPC $\Gamma'' = \Gamma_{\psi}''^{\mathbf{X}|\mathbf{O}} = \psi_X\left(\widehat{\lambda}_1^{\mathbf{X}|\mathbf{O}}, \ldots, \widehat{\lambda}_D^{\mathbf{X}|\mathbf{O}}\right)$, where

$$\widehat{\lambda}_i = \frac{\frac{1}{2} + \sum_{b=1}^{B} \mathbb{I}\left\{\Gamma_i^{*b} \geq \Gamma_i^O\right\}}{B+1}$$

is the p-value function estimate for each Γ_i^*, Γ_i^O is the observed value, and $\psi_X \psi \in \mathcal{C}$.

Note that according to Rubin (1976), we may ignore the variable Δ because in this context we have assumed that Δ does not provide any information about symbolic treatment effects (since we are dealing with treatment independent censoring). Thus, we can ignore the process that causes censoring data and analysis can be carried out conditionally on the actually observed data.

9.4.4 NPC Test for Treatment-Dependent Censoring

In this section we consider a multidimensional permutation test in the case of treatment-dependent censoring (TDC-NPC). This test, initially due to Bonnini et al. (2005), is a conditional test based on the probability of failure and on the distribution of observed time to failure. Note that in the case of treatment-dependent censoring, it is assumed that in the alternative hypothesis the symbolic treatment may also influence the underlying censoring process. In fact, the treatment may affect both the distributions of variable X and censoring indicator Δ.

In the present setting, we are interested in testing the global null hypothesis,

$$\begin{aligned} H_0^G &: \left\{\left[S_1(t_i) = S_2(t_i),\ \forall t_i, i = 1, \ldots, D\right] \bigcap \left[\Delta_1 \stackrel{d}{=} \Delta_2\right]\right\} \\ &= \left\{\left[X_1 \stackrel{d}{=} X_2\right] \bigcap \left[\Delta_1 \stackrel{d}{=} \Delta_2\right]\right\}, \end{aligned}$$

against the overall alternative,

$$\begin{aligned} H_1^G &: \left\{\left[S_1(t_i) \leq\neq\geq S_2(t_i),\ \forall t_i, \exists t_i : S_1(t_i) <\neq> S_2(t_i)\right] \bigcap \left[\Delta_1 \stackrel{d}{\neq} \Delta_2\right]\right\} \\ &= \left\{\left[X_1 \stackrel{d}{\leq\neq\geq} X_2\right] \bigcap \left[\Delta_1 \stackrel{d}{\neq} \Delta_2\right]\right\}. \end{aligned}$$

In the case of treatment-dependent censoring, H_0^G must take account of homogeneity in distribution with respect to the two samples of the actually observed and collected data $\mathbf{X}$ jointly with that associated with the censored data process $\mathbf{O}$, because in this setting it is assumed that, in the alternative, treatment may also influence the censoring process. In fact, the treatment may affect both the distributions of variables X and the censoring indicator O. Hence, in the case of the treatment-dependent censoring data model, the null hypothesis requires the joint distributional equality of the censored data processes in the two groups, giving rise to $\mathbf{O}$, and of response data $\mathbf{X}$ conditional on $\mathbf{O}$, i.e.

$$H_0^G : \left\{\left[\mathbf{O}_1 \stackrel{d}{=} \mathbf{O}_2\right]\bigcap\left[\left(\mathbf{X}_1 \stackrel{d}{=} \mathbf{X}_2\right)|\mathbf{O}\right]\right\}.$$

The assumed exchangeability, in the null hypothesis, of the n individual data vectors in $(\mathbf{X}, \mathbf{O})$, with respect to the two groups, implies that the treatment effects are null on *all* observed and unobserved variables. In other words, we assume that there is no difference in distribution for the multivariate censoring indicator variables $\mathbf{O}_j$, $j = 1, 2$, and, conditionally on $\mathbf{O}$, for actually observed data $\mathbf{X}$. As a consequence, it is not necessary to specify both the censored data process and the data distribution, provided that marginally unbiased permutation tests are available. In particular, it is not necessary to specify the dependence relation structure in $(\mathbf{X}, \mathbf{O})$ because it is nonparametrically processed.

In this framework, the global null and alternative hypotheses may be broken down into the $2 \times D$ sub-hypotheses

$$\begin{aligned} H_0^G &: \left\{\left[\bigcap_{i=1}^{D}\left(O_{i1} \stackrel{d}{=} O_{i2}\right)\right]\bigcap\left[\bigcap_{i=1}^{D}\left(X_{i1} \stackrel{d}{=} X_{i2}\right)|\mathbf{O}\right]\right\} \\ &= \left\{H_0^{\mathbf{O}} \bigcap H_0^{\mathbf{X}|\mathbf{O}}\right\} = \left\{\left(\bigcap_{i=1}^{D} H_{0i}^{\mathbf{O}}\right)\bigcap\left(\bigcap_{i=1}^{D} H_{0i}^{\mathbf{X}|\mathbf{O}}\right)\right\} \end{aligned}$$

against

$$H_1^G : \left\{\left(\bigcup_{i=1}^{D} H_{1i}^{\mathbf{O}}\right)\bigcup\left(\bigcup_{i=1}^{D} H_{1i}^{\mathbf{X}|\mathbf{O}}\right)\right\},$$

where $H_{0i}^{\mathbf{O}}$ refers to the equality in distribution among the two levels of the ith marginal component of the censoring indicator (censoring) process, and $H_{0i}^{\mathbf{X}|\mathbf{O}}$ refers to the equality in distribution of the ith component of $\mathbf{X}$, conditional on $\mathbf{O}$.

For treatment-dependent censoring models we must also combine the D test statistics on the components of the censoring indicator $\mathbf{O}$, provided that all partial tests are marginally unbiased. More specifically, to test $H_0^G : \{[\bigcap_i H_{0l}^{\mathbf{O}}]\bigcap[\bigcap_i H_{0l}^{\mathbf{X}|\mathbf{O}}]\}$ against $H_1^G : \{[\bigcup_i H_{1l}^{\mathbf{O}}]\bigcup[\bigcup_i H_{1l}^{\mathbf{X}|\mathbf{O}}]\}$ we must now combine D tests $\Gamma_i^{*\mathbf{O}}$ and D tests $\Gamma_i^{*\mathbf{X}|\mathbf{O}}$, $i = 1, \ldots, D$. Hence,

$$\Gamma'' = \psi(\widehat{\lambda}_1^{\mathbf{O}}, \ldots, \widehat{\lambda}_D^{\mathbf{O}}; \widehat{\lambda}_1^{\mathbf{X}|\mathbf{O}}, \ldots, \widehat{\lambda}_D^{\mathbf{X}|\mathbf{O}}).$$

Remark 1. Notice that the NPC of $2 \times D$ partial tests may be done in at least three different ways:

(a) by taking one single combining function on all $2 \times D$ partial tests such as $\Gamma_a'' = \psi(\lambda_1^{\mathbf{O}}, \ldots, \lambda_D^{\mathbf{O}}; \lambda_1^{X|\mathbf{O}}, \ldots, \lambda_D^{X|\mathbf{O}})$;

(b) by taking D second-order combinations, one for each component variable, $\Gamma_{bi}'' = \psi_i(\lambda_i^{\mathbf{O}}; \lambda_i^{X|\mathbf{O}})$, $i = 1, \ldots, D$, followed by a third-order combination $T_b''' = \psi(\lambda_{b1}'', \ldots, \lambda_{bD}'')$ – this is the procedure presented above.

(c) by taking two second-order combinations, $\Gamma''_{c\mathbf{O}} = \psi_{\mathbf{O}}(\lambda_1^{\mathbf{O}}, \ldots, \lambda_D^{\mathbf{O}})$ and $\Gamma''_{c\mathbf{X}|\mathbf{O}} = \psi_X(\lambda_1^{X|\mathbf{O}}, \ldots, \lambda_D^{X|\mathbf{O}})$ respectively on the censoring indicator $\mathbf{O}$ and on the actually observed $(\mathbf{X}|\mathbf{O})$, followed by a third-order combination $\Gamma'''_c = \psi(\lambda''_{c\mathbf{O}}; \lambda''_{c\mathbf{X}|\mathbf{O}})$.

If in all phases and with each of the three methods of combination the same combining function ψ is used, then Γ''_a, Γ'''_b and Γ'''_c are almost permutationally equivalent, except for approximations due to finite simulations and nonlinearity of combining functions.

In addition, due to assumptions on partial tests, the second-level partial test $\Gamma''_{c\mathbf{X}|\mathbf{O}}$ is (approximately) marginally unbiased for $H_0^{\mathbf{X}|\mathbf{O}} : \left\{\left[\left(X_1 \overset{d}{=} \ldots \overset{d}{=} X_C\right) |\mathbf{O}\right]\right\}$ and so allows for a form of separate testing on actually observed data, conditional on $\mathbf{O}$, even under treatment-dependent censoring. This is useful in many circumstances, especially when interest centres on actually observed data.

For each of the D sub-hypotheses $H_{0i}^{\mathbf{O}}$ against $H_{1i}^{\mathbf{O}}$, a permutation statistic such as Fisher's exact probability test or any other suitable test statistic for proper testing of binary data is generally appropriate (see Section 2.8 and Chapter 6). Hence, the partial permutation test statistics for testing the sub-hypothesis $H_{0i}^{\mathbf{O}}$ against the sub-alternative $H_{1i}^{\mathbf{O}}$ take the form

$$\Gamma_i^{\mathbf{O}*} = \sum_{m=1}^{n_2} O_{m2i}^*.$$

This partial test is permutationally equivalent to Fisher's exact probability test. For each of the D sub-hypotheses $H_{0i}^{\mathbf{X}|\mathbf{O}}$, $\mathbf{O}$ is fixed at its observed value, so that we may proceed conditionally. Proper partial tests are also discussed in Chapter 7.

Remark 2. Partial tests $\Gamma_i^{\mathbf{O}*}$ on the components of $\mathbf{O}$ are exact, unbiased and consistent, whereas $\Gamma_i^{\mathbf{X}|\mathbf{O}*}$ on the components of $\mathbf{X}$ are consistent but approximately exact and unbiased. Thus the combined test T'' is consistent and approximately exact and unbiased for all $\psi \in C$ (see also Sections 7.9 and 7.10).

A comprehensive simulation study on these tests can be found in Campigotto (2009).

9.5 An Application to a Biomedical Study

Tricuspid valve replacement (TVR) has historically been associated with high mortality and morbidity, and current knowledge of long-term results of TVR is limited. We report the results of an experiment (Garatti et al., 2009) in a high-risk patient population with an emphasis on postoperative or in-hospital mortality and long-term survival following a well-known risk factor known as the New York Heart Association (NYHA) functional classification before surgery. This provides a simple way of classifying the extent of heart failure. It is a severity index which places patients into one of four categories based on how limited they are during physical activities, with limitations or symptoms related to normal breathing and varying degrees of shortness of breath and/or angina pain. In this study 16 patients were in the class II functional capacity group (28%), 30 were in class III (54%) and 10 were in class IV (18%). We compared the group of individuals in NYHA class III or IV (*group of interest*) and the group of patients in class II (*control group*).

The aim of a study carried out at the San Donato Hospital in Milan (Italy) was survival analysis related to overall mortality (either perioperative mortality or long-term follow-up (FU) mortality after discharge). Perioperative mortality was classed as post-operative and in-hospital mortality (all deaths occurring at the hospital) and 30-day mortality (all deaths occurring within 30 days of surgery).

This was a retrospective study where 56 patients underwent TVR at the San Donato Hospital over a 15-year period from June 1990 to December 2005. During the study period, 21 (37.5%) deaths occurred, 8 (38%) at the hospital and 13 (62%) during the follow-up. Figure 9.1 shows the survival curves for the two samples.

The data set is relatively small (containing 56 subjects with approximately balanced sample sizes) with moderate to heavy treatment-dependent censoring. In this application the hazard rates seem to be proportional.

In order to find possible significant risk factors affecting survival analysis, we performed both the traditional asymptotic log-rank test and the four proposed permutation tests (see Table 9.2). A p-value < 0.05 was considered statistically significant. As regards the four permutation solutions, we used $B = 10\,000$ conditional Monte Carlo iterations (CMC).

We refer the reader to the NHYA folder on the book's website for MATLAB code that can be used to carry out the analysis.

Table 9.2 Results of survival analysis for NYHA. Department of Cardiovascular Disease, Policlinico S. Donato, Milan, June 1990 to December 2005

Alternatives	Permutation tests TIC-NPC	Asymptotic tests WKN
Two-sided	0.020	0.027

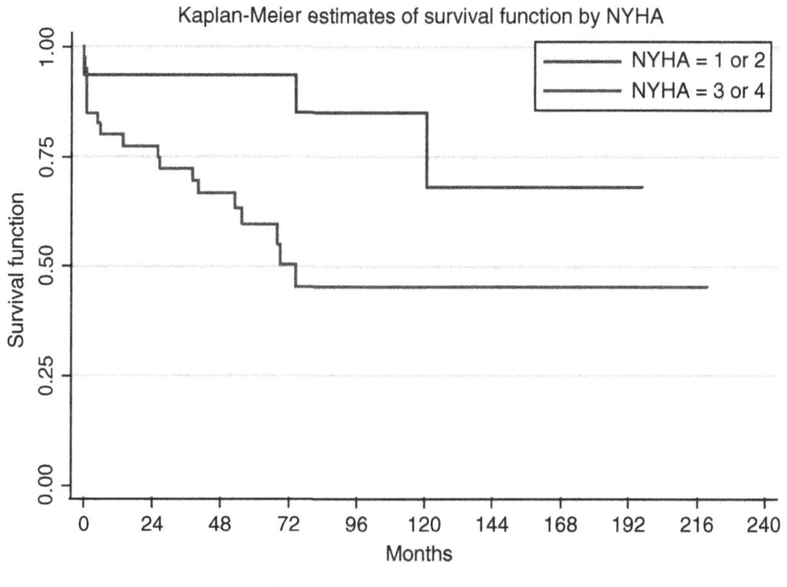

Figure 9.1 Kaplan–Meier survival estimates by NYHA. Department of Cardiovascular Disease, Policlinico S.Donato, Milan, June 1990 to December 2005

10

NPC Tests in Shape Analysis

10.1 Introduction

Statistical shape analysis is a cross-disciplinary field characterized by flexible theory and techniques, potentially adaptable to any appropriate configuration matrix. Specific applications of shape analysis may be found in archaeology, architecture, biology, geography, geology, agriculture, genetics, medical imaging, security applications such as face recognition, the entertainment industry (movies, games), computer-aided design and manufacturing, and so on. David Kendall and Fred Bookstein are without doubt the pioneers in this field. The statistical community has shown increasing interest in shape analysis in the last decade, and particular efforts have been made to develop powerful statistical methods based on models for shape variation of entire configurations of points corresponding to the locations of morphological landmarks.

Inferential methods known in the shape analysis literature make use of configurations of landmarks optimally superimposed using a least-squares procedure, or analyse matrices of inter-landmark distances, for example by means of Euclidean distance matrix analysis (EDMA). In the two-sample case, a practical method for comparing the mean shapes in the two groups is to use Procrustes tangent space coordinates and, if data are concentrated (i.e. close in shape or showing only small variations in shape), calculate the Mahalanobis distance and then Hotelling's T^2 test statistic. Under the assumption of isotropy, another simple approach is to work with statistics based on the squared Procrustes distance and then calculate Goodall's F test statistic (Goodall, 1991).

These tests are based on quite stringent assumptions, such as the equality of covariance matrices, the independence of variation within and among landmarks and the multinormality of the model describing landmarks.

As pointed out in Good (2000), the assumption of equal covariance matrices may be unreasonable in certain applications, the multinormal model in the tangent space may be doubted, and sometimes there may be fewer individuals than landmarks, implying over-dimensioned spaces (curse of dimensionality) and loss of power for Hotelling's T^2 test (Blair et al., 1994). Hence, an alternative is to consider a permutation approach. Further limitations of traditional inferential procedures have been highlighted in Terriberry et al. (2005). Actually, useful shape models contain parameters lying in non-Euclidean spaces (in this context the term 'parameters' is used in place of 'variables'). More specifically, in morphometrics, four kinds of parameters should be taken into account: nuisance parameters (e.g. translation and rotation); geometric parameters, such as shape coordinates; statistical parameters, such as mean differences or correlations; and finally, another set of geometric parameters, such as partial warp scores or Procrustes residuals (Slice et al., 1996). Some of

Permutation Tests for Complex Data: Theory, Applications and Software Fortunato Pesarin, Luigi Salmaso

these parameters may have a large variance, may be highly correlated, or have completely different scales, thus invalidating analyses and final results. Hence traditional statistical tools designed for Euclidean spaces must be used with particular care if they are applicable. In contrast, permutation tests are appealing because they make no distributional assumptions, requiring only that the data in each group are exchangeable under the null hypothesis. Terriberry et al. (2005) presented an application of NPC methodology in shape analysis, but the properties of the method itself are not investigated further. On the strength of these considerations, we suggest an extension of the NPC methodology to shape analysis. Under very mild and reasonable conditions, the NPC method is found to be consistent, unbiased and extendable to unconditional inferences. We remark that in the parametric approach, this extension is possible when the data set is randomly selected by well-designed sampling procedures on well-defined population distributions. When similarity and conditional unbiasedness properties are jointly satisfied, and if correctly applicable, permutation tests allow for inferential extensions at least in a weak sense (Pesarin, 2002; Ludbrook and Dudley, 1998; see also Section 3.5 above).

We emphasize that permutation tests require homogeneous covariance matrices in order to guarantee exchangeability only in H_0, thus relaxing more stringent assumptions required by parametric tests since they do not require homoscedasticity in the alternative.

We begin with a review of inferential methods from the shape analysis literature, highlighting some drawbacks to the use of Hotelling's T^2 test. Then, focusing on the two-sample case, through an exhaustive comparative simulation study, we evaluate the behaviour of traditional tests along with nonparametric permutation tests using multi-aspect (MA) procedures (see Example 3, 4.6) and domain combinations. In this nonparametric framework we also analyse the case of heterogeneous and dependent variation at each landmark. Furthermore, we examine the effect of superimposition on the power of NPC tests. Due to their nonparametric nature, we may assert that the suggested tests provide efficient solutions which allow us to deal with data sets including many informative landmarks and few specimens.

10.2 A Brief Overview of Statistical Shape Analysis

Statistical shape analysis relates to the study of random objects, the concept of shape corresponding to some geometrical information that is invariant under translation, rotation and scale effects. An intuitive definition of shape is given by Kendall (1977).

Definition 1. *Shape is all the geometrical information that remains when location, scale and rotational effects are filtered out from an object.*

Hence two objects have the same shape if they are invariant under the Euclidean similarity transformations of translation, scaling and rotation (Dryden and Mardia, 1998).

Definition 2. *Size and shape represent all the geometrical information that remains when location and rotational effects are filtered out from an object.*

Thus, in order to have the same size and shape, two objects are required to be rigid body transformations of each other (Dryden and Mardia, 1998).

10.2.1 How to Describe Shapes

'Landmark-based' analysis, where shapes are represented by a discrete sampling of the object contours (Dryden and Mardia, 1998; Small, 1996), has played a substantial role in shape analysis research. Bookstein and his colleagues recommended the use of landmarks for the analysis of biological features and constrained the choice of landmarks to prominent features of the organism or biological structure (Dryden and Mardia, 1998). Hence, these points are biologically active sites on organisms and are defined in Dryden and Mardia (1998) as follows.

Definition 3. *A landmark is a point (locus) of correspondence on each object that matches between and within populations.*

These loci have the same name (they are homologous), as well as Cartesian coordinates, and correspond in some sensible way over the forms of a data set. We recall that in geometric morphometrics the term 'homologous' has no meaning other than that the same name is used for corresponding parts in different species or developmental stages (Slice et al., 1996). Moreover, these points represent a foundation for the explanations of the biological processes, and still today many of the explanations of form accepted as epigenetically valid adduce deformations of the locations of landmarks (Bookstein, 1986). Srivastava et al. (2005) emphasized some limitations to the landmark-based representations. Despite the effectiveness of this approach in applications where landmarks are readily available (e.g. physician-assisted medical image analysis), automatic detection of landmarks is not straightforward and the resulting shape analysis is greatly determined by the choice of landmarks. In addition, shape interpolation with geodesics in this framework lacks a physical interpretation.

Landmarks can basically be classified into three groups: anatomical, mathematical, and pseudo-landmarks.

- An *anatomical landmark* is a point assigned by an expert that corresponds between organisms in some biologically meaningful way (e.g. the corner of an eye or the meeting of two sutures on a skull).
- *Mathematical landmarks* are points located on an object in accordance with some mathematical or geometrical property of the figure (e.g. at a point of high curvature or at an extreme point). Mathematical landmarks are particularly useful in automatic recognition and analysis.
- *Pseudo-landmarks* are constructed points on an organism, located either around the outline or in between anatomical or mathematical landmarks. Continuous curves can be approximated by a large number of pseudo-landmarks along the curve. Also, pseudo-landmarks are useful in matching surfaces when points can be located on a regular grid over each surface.

They can be grouped into three further types (Dryden and Mardia, 1998).

- *Type I landmarks* (usually the easiest and the most reliable to locate) are mathematical points whose homology is reinforced by the strongest evidence, such as a local pattern of juxtaposition of tissue types or a small patch of some unusual histology.
- *Type II landmarks* are defined by local properties such as maximal curvatures (i.e. they are mathematical points whose homology is strengthened only by geometric, not histological, evidence), for instance the sharpest curvature of a tooth.
- *Type III landmarks* are the most difficult and the least reliable to locate. They occur at extremal points or constructed landmarks (e.g. maximal diameters and centroids) and have at least one deficient coordinate, such as either end of a longest diameter, or the bottom of a concavity. They characterize more than one region of the form and could be treated by geometric morphometrics as landmark points, even if they can be tricky because of the deficiency they embody.

Anatomical landmarks are usually of type I or II and mathematical landmarks are usually of type II or III. Pseudo-landmarks are commonly taken as equi-spaced along outlines between pairs of landmarks of type I or II, and in this case the pseudo-landmarks are of type III.

Along with landmarks, it is possible to collect semi-landmark points. These are located on a curve and allowed to slip a small distance with respect to another corresponding curve. The term 'semi' is used because the landmarks lie in a lower number of dimensions than other types of landmarks, for example along a one-dimensional curve in a two-dimensional image (Dryden and Mardia, 1998). Semi-landmarks are defined in relation to other landmarks (e.g. 'midway between landmarks 1 and 2'). Indeed, they have no anatomical identifiers but remain corresponding points in a sense satisfactory for subsequent morphometric interpretation (Bookstein, 1997). Hence, these loci fail to be true landmarks since they do not enjoy the homology property in its traditional sense (i.e. they lie on homologous curves while their exact position along these usually smooth regions or curves is unclear). Defining semi-landmarks may be useful in studying substantial regions in an object that cannot be defined by simply using anatomical or mathematical landmarks, or a region which consists of two or more real landmark points (Adams et al., 2004).

On the basis of these considerations, Katina et al. (2007) give another landmark classification, including the information carried by semi-landmarks on curves and surfaces. In particular, it is possible to define the following landmark types:

- *Type 1*: discrete juxtaposition of tissues.
- *Type 2*: extreme of curvature characterizing a single structure.
- *Type 3*: landmark points characterized locally by information from multiple curves and surfaces and by symmetry:
  - *Type 3a*: intersection of a ridge curve and the midcurve on the same surface;
  - *Type 3b*: intersection of an observed curve and the midcurve;
  - *Type 3c*: intersection of a ridge curve and an observed curve on the same surface.
- *Type 4*: semi-landmarks on ridge curves and symmetric curves (midsagittal curve).
- *Type 5*: semi-landmarks on surfaces.
- *Type 6*: constructed semi-landmarks.

10.2.2 Multivariate Morphometrics

Morphometrics is the study of shape variation and its covariation with other variables and represents an integral part of the biology of organisms (Adams, 1999). Its goal is the objective description of the changes in the form of an organism – its shape and size – during ontogeny or during the course of evolution (Bookstein, 1986). Databases of landmark locations are usually processed using techniques such as multivariate morphometrics and deformation analysis. In fact we can evaluate configurations of landmark points by means of variables expressing aspects of the size or shape of single specimens, such as distances or ratios of distances, or can directly measure the relation between one form and another as a deformation (Dryden and Mardia, 1998). Both strategies are useful tools for examining group differences in size and shape or between size change and shape change (Bookstein, 1986). With reference to multivariate morphometrics, this approach is often applied without regard to homology, that is, it does not require size or shape measures to derive from the locations of homologous landmarks. As a consequence, the homology of linear distances is difficult to assess because many distances (e.g. maximum width) are not defined by homologous points. The large amount of measurements obtained through this method are analysed using standard multivariate statistical methods (canonical variates analysis, principal components analysis, factor analysis, linear modelling, discriminatory analysis, component extraction) and any findings are separately interpreted coefficient by coefficient. However, the geometric origin of the variables measured is generally not exploited further. Moreover, visualizing results through graphical representations of shape is very demanding because the geometric relationships among the variables (linear distances) are not preserved, thus losing some aspects of shape.

On the other hand, deformation analysis has been introduced into descriptive biology by Thompson (1961) under the label of 'Cartesian transformation'. We recall the notion of deformation as given in Dryden and Mardia (1998).

Definition 4. *A deformation is a mapping which takes neighbouring points to neighbouring points and which alters lengths of little segments by factors which never get too large or too small. It is an informal version of what mathematicians call a diffeomorphism, a one-to-one transformation which, along with its inverse, has a derivative at every point of a region and its image.*

Thompson (1961) suggested directly observing a comparison of biological forms as a geometric object of measurement in its own right, rather than as the mere numerical difference of measures taken separately upon forms. In particular he proposed to represent the form change as a deformation of the picture plane corresponding closely to what biologists already knew as homology: the smooth mapping of one form onto the other sending landmarks onto their homologues and interpolated suitably in between (Bookstein, 1986).

But the field of morphometrics has lately experienced a revolution. During the 1980s various authors, among them Fred Bookstein and James Rohlf, proposed to combine traditional multivariate morphometrics and deformation analysis, calling this synthesis *geometric morphometrics*. The term 'geometric' relates to the geometry of Kendall's shape space, that is, the estimation of mean shapes and the description of sample variation of shape using the geometry of Procrustes distance. Multivariate morphometrics is usually carried out in a linear tangent space to the non-Euclidean shape space in the vicinity of the mean shape. It could be defined as a collection of approaches for the multivariate statistical analysis of Cartesian coordinate data, often limited to landmark point locations. More directly, it is described as the class of morphometric methods that capture the geometry of the morphological structures of interest and preserve complete information about the relative spatial arrangements of the data throughout the analyses. As a consequence, results of high-dimensional multivariate analyses can be mapped back into physical space to achieve appealing and informative visualizations, contrary to alternative traditional methods (Slice, 2005).

The direct analysis of databases of landmark locations is not convenient because of the presence of nuisance parameters, such as position, orientation and size. In order to carry out a useful statistical shape analysis, a generalized least-squares (GLS) or generalized Procrustes analysis (GPA) superimposition is performed to eliminate non-shape variation in configurations of landmarks and to align the specimens to a common coordinate system (Rohlf and Slice, 1990). Along with GPA, we mention another registration method, the two-point registration, that provides Bookstein's shape coordinates. The aligned specimens identify points in a non-Euclidean space, which is approximated by a Euclidean tangent space for standard multivariate statistical analyses (Slice et al., 1996; Rohlf, 1999). With reference to the GPA superimposition method, first the centroid of each configuration is translated to the origin, and configurations are scaled to a common unit size (by dividing by centroid size; see Bookstein, 1986). Finally, the configurations are optimally rotated to minimize the squared differences between corresponding landmarks (Gower, 1975; Rohlf and Slice, 1990). This is an iterative process, useful for computing the mean shape.

Generalized resistant-fit (GRF) procedures, making use of median and repeated median-based estimates of fitting parameters rather than least-squares estimates, are also available (Slice et al., 1996). In particular, they are more efficient for revealing differences between two objects when the major differences are mostly in the relative positions of a few landmarks (Rohlf and Slice, 1990). Even if they lack the well-developed distributional theory associated with the least-squares fitting techniques, being robust, these methods seem to be protected against departure from the assumptions of the analysis (e.g. independent, identically and normally distributed errors) and seem to be unresponsive to the potentially strong influences of atypical or incorrect data values (Siegel and Benson, 1982).

In the presence of semi-landmarks, a method worthy of note is that of 'sliding semi-landmarks', allowing outlines to be combined with landmark data in one analysis, providing a richer description of the shapes. The iterative procedure involves first sliding the semi-landmarks to the left or right along a curve during the GPA superimposition in an attempt to minimize the distance between the adjusted position and the corresponding point in the consensus or to reduce the overall bending energy required to fit the specimens to the sample average configuration. Computations are iterative and the algorithm provides smooth and interpretable deformation grids among the forms. For details, see Bookstein (1997), Adams et al. (2004), Slice et al. (1996) and the TpsRelw software guide by Rohlf (2008a).

After superimposition, differences in shape can be described either in terms of differences in coordinates of corresponding landmarks between objects (Bookstein, 1996) or in terms of differences in the deformation grids representing the objects, for example using the thin-plate spline method (Bookstein, 1991). The thin-plate spline is a global interpolating function that maps the landmark coordinates of one specimen to the coordinates of the landmarks in another specimen; it is a mathematically rigorous realization of Thompson's (1961) idea of transformation grids, where one object is deformed or 'warped' into another. The parameters describing these deformations (partial warp scores) can be used as shape variables for statistical comparisons of variation in shape within and between populations (Adams, 1999; Rohlf, 1993). As a result, thin-plate splines can be interpreted as one method for generating a coordinate system for the above-mentioned tangent space.

Along with the superimposition methods, several alternative procedures for obtaining shape information from landmark data have been proposed (Adams et al., 2004). Here we mention EDMA methods proposed by Lele and Richtsmeier (1991), a related approach using standard multivariate methods on logs of size-scaled inter-landmark distances (Rao and Suryawanshi, 1996) and methods based on interior angles (Rao and Suryawanshi, 1998).

10.3 Inference with Shape Data

Rohlf (2000) reviews the main tests used in the field of shape analysis and compares the statistical power of the various tests that have been proposed to test for equality of shape in two populations. Although his work is limited to the simplest case of homogeneous, independent, spherical variation at each landmark and the sampling experiments emphasize the case of triangular shapes, it allows practitioners to choose the method with the highest statistical power under a set of assumptions appropriate for the data. By means of a simulation study, he found that Goodall's F test had the highest power followed by the T^2 test using Kendall tangent space coordinates. Power for T^2 tests using Bookstein shape coordinates was good if the baseline was not the shortest side of the triangle. The Rao and Suryawanshi shape variables had much lower power when triangles were not close to being equilateral. Power surfaces for the EDMA-I T statistic revealed very low power for many shape comparisons, including those between very different shapes. Power surfaces for the EDMA-II Z statistic depended strongly on the choice of baseline used for size scaling (Rohlf, 2000).

All these tests are based on quite stringent assumptions. In particular, the tests based on the T^2 statistic – for example, T^2 tests using Bookstein, Kendall tangent space coordinates, and Rao and Suryawanshi (1996, 1998) shape variables – require independent samples and homogeneous covariance matrices in both H_0 and H_1, and shape coordinates distributed according to the multivariate normal distribution. We remark that Hotelling's T^2 is derived under the assumption of population multivariate normality and it is not recommended unless the number of subjects is much larger than that of landmarks (Dryden and Mardia, 1998). It is well known that Hotelling's T^2 test is formulated to detect any departures from the null hypothesis and therefore often lacks power to detect specific forms of departures that may arise in practice – that is, the T^2 test fails to provide an easily implemented one-sided (directional) hypothesis test (Blair et al., 1994).

Goodall's F test requires a restrictive isotropic model in which configurations are isotropic normal perturbations from mean configurations, and it assumes that the distributions of the squared Procrustes distances are approximately chi-squared distributed.

Turning to methods based on inter-landmark distances, EDMA-I T assumes independent samples and the equality of the covariance matrices in the two populations being compared (Lele and Cole, 1996), while EDMA-II Z assumes only independent samples and normally distributed variation at each landmark.

In order to complete this review on main tests used in shape analysis, we recall the pivotal bootstrap methods for k-sample problems, in which each sample consists of a set of real (the directional case) or complex (the two-dimensional shape case) unit vectors, proposed in the paper by Amaral et al. (2007). The basic assumption here is that the distribution of the sample mean shape (or direction or axis) is highly concentrated. This is a substantially weaker assumption than is entailed in tangent space inference (Dryden and Mardia, 1998) where observations are presumed to be highly concentrated. For mathematical, statistical and computational details we refer to Amaral et al. (2007). We have already drawn attention to Good's (2000) observation that the assumption of equal covariance matrices may be unreasonable, especially under the alternative, the multinormal model in the tangent space may be doubted and sometimes there are few individuals and many more landmarks, implying over-dimensioned spaces and loss of power for Hotelling's T^2 test. Hence, when sample sizes are too small, or the number of landmarks is too large, it is essentially inefficient to assume that observations are normally distributed. An alternative procedure is to consider a permutation version of the test (see Good, 2000; Dryden and Mardia, 1993; Bookstein, 1997; Terriberry et al., 2005). Permutation methods are distribution-free and allow for quite efficient solutions which may be tailored for sensitivity to specific treatment alternatives providing one-sided as well as two-sided tests of hypotheses (Blair et al., 1994).

In the wake of these considerations, we propose an extension of the NPC methodology in Chapter 4 to shape data. Generally, permutation tests require homogeneous covariance matrices under H_0 in order to guarantee exchangeability, thus relaxing the stringent assumptions of parametric tests. This is consistent with the notion that the true H_0 implies the equality in multivariate distribution of observed variables.

10.4 NPC Approach to Shape Analysis

10.4.1 Notation

Let $\mathbf{X}_1$ be the $n_1 \times (k \times m)$ matrix of aligned data (e.g. specimens or individuals) in the tangent space in the first group. By 'aligned' we mean that we are considering the shape coordinates obtained through GPA. We recall that the direct analysis of databases of landmark locations is not convenient because of the presence of nuisance parameters, such as position, orientation and size. Usually, in order to carry out a valuable statistical shape analysis, a GLS or GPA superimposition is performed to eliminate non-shape variation in configurations of landmarks and to align the specimens to a common coordinate system (Rohlf and Slice, 1990). Hence the GPA procedure is performed to estimate a mean shape and to align the specimens to it.

Similarly, $\mathbf{X}_2$ is the $n_2 \times (k \times m)$ matrix of aligned specimens in the tangent space, that is, the second group of subjects. Let $\mathbf{X} = \begin{pmatrix} \mathbf{X}_1 \\ \mathbf{X}_2 \end{pmatrix}$ be the $n \times (k \times m)$ matrix of aligned specimens in the tangent space, that is, our data set may be written as $\mathbf{X} = \mathbf{X}_1 \biguplus \mathbf{X}_2$, where $n = n_1 + n_2$. Hence, $\mathbf{X}$ is a matrix of data with specimens in rows and landmark coordinates in columns. The unit-by-unit representation of $\mathbf{X}$ is then $\{X_{hji},\ i = 1, \ldots, n,\ j = 1, 2,\ h = 1, \ldots, km;\ n_1, n_2\}$.

For simplicity, we may assume that the landmark coordinates in the tangent space behave according to the model $X_{hji} = \mu_h + \delta_{hj} + \sigma_h Z_{hji}$, $i = 1, \ldots, n$, $j = 1, 2$, $h = 1, \ldots, km$, where:

- k is the number of landmarks in m dimensions;
- μ_h is a population constant for the hth variable;
- δ_{hj} represents treatment effect (i.e. the non-centrality parameter) in the jth group on the hth variable which, without loss of generality, is assumed to be $\delta_{h1} = 0$, $\delta_{h2} \geq$ (or $\leq$) 0;
- σ_h is the scale coefficient specific to the hth variable;
- Z_{hji} are random errors assumed to be exchangeable with respect to treatment levels, independent with respect to units, with null mean vector ($\mathbb{E}(\mathbf{Z}) = 0$) and finite second moment.

With reference to the scale coefficients σ_h, we observe that these parameters may be very useful since they reflect the 'intrinsic' biases in the registration of landmarks. There are in fact landmark points readily available, hence easier to capture than others by the operator or machine. As a consequence, they are less variable in their location. Hence, landmark coordinates in the first group differ from those in the second group by a 'quantity' $\boldsymbol{\delta}$, where $\boldsymbol{\delta}$ is the km-dimensional vector of effects. $\left\{X^*_{hji},\ i = 1, \ldots, n,\ j = 1, 2,\ h = 1, \ldots, km;\ n_1, n_2\right\}$ denotes a permutation of the original data.

Therefore, the specific hypotheses may be expressed as

$$H_0 : \left\{\bigcap_{h=1}^{km}\left[X_{h1} \overset{d}{=} X_{h2}\right]\right\} \text{ against } H_1 : \left\{\bigcup_{h=1}^{km}\left[X_{h1} \overset{d}{>} (<) X_{h2}\right]\right\},$$

where $\overset{d}{>}$ stands for distribution (or stochastic) dominance.

Let $T^*_h = \mathrm{Sg}(\delta_h)\sum_{i \leq n_1} X^*_{h1i}$, $h = 1, \ldots, km$, be the km partial tests where $\mathrm{Sg} = +1$ if $\delta_h > 0$ and -1 otherwise. All these tests are exact, marginally unbiased and consistent, and significant for large values, so that the NPC theory properly applies. The hypothesis testing problem is broken down into two stages, considering both the coordinate and the landmark level (and, if present, the domain level). We actually refer to domains as subgroups of landmarks sharing anatomical, biological or locational features. However, we remark that a domain may be seen as a latent variable.

Hence, we formulate partial test statistics for one-sided hypotheses and then consider the global test T'' obtained after combining at the first stage with respect to m, then with respect to k (of course, this sequence may be reversed). We may also apply the MA procedure to each coordinate of a single landmark and then consider their combination. For example, if we consider four landmarks, first of all we can consider a test for each coordinate (x and y coordinates in the 2D case) of each landmark. Once the aspects of interest have been decided (e.g. the first two moments), we can focus on the coordinate level, or on the landmark level after combining coordinates, or on the domain level as well, and finally on the global test (see Figure 10.1).

One of the main features and advantages of the proposed approach is that by using the MA procedure and the information about domains, we are able to obtain not only a global p-value, as in traditional tests, but also a p-value for each of the defined aspects or domains. Hence, our procedure makes it possible to construct a hierarchical tree, allowing for testing at different levels of the tree (see Figure 10.1). On the one hand, partial tests may provide marginal information for each specific aspect; on the other, they jointly provide information on the global hypothesis. In this way, if we find a significant departure from H_0, we can investigate the nature of this departure in detail. Furthermore, we can move from the top of the tree to the bottom and, for interpreting results in a hierarchical way, from the bottom to the top. It is worth noting that "intermediate" level p-values need to be adjusted for multiplicity.

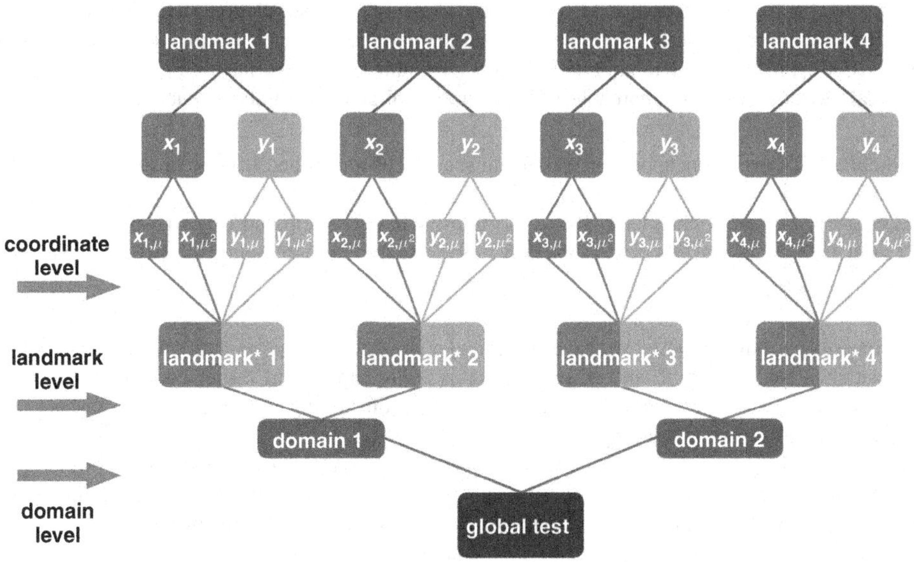

Figure 10.1 Different levels of combination

10.4.2 Comparative Simulation Study

Let us assume that our samples consist of configurations of $k = 8$ landmarks in $m = 2$ dimensions characterized by slightly different means. Suppose that we are dealing with male and female skull configurations of a particular animal created *ad hoc*, representing the two independent samples (see Table 10.1). Since sample means differ slightly from each other in 6 out of 16 coordinates (highlighted in bold in Table 10.1), we have generated data using such a configuration in order to evaluate the power of the competing tests.

In all simulations we have $B = 1000$ CMC iterations and $MC = 1000$ Monte Carlo runs. Throughout the simulation study, we have selected both Liptak and Fisher combining functions.

Table 10.1 Hypothetical mean configurations

Domain	No.	Lnd. name	Male		Female	
			x	y	x	y
1	1	nasion	65.00	223.00	65.00	**222.85**
1	2	basion	54.00	−40.00	**53.75**	−40.00
2	3	staphylion	0.00	0.00	0.00	0.00
2	4	prosthion	0.00	35.00	0.00	**34.50**
2	5	nariale	19.00	121.00	**18.90**	121.00
3	6	bregma	70.00	203.00	70.00	203.00
3	7	lambda	110.00	112.00	**109.95**	112.00
3	8	opisthion	104.00	17.00	104.00	**16.88**

Focusing on the two-sample case, we have carried out the simulation study under the same conditions of homogeneous, independent, spherical variation at each landmark, as described in Rohlf (2000). In the complete simulation study (Brombin, 2009) we have compared, in terms of statistical power, traditional approaches for the statistical analysis of shape, such as

- Hotelling's T^2 test using approximate tangent space coordinates and Bookstein shape coordinates,
- Goodall's F test,
- EDMA-I (Lele and Richtsmeier, 1991) and EDMA-II tests (Lele and Cole, 1995; 1996),
- T^2 test using Rao and Suryawanshi (1996, 1998) shape variables only for large sample sizes, e.g. $n_1 = n_2 = 50$.

As regards permutation tests, we have in particular considered (i) permutation Hotelling's T^2, (ii) permutation global tests with and without domains using Liptak and Fisher combining functions, (iii) permutation MA tests with and without domains, considering location and scale aspects and using Liptak and Fisher combining functions.

Let n_j, $j = 1, 2$, be the sample sizes of the two samples and let the landmark coordinates be multivariate normally distributed. Three domains have been considered, i.e. baseline (nasion and basion), face (staphylion, prosthion and nariale), and braincase (bregma, lambda and opisthion). We use G to denote the global test obtained after combining all partial tests, and G_d to denote the global test that takes into account the information about domains (hence obtained after combining partial tests on chosen domains). MA, if present, indicates that the MA procedure has been applied. $T^{2,perm}$ indicates the nonparametric permutation counterpart of Hotelling's T^2.

We wish to point out that in this simulation study we only show global p-values. Hence, in this case, we do not need to cope with the multiplicity problem. For the sake of space, we report only the results of NPC tests with small sample sizes ($n_1 = n_2 = 10$, $\sigma^2 = 0.25$). In Table 10.2, the better-performing tests are highlighted in bold. In particular, we show that the global nonparametric test using Fisher's combining function, in its standard, domain and MA versions, has better power in almost all situations.

Our simulation study emphasizes the good behaviour in terms of power and the flexibility of the nonparametric permutation solution in shape analysis, since it allows us to carry out a shape analysis even in the presence of small sample sizes and a large number of shape variables. We wish to highlight that, under the null hypothesis, the proportion of test rejections is very close to the given significance level (see Table 10.2).

10.5 NPC Analysis with Correlated Landmarks

In this framework we analyse the case of heterogeneous and dependent variation at each landmark, and evaluate the power and attained α-level. The superimposition step has been included in the routine since GPA superimposition may modify dependency structures among landmarks (Rohlf, 2008b). In order to obtain a non-singular covariance matrix, we have performed an eigenvalue decomposition of the original variance–covariance matrix and transformed the original eigenvalues λ. We have considered transformations such as $\lambda^{1/3}$ and $\lambda^{1/10}$, rescaled by their trace (see the effect of transforming eigenvalues on the scatterplot in Figure 10.2). Then we have recalculated the covariance matrix $\mathbf{\Sigma}^{\dagger}$, using the relation $\mathbf{\Sigma}^{\dagger} = \mathbf{V}\mathbf{\Lambda}^{\dagger}\mathbf{V}^{\top}$, where $\mathbf{\Lambda}^{\dagger}$ is a diagonal matrix with the transformed eigenvalues, $\mathbf{V}$ is an orthogonal matrix containing the corresponding eigenvectors, and $\mathbf{V}^{\top}$ is $\mathbf{V}$ transposed. Under the alternative, data have been generated using different means and the same covariance matrix $\mathbf{\Sigma}^{\dagger}$. In Table 10.3 we display hypothetical mean configurations, representing 3D male and female *Macaca fascicularis* monkey skulls (for details, see Frost et al., 2003).

Table 10.2 Attained α-level and power behaviour: $n_1 = n_2 = 10$, $B = \text{CMC} = 1000$, $\sigma^2 = 0.25$

	Attained α-level					
	$\alpha = 0.01$	$\alpha = 0.05$	$\alpha = 0.10$	$\alpha = 0.20$	$\alpha = 0.30$	$\alpha = 0.50$
$T^{2,perm}$	0.011	0.056	0.112	0.211	0.300	0.506
G (Liptak)	0.017	0.054	0.107	0.215	0.307	0.512
G_d (Liptak)	0.014	0.055	0.115	0.213	0.304	0.512
G (Fisher)	0.014	0.058	0.120	0.208	0.307	0.506
G_d (Fisher)	0.013	0.057	0.121	0.205	0.313	0.490
G (Liptak, MA)	0.016	0.051	0.109	0.217	0.311	0.516
G_d (Liptak, MA)	0.013	0.046	0.116	0.218	0.310	0.514
G (Fisher, MA)	0.010	0.058	0.118	0.214	0.311	0.504
G_d (Fisher, MA)	0.012	0.061	0.115	0.210	0.317	0.484
	Power					
	$\alpha = 0.01$	$\alpha = 0.05$	$\alpha = 0.10$	$\alpha = 0.20$	$\alpha = 0.30$	$\alpha = 0.50$
$\boldsymbol{T^{2,perm}}$	**0.087**	**0.229**	**0.320**	**0.496**	**0.611**	**0.775**
G (Liptak)	0.054	0.172	0.279	0.434	0.549	0.741
G_d (Liptak)	0.048	0.165	0.265	0.419	0.526	0.724
G (Fisher)	**0.079**	**0.214**	**0.326**	**0.480**	**0.615**	**0.769**
G_d (Fisher)	0.067	0.203	0.302	0.452	0.581	0.753
G (Liptak, MA)	0.056	0.175	0.275	0.440	0.549	0.748
G_d (Liptak, MA)	0.052	0.170	0.265	0.419	0.527	0.723
G (Fisher, MA)	**0.075**	**0.218**	**0.312**	**0.477**	**0.604**	**0.767**
G_d (Fisher, MA)	0.071	0.203	0.290	0.447	0.568	0.755

Table 10.3 Configurations

	Landmark	Male			Female		
No.	Lnd. name	x_M	y_M	z_M	x_F	y_F	z_F
1	Inion	17.7752	18.9981	6.9585	17.5252	18.9981	6.9585
2	Bregma	15.9101	16.3499	9.2159	15.9101	16.4499	9.2159
3	Glabella	13.6833	12.7086	7.6433	13.6833	12.7086	7.6433
4	Nasion	13.6799	12.6892	7.5628	13.8299	12.6892	7.5628
5	Rhinion	12.9273	11.2649	5.1792	12.9273	11.2149	5.1792
6	Nasospinale	12.6114	10.5523	3.6257	12.6114	10.5523	3.6257
7	Prosthion	12.4725	10.233	2.8531	12.4725	10.2330	2.8531
8	Opisthion	17.1882	17.8852	5.0014	17.1882	17.8852	5.1514
9	Basion	16.5070	16.7665	4.4799	16.5070	16.7165	4.4799
10	Staphylion	14.6975	13.8755	4.1783	14.6075	13.8755	4.1783
11	Incisivion	13.2442	11.4665	3.5466	13.2442	11.4665	3.5166

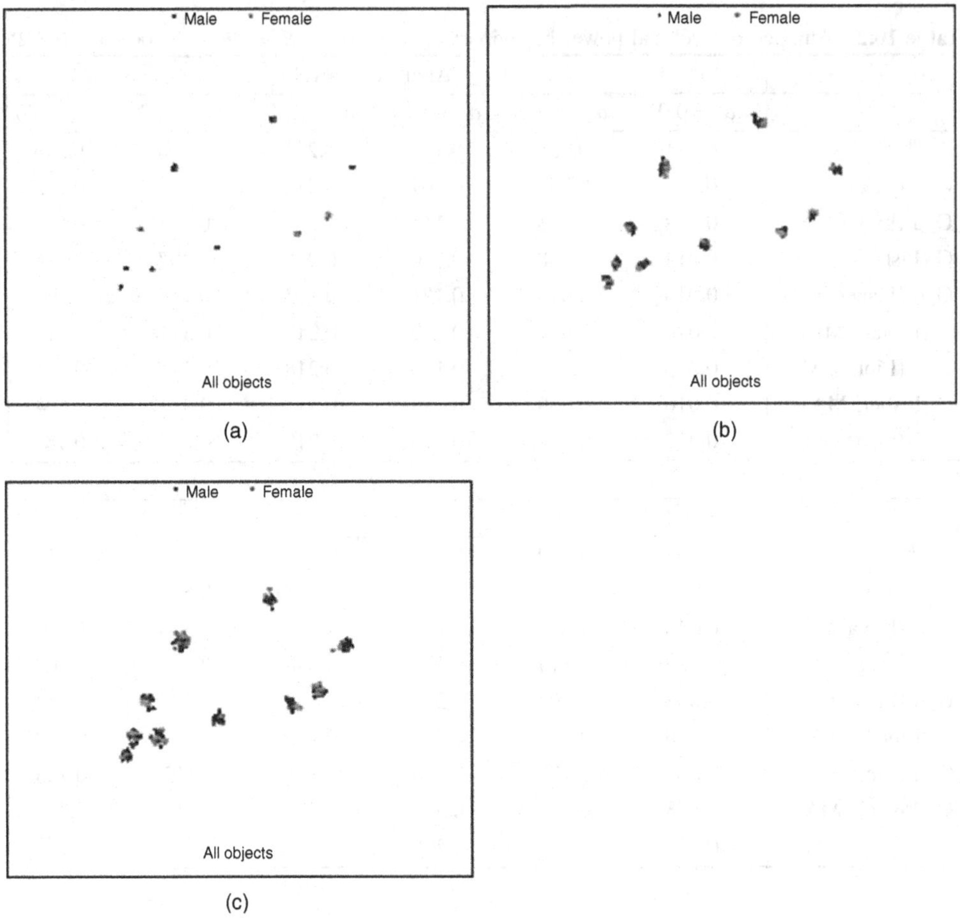

Figure 10.2 Original eigenvalues (a) λ, (b) $\lambda^{1/3}$, and (c) $\lambda^{1/10}$

Let n_j, $j = 1, 2$, denote the sample size in the two groups. In particular, we have considered the settings $n_1 = n_2 = 5$, $n_1 = n_2 = 10$, $n_1 = 5$, $n_2 = 10$. In the simulation study we have evaluated the power and attained α-level when the number of 3D landmarks k was, in turn, equal to 3, 6, 9, 11. Three domains have been considered: the first includes landmarks 1, 2 and 11; the second includes landmarks 3–7; the third includes landmarks 8–10.

We use $T^{2,perm}$ to denote Hotelling's T^2 permutation counterpart, with G the combination of all partial tests, and G_d the combination using domains. For the sake of space, we present simulation results only for the case in which the number of 3D landmarks k is equal to 6, $n_1 = n_2 = 10$ and transformed λ is $\lambda^{1/3}$(see Table 10.4).

In all the simulations under H_0, when using a global test with Fisher's combining function, MA procedure and domain information, the type I error rate was too large, thus invalidating inferential conclusions. For example, in Table 10.4, focusing on $\alpha = 0.05$, $G_{MA,F}$ has a corresponding attained $\alpha = 0.112$ and $G_{d,MA,F}$ has a corresponding attained $\alpha = 0.122$. Recall that at the beginning we

Table 10.4 Attained α-level and power behaviour: $n_1 = n_2 = 10$, $\lambda^{1/3}$, $k = 6$, $m = 3$, $B = CMC = 1000$

	Attained α-level					
Test	0.01	0.05	0.10	0.20	0.30	0.50
$T^{2,perm}$	0.012	0.056	0.102	0.204	0.295	0.493
G_T	0.000	0.052	0.088	0.195	0.290	0.489
$G_{d,T}$	0.004	0.038	0.103	0.195	0.288	0.475
G_F	0.011	0.055	0.099	0.197	0.292	0.504
$G_{d,F}$	0.011	0.050	0.098	0.196	0.293	0.496
$G_{MA,T}$	0.008	0.031	0.065	0.136	0.206	0.350
$G_{d,MA,T}$	0.008	0.027	0.073	0.143	0.216	0.345
$G_{MA,F}$	0.041	0.112	0.164	0.266	0.340	0.508
$G_{d,MA,F}$	0.043	0.122	0.185	0.305	0.402	0.550
$G_{(\mu,\mathbb{M}d),F}$	0.012	0.055	0.099	0.190	0.288	0.466
$G_{d,(\mu,\mathbb{M}d),F}$	0.011	0.054	0.098	0.191	0.300	0.479
$G_{(\mu,\mu^2),F}$	0.010	0.048	0.109	0.195	0.286	0.509
$G_{d,(\mu,\mu^2),F}$	0.009	0.043	0.096	0.192	0.287	0.485
	Power					
Test	0.01	0.05	0.10	0.20	0.30	0.50
$T^{2,perm}$	0.337	0.621	0.757	0.888	0.930	0.981
G_T	0.000	0.538	0.709	0.855	0.923	0.978
$G_{d,T}$	0.123	0.478	0.683	0.828	0.902	0.971
G_F	0.299	0.564	0.709	0.858	0.916	0.973
$G_{d,F}$	0.286	0.547	0.679	0.819	0.898	0.962
$G_{MA,T}$	0.234	0.456	0.636	0.754	0.851	0.945
$G_{d,MA,T}$	0.255	0.401	0.600	0.755	0.824	0.923
$G_{(\mu,\mathbb{M}d),F}$	0.284	0.571	0.711	0.853	0.918	0.969
$G_{d,(\mu,\mathbb{M}d),F}$	0.164	0.431	0.596	0.797	0.880	0.952
$G_{(\mu,\mu^2),F}$	0.224	0.467	0.612	0.740	0.833	0.922
$G_{d,(\mu,\mu^2),F}$	0.137	0.369	0.495	0.663	0.782	0.903

decided to evaluate location (mean) and distributional (Anderson–Darling statistic) aspects. Hence, GPA superimposition may modify dependency structures, thus altering the final distribution. Fisher's combining function is more sensitive with MA procedures than Tippett's combining function. However, if we change the aspects (e.g. if we consider mean μ and median $\mathbb{M}d$ or mean and second moment μ^2), the proportion of rejection of the tests $G_{MA,F}$ and $G_{d,MA,F}$ now achieves the given significance α-level. These results highlight the fact that GPA affects the initial distribution of the data, hence particular care is needed when an MA procedure is carried out.

10.6 An Application to Mediterranean Monk Seal Skulls

10.6.1 The Case Study

The Mediterranean monk seal (*Monachus monachus*) is believed to be the world's rarest pinniped and one of the most endangered mammals in the world, with fewer than 600 individuals currently surviving (Johnson et al., 2006).

The data we have at hand refer to 17 *M. monachus* skulls. Information about sex and age category has been collected by Mo, in collaboration with the staff of the Department of Experimental Veterinary Sciences of the University of Padua (Mo, 2005). In particular, 4 of the seals are male (1 young, 2 old and 1 with unknown age) and 5 are female (1 young, 4 old). For 8 of them we do not have information about sex (and for 5 of them we do not have information about age). For technical details on the methods for the biological determination of age category used by veterinarians, see Mo (2005). However, to give an idea, age categories up to 4 include specimens aged between 0 and 4 months (from newborns to fat short pups), and age categories 5–9 include specimens older than 5 months (from weaned youngsters to black adult males).

Left-lateral, frontal, posterior, dorsal and ventral views of the skull are available for each subject. It is of interest to assess whether there is a shape difference between the sexes and the ages, which parts determine these possible differences and how shape relates to size and other covariates. It is worth noting that combining shape analysis with nonparametric permutation techniques may be useful when dealing with small samples.

Initially we tried to find differences between male and female groups. Given the low sample size and the fact that using information on sex we have only two young specimens out of nine whose sex is known, it is very hard to detect a sexual dimorphism and if there is one, it is masked by age effects. This confirms claims by Mo (2005) and Marchessaux (1989) that there is no apparent sexual dimorphism although it is likely that the number of adult males reaching their maximum size is higher than that of adult females. Also, where it is found, dimorphism is related more to the length of body or to the coloration of the pelage than to cranial differences (Brunner et al., 2003). For all these reasons, we apply the nonparametric methodology previously discussed in order to detect differences between different age categories. We can thus obtain a sample of 12 specimens, 4 of which are considered young (i.e. of age category less than or equal to 4) and 8 are considered adult (i.e. of age category greater than 4). We have chosen eight anatomical landmarks in 2D – auricular (au), mastoidal (ms), bregma (b), rhinion (rhi), nasospinale (ns), prosthion (pr), canine base (cb) and canine tip (ct), as shown in Figure 10.3. We have also defined two domains: the *midsagittal and nasal* (b, rhi, ns, pr) and the *canine tooth* (cb, ct). Landmarks have been digitized using tpsDig2 software (Rohlf, 2007).

Using the NPC methodology (with $B = 10\,000$) we obtain the following p-values: $p = 0.0080$ (adjusted $p = 0.0158$) for the auricular landmark point, $p = 0.0014$ for the mastoidal (adjusted $p = 0.0038$), $p = 0.0011$ (adjusted $p = 0.0026$) for the first domain and $p = 0.0311$ (adjusted $p = 0.0311$) for the second domain. The global p-value is $p = 0.0026$. Further results are shown in Tables 10.5 and 10.6.

Hence, even if we cannot find sexual dimorphism in our data, we are able to find a significant difference between young and adult specimens studying landmarks and combinations with domains other than evaluating the global test.

To complicate the analysis a little, thus illustrating how the MA procedure works, we carried out a similar analysis to that presented above, introducing a step allowing us to perform an MA analysis. In particular, we examined three aspects: scatter (by means of second moment, denoted by μ^2), location (median, $\mathbb{M}d$) and distribution (by means of a Kolmogorov–Smirnov statistic, KS). The results of this further analysis are displayed in Tables 10.7–10.9. We emphasize that when adjusting p-values at coordinate level, the correction (adjustment) is more 'severe'. Hence, the tree

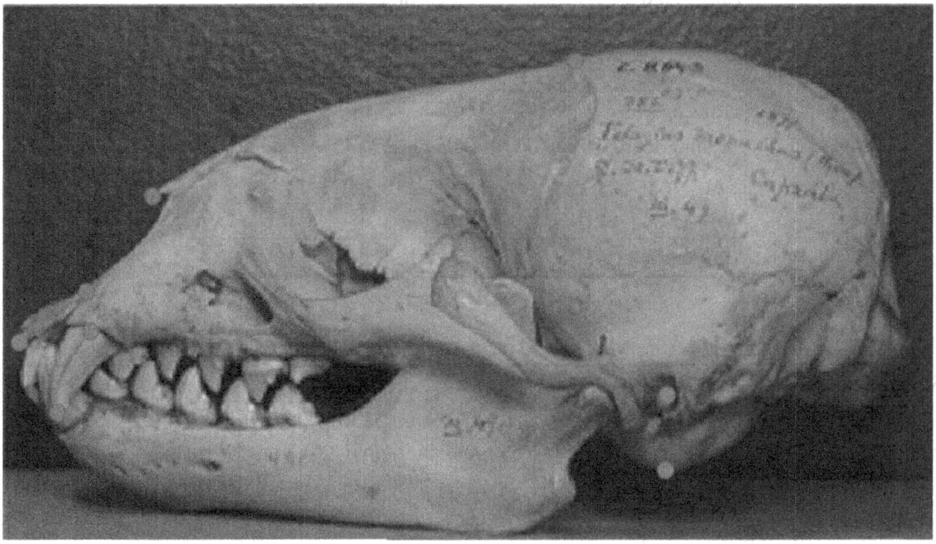

Figure 10.3 Landmarks: au, ms, b, rhi, ns, pr, cb and ct

Table 10.5 Controlling FWE at different levels of the analysis

Landmarks	Coordinate	NPC p-value coordinate	p-value FWE coordinate	NPC p-value landmark	p-value FWE landmark
bregma	x1	0.5624	0.8903		
	y1	0.0020	0.0198	0.0040	0.0202
rhinion	x2	0.0438	0.2769		
	y2	0.9602	0.9916	0.0876	0.0876
ns	x3	0.0007	0.0041		
	y3	0.9291	0.9916	0.0008	0.0026
pr	x4	0.3360	0.7938		
	y4	0.0046	0.0453	0.0089	0.0353
cb	x5	0.0079	0.0733		
	y5	0.5042	0.8903	0.0157	0.0432
ct	x6	0.1709	0.5980		
	y6	0.0246	0.1895	0.0482	0.0855
au	x7	0.1044	0.4720		
	y7	0.0042	0.0449	0.0080	0.0353
ms	x8	0.5226	0.8903		
	y8	0.0007	0.0041	0.0014	0.0065
Global			0.0041		0.0026

Table 10.6 Controlling FWE at domain level

	NPC p-value domain	p-value FWE domain
midsagittal & nasal	0.0011	0.0026
canine teeth	0.0311	0.0311
au	0.0080	0.0158
ms	0.0014	0.0038
Global		0.0026

Table 10.7 An example of the MA procedure using the Monachus data set

Statistic	bregma x	bregma y	rhinion x	rhinion y	ns x	ns y	pr x	pr y
μ^2	0.5014	0.0028	0.0413	0.9230	0.0010	0.8897	0.3308	0.0058
$\mathbb{M}d$	0.5362	0.2087	0.0679	0.8030	0.2003	1.0000	0.7171	0.0716
KS	0.5546	0.0454	0.0101	0.2651	0.0419	1.0000	0.9721	0.0445
Global (over aspects)	0.7176	0.0044	0.0200	0.5248	0.0010	0.9916	0.6518	0.0078
p-FWE (adjusted global p-values)	0.9931	0.0511	0.1562	0.9716	0.0079	0.9931	0.9931	0.0760
Statistic	cb x	cb y	ct x	ct y	au x	au y	ms x	ms y
μ^2	0.0106	0.5536	0.1579	0.0295	0.1101	0.0041	0.5067	0.0021
$\mathbb{M}d$	0.0661	0.7209	0.1755	0.4479	0.0703	0.1400	0.7662	0.4461
KS	0.1167	0.5570	0.5502	0.2631	0.1262	0.0104	0.5560	0.0028
Global (over aspects)	0.0203	0.8255	0.2988	0.0388	0.1550	0.0061	0.7527	0.0021
p-FWE (adjusted global p-values)	0.1562	0.9931	0.8631	0.2742	0.6287	0.0677	0.9931	0.0209

Table 10.8 Obtaining information on landmarks (raw and adjusted p-values)

	p-value	p-FWE
bregma	0.0087	0.0488
rhinion	0.0318	0.0823
ns	0.0010	0.0041
pr	0.0156	0.0570
cb	0.0397	0.0823
ct	0.0748	0.0823
au	0.0121	0.0570
ms	0.0028	0.0148
Global		0.0041

Table 10.9 Obtaining information on domains (raw and adjusted p-values)

	NPC p-value domain	p-value FWE domain
midsagittal & nasal	0.0028	0.0105
canine teeth	0.0734	0.0734
au	0.0121	0.0228
ms	0.0028	0.0105
Global		0.0105

level at which to apply the control for multiplicity should be conveniently chosen, depending on the goal of the analysis.

10.6.2 Some Remarks

The case study has shown that NPC tests, due to their nonparametric nature, may be computed even when the number of covariates exceeds the number of cases. With reference to the problem of small sample sizes, we recall that the results obtained within the NPC framework can be extended to the corresponding reference population. In Section 3.5 it is proved that it is possible to extend the permutation conditional to unconditional or population inferences since permutation tests are provided with similarity and conditional unbiasedness properties. In the parametric field, this extension is in fact possible when the data set is randomly selected by well-designed sampling procedures on well-defined population distributions, provided that their nuisance parameters have boundedly complete statistics in the null hypothesis or are provided with invariant statistics. In practice, this situation does not always occur and parametric inferential extensions might be incorrect or even misleading. Permutation tests allow for such extensions, at least in a weak sense, requiring that the similarity and conditional unbiasedness properties are jointly satisfied. Moreover, we have shown how the NPC methodology enables the researcher to give a local assessment using a combination within domains. We feel confident that developing geometric morphometrics techniques in a nonparametric permutation framework makes it possible to obtain valid solutions for high-dimensional and small sample size problems.

10.6.3 Shape Analysis Using MATLAB

In order to carry out the two applications discussed in Section 10.6.1, we first used some freely distributed dedicated software by James Rohlf (e.g. tpsUtil, tpsDIG2, tpsRelw, available at `http://life.bio.sunysb.edu/morph/`) to perform various utility operations, useful for data acquisition or providing statistical analyses using partial warp scores as shape variables and/or expressing the results of a morphometric analysis as a shape deformation.

tpsUtil is a utility program for creating an empty TPS file from a directory with image files. Having obtained this file containing the images, it is necessary to get landmark and semi-landmark coordinates. The coordinate locations of landmarks and semi-landmarks, representing the raw data, were digitized using tpsDIG2 software (Rohlf, 2007). However, since the direct analysis of databases of landmark locations is not appropriate because of the presence of nuisance parameters, such as location, rotation and scale, to carry out a valid statistical shape analysis a generalized least-squares superimposition (or generalized Procrustes analysis) was carried out to eliminate non-shape variation in configurations of landmarks and to align the specimens to a common coordinate system. The data we used for our analyses are the coordinates of landmarks and semi-landmarks, after sliding semi-landmarks and performing superimposition (hence we have considered shape variables). tpsRelw (Rohlf, 2008a) allows the sliding of semi-landmarks, thus obtaining shape variables. The procedure involves first sliding the semi-landmarks to the left or right along a curve so as to minimize the amount of shape change between each specimen and the Procrustes average of all the specimens.

Inferential analyses were carried out using MATLAB routines. Since in both cases we simply deal with a two independent sample problem, we have used four MATLAB functions:

- `np_2s`, which tests the equality in distribution in two-sample designs, whenever the data set contains continuous or dichotomous variables and even in the presence of missing values;
- `np_2s_MA`, which performs a one-way multi-aspect ANOVA for continuous variables, testing the equality in distribution in two-sample designs, whenever the data set contains continuous or dichotomous variables and even in the presence of missing values;
- `NPC`, allowing us to apply the NPC methodology, thus combining the hypotheses of interest and producing a multivariate test;
- `NPC_FWE`, performing a closed testing procedure that controls the FWE (i.e. the probability of finding at least one type I error in all the tests carried out).

Input data for performing the analyses are provided in a standard Excel format, in a file called `monachus.xls`. MATLAB code is given below.

```
[D,data,code]=xlsimport('monachus');
reminD(D)
[D.name]
[no Ts]=v(2:size(D,2));
[P T options]=NP_2s(Ts,'group',10000,0);
[bregma p_glob_bregma]=NPC(P(:,1:2),'T');
[rhinion p_glob_rhinion]=NPC(P(:,3:4),'T');
[ns p_glob_ns]=NPC(P(:,5:6),'T');
[pr p_glob_pr]=NPC(P(:,7:8),'T');
[cb p_glob_cb]=NPC(P(:,9:10),'T');
[ct p_glob_ct]=NPC(P(:,11:12),'T');
[au p_glob_au]=NPC(P(:,13:14),'T');
[ms p_glob_ms]=NPC(P(:,15:16),'T');

%%%%
% code useful when performing MA analysis
stats={':Y.^2','Me','KS'}
[P_MA T_MA options]=NP_2s_MA(Ts,'group',10000,stats,'T',0);
[bregma p_glob_bregma]=NPC(P_MA(:,1:2,4),'T');
[rhinion p_glob_rhinion]=NPC(P_MA(:,3:4,4),'T');
[ns p_glob_ns]=NPC(P_MA(:,5:6,4),'T');
[pr p_glob_pr]=NPC(P_MA(:,7:8,4),'T');
[cb p_glob_cb]=NPC(P_MA(:,9:10,4),'T');
[ct p_glob_ct]=NPC(P_MA(:,11:12,4),'T');
[au p_glob_au]=NPC(P_MA(:,13:14,4),'T');
[ms p_glob_ms]=NPC(P_MA(:,15:16,4),'T');
%%%%
```

```
midsagittal=NPC([bregma,rhinion,ns,pr],'T');
canine=NPC([cb,ct],'T');
globalAN1_2=NPC_FWE([midsagittal,canine,au,ms],'T');
global2AN1_2=NPC_FWE(P,'T');
global3AN1_2=NPC_FWE([bregma,rhinion,ns,pr,cb,ct,au,ms],'T');
```

10.6.4 Shape Analysis Using R

The same analysis was performed in R. As already stated, this data set comprises some anthropological measurements taken on the skulls of two groups of seals. Each variable consists of two measurements, with respect to the x and y axes. This is an example of a two-sample multivariate test, where we have to combine the information within each variable first (i.e. combine the tests on the x and y axes), and there is a further requirement that the tests on the variables must be combined according to some pre-specified domains. We consider the difference in means as the test statistic in each aspect/variable.

```
setwd("C:/path")
data<-read.csv("monachus.csv",header=TRUE)
p = dim(data)[2]-1
n = table(data[,1]) ; n

0 1
8 4

contr = rep(c(1/sum(n[2]),-1/sum(n[1])),n[2:1])
contr

 [1]  0.250  0.250  0.250  0.250 -0.125 -0.125 -0.125 -0.125 -0.125
[10] -0.125 -0.125 -0.125

B = 5000
T=array(0,dim=c((B+1),p))

g = data[,1]; data = data[,-1]

T[1,] = t(data)%*%contr

for(bb in 2:(B+1)){
data.star=data[sample(1:sum(n)),]
T[bb,] = t(data.star)%*%contr
}

source("t2p.r")
P = t2p(abs(T)); colnames(P)=colnames(data);
cat("Rough p-values \n");t(t(P[1,]))

Rough p-values
           [,1]
bregma_x  0.5694
```

```
bregma_y   0.0010
rhinion_x  0.0358
rhinion_y  0.9596
ns_x       0.0016
ns_y       0.9172
pr_x       0.3480
pr_y       0.0096
cb_x       0.0046
cb_y       0.4970
ct_x       0.1694
ct_y       0.0262
au_x       0.1058
au_y       0.0052
ms_x       0.5244
ms_y       0.0012
```

The results above are the p-values related to each aspect (X or Y) in each variable. Now we wish to combine the X and Y information, using Fisher's combining function. The null hypothesis to be assessed is $H_{0j}\colon \{[X_{j1} \stackrel{d}{=} X_{j2}] \bigcap [Y_{j1} \stackrel{d}{=} Y_{j2}]\}$, where X_{j1} and Y_{j2} respectively denote the X and Y measurement of the jth variable, $j = 1, \ldots, p$. The alternative hypothesis is $H_{1j}\colon \{[X_{j1} \stackrel{d}{>} X_{j2}] \bigcup [Y_{j1} \stackrel{d}{>} Y_{j2}]\}$. The test statistics are in accordance with the rule *large is significant*.

```
p=p/2 ; T.xy = array(0,dim=c((B+1),p))

for(j in 1:p){
T.xy[,j] = -2*log(P[,((2*j)-1)]*P[,(2*j)])
}

P.xy = t2p(T.xy) ;
colnames(P.xy)=c('bregma','rhinion','ns','pr','cb','ct','au','ms')
cat("Combined tests on each variable: \n");P.xy[1,]

Combined tests on each variable:
 bregma rhinion      ns      pr      cb      ct      au      ms
 0.0036  0.1858  0.0072  0.0232  0.0188  0.0180  0.0022  0.0024
```

Now we wish to combine the information of the partial tests on each variable into four domains: the first domain D_1 covers the variables `bregma`, `rhinion`, `ns`, and `pr`; the second domain D_2 covers the variables `cb` and `ct`; the third domain is just the variable `au` and the fourth is the variable `ms`. We apply Fisher's combining function to the p-values within each domain.

```
dom=c(rep(1,4),c(2,2),3,4)
D = length(table(dom))
T.dom = array(0,dim=c((B+1),D))

for(d in 1:D){
ID = dom%in%d
if(sum(ID)>1){T.dom[,d] = apply(P.xy[,ID],1,function(x){-2*log(prod(x))}) }
if(sum(ID)==1){T.dom[,d] = -2*log(P.xy[,d]) }
}
```

```
P.dom = t2p(T.dom) ; colnames(P.dom) = c('D1','D2','D3','D4')
P.dom[1,]

    D1     D2     D3     D4
0.0008 0.0086 0.0050 0.0246
```

The partial p-values show that the greater differences between the two groups are related to D_1 and D_2. If multiplicity control is required, apply the function `FWE.minP`.

```
source("FWEminP.r")
FWE.minP(P.dom)

D1 0.0022
D2 0.0172
D3 0.0144
D4 0.0246
```

The data set and the corresponding software codes can be found in the `monachus` folder on the book's website.

11

Multivariate Correlation Analysis and Two-Way ANOVA

In this chapter we present two real case studies in ophthalmology, concerning complex repeated measures problems, carried out at the Department of Ophthalmology, University of Padua (see Midena et al., 2007a, 2007b). For each data set, different analyses have been suggested in order to emphasize, from time to time, specific aspects of the data structure itself. In this way we enable the reader to choose the most appropriate analyses for his/her research purposes.

The first case study concerns a clinical trial in which 13 patients with bilateral age-related macular degeneration were evaluated. Their eyes were observed at $k = 66$ different and fixed positions. Hence repeated measures issues arise. Five outcome variables were recorded and analysed.

The second case study concerns a clinical trial with a five-year follow-up period, the aim of which is to evaluate the long-term side-effects of mitomycin C (MMC) assisted photorefractive keratectomy (PRK) on corneal keratocytes of highly myopic eyes. At baseline, one eye was randomly assigned to receive topical MMC treatment (treated eye, T) and the other eye was treated without MMC, using standard corticosteroid treatment (non-treated eye, NT). Corneal keratocyte density was quantified by corneal confocal microscopy at baseline and 5 years after surgery. Fourteen variables and four domains in total were analysed.

MATLAB and R codes, as well as raw data (.xls and .csv files), for carrying out both the analyses can be found on the book's website.

11.1 Autofluorescence Case Study

We consider a real case study in ophthalmology concerning a clinical trial in which 13 patients (3 men and 10 women) with bilateral age-related macular degeneration (AMD) were evaluated. Their eyes (7 left and 6 right eyes) were observed at $k = 66$ different and fixed positions each. Hence, repeated measures issues arise. Responses are in fact considered consecutive with respect to the conventional ordering: from the centre to the boundary and clockwise (see Figure 11.1) in order to maintain correspondence of responses to positions. Since these positions are *a priori* well defined, they may be seen as a special case of landmarks (see Chapter 10 for details of landmark definitions).

More precisely, 61 repeated measures of each variable at different locations were collected; 66 measures in total, 5 of which overlap in correspondence with the central point 255 (see Figure 11.1).

Early AMD has been correlated with different functional alterations, but it remains unclear how fundus lesions and overlying sensitivity are connected to each other. We observe that this data set was previously analysed in Midena et al. (2007a), in order to compare fundus-related sensitivity (microperimetry) and fundus autofluorescence of the macular area with drusen and pigment abnormalities in early AMD. In particular, all the microperimetry data were analysed

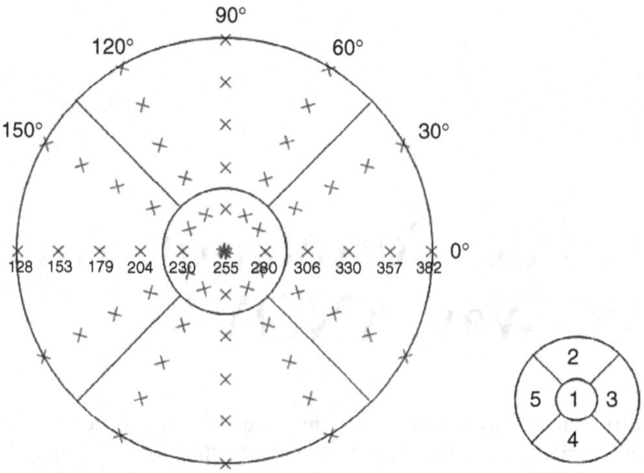

Figure 11.1 A microperimetric customized radial grid pattern of 61 positions. The positions are projected randomly onto the 0°, 30°, 60°, 90°, 120°, 150° axes, 1° apart

using a traditional ANOVA test for repeated measures and a Bonferroni test. On the other hand, the autofluorescence data were examined using the chi-square test and, in order to assess inter-observer variability, Cohen's kappa statistic (Altman, 1991) was computed. Here we deal with this multivariate problem by using a nonparametric permutation approach instead of traditional methods. In particular, the primary goal of the study is to assess the relationships (correlations) among some variables with repeated observations (Roy, 2006). When dealing with this kind of study, it may be of interest not only to characterize changes in the repeated values of the response variable, but also to determine the explanatory variables most associated with any change (Der and Everitt, 2006). Moreover, it is interesting to establish whether there is a monotonic stochastic ordering of the response variables as the categories of each explanatory variable increases. In particular, the following outcome variables were recorded and analysed:

- *OCT retinal thickness* (denoted by O), expressed in micrometers (μm), was quantitatively measured by optical coherence tomography (OCT);
- *retinal sensitivity* (M), expressed in decibels (dB) and measured by microperimetry examination;
- *autofluorescence* (A), an ordinal covariate with three levels, 0 = not present, 1 = hypo (less than normal), 2 = hyper (more than normal);
- *pigment abnormalities* (P), a binary covariate with values 0 = absent, 1 = present.
- *drusen*(D), an ordinal covariate with four levels, 0 = absent, 1 = small (<63 μm), 2 = intermediate (63–125 μm), 3 = large (>125 μm). Drusen are subretinal pigment epithelial deposits that are characteristic of but not uniquely associated with AMD.

O and M are the variables of interest (according to the standard terminology for multivariate linear regression, they are analogous to independent variables). Atrophy has not been considered.

11.1.1 A Permutation Solution

Each eye has been observed at $k = 66$ different and fixed positions. Three covariates $\mathbf{X}$ (i.e. A, P and D) and two responses $\mathbf{Y}$ (i.e. O and M) have been collected. The set of $\mathbf{X}$ and $\mathbf{Y}$ variables may be denoted by

$$\mathbf{Y} = \{Y_{wit},\ i = 1 \ldots, n,\ w = 1, 2,\ t = 1, \ldots, k\},$$

$$\mathbf{X} = \{X_{hit},\ i = 1 \ldots, n,\ h = 1, \ldots, 3,\ t = 1, \ldots, k\}.$$

We recall that we are dealing with repeated measures in space (different positions/landmarks in the eye).

The first analysis we carried out is illustrated in Figure 11.2. In particular, for each t we calculated Pearson's correlation coefficient ρ between A, P and D (**X** variables) and O and M (**Y** variables). Hence, for each position, we compute six correlation coefficients, $\rho_t(O, A)$, $\rho_t(O, P)$, $\rho_t(O, D)$, $\rho_t(M, A)$, $\rho_t(M, P)$ and $\rho_t(M, D)$. We have

$$\rho_t(w, h) = \frac{\sum(Y_{hi} - \overline{Y}_h)(X_{wi} - \overline{X}_w)}{\sqrt{\sum(Y_{hi} - \overline{Y}_h)^2 \sum(X_{wi} - \overline{X}_w)^2}},$$

where $i = 1 \ldots, n$, $w = O, M$, $h = A, P, D$, $t = 1, \ldots, k$. The hypotheses of interest are

$$H_{0(w,h)} = \left\{ \bigcap_{t=1}^{k} [\rho_t(w, h) = 0] \right\},$$

$$H_{1(w,h)} = \left\{ \bigcup_{t=1}^{k} [\rho_t(w, h) < 0] \right\},$$

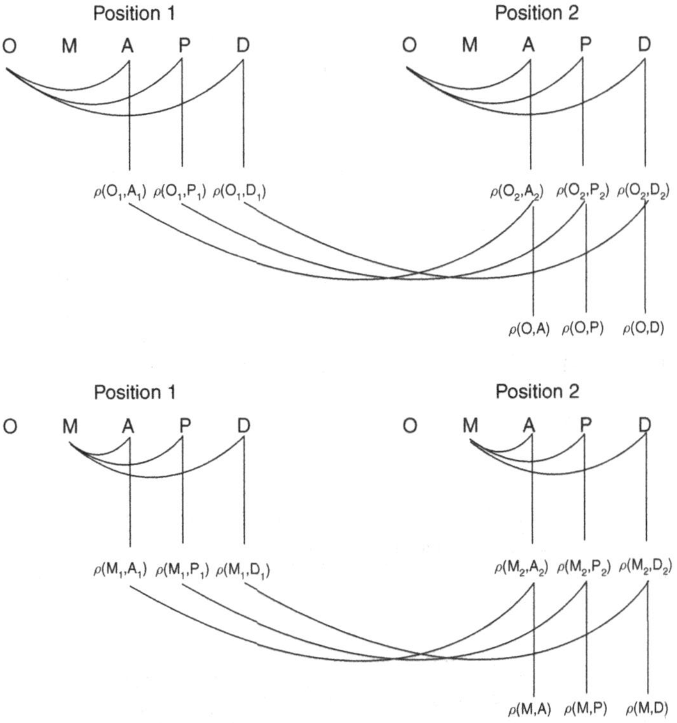

Figure 11.2 Sketch of the first analysis, in the simplest case with only two positions

Table 11.1 Retinal thickness and retinal sensibility p-values, controlling FWE. $B = 1000$

		A	P	D
O	p-value	0.0230	0.0375	0.0090
	p-FWE	0.0435	0.0375	0.0235
M	p-value	0.0005	0.0610	0.0005
	p-FWE	0.0005	0.0610	0.0005

where $w = O, M,\ h = A, P, D$. We remark that when working in a parametric framework, it is notoriously difficult to specify the direction of the alternative so as to obtain one-sided tests, and only two-sided tests (testing the alternative $H_1 : \rho_t(w, h) \neq 0$) are usually carried out. To save space, instead of summarizing the results obtained separately for A, P and D, thus showing partial p-values for each position t, we display only global raw and adjusted p-values for the correlations $\rho(O, A)$, $\rho(O, P)$, $\rho(O, D)$, $\rho(M, A)$, $\rho(M, P)$ and $\rho(O, D)$, obtained after testing the above system of hypotheses (see Table 11.1). Note, finally, that even if this problem is multivariate in nature, by means of NPC methodology we are able to evaluate the relationships between $\mathbf{X}$ and $\mathbf{Y}$ variables, processing them jointly.

In the second analysis we wish to establish if there is a monotonic stochastic ordering of the response variables as the categories of each explanatory variable increase. Hence, the sub-hypotheses of interest are

$$H_{0(w,h)} : \left\{ \bigcap_{t=1}^{66} \left(Y_{1_h t} \stackrel{d}{=} \ldots \stackrel{d}{=} Y_{s_h t} \right) \right\}, \quad w = O, M,\ h = A, P, D,$$

against

$$H_{1(w,h)} : \left\{ \bigcup_{t=1}^{66} \left(Y_{1_h t} \stackrel{d}{\geq} \ldots \stackrel{d}{\geq} Y_{s_h t} \right) \right\}, \quad w = O, M,\ h = A, P, D,$$

where s_h indicates the level of hth variable $\mathbf{X}$. In fact, $s = 3$ when considering autofluorescence (A), $s = 2$ in retinal pigmentation (P) and, finally, $s = 4$ when considering drusen (D). Under the alternative, the monotonic stochastic ordering of O or M is emphasized. As in the previous analysis, we display only global raw and adjusted p-values for the 'intersections' (O, A), (O, P), (O, D), (M, A), (M, P) and (O, D) (see Table 11.2). For details of stochastic ordering issues, refer to the extensive discussion in Chapter 8.

Table 11.2 NPC procedure for monotonic stochastic ordering. $B = 1000$

		A	P	D
O	p-value	0.0005	0.0005	0.0005
	p-FWE	0.0000	0.0000	0.0005
M	p-value	0.0005	0.0005	0.0005
	p-FWE	0.0000	0.0000	0.0005

11.1.2 Analysis Using MATLAB

All the nonparametric analyses have been carried out using MATLAB and R routines. Here we give a brief description of the functions we used in MATLAB. We refer the reader to the book's website for details.

The function `np_rho` computes the nonparametric correlation coefficient (providing both the test statistic and the associated p-value) and requires as input the following arguments:

- **Y**, the data matrix of categorical responses;
- **X**, the indicator variable,
- **Z**, the vector representing the stratification variable (connected, as in repeated measures);
- `type`, allowing for Pearson's linear correlation, or Kendall's τ, or Spearman's ρ;
- `connected`, a binary variable indicating whether independent permutations or the same permutations should be performed on each of the strata, and taking value 0 in the former case and 1 in the latter (we have set this parameter equal to 1 since we are dealing with repeated measures).

We have the variables O and M as a matrix of responses **Y**, variables A, P and D as **X**, and the variable 'location/time' as **Z** (i.e. measurement number). We then connected it, since we are dealing with repeated measures, thus performing the same permutations on each strata. A right-tailed test was considered, checking the alternative hypothesis stating the presence of negative correlation.

The function `NPC` allowed us to apply the NPC methodology, thus combining the hypotheses of interest and producing multivariate tests.

Using `NPC_FWE`, we carried out a closed testing procedure that controls the FWE (i.e. probability of finding at least one type I error in all the tests performed). We emphasize that, when carrying out closed testing, we chose Tippett's combining function which corresponds to a Min$P$ step-down procedure (see details in Chapter 5).

Finally, we used the `xlsimport` function to import the relevant Excel data files of (not shown in the code below).

We carried out $B = 1000$ CMC permutations. Note that in the following analyses the terms 'location' and 'time' are used interchangeably, and can be taken to have the same meaning.

```
B=1000
[P, T, options] = NP_rho({'OCT' 'MICRO'} ,{'AF' 'PIG' 'DRU'},
   'Time',B,-1,'Pearson',1);
options.Combdims=4;
P2=NPC(P,'F',options);
options.Combdims=2;
P3fwe=NPC_FWE(P2,'T',options);
B=space_perm(13,1000);
[P, T] = by_strata(lD(1),'NP_stord',lD(7:8) ,lD(9:11),B,-1);
options.Combdims=4;
P2_F=NPC(P,'F',options);
P2_D=NPC(T,'D',options);
options.Combdims=2;
P3fweF=NPC_FWE(P2_F,'T',options);
P3fweD=NPC_FWE(P2_D,'T',options);
```

11.1.3 Analysis Using R

We now present the analysis carried out using routines developed in R.

We used the `read.csv` function to read the data file in table format, thus creating a data frame from it, with subjects corresponding to lines, and variables to fields in the file. This function is in fact intended for reading 'comma separated value' files ('.csv'). We refer the reader to the book's website for details on R code.

As previously mentioned, we have three explanatory variables (`X1` = Autofl, `X2` = Pigment, `X3` = Drusen) and two responses (`Y1` = Oct, `Y2` = Micro). After loading data set of each variable, we simplify the data structure with an explanatory matrix `X` of dimensions $n \times p \times 3$ and a response matrix `Y` with dimensions $n \times p \times 2$, where n is the number of subjects and p the number of variables (positions/times). We will investigate whether there is a significant negative correlation between each response and each explanatory variable at each time $j = 1, \ldots, p$. The test statistic is the covariance (not the correlation coefficient ρ as in the previous analysis performed in MATLAB). The matrix `R` contains the permutation distribution of the partial test statistic. The function `COV` computes the covariances of each pair of columns of the two matrices.

```
setwd("C:/path") ; source("t2p.r") ; source("FWEminP.r")

COV<-function(x,y){
n=dim(x)[1] ; p=dim(x)[2];
r=apply(x*y,2,mean)-apply(x,2,mean)*apply(y,2,mean)
return(r)
}

Y1<-read.csv("Oct.csv",header=TRUE)
Y2<-read.csv("Micro.csv",header=TRUE)
X1<-read.csv("Autofl.csv",header=TRUE)
X2<-read.csv("Pigment.csv",header=TRUE)
X3<-read.csv("Drusen.csv",header=TRUE)

n=dim(Y1)[1] ; p=dim(Y1)[2]
Y=array(0,dim=c(n,p,2))
Y[,,1] = as.matrix(Y1) ; Y[,,2] = as.matrix(Y2)

X=array(0,dim=c(n,p,3))
X[,,1] = as.matrix(X1)
X[,,2] = as.matrix(X2)
X[,,3] = as.matrix(X3)

B=2000
R<-array(0,dim=c((B+1),p,3,2))

for(i in 1:2){ for(j in 1:3){ R[1,,j,i]=COV(Y[,,i],X[,,j]) } }

for(bb in 2:(B+1)){
Y.star = Y[sample(1:n),,]
for(i in 1:2){ for(j in 1:3){ R[bb,,j,i]=COV(Y.star[,,i],X[,,j]) } }
}
P=t2p(-R)

AF.res = data.frame(colnames(X1),P[1,,1,1],P[1,,1,2])
colnames(AF.res) = c("Time","Oct","Micro")
PIG.res = data.frame(colnames(X2),P[1,,2,1],P[1,,2,2])
```

```
colnames(PIG.res) = c("Time","Oct","Micro")
DRU.res = data.frame(colnames(X3),P[1,,3,1],P[1,,3,2])
colnames(DRU.res) = c("Time","Oct","Micro")
```

The rows of matrices `AF.res`, `PIG.res` and `DRU.res` contain the raw partial p-values assessing a significant negative correlation between each response and the related explicative variable at each time/location $t = 1, \ldots, p$. In order to obtain a global test for each combination of response/explanatory variable, we may combine the partial test with respect to time (with Fisher's function).

```
T.glob = apply(P,c(1,3,4),function(x){-2*log(prod(x))})
P.glob = t2p(T.glob)[1,,] ; colnames(P.glob)=c("Oct","Micro")
rownames(P.glob) = c("Auto","PIG","DRU") ; P.glob

        Oct  Micro
Auto 0.0135 0.0000
PIG  0.0465 0.0575
DRU  0.0045 0.0000
```

A control for multiplicity is obtained by running the function `FWE.minP` on the global p-values computed on `T.glob`:

```
P.Oct= t2p(T.glob)[,,1] ; P.Mic= t2p(T.glob)[,,2]
FWE.res<-cbind(FWE.minP(P.Oct),FWE.minP(P.Mic))
rownames(FWE.res) = c("Auto","PIG","DRU")
colnames(FWE.res)=c("Oct","Micro") ; FWE.res

        Oct  Micro
Auto 0.0255 0.0000
PIG  0.0465 0.0575
DRU  0.0115 0.0000
```

Finally, if a unique global test is required for each explanatory variable, we can combine the elements of `P.Oct` and `P.Mic`:

```
T.APD = apply(t2p(T.glob),c(1,2),function(x){-2*log(prod(x))})
P.APD = t2p(T.APD) ; colnames(P.APD)=c("Auto","PIG","DRU")
P.APD[1,]

  Auto    PIG    DRU
0.0000 0.0165 0.0000
```

In the second part of this analysis we wish to establish whether there is a decreasing trend on the response variables as the categories of each explanatory variable increase. This is a stochastic ordering problem for continuous variables that can be solved by the `stoch.ord` function. The arguments of this function are a vector of continuous responses (`y`), a categorical variable (`x`) specifying the time/group labels, and the direction of the alternative hypothesis (-1 meaning 'decreasing' and 1 meaning 'increasing'). The function returns a vector of p-values with its first element being the p-value of the analysis and the latter elements being the null distribution of the global p-value. The integers printed refer to the time-points. The function can process one variable at time; in order to maintain the dependence among the response variables the function is always run with a common seed, which will be the input to the set.seed function (thus providing exactly the same permutations each time).

```
source("stoch_ord.r")
P = array(0,dim=c((B+1),p,3,2))
for(i in 1:p){
for(j in 1:3){
for(k in 1:2){
P[ ,i,j,k] = stoch.ord(Y[,i,k],X[,i,j],alt=-1,B=B)
}
}
print(i)
}
```

The matrix P has dimension $(B+1) \times p \times 3 \times 2$. The raw p-values realted to the stochastic ordering problem at each time for each combination of response/explanatory variables can be found in the first element of the first dimension of P.

```
RES = P[1,,,] ;
rownames(RES)<-paste("T",seq(1,p),sep="")
colnames(RES)=c("Auto","PIG","DRU")
```

Now the matrix of partial p-values RES has dimension $p \times 3 \times 2$. Therefore, to see the partial p-values related to Oct and Micro for the first 10 times, type RES[1:10,,1] and RES[1:10,,2] respectively. If a global test is required, we can combine the elements of P with respect to the second dimension (time).

```
T.glob <- apply(P,c(1,3,4),function(x){-2*log(prod(x))})
P.glob <- t2p(T.glob)[1,,]
rownames(P.glob)=c("Auto","PIG","DRU")
colnames(P.glob) = c("Oct","Micro"); P.glob

     Oct Micro
Auto   0     0
PIG    0     0
DRU    0     0
```

If a multiplicity correction is required, run the FWE.minP function again:

```
T.FWE = t2p(T.glob)
P.FWE = cbind(FWE.minP(T.FWE[,,1]),FWE.minP(T.FWE[,,2]))
rownames(P.FWE) = rownames(P.glob);
colnames(P.FWE) = colnames(P.glob); P.FWE

        Oct Micro
Auto    0     0
PIG     0     0
DRU     0     0
```

The data set and the corresponding software codes can be found in the autofluorescence folder on the book's website.

11.2 Confocal Case Study

We now turn to a real case study in ophthalmology related to a clinical trial with a five-year follow-up period, where the aim was to evaluate the long-term side-effects of mitomycin C (MMC) assisted photorefractive keratectomy (PRK) on corneal keratocytes of highly myopic eyes. In this prospective, randomized, double-masked study, 28 patients with high bilateral myopia (i.e. myopia from −7.00 to −14.25 diopters), who underwent PRK on both eyes, were examined.

At baseline, one eye was randomly assigned to receive topical MMC treatment (treated eye, T) while the other eye was treated without MMC, using standard corticosteroid treatment (non-treated eye, NT). Corneal keratocyte density was quantified by corneal confocal microscopy at baseline and 5 years after surgery. Confocal microscopy was used to evaluate the thickness of the whole cornea, corneal epithelium, the density of subbasal and stromal nerves, and keratocytes both before and after surgery.

Four corneal parameters (i.e. groups of outcome variables, also called *domains*), were analysed: corneal endothelium, corneal nerves, corneal epithelium and corneal keratocytes. The *corneal endothelium* domain includes variables such as:

1. cellular density (cells/mm^2);
2. pleomorphism (expressed as a percentage);
3. polymegathism (expressed as a percentage).

The *corneal nerves* domain includes the variables:

1. density (defined as the total length of the nerve fibres existing in one image, expressed in micrometres of nerve fibre within an area of 74 340 mm^2);
2. P_F, the number of fibres present in one image;
3. P_{BE}, defined as the number of beadings present in 100 μm of nerve fibre;
4. P_{BI}, bifurcation order;
5. P_T, tortuosity grade, classified into four grades according to a scale (qualitative variable, score range 0 to 4).

In the *corneal epithelium* domain the variable of interest is epithelium thickness (expressed in μm). *Corneal keratocytes* constitute the last domain. These cells are also known as fibroblasts, the basic cell type found in the corneal stroma. In particular, in order to analyse corneal keratocyte density at baseline and 5 years after PRK, five stromal layers were considered: 0–10% (anterior); 11–33%; 34–66%; 67–90 and 91–100% (posterior) depth (Erie et al., 2005). Once again we refer to Midena et al. (2007b) for details of methods for image selection and keratocyte nucleus (cell) identification.

11.2.1 A Permutation Solution

Here we are concerned with paired data and repeated measures. A V-dimensional non-degenerate variable is observed on k different time occasions in n patients in two experimental situations, corresponding to two levels of the treatment. In our study, $V = 14$ variables have been assessed on $k = 2$ different time occasions (baseline and five-year follow-up) in $n = 28$ patients in two experimental situations, corresponding to two levels of the treatment (two treatment groups, MMC and steroid therapy). The whole data set is represented by

$$\mathbf{X} = \{X_{hjit},\ t = 1, \ldots, k,\ i = 1, \ldots, n,\ j = 1, 2,\ h = 1, \ldots, V\}$$

$$= \{\mathbf{X}_{jit},\ t = 1, \ldots, k,\ i = 1, \ldots, n,\ j = 1, 2\}$$
$$= \{\mathbf{X}_{ji},\ i = 1, \ldots, n,\ j = 1, 2\},$$

where the same symbol **X** is used to represent the V-variate response variable $(X_1, \ldots, X_V)$, the k-variate time profile $(\mathbf{X}_1, \ldots, \mathbf{X}_k)$ and the whole data set. We assume that the response variables behave according to the model

$$X_{hjit} = \mu_h + \mu_{hit} + \delta_{hjt} + \sigma_{ht}(\delta_{hjt}) \cdot Z_{hjit},$$

$t = 1, \ldots, k$, $i = 1, \ldots, n$, $j = 1, 2$ (also denoted below by T and NT, representing the treatment groups), $h = 1, \ldots, V$, where μ_h represents a population constant for the hth variable; μ_{hit} represents a time effect on the hth variable at time t and specific to the ith individual; δ_{hjt} represents treatment time effect at level j on the hth variable which, without loss of generality, is assumed to be $\delta_{h1t} = 0$, $\delta_{h2t} \leq$ (or $\geq$) 0, for all (h, t); $\sigma_{ht}(\delta_{hjt}) > 0$ represent population scale coefficients for variable h at time t, which are assumed to be invariant with respect to units but which may depend on treatment levels through effects δ_{hjt}, provided that, when $\delta_{h2t} \neq 0$, stochastic dominance relationships $X_{h1} \overset{d}{<}$ (or $\overset{d}{>}$)X_{h2}, $h = 1, \ldots, V$, are satisfied; Z_{hjit} are V-variate random errors, which are assumed to be exchangeable with respect to treatment levels, independent with respect to units, with null mean vector, $\mathbb{E}(\mathbf{Z}) = \mathbf{0}$, and with unspecified distribution $P \in \mathcal{P}$; in particular, these multivariate errors may be dependent with respect to component variables and time through any kind of monotonic regression.

The following problems may be of interest and should therefore be formalized into hypothesis systems and then tested.

(a) We perform a one-sample paired data analysis and, by using time as a stratification variable and treatment as a group variable, we test whether or not treated and untreated eyes are equal in distribution respectively at baseline time (t_0) and at the five-year follow-up (t_1). We observe that two patients in the t_0 stratum (patients 16 and 26) have missing observations. These missing data will be processed when using MATLAB routines, while observations referring to patients 16 and 26 will be omitted when using R routines. Also worthy of note is that the permutations are also paired and there are two observations for each patient. In short, we are interested in evaluating *treatment effect* separately for each time (t_0 and t_1). The V-dimensional vector of test statistics T^*_{ht}, $h = 1, \ldots, V$, is obtained by considering the test statistics at t_0 and t_1,

$$T^*_{ht} = \sum_{i=1}^{n} [X_{h2it} - X_{h1it}] \cdot S^*_i,\ \ h = 1, \ldots, V,\ \ t = t_0, t_1,$$

where S^*_i are i.i.d. and the $\pm$ signs, common to all variables, are random realizations of the random variable $1 - 2\mathcal{B}n(1, 1/2)$. Hence, the partial and global hypotheses

$$H_{0ht} : \left\{X_{h1t} \overset{d}{=} X_{h2t}\right\},\quad h = 1, \ldots, V,\ t = t_0, t_1$$

should be tested and then suitably combined to test the global null hypothesis

$$H_{0t} : \left\{\bigcap_{h=1}^{V} H_{0ht}\right\},\quad t = t_0, t_1,$$

against the partial alternatives

$$H_{1ht} : \left\{X_{h1t} \overset{d}{<\neq>} X_{h2t}\right\},\ h = 1, \ldots, V,\ t = t_0, t_1,$$

suitably combined in a global test, in which at least one among the H_{0h} is not true,

$$H_{1t}:\left\{\bigcup_{h=1}^{V} H_{1ht}\right\},\quad t=t_0,t_1.$$

(b) It is of interest to evaluate the *time effect*. After calculating the differences between the two times, for each treatment group we obtain the V-dimensional vector of test statistics T^*_{hj}, $h = 1,\ldots,V$, by considering the test statistics for treated and untreated eyes,

$$T^*_{hj}=\sum_{i=1}^{n}\left[X_{hjit_0}-X_{hjit_1}\right]\cdot S^*_i,\quad h=1,\ldots,V,\ j=1,2,$$

where S^*_i is as previously defined. Hence, the partial and global hypotheses

$$H_{0hj}\colon\left\{X_{hit_0}\overset{d}{=}X_{hjt_1}\right\},\quad h=1,\ldots,V,\ j=1,2,$$

should be tested and then suitably combined to test the global null hypothesis

$$H_{0j}:\left\{\bigcap_{h=1}^{V} H_{0hj}\right\},\quad j=1,2,$$

against the partial alternatives

$$H_{1hj}\colon\left\{X_{hjt_0}<\overset{d}{\neq}>X_{hjt_1}\right\},\quad h=1,\ldots,V,\ j=1,2,$$

suitably combined in a global test, in which at least one among the H_{0h} is not true,

$$H_{1t}:\left\{\bigcup_{h=1}^{V} H_{1hj}\right\},\quad j=1,2.$$

Moreover, by taking into account $c < V$ predefined domains of the outcome variables, we may further decompose the global null hypothesis, thus combining by means of the NPC methodology partial p-values corresponding to the variables belonging to the same domain. We recall that in our study $c = 4$ (i.e. corneal endothelium, corneal nerves, corneal epithelium and keratocytes). It is worth noting that in this case the NPC may be conveniently processed in two stages: first we combine all variables belonging to the same domain into a domain-based test and then we combine all the domains to obtain a global test.

11.2.2 MATLAB and R Codes

Evaluation of Treatment Effect

In this analysis, we evaluate treatment effect at baseline time (t_0) and after 5 years (t_1). The results obtained, displayed in Tables 11.3 and 11.4, allow us to state that there is no strong evidence against the null hypothesis of equality in distribution of treated (T) and not treated (NT) patients respectively at t_0 and t_1.

Table 11.3 Evaluating treatment effect

	Baseline		Follow-up	
Variable	p-value	p-FWE	p-value	p-FWE
Cellular_density	0.1570	0.8368	0.4115	0.9920
Polymegathism	0.2268	0.8592	0.4695	0.9920
Pleomorphism	0.9678	1.0000	0.7319	0.9958
Density	0.0594	0.4847	0.2753	0.9694
P_F	0.4111	0.9710	0.8494	0.9994
P_{BE}	0.2038	0.8564	0.4733	0.9920
P_{BI}	1.0000	1.0000	1.0000	1.0000
P_T	0.9888	1.0000	1.0000	1.0000
Epithelium_thickness	0.6269	0.9942	0.2178	0.9524
K_1_10	0.9446	1.0000	0.8900	0.9994
K_11_33	0.6613	0.9942	0.1942	0.9342
K_34_66	0.4949	0.9800	0.4407	0.9920
K_67_90	0.5059	0.9802	0.6547	0.9958
K_91_100	0.1804	0.8518	0.3987	0.9920
p-Global		0.4847		0.9342

Table 11.4 Treatment effect over domains

	Baseline		Follow-up	
Variable	p-value	p-FWE	p-value	p-FWE
Domain 1	0.5059	0.9438	0.5529	0.9128
Domain 2	0.6899	0.9438	0.9484	0.9484
Domain 3	0.6269	0.9438	0.2178	0.6327
Domain 4	0.5595	0.9438	0.5557	0.9128
p-Global		0.9438		0.6327

The MATLAB code used to carry out the analysis is as follows.

```
B=5000
[D,data,code]=xlsimport('confocal');
reminD(D)
T_NT=(v('T_NT')-1)*2-1;
FU_BA=(v('FU_BA')-1)*2-1;
Y=D(4:end);

%Treated vs Untreated
%Evaluating TREATMENT effect (separately for Base-
line (t0) and 5 years FU (t1))
```

```
P_T_NT_byStrata=by_strata(FU_BA,'NP_ReM',Y,T_NT,'all',B,0,0);
options.labels.dims{2}=[Y.name];
options.labels.dimslabel{2}='response';
options.labels.dims{3}={'Baseline' 'FU'};
options.labels.dimslabel{3}='BASELINE vs FU';
pmat_show(P_T_NT_byStrata,.05,options);


options.Combdims=2;
P2_T_NT=NPC_FWE(P_T_NT_byStrata,'T',options);
pmat_show(P2_T_NT,.05,options);

options.Combdims=2;

options.OUT=0;
P_T_NT_dom(:,1,:)=NPC(P_T_NT_byStrata(:,1:3,:),'L',options);
P_T_NT_dom(:,2,:)=NPC(P_T_NT_byStrata(:,4:8,:),'L',options);
P_T_NT_dom(:,3,:)=P_T_NT_byStrata(:,9,:);
P_T_NT_dom(:,4,:)=NPC(P_T_NT_byStrata(:,10:14,:),'L',options);

options.OUT=1;
optdom=options;
optdom.labels.dims{2}={'domain 1' 'domain 2' 'domain 3' 'domain 4'}
P2_T_NT_dom=NPC_FWE(P_T_NT_dom,'T',optdom);
pmat_show(P2_T_NT_dom,.05,optdom)
```

We focus on two routines used in both this analysis and a later one, namely `NPC_ReM` and `by_strata`. The former,

```
function [P, T,options] = NP_ReM(Y,X,DES,B,tail,options)
```

requires the following input arguments:

- `Y`, the $nk \times V$ matrix, where n denotes the sample size, measured k times over V variables, containing all the repeated measures for each variable.
- `X`, an indicator vector nk. `X` values corresponding to the same value of `X` are placed in the same sample.
- `DES` is the design matrix. Several choices for this parameter are available. Throughout the analysis the (default) option `'Seq'` for sequential comparisons is used. However, it is possible to use the option `'All'` (which considers all the possible comparisons) or `'Bal'` (comparing the baseline to the other measures) or `'Trd'` (evaluating a possible trend in the responses).
- `B` denotes the number of permutations.
- `tail` is the vector specifying the direction of the alternatives for the variables under study.
- The parameter `options` refer to the possibility of printing the observed p-values.

The function `by_strata`,

```
[P T options]=by_strata(strata_vector, function_name, Y,X,
arguments);
```

may be found in the `utilities` folder on the book's website and requires the following input arguments:

- strata, i.e. the stratification variable.
- function_name, allowing us to choose any 'NP_*' function.
- Y and X, denoting respectively the matrix containing the variables under study and the group-indicator vector. The same considerations as given for the previous function hold here.
- arguments, i.e. all the arguments required by the function chosen at the second step.

Note that this function uses independent permutations in each stratum.

We now present the analysis carried out using routines developed in R. We refer the reader to the book's website for details of the R code. As already stated, this first analysis evaluating treatment effects is a paired one-sample analysis, and there are two patients in the CASI group (16 and 26) who should be omitted because they have missing observations (note that the permutations are also paired, and there are only two observations for each patient):

```
setwd("C:/path")
data<-read.csv("CONFOCAL.csv",header=TRUE)
ID<-data$FU_BA=='FU' & (data$PAT_ID==16 | data$PAT_ID==26);
data = data[ID==FALSE,]
```

Let d0 and d5 be the strata defined at t_0 and t_1, nT0 and nT5 their sample sizes. D0 and D5 are matrices of the observed differences D_{ji}, $i = 1, \ldots, n$, $j = 1, \ldots, p$.

```
d0<-data[data$FU_BA=='BASELINE',]
d5<-data[data$FU_BA=='FU',]

nT0  <-dim(d0)[1]/2
nT5  <-dim(d5)[1]/2

#### TEST T vs. NT  ####

D0= as.matrix(d0[1:nT0,-c(1:3)]-d0[-c(1:nT0),-c(1:3)])
D5= as.matrix(d5[1:nT5,-c(1:3)]-d5[-c(1:nT5),-c(1:3)])
```

Data columns 1 and 3 (variables Patient ID and T_NT) have been removed because they are non-informative. In order to perform a partial test on each variable we consider $B = 2000$ random independent permutations of n $\pm$ signs for each variable (the matrices S0 and S5). The vectors T0 and T5 contain the null distribution of the test statistics T_{ht}^*, $h = 1, \ldots, V$. They are obtained by multiplying the matrices of observed differences and the matrices of random $\pm$ signs, therefore we are considering the following test statistics at t_0 and t_1:

$$T_{ht}^* = \sum_{i=1}^{n} [X_{h2it} - X_{h1it}]\, S_i^*, \;\; t = t_0, t_1,$$

where $S_i^* = 1 - 2\mathcal{B}n(1, 1/2)$ are i.i.d. random signs.

```
B=2000
p=dim(D0)[2]
T0<-array(0,dim=c((B+1),p))
T5<-array(0,dim=c((B+1),p))

T0[1,] = apply(D0,2,sum)
T5[1,] = apply(D5,2,sum)
```

```
for(bb in 2:(B+1)){## independently within variables

S0=array(1-2*rbinom(nT0*p,1,.5),dim=dim(D0))
S5=array(1-2*rbinom(nT5*p,1,.5),dim=dim(D5))

for(j in 1:p){
T0[bb,j] = D0[,j]%*%S0[,j]
T5[bb,j] = D5[,j]%*%S5[,j]
}
}
```

In order to obtain the partial p-values, we load the `t2p` function, which returns a matrix with the null distribution of p-values from a matrix containing the permutation distribution of the test statistic. The p-values are obtained according to the rule *large is significant*, therefore the inputs of the function are the absolute values of the matrices `T0` and `T5`.

```
source("t2p.r")
P0<-t2p(abs(T0))  ;  colnames(P0)=colnames(D0)
P5<-t2p(abs(T5))  ;  colnames(P5)=colnames(D5)
```

The p-values of the partial tests are the elements of the first rows of `P0` and `P5`.

```
par.res = cbind(P0[1,],P5[1,]); colnames(par.res) = c("T0","T5")
par.res
                          T0      T5
CELLULAR_DENSITY      0.1850  0.3140
POLYMEGATHISM         0.2245  0.4740
PLEOMORPHISM          0.9700  0.9215
dens                  0.0595  0.2765
P.F                   0.4230  0.8495
PBE                   0.2065  0.5220
PBI                   1.0000  1.0000
PT                    1.0000  1.0000
EPITHELIUM_THICKNESS  0.6905  0.2545
X1to10                0.9550  0.8855
X11to33               0.6720  0.2070
X34to66               0.4995  0.5630
X67to90               0.5290  0.6410
X91to100              0.1995  0.1380
```

The partial p-values do not provide any evidence of treated–untreated differences in any variable. In order to combine the partial tests into four domains, we create a categorical variable `dom`, and combine the columns of `P0, P5` for the kth domain with *Fisher's* function:

```
dom<-c(rep(1,3),rep(2,5),3,rep(4,5))
dom
[1] 1 1 1 2 2 2 2 2 3 4 4 4 4 4

TD0<-array(0,dim=c((B+1),4))
TD5<-array(0,dim=c((B+1),4))
```

```
for(k in 1:4){

TD0[,k] = apply(as.matrix(P0[,dom==k]),1,function(x){-2*log(prod(x))})
TD5[,k] = apply(as.matrix(P5[,dom==k]),1,function(x){-2*log(prod(x))})

}
PD0<-t2p(TD0); PD5<-t2p(TD5)
colnames(PD0) = colnames(PD5) = paste("dom",seq(1,4),sep="")
dom.res=cbind(PD0[1,],PD5[1,]);colnames(dom.res)=c("T0","T5")
dom.res

         T0     T5
dom1 0.3750 0.6800
dom2 0.2755 0.8495
dom3 0.6905 0.2545
dom4 0.7570 0.4945
```

None of the combined tests of each domain are significant.

Evaluation of Time Effect

In this analysis, the time effect is examined separately for MMC and standard steroid therapy. The results obtained using the NPC methodology are displayed in Tables 11.5 and 11.6 and in Figures 11.3 and 11.4, where possible significances of the partial and global p-values are displayed

Table 11.5 Evaluating time effect

	Untreated		Treated	
Variable	p-value	p-FWE	p-value	p-FWE
Cellular_density	0.4567	0.9310	0.0478	0.2983
Polymegathism	0.0074	0.0468	0.0082	0.0694
Pleomorphism	0.3813	0.9290	0.5613	0.9216
Density	0.0002	0.0022	0.0014	0.0130
P_F	0.7047	0.9464	0.0498	0.3059
P_{BE}	0.0006	0.0054	0.0002	0.0022
P_{BI}	0.5109	0.9322	0.7623	0.9216
P_T	0.5171	0.9322	0.5651	0.9216
Epithelium_thickness	0.0028	0.0238	0.0842	0.3917
K_1_10	0.0790	0.4089	0.0768	0.3529
K_11_33	0.7145	0.9464	0.2750	0.7758
K_34_66	0.9136	0.9464	0.5185	0.9216
K_67_90	0.3363	0.9224	0.3833	0.8680
K_91_100	0.0002	0.0022	0.0002	0.0022
p-Global		0.0022		0.0022

using a palette of grey colours. A colour key is shown to the right of the plot. In order to control the familywise error rate and compute adjusted p-values, a closed testing procedure is used.

We find that three variables (density, P_{BE} and K_91_100) are significant with and without controlling the FWE in both treated and untreated groups, as well as global p-values obtained after combining all the partial p-values using Tippett's combining function. The variables polymegathism and epithelium thickness are significant only in the untreated group, before and after controlling the FWE, while the variables cellular density, polymegathism and P_F are significant in the treated group only before controlling the FWE. Domain 2 is significant with and without controlling the FWE in both treated and untreated groups, while domain 3 is significant only in the untreated group, before and after controlling the FWE. Raw p-values corresponding to domains 1 and 4 are significant in the treated group.

```
%Evaluating differences T_1-T_0, separately for treated
%and untreated patients
%Evaluating TIME effect

P_FU_BA_byStrata=by_strata(T_NT,'NP_ReM',Y,FU_BA,'all',B,0,0);
options.labels.dims{2}=[Y.name];
options.labels.dims{3}={'Untreated' 'Treated'}
options.labels.dimslabel{3}='Untreated vs Treated';
pmat_show(P_FU_BA_byStrata,.05,options)

P2_FU_BA=NPC_FWE(P_FU_BA_byStrata,'T',options);
pmat_show(P2_FU_BA,.05,options)

options.OUT=0;
options.Combdims=2;
P_FU_BA_dom(:,1,:)=NPC(P_FU_BA_byStrata(:,1:3,:),'L',options);
P_FU_BA_dom(:,2,:)=NPC(P_FU_BA_byStrata(:,4:8,:),'L',options);
P_FU_BA_dom(:,3,:)=P_FU_BA_byStrata(:,9,:);
P_FU_BA_dom(:,4,:)=NPC(P_FU_BA_byStrata(:,10:14,:),'L',options);

options.OUT=1;
optdom=options;
optdom.labels.dims{2}={'domain 1' 'domain 2' 'domain 3' 'domain 4'}
P2_FU_BA_dom=NPC_FWE(P_FU_BA_dom,'T',optdom);
pmat_show(P_FU_BA_dom,.05,optdom)
pmat_show(P2_FU_BA_dom,.05,optdom)
```

We now present the analysis carried out using R. Once more we refer the reader to the book's website for details of R code. In the `CASI` v. `BASE` comparisons, we exclude two patients with some missing observations. Note that this is again a two-sample problem with paired data, so we repeat the previous testing procedure here, although we carry out separate analyses for treated/untreated observations and compare paired observations related to time 0 (group `BASE`) and time 5 (group `CASI`).

```
data<-read.csv("CONFOCAL.csv",header=TRUE)

dT<-data[data$T_NT=='T',] ; dNT<-data[data$T_NT=='NT',]

# IDT<-dT$FU_BA=='FU' & (dT$PAT_ID == 16 | dT$PAT_ID == 26)
```

Table 11.6 Time effect over domains

	Untreated		Treated	
Variable	p-value	p-FWE	p-value	p-FWE
Domain 1	0.0916	0.1778	0.0314	0.0890
Domain 2	0.0058	0.0174	0.0038	0.0146
Domain 3	0.0028	0.0112	0.0842	0.0904
Domain 4	0.1548	0.1778	0.0456	0.0904
p-Global		0.0112		0.0146

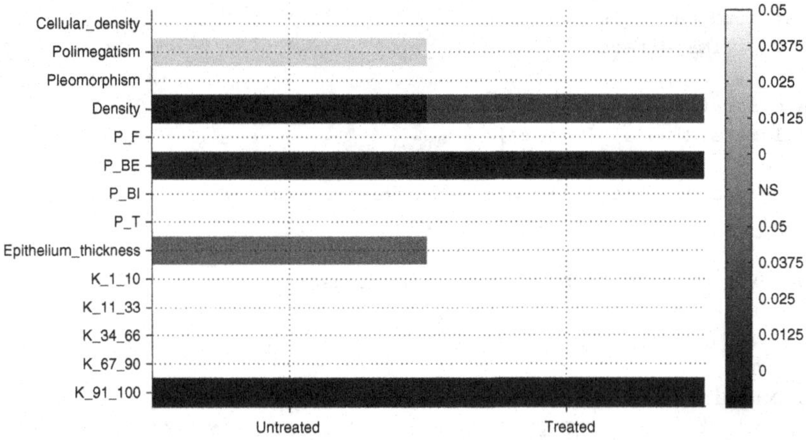

Figure 11.3 Output of `pmat_show` function for the time effect analysis showing corresponding p-values adjusted for multiplicity (i.e. controlling the FWE)

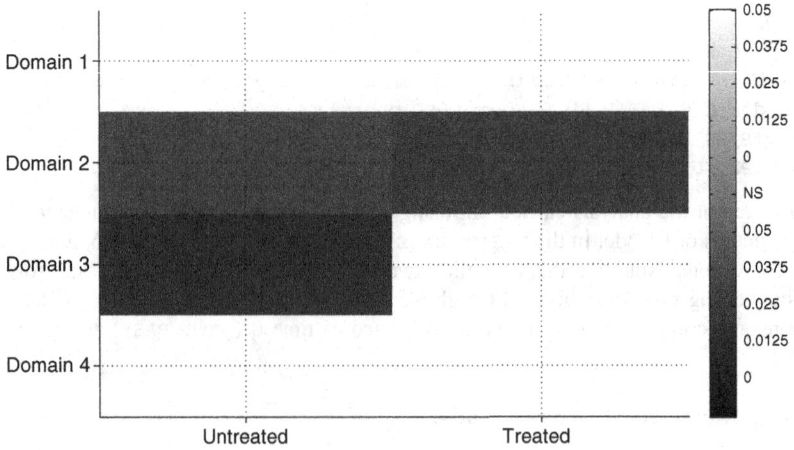

Figure 11.4 Output of `pmat_show` function for the time effect analysis when information on domains is taken into account and after adjusting p-values for multiplicity

```
# IDNT<-dNT$FU_BA=='FU' & (dNT$PAT_ID == 16 | dNT$PAT_ID == 26)

IDT <- dT$PAT_ID == 16 | dT$PAT_ID == 26
IDNT<- dNT$PAT_ID == 16 | dNT$PAT_ID == 26

dT<-dT[IDT==FALSE,] ; dNT<-dNT[IDNT==FALSE,]


# nT<-table(dT$FU_BA)[2:1]; nNT<-table(dNT$FU_BA)[2:1]
# cT<-rep(c(1,-1),nT) ; cNT<-rep(c(1,-1),nNT)

dT=as.matrix(dT[,-c(1:3)]) ; dNT=as.matrix(dNT[,-c(1:3)])
p=dim(dT)[2]
n=dim(dT)[1]/2


DT= dT[1:n,]-dT[-c(1:n),]
DNT=dNT[1:n,]-dNT[-c(1:n),]

TT<-array(0,dim=c((B+1),p)) ; TNT<-array(0,dim=c((B+1),p))
TT[1,] = apply(DT,2,sum)
TNT[1,] = apply(DNT,2,sum)
```

As in the previous example, the matrices DT and DNT contain the observed differences CASI versus BASE) related to each variable and each statistical unit (treated/untreated eyes, in this example).

```
for(bb in 2:(B+1)){## independently within variables

ST=array(1-2*rbinom(n*p,1,.5),dim=dim(DT))
SNT=array(1-2*rbinom(n*p,1,.5),dim=dim(DNT))

for(j in 1:p){
TT[bb,j] = DT[,j]%*%ST[,j]
TNT[bb,j] = DNT[,j]%*%SNT[,j]
}
}

source("t2p.r")
PT=t2p(abs(TT)) ; PNT=t2p(abs(TNT))
colnames(PT)=colnames(dT) ; colnames(PNT)=colnames(PT)
TNT.res = cbind(PT[1,],PNT[1,]) ; colnames(TNT.res) = c("T","NT")
TNT.res
                       T     NT
CELLULAR_DENSITY  0.0620 0.4840
POLYMEGATHISM     0.0100 0.0065
PLEOMORPHISM      0.6760 0.6835
dens              0.0010 0.0000
P.F               0.0490 0.8955
PBE               0.0000 0.0010
PBI               0.7995 0.8080
PT                0.5490 0.5070
```

```
EPITHELIUM_THICKNESS 0.1135 0.0025
X1to10               0.2095 0.1845
X11to33              0.5300 0.9600
X34to66              0.8935 0.7010
X67to90              0.0925 0.1750
X91to100             0.0000 0.0000
```

In order to control FWE, we use the `FWEminP` function to apply a Bonferroni–Holms adjustment to a vector of raw p-values through a step-down application of Tippett's combining function.

```
source("FWEminP.r")
FWE.res = cbind(FWE.minP(PT),FWE.minP(PNT))
colnames(FWE.res)=c("T","NT") ; rownames(FWE.res)=colnames(dT)
FWE.res
```

```
                       T      NT
CELLULAR_DENSITY      0.4055 0.9805
POLYMEGATHISM         0.0885 0.0500
PLEOMORPHISM          0.9630 0.9915
dens                  0.0095 0.0000
P.F                   0.3470 0.9915
PBE                   0.0000 0.0105
PBI                   0.9630 0.9915
PT                    0.9630 0.9810
EPITHELIUM_THICKNESS  0.5215 0.0175
X1to10                0.7355 0.7920
X11to33               0.9630 0.9600
X34to66               0.8935 0.9915
X67to90               0.5085 0.7920
X91to100              0.0000 0.0000
```

11.3 Two-Way (M)ANOVA

In the previous section we evaluated both treatment effect (first analysis) and time effect (second analysis). To analyse the data set, we may also consider a two-way MANOVA, according to possible different goals. This example can also be analysed by applying the synchronized permutations introduced in Example 8, 2.7 and in Section 11.3.1. We consider the variables time and treatment as the main factors with two levels each. We have treated and untreated eyes, at baseline time ($t = 0$) and at 5 years after surgery ($t = 1$). Synchronized permutation testing requires the same number of observations for each treatment. Thus, we do not consider patients 16 and 26, since they have missing observations.

Along with time and treatment factors, we may consider their interaction, addressing whether time effects depend on a given treatment.

11.3.1 Brief Overview of Permutation Tests in Two-Way ANOVA

In the MANOVA analysis previously discussed, a 2^2 factorial design was used. In an $I \times J$ replicated factorial design, data are assumed to behave according to the linear model (see also Example 8, 2.7)

$$X_{ijk} = \mu + \alpha_i + \beta_j + (\alpha\beta)_{ij} + Z_{ijk}$$

with $i = 1, \ldots, I$, $j = 1, \ldots, J$, $k = 1, \ldots, n$, where y_{ijk} are the experimental responses, μ is the general mean, α_i, β_j are the effects of the main factors, $(\alpha\beta)_{ij}$ are interaction effects, ε_{ijk} are exchangeable experimental errors, with zero mean, from an unknown continuous distribution P, and n is the number of replicates in each cell. Assuming the usual side-conditions, the following relations hold:

$$\sum_i \alpha_i = 0, \sum_j \beta_j = 0, \sum_i (\alpha\beta)_{ij} = 0 \ \forall\, j, \sum_j (\alpha\beta)_{ij} = 0 \ \forall\, i.$$

The hypotheses of interest are $H_{0A}\colon \{\alpha_i = 0 \ \forall\, i\}$, against $H_{1A}\colon \{\exists\, i : \alpha_i \neq 0\}$, $H_{0B}\colon \{\beta_j = 0 \ \forall\, j\}$ against $H_{1B}\colon \{\exists\, j : \beta_j \neq 0\}$, $H_{0AB}\colon \{(\alpha\beta)_{ij} = 0 \ \forall i, j\}$ against $H_{1AB}\colon \{\exists i, j : (\alpha\beta)_{ij} \neq 0\}$. Since observations from different blocks have different means, they are not exchangeable. To carry out nonparametric testing on main effects, a restricted kind of randomization should be introduced. Assume that it is of interest to test for H_{0A} and that ν^* units are randomly exchanged between blocks A_iB_j and A_sB_j, $j = 1, \ldots, J$. Let us consider the permutation structure (Remark 1, 2.7) of intermediate statistics to separately compare levels i and s of factor A at level j of factor B:

$${}^aT^*_{is|j} = \sum_k X^*_{ijk} - \sum_k X^*_{sjk} = (n - 2\nu^*)[\alpha_i - \alpha_s + (\alpha\beta)_{ij} - (\alpha\beta)_{sj}] + \bar{Z}^*_{is|j},$$

where an asterisk means that we have performed a random synchronized permutation (since ν^* is invariant with respect to the levels of factor B) among units in blocks A_iB_j and A_sB_j, $\bar{Z}^*_{is|j} = (n - \nu^*)[\bar{Z}^*_{ij} - \bar{Z}^*_{sj}] + \nu^*[\tilde{Z}^*_{ij} - \tilde{Z}^*_{sj}]$ is the permutation error term, and $\bar{Z}^*_{ij}, \bar{Z}^*_{sj}, \tilde{Z}^*_{ij}, \tilde{Z}^*_{sj}$ are random sample means of respectively $n - \nu^*$ and ν^* errors from different pairs of blocks. Due to the side-conditions, the test statistic

$${}^aT^*_A = \sum_{i<s} \left[\sum_j {}^aT^*_{is|j} \right]^2,$$

only depends on levels of factor A and on exchangeable errors $(\bar{\varepsilon}^*_{is|j})$, hence it is a separate exact test on factor A. ${}^aT^*_A$ is a sum of all possible pairs of cells of the intermediate statistics ${}^aT^*_{is|j}$, $j = 1, \ldots, J$. Similarly, it is possible to define ${}^bT^*_{jh|i} = \sum_k X^*_{ijk} - \sum_k X^*_{ihk}$, that is, the intermediate statistic for comparing levels j and h of factor B at level i of factor A. Then, the test statistic

$${}^bT^*_A = \sum_{j<h} \left[\sum_i {}^bT^*_{jh|i} \right]^2,$$

only depends on levels of factor B and on the exchangeable error, hence it is a separate exact test on factor B. Finally, it is also possible to define two test statistics for the interaction

$${}^aT^*_{AB} = \sum_{i<s} \sum_{j<h} [{}^aT^*_{is|j} -{}^a T^*_{is|h}]^2, \quad {}^bT^*_{AB} = \sum_{j<h} \sum_{i<s} [{}^bT^*_{jh|i} -{}^b T^*_{jh|s}]^2.$$

Note that ${}^aT^*_{AB}$ is obtained from synchronized permutations involving the factor A row, whereas ${}^bT^*_{AB}$ is obtained from permutations involving the factor B column. Each statistic for the interaction depends only on interaction levels and on exchangeable errors, hence there is a separate exact test on interaction AB. To obtain only one test, the NPC procedure may be applied. Once we have defined the test statistics for main effects and interaction, in order to apply these tests, the synchronized permutation strategy should be defined. The basic concept of synchronized permutations is to exchange the same number ν^* of units within each pair of considered blocks. There are two ways to obtain a synchronized permutation: by exchanging units in the same original positions within each block (constrained synchronized permutations, CSPs) or by exchanging units independently of their original position (unconstrained synchronized permutations, USPs). The main point to remember is that in both cases the same number of units must be exchanged within all pairs of cells in the pair of rows/columns considered. CSPs are defined by the number of units being exchanged (ν^*) and their original positions within each block. As a consequence, the total number of CSPs depends only on which exchange has been made in the first pair of blocks. Since there are

$$C_{CSP} = \binom{2n}{n}$$

possible ways to exchange units in the first pair of blocks, C_{CSP} is the cardinality of the CSPs. Another point to take into account is the cardinality of distinct permutation test statistics (e.g. the number of distinct ${}^aT_A^*$s if we are testing for factor A): the square in formulas of test statistics produces a symmetry, that is, there are two distinct permutations generating the same value of ${}^aT_A^*$, hence the total number of distinct ${}^aT_A^*$s is $C_{CSP}/2$. USPs do not require the exchanged units to be in the same original position within the blocks. Thus from a naive point of view, we could apply the same algorithm with an initial random shuffling within each single block in order to obtain the USPs. For further details we refer the reader to Basso et al. (2009a).

11.3.2 MANOVA Using MATLAB and R Codes

Below we show the results obtained for the MANOVA analysis in the nonparametric permutation framework. In order to control the FWE and compute adjusted p-values, a closed testing procedure is used. In particular, we perform the strong FWE over the variables, through and within each factor (see Tables 11.7 and 11.8), including also the information on domains (Table 11.9). Similar

Table 11.7 Two-way MANOVA and p-value adjustment over variables through factors (p-FWE)

	Baseline v. FU		T v. NT		Interaction		
Variable	p-value	p-FWE	p-value	p-FWE	p-value	p-FWE	p-Global
Cellular_density	0.1274	0.2324	0.0990	0.2324	0.2558	0.2558	0.2324
Polymegathism	0.1956	0.3361	0.0014	0.0042	0.5211	0.5211	0.0042
Pleomorphism	0.8388	0.9696	0.4027	0.7780	0.8742	0.9696	0.7780
Density	0.4055	0.4055	0.0002	0.0006	0.0054	0.0102	0.0006
P_F	0.5931	0.8226	0.1898	0.4717	0.8994	0.8994	0.4717
P_{BE}	0.0948	0.1784	0.0002	0.0006	0.2010	0.2010	0.0006
P_{BI}	0.8374	0.9894	0.4165	0.7409	1.0000	1.0000	0.7409
P_T	1.0000	1.0000	0.3719	0.7473	1.0000	1.0000	0.7473
Epithelium_thickness	0.2090	0.3765	0.0040	0.0118	0.5587	0.5587	0.0118
K_1_10	0.8834	0.9870	0.0462	0.1332	0.9474	0.9870	0.1332
K_11_33	0.6557	0.6659	0.4177	0.6659	0.3069	0.6659	0.6659
K_34_66	0.9276	0.9372	0.7465	0.9372	0.3407	0.7155	0.7155
K_67_90	0.4855	0.7281	0.2949	0.6551	0.9130	0.9130	0.6551
K_91_100	0.1330	0.2436	0.0002	0.0006	0.5999	0.5999	0.0006

Table 11.8 Two-way MANOVA and p-value adjustment over variables within each factor (p-FWE)

	Baseline v. FU		T v. NT		cInteraction	
Variable	p-value	p-FWE	p-value	p-FWE	p-value	p-FWE
Cellular_density	0.1274	0.7643	0.0990	0.4811	0.2558	0.9546
Polymegathism	0.1956	0.8728	0.0014	0.0114	0.5211	0.9978
Pleomorphism	0.8388	0.9998	0.4027	0.8722	0.8742	1.0000
Density	0.4055	0.9776	0.0002	0.0020	0.0054	0.0646
P_F	0.5931	0.9976	0.1898	0.6845	0.8994	1.0000
P_{BE}	0.0948	0.6919	0.0002	0.0020	0.2010	0.9204
P_{BI}	0.8374	0.9998	0.4165	0.8722	1.0000	1.0000
P_T	1.0000	1.0000	0.3719	0.8722	1.0000	1.0000
Epithelium_thickness	0.2090	0.8728	0.0040	0.0280	0.5587	0.9994
K_1_10	0.8834	0.9998	0.0462	0.2662	0.9474	1.0000
K_11_33	0.6557	0.9978	0.4177	0.8722	0.3069	0.9696
K_34_66	0.9276	0.9998	0.7465	0.8722	0.3407	0.9754
K_67_90	0.4855	0.9908	0.2949	0.7996	0.9130	1.0000
K_91_100	0.1330	0.7643	0.0002	0.0020	0.5999	0.9994
p-Global		0.6919		0.0020		0.0646

Table 11.9 Two-way MANOVA and strong FWE control

	Baseline v. FU		T v. NT		Interaction	
Variable	p-value	p-FWE	p-value	p-FWE	p-value	p-FWE
Domain 1	0.2809	0.6305	0.0258	0.0516	0.6045	0.9686
Domain 2	0.8052	0.9282	0.0002	0.0008	0.9248	0.9686
Domain 3	0.2090	0.6091	0.0040	0.0116	0.5587	0.9686
Domain 4	0.7439	0.9282	0.0696	0.0696	0.7824	0.9686
		0.6091		0.0008		0.9686

information is given in Figures 11.5–11.7, where significant variables are associated with different shades of grey. We find that raw p-values corresponding to polymegathism, density, P_{BE}, epithelium thickness, K_1_10 and K_91_100 are significantly different between treatment levels and the density variable is significant in the interaction term. When controlling the FWE, K_1_10 is no more significant, and density is no more significant in the interaction term. When including domains, we find that domains 1, 2 and 3 are significantly different between treatment levels.

```
%Factorial design 2×2 for repeated measurements
%% Effects: Treated VS Untreated,  Baseline VS FU, interaction
```

```
DES=[ 1  1  1; 1 -1 -1; -1 1 -1;-1 -1 1]

levs= [FU_BA T_NT]*[1 2]';
options.labels.dims{3}={'FU_BA', 'T_NT', 'INTERACT'};
[P t options]=NP_ReM(Y,levs,DES,B,0,options);
options.labels.dimslabel{3}='Two-way MANOVA';
pmat_show(P,.05,options)

%combining effects over variables
options.Combdims=3;
P_fact=NPC(P,'F',options);
%adjusting p-values
P2_fact=NPC_FWE(P,'T',options);
pmat_show(P2_fact,.05,options)

%combining variables
options.Combdims=2;
P_var=NPC_FWE(P,'T',options);
Pdom(:,1,:)=NPC(P(:,1:3,:),'L',options);
Pdom(:,2,:)=NPC(P(:,4:8,:),'L',options);
Pdom(:,3,:)=P(:,9,:);
Pdom(:,4,:)=NPC(P(:,10:14,:),'L',options);

optdom=options;
optdom.labels.dims{2}={'domain 1' 'domain 2' 'domain 3' 'domain 4'}
P2dom=NPC_FWE(Pdom,'T',optdom);
pmat_show(Pdom,.05,optdom)
pmat_show(P2dom,.05,optdom)
```

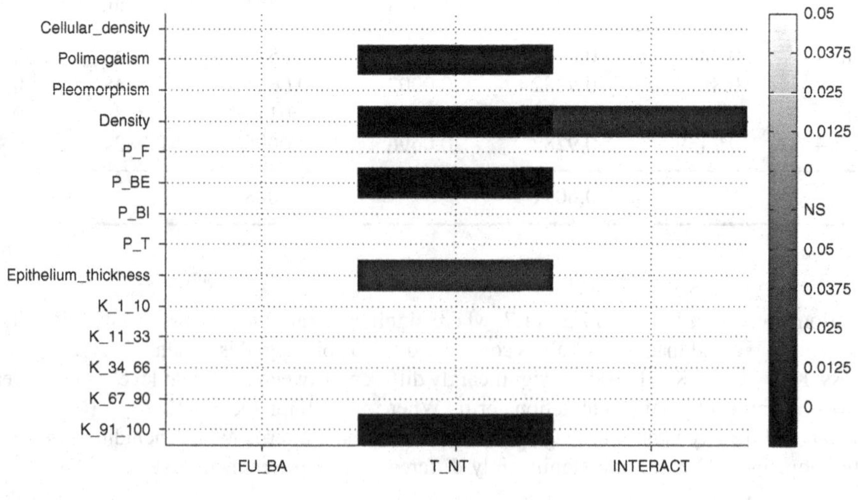

Figure 11.5 Output of `pmat_show` function for the MANOVA analysis showing adjusted *p*-values (adjustment is performed over variables through factors)

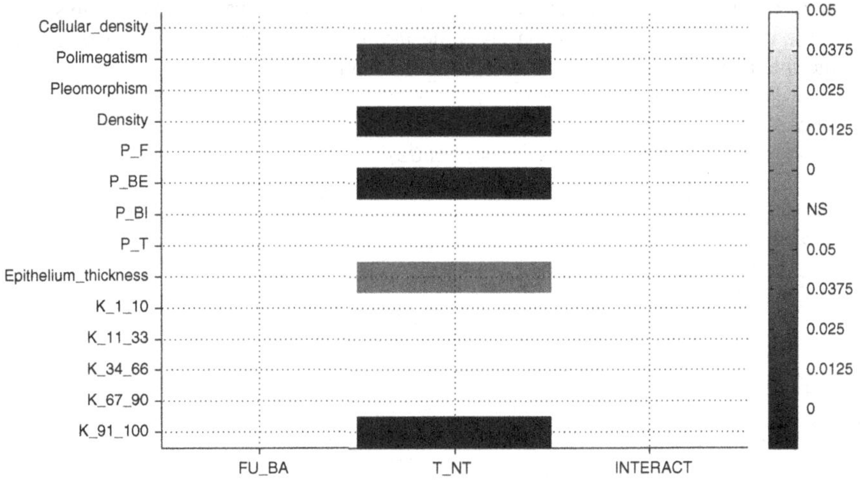

Figure 11.6 Output of `pmat_show` function for the MANOVA analysis showing corresponding p-values adjusted for multiplicity (adjustment is done over variables within each factor)

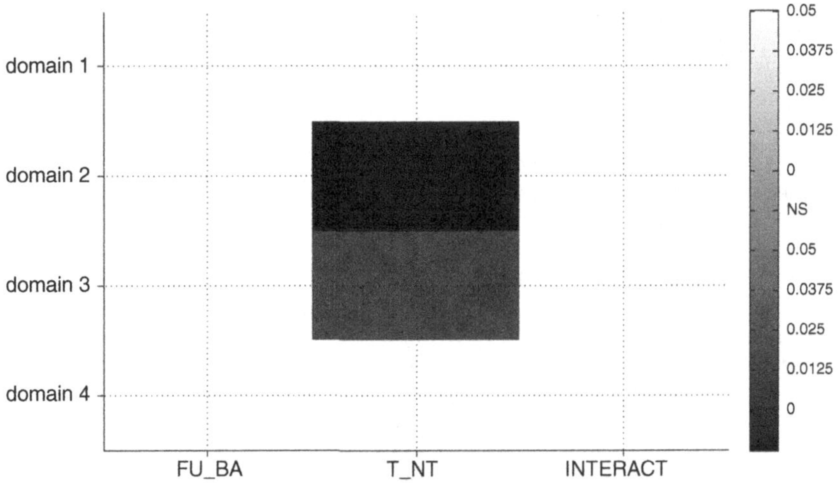

Figure 11.7 Output of `pmat_show` function for the MANOVA analysis when information on domains is taken into account and after adjusting p-values for multiplicity

We now turn to the analysis in R. We consider the variables `tempo` and `T_NT` as main factors with two levels each. Since synchronized permutation testing requires the same number of observations for each treatment, patients 16 and 26 were not considered, since they have missing observations. Again, we refer the reader to the book's website for details of R code.

The constrained synchronized permutation test is done by the function `CSP`, which requires as argments a vector of data `y` and a matrix of labels `x`. The first column of `x` contains the levels of `time`, whereas the second contains the level of `T_NT`. A further option is the desired number of permutations C, which here is set equal to 2000. We perform a two-way ANOVA analysis for each variable. The output of the CSP function gives the p-values assessing the null hypothesis on main effects and interaction. These results are stored in the matrix P.

```
source("CSP.n")
data<-read.csv("CONFOCAL.csv",header=TRUE)
data = data[data$PAT_ID != 16 & data$PAT_ID != 26,]

x=data[,2:3]
x[,1]=ifelse(x[,1]=='FU',1,2)
x[,2]=ifelse(x[,2]=='T',1,2)


p=dim(data)[2]-3
P = array(0,dim=c(p,3))

C = 2000

for(j in 4:(3+p)){

y = data[,j]
t = CSP(y,x,C=C)
P[(j-3),] = c(t$pa,t$pb,t$pab)

print((j-3))
}

rownames(P) = colnames(data[,-c(1:3)])
colnames(P) = c("FU/BASELINE","T/NT","Interaction")

P
              FU/BASELINE  T/NT Interaction
CELLULAR_DENSITY                0.105 0.154       0.341
POLYMEGATHISM      0.001 0.303        0.900
PLEOMORPHISM       0.633 0.988        0.986
dens               0.000 0.412        0.019
P.F                0.283 0.556        0.959
PBE                0.000 0.195        0.410
PBI                0.638 0.872        1.000
PT                 0.319 1.000        1.000
EPITHELIUM_THICKNESS            0.002 0.357       0.545
X1to10             0.114 0.787        0.997
X11to33            0.666 0.636        0.730
X34to66            0.856 0.954        0.790
X67to90            0.058 0.649        0.981
X91to100           0.000 0.074        0.995
```

The data set and the corresponding software codes can be found in the confocal folder on the book's website.

12

Some Case Studies Using NPC Test R10 and SAS Macros

In this chapter we present some complex data sets analysed using standalone software specifically implemented for combination-based permutation tests (NPC Test R10) and also using SAS macros. The NPC Test R10 software the SAS macros can be downloaded along with a user manual from the book's website.

12.1 An Integrated Approach to Survival Analysis in Observational Studies

In the framework of survival analysis, we present an integrated procedure combining post-stratification with a propensity score and combination-based permutation tests, based on an observational study.

In this section we first introduce the application problem and perform the survival analysis based on the whole sample. Then we stratify the sample using an estimated propensity score to control for overt biases based on the theory of covariance adjustment in observational studies (Rosenbaum, 2002). Finally, we use the NPC approach of within-strata partial results to test the global hypothesis related to the whole sample.

12.1.1 A Case Study on Oesophageal Cancer

We consider a survival observational study on times to death for oesophageal cancer patients undergoing oesophagectomy. An ageing population and longer life expectancy have led to increasing numbers of elderly patients with oesophageal carcinoma being referred for surgical treatment. The effect of advanced age on the outcome of oesophagectomy is a matter of some controversy. Indeed, there is an apparent contradiction between studies where the individual risk of mortality after oesophagectomy is strongly linked to patient age and performance status, with lower long-term survival among elderly patients, and studies confirming improvements in the results of oesophagectomy in patients over 70 with mortality and morbidity rates comparable to their younger counterparts. A study performed by the Department of Medical and Surgical Science at the University of Padua (Ruol et al., 2007) was carried out to assess the effects of age on the long-term survival of patients undergoing oesophagectomy for oesophageal cancer.

The long-term survival rate of patients at least 70 years old, undergoing oesophagectomy between January 1992 and June 2005, for cancer of the oesophagus or oesophagogastric junction, was compared with that of younger patients. The analysis involved a total of $n = 664$ patients with cancer

Permutation Tests for Complex Data: Theory, Applications and Software Fortunato Pesarin, Luigi Salmaso

Table 12.1 Sample sizes and distribution of events

	Subjects	Deaths
Age group	*n*	*d*
under 70	529 (79.7%)	329 (62.2%)
70 and above	135 (20.3%)	82 (60.1%)
Total	664	411

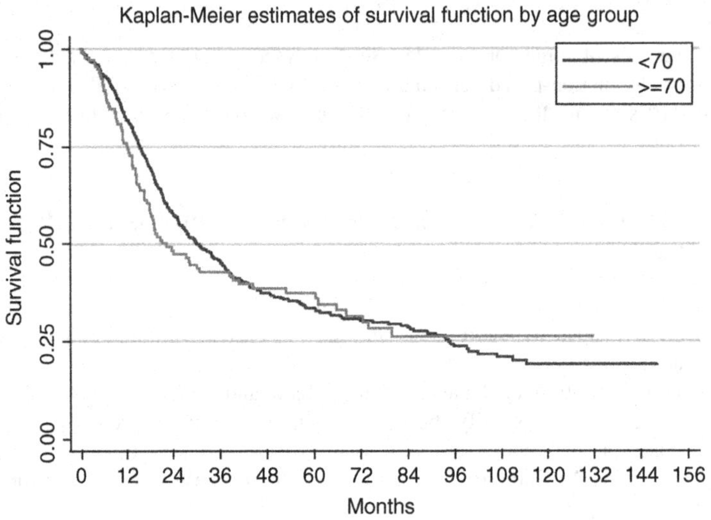

Figure 12.1 Kaplan-Meier estimates of survival function by age group

of the oesophagus or oesophagogastric junction (*Siewert type I-II*): $n_1 = 135$ (20.3%) patients were at least 70 years old (group of interest) and $n_2 = 529$ (79.7%) were under 70 (control group); see Table 12.1. In the elderly group 92 (68.2%) patients were aged between 70 and 74, 35 (25.9%) were aged 75–79, and 8 (5.9%) were in their 80s. Patients were routinely followed up by their surgeons 1, 3, 6 and 12 months after the operation and every 6–12 months thereafter. Complete tumour resection was defined as $R0$, and incomplete resection with microscopic or macroscopic residual disease was defined as $R1$ and $R2$ respectively. Overall, 594 (89.5%) patients were considered to have curative resection ($R0$) : 124 (91.9%) out of 135 in the group of interest and 470 (88.8%) out of 529 in the control group. During the follow-up period, the overall number of deaths was 82 (60.1%) among patients over 70 and 329 (62.2%) among the younger patients.

This case study is characterized by a fairly large number of observations with a left unbalanced sample size, a light to moderate equal censoring model and, as can be seen in Figure 12.1, a clearly visible pattern of crossing hazard rates. Figure 12.1 shows the raw survival curves for the two samples.

In Chapter 9 we saw that the weighted Kaplan–Meier (WKM) test has good power behaviour in this situation. As regards TIC-NPC, it also behaves well in terms of power and represents an effective and robust combination-based permutation procedure to detect differences among groups.

12.1.2 A Permutation Solution

In this application it is of interest to determine whether or not the two survival time distributions have arisen from an identical survival function based on the two independent samples.

Using the same notation as in Chapter 9, the hypotheses of interest are

$$H_0^G : \left\{[S_1(t_i) = S_2(t_i)\ \forall t_i,\ i = 1, \ldots, D] \bigcap \left[\Delta_1 \stackrel{d}{=} \Delta_2\right]\right\}$$
$$= \left\{\left[X_1 \stackrel{d}{=} X_2\right] \bigcap \left[\Delta_1 \stackrel{d}{=} \Delta_2\right]\right\},$$

against

$$H_1^G : \{S_1(t_i) \leq\neq\geq S_2(t_i)\ \forall t_i,\ \exists t_i : S_1(t_i) <\neq> S_2(t_i)\}$$
$$= \left\{X_1 \stackrel{d}{\leq\neq\geq} X_2\right\}.$$

The global null and alternative hypotheses are broken down into D sub-hypotheses, one for each of the D observed times $t_1^o < \ldots < t_D^o$

$$H_0^G = H_0^{\mathbf{X}|\mathbf{O}} : \left\{\bigcap_{i=1}^{D} \left(X_{1i} \stackrel{d}{=} X_{2i}\right) |\mathbf{O}\right\} = \left\{\bigcap_{i=1}^{D} H_{0i}^{\mathbf{X}|\mathbf{O}}\right\}$$

against $H_1^G : \left\{\bigcup_{i=1}^{D} H_{1i}^{\mathbf{X}|\mathbf{O}}\right\}$.

In Table 12.2, TIC-NPC refers to the multidimensional permutation test in the case of treatment-independent censoring (Callegaro et al., 2003) and WKM indicates the asymptotic weighted Kaplan–Meier test (Pepe and Fleming, 1989).

12.1.3 Survival Analysis with Stratification by Propensity Score

Possible effects of treatments and risk factors would preferably be investigated in randomized experiments (or designed experiments) which randomly assign experimental units to the different groups under study. In such an experiment, the random assignment of individuals to the groups is under the control of the experimenter, who ensures that the groups are comparable in terms of all covariates, relevant and irrelevant, known and unknown, measured and unmeasured. Observational studies (sometimes referred to as quasi-experiments or non-experimental studies) are different from experiments in that the investigator lacks experimental control and, in particular, cannot manage the random assignment of subjects to treatments, so the groups may not have been comparable and may therefore reproduce these differences rather than effects of treatments and risk factors under study. For instance, when treated patients are compared to controls, differing outcomes may reflect either effects caused by the treatment or differences in prognosis before treatment. Nevertheless, in some scenarios, an observational study aims to estimate treatment and risk factor effects in a

Table 12.2 Comparison of two survival curves

Alternatives	Permutation Tests	Asymptotic Tests
	TIC-NPC	WKM
Two-sided	0.022	0.333

context in which it is impractical or unethical to perform randomized experiments (see Cochran, 1968; Berger, 2005).

An observational study is therefore biased if the various groups differ in ways that matter in relation to the outcomes under study. Biases may refer to systematic differences between groups with respect to one or more prognostic variables. Fortunately, differences in observed covariates can be reduced if not removed. Units belonging to different groups may be seen to vary in terms of some observed and accurately measured characteristics, but these noticeable differences may be controlled by comparing individuals belonging to different groups who have the same values of the observed covariates, that is, subjects should belong to the same matched set or stratum.

Techniques such as matching, stratification and model-based adjustments, along with combinations of these procedures, are generally applied to eliminate conspicuous biases 'accurately recorded in the data at hand, that is, biases visible in imbalances in observed covariates' (Rosenbaum, 2005). Covariance adjustment with propensity score (Rosenbaum and Rubin, 1983; Rosenbaum, 2002, 2005) is another procedure which can be used to accomplish this goal, combining good quality matching and stratification with the properties of model-based adjustments. The propensity score is often applied directly as the basis for inference. In addition, matching or stratifying propensity scores is often used in conjunction with further model-based adjustments within matched pairs or strata using regression or generalized linear models. The propensity score is a procedure for constructing matched pairs or matched sets or strata that balance a large number of observed covariates. In fact, in an observational study, adjustment for an estimated propensity score, that is, matching or stratifying treated and control subjects in a single score, tends to balance all of the observed covariates that were used to construct it. The resulting matched sets or strata are heterogeneous in the covariates, but the covariates tend to have similar distributions in the treated and control groups; therefore, the groups as a whole appear to be comparable.

A propensity score is defined as the probability of belonging to a group of interest, given the observed covariates. This score, which can also be given a prognostic interpretation, summarizes all the information required to balance the distribution of confounding variables between groups, in order to have sets or strata with homogeneous units. In practice, the propensity score is unknown and must be estimated, for instance, using a logistic regression model where the binary response variable represents the indicator of the group to which the units belong, and the vector of the independent variables is the vector of the observed covariates, which could record a great number of pretreatment measurements describing the characteristics of the subjects.

Clearly, propensity scores balance only observed covariates used to construct the score, but it is important to note that, unlike randomization, they cannot be expected to remove biases due to unobserved or unrecorded differences before treatment between treated and control individuals. Indeed, relevant unobserved or unknown pretreatment differences cannot be removed by covariance adjustments, see also Example 11, 2.7. These biases are addressed partly in the design of an observational study and partly in the analysis of the study by additional procedures, such as sensitivity analysis (Rosenbaum, 2002). In an observational study, hidden bias is a serious and crucial concern that can affect the study's conclusions. For that reason, statistical methodologies of adjustment for confounding factors, such as covariance adjustment with propensity score, should work well when the study is free of hidden bias.

12.2 Integrating Propensity Score and NPC Testing

From the paper by Ruol et al. (2007), we use six binary covariates as predictors of the propensity to belong to the treated group as a toy example.

In order to estimate the propensity score, we place these covariates in a stepwise exact logistic regression model (Mehta et al., 2000) using the nonparametric procedure (Finos et al., 2009) presented in Chapter 5 to adjust the p-values of the selected model.

Having estimated the propensity score, following the indications of Cochran (1968) and Rosenbaum (2005), we decided to divide the whole sample into five strata based on the values of the score. Indeed, Cochran showed that five strata formed from a single continuous covariate can remove about 90% of the bias. Table 12.3 shows the resulting sample sizes and the distribution of the events of interest within each stratum. Figures 12.2–12.6 present the survival curves for the two samples within each stratum.

In stratum 1, the two survival functions cross at several failure time points, and a crossing hazard rates model is clearly visible.

Table 12.3 Sample sizes and distribution of events

		Age group		
Stratum	Sample size/Events	Less than 70	70 or older	Total
1	n	116(91.3%)	11(8.7%)	127
	Events	70(90.9%)	7(9.1%)	77
2	n	120(85.1%)	21(14.9%)	141
	Events	78(90.7%)	8(9.3%)	86
3	n	104(78.8%)	28(21.2%)	132
	Events	81(79.4%)	21(20.6%)	102
4	n	105(74.5%)	36(25.5%)	141
	Events	43(74.1%)	15(25.9%)	58
5	n	84(68.3%)	39(31.7%)	123
	Events	57(64.8%)	31(35.2%)	88

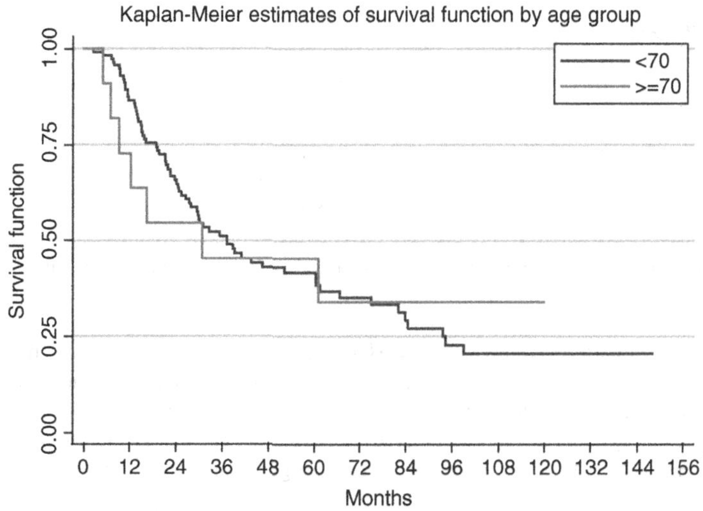

Figure 12.2 Kaplan–Meier survival estimates by group in stratum 1

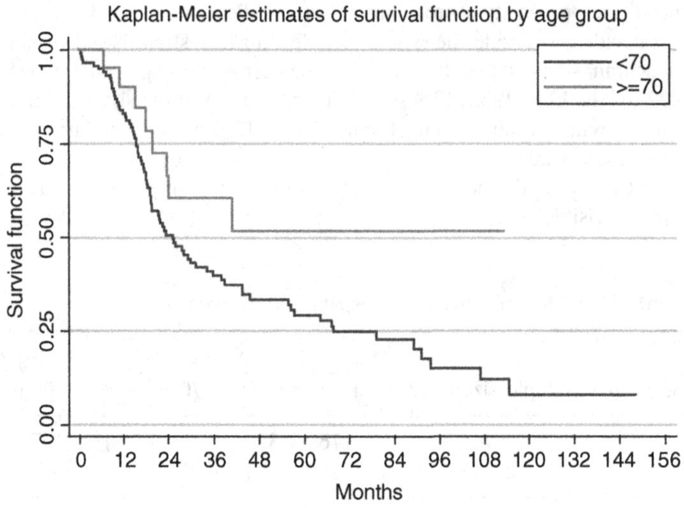

Figure 12.3 Kaplan–Meier survival estimates by group in stratum 2

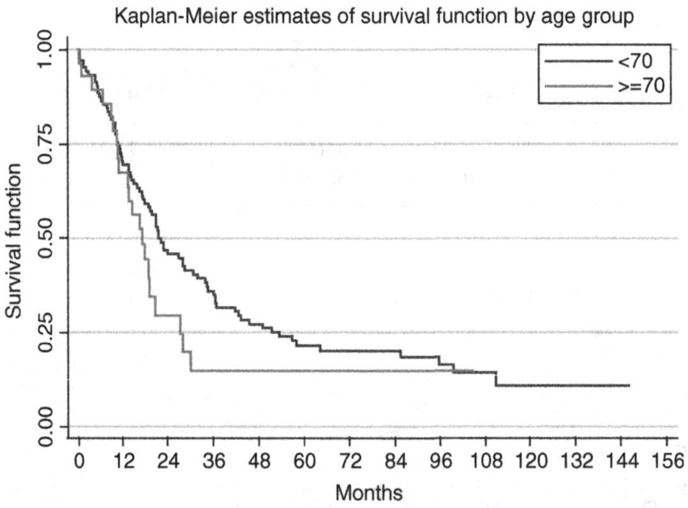

Figure 12.4 Kaplan–Meier survival estimates by group in stratum 3

In stratum 2, the situation is quite different. Here the two survival functions are moderately close for early failure times. Subsequently the differences between the two curves increase considerably. In this configuration there is evidence of a model with late hazard rate differences.

In stratum 3, the two survival functions overlap in the early and late parts of the curves. There are noticeable differences in between. Here the figure suggests a model with middle hazard rate differences.

In stratum 4, the two survival curves are relatively close to each other and the functions cross at several failure time points.

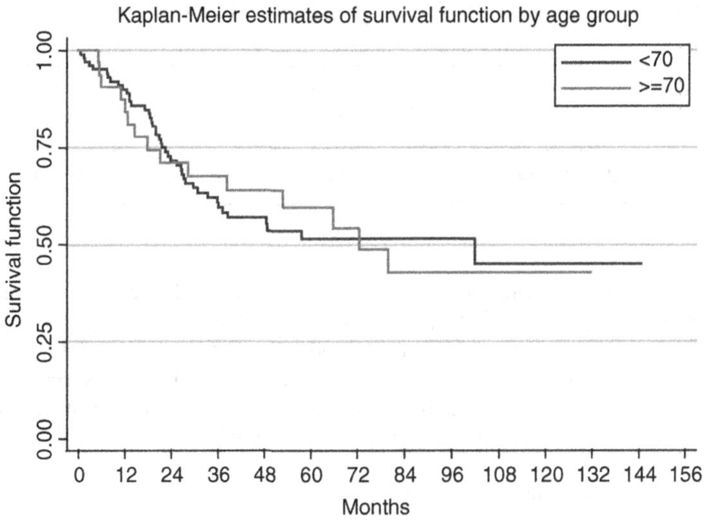

Figure 12.5 Kaplan–Meier survival estimates by group in stratum 4

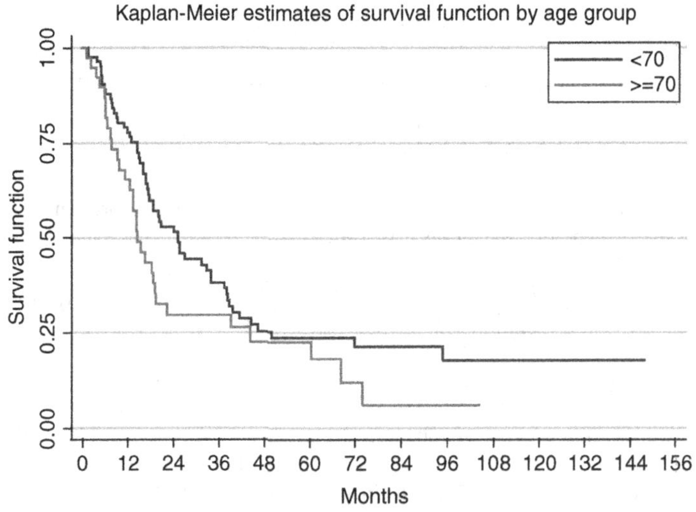

Figure 12.6 Kaplan–Meier survival estimates by group in stratum 5

In stratum 5, for early and late failure times the survival function of the older subjects is always significantly lower than the function of their younger counterparts. In between these points, the survival curves are very close and the two functions cross.

In this case a more complex data configuration is present because our testing problem requires that we take into consideration the stratification variable introduced by the estimated propensity score to reduce possible biases caused by the observed and known covariates used to construct the score.

Table 12.4 p-values obtained in the survival analysis carried out using the TIC-NPC procedure

Alternative	Stratum 1	2	3	4	5	Global
Two-sided	0.3546	0.2624	0.1843	0.5851	0.1560	0.081

Therefore, the NPC may be viewed as a multi-phase procedure characterized by an intermediate combination, where we first combine the partial tests with respect to D times within each s stratum (with $s = 1, \ldots, 5$), and then we combine the second-order tests with respect to strata in a single third-order combined test (see Table 12.4). It should be noted that this multivariate and multistrata hypothesis testing problem cannot be solved using a standard asymptotic test procedure such as the weighted Kaplan–Meier test, as we did for the unstratified survival analysis based on the whole sample.

The overall hypotheses of interest remain

$$H_0^G : \left\{[S_1(t_i) = S_2(t_i)\ \forall t_i,\ i = 1, \ldots, D] \bigcap \left[\Delta_1 \stackrel{d}{=} \Delta_2\right]\right\}$$
$$= \left\{\left[X_1 \stackrel{d}{=} X_2\right] \bigcap \left[\Delta_1 \stackrel{d}{=} \Delta_2\right]\right\},$$

against

$$H_1^G : \{S_1(t_i) \leq\neq\geq S_2(t_i)\ \forall t_i,\ \exists t_i : S_1(t_i) < \neq > S_2(t_i)\}$$
$$= \left\{X_1 \stackrel{d}{\leq\neq\geq} X_2\right\}.$$

However, in this way, the global null and alternative hypotheses are broken down into $D \times S$ sub-hypotheses, one for each of the D observed times $t_1^o < \ldots < t_D^o$ and each stratum

$$H_0^G = H_0^{\mathbf{X}|\mathbf{O}} : \left\{\bigcap_{s=1}^{S}\left[\bigcap_{i=1}^{D}\left(X_{1is} \stackrel{d}{=} X_{2is}\right)|\mathbf{O}\right]\right\} = \left\{\bigcap_{s=1}^{S}\left[\bigcap_{i=1}^{D} H_{0is}^{\mathbf{X}|\mathbf{O}}\right]\right\}$$

against $H_1^G : \left\{\bigcup_{s=1}^{S}\left[\bigcup_{i=1}^{D} H_{1is}^{\mathbf{X}|\mathbf{O}}\right]\right\}$.

Multivariate stratification using the propensity score has an effect on the TIC-NPC procedure. In fact, its inferential results obviously change and lead to statistical conclusions that differ from those in the survival analysis without stratification. In particular, the TIC-NPC test induces rejection of the null hypothesis of no difference in survival between the two groups before the stratification, whereas it leads to acceptance of the null hypothesis after the survival analysis with stratification (see Table 12.4).

12.2.1 Analysis Using MATLAB

We refer the reader to the `survival` folder on the book's website for details of the MATLAB code we used to perform the survival analysis before and after the stratification of the whole data set into five strata using the estimated propensity score obtained through the use of a stepwise exact logistic regression model.

12.3 Further Applications with NPC Test R10 and SAS Macros

12.3.1 A Two-Sample Epidemiological Survey: Problem Description

A prospective multicentre epidemiological study known as SETIG involving the surveillance of treatments in severe infections took place between 1995 and 1997 (Arboretti et al., 2000). Its objective was to describe the natural history of clinically relevant infections, and to compare different diagnostic and therapeutic approaches. A network of 79 Italian hospitals was established, with 141 participating wards. Data on 1159 cases of different infectious diseases were collected. A form was designed to collect patient information on: age and gender; admission ward; presumed diagnosis; type of infection; concomitant diseases; diagnostic data; antibiotics prescribed; dosage and duration of antibiotic therapy; and clinical outcomes (death, resolution of infection, length of stay in hospital, duration of treatment). An important aim of the study was to compare different therapeutic approaches in terms of clinical outcomes. The observational nature of the study requires particular care in the analysis because the incorrect use of observational data for evaluating therapy causal effects can produce biased treatment comparisons. In order to improve the validity of causal inferences in the comparison of several outcomes from different therapies, an approach based on testing for *coherent alternatives* (Rosenbaum, 2002) is used. According to this methodology, we should define a complex pattern of responses as being coherent with the hypothesis of a causal treatment effect. The evidence in favour of the alternative hypothesis implies that the whole causal pattern is confirmed by the data. In the SETIG survey, the hypothesis of interest concerns the comparative effect of the *specific therapeutic approach* versus an *empirical approach* with respect to several outcomes which constitute the pattern of interest, where 'specific therapeutic approach' means that an antibiotic specific to the particular infection is used, and 'empirical approach' means that a generic wide-spectrum antibiotic is used. The causal pattern of coherent alternatives is the following: compared to patients treated with the empirical therapy, patients receiving the specific therapy should show a reduced death rate, a shorter duration of treatment, a shorter length of stay in hospital, and a higher rate of cured infections. It is clear that in statistical terms such a pattern can be reduced to a two-sample multivariate comparison with restricted alternatives and mixed variables with a possible presence of missing values.

The analysis was carried out on 334 patients with sepsis, 154 of whom were treated with empirical therapy and 180 with a specific therapy. These 334 patients were selected from an entire set of 1159 cases on the basis of their homogeneity with respect to the kind of sepsis, seriousness of infection, etc. The 334 patients were first stratified into four homogenous strata with respect to possible confounding factors (age, number of concomitant factors, presence of diabetes, presence of surgical intervention, presence of a tumour, type of unit where patient is admitted, geographic area, etc.). For the construction of the strata $s = 1, 2, 3, 4$, a stratification by *propensity score* was used following Rosenbaum and Rubin (1983). The propensity score is defined as the probability of being assigned to a particular treatment, given a vector of concomitant variables (i.e. the confounding factors). This score, which may also have a prognostic interpretation, summarizes all the information required to balance the distribution of confounding variables between treatment groups, in order to have strata with homogeneous units. In this study, the propensity score was evaluated by a logistic model (see Arboretti et al., 1999, 2000).

Four variables were taken into consideration: death (D) and clinical resolution (R), both binary variables with 1 denoting 'yes' and 0 'no'; duration of treatment (U), a binary variable with 1 denoting 'more than 15 days' and 0 denoting '15 days or less'; and length of stay (L), a positive integer variable. In the analysis, group 1 contains all subjects treated with a specific therapy and group 2 those treated with an empirical therapy. The multidimensional system of overall hypotheses defining the causal pattern can be written, and correspondingly analysed, either within strata with respect to variables, as

$$H_0 : \left\{\bigcap_s \left[(D_{1s} \stackrel{d}{=} D_{2s}) \bigcap (R_{1s} \stackrel{d}{=} R_{2s}) \bigcap (U_{1s} \stackrel{d}{=} U_{2s}) \bigcap (L_{1s} \stackrel{d}{=} L_{2s})\right]\right\},$$

against

$$H_1 : \left\{\bigcup_s \left[(D_{1s} \stackrel{d}{>} D_{2s}) \bigcup (R_{1s} \stackrel{d}{<} R_{2s}) \bigcup (U_{1s} \stackrel{d}{>} U_{2s}) \bigcup (L_{1s} \stackrel{d}{>} L_{2s})\right]\right\},$$

or within variables with respect to strata, as

$$H_0 : \left\{\left[\bigcap_s (D_{1s} \stackrel{d}{=} D_{2s})\right] \bigcap \left[\bigcap_s (R_{1s} \stackrel{d}{=} R_{2s})\right] \bigcap \left[\bigcap_s (U_{1s} \stackrel{d}{=} U_{2s})\right] \bigcap \left[\bigcap_s (L_{1s} \stackrel{d}{=} L_{2s})\right]\right\}$$

against

$$H_1 : \left\{\left[\bigcup_s (D_{1s} \stackrel{d}{>} D_{2s})\right] \bigcup \left[\bigcup_s (R_{1s} \stackrel{d}{<} R_{2s})\right] \bigcup \left[\bigcup_s (U_{1s} \stackrel{d}{>} U_{2s})\right] \bigcup \left[\bigcup_s (L_{1s} \stackrel{d}{>} L_{2s})\right]\right\}.$$

We note that in each stratum this problem presents three binary variables and one quantitative; moreover, all sub-alternatives are one-sided, three in a positive direction and one negative. Variables U and L present some missing values which may be missing not completely at random. However, in this respect, our analysis is performed conditionally since we are mainly interested in the direct effects of two treatments.

As regards the analysis, this is carried out first within each stratum and variable and then between strata or between variables. As variables are either binary or quantitative and there are missing values (see Chapter 7), all partial tests have the form

$$T^*_{hs} = \varphi_h \left(\sum_i X^*_{h1si} \cdot \gamma^*_{h1s} - \sum_i X^*_{h2si} \cdot \gamma^*_{h2s}\right),$$

$s = 1, \ldots, 4,\ h = D, R, U, L$, where $\gamma^*_{hjs} = (\nu^*_{hks}/\nu^*_{hjs}),\ k \neq j = 1, 2$, and the function $\varphi_h(\cdot)$ is $-(\cdot)$ or $+(\cdot)$ according to whether the hth sub-alternative is $\stackrel{d}{<}$ or $\stackrel{d}{>}$.

This analysis was carried out using MATLAB, R, NPC Test R10, and SAS 9.0, all with $B =$ 10 000 CMC iterations.

12.3.2 Analysing SETIG Data Using MATLAB

We refer the reader to the book's website for MATLAB code. The results of raw and adjusted partial tests are reported in Tables 12.5 and 12.6.

Table 12.7 displays the results of the NPC obtained using Fisher's combining function to combine second-order p-values for the four variables within strata. The third-order combined p-value, again using Fisher's combining function, for the overall hypothesis on second-order p-values gives $\hat{\lambda}''' = 0.0000$.

Table 12.5 Results of partial raw tests on the SETIG data

Stratum	D	R	U	L
1	0.0241	0.0520	0.4597	0.2162
2	0.2161	0.0768	0.1548	0.0968
3	0.4028	0.3279	0.8305	0.5841
4	0.0044	0.0056	0.0331	0.0080

Table 12.6 Results of partial adjusted tests on the SETIG data

Stratum	D	R	U	L
1	0.0565	0.1384	0.4597	0.2984
2	0.2335	0.1997	0.2335	0.2075
3	0.6763	0.6763	0.8305	0.7409
4	0.0114	0.0117	0.0331	0.0154

Table 12.7 Results of within-strata analysis

Stratum	$\hat{\lambda}''_s$
1	0.0243
2	0.0499
3	0.5655
4	0.0000

The results of our analysis are self-evident. However, from the partial p-values we can see that the subjects of the third stratum react substantially in the same way to both antibiotic treatments. This may induce researchers to study the problem in greater depth, possibly by means of a different analysis, or by collecting more individuals, or by considering more informative concomitant variables for use in determining the propensity score, etc.

Investigation of results with regard to variable L may also suggest a more in-depth analysis, possibly by taking into consideration a form of competing behaviour of the death variable with respect to length of stay in hospital. In fact, given that dead individuals leave hospital before treatment is completed, their length of stay is shorter than it would otherwise have been. If empirically treated subjects have a higher death rate than specifically treated subjects, the latter may have a greater mean value than the former when comparing the L of two groups. In other words, death behaves as a confounding variable for length of stay.

As a general comment we may say that the joint use of multivariate permutation tests under order restrictions and stratification methods to control confounding factors gives us an effective tool to test for possible significance of quite a complex multivariate causal pattern in both experimental and observational studies, provided that in the null hypothesis the exchangeability of data with respect to groups is assumed. We note that the exchangeability condition is quite natural in experimental studies, but it is also generally easy to justify in most observational studies – especially when units enrolled in the study are homogeneous, at least within strata, with respect to the most important covariates.

The NPC approach also allows us to analyse the causal pattern: (a) at the univariate level by means of partial tests; (b) at the multivariate level, either within strata or within variables, by means of second-order combined tests; and (c) at the global level by means of a third order of combination. Hence, researchers are able to examine the contribution of each component variable to the possible global significance. This is not possible with Rosenbaum's (2002) POSET method, which focuses only on the possible global significance. Furthermore, the NPC approach is much more flexible when dealing with either ignorable or non-ignorable missing values.

12.3.3 Analysing the SETIG Data Using R

We consider the test statistic for difference in means which takes into account missing observations (described in Chapter 7), and which requires us to compute the number of non-missing observations ν_1^* and ν_2^* in the two samples, at each permutation. Here `O` is and indicator vector of the non-missing observations, and `nu` is the vector with the observed number of non-missing observations in the two samples.

```
setwd("C:\\path")
data<-read.csv("setig.csv",header=TRUE,na.strings='NaN')
alternative = c(1,-1,1,1)

strata = data[,1] ; ID = data[,2]
data = data[,-c(1,2)]
p=dim(data)[2] ; nstr = length(table(strata))

B=10000
T=array(0,dim=c((B+1),p,nstr))

for(s in 1:nstr){

Y=data[strata==s,]

for(j in 1:p){

paz = ID[strata==s]
O = ifelse(is.na(Y[,j])==TRUE,0,1)
y = ifelse(is.na(Y[,j])==TRUE,0,Y[,j])
nu = table(O,paz)
if(dim(nu)[1]>1){nu=nu[2,]}

D1 = sum(y[paz==1]) ;  D2 = sum(y[paz==2])
T[1,j,s] = D1*sqrt(nu[2]/nu[1])-D2*sqrt(nu[1]/nu[2])

}## end j
}## end s

for(bb in 2:(B+1)){
```

```
for(s in 1:nstr){

Y=data[strata==s,]
n=dim(Y)[1]

Y.star = Y[sample(1:n),]

for(j in 1:p){

paz = ID[strata==s]
O = ifelse(is.na(Y.star[,j])==TRUE,0,1)
y = ifelse(is.na(Y.star[,j])==TRUE,0,Y.star[,j])
nu = table(O,paz)
if(dim(nu)[1]>1){nu=nu[2,]}

D1 = sum(y[paz==1]) ;  D2 = sum(y[paz==2])

T[bb,j,s] =  D1*sqrt(nu[2]/nu[1])-D2*sqrt(nu[1]/nu[2])

if(T[bb,j,s]=='NaN'){T[bb,j,s] = Inf}
}## end j
}## end s
print(bb)
}## end bb

source("t2p.r")
T1=T

for(j in 1:4){
T1[,j,]=T[,j,]*alternative[j]
}

P=t2p(T1)
P=ifelse(is.na(P)==TRUE,1,P)

res = t(P[1,,])
colnames(res) = colnames(data)
rownames(res) = seq(1,4)
res

       D      R      U      L
1 0.0241 0.0483 0.4504 0.2099
2 0.2180 0.0733 0.1620 0.0940
3 0.4139 0.3230 0.8362 0.5896
4 0.0030 0.0044 0.0392 0.0062
```

Some problems may be arise when testing the variable R on the first stratum: here there are 54 missing observations out of 83; moreover, there are 15 patients treated with empirical therapy in the first stratum, therefore it might happen that some permutations lead to obtain missing observations only in the first group, and therefore crude $\sqrt{\nu_2^*/\nu_1^*}$ coefficients do not exist and $S_1^* = 0$. This apparent drawback is solved by simply ignoring all elements of the partition of the permutation sample space $\mathcal{X}_{/\mathbf{X}}$ (see Chapter 7) with $\min(\nu_1^*, \nu_2^*) < 3$ because all subdistributions have at least the same mean and variance.

The matrix `res` contains the partial p-values related to each variable in each stratum. Significant differences in the two therapies can be found especially in the first and fourth strata. In order to obtain an adjustment of p-values for multiplicity, we run the `FWE.minP` function independently stratum by stratum:

```
source("FWEminP.r")

p.fwe = array(0,dim=c(nstr,p))

for(s in 1:nstr){
p.fwe[s,] = FWE.minP(P[,,s])
}
colnames(p.fwe) = colnames(res)
rownames(p.fwe) = rownames(res)
p.fwe

       D      R      U      L
1 0.0572 0.1239 0.4504 0.2940
2 0.2180 0.1905 0.2071 0.2071
3 0.7300 0.6770 0.8362 0.7458
4 0.0069 0.0099 0.0392 0.0099
```

We now combine the partial tests within each stratum with Fisher's combining function.

```
T.glob.s = array(0,dim=c((B+1),s))

for(k in 1:nstr){
T.glob.s[,k] = apply(P[,,k],1,function(x){-2*log(prod(x))})
}

P.glob.s = t2p(T.glob.s)
res.str = P.glob.s[1,]
names(res.str) = paste("Stratum",seq(1,s),sep=" ")
res.str

Stratum 1 Stratum 2 Stratum 3 Stratum 4
   0.0228    0.0496    0.5801    0.0000
```

Finally, to obtain a unique global test we can compare the partial tests on each stratum:

```
T.glob = apply(P.glob.s,1,function(x){-2*log(prod(x))})
p.glob = t2p(T.glob)[1]
p.glob
```

12.3.4 *Analysing the SETIG Data Using NPC Test*

The previous analysis was also carried out with the NPC Test R10 software. On the basis of the outcomes previously considered, we define a binary group variable which denotes the two groups of patients to be compared. The 334 patients are stratified into four homogenous strata by defining a stratification variable (taking values $s = 1, 2, 3, 4$) as obtained using the propensity score technique. The four response variables of interest are death (D), clinical resolution (R), duration of treatment (U), and length of stay (L).

Hence, in each stratum we have one quantitative and three binary variables. Additionally, variables U and L have some missing values which may be missing not completely at random. Figure 12.7 displays part of the data set from the NPC Test worksheet, where missing values are denoted by '?'. At this stage, the user is asked to specify the *type of analysis*: two-sample testing, C-sample testing or one-way ANOVA, C related samples, stochastic ordering or correlation analysis.

NPC Test allows the user to constructively specify the testing analysis in full, from the specification of the system of hypotheses of interest to the construction of partial and combined tests. This is achieved by a sequence of five dialogue boxes. Once the type of analysis has been selected, the user starts constructing the hypotheses to test from the definition of samples and strata in the first dialogue box. Figure 12.8 displays the first dialogue box for the SETIG case study. In the present instance, we select two-sample testing.

The second dialogue box (see Figure 12.9) allows the user to specify the multidimensional system of hypotheses for each variable within each stratum, to choose the appropriate test statistic and to set the number of CMC iterations. In the SETIG study, the system of hypotheses of interest, within each stratum, is

$$H_{0s} : \left\{ (D_{1s} \stackrel{d}{=} D_{2s}) \bigcap (R_{1s} \stackrel{d}{=} R_{2s}) \bigcap (U_{1s} \stackrel{d}{=} U_{2s}) \bigcap (L_{1s} \stackrel{d}{=} L_{2s}) \right\}, \quad s = 1, \ldots, 4,$$

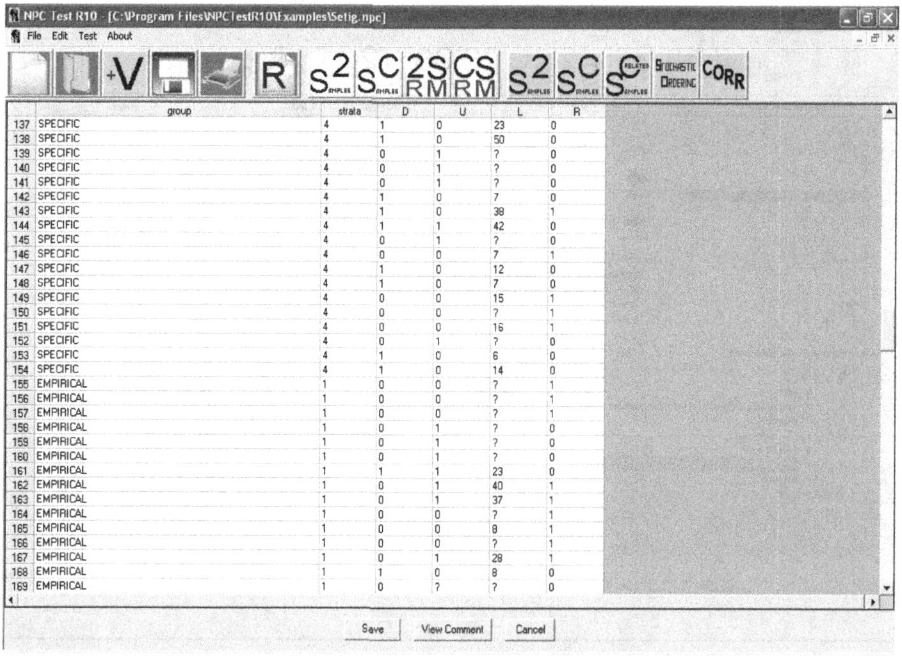

	group	strata	D	U	L	R
137	SPECIFIC	4	1	0	23	0
138	SPECIFIC	4	1	0	50	0
139	SPECIFIC	4	0	1	?	0
140	SPECIFIC	4	0	1	?	0
141	SPECIFIC	4	0	1	?	0
142	SPECIFIC	4	1	0	7	0
143	SPECIFIC	4	1	0	38	1
144	SPECIFIC	4	1	1	42	0
145	SPECIFIC	4	0	1	?	0
146	SPECIFIC	4	0	0	7	1
147	SPECIFIC	4	1	0	12	0
148	SPECIFIC	4	1	0	7	0
149	SPECIFIC	4	0	0	15	1
150	SPECIFIC	4	0	0	?	1
151	SPECIFIC	4	0	0	16	1
152	SPECIFIC	4	0	1	?	0
153	SPECIFIC	4	1	0	6	0
154	SPECIFIC	4	1	0	14	0
155	EMPIRICAL	1	0	0	?	1
156	EMPIRICAL	1	0	0	?	1
157	EMPIRICAL	1	0	0	?	1
158	EMPIRICAL	1	0	1	?	0
159	EMPIRICAL	1	0	1	?	0
160	EMPIRICAL	1	0	1	?	0
161	EMPIRICAL	1	1	1	23	0
162	EMPIRICAL	1	0	1	40	1
163	EMPIRICAL	1	0	1	37	1
164	EMPIRICAL	1	0	0	?	1
165	EMPIRICAL	1	0	0	8	1
166	EMPIRICAL	1	0	0	?	1
167	EMPIRICAL	1	0	1	28	1
168	EMPIRICAL	1	1	0	8	0
169	EMPIRICAL	1	0	?	?	0

Figure 12.7 NPC Test R10 worksheet with SETIG data

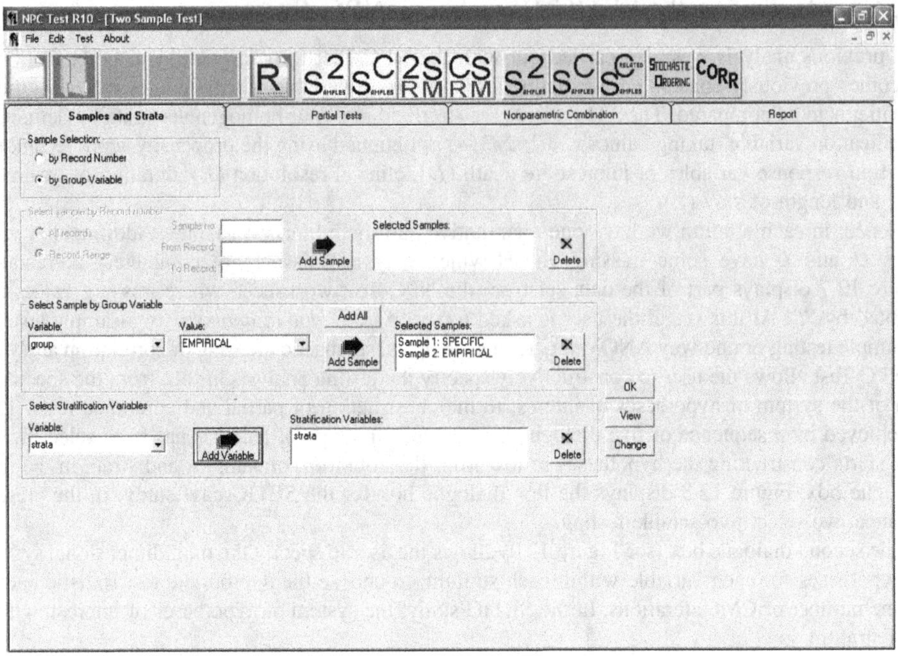

Figure 12.8 NPC Test R10 samples and strata dialogue box

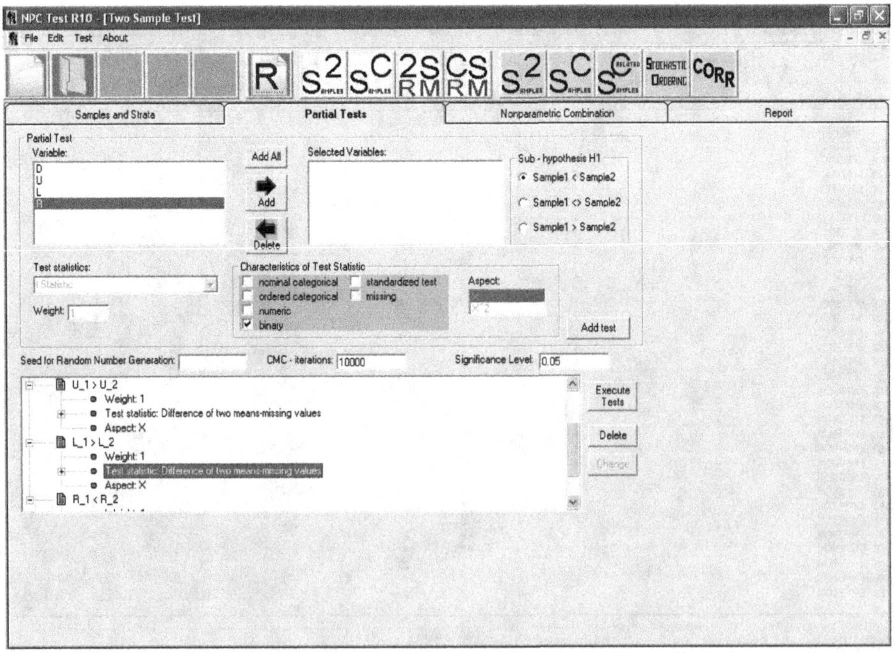

Figure 12.9 NPC Test R10 partial tests dialogue box

against

$$H_{1s}: \left\{(D_{1s} \overset{d}{>} D_{2s}) \bigcup (R_{1s} \overset{d}{<} R_{2s}) \bigcup (U_{1s} \overset{d}{>} U_{2s}) \bigcup (L_{1s} \overset{d}{>} L_{2s})\right\}, \quad s = 1, \ldots, 4.$$

For variables D and R, suitable test statistics are the *difference of two means* (or alternatively a t statistic). For variables L and U, on the other hand, we select the permutation test statistic for missing values, as discussed in Chapter 7.

Once all partial tests for each stratum have been computed, in the third dialogue box we wish to perform the closed testing procedure within each stratum. This is done by selecting the four variables of each stratum and then clicking the NPC_FWE button (we have chosen the Tippett combination function and the 'by Selection' option in the 'Combination of Tests' box). Figure 12.10 displays the results of the closed testing procedures and in particular the result of the first of the 16 adjusted tests.

Finally, once all partial tests for each stratum have been computed and adjusted within strata, in the same dialogue box the user can choose the combining function and accordingly the multivariate and multistrata permutation tests to perform. There are two main default choices: combination of partial tests *within strata*, for

$$H_0: \left\{\bigcap_s \left[(D_{1s} \overset{d}{=} D_{2s}) \bigcap (R_{1s} \overset{d}{=} R_{2s}) \bigcap (U_{1s} \overset{d}{=} U_{2s}) \bigcap (L_{1s} \overset{d}{=} L_{2s})\right]\right\}$$

against

$$H_1: \left\{\bigcup_s \left[(D_{1s} \overset{d}{>} D_{2s}) \bigcup (R_{1s} \overset{d}{<} R_{2s}) \bigcup (U_{1s} \overset{d}{>} U_{2s}) \bigcup (L_{1s} \overset{d}{>} L_{2s})\right]\right\},$$

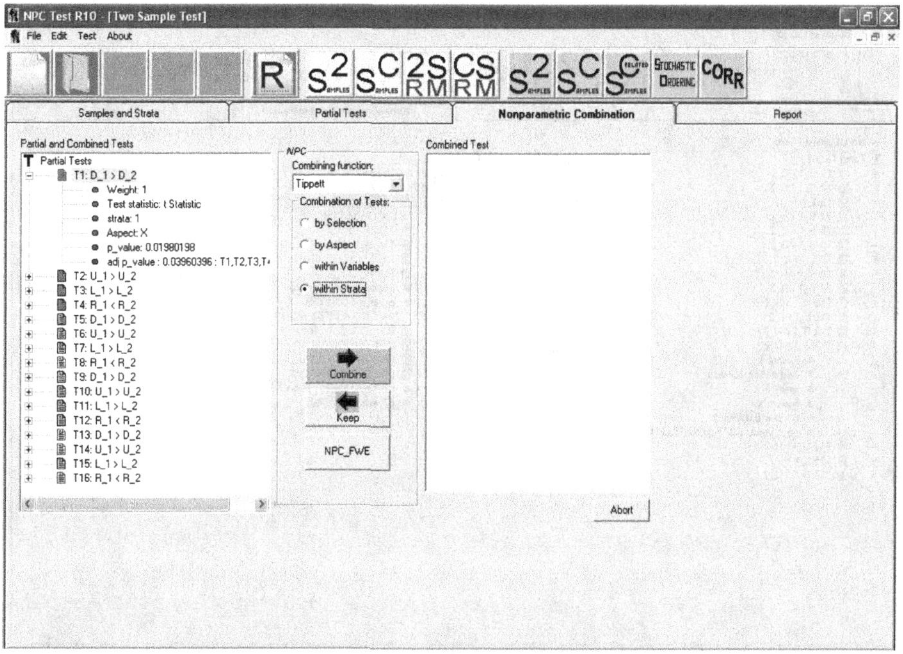

Figure 12.10 NPC Test R10 combination worksheet

and combination of partial tests *within variables* with respect to strata, for

$$H_0 : \left\{\left[\bigcap_s (D_{1s} \stackrel{d}{=} D_{2s})\right] \bigcap \left[\bigcap_s (R_{1s} \stackrel{d}{=} R_{2s})\right] \bigcap \left[\bigcap_s (U_{1s} \stackrel{d}{=} U_{2s})\right] \bigcap \left[\bigcap_s (L_{1s} \stackrel{d}{=} L_{2s})\right]\right\}$$

against

$$H_1 : \left\{\left[\bigcup_s (D_{1s} \stackrel{d}{>} D_{2s})\right] \bigcup \left[\bigcup_s (R_{1s} \stackrel{d}{<} R_{2s})\right] \bigcup \left[\bigcup_s (U_{1s} \stackrel{d}{>} U_{2s})\right] \bigcup \left[\bigcup_s (L_{1s} \stackrel{d}{>} L_{2s})\right]\right\}.$$

Figure 12.11 displays the option *test combination within strata*, using the Fisher combining function. Of course, the user can also construct any *test combination* using the option 'by Selection' in the 'Combination of Tests' box.

Finally, a fourth dialogue box allows the user to generate a report (Figure 12.12), which contains the descriptive statistics (which are automatically generated) and test results for the data examined. The user can add comments to the report if required.

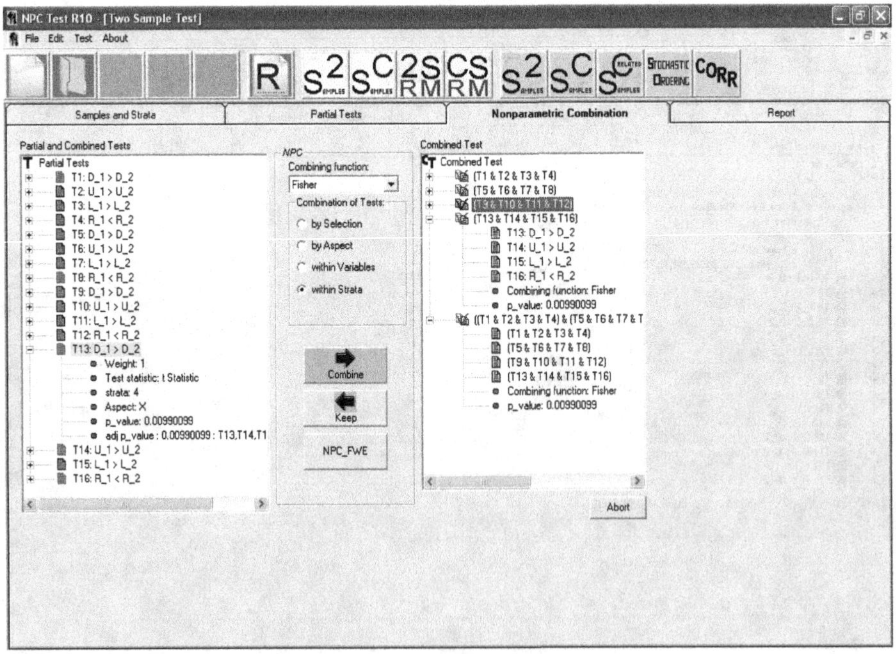

Figure 12.11 NPC Test R10 within-strata test combination worksheet

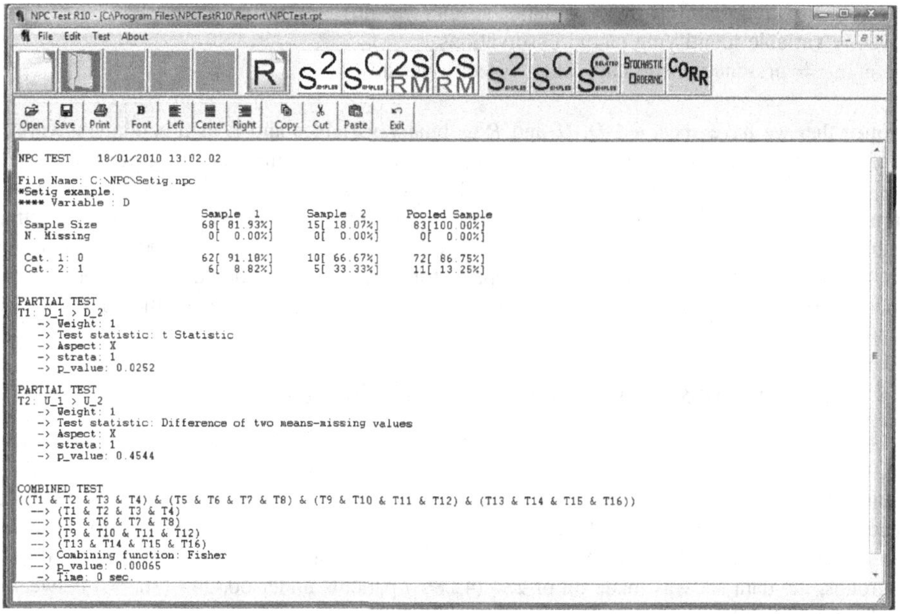

Figure 12.12 NPC Test R10 report worksheet

12.3.5 Analysis of the SETIG Data Using SAS

In order to carry out the analysis in SAS, the `NPC_2samples` macro available from the book's website can be used along with PROC MULTTEST to compute the closed testing procedure within strata, using the exact permutation Min$P$ tests (Westfall and Young, 1993). The main instructions to run the macro are as follows:

```
filename mac_npc '...\npc.sas';
%include mac_npc;

%npc(data=setig, var_byn= D U R, var_con=L,
dom_byn=great great less, dom_con=great,
clas=group, nsample=10000, strato=strata, paired=no, missing=yes);
```

The following input parameters are required:

- `data` = name of the data set;
- `var_byn` = list of binary variables;
- `var_cat` = list of categorical variables, non-binary;
- `var_con` = list of continuous variables;
- `dom_byn` = list of directional marginal sub-hypotheses for binary variables – if $X_{Au} < X_{Bu}$, specify LESS; if $X_{Au} > X_{Bu}$, specify GREAT; if $X_{Au} \neq X_{Bu}$, specify NOTEQ;
- `dom_con` = list of directional marginal sub-hypotheses for continuous variables, see above;
- `weights` = list of weights for the variables – first specify weights for binary variables, then weights for categorical variables and for continuous variables;
- `clas` = variable defining the two groups (character variable);
- `nsample` = number of conditional resamplings;
- `strato` = variable defining strata (character variable);

- `paired` = paired data (character variable, specify yes/no);
- `unit` = variable identifying paired observations;
- `missing` = presence of missing values (specify yes/no).

Notice that we have specified *D*, *U* and *R* as binary variables and *L* as a continuous variable. Furthermore, we have considered *Group* and *Strata* as variables defining the two groups and the four strata, respectively. In order to specify the multidimensional system of hypotheses for each variable within each stratum, a left-tailed test for variable *R* and three right-tailed tests for the other three variables were considered. This analysis was carried out using `nsample`= 10 000 CMC iterations. Since we are dealing with missing data, the option for missing values should be activated. The data set and the corresponding software codes can be found in the `setig` folder on the book's website.

12.4 A Comparison of Three Survival Curves

In this section we extend the procedure illustrated in Chapter 9 to the case of $C > 2$ survival curves using the example presented in Section 12.1 on oesophagectomy for oesophageal cancer in elderly patients (Ruol et al., 2007). In this application, it is of interest to separate subjects under 70 years of age into two different groups. Therefore, the three groups to be compared are: (a) individuals under 60; (b) individuals between 60 and 69; and (c) individuals 70 and over. Having defined the age groups, the data set was made up of 284 (42.8%) patients under 60, 245 (36.9%) between 60 and 69, and 135 (20.3%) aged 70 or over.

Remember that the aim of the study was to determine whether or not the three survival time distributions arose from a common survival function based on the three independent samples. Therefore, the hypothesis of interest concerns the effects of patient ageing with respect to long-term survival for oesophageal cancer patients undergoing oesophagectomy. The case study is characterized by a fairly large number of observations with a light to moderate equal censoring model and, as can be seen in Figure 12.13, a crossing hazard rates pattern is clearly visible.

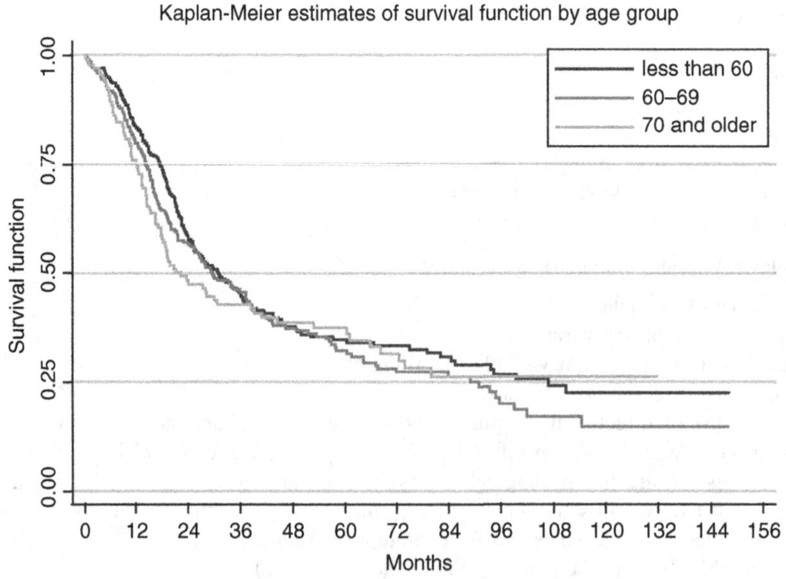

Figure 12.13 Kaplan–Meier survival estimates by group

First, we carry out the survival analysis based on the whole sample. Next, we stratify the sample, using the propensity score, as a toy example. Then, we perform the multidimensional nonparametric permutation test within strata.

For the present case study and using the same notation as in Chapter 9, the system of hypotheses of interest is

$$H_0^G : \left\{[S_1(t_i) = S_2(t_i) = S_3(t_i) = S(t_i)\,\forall t_i,\ i = 1, \ldots, D]\bigcap\left[\Delta_1 \stackrel{d}{=} \Delta_2 \stackrel{d}{=} \Delta_3 = 0\right]\right\}$$

$$= \left\{\left[X_1 \stackrel{d}{=} X_2 \stackrel{d}{=} X_3\right]\bigcap\left[\Delta_1 \stackrel{d}{=} \Delta_2 \stackrel{d}{=} \Delta_3 = 0\right]\right\}$$

against the overall alternative

$$H_1^G : \left\{\text{at least one of the three } S_j(t_i) \text{ is different for some } t_i\right\}$$

$$= \left\{\text{at least one of the three } X_j \text{ arises from a different distribution}\right\}.$$

The global null and alternative hypotheses are broken down into D sub-hypotheses, one for each of the D times $t_1 < \ldots < t_D$,

$$H_0^G = H_0^{\mathbf{X}|\mathbf{O}} : \left\{\bigcap_{i=1}^{D}\left(X_{1i} \stackrel{d}{=} X_{2i} \stackrel{d}{=} X_{3i}\right)|\mathbf{O}\right\} = \left\{\bigcap_{i=1}^{D} H_{0i}^{\mathbf{X}|\mathbf{O}}\right\}$$

against

$$H_1^G : \left\{\bigcup_{i=1}^{D} H_{1i}^{\mathbf{X}|\mathbf{O}}\right\}.$$

Hence, in the notation of Chapter 9, the partial permutation test statistics for testing the sub-hypothesis $H_{0i}^{\mathbf{X}|\mathbf{O}}$ against the sub-alternative $H_{1i}^{\mathbf{X}|\mathbf{O}}$ take the form

$$\Gamma_i^{\mathbf{X}|\mathbf{O}*} = S_2^*(t_i)\sqrt{\frac{\nu_{1i}^*}{\nu_{2i}^*}} - S_1^*(t_i)\sqrt{\frac{\nu_{2i}^*}{\nu_{1i}^*}},$$

after which an NPC follows with a suitable combining function.

12.4.1 Unstratified Survival Analysis

TIC-NPC is the multidimensional permutation test to use when there is equal censoring (Callegaro et al., 2003). After carrying out a two-stage NPC procedure using Fisher's combining function to combine the partial p-values for the D observed failure times, we obtained a global p-value, for the overall hypothesis, of 0.1900. This suggests there are no differences between the three age groups in the distribution of the time to death for oesophageal cancer patients undergoing oesophagectomy. This analysis was carried out using NPC Test R10, SAS 9.0 and MATLAB with $B = 10\,000$ CMC iterations.

12.4.2 Survival Analysis with Stratification by Propensity Score

In this three-sample case, the estimation of the propensity score $\widehat{e}(A)$ is carried out using an asymptotic ordinal logistic regression model. As in the two-sample application, we partitioned the whole sample into five strata based on the values of this estimated propensity score. Figures 12.14–12.18 present the survival curves for the three samples within each stratum.

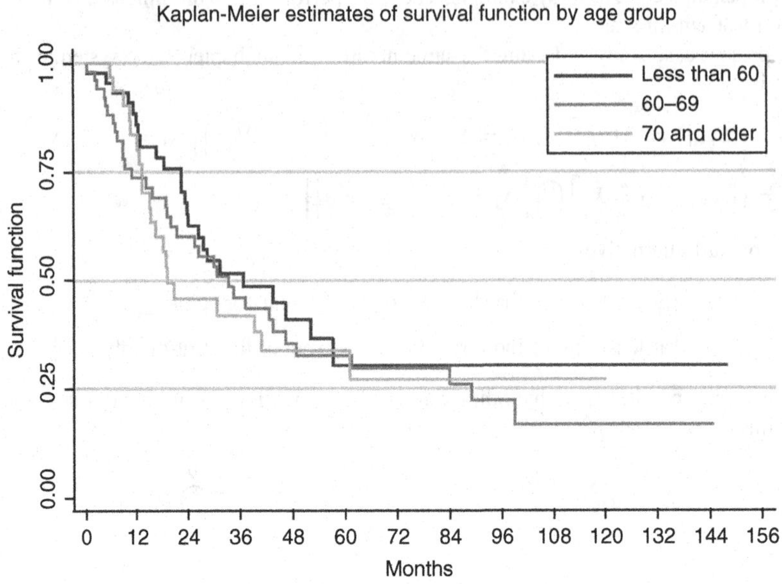

Figure 12.14 Kaplan–Meier survival estimates by group in stratum 1

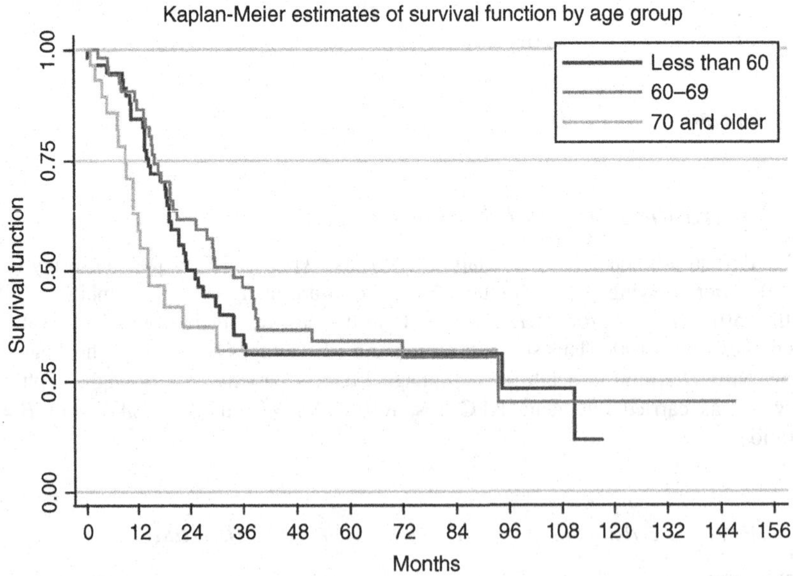

Figure 12.15 Kaplan–Meier survival estimates by group in stratum 2

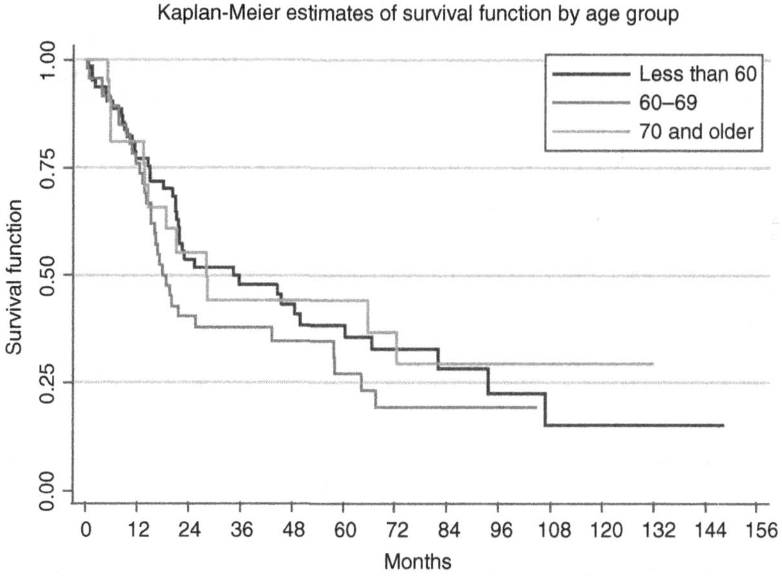

Figure 12.16 Kaplan–Meier survival estimates by group in stratum 3

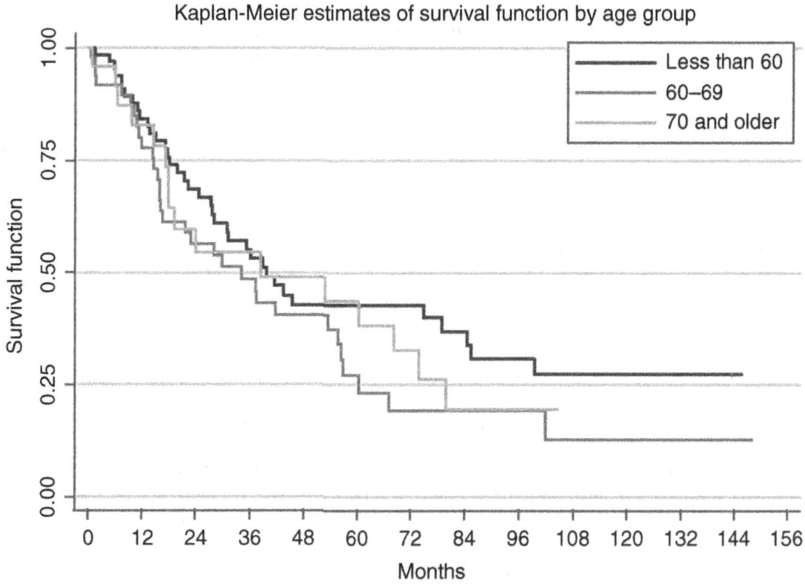

Figure 12.17 Kaplan–Meier survival estimates by group in stratum 4

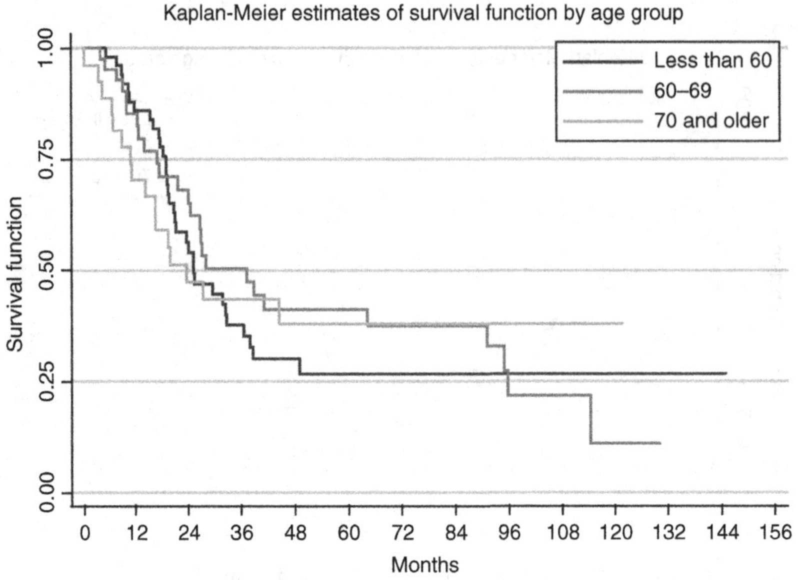

Figure 12.18 Kaplan–Meier survival estimates by group in stratum 5

In this case there is a more complex data configuration because in our hypothesis testing problem we also have to consider the stratification variable, introduced by the estimated propensity score in order to remove or control the possible biases due to the observed and known covariates used to construct the score.

The NPC may therefore be like a multi-phase procedure characterized by an intermediate combination, where we first combine the partial tests with respect to D times within each stratum $s = 1, \ldots, 5$, and then combine the second-order tests with respect to strata into a single third-order combined test. It should be noted that this multitesting and multistrata hypothesis testing problem cannot be solved using a standard asymptotic test procedure such as the weighted Kaplan–Meier test.

The overall hypotheses of interest remain

$$H_0^G : \left\{[S_1(t_i) = S_2(t_i) = S_3(t_i) = S(t_i) \ \forall t_i, \ i = 1, \ldots, D] \bigcap \left[\Delta_1 \overset{d}{=} \Delta_2 \overset{d}{=} \Delta_3 \overset{d}{=} 0\right]\right\}$$
$$= \left\{\left[X_1 \overset{d}{=} X_2 \overset{d}{=} X_3\right] \bigcap \left[\Delta_1 \overset{d}{=} \Delta_2 \overset{d}{=} \Delta_3 \overset{d}{=} 0\right]\right\},$$

against

$$H_1^G : \left\{\text{at least one of the three } S_j(t_i) \text{ is different for some } t_i\right\}$$
$$= \left\{\text{at least one of the three } X_j \text{ arises from a different distribution}\right\}.$$

However, now the global null and alternative hypotheses are broken down into $D \times S$ sub-hypotheses, one for each of the D times $t_1 < \ldots < t_D$ and for each stratum

$$H_0^G = H_0^{\mathbf{X}|\mathbf{O}} : \left\{\bigcap_{s=1}^{S} \left[\bigcap_{i=1}^{D} \left(X_{1is} \overset{d}{=} X_{2is} \overset{d}{=} X_{3is}\right) |\mathbf{O}\right]\right\} = \left\{\bigcap_{s} \left[\bigcap_{i} H_{0is}^{\mathbf{X}|\mathbf{O}}\right]\right\}$$

Table 12.8 Survival analysis results using the TIC-NPC procedure

p-values	Stratum					Global
	1	2	3	4	5	
Two-sided alternative	0.143	0.024	0.410	0.567	0.109	0.034

against

$$H_1^G : \left\{ \bigcup_s \left[\bigcup_i H_{1is}^{\mathbf{X}|\mathbf{O}} \right] \right\}.$$

Here again we use the TIC-NPC test as there is treatment-independent censoring (Callegaro et al., 2003). Table 12.8 displays the results of the three-phase NPC procedure using Fisher's combining function to combine the $D \times S$ partial p-values.

The data set and the corresponding software codes can be found in the `survival` folder on the book's website.

12.5 Survival Analysis Using NPC Test and SAS

We defined a categorical variable, *group*, to represent the three groups of patients to be compared – 0 being 'patients under 60 years of age'; 1 being 'patients aged between 60 and 69'; and 2 being 'patients aged 70 and over'. The 664 patients were stratified into five homogenous strata by defining the *strata* variable (taking values $s = 1, \ldots, 5$) as obtained by the estimated propensity score procedure. The response variables of interest are the D binary variables $V_{mji}(t_i)$ as previously described. Recall that all the variables are binary variables with or without missing values. This analysis was carried out using `nsample` = 10 000 CMC iterations.

12.5.1 Survival Analysis Using NPC Test

For the present case study, we select the C-sample testing (also known as one-way ANOVA) analysis. We start by constructing the hypotheses to test using the definition of samples and strata (the latter necessary only in the case of the stratified analysis). Then, we specify the multidimensional system of hypotheses to choose the appropriate test statistics and to set the number of CMC iterations. Figure 12.19 shows the dialogue box used to select all the necessary single components of the NPC procedure. For binary variables without missing values, a suitable test statistic is the *difference of two means* (or alternatively a t statistic), while for variables with missing data we select the permutation test statistic for missing values discussed in Chapter 7.

With respect to the unstratified analysis, once all partial tests have been selected and computed in the dialogue box in Figure 12.19, we combine all the D partial p-values by choosing the combining function and, accordingly, the multivariate permutation tests to perform. The NPC procedure is performed by clicking on the 'Combine' button (in this case using Fisher's combining function and choosing the combination of partial tests by clicking on the 'by Selection' option in the 'Combination of Tests' box). Figure 12.20 displays the NPC of all the D binary variables, using Fisher's combination function.

With respect to the stratified analysis, once all partial tests for each stratum have been selected and computed in the dialogue box in Figure 12.21, we combine all the D partial p-values within strata (second-order p-values) and finally obtain the global (third-order) p-value. The NPC procedure is carried out by clicking on the 'Combine' button (in this case using Fisher's combining function

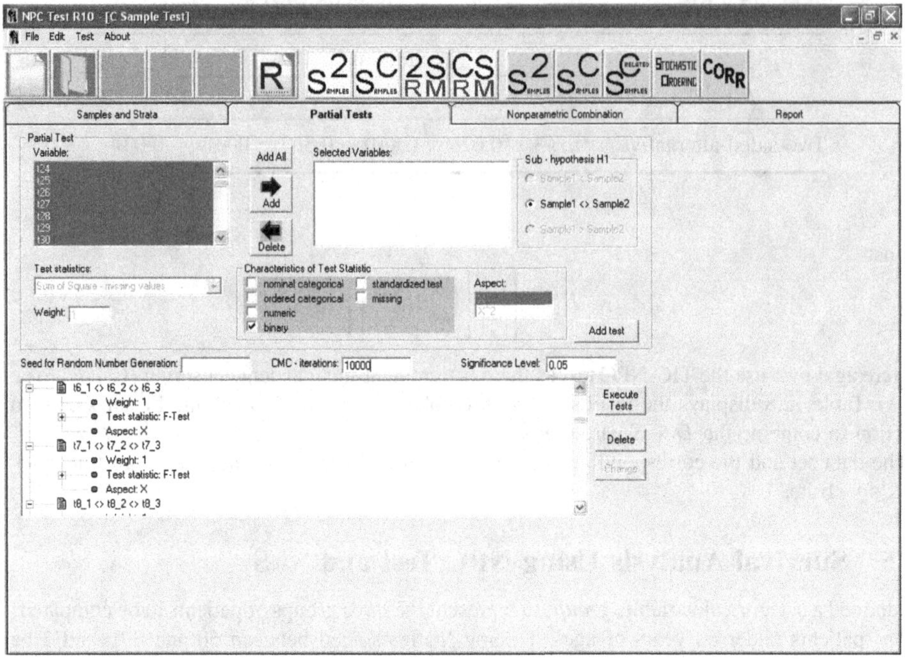

Figure 12.19 NPC Test R10 partial tests dialogue box

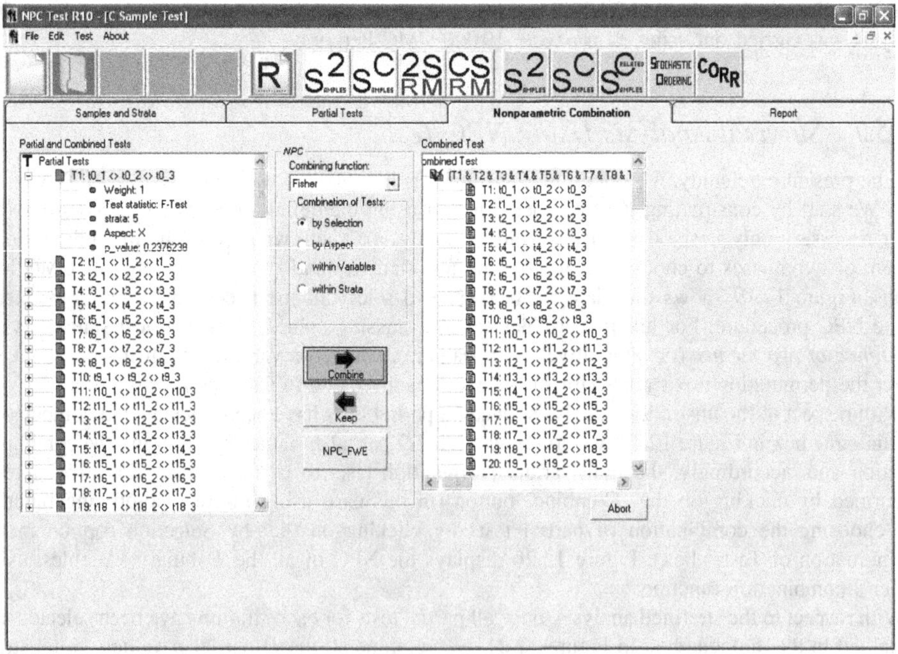

Figure 12.20 NPC Test R10 combination worksheet (by selection)

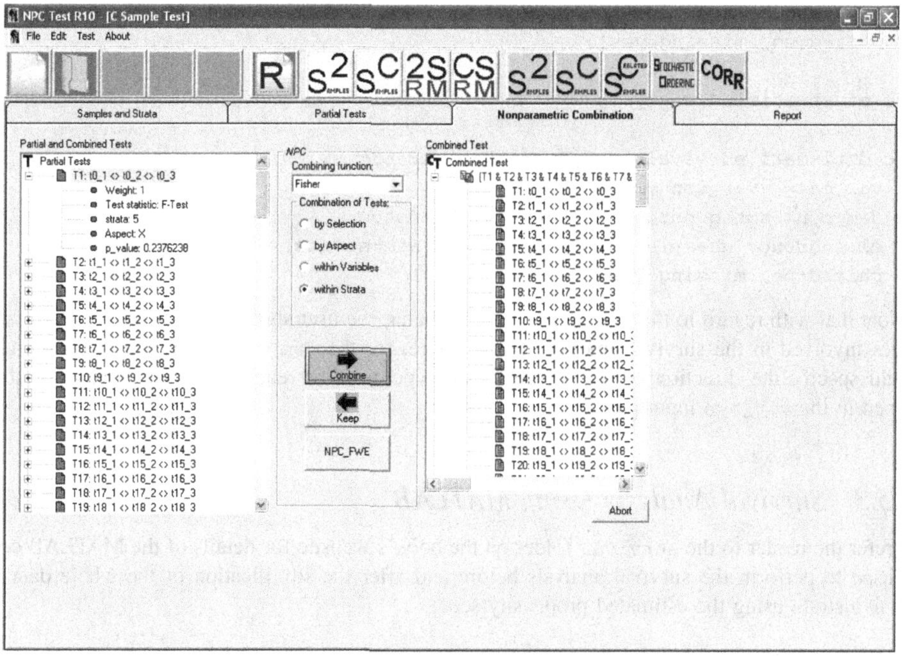

Figure 12.21 NPC Test R10 combination worksheet (within-strata)

and choosing the combination of partial tests within strata by clicking on the 'within Strata' option in the 'Combination of Tests' box), for

$$H_0 : \left\{ \bigcap_{s=1}^{S} \left[\bigcap_{i=1}^{D} H_{0is}^{\mathbf{X}|\mathbf{O}} \right] \right\}$$

against

$$H_1 : \left\{ \bigcup_{s} \left[\bigcup_{i} H_{1is}^{\mathbf{X}|\mathbf{O}} \right] \right\}.$$

12.5.2 Survival Analysis Using SAS

In order to carry out the analysis in SAS 9.0, the code available from the book's website can be used. The main instructions for running the `NPC_Csamples` macro are as follows:

```
filename mac_npc '...\npc.sas';
%include mac_npc;

case of unstratified analysis

%npc(dati=dati.survival, var_byn= t0 t1 t2 t3 ... tD,
    var_cat=, var_con=,
    dom_byn= noteq noteq noteq noteq ... noteq,
```

```
    clas=group, nsample=10000, unit=, strato=,
    paired=no, missing=yes);

case of stratified analysis

%npc(dati=dati.survival, var_byn= t0 t1 t2 t3 ... tD,
    var_cat=, var_con=,
    dom_byn= noteq noteq noteq noteq ... noteq,
    clas=group, nsample=10000, unit=, strato=strata,
    paired=no, missing=yes);
```

Note that with regard to the `var_byn` input parameter, the user should include all the D variable names involved in the survival analysis. Furthermore, for the `dom_byn` input parameter, the user should specify the direction of the marginal sub-hypothesis for each of the D binary variables defined in the `var_byn` input parameter.

12.5.3 Survival Analysis Using MATLAB

We refer the reader to the `survival` folder on the book's website for details of the MATLAB code we used to perform the survival analysis before and after the stratification of the whole data set into five strata using the estimated propensity score.

12.6 Logistic Regression and NPC Test for Multivariate Analysis

In the framework of exploratory studies, where the aim is to investigate possible significant risk factors related to the outcome of analysis, it is often of interest to indicate the relative importance of the various independent explanatory variables inspected in the analysis.

When the outcome is a binary variable, representing for instance two levels of a treatment or two groups of individuals with different characteristics, in order to find possible significant factors affecting the outcome of interest, we can perform a multivariate analysis with two alternative procedures:

(i) a multivariate logistic regression of the binary category representing the outcome, on the observed explanatory variables;
(ii) an NPC procedure of dependent componentwise tests within a two-sample design.

The NPC testing procedure may detect more possible significant effects than the usual logistic regression, hence we emphasize the importance of using both procedures when dealing with an exploratory study. Let us consider two examples.

12.6.1 Application to Lymph Node Metastases

The presence of lymph node metastases is one of the most important predictors of survival in patients with oesophageal cancer. The international classification focuses on the presence/absence of metastatic lymph nodes; however, it would be of interest to check for specific metastatic lymph node sites as they may lead to different outcomes.

The aim here was to investigate possible associations of the metastatic lymph node sites with respect to outcomes, beyond the main classification (presence/absence).

Table 12.9 Sample sizes and distribution of events of interest (DOD patients only)

Type of analysis	subjects at risk (n)	events
Only DOD patients	136	74 (54.4%)

The analysis involved a total of $n = 136$ patients with cancer of the oesophagogastric junction seen at the Department of Medical and Surgical Science, University of Padua, over an 11-year period from January 1992 to December 2002. Other conditions were: complete resection ($R0$), adenocarcinoma histotype, and no neoadjuvant therapy. Patients were routinely followed up (FU) by their surgeons 1, 3, 6 and 12 months after the operation and every 6–12 months thereafter.

Of the 136 patients, 116 (85.3%) were male and 20 (14.7%) female; 4 (2.9%) patients were less than 40 years old, 41 (30.2%) were between 40 and 60, 91 (66.9%) were over 60; 55 (40.4%) patients had negative lymph nodes and 81 (59.6%) had positive lymph nodes (see Table 12.9).

Patient outcome (at five-years of FU) was encoded as ALIVE (patient alive), DOD (died of disease), or DOC (died of other causes). The metastatic lymph node presence percentage was higher (85.1%) in DOD patients and lower (29.0%) in ALIVE patients, so a worse outcome seemed to be related to the presence of metastatic lymph nodes. In DOC patients the percentage was 58.1%, so we can say that DOC patients had no strong prevalence of presence or absence.

Therefore, the statistical analysis was focused on the study of significant risk factors related to patient survival in terms of mortality. Since the main task is related to prognostic factors with respect to oesophageal cancer, DOC patients were excluded and the five-year outcome (ALIVE against DOD) was used as a response variable. We took as risk factors the explanatory variables related to the six sites of metastatic lymph nodes (paraoesophageal, paracardial, perigastric, subcarinal, tripod, other lymph node sites) that were significant in the univariate analysis and the pathological stage (tumour stage, presence of positive lymph nodes, presence of metastatic lymph nodes), the number of lymph nodes examined and the number of metastatic lymph nodes (see Table 12.10). We took as confounding factors gender, age group, degree of tumour differentiation (well/moderate against poor differentiation) and post-therapy (yes/no).

In order to find possible significant risk factors affecting the survival of individuals at the five-year follow-up stage, we carried out a multivariate analysis with two alternative procedures:

- a multivariate logistic regression of the binary category ALIVE/DOD on the observed explanatory variables;
- the NPC testing procedure for a two-sample problem, where we compared the alive group with the group of deaths during the five-year follow-up period.

Logistic regression was carried out using SAS 9.0 and the NPC procedure was performed using NPC Test R10. The results obtained using SAS are shown in Table 12.11, while those obtained using NPC Test R10 are displayed in Table 12.12.

Logistic regression detected significant effects of paracardial lymph nodes (p-value 0.0078) and presence of positive lymph nodes (0.0056); the NPC Test (without controlling for multiplicity) detected significant effects of paraoesophageal (p-value 0.0006), paracardial (0.00015), perigastric (0.0035), tripod (0.01), other site lymph nodes (0.017), tumour stage (0.00015), presence of positive lymph nodes (0.00015), number of metastatic lymph nodes (0.00015), post-therapy (0.03), and degree of tumour differentiation (0.026) and global $p = 0.0002$.

The data set can be found in the `lymphnodes` folder on the book's website.

Table 12.10 Risk factors and coding

Risk factor	*Group coding*
Paraoesophageal	Yes; No
Paracardial	Yes; No
Perigastric	Yes; No
Subcarinal	Yes; No
Tripod	Yes; No
Other lymph node sites	Yes; No
Tumour stage	$0, 1, \ldots, 4$
Presence of positive lymph nodes	Yes; No
Presence of metastatic lymph nodes	Yes; No
No. of examined lymph nodes	
No. of metastatic lymph nodes	
Gender	Male; Female
Age group	< 40, [40, 60], > 60
Degree of tumour differentiation	well/moderate/poor
Post therapy	Yes; No

Table 12.11 Multivariate logistic regression (using SAS) of predictors of survival (ALIVE v. DOD)

Risk factor	χ^2	p-value	Odds ratio
Paraoesophageal	0.04	0.8500	
Paracardial	7.08	0.0078	0.199 (0.06–0.65)
Perigastric	1.15	0.2800	
Subcarinal	0.36	0.5500	
Tripod	0.11	0.7400	
Other lymph node sites	0.06	0.8100	
Tumour stage	4.94	0.4200	
Presence of positive lymph nodes	7.66	0.0056	0.199 (0.06–0.62)
Metastatic lymph nodes	0.34	0.5600	
No. of examined lymph nodes	0.05	0.8300	
No. of metastatic lymph nodes	1.40	0.2400	
Gender	0.14	0.7100	
Age group	0.23	0.6300	
Degree of tumour differentiation	0.39	0.5300	
Post therapy	1.20	0.2700	

12.6.2 Application to Bladder Cancer

The multicentre retrospective study discussed here refers to 1312 patients (1149 (87.6%) male and 163 (12.4%) female included in a comprehensive database of anamnestic, clinical and pathological data) who underwent radical cystectomy, bilateral iliac and obturator node dissection, and urinary diversion for bladder cancer in the period 1982–2002.

Bladder cancer was diagnosed or confirmed by transurethral resection. Physical examination, chest X-ray, excretory urography and whole-bone scans were carried out on all patients. Computed

Table 12.12 Multivariate analysis (using NPC) of predictors of survival (ALIVE v. DOD)

Risk factor	p-value
Paraoesophageal	0.0006
Paracardial	0.0002
Perigastric	0.0035
Subcarinal	0.0570
Tripod	0.0100
Other lymph node sites	0.0170
Tumour stage	0.0002
Presence of positive lymph nodes	0.0002
Presence of metastatic lymph nodes	0.2300
No. of examined lymph nodes	0.4500
No. of metastatic lymph nodes	0.0002
Gender	0.6300
Age group	0.3500
Degree of tumour differentiation	0.0260
Post therapy	0.0300
Global test	0.0002

tomography or magnetic resonance imaging and abdominal ultrasonography were used for clinical staging. The tumour was staged according to the 2002 TNM classification and graded according to the 1997 WHO system. The indications for radical cystectomy included muscle invasive bladder cancer, superficial bladder cancer refractory to intravesical therapy, and multifocal stage T1G3 tumour associated with diffuse carcinoma *in situ*. No patient had clinically distant metastases at the time of operation. The patients were initially seen 2 months after surgery, every 4 months in the first year after surgery and thereafter every 6 months until disease progression or death.

The aim of the study was to identify the variables that best discriminate between dead and alive patients with a bladder cancer diagnosis. Three phases were identified (which may be seen as domains in subsequent analyses):

- *Phase I*: anamnestic findings (i.e. patient's state of health at first medical visit; this domain includes the variables measured in the time interval between first symptoms and diagnosis);
- *Phase II*: diagnostic findings (i.e. patient's condition after bladder cancer diagnosis; this domain comprises the variables concerning clinical stadiation or clinical findings);
- *Phase III*: pathological stadiation and post-surgical status (i.e. patient's state after surgery).

Patients lost at follow-up or who died from causes other than bladder cancer were excluded. The total remaining sample size (1003 individuals) was divided into two groups. The first group (469 dead patients) included DOD (dead of disease) and AWD (alive with disease) patients; the second group (534 alive patients) included NED (no evidence of disease) patients.

Overall survival 5 years after radical cystectomy was chosen as the dependent variable. Patients' sex and age were not considered, therefore no stratification was performed on the basis of this information.

Because of the use of the TNM classification of bladder cancer, the original variables were transformed into categorical (ordinal) variables, giving 30 endpoints.

12.6.3 NPC Results

Staging systems are the simplest example of a prediction tool, categorizing the disease based on outcome. The statistical interpretation of medical data represents an alternative solution to the problem of accurately predicting the behaviour of cancer, but traditional models such as logistic regression are limited because of their inadequate capacity to handle the variability and complexity of data (Shabsigh and Bochner, 2006). Recently, newer statistical techniques, such as risk stratification groups, risk tables, nomograms and artificial intelligence, have been designed. We propose the use

Table 12.13 Results obtained using NPC methodology. Raw and adjusted p-values are shown. Tippett's combining function was used and 10 000 permutations were carried out

Domain	Variable	Code	p-value	p-FWE
Phase I	Previous superficial TCC	TCC1	0.8217	0.9973
	Focality	FOC1	0.0523	0.5392
	Tumor	T1	0.9488	0.9973
	Grading	G1	0.7321	0.9947
	Carcinoma in situ (CIS)	CIS1	0.1803	0.8691
Phase II	Focality	FOC2	0.8293	0.9973
	Tumor	T2	0.0010	**0.0165**
	Carcinoma in situ (CIS)	CIS2	0.9970	0.9973
	Grading	G2	0.4876	0.9816
	Regional lymph nodes	N2	0.0075	0.1085
	Metastases	M2	0.0001	**0.0014**
	Highway urinary obstruction	HighOBSTR2	0.0002	**0.0030**
Phase III (1)	Tumor	T3	0.0003	**0.0051**
	Carcinoma in situ (CIS)	CIS3	0.0746	0.6302
	Grading	G3	0.0008	**0.0135**
	Regional lymph nodes	N3	0.0002	**0.0030**
	Metastases	M3	0.0001	**0.0014**
	Histology	Histol3	0.1065	0.7370
Phase III (2)	Vesical trigone infiltration	infTRIG3	0.1484	0.8362
	Corpus invasion	infCORPUS3	0.2918	0.9524
	Urethral involvement	invURETH3	0.0007	**0.0116**
	Vascular invasion	invVASC3	0.0616	0.5811
	Lymphonodal invasion	invLYMPH3	0.0001	**0.0014**
	Prostatic invasion	invPROST3	0.0001	**0.0014**
Phase III (3)	Adenocarcinoma of the prostate	ADENcrPR3	0.2228	0.9189
	Highway TCC	HighTCC3	0.0029	**0.0460**
	Desease restarting	desease3	0.3785	0.9641
	Chemotherapy pre-surgery	CHTpreOP3	0.6502	0.9940
	Chemotherapy post-surgery	CHTpostOP3	0.0001	**0.0014**
	Therapy restarting	therapy3	0.3142	0.9524
Global p-value				**0.0014**

Table 12.14 Combining variables within domains. Raw and adjusted p-values are shown, as well as global p-value

Domain	p-value	p-FWE
Combining anamnestic variables (Phase I)	0.2366	0.2366
Combining diagnostic variables (Phase II)	0.0019	**0.0032**
Combining variables 13–18 (Phase III (1))	0.0012	**0.0012**
Combining variables 19–24 (Phase III (2))	0.0005	**0.0005**
Combining variables 25–30 (Phase III (3))	0.0006	**0.0006**
Global p-value		**0.0005**

of the NPC of dependent permutation tests methodology in order to identify prognostic factors in patients with a diagnosis of bladder cancer. Moreover, we adjust p-values using a closed testing procedure, controlling the FWE rate.

With reference to the clinical analysis, we did not find significant differences between the two groups (dead and alive patients) in the first phase of anamnestic findings. However, in the second phase – the diagnostic findings – we found differences in terms of tumour clinical staging (T2), metastases (M2) and hydronephrosis (HighOBSTR2). As regards the third phase – pathological stadiation and post-surgical status – we found significant differences between the two groups in terms of tumour pathological staging (T3), grading (G3), regional lymph node involvement (N3), metastases (M3), urethral involvement (invURETH3), lymphatic invasion (invLYMPH3), prostatic invasion (invPROST3), highway TCC (HighTCC3) and post-surgery chemotherapy (CHTpostOP3). Moreover, p-values referring to the second and third domains (i.e. the combination of the variables involved in the second and third phases) as well as the global p-values were significant. Results are shown in Tables 12.13 and 12.14. All the analyses were done using MATLAB software and $B = 10\,000$ permutations were carried out to estimate the permutation distribution.

The corresponding MATLAB code used is as follows:

```
[D,data]=xlsimport('Urology.xls');
reminD(D)
B=10000;
Y=data(:,6:35);
X=data(:,1);
dati=expand_categ(Y,2);
p_value=NP_2s(dati,X,B,0,1);
TCC1=NPC(p_value(:,1),'T');
FOC1=NPC(p_value(:,2:3),'T');
T1=NPC(p_value(:,4:14),'T');
G1=NPC(p_value(:,15:18),'T');
CIS1=NPC(p_value(:,19:20),'T');
FOC2=NPC(p_value(:,21),'T');
T2=NPC(p_value(:,22:33),'T');
CIS2=NPC(p_value(:,34),'T');
G2=NPC(p_value(:,35:37),'T');
N2=NPC(p_value(:,38:40),'T');
M2=NPC(p_value(:,41),'T');
HighOBSTR2=NPC(p_value(:,42:43),'T');
T3=NPC(p_value(:,44:55),'T');
```

```
CIS3=NPC(p_value(:,56),'T');
G3=NPC(p_value(:,57:59),'T');
N3=NPC(p_value(:,60:62),'T');
M3=NPC(p_value(:,63),'T');
Histol3=NPC(p_value(:,64:68),'T');
infTRIG3=NPC(p_value(:,69),'T');
infCORPUS3=NPC(p_value(:,70),'T');
invURETH3=NPC(p_value(:,71),'T');
invVASC3=NPC(p_value(:,72),'T');
invLYMPH3=NPC(p_value(:,73),'T');
invPROST3=NPC(p_value(:,74),'T');
ADENcrPR3=NPC(p_value(:,75),'T');
HighTCC3=NPC(p_value(:,76),'T');
desease3=NPC(p_value(:,77),'T');
CHTpreOP3=NPC(p_value(:,78),'T');
CHTpostOP3=NPC(p_value(:,79),'T');
therapy3=NPC(p_value(:,80),'T');

phase1=NPC(p_value(:,1:20),'T');

phase2=NPC(p_value(:,21:43),'T');

phase3_1=NPC(p_value(:,44:68),'T');

phase3_2=NPC(p_value(:,69:74),'T');
phase3_3=NPC(p_value(:,75:80),'T');


domain=[phase1 phase2 phase3_1 phase3_2 phase3_3];
P_adj=NPC_FWE(domain,'T');

variable=[TCC1 FOC1 T1 G1 CIS1 FOC2 T2 CIS2 G2 N2 M2
HighOBSTR2 T3 CIS3 G3 N3 M3 Histol3 infTRIG3 infCORPUS3
invURETH3 invVASC3 invLYMPH3 invPROST3 ADENcrPR3 HighTCC3
desease3 CHTpreOP3 CHTpostOP3 therapy3];
P_adj=NPC_FWE(variable,'T');
```

12.6.4 Analysis by Logistic Regression

Here we present the multivariate analysis carried out using standard logistic regression. It is beyond the scope of this example to discuss alternative statistical techniques (e.g. Cox's proportional hazards, logistic regression, artificial intelligence, artificial neural networks and neuro-fuzzy modelling) employed when analysing prognostic information (see Catto et al., 2003; Qureshi et al., 2000). We use one of the standard methods for processing this kind of data set, emphasizing some of the difficulties arising when handling missing data.

If we perform a standard logistic regression, we have to invoke the principle of deletion or imputation in order to deal with missing data. However, following the previous nonparametric solution, we need not delete/impute. Therefore, if we delete incomplete patient files, our analyses

Table 12.15 Logistic regression: including all the variables

	Estimate	Std. Error	z-value	$\Pr(\cdot > \|z\|)$
(Intercept)	−2.7428	0.9973	−2.75	0.0060
TCC1	1.1855	1.1152	1.06	0.2877
FOC1	0.9105	0.4795	1.90	0.0576
T1	−0.1259	0.1963	−0.64	0.5212
G1	−0.3447	0.2607	−1.32	0.1861
CIS1	−0.5648	0.7430	−0.76	0.4472
FOC2	−0.0980	0.3969	−0.25	0.8050
T2	0.1288	0.0577	2.23	0.0256
CIS2	0.3809	0.5312	0.72	0.4733
G2	−0.1321	0.2361	−0.56	0.5758
N2	−0.7535	0.7490	−1.01	0.3144
M2	0.5308	7174.9827	0.00	0.9999
HighOBSTR2	0.4451	0.2269	1.96	0.0498
T3	0.1087	0.0516	2.11	0.0351
CIS3	0.2799	0.3163	0.88	0.3763
G3	0.2569	0.2759	0.93	0.3517
N3	1.0088	0.5815	1.73	0.0828
M3	18.3200	2989.2411	0.01	0.9951
Histol3	0.2088	0.1389	1.50	0.1329
infTRIG3	−0.3609	0.2554	−1.41	0.1575
infCORPUS3	−0.4591	0.3076	−1.49	0.1355
invURETH3	−0.9715	0.5895	−1.65	0.0994
invVASC3	0.5834	0.4128	1.41	0.1576
invLYMPH3	0.4656	0.2617	1.78	0.0752
invPROST3	0.5103	0.3812	1.34	0.1807
ADENcrPR3	0.1146	0.2915	0.39	0.6942
HighTCC3	0.4137	0.5365	0.77	0.4406
desease3	35.7373	1179.0638	0.03	0.9758
CHTpreOP3	−1.5869	1.1331	−1.40	0.1614
CHTpostOP3	−0.9515	0.7094	−1.34	0.1798
therapy3	−17.2044	987.9492	−0.02	0.9861

are based on a sample of 481 subjects (231 dead patients in total). The results of the stepwise regression model are summarized in Tables 12.15 and 12.16.

12.6.5 Some Comments

Using the NPC methodology, we are able to obtain more information which is qualitatively and quantitatively different from that provided by standard statistical methods, such as logistic regression. Note that, using logistic regression, five significant variables may be identified (`FOC1`, `T1`, `HighOBSTR2`, `T3` and `N3`). Using the NPC methodology, 12 prognostic factors may be identified.

The NPC methodology allows for the evaluation of new treatment strategies and may be useful in observational studies. Through application to a real case study, we have shown that the NPC methodology can offer important and substantial contributious to successful research in biomedical studies with several endpoints. The advantages of this approach are its flexibility in handling any

Table 12.16 GLM Stepwise regression (second model)

	Estimate	Std. Error	z-value	$\Pr(\cdot > \|z\|)$
(Intercept)	−2.5134	0.3816	−6.59	0.0000
FOC1	0.7849	0.2898	2.71	0.0068
T1	−0.2250	0.1078	−2.09	0.0369
T2	0.0802	0.0472	1.70	0.0894
HighOBSTR2	0.4056	0.1939	2.09	0.0365
T3	0.1750	0.0355	4.92	0.0000
N3	0.5499	0.2033	2.70	0.0068
CHTpreOP3	−1.3631	0.9652	−1.41	0.1579

type of variable (categorical and quantitative, with or without missing values) while at the same time taking dependencies among those variables into account without the need to model them.

The data set and the corresponding software codes can be found in the `Urology` folder on the book's website.

References

Aalen, O. O. (1978). Nonparametric inference for a family of counting processes. *Annals of Statistics* **6**, 701–726.

Abbate, C., Giorgianni, F., Munaò, F., Pesarin, F. and Salmaso, L. (2001). Neurobehavioural evaluation in humans exposed to hydrocarbons: a new statistical approach. *Psychotherapy and Psychosomatics* **70**, 44–49.

Abbate, C., Micali, E., Giorgianni, C., Munaò, F., Brecciaroli, R., Salmaso, L. and Germanò D. (2004). Affective correlates of occupational exposure to whole-body vibration. A case control study. *Psychotherapy and Psychosomatics* **73**, 375–379.

Adams, D. C. (1999). Methods for shape analysis of landmark data from articulated structures. *Evolutionary Ecology Research* **1**, 959–970.

Adams, D. C., Rohlf, F. J. and Slice, D. E. (2004). Geometric morphometrics: ten years of progress following the 'revolution'. *Italian Journal of Zoology* **71**, 5–16.

Agresti, A. (2002). *Categorical Data Analysis*. New York: John Wiley & Sons, Inc.

Albers, W., Bickel, P. J. and Van Zwet, W. R. (1976). Asymptotic expansions for the power of distribution free tests in the one-sample problem. *Annals of Statistics* **4**, 108–156.

Ali, M. W. (1990). Exact versus asymptotic tests of trend of tumor prevalence in tumorigenicity experiments. A comparison of p-values for small frequency of tumors. *Drug information Journal* **24**, 727–737.

Altman, D. (1991). *Practical Statistics for Medical Research*. London: Chapman & Hall.

Amaral, G., Dryden, I. and Wood, A. (2007). Pivotal bootstrap methods for k-sample problems in directional statistics and shape analysis. *Journal of the American Statistical Association* **102**, 695–707.

Andersen, P. K., Borgan, Ø., Gill, R. and Keiding, N. (1982). Linear nonparametric tests for comparison of counting process with application to censored survival data (with discussion). *International Statistical Review* **50**, 219–258.

Ansari, A. R. and Bradley, R. A. (1960). Rank-sum tests for dispersions. *Annals of Mathematical Statistics* **31**, 1174–1189.

Arboretti, R. (2008). A permutation approach for multivariate repeated measures with application to a complex randomized clinical trial. *Advances and Applications in Statistics* **9**, 261–282.

Arboretti Giancristofaro, R. and Bonnini, S. (2008). Moment-based multivariate permutation tests for ordinal categorical data. *Journal of Nonparametric Statistics* **20**, 383–393.

Arboretti Giancristofaro, R. and Bonnini, S. (2009). Some new results on univariate and multivariate permutation tests for ordinal categorical variables under restricted alternatives. *Statistical Methods and Applications* **18**, 221–236.

Arboretti Giancristofaro, R. and Salmaso, L. (2003). Model performance analysis and model validation in logistic regression. *Statistica* **LXIII**, 375–396.

Arboretti, R., Pesarin, F., Salmaso, L. and Tognoni, G. (1997). Statistical analysis of repeated measurements: an application to a comparative epidemiological study. In C.P. Kitsos (ed.), *Volume of Abstracts of the ISI Satellite Conference on Industrial Statistics: Aims and Computational Aspects*, Athens University of Economics and Business, 9–10, Athens.

Arboretti, R., Pesarin, F., Romero, M. and Salmaso, L. (1999). SAS macro for multivariate and multistrata permutation tests. *Proceedings of SUGItalia 99*, 439–451. SAS Institute, Milan.

Permutation Tests for Complex Data: Theory, Applications and Software Fortunato Pesarin, Luigi Salmaso

Arboretti, R., Pesarin, F., Romero, M. and Salmaso, L. (2000a). Il progetto SETIG e la valutazione comparativa delle strategie terapeutiche adottate: metodologia statistica e applicazione. *Italian Journal of Clinical Pharmacy* **14**, 26–36.

Arboretti, R., Pesarin, F. and Salmaso, L. (2000b). Metodi non parametrici per il confronto tra $C \geq 2$ campioni multivariati mediante test di permutazione. *Proceedings of SUGItalia 2000*, 591–596. SAS Institute, Milan.

Arboretti Giancristofaro, R., Marozzi, M. and Salmaso, L. (2005a). Repeated measures designs: a permutation approach for testing for active effects. *Far East Journal of Theoretical Statistics* (Special Volume on Biostatistics) **16**, 303–325.

Arboretti Giancristofaro, R., Marozzi, M. and Salmaso, L. (2005b). Nonparametric pooling and testing of preference ratings for full-profile conjoint analysis experiments. *Journal of Modern Applied Statistical Methods* **4**, 545–552.

Arboretti Giancristofaro, R., Bonnini, S. and Salmaso, L. (2007a). A performance indicator for multivariate data. *Quaderni di Statistica* **9**, 1–29.

Arboretti Giancristofaro, R., Pesarin, F. and Salmaso, L. (2007b). Permutation Anderson–Darling type and moment based test statistics for univariate ordered categorical data. *Communications in Statistics – Simulation and Computation* **36**, 139–150.

Arboretti Giancristofaro, R., Pesarin, F., Salmaso, L. and Solari, A. (2007c). Nonparametric procedures for testing for dropout rates on university courses with application to an Italian case study. In S. Sawilowsky (ed.), *Real Data Analysis*. Charlotte, NC: Information Age Publishing, pp. 335–353.

Arboretti Giancristofaro, R., Pesarin, F. and Salmaso, L. (2007d). Nonparametric approaches for multivariate testing with mixed variables and for ranking on ordered categorical variables with an application to the evaluation of PhD programs. In S. Sawilowsky (ed.), *Real Data Analysis*. Charlotte, NC: Information Age Publishing, pp. 355–385.

Arboretti Giancristofaro, R., Babbo, G. L., Corain, L. and Salmaso, L. (2007e). Advantages of the nonparametric combination (NPC) multivariate testing method with application to the study of ovarian tumour mass malignity. *JP Journal of Biostatistics* **1**, 39–52.

Arboretti, R., Solmi, F., Salmaso, L., Grego, F. and Maturi, F. (2007f). Nonparametric methods for the comparison of two techniques of surgical operation to repair abdominal aortic aneurysms. *JP Journal of Biostatistics* **1**, 109–122.

Arboretti Giancristofaro, R., Bonnini, S. and Salmaso, L. (2007g). A performance indicator for multivariate data. *Quaderni di Statistica* **9**, 1–29.

Arboretti Giancristofaro, R., Bonnini, S. and Pesarin, F. (2008a). A permutation approach for testing heterogeneity in two-sample categorical variables. *Statistics and Computing* **19**, 209–216.

Arboretti, R., Brombin, C., Pellizzari, S., Salmaso, L. and Mozzanega, B. (2008b). Nonparametric methods applied to nuchal translucency and fetal macrosomia. *Journal of Biostatistics* **2**, 19–36.

Arboretti Giancristofaro, R., Bonnini, S. and Pesarin, F. (2009a). A permutation approach for testing heterogeneity in two-sample problems. *Statistics and Computing* **19**, 209–216.

Arboretti Giancristofaro, R., Bonnini, S. and Salmaso L. (2009b). Employment status and education/employment relationship of PhD graduates from the University of Ferrara. *Journal of Applied Statistics*. http://dx.doi.org/10.1080/02664760802638108.

Arboretti Giancristofaro, R., Corain, L., Salmaso, L. and Tempesta, T. (2009c). An integrated approach using choice-based conjoint analysis and combination-based permutation tests with application to wine quality perception and landscape. *Proceedings of Eursibis'09 Conference*, Cagliari, 30 May – 3 June, pp. 183–184.

Arboretti Giancristofaro, R., Bonnini, S. and Salmaso, L. (2009d). Ordinal classification in multidimensional problems. In S. M. Ermakov, V. B. Melas and A. N. Pepelyshev (eds), *Proceedings of the 6th St. Petersburg Workshop on Simulation*. St. Petersburg, June 28–July 4, pp. 504–550. ISBN 978-5-9651-0354-6.

Azzalini, A. (1984). Estimation and hypothesis testing for collection of autoregressive time series. *Biometrika* **71**, 85–90.

Bacelli, F. and Makowski, A. M. (1989). Multidimensional stochastic ordering and associated random variables. *Operations Research* **37**, 478–487.

Baker, S. G. and Laird, N. M. (1988). Regression analysis for categorical variables with outcome subject to nonignorable nonresponse. *Journal of the American Statistical Association* **83**, 62–69.

Ballin, M. and Pesarin, F. (1990). Una procedura di ricampionamento e di combinazione non parametrica per il problema di Behrens–Fisher multivariato. *Atti della Società Italiana di Statistica* **2**, 351–358.

Barabesi, L. (1998). The computation of the distribution of the sign test statistic for ranked-set sampling. *Communications in Statistics – Simulation and Computation* **27**, 833–842.

Barabesi, L. (2000). On the exact bootstrap distribution of linear statistics. *Metron* **58**, 53–64.

Barabesi, L. (2001). The unbalanced ranked-set sample sign test. *Journal of Nonparametric Statistics* **13**, 279–289.

Barlow, R. E., Bartholomiew, D. J., Brummer, J. M. and Brunk, H. D. (1972). *Statistical Inference under Order Restrictions*. New York: John Wiley & Sons, Inc.

Barndorff-Nielsen, O. E. and Cox, D. R. (1979). Edgeworth and saddlepoint approximations with statistical applications. *Journal of the Royal Statistical Society, Series B* **41**, 279–312.

Barton, C. N. and Cramer, E. C. (1989). Hypothesis testing in multivariate linear models with randomly missing data. *Communications in Statistics – Simulation and Computation* **18**, 875–895.

Barton, D. E. and David, F. N. (1961). Randomization bases for multivariate tests I: the bivariate case, randomness of N points in a plane. In *Bulletin of the International Statistical Institute: Proceedings of the 33rd Session (Paris)*, pp. 455–467. ISI, The Hague.

Barzi, F., Celant, G., Di Castelnuovo, A., Pesarin, F. and Salmaso, L. (1999a). Multivariate permutation tests for repeated measures: an application to a tumor-growth case study. In *Volume of Abstracts of 3rd Congress of the Italian Biometric Society-International Biometric Society Italian Region*, pp. 53–55, CNR, Rome.

Barzi, F., Celant, G., Di Castelnuovo, A., Pesarin, F. and Salmaso, L. (1999b). Permutation tests for repeated measures when the number of measures is greater than the number of observations. In H. Friedl, A. Berghold and G. Kauermann (eds), *Proceedings of the 14th International Workshop on Statistical Modelling*, pp. 436–439. Technical University, Graz.

Bassi, P. F., Brombin, C., Salmaso, L. and Racioppi, M. (2007). Nonparametric methods in observational studies for testing for association in bladder cancer. *Atti del IV Congresso Nazionale della Società Italiana di Statistica Medica ed Epidemiologia Clinica*, Monreale, 19–22 September, pp. 347–352.

Basso, D. and Salmaso, L. (2006). A discussion of permutation tests conditional to observed responses in unreplicated 2^M full factorial designs. *Communications in Statistics – Theory and Methods* **35**, 83–97.

Basso, D. and Salmaso, L. (2007). A comparative study of permutation tests for unreplicated two-level experiments. *Journal of Applied Statistical Science* **15**, 87–98.

Basso, D. and Salmaso, L. (2009a). A permutation test for umbrella alternatives. *Statistics and Computing*. To appear.

Basso, D. and Salmaso, L. (2009). A permutation test for umbrella alternatives. In M. Bini, P. Monari, D. Piccolo and L. Salmaso (eds), *Statistical Methods for the Evaluation of Educational Services and Quality of Products*. Heidelberg: Physica-Verlag. To appear.

Basso, D., Salmaso, L., Hevangeralas, H. and Koukouvinos, C. (2004). Nonparametric testing for main effects on inequivalent designs. In A. Di Bucchianico, H. Lauter, and H. P. Wynn (eds), *MODA 7 – Advances in Model-Oriented Design and Analysis*. Heldelberg: Physica, pp. 33–40.

Basso, D., Chiarandini, M. and Salmaso, L. (2007a). Synchronized permutation tests in replicated $I \times J$ designs. *Journal of Statistical Planning and Inference* **137**, 2564–2578.

Basso, D., Finos, L. and Salmaso, L. (2007b). A new association test in linear models with application to experimental designs. *Journal of Applied Statistical Science* **15**, 99–109.

Basso, D., Salmaso, L. and Pesarin, F. (2007c). On synchronized permutation tests in two-way ANOVA, *Atti del convegno internazionale Moda 8*, Castilla-La Mancha, 4–8 June, pp. 25–31.

Basso, D., Pesarin, F. and Salmaso, L. (2008). Testing effects in ANOVA experiments: direct combination of all pair-wise comparisons using constrained synchronized permutations (invited paper). *Proceedings of COMPSTAT 2008, International Conference on Computational Statistics*. Porto, 24–29 August, vol. I. Heidelberg: Physica-Verlag, pp. 411–422.

Basso, D., Pesarin, F., Salmaso, L. and Solari, A. (2009a). *Permutation Tests for Stochastic Ordering and ANOVA: Theory and Applications in R*. New York: Springer.

Basso, D., Corain, L., Salmaso, L., Spadoni, L., Tveit, C. (2009b). Multivariate Ranking Methods for Design and Analysis of Experiments. In S. M. Ermakov, V. B. Melas and A. N. Pepelyshev (eds), *Proceedings of the 6th St. Petersburg Workshop on Simulation*. St. Petersburg, June 28 – July 4, pp. 472–476. ISBN 978-5-9651-0354-6.

Basawa, I. V. and Prakasa-Rao, B. L. S. (1980). *Statistical Inference for Stochastic Processes*. New York: Academic Press.

Bassukas, I. D. (1994). Comparative Gompertzian analysis of tumour growth patterns. *Cancer Research* **54**, 4385–4392.

Bassukas, I. D. and Maurer-Schulze, B. (1988). The recursion formula of the Gompertz function: a simple method for the estimation and comparison of tumour growth curves. *Growth, Development and Ageing* **52**, 113–122.

Basu, D. (1978). Relevance of randomization in data analysis. In N. K. Namboodiri (ed.), *Survey Sampling and Measurement*, pp. 267–292. New York: Academic Press.

Basu, D. (1980). Randomization analysis of experimental data: the Fisher randomization test. *Journal of the American Statistical Association* **75**, 575–582.

Bell, C. B. and Doksum, K. A. (1967). Distribution-free tests of independence. *Annals of Mathematical Statistics* **38**, 429–446.

Bell, C. B. and Sen, P. K. (1984). Randomization procedures. In P. R. Krishnaiah and P. K. Sen (eds), *Handbook of Statistics* **4**. Amsterdam: North Holland, pp. 1–29.

Bell, C. B., Woodroofe, M. and Avadhani, T. V. (1970). Some nonparametric tests for stochastic processes. In M. L. Puri (ed.), *Nonparametric Techiques in Statistical Inference*. Cambridge: Cambridge University Press.

Benjamini, Y. and Hochberg, Y. (1995). Controlling the false discovery rate: a practical and powerful approach to multiple testing. *Journal of the Royal Statistical Society B* **57**, 289–300.

Benjamini, Y. and Hochberg, Y. (1997). Multiple hypotheses testing with weights. *Scandinavian Journal of Statistics* **24**, 407–418.

Berger, J. O. and Wolpert, R. L. (1988). *The Likelihood Principle*. Hayward, CA: Institute of Mathematical Statistics.

Berger, V. W. (2000). Pros and cons of permutation tests in clinical trials. *Statistics in Medicine* **19**, 1319–1328.

Berger, V. W. (2005). *Selection bias and covariate inbalances in randomized clinical trials*. Hoboken, NJ: John Wiley & Sons, Inc.

Berger, V. W., Permutt, P. and Ivanova, A. (1998). Convex hull test for ordered categorical data. *Biometrics* **54**, 1541–1550.

Berk, R. H. and Jones, D. H. (1978). Relatively optimal combination of tests. *Scandinavian Journal of Statistics* **15**, 158–162.

Berry, K. J. (1982). Algorithm AS 179: enumeration of all permutations of multi-sets with fixed repetition numbers. *Applied Statistics* **31**, 169–173.

Berry, K. J. and Mielke, P. W. (1983). Moment approximations as an alternative to the F test in analysis of variance. *British Journal of Mathematical and Statistical Psychology* **36**, 202–206.

Berry, K. J. and Mielke, P. W. (1984). Computation of exact probability values for multiresponse permutation procedures (MRPP). *Communications in Statistics – Simulation and Computation* **13**, 417–432.

Berry, K. J. and Mielke, P. W. (1985). Computation of exact and approximate probability values for a matched pairs permutation test. *Communications in Statistics – Simulation and Computation* **14**, 229–248.

Berry, K. J., Mielke, P. W. and Wary, R. (1986). Approximate MRPP p-values obtained from four exact moments. *Communications in Statistics – Simulation and Computation* **15**, 581–589.

Bertacche, R. and Pesarin, F. (1997). Treatment of missing data in multidimensional testing problems for categorical variables. *Metron* **55**, 135–149.

Berti, G. A., Corain, L., Monti, M. and Salmaso, L. (2006). Nonparametric methods and modelling applied to the numerical simulation of hot forging. *Statistica Applicata – Italian Journal of Applied Statistics* **18**, 51–70.

Bertoluzzo, F., Pesarin, F. and Salmaso, L. (2009). Nonparametric weighted step down Holm method with heteroscedastic variables. In S. M. Ermakov, V. B. Melas and A. N. Pepelyshev (eds), *Proceedings of the 6th St. Petersburg Workshop on Simulation*. St. Petersburg, June 28 – July 4, pp. 477–481. ISBN 978-5-9651-0354-6.

Bickel, P. J. and Van Zwet, W. R. (1978). Asympototic expansions for the power of distribution free tests in the two-sample problem. *Annals of Statistics* **6**, 987–1004.

Bird, S. M., Cox, D., Farewell, D. T., Goldstein, H., Holt, T. and Smith, P. C. (2005). Performance indicators: good, bad and ugly. *Journal of the Royal Statistical Society A* **168**, 1–27.

Birnbaum, A. (1954). Combining independent tests of significance. *Journal of the American Statistical Association* **49**, 559–574.

Birnbaum, A. (1955). Characteriazations of complete classes of tests of some multiparametric hypotheses, with applications to likelihood ratio tests. *Annals of Mathematical Statistics* **26**, 21–36.

Blair, R. C., Higgins, J. J., Karniski, W. and Kromrey, J. D. (1994). A study of multivariate permutation tests which may replace Hotelling's t^2 test in prescribed circumstances. *Multivariate Behavioral Research* **29**, 141–163.

Blair, R. C., Troendle, J. F. and Beck, R. W. (1996). Control of familywise errors in multiple endpoint assessments via stepwise permutation tests. *Statistics in Medicine* **15**, 1107–1121.

Bonnini, S., Salmaso, L. and Solari, A. (2005). Multivariate permutation tests for evaluating effectiveness of universities through the analysis of student dropouts. *Statistica & Applicazioni* **3**, 37–44.

Bonnini, S., Corain, L., Munaò, F. and Salmaso, L. (2006a). Neurocognitive effects in welders exposed to aluminium: an application of the NPC test and NPC ranking methods. *Statistical Methods & Applications – Journal of the Italian Statistical Society* **15**, 191–208.

Bonnini, S., Corain, L. and Salmaso, L. (2006b). A new statistical procedure to support industrial research into new product development. *Quality and Reliability Engineering International* **22**, 555–566.

Bonnini, S., Corain, L., Cordellina, A., Crestana, A., Musci, R. and Salmaso, L. (2009). A novel global performance score with application to the evaluation of new detergents. In M. Bini, P. Monari, D. Piccolo and L. Salmaso (eds), *Statistical Methods for the Evaluation of Educational Services and Quality of products*. Heidelberg: Physica-Verlag. To appear.

Bookstein, F. L. (1986). Size and shape spaces for landmark data in two dimensions. *Statistical Science* **1**, 181–242.

Bookstein, F. L. (1991). *Morphometric Tools For Landmark Data: Geometry and Biology*. Cambridge: Cambridge University Press.

Bookstein, F. L. (1996). Combining the tools of geometric morphometrics. In *Advances in Morphometrics*, vol. 284. New York: Plenum Press, pp. 131–151.

Bookstein, F. L. (1997). Shape and the information in medical images: a decade of the morphometric synthesis. *Computer Vision and Image Understanding* **66**, 97–118.

Boos, D. D. and Brownie, C. (1989). Bootstrap methods for testing homogeneity of variances. *Technometrics* **31**, 69–82.

Borovkov, A. (1987). *Statistique Mathématique*. Moscow: Mir.

Bosq, D. (2000). *Linear Processes in Function Spaces: Theory and Applications*. New York: Springer-Verla.

Boyett, J. M. and Shuster, J. J. (1977). Nonparametric one-sided tests in multivariate analysis with medical applications. *Journal of the American Statistical Association* **72**, 665–668.

Braun, T. and Feng, Z. (2001). Optimal permutation tests for the analysis of group randomized trials. *Journal of the American Statistical Association* **96**, 1424–1432.

Breslow, N. (1970). A generalized Kruskal–Wallis test for comparing K samples subject to unequal patterns of censorships. *Biometrika* **57**, 579–594.

Breslow, N. E., Edler, L. and Berger, J. (1984). A two-sample censored-data rank test for acceleration. *Biometrics* **40**, 1049–1062.

Bretz, F., Maurer, W., Brannath, W. and Posch, M. (2009). A graphical approach to sequentially rejective multiple test procedures. *Statistics in Medicine* **28**, 586–604.

Brombin, C. (2009). A nonparametric permutation approach to statistical shape analysis. Ph.D. thesis, University of Padua.

Brombin, C. and Salmaso, L. (2008). Searching for powerful tests in shape analysis (contributed paper). In *Proceedings of COMPSTAT 2008, International Conference on Computational Statistics*, Porto, 24–29 August, vol. II. Heidelbrg: Physica-Verlag, pp. 3–10.

Brombin, C. and Salmaso, L. (2009). Multi-aspect permutation tests in shape analysis with small sample size. *Computational Statistics & Data Analysis* **53**(12), 3921–3931.

Brombin, C., Mo, G., Zotti, A., Giurisato, M., Cozzi, B. and Salmaso, L. (2009). A landmark analysis-based approach to age and sex classification of the skull of the Mediterranean monk seal (*Monachus monachus*) (Hermann, 1779). *Anatomia, Histologia, Embryologia*. To appear.

Brookmeyer, R. and Crowley, J. J. (1982). A confidence interval for the median survival time. *Biometrics* **38**, 29–41.

Brown, B., Hollander, M. and Korwar, R. M. (1974). Nonparametric tests of independence for censored data, with applications to heart transplant studies. In F. Proschan and R. J. Serfling (eds), *Reliability and Biometry*. Philadelphia: SIAM, pp. 327–354.

Brown, B. M. and Hettmansperger, T. P. (1989). An affine bivariate version of the sign test. *Journal of the Royal Statististical Society B* **51**, 117–125.

Brunner, E. and Munzel, U. (2000). The nonparametric Behrens–Fisher problem: asymptotic theory and small-sample approximation. *Biometrical Journal* **42**, 17–25.

Brunner, E., Puri, M. L. and Sun, S. (1995). Nonparametric methods for stratified two-sample design with application to multiclinic trials. *Journal of the American Statistical Association* **90**, 1004–1014.

Brunner, S., Bryden, M. M. and Shaughnessy, P. D. (2003). Cranial ontogeny of otariid seals. *Systematics and Biodiversity* **2**, 83–110.

Calian, V., Li, D. and Hsu, J. C. (2008). Partitioning to uncover conditions for permutation tests to control multiple testing error rates. *Biometrical Journal* **50**, 756–766.

Callegaro, A., Pesarin, F. and Salmaso, L. (2003). Test di permutazione per il confronto di curve di sopravvivenza. *Statistica Applicata* **15**, 241–261.

Campigotto, F. (2009). Comparison of nonparametric tests for survival analysis: applications to biomedical studies. Bachelor's thesis, University of Padua, Italy.

Catto, J. W. F., Linkens, D. A., Abbod, M. F., Chen, M., Burton, J. L., Feeley, K. M. and Hamdy, F. C. (2003). Artificial intelligence in predicting bladder cancer outcome: a comparison of neuro-fuzzy modeling and artificial neural networks. *Clinical Cancer Research* **9**, 4172–4177.

Celant, G. and Pesarin, F. (2000). Alcune osservazioni critiche riguardanti l'analisi bayesiana condizionata. *Statistica* **60**, 25–37.

Celant, G. and Pesarin, F. (2001). Sulla definizione di analisi condizionata. *Statistica* **61**, 185–194.

Celant, G., Pesarin, F. and Salmaso, L. (2000a). Some comparisons between a parametric and a nonparametric solution for tests with repeated measures. *Metron* **58**, 64–79.

Celant, G., Pesarin, F., Salmaso, L. (2000). Two sample permutation tests for repeated measures with missing values. *Journal of Applied Statistical Science* **9**, 291–304.

Chakraborti, S. and Shaafsma, W. (1996). On the choice of scores in contingency tables. In *Proceedings of the Biometric Section of the American Statistical Association*, pp. 329–333. American Statistical Association, Washington, DC.

Chatterjee, S. K. (1984). Restricted alternatives. In P. R. Krishnaiah and P. K. Sen (eds), *Handbook of Statistics* **4**. Amsterdam: North Holland, pp. 327–345.

Chen, J. J. and Gaylor, D. W. (1986). The upper percentiles of the distribution of the log-rank statistics for small number of tumors. *Communications in Statistics – Simulation and Computation* **15**, 991–1002.

Chiano, M. N. and Clayton, D. G. (1998). Genotypic relative risks under ordered restriction. *Genetic Epidemiology* **15**, 135–146.

Chung, J. H. and Fraser, D. A. S. (1958). Randomization tests for a multivariate two-sample problem. *Journal of the American Statistical Association* **53**, 729–735.

Cochran, W. G. (1952). The χ^2-goodness of fit. *Annals of Mathematical Statistics* **23**, 493–507.

Cochran, W. G. (1968). The effectiveness of adjustment by subclassification in removing bias in observational studies. *Biometrics* **24**, 295–313.

Cohen, A. and Sackrowitz, H. B. (1998). Directional tests for one-sided alternatives in multivariate models. *Annals of Statistics* **26**, 2321–2338.

Commenges, D. (1996). Permutation tests based on transformation towards exchangeability. In *Proceedings of the 4th World Congress of the Bernoulli Society*, 157. Bernoulli Society, Vienna.

Conaway, M. R. (1994). Casual nonresponse models for repeated categorical measurements. *Biometrics* **50**, 1102–1116.

Copas, J. B. and Long, T. (1991). Estimating the residual variance in orthogonal regression with variable selection. *The Statistician* **40**, 51–59.

Corain, L. and Salmaso, L. (2003). An empirical study on new product development process by nonparametric combination (NPC) testing methodology and post-stratification. *Statistica* **63**, 335–357.

Corain, L. and Salmaso, L. (2004). Multivariate and multistrata nonparametric tests: the nonparametric combination method. *Journal of Modern Applied Statistical Methods* **3**, 443–461.

Corain, L. and Salmaso, L. (2007a). A critical review and a comparative study on conditional permutation tests for two-way ANOVA. *Communications in Statistics – Simulations and Computation* **36**, 791–805.

Corain, L. and Salmaso, L. (2007b). A nonparametric method for defining a global preference ranking of industrial products. *Journal of Applied Statistics* **34**, 203–216.

Corain, L. and Salmaso, L. (2009a). Future trends on global performance indicators in industrial research. In G. I. Hayworth (ed.), *Reliability Engineering Advances*. Hauppauge, NY: Nova. To appear.

Corain, L. and Salmaso, L. (2009b). Nonparametric tests for the randomized complete block design with ordered categorical variables. In M. Bini, P. Monari, D. Piccolo and L. Salmaso (eds), *Statistical Methods for the Evaluation of Educational Services and Quality of Products*. Heidelberg: Physica-Verlag. To appear.

Corain, L., Salmaso, L. and Tessarolo, P. (2002). Developing successful products: a multivariate permutation analysis. *Statistica Applicata* **14**, 421–442.

Corain, L., Melas, V. B. and Salmaso, L. (2009a). A discussion on nonlinear models for price decision in rating based product preference models. *Communications in Statistics – Simulations and Computation* **38**, 1178–1201.

Corain, L., Salmaso, L., Ragazzi, S. (2009b). Permutation tests for the multivariate randomized complete block design with application to sensorial testing. In S. M. Ermakov, V. B. Melas and A. N. Pepelyshev (eds), *Proceedings of the 6th St. Petersburg Workshop on Simulation*. St. Petersburg, June 28 – July 4, pp. 465–471. ISBN 978-5-9651-0354-6.

Corcoran, C., Ryan, L., Mehta, C. R., Senchaudhuri, P., Patel, N. P. and Molenberghs, G. (2001). An exact trend test for correlated binary data. *Biometrics* **57**, 941–948.

Corrain, C., Mezzavilla, F., Pesarin, F. and Scardellato, U. (1977). Il valore discriminativo di alcuni fattori Gm, tra le poplazioni pastorali del Kenya. In *Atti e Memorie dell'Accademia Patavina di Scienze, Lettere ed Arti, LXXXIX, Parte II: Classe di Scienze Matematiche e Naturali*, pp. 55–63.

Cox, D. R. (1972). Regression models and life tables (with discussion). *Journal of the Royal Statistical Society B* **34**, 187–220.

Cox, D. R. and Hinkley, D. V. (1974). *Theoretical Statistics*. London: Chapman & Hall.

Cressie, N. A. C. and Read, T. R. C. (1988). *Goodness of Fit Statistics for Discrete Multivariate Data*. New York: Springer-Verlag.

Crowder, M. J. and Hand, D. J. (1990). *Analysis of Repeated Measures*. London: Chapman & Hall.

D'Agostino, R. B. and Stephens, M. A. (1986). *Goodness of Fit Techniques*. New York: Marcel Dekker.

Dalla Valle, F., Pesarin, F. and Salmaso, L. (2000). A comparative simulation study on permutation tests for effects in two-level factorial designs. *Metron* **58**, 147–161.

Dalla Valle, F., Pesarin, F. and Salmaso, L. (2003). Paired permutation testing in 2k unreplicated factorials. *Journal of the Italian Statistical Society – Statistical Methods and Applications* **11**, 265–276.

Davidian, M. and Giltinan, D. M. (1995). *Nonlinear Models for Repeated Measurement Data*. London: Chapman & Hall.

Davison, A. C. and Hinkley, D. V. (1988). Saddlepoint approximations in resampling methods. *Biometrika* **75**, 417–431.

De Martini, D. (1998). Metodi per il calcolo della potenza e la determinazione della numerosità campionaria per test non parametrici – con realizzazione di un programma di calcolo. Ph.D. thesis, Università di Genova, Genova.

Dempster, A. P., Laird, N. M. and Rubin, D. B. (1977). Maximum likelihood from incomplete data via the EM algorithm. *Journal of the Royal Statistical Society B* **39**, 1–38.

Der, G. and Everitt, B. S. (2006). *Statistical Analysis of Medical Data Using SAS*. Boca Raton, FL: Chapman & Hall/CRC.

Dharmadhikari, S. and, Joag-Dev, K. (1988). *Unimodality, Convexity, and Applications*. Boston: Academic Press.

Diggle, P. J., Liang, K. Y. and Zeger, S. L. (2002). *Analysis of Longitudinal Data*. Oxford: Oxford University Press.

DiRienzo, A. G. (2003). Nonparametric comparison of two survival-time distributions in the presence of dependent censoring. *Biometrics* **59**, 497–504.

Dmitrienko, A., Offen, W. W. and Westfall, P. H. (2003). Gatekeeping strategies for clinical trials that do not require all primary effects to be significant. *Statistics in Medicine* **22**, 2387–2400.

Dryden, I. L. and Mardia, K. V. (1993). Multivariate shape analysis. *Sankhyā A* **55**, 460–480.

Dryden, I. L. and Mardia, K. V. (1998). *Statistical Shape Analysis*. Chichester: John Wiley & Sons, Ltd.

Dudoit, S. and Van der Laan, M. J. (2008). *Multiple Testing Procedures with Applications to Genomics*. New York: Springer-Verlag.

Dunnett, C. W. and Tamhane, A. C. (1992). A step-up multiple test procedure. *Journal of the American Statistical Association* **87**, 162–170.

Dwass, M. (1957). Modified randomization tests for nonparametric hypotheses. *Annals of Mathematical Statistics* **28**, 181–187.

Dykstra, R. L., Kochar, S. and Robertson, T. (1995). Inference for likelihood ratio ordering in the two-sample problem. *Journal of the American Statistical Association* **90**, 1039–1040.

Eden, T. and Yates, B. A. (1933). On the validity of Fisher's z test when applied to an actual example of non-normal data. *Journal of Agricultural Science* **23**, 6–17.

Edgington, E. S. (1995). *Randomization Tests*, 3rd edn. New York: Marcel Dekker.

Edgington, E. S. and Onghena, P. (2007). *Randomization Tests*, 4th edn. Boca Raton, FL: Chapman & Hall/CRC.

Efron, B. (1992). Missing data, imputation and the bootstrap. Technical report no. 397, Department of Statistics, Stanford University, CA.

Efron, B. and Tibshirani, R. J. (1993). *An Introduction to the Bootstrap*. New York: Chapman & Hall.

El Barmi, H. and Dykstra, R. (1995). Testing for and against a set of linear inequality constraints in a multinomial setting. *Canadian Journal of Statistics* **23**, 131–143.

El Barmi, H. and Mukerjee, H. (2005). Inferences under stochastic ordering constraint: the k-sample case. *Journal of the American Statistical Association* **100**, 252–261.

Entsuah, A. R. (1990). Randomization procedures for analyzing clinical trial data with treatment related withdrawls. *Communications in Statistics – Theory and Methods* **19**, 3859–3880.

Erie, J. C., McLaren, J. W., Hodge, D. O., Bourne, W. M. (2005). Long-term corneal keratocyte deficits after photorefractive keratectomy and laser in situ keratomitensis. *Transactions of the American Ophthalmological Society* **103**, 56–66.

Farrar, D. B. and Crump, K. S. (1988). Exact tests for any carcinogenic effect in animal bioassays. *Fundamental and Applied Toxicology* **11**, 652–663.

Fassò, A. and Pesarin, F. (1986). A new nonparametric independence test. *Serdika* **12**, 65–72.

Fattorini, L. (1996). Il bambino, la famiglia, la contrada: analisi statistica di un questionario. In *Atti del Convegno 'Bambini e territorio nelle contrade di Siena'*, Siena, Italy. Dipartimento di Metodi Quantitativi, Università di Siena, Siena.

Fava, G., Ruini, C., Rafanelli, C., Finos, L., Salmaso, L., Mangelli, L. and Sirigatti, S. (2005). Well-being therapy of generalized anxiety disorder. *Psychotherapy and Psychosomatics* **74**, 26–30.

Fay, R. E. (1986). Causal models for patterns of nonresponse. *Journal of the American Statistical Association* **81**, 354–365.

Ferguson, T. S. (1973). A Bayesian analysis of some nonparametric problems. *Annals of Statistics* **1**, 209–230.

Ferraty, F. and Vieu, P. (2006). *Nonparametric Functional Data Analysis: Theory and Practice*. New York: Springer.

Filippini, R., Salmaso, L. and Tessarolo, P. (2004). Product development time performance: investigating the effect of interactions between drivers. *Journal of Product Innovation Management* **21**, 199–214.

Finos, L. and Salmaso, L. (2004). Nonparametric multi-focus analysis for categorical variables. *Communications in Statistics – Theory and Methods* **33**, 1931–1941.

Finos, L. and Salmaso, L. (2005). A new nonparametric approach for multiplicity control: optimal subset procedures. *Computational Statistics* **20**, 643–654.

Finos, L. and Salmaso, L. (2006). Weighted methods controlling the multiplicity when the number of variables is much higher than the number of observations. *Journal of Nonparametric Statistics* **18**, 245–261.

Finos, L. and Salmaso, L (2007). FDR- and FWE-controlling methods using data-driven weights. *Journal of Statistical Planning and Inference* **137**, 3859–3870.

Finos, L., Pesarin, F. and Salmaso, L. (2000a). Test combinati per il controllo della molteplicità mediante procedure di closed testing. Working Paper no. 2000.8, Department of Statistics, University of Padua.

Finos, L., Pesarin, F. and Salmaso, L. (2000b). Confronti multipli non parametrici: nuovi sviluppi. Working Paper no. 2000.9, Department of Statistics, University of Padua.

Finos, L., Pesarin, F. and Salmaso, L. (2003). Test combinati per il controllo della molteplicità mediante procedure di closed testing. *Statistica Applicata* **15**, 301–329.

Finos, L., Salmaso, L. and Solari, A. (2007). Conditional inference under simultaneous stochastic ordering constraints. *Journal of Statistical Planning and Inference* **137**, 2633–2641.

Finos, L., Pesarin, F., Salmaso, L. and Solari, A. (2008). Exact inference for multivariate ordered alternatives. *Statistical Methods & Applications* **17**, 195–208.

Finos, L., Brombin, C. and Salmaso, L. (2010). Adjusting stepwise p-values in generalized linear models. Accepted for publication in *Communications in Statistics – Theory and Methods*.

Fisher, R. A. (1925). *Statistical Methods for Research Workers*. Edinburgh: Oliver & Boyd.

Fisher, R. A. (1935). The fiducial argument in statistical inference. *Annals of Eugenics* **49**, 174–180.

Fleming, T. R. and Harrington, D. P. (1981). A class of hypothesis tests for one and two sample censored survival data. *Communications in Statistics – Theory and Methods* **10**, 763–794.

Fleming, T. R. and Harrington, D. P. (1991). *Counting Processes and Survival Analysis*. New York: John Wiley & Sons, Inc.

Fleming, T. R., Harrington, D. P. and O'Sullivan, M. (1987). Supremum versions of the log-rank and generalized Wilcoxon statistics. *Journal of the American Statistical Association* **82**, 4312–4320.

Folks, J. L. (1984). Combinations of independent tests. In P. R. Krishnaiah and P. K. Sen (eds), *Handbook of Statistics* **4**. Amsterdam: North-Holland, pp. 113–121.

Fraser, D. A. S. (1957). *Nonparametric Methods in Statistics*. New York: John Wiley & Sons, Inc.

Friedman, M. (1937). The use of ranks to avoid the assumption of normality implicit in the analysis of variance. *Journal of the American Statistical Association* **32**, 675–701.

Friman, A. and Westin, C. F. (2005). Resampling fMRI time series. *NeuroImage* **25**, 859-867.

Frost, S. R., Marcus, L. F., Bookstein, F. L., Reddy, D. P. and Delson, E. (2003). Cranial allometry, phylogeography, and systematics of large-bodied papionins (primates: Cercopithecinae) inferred from geometric morphometric analysis of landmark data. *The Anatomical Record (Part A)* **275A**, 1048–1072.

Fuchs, C. (1982). Maximum likelihood estimation and model selection in contingency tables with missing data. *Journal of the American Statistical Association* **77**, 270–278.

Gabriel, K. R. and Hall, W. J. (1983). Rerandomization inference on regression and shift effects: computationally feasible methods. *Journal of the American Statistical Association* **78**, 827–836.

Gail, M. and Mantel, N. (1977). Counting the number of $r \times c$ contingency tables with fixed marginals. *Journal of the American Statistical Association* **72**, 859–862.

Galambos, J. and Simonelli, I. (1996). *Bonferroni-Type Inequality with Applications*. New York: Springer-Verlag.

Galimberti, S. and Valsecchi, M. G. (2002). *Permutation Tests to Compare Survival Curves for Matched Data*. XLI SIS Scientific Meeting, Milan, 5–7 June, SIS – The Italian Statistical Society.

Garatti, A., Canziani, A., Mossuto, E., Gagliardotto, P., Innocente, F., Pusceddu, G., Frigiola, A. and Menicanti, L. (2009). Long-term results of tricuspid valve replacement: fifteen-years experience at a single-institution. Submitted for publication in *Journal of Thoracic and Cardiovascular Surgery*.

Gaugler, T., Kim, D. and Liao, S. (2007). Comparing two survival time distributions: an investigation of several weight functions for the weighted logrank statistic. *Communications in Statistics – Simulation and Computation* **36**, 423–435.

Gautman, S. (1997). Test for linear trend in $2 \times k$ ordered tables with open-ended categories. *Biometrics* **53**, 1163–1169.

Gehan, E. A. (1965). A generalized Wilcoxon test for comparing aarbitrarily singly censored samples. *Biometrika* **52**, 203–223.

Gilbert, J. P. (1962), *Random censorship*. Ph.D. dissertation, University of Chicago, Chicago.

Gill, R. D. (1980). *Censoring and Stochastic Integrals*. Amsterdam: Mathematisch Centrum.

Gini, C. (1912). Variabilità e mutabilità. In *Studi economico-giuridici della Facoltà di Giurisprudenza*. Università di Cagliari.

Giraldo, A. and Pesarin, F. (1992). Verifica d'ipotesi in presenza di dati mancanti e tecniche di ricampionamento. In *Atti della XXXVI Riunione Scientifica della Società Italiana di Statistica* **2**, pp. 271–278.

Giraldo, A. and Pesarin, F. (1993). A resampling procedure for missing data in testing problems. In G. Diana, L. Pace and A. Salvan (eds), *Atti della Conferenza 'Due Temi di Statistica Metodologica'*, pp. 115–122. Curto, Napoli.

Giraldo, A., Pallini, A. and Pesarin, F. (1994). A nonparametric testing procedure for missing data. In R. Dutter and W. Grossmann (eds), *COMPSTAT 1994: Proceedings in Computational Statistics*. Heidelberg: Physica-Verlag, pp. 503–508.

Glynn, R. J., Laird, N. M. and Rubin, D. B. (1986). Selection modeling versus mixed with nonignorable response. In H. Wainer (ed), *Drawing Inferences from Self- Selected Samples*. Berlin: Springer-Verlag, pp. 115–142.

Goeman, J. J. and Mansman, U. (2008). Multiple testing on the directed acyclic graph of gene ontology. *Bioinformatics* **24**, 537–544.

Goggin, M. L. (1986). The 'too few cases/too many variables' problem in implementation research. *Western Political Quarterly* **39**, 328–347.

Good, P. (1991). Most powerful tests for use in matched pair experiments when data may be censored. *Journal of Statistical Computation and Simulation* **38**, 57–63.

Good, P. (2000). *Permutation Tests: A Practical Guide to Resampling Methods for Testing Hypotheses*. New York: Springer-Verlag.

Good, P. (2005). *Introduction to Statistics through Resampling Methods and R/S-PLUS*. Hoboken, NJ: Wiley-Interscience.

Good, P. (2006). *Resampling Methods: A Practical Guide to Data Analysis*. Boston: Birkhäuser.

Goodall, C. R. (1991). Procrustes methods in the statistical analysis of shape. *Journal of the Royal Statistical Society B* **53**, 285–339.

Goutis, C., Casella, G. and Wells, M. T. (1996). Assessing evidence in multiple hypotheses. *Journal of the American Statistical Association* **91**, 1268–1277.

Gower, J. C. (1975). Generalized Procrustes analysis. *Psychometrika* **40**, 33–50.

Grechanovsky, E. and Pinsker, I. (1995). Conditional p-values for the F-statistic in a forward selection procedure. *Computational Statistics & Data Analysis* **20**, 239–263.

Greenberg, B. G. (1951). Why randomize? *Biometrics* **7**, 309–322.

Greenland, S. (1991). On the logical justification of conditonal tests for two-by-two contingency tables. *American Statistician* **45**, 248–251.

Greenlees, J. S., Reece, W. S. and Zieschang, K. D. (1982). Imputation of missing values when the probability of responce depends on the variable being imputed. *Journal of the American Statistical Association* **77**, 251–261.

Guarda-Nardini, L., Stifano, M., Brombin, C., Salmaso, L. and Manfredini, D. (2007). A one-year case series of arthrocentesis with hyaluronic acid injections for temporomandibular joint osteoarthritis. *Oral Surgery, Oral Medicine, Oral Pathology, Oral Radiology & Endodontics* **103**, e14–e22.

Guarda-Nardini, L., Manfredini, D., Salamone, M., Salmaso, L., Tonello, S. and Ferronato, G. (2008). Efficacy of botulinum toxin in treating myofascial pain in bruxers: a controlled placebo pilot study. *CRANIO: The Journal of Craniomandibular Practice* **26**, 126–135.

Hájek, J. (1961). Some extensions of the Wald–Wolfowitz–Noether theorem. *Annals of Mathematical Statistics* **32**, 506–523.

Hájek, J. and Šidák, Z. (1967). *Theory of Rank Tests*. New York: Academic Press.

Hájek, J., Šidák, Z. and Sen, P. K. (1999). *Theory of Rank Tests*. San Diego, CA: Academic Press.

Han, K. E., Catalano, P. J., Senchaudhuri, P. and Metha, C. (2004). Exact analysis of dose response for multiple correlated binary outcomes (2004) *Biometrics* **60**, 216–224.

Harrington, D. P. and Fleming, T. R. (1982), A class of rank test procedures for censored survival data. *Biometrika* **69**, 553–566.

Harshman, R. A. and Lundy, M. E. (2006). A randomization method of obtaining valid p-values for model changes selected 'post hoc'. Poster presented at the *Seventy-first Annual Meeting of the Psychometric Society*. Montreal, Canada, June. Available at http://publish.uwo.ca/~harshman/imps2006.pdf.

Heimann, G. and Neuhaus, G. (1998). Permutational distribution of the log-rank statistic under random censorship with applications to carcinogenicity assays. *Biometrics* **54**, 168–184.

Heinze, G., Guant, M. and Schemper, M. (2003). Exact log-rank tests for unequal follow-up. *Biometrics* **59**, 1151–1157.

Heyse, J. F. and Rom, D. (1988). Adjusting for multiplicity of statistical tests in the analysis of carcinogenicity studies. *Biometrical Journal* **30**, 883–896.

Heller, G. and Venkatraman, E. S. (2004). A nonparametric test to compare survival distributions with covariate adjustment. *Journal of the Royal Statistical Society B* **66**, 719–733.

Heritier, S. R., Gebski, V. J. and Keech, A. C. (2003). Inclusion of patients in clinical trial analysis: the intention-to-treat principle. *Medical Journal of Australia* **179**, 438–440.

Hettmansperger, T. P. (1984). *Statistical Inference Based on Ranks*. New York: John Wiley & Sons, Inc.

Hettmansperger, T. P. and McKean, J. W. (1998). *Robust Nonparametric Statistical Methods*. London: Arnold.

Higgins, J. J. and Noble, W. (1993). A permutation test for a repeated measures design. In D. Johnson (ed.), *Proceedings of the 5th Annual Conference on Applied Statistics in Agriculture*. Kansas State University, Manhattan, KS, pp. 240–254.

Hinkley, D. V. (1975). On power transformations to symmetry. *Biometrika* **62**, 101–111.

Hinkley, D. V. (1989). Bootstrap significance tests. In *Bulletin of the International Statistical Institute: Proceedings of the 47th Session (Paris)*, Invited Papers, 2, pp. 65–74. ISI, Voorburg, Netherlands.

Hirotsu, C. (1986). Cumulative chi-squared statistic or a tool for testing goodness of fit. *Biometrika* **73**, 165–173.

Hirotsu, C. (1998a). Max t test for analysing a dose-response relationship – an efficient algorithm for p value calculation. In L. Pronzato (ed.), *Volume of Abstracts of MODA-5 (5th International Conference on Advances in Model Oriented Data Analysis and Experimental Design)*. CIRM, Marseille.

Hirotsu, C. (1998b). Isotonic inference. In *Encyclopedia of Biostatistics*. New York: John Wiley & Sons, Inc, pp. 2107–2115.

Hjorth, J. S. U. (1994). *Computer intensive statistical methods: validation model selection and bootstrap*. New York: Chapman & Hall.

Hochberg, Y. and Tamhane, A. C. (1987). *Multiple Comparison Procedures*. New York: John Wiley & Sons, Inc.

Hocking, R. R. (1976). The analysis and selection of variables in linear regression. *Biometrics* **32**, 1–49.

Hoeffding, W. (1951a). Optimum nonparametric tests. In J. Neyman (ed.), *Proceedings of the Second Berkeley Symposium on Mathematical Statististics and Probabability*. University of California Press, Berkeley, pp. 83–92.

Hoeffding, W. (1951b). A combinatorial central limit theorem. *Annals of Statistics* **22**, 558–566.

Hoeffding, W. (1952). The large-sample power of tests based on permutations of observations. *Annals of Mathematical Statistics* **23**, 169–192.

Hogg, R. (1962). Iterated tests of equality of several distributions. *Journal of the American Statistical Association* **57**, 579–585.

Hoh, J. and Ott, J. (2000). Scan statistics to scan markers for susceptibility genes. *Proceedings of the National Academy of Sciences USA* **96**, 9615–9617.

Hollander, M. and Wolfe, D. A. (1999). *Nonparametric Statistical Methods*, 2nd edn. New York: John Wiley & Sons, Inc.

Holm, S. (1979). A simple sequentially rejective multiple test procedure. *Scandinavian Journal of Statistics* **6**, 65–70.

Horn, M. and Dunnet, C. W. (2004). Power and sample size comparisons of stepwise FWE and FDR controlling test procedures in the normal many-one case. In Y. Benjamini, F. Bretz and S. Sarkar (eds), *Recent Developments in Multiple Comparison Procedures*. Beachwood, OH: Institute of Mathematical Statistics, pp. 48–64.

Hossein-Zadeh, G. A., Soltanian-Zadeh, H. and Ardekani, B. A. (2003). Multiresolution fMRI activation detection using translation invariant wavelet transform and statistical analysis based on resampling. *IEEE Transactions on Medical Imaging* **22**, 302–314.

Hsu, J. C. (1996). *Multiple Comparisons. Theory and Methods*. London: Chapman & Hall.

Hudgens, M. and Satten, G. (2002). Midrank unification of rank tests for exact, tied and censored data. *Journal of Nonparametric Statistics* **14**, 569–581.

Hung, H. M. J., Wang, S. J., Tsong, Y., Lawrence, J. and O'Neil, R. T. (2003). Some fundamental issues with non-inferiority testing in active controlled trial. *Statistics in Medicine* **22**, 213–225.

Huque, M. F. and Alosh, M. (2008). A flexible fixed-sequence testing method for hierarchically ordered correlated multiple endpoints in clinical trials. *Journal of Statistical Planning and Inference* **138**, 321–335.

Ives, F. M. (1976). Permutation enumeration: four new permutation algorithms. *Communications of the Association for Computing Machinery* **19**, 68–70.

Jacod, J. and Protter, P. (2000). *Probability Essentials*. Berlin: Springer-Verlag.

James, B., James, K. L. and Siegmund, D. (1987). Tests for a change-point. *Biometrika* **74**, 71–83.

Janssen, A. J. and Pauls, T. (2003). How do bootstrap and permutation tests work? *Annals of Statistics* **31**, 768–806.

Joe, H. (1997). *Multivariate Models and Dependence Concepts*. London: Chapman & Hall.

Jogdeo, K. (1968). Asymptotic normality in nonparametric methods. *Annals of Mathematical Statistics* **39**, 905–922.

John, R. D. and Robinson, J. (1983a). Edgeworth expansions for the power of permutation tests. *Annals of Statistics* **11**, 625–631.

John, R. D. and Robinson, J. (1983b). Significance levels and confidence intervals for permutation tests. *Journal of Statistical Computation and Simulation* **16**, 161–173.

Johnson, N. L. and Leone, F. C. (1964). *Statistical and Experimental Design in Engineering and the Physical Sciences*. New York: John Wiley & Sons, Inc.

Johnson, R. A. and Wichern, D. W. (2007). *Applied Multivariate Statistical Analysis*, 6th edn. Upper Saddle River, NJ: Pearson Prentice Hall.

Johnson, W. M., Karamanlidis, A. A., Dendrinos, P., Fernández de Larrinoa, P., Gazo, G., González, L. M., Güçlüsoy, H., Pores, R. and Schnellmann, M. (2006). Monk seal fact files. Biology, behaviour, status and conservation of the foca monje del mediterráneo, Monachus monachus. *The Monachus Guardian* http://www.monachus-guardian.org.

Jonckheere, A. (1954). A distribution free k-sample test against ordered alternatives. *Biometrica* **41**, 133–145.

Jones, M. P. and Crowley, J. (1989). A general class of nonparametric tests for survival analysis. *Biometrics* **45**, 157–170.

Kalbfleisch, J. D. and Prentice, R. L. (1980). *The Statistical Analysis of Failure Time Data*. Wiley, New York.

Katina, S., Bookstein, F. L., Gunz, P. and Schaefer, K. (2007). Was it worth digitizing all those curves? A worked example from craniofacial primatology. *American Journal of Physical Anthropology, Abstracts of American Academy of Physician Assistants (AAPA) poster and podium presentations* **132**, S44, 140.

Kellerer, A. M. and Chmelevsky, D. (1983). Small-sample properties of censored-data rank tests. *Biometrics* **39**, 675–682.

Kempthorne, O. (1955). The randomization theory of experimental inference. *Journal of the American Statistical Association* **50**, 964–967.

Kempthorne, O. (1966). Some aspects of experimental inference. *Journal of the American Statistical Association* **61**, 11–34.

Kempthorne, O. (1977). Why randomize? *Journal of Statistical Planning and Inference* **1**, 1–25.

Kempthorne, O. (1979). Sampling inference, experimental inference and observation inference. *Sankhyā* **40**, 115–145.

Kempthorne, O. (1982). Review of randomization tests by E.S. Edgington. *Biometrics* **38**, 864–867.

Kendall, D. G. (1977). The diffusion of shape. *Advances in Applied Probability* **9**, 428–430.

Klingenberg, B. and Agresti, A. (2006). Multivariate extensions of McNemar's test. *Biometrics* **62**, 921–928.

Klingenberg, B., Solari, A., Salmaso, L. and Pesarin, F. (2008). Testing marginal homogeneity against stochastic order in multivariate ordinal data. *Biometrics* **65**, 452–462.

Kosorok, M. R. and Lin, C.-Y. (1999). The versatility of function-indexed weighted log-rank statistics. *Journal of the American Statistical Association* **94**, 320–332.

Kounias, S. (1995). Poisson approximation and Bonferroni bounds for the probability of the union of events. Recent developments in mathematical statistics. *International Journal of Mathematical and Statistical Sciences* **4**, 43–52.

Kounias, S. and Sotirakoglou, K. (1989). Bonferroni bounds revisited. *Journal of Applied Probability* **26**, 233–241.

Koziol, J. A. (1978). A two-sample Cramér–von Mises test for randomly censored data. *Biometrical Journal* **20**, 603–608.

Kramer, C. Y. (1956). Extension of multiple range tests to group means with unequal numbers of replications. *Biometrics* **12**, 307–310.

Kudo, A. (1963). A multivariate analogue of the one-sided test. *Biometrika* **50**, 403–418.

Lago, A. and Pesarin, F. (2000). Nonparametric combination of dependent rankings with application to the quality assessment of industrial products. *Metron* **58**, 39–52.

Laird, N. M. (1988). Missing data in longitudinal studies. *Statics in Medicine* **7**, 305–315.

Laird, N. M. and Ware, J. H. (1982). Random effects models for longitudinal data. *Biometrics* **38**, 963–974.

Lambert, D. (1985). Robust two-sample permutation tests. *Annals of Statistics* **13**, 606–625.

Landenna, G. and Marasini, D. (1990). *Metodi Statistici Non Parametrici*. Il Mulino, Bologna.

Lee, J. W. (1996). Some versatile tests based on the simultaneous use of weighted log-rank statistics. *Biometrics* **52**, 721–725.

Lee, S. H., Lee, E. J. and Omolo, B. O. (2008). Using integrated weighted survival difference for the two-sample censored data problem. *Computational Statistics and Data Analysis* **52**, 4410–4416.

Lehmann, E. L. (1986). *Testing Statistical Hypotheses*, 2nd edn. New York: John Wiley & Sons, Inc.

Lehmann, E. L. (2006). *Nonparametrics: Statistical Methods Based on Ranks*. New York: Springer-Verlag.

Lehmann, E. L. (2009). Parametric versus nonparametrics: two alternative methodologies. *Journal of Nonparametric Statistics* **21**, 397–405.

Lehmann, E. L. and Romano, J. P. (2005). *Testing Statistical Hypotheses*, 3rd edn. New York: Springer-Verlag.

Lehmann, E. L. and Scheffé, H. (1950). Completeness similar regions, and unbiased estimation. *Sankhyā* **10**, 305–340.

Lehmann, E. L. and Scheffé, H. (1955). Completeness similar regions, and unbiased estimation – part II. *Sankhyā* **15**, 219–236.

Lehmann, E. L. and Stein, C. (1949). On the theory of some nonparametric hypotheses. *Annals of Mathematical Statistics* **20**, 28–45.

Lele, S. and Cole, T. M. (1995). Euclidean distance matrix analysis: a statistical review. *Current Issues in Statistical Shape Analysis* **3**, 49–53.

Lele, S. and Cole, T. M. (1996). A new test for shape differences when variancecovariance matrices are unequal. *Journal of Human Evolution* **31**, 193–212.

Lele, S. and Richtsmeier, J. T. (1991). Euclidean distance matrix analysis: a coordinate free approach for comparing biological shapes using landmark data. *American Journal of Physical Anthropology* **86**, 415–427.

Lindsay, J. K. (1994). *Models for Repeated Measures*. Oxford: Oxford University Press.

Liptak, I. (1958). On the combination of independent tests. *Magyar Tudományos Akadémia Matematikai Kutató Intézének Közleményei* **3**, 127–141.

Littell, R. C. and Folks, J. L. (1971). Asymptotic optimality of Fisher's method of combining independent tests. *Journal of the American Statistical Association* **66**, 802–806.

Littell, R. C. and Folks, J. L. (1973). Asymptotic optimality of Fisher's method of combining independent tests II. *Journal of the American Statistical Association* **68**, 193–194.

Little, R. J. A. (1982). Models for nonresponse in sample survey. *Journal of the American Statistical Association* **77**, 237–250.

Little, R. J. A. and Rubin, D. B. (1987). *Statistical Analysis with Missing Data*. New York: John Wiley & Sons, Inc.

Lock, R. H. (1986). Using the computer to approximate permutation tests. *Proceedings of the Statistical Computing Section of the American Statististical Association*, pp. 349–352. American Statistical Assiocation, Alexandria, VA.

Lock, R. H. (1991). A sequential approximation to a permutation test. *Communications in Statistics – Simulation and Computation* **20**, 341–363.

Ludbrook, J. (1998). Multiple comparison procedures updated. *Clinical and Experimental Pharmacology and Physiology* **25**, 1032–1037.

Ludbrook, J. and Dudley, H. (1998). Why permutation tests are superior to t and F tests in biomedical research. *American Statistician* **52**, 127–132.

Lunneborg, C. (1999). *Data Analysis by Resampling: Concepts and Applications*. Belmont, CA: Duxbury Press.

Mack, G. A. and Wolfe, A. D. (1981). K-sample rank tests for umbrella alternatives. *Journal of the American Statistical Association* **76**, 175–181.

Mandel, J. (1993). The analysis of two-way tables with missing values. *Applied Statistics* **42**, 85–93.

Manfredini, D., Basso, D., Salmaso, L. and Guarda-Nardini, L. (2008). Temporomandibular joint click sound and magnetic resonance-depicted disk position: which relationship? *Journal of Dentistry* **36**, 256–260.

Mann, H. B. and Whitney, D. R. (1947). On a test of whether one of two random variables is stochastically larger than the other. *Annals of mathematical statistics* **18**, 50–60.

Manly, B. F. J. (1997). *Randomization, Bootstrap and Monte Carlo Methods in Biology*, 2nd edn. London: Chapman & Hall.

Mansouri, H. (1990). Rank tests for ordered alternatives in analysis of variance. *Journal of Statistical Planning and Inference* **24**, 107–117.

Mantel, N. (1966). Evaluation of survival data and two new rank order statistics arising in its consideration. *Cancer Chemotherapy Reports* **50**, 163–170.

Mantel, N. and Valand, R. S. (1970). A technique of nonparametric multivariate analysis. *Biometrics* **26**, 547–558.

Marchessaux, D. (1989). *Recherches sur la Biologie, l'Ecologie et le Statut du Phoque Moine, Monachus monachus*. GIS Posidonie Publ., Marseille, France.

Marcus, R., Peritz, E. and Gabriel, K. R. (1976). On closed testing procedures with special reference to ordered analysis of variance. *Biometrika* **63**, 655–660.

Maritz, J. S. (1995). A permutation test allowing for missing values. *Australian Journal of Statistics* **37**, 153–159.

Marozzi, M. and Salmaso, L. (2006). Multivariate bi-aspect testing for the two-sample location problem. *Communication in Statistics – Theory and Methods* **35**, 477–488.

Marshall, A. W. and Olkin, I. (1979). *Inequalities: Theory of Majorization and Its Applications*. New York: Academic Press.

Massaro, M. and Blair, D. (2003). Comparison of population numbers of yellow-eyed penguins, *Megadyptes antipodes*, on Stewart Island and on adjacent cat-free islands. *New Zealand Journal of Ecology* **27**, 107–113.

Mazzaro, D., Pesarin, F. and Salmaso, L. (2001). A discussion on multi-way ANOVA using a permutation approach. *Statistica* **61**, 15–26.

McDaniel, K. L. and Moser, V. C. (1993). Utility of a neurobehavioral screening battery for differentiating the effects of two pyrethroids, permethrin and cypermethrin. *Neurotoxicology and Teratology* **15**, 71–83.

Mehta, C. R. and Patel, N. R. (1980). A network algorithm for the exact treatment of the $2 \times K$ contingency table. *Communications in Statistics – Simulation and Computation* **9**, 649–664.

Mehta, C. R. and Patel, N. R. (1983). A network algorithm for performing Fisher's exact test in $r \times c$ contingency tables. *Journal of the American Statistical Association* **78**, 427–434.

Mehta, C. and Patel, N. (1999). *StatXact 4 for Windows, User's Manual*. Cambridge, MA: Cytel Software Corporation.

Mehta, C. R., Patel, N. R. and Gray, R. (1985). On computing an exact confidence interval for the common odds ratio in several 2×2 contingency tables. *Journal of the American Statistical Association* **80**, 969–973.

Mehta, C. R., Patel, N. R. and Senchauduri, P. (1988a). Importance sampling for estimating exact probabilities in permutational inference. *Journal of the American Statistical Association* **83**, 999–1005.

Mehta, C. R., Patel, N. R. and Wei, L. J. (1988b). Computing exact permutational distributions with restricted randomization designs. *Biometrika* **75**, 295–302.

Mehta, C. R., Patel, N. R. and Senchaudhuri, P. (2000). Efficient Monte Carlo methods for conditional logistic regression. *Journal of the American Statistical Association* **95**, 99–108.

Midena, E., Vujosevic, S., Convento, E., Manfrè, A., Cavarzeran, F. and Pilotto, E. (2007a). Microperimetry and fundus autofluorescence in patients with early age-related macular degeneration. *British Journal of Ophthalmology* **91**, 1499–1503.

Midena, E., Gambato, C., Miotto, S., Cortese, M., Salvi, R. and Ghirlando, A. (2007b). Long-term effects on corneal keratocytes of mitomycin C during photorefractive keratectomy: a randomized contralateral eye confocal microscopy study. *Journal of Refractive Surgery* **23**, 1011–1014.

Mielke, P. W. and Berry, K. J. (2007). *Permutation Methods, A Distance Function Approach*, 2nd edn. New York: Springer-Verlag.

Mielke, P. W. and Iyer, H. K. (1982). Permutation techniques for analyzing multi-response data from randomized block experiments. *Communications in Statistics–Theory and Methods* **3**, 1427–1437.

Miller, A. J. (1984). Selection of subsets of regression variables. *Journal of the Royal Statistical Society A* **147**, 389–425.

Miller, R. (1981). *Simultaneous Statistical Inference*, 2nd edn. New York: Springer-Verlag.

Mo, G. (2005). Anatomy of the skull of Mediterranean monk seal, *Monachus monachus*, functional morphology, densitometry, morphometrics and notes on natural history. Ph.D. thesis, University of Padua.

Molenberghs, G. and Ryan, L. M. (1999). An exponential family model for clustered multivariate binary data. *Environmetrics* **10**, 279–300.

Motoo, M. (1957). On the Hoeffding's combinatorial central limit theorem. *Annals of the Instute of Statistical Mathematics* **8**, 145–154.

Moser, V. C. (1989). Screening approaches to neurotoxicity: a functional observational battery. *Journal of the American College of Toxicology* **8**, 85–93.

Moser, V. C. (1992). Applications of a neurobehavioral screening battery. *Journal of the American College of Toxicology* **10**, 661–669.

Moses, L. E. (1963). Rank test for dispersion. *Annals of Mathematical Statistics* **34**, 973–983.

Mukerjee, H., Robertson, T. and Wright, F. T. (1986). Multiple contrast tests for testing against a simple free ordering. In R. Dykstra, T. Robertson and F. T. Wright (eds), *Advances in Order Restricted Statistical Inference*. Berlin: Springer-Verlag.

Munzel, U. (1999). Nonparametric methods for paired samples. *Statistica Neerlandica* **53**, 277–286.

Nelson, L. S. (1992). A randomization test for ordered alternatives. *Journal of Quality Technology* **24**, 51–53.

Nelson, W. (1972). Theory and applications of hazard plotting for censored failure data. *Technometrics* **14**, 945–965.

Neuhäuser, M. (2007). A comparative study of nonparametric two-sample tests after Levene's transformation. *Journal of Statistical Computation and Simulation* **77**, 517–526.

Neuhäuser, M., Leisler, B. and Hothorn, L. A. (2003). A trend test for the analysis of multiple paternity. *Journal of Agricultural, Biological, and Environmental Statistics* **8**, 29–35.

Nikitin, Y. Y. (1995). *Asymptotic Efficiency of Nonparametric Tests*. Cambridge: Cambridge University Press.

Noether, G. E. (1949). On a theorem by Wald and Wolfowitz. *Annals of Statistics* **20**, 455–458.

Noether, G. E. (1978). Distribution-free confidence intervals. *Statistica Neerlandica* **32**, 109–122.

Nogales, A. G., Oyola, J. A. and Pérez, P. (2000). Invariance, almost invariance and sufficiency. *Statistica* **60**, 277–286.

Nordheim, E. (1984). Inference from nonrandomly missing categorical data: an example from a genetic study on Turner's syndrome. *Journal of the American Statistical Association* **79**, 772–780.

Nüesch, P. E. (1966). On the problem of testing location in multivariate populations for restricted alternatives. *Annals of Mathematical Statistics* **37**, 113–119.

O'Brien, P. C. (1978). A nonparametric test for association with censored data. *Biometrics* **34**, 243–250.

Odén, A. and Wedel, H. (1975). Arguments for Fisher's permutation test. *Annals of Statistics* **3**, 510–520.

Oosterhoff, J. (1969). *Combinations of One-Sided Statistical Tests*, Mathematical Centre Tracts, 28. Amsterdam: Mathematische Centrum.

Oshungade, I. (1989). Some methods of handling item non-response in categorical data. *The Statistician* **38**, 281–296.

Owen, A. B. (1988). Empirical likelihood ratio confidence intervals for a single functional. *Biometrika* **75**, 237–249.

Pagano, M. and Tritchler, D. (1983). On obtaining permutation distributions in polynomial time. *Journal of the American Statistical Association* **78**, 435–440.

Page, E. B. (1963). Ordered hypotheses for multiple treatments: a significance test for linear ranks. *Journal of the American Statistical Association* **58**, 216–230.

Pallini, A. (1990). Il campionamento per importanza nella stima del valore critico di un test di permutazione. *Quaderni di Statistica ed Econometria* **12**, 211–220.

Pallini, A. (1991). Semiparametric combinations of dependent tests using Latin hypercube resampling. *Statistica Applicata* **3**, 38–45.

Pallini, A. (1992a). Test bootstrap e di permutazione per funzioni di matrici di covarianza. In *Atti della XXXVI Riunione Scientifica della Società Italiana di Statistica* **2**, pp. 263–270. SIS, Rome.

Pallini, A. (1992b). On characteristic function-based bootstrap tests. *Journal of the Italian Statistical Society* **1**, 77–86.

Pallini, A. (1992c). Semiparametric combinations of two dependent tests. *Statistica* **52**, 263–270.

Pallini, A. (1994). Bahadur exact slopes for a class of combinations of dependent tests. *Metron* **52**, 53–65.

Pallini, A. and Pesarin, F. (1990). Combinazione di test dipendenti: una soluzione basata su tecniche di ricampionamento. In *Atti della XXXV Riunione Scientifica della Società Italiana di Statistica* **2**, pp. 367–374. SIS, Rome.

Pallini, A. and Pesarin, F. (1992a). Test di permutazione. *SIS Formazione* Pescara.

Pallini, A. and Pesarin, F. (1992b). A class of combinations of dependent tests by a resampling procedure. In K. H. Jöckel, G. Rothe and W. Sendler (eds), *Bootstrapping and Related Techniques*, Lecture Notes in Economics and Mathematical Systems, 376. Springer-Verlag, Berlin, pp. 93–97.

Pallini, A. and Pesarin, F. (1994). Ricampionamento via calibrazione per il bootstrap condizionato. In *Atti della XXXVII Riunione Scientifica della Società Italiana di Statistica*, pp. 473–480. SIS, Rome.

Park, T. and Brown, M. B. (1994). Models for categorical data with nonignorable nonresponse. *Journal of the American Statistical Association* **89**, 44–52.

Parthasarathy, K. R. (1977). *Introduction to Probability and Measure*. London: Macmillan.

Patil, G. P. and Rao, C. R. (1977). The weighted distributions: a survey of their applications. In P. R. Krishnaiah (ed.), *Applications of Statistics*. Amsterdam: North-Holland, pp. 383–405.

Pepe, M. S. and Fleming, T. R. (1989). Weighted Kaplan–Meier statistics: a class of distance tests for censored survival data. *Biometrics* **45**, 497–507.

Pepe, M. S. and Fleming, T. R. (1991). Weighted Kaplan–Meier statistics: large sample and optimality considerations. *Journal of the Royal Statistical Society B* **53**, 341–352.

Perlman, M. D. and Wu, L. (1998). A class of conditional tests for a multivariate one-sided alternative. Technical report 328, Department of Statistics, University of Washington, Seattle.

Pesarin, F. (1984). On randomized statistical procedures. In M. Iosifescu (ed.), *Proc. of the Seventh Conference on Probability Theory*. Acad. Rep. Socialiste Romania, Bucharest, pp. 295–306.

Pesarin, F. (1988). Combinazione non parametrica di test dipendenti. In G. Diana, C. Provasi and R. Vedaldi (eds), *Metodi Statistici per la Tecnologia e l'Analisi Multivariata*, pp. 241–248. Department of Statistics, University of Padua.

Pesarin, F. (1989). The analysis of spatial point patterns combining a set of indicators. In *Atti del III Congresso della Società Italiana di Ecologia*, pp. 805–808. SIE, Siena.

Pesarin, F. (1990a). Tecniche di ricampionamento e verifica multidimensionale delle ipotesi. *Quaderni di Statistica e Econometria* **12**, 39–60.

Pesarin, F. (1990b). On a nonparametric combination method for dependent permutation tests with applications. *Psychotherapy and Psychosomatics* **54**, 172–179.

Pesarin, F. (1990c). Multidimensional testing for mixed variables by resampling procedures. *Statistica Applicata* **2**, 395–406.

Pesarin, F. (1991a). Tecniche di ricampionamento e verifica d'ipotesi multidimensionale. *Statistica* **50**, 483–501.

Pesarin, F. (1991b). Some multidimensional testing problems for missing values via a resampling procedure. *Statistica Applicata* **3**, 567–577.

Pesarin, F. (1992). A resampling procedure for nonparametric combination of several dependent tests. *Journal of the Italian Statistical Society* **1**, 87–101.

Pesarin, F. (1993). Analisi di tabelle di contingenza multidimensionali tramite tecniche di ricampionamento. *Archivio per l'Antropologia e la Etnologia* **123**, 575–588.

Pesarin, F. (1994). Goodness of fit testing for ordered discrete distributions by resampling techniques. *Metron* **52**, 57–71.

Pesarin, F. (1995). A new solution for the generalized Behrens–Fisher problem. *Statistica* **55**, 131–146.

Pesarin, F. (1996a). An almost exact solution for the multivariate Behrens–Fisher problem. In A. Azzalini and F. Parpinel (eds), *Studi in Onore di Oliviero Lessi*, pp. 109–126. Department of Statistics, University of Padua.

Pesarin, F. (1996b). Il problema di Behrens–Fisher dal punto di vista dei test di permutazione. In D. Marasini and P. Ferrari (eds), *Studi in Onore di Giampiero Landenna*, pp. 435–456. Giuffré, Milan.

Pesarin, F. (1996c). Some testing problems with repeated measures by using a nonparametric combination method. In G. Zanardi (ed.), *Atti della Giornata di Studio su Qualità dei Dati, Campionamento, Inferenza*, pp. 157–167. Università Cà Foscari, Venezia.

Pesarin, F. (1997a). An almost exact solution for the multivariate Behrens–Fisher problem. *Metron* **55**, 85–100.

Pesarin, F. (1997b). A nonparametric combination method for dependent permutation tests with application to some problems with repeated measures. In C. P. Kitsos and L. Edler (eds), *Proceedings of the ISI Satellite Meeting on Industrial Statistics*, pp. 259–268. Heidelberg: Physica-Verlag.

Pesarin, F. (1999a). Exact permutation testing for effects in replicated two-way factorial designs. Working Paper no. 1999.11, Department of Statistics, University of Padua.

Pesarin, F. (1999b). *Permutation Testing of Multidimensional Hypotheses by Nonparametric Combination of Dependent Tests*. CLEUP, Padua.

Pesarin, F. (2001). *Multivariate Permutation Tests: With Application in Biostatistics*. Chichester: John Wiley & Sons, Ltd.

Pesarin, F. (2002). Extending permutation conditional inference to unconditional one. *Statistical Methods and Applications* **11**, 161–173.

Pesarin, F. and Salmaso, L. (1998a). Introduzione a CR-test: conditional resampling test. Working Paper no. 1998.9, Department of Statistics, Unversity of Padua.

Pesarin, F. and Salmaso, L. (1998b). Conditional resampling test – statistical software to perform multivariate resampling techniques for testing nonparametric hypotheses. *Computational Statistics & Data Analysis* **28**, 327.

Pesarin, F. and Salmaso, L. (1999). Exact permutation testing for effects in replicated 2^k factorial designs. Working Paper no. 1999.12, Department of Statistics, University of Padua.

Pesarin, F. and Salmaso, L. (2000a). Permutation tests for effects in unreplicated factorial designs. *Proceedings of the XII Conference of the Greek Statistical Institute*, pp. 465–476. GSI, Athens.

Pesarin, F. and Salmaso, L. (2000b). Goodness of fit testing for ordered categorical variables. In *Abstracts Book of the International Workshop GOF2000 on Goodness of fit Tests and Validity of Models*, pp. 118–120. ISUP, Paris.

Pesarin, F. and Salmaso, L. (2002). Exact permutation tests for unreplicated factorials. *Journal of Applied Stochastic Models in Business and Industry* **18**, 287–299.

Pesarin, F. and Salmaso, L. (2006). Permutation tests for univariate and multivariate ordered categorical data. *Austrian Journal of Statistics* **35**, 315–324.

Pesarin, F. and Salmaso, L. (2009). Finite-sample consistency of combination-based permutation tests with application to repeated measures designs. *Journal of Nonparametric Statistics*. http://dx.doi.org/10.1080/10485250902807407.

Peto, R. and Peto, J. (1972). Asymptotically efficient rank invariant test procedures (with discussion). *Journal of the Royal Statistical Society A* **135**, 185–207.

Petrondas, D. A. and Gabriel, K. R. (1983). Multiple comparisons by rerandomization tests. *Journal of the American Statistical Association* **78**, 949–957.

Pfanzagl, J. (1974). On the Behrens–Fisher problem. *Biometrika* **61**, 39–47.

Phillips, M. J. (1993). Contingency table with missing data. *The Statistician* **42**, 9–18.

Piccolo, D. (2000). *Statistica*, 2nd edn. Il Mulino, Bologna.

Pitman, E. J. C. (1937a). Significance tests which may be applied to samples from any populations. *Supplement to the Journal of the Royal Statististical Society* **4**, 119–130.

Pitman, E. J. C. (1937b). Significance tests which may be applied to samples from any populations II: the correlation coefficient test. *Supplement to the Journal of the Royal Statististical Society* **4**, 225–232.

Pollard, J. H. (1977). *A Handbook of Numerical Statistical Techniques*. Cambridge: Cambridge University Press.

Prentice, R. L. (1978). Linear rank tests with right censored data. *Biometrika* **65**, 167–179.

Puri, M. L. and Sen, P. K. (1971). *Nonparametric Methods in Multivariate Analysis*. New York: John Wiley & Sons, Inc.

Qureshi, K. N., Naguib, R. N., Hamdy, F. C., Neal, D. E. and Mellon, J. K. (2000). Neural network analysis of clinicopathological and molecular markers in bladder cancer. *Journal of Urology* **163**, 630–633.

Ramsay, J. O. and Silverman, B. W. (1997). *Functional Data Analysis*. New York: Springer-Verlag.

Ramsay, J. O. and Silverman, B. W. (2002). *Applied Functional Data Analysis*. New York: Springer-Verlag.

Randles, R. H. and Wolfe, D. A. (1979). *Introduction to the Theory of Nonparametric Statistics*. New York: John Wiley & Sons, Inc.

Randles, R. H. (2001). On neutral responses (zeros) in the sign test and ties in Wilcoxon–Mann–Whitney test. *American Statistician* **55**, 96–101.

Rao, C. and Suryawanshi, S. (1996). Statistical analysis of shape of objects based on landmark data. *Proceedings of the National Academy of Sciences of the United States of America* **93**, 12132–12136.

Rao, C. and Suryawanshi, S. (1998). Statistical analysis of shape through triangulation of landmarks: a study of sexual dimorphism in hominids. *Proceedings of the National Academy of Sciences of the United States of America* **95**, 4121–4125.

Reboussin, D. M. and DeMets, D. L. (1996). Exact permutation inference for two sample repeated measures data. *Communications in Statistics – Theory and Methods* **25**, 2223–2238.

Rényi, A. (1996). *Calcul des probabilités*. Paris: Dunod.

Robertson, T., Wright, F. T. and Dykstra, R. L. (1988). *Order Restricted Statistical Inference*. Chichester: John Wiley & Sons, Ltd.

Robinson, J. (1980). An asymptotic expansion for permutation tests with several samples. *Annals of Mathematical Statistics* **8**, 851–864.

Robinson, J. (1982). Saddlepoint approximations for permutation tests and confidence intervals. *Journal of the Royal Statistical B* **44**, 91–101.

Robinson, J. (1987). Nonparametric confidence intervals in regression: the bootstrap and randomization methods. In M. L. Puri, J. P. Vilaplana and W. Wertz (eds), *New Perspectives in Theoretical and Applied Statistics*, 243–255. Wiley, New York.

Rohlf, F. J. (1993). Relative warp analysis and an example of its application to mosquito wings. In L. F. Marcus, E. Bello and A. Garcia-Valdecasas (eds), *Contributions to Morphometrics*, vol. 8. Mardrid: Monografias del Museo Nacional de Ciencias Naturales, pp. 131–159.

Rohlf, F. J. (1999). Shape statistics: Procrustes superimpositions and tangent spaces. *Journal of Classification* **16**, 197–225.

Rohlf, F. J. (2000). Statistical power comparisons among alternative morphometric methods. *American Journal of Physical Anthropology* **111**, 463–478.

Rohlf, F. J. (2007). *tpsDig2, digitize landmarks and outlines, version 2.12*. Department of Ecology and Evolution, State University of New York at Stony Brook.

Rohlf, F. J. (2008a). *tpsRelw, Relative warps analysis, version 1.46*. Department of Ecology and Evolution, State University of New York at Stony Brook.

Rohlf, F. J. (2008b). *TPStri, Explore shapes of triangles, version 1.25*. Department of Ecology and Evolution, State University of New York at Stony Brook.

Rohlf, F. J. and Slice, D. E. (1990). Extensions of the Procrustes method for the optimal superimposition of landmarks. *Systematic Zoology* **39**, 40–59.

Romano, J. P. (1988). A bootstrap revival of some nonparametric distance tests. *Journal of the American Statistical Association* **83**, 698–708.

Romano, J. P. (1989). Bootstrap and randomization tests of some nonparametric hypotheses. *Annals of Statistics* **17**, 141–159.

Romano, J. P. (1990). On the behaviour of randomization tests without group variance assumption. *Journal of the American Statistical Association* **85**, 686–692.

Rosenbaum, P. R. (2002). *Observational Studies*. New York: Springer-Verlag.

Rosenbaum, P. R. (2005). Observational study. In *Encyclopedia of Statistics in Behavioral Science*, vol. 3. Chichester: John Wiley & Sons, Ltd, pp. 1451–1462.

Rosenbaum, P. R. and Rubin, D. B. (1983). The central role of the propensity score in observational studies for causal effects. *Biometrika* **70**, 41–55.

Rosenthal, R. and Rubin, D. B. (1983). Ensemble-adjusted p values. *Psychological Bulletin* **94**, 540–541.

Ross, S. M. (1983). *Stochastic Processes*. New York: John Wiley & Sons, Inc.

Roy, A. (2006). Estimating correlation coefficient between two variables with repeated observations using mixed effects model. *Biometrical Journal* **48**, 286–301.

Rubin, D. B. (1976). Inference and missing data. *Biometrika* **63**, 581–592.

Runger, G. C. and Eaton, M. L. (1992). Most powerful invariant permutation tests. *Journal of Multivariate Analysis* **42**, 202–209.

Ruol, A., Portale, G., Zaninotto, G., Cagol, M., Cavallin, F., Castoro, C, Chiarion Sileni, V., Alfieri, R., Rampado, S. and Ancona, E. (2007). Results of esophagectomy for esophageal cancer in elderly patients: age has a little influence on outcome and survival. *Journal of Thoracic and Cardiovascular Surgery* **133**, 1186–1192.

Ruymgaart, F. H. (1980). A unified approach to the asymptotic distribution theory of certain midrank statistics. In J. P. Raoult (ed.), *Statistique non Paramétrique Asymptotique*, Lecture Notes in Mathematics, 821. Berlin: Springer-Verlag, pp. 1–18.

Salmaso, L. (2003). Synchronized permutation tests in factorial designs. *Communications in Statistics – Theory and Methods* **32**, 1419–1437.

Salmaso, L. (2005). Permutation tests in screening two-level factorial experiments. *Advances and Applications in Statistics* **5**, 91–110.

Salmaso, L. and Solari, A. (2005). Multiple aspect testing for case–control designs. *Metrika* **12**, 1–10.

Salmaso, L. and Solari, A. (2006). Nonparametric iterated combined tests for genetic differentiation. *Computational Statistics & Data analysis* **50**, 1105–1112.

Sampson, A. R. and Whitaker, L. R. (1989). Estimation of multivariate distributions under stochastic ordering. *Journal of the American Statistical Association* **84**, 541–548.

Savage, I. R. (1956). Contributions to the theory of rank order statistics. The two sample case. *Annals of Mathematical Statistics* **27**, 590–615.

Scheffé, H. (1943a). On a measure problem arising in the theory of non-parametric tests. *Annals of Mathematical Statistics* **14**, 227–233.

Scheffé, H. (1943b). Statistical inference in the non-parametric case. *Annals of Mathematical Statistics* **14**, 305–332.

Scheffé, H. (1943c). On solution of the Behrens–Fisher problem based on the t-distribution. *Annals of Mathematical Statistics* **14**, 1430–1432.

Scheffé, H. (1944). A note on the Behrens–Fisher problem. *Annals of Mathematical Statistics* **14**, 35–44.

Schemper, M. A. (1984). A survey of permutation tests for censored survival data. *Communications in Statistics. Theory and Methods* **13**, 433–448.

Sen, P. K. (1983). On permutational central limit theorems for general multivariate linear statistics. *Sankhyā A* **45**, 141–149.

Sen, P. K. (1985). Permutational central limit theorems. In S. Kotz, N. L. Johnson and C. B. Read (eds), *Encyclopedia of Statistical Sciences* **6**. New York: John Wiley & Sons, Inc, pp. 683–687.

Servy, E. C. and Sen, P. K. (1987). Missing values in multi-sample rank permutation tests for MANOVA and MANCOVA. *Sankhyā A* **49**, 78–95.

Shabsigh, A. and Bochner, B. H. (2006). Use of nomograms as predictive tools in bladder cancer. *World Journal of Urology* **24**, 489–498.

Shafer, G. and Olkin, I. (1983). Adjusting p values to account for selection over dichotomies. *Journal of the American Statistical Association* **78**, 674–678.

Sham, P. C. and Curtis, D. (1995). Monte Carlo tests for associations between disease and alleles at highly polymorphic loci. *Annals of Human Genetics* **59**, 97–105.

Shannon, C. E. (1948). A mathematical theory of communication. *AT&T Technology Journal* **27**, 379–423.

Shapiro, C. P. and Hubert, L. (1979). Asymptotic normality of permutation statistics derived from weighted sums of bivariate functions. *Annals of Statistics* **7**, 788–794.

Shen, Y. and Cai, J. (2001). Maximum of the weighted Kaplan–Meier tests with application to cancer prevention and screening trials. *Biometrics* **57**, 837–843.

Shorack, G. R. (1967). Testing against ordered alternative in model I analysis of variance: normal theory and non-parametrics. *Annals of Mathematical Statistics* **38**, 1740–1753.

Shorack, G. R., Wellner, J. A. (1986). *Empirical Processes with Applications to Statistics*. New York: John Wiley & Sons, Inc.

Shumacher, M. (1984). Two sample tests of Cramér-vON Mises and Kolmogorov–Smirnov type for randomly censored data. *International Statistical Review* **52**, 263–281.

Shuster, J. J. and Boyett, J. M. (1979). Nonparametric multiple comparison procedures. *Journal of the American Statistical Association* **74**, 379–382.

Siegel, A. F. and Benson, R. H. (1982). A robust comparison of biological shapes. *Biometrics* **38**, 341–350.

Silvapulle, J. S. and Sen, P. K. (2005). *Constrained Statistical Inference. Inequality, Order and Shape Restrictions*. Hoboken, NJ: John Wiley & Sons, Inc.

Simes, R. J. (1986). An improved Bonferroni procedure for multiple tests of significance. *Biometrika* **73**, 751–754.

Slice, D. E. (2005). *Modern Morphometrics in Physical Anthropology*. New York: Springer-Verlag.

Slice, D. E., Bookstein, F. L., Marcus, L. F. and Rohlf, F. J. (1996). A glossary for geometric morphometrics. *Advances in Morphometrics* **284**, 531–551.

Small, C. G. (1996). *The Statistical Theory of Shape*. New York: Springer-Verlag.

Solari, A., Salmaso, L., El Barmi, H. and Pesarin, F. (2008). Conditional tests in a competing risks model. *Lifetime Data Analysis* **14**, 154–166.

Soper, K. A. and Tonkonoh, N. (1993). The discrete distribution used for the log-rank test can be inaccurate. *Biometrical Journal* **35**, 291–298.

Sonnemann, E. (2008). General solutions to multiple testing problems. *Biometrical Journal* **50**, 641–656.

Spino, C. and Pagano, M. (1991). Efficient calculation of the permutation distribution of trimmed means. *Journal of the American Statistical Association* **86**, 729–737.

Spjø tvoll, E. (1972). On the optimality of some multiple comparison procedures. *Annals of Mathematical Statistics* **43**, 398–411.

Srivastava, A., Joshi, S., Mio, W. and Liu, X. (2005). Statistical shape analysis: clustering, learning and testing. *IEEE Transactions on Pattern Analysis and Machine Intelligence* **27**, 590–602.

Stoyan, D. (1983). *Comparison Methods for Queues and Other Stochastic Models*. Chichester: John Wiley & Sons, Ltd.

Tarone, R. E. (1975). Tests for trend in life table analysis. *Biometrika* **62**, 679–690.

Tarone, R. E. and Ware, J. H. (1977). On distribution-free tests for equality of survival distributions. *Biometrika* **64**, 156–160.

Terriberry, T. B., Joshi, S. and Garig, G. (2005). Hypothesis testing with nonlinear shape models. In *Information Processing in Medical Imaging*. Berlin: Springer-Verlag, pp. 15–26.

Thompson, D. W. (1961). *On Growth and Form*. Cambridge: Cambridge University Press.

Thornett, M. L. (1982). The role of randomization in model-based inference. *Australian Journal of Statistics* **24**, 137–145.

Tsui, K. and Weerhandi, S. (1989). Generalized p-values in significance testing of hypotheses in presence of nuisance parameters. *Journal of the American Statistical Association* **84**, 602–607.

Vadiveloo, J. (1983). On the theory of modified randomization tests for nonparametric hypotheses. *Communications in Statistics – Theory and Methods* **11**, 1581–1596.

Van den Brink, W. P. and Van den Brink, S. G. J. (1990). A modified approximate permutation test procedure. *Computational Statistics Quarterly* **3**, 241–247.

Wald, A. and Wolfowitz, J. (1944). Statistical tests based on permutations of the observations. *Annals of Mathematical Statistics* **15**, 358–372.

Wang, Y. (1996). A likelihood ratio test against stochastic ordering in several populations, *Journal of the American Statistical Association* **91**, 1676–1683.

Ware, J. H. (1985). Linear models for the analysis of longitudinal studies. *American Statistician* **39**, 95–101.

Wasserman, L. (2006). *All of Nonparametric Statistics*. New York: Springer.

Watson, G. S. (1957). Sufficient statistics, similar regions and distribution-free tests. *Journal of the Royal Statistical Society B* **19**, 262–267.

Wei, L. J. and Johnson, W. E. (1985). Combining dependent tests with incomplete repeated measurements. *Biometrika* **72**, 359–364.

Wei, L. J. and Lachin, J. M. (1984). Two-sample asymptotically distribution-free tests for incomplete multivariate observations. *Journal of the American Statistical Association* **79**, 653–661.

Welch, W. J. and Gutierrez, L. G. (1988). Robust permutation tests for matched-pairs designs. *Journal of the American Statistical Association* **83**, 450–455.

Westberg, M. (1986). Tippett-adaptive method of combining independent statistical tests. Reaserch Report 3, Department of Statistics, Gothenburg, Sweden.

Westfall, P. H. (1997). Multiple testing of general contrasts using logical constraints and correlations. *Journal of the American Statistical Association* **92**, 299–306.

Westfall, P. H. and Krishen, A. (2001). Optimally weighted, fixed sequence, and gatekeeping multiple testing procedures. *Journal of Statistical Planning and Inference* **99**, 25–40.

Westfall, P. H. and Troendle, J. F. (2008). Multiple testing with minimal assumptions. *Biometrical Journal* **50**, 745–755.

Westfall, P. H. and Wolfinger, R. D. (2000). *Closed Multiple Testing Procedures and PROC MULTTEST*. Observations SAS Institute Inc.

Westfall, P. H. and Young, S. S. (1989). *P*-value adjustments for multiple tests in multivariate binomial model. *Journal of the American Statistical Association* **84**, 780–786.

Westfall, P. H. and Young, S. S. (1993). *Resampling-Based Multiple Testing: Examples and Methods for p-Value Adjustment*. New York: John Wiley & Sons, Inc.

Westfall, P. H., Tobias, R. D., Rom, D., Wolfinger, R. D. and Hochberg, Y. (1999). *Multiple Comparisons and Multiple Tests Using the SAS System*. Cary, NC: SAS Institute.

Wilcoxon, F. (1945). Individual comparisons by ranking method. *Biometrics* **1**, 80–83.

Witting, H. (1995). Rank tests for scale: Hájek's influence and recent developments. *Kybernetika* **31**, 269–291.

Wu, L. and Gilbert, P. (2002). Flexible weighted log-rank tests optimal for detecting early and/or late survival differences. *Biometrics* **58**, 997–1004.

Zanella, A. (1973). Sulle procedure di classificazione simultanea. *Statistica* **33**, 63–120.

Zhang, Y. and Rosenberger, W. F. (2005). On asymptotic normality of the randomization-based logrank test. *Journal of Nonparametric Statistics* **17**, 833–839.

Zimmerman, H. (1985a). Exact calculations of permutation distributions for 2 dependent samples. *Biometrical Journal* **27**, 349–352.

Zimmerman, H. (1985b). Exact calculations for permutation distribution for 2 independent samples. *Biometrical Journal* **27**, 431–443.

Index

Permutation Tests for Complex Data: Theory, Applications and Software Fortunato Pesarin, Luigi Salmaso

WILEY SERIES IN PROBABILITY AND STATISTICS

Established by WALTER A. SHEWHART and SAMUEL S. WILKS

Editors
David J. Balding, Noel A. C. Cressie, Garrett M. Fitzmaurice, Harvey Goldstein, Geert Molenberghs, David W. Scott, Adrian F.M. Smith, Ruey S. Tsay, Sanford Weisberg

Editors Emeriti
Vic Barnett, J. Stuart Hunter, David G. Kendall, Jozef L. Teugels

The Wiley Series in Probability and Statistics is well established and authoritative. It covers many topics of current research interest in both pure and applied statistics and probability theory. Written by leading statisticians and institutions, the titles span both state-of-the-art developments in the field and classical methods.

Reflecting the wide range of current research in statistics, the series encompasses applied, methodological and theoretical statistics, ranging from applications and new techniques made possible by advances in computerized practice to rigorous treatment of theoretical approaches.

This series provides essential and invaluable reading for all statisticians, whether in academia, industry, government, or research.

ABRAHAM and LEDOLTER · Statistical Methods for Forecasting
AGRESTI · Analysis of Ordinal Categorical Data
AGRESTI · An Introduction to Categorical Data Analysis
AGRESTI · Categorical Data Analysis, Second Edition
ALTMAN, GILL and McDONALD · Numerical Issues in Statistical Computing for the Social Scientist
AMARATUNGA and CABRERA · Exploration and Analysis of DNA Microarray and Protein Array Data
ANDĚL · Mathematics of Chance
ANDERSON · An Introduction to Multivariate Statistical Analysis, Third Edition
* ANDERSON · The Statistical Analysis of Time Series
ANDERSON, AUQUIER, HAUCK, OAKES, VANDAELE and WEISBERG · Statistical Methods for Comparative Studies
ANDERSON and LOYNES · The Teaching of Practical Statistics
ARMITAGE and DAVID (editors) · Advances in Biometry
ARNOLD, BALAKRISHNAN and NAGARAJA · Records
* ARTHANARI and DODGE · Mathematical Programming in Statistics
* BAILEY · The Elements of Stochastic Processes with Applications to the Natural Sciences
BALAKRISHNAN and KOUTRAS · Runs and Scans with Applications
BALAKRISHNAN and NG · Precedence-Type Tests and Applications
BARNETT · Comparative Statistical Inference, Third Edition
BARNETT · Environmental Statistics: Methods & Applications
BARNETT and LEWIS · Outliers in Statistical Data, Third Edition
BARTOSZYNSKI and NIEWIADOMSKA-BUGAJ · Probability and Statistical Inference
BASILEVSKY · Statistical Factor Analysis and Related Methods: Theory and Applications
BASU and RIGDON · Statistical Methods for the Reliability of Repairable Systems
BATES and WATTS · Nonlinear Regression Analysis and Its Applications
BECHHOFER, SANTNER and GOLDSMAN · Design and Analysis of Experiments for Statistical Selection, Screening and Multiple Comparisons
BEIRLANT, GOEGEBEUR, SEGERS, TEUGELS and DE WAAL · Statistics of Extremes: Theory and Applications
BELSLEY · Conditioning Diagnostics: Collinearity and Weak Data in Regression
BELSLEY, KUH and WELSCH · Regression Diagnostics: Identifying Influential Data and Sources of Collinearity

*Now available in a lower priced paperback edition in the Wiley Classics Library.

BENDAT and PIERSOL · Random Data: Analysis and Measurement Procedures, Third Edition
BERNARDO and SMITH · Bayesian Theory
BERRY, CHALONER and GEWEKE · Bayesian Analysis in Statistics and Econometrics: Essays in Honor of Arnold Zellner
BHAT and MILLER · Elements of Applied Stochastic Processes, Third Edition
BHATTACHARYA and JOHNSON · Statistical Concepts and Methods
BHATTACHARYA and WAYMIRE · Stochastic Processes with Applications
BIEMER, GROVES, LYBERG, MATHIOWETZ and SUDMAN · Measurement Errors in Surveys
BILLINGSLEY · Convergence of Probability Measures, Second Edition
BILLINGSLEY · Probability and Measure, Third Edition
BIRKES and DODGE · Alternative Methods of Regression
BISWAS, DATTA, FINE and SEGAL · Statistical Advances in the Biomedical Sciences: Clinical Trials, Epidemiology, Survival Analysis, and Bioinformatics
BLISCHKE and MURTHY (editors) · Case Studies in Reliability and Maintenance
BLISCHKE and MURTHY · Reliability: Modeling, Prediction and Optimization
BLOOMFIELD · Fourier Analysis of Time Series: An Introduction, Second Edition
BOLLEN · Structural Equations with Latent Variables
BOLLEN and CURRAN · Latent Curve Models: A Structural Equation Perspective
BOROVKOV · Ergodicity and Stability of Stochastic Processes
BOSQ and BLANKE · Inference and Prediction in Large Dimensions
BOULEAU · Numerical Methods for Stochastic Processes
BOX · Bayesian Inference in Statistical Analysis
BOX · R. A. Fisher, the Life of a Scientist
BOX and DRAPER · Empirical Model-Building and Response Surfaces
* BOX and DRAPER · Evolutionary Operation: A Statistical Method for Process Improvement
BOX · Improving Almost Anything *Revised Edition*
BOX, HUNTER and HUNTER · Statistics for Experimenters: An Introduction to Design, Data Analysis and Model Building
BOX, HUNTER and HUNTER · Statistics for Experimenters: Design, Innovation and Discovery, Second Edition
BOX and LUCEÑO · Statistical Control by Monitoring and Feedback Adjustment
BRANDIMARTE · Numerical Methods in Finance: A MATLAB-Based Introduction
BROWN and HOLLANDER · Statistics: A Biomedical Introduction
BRUNNER, DOMHOF and LANGER · Nonparametric Analysis of Longitudinal Data in Factorial Experiments
BUCKLEW · Large Deviation Techniques in Decision, Simulation and Estimation
CAIROLI and DALANG · Sequential Stochastic Optimization
CASTILLO, HADI, BALAKRISHNAN and SARABIA · Extreme Value and Related Models with Applications in Engineering and Science
CHAN · Time Series: Applications to Finance
CHARALAMBIDES · Combinatorial Methods in Discrete Distributions
CHATTERJEE and HADI · Regression Analysis by Example, Fourth Edition
CHATTERJEE and HADI · Sensitivity Analysis in Linear Regression
CHERNICK · Bootstrap Methods: A Practitioner's Guide
CHERNICK and FRIIS · Introductory Biostatistics for the Health Sciences
CHILÉS and DELFINER · Geostatistics: Modeling Spatial Uncertainty
CHOW and LIU · Design and Analysis of Clinical Trials: Concepts and Methodologies, Second Edition
CLARKE · Linear Models: The Theory and Application of Analysis of Variance
CLARKE and DISNEY · Probability and Random Processes: A First Course with Applications, Second Edition
* COCHRAN and COX · Experimental Designs, Second Edition
CONGDON · Applied Bayesian Modelling

*Now available in a lower priced paperback edition in the Wiley Classics Library.

CONGDON · Bayesian Models for Categorical Data
CONGDON · Bayesian Statistical Modelling, Second Edition
CONOVER · Practical Nonparametric Statistics, Second Edition
COOK · Regression Graphics
COOK and WEISBERG · An Introduction to Regression Graphics
COOK and WEISBERG · Applied Regression Including Computing and Graphics
CORNELL · Experiments with Mixtures, Designs, Models and the Analysis of Mixture Data, Third Edition
COVER and THOMAS · Elements of Information Theory
COX · A Handbook of Introductory Statistical Methods
* COX · Planning of Experiments
CRESSIE · Statistics for Spatial Data, Revised Edition
CSÖRGÖ and HORVÁTH · Limit Theorems in Change Point Analysis
DANIEL · Applications of Statistics to Industrial Experimentation
DANIEL · Biostatistics: A Foundation for Analysis in the Health Sciences, Sixth Edition
* DANIEL · Fitting Equations to Data: Computer Analysis of Multifactor Data, Second Edition
DASU and JOHNSON · Exploratory Data Mining and Data Cleaning
DAVID and NAGARAJA · Order Statistics, Third Edition
* DEGROOT, FIENBERG and KADANE · Statistics and the Law
DEL CASTILLO · Statistical Process Adjustment for Quality Control
DEMARIS · Regression with Social Data: Modeling Continuous and Limited Response Variables
DEMIDENKO · Mixed Models: Theory and Applications
DENISON, HOLMES, MALLICK and SMITH · Bayesian Methods for Nonlinear Classification and Regression
DETTE and STUDDEN · The Theory of Canonical Moments with Applications in Statistics, Probability and Analysis
DEY and MUKERJEE · Fractional Factorial Plans
DILLON and GOLDSTEIN · Multivariate Analysis: Methods and Applications
DODGE · Alternative Methods of Regression
* DODGE and ROMIG · Sampling Inspection Tables, Second Edition
* DOOB · Stochastic Processes
DOWDY, WEARDEN and CHILKO · Statistics for Research, Third Edition
DRAPER and SMITH · Applied Regression Analysis, Third Edition
DRYDEN and MARDIA · Statistical Shape Analysis
DUDEWICZ and MISHRA · Modern Mathematical Statistics
DUNN and CLARK · Applied Statistics: Analysis of Variance and Regression, Second Edition
DUNN and CLARK · Basic Statistics: A Primer for the Biomedical Sciences, Third Edition
DUPUIS and ELLIS · A Weak Convergence Approach to the Theory of Large Deviations
EDLER and KITSOS (editors) · Recent Advances in Quantitative Methods in Cancer and Human Health Risk Assessment
* ELANDT-JOHNSON and JOHNSON · Survival Models and Data Analysis
ENDERS · Applied Econometric Time Series
ETHIER and KURTZ · Markov Processes: Characterization and Convergence
EVANS, HASTINGS and PEACOCK · Statistical Distribution, Third Edition
FELLER · An Introduction to Probability Theory and Its Applications, Volume I, Third Edition, Revised; Volume II, Second Edition
FISHER and VAN BELLE · Biostatistics: A Methodology for the Health Sciences
FITZMAURICE, LAIRD and WARE · Applied Longitudinal Analysis
* FLEISS · The Design and Analysis of Clinical Experiments
FLEISS · Statistical Methods for Rates and Proportions, Second Edition
FLEMING and HARRINGTON · Counting Processes and Survival Analysis
FULLER · Introduction to Statistical Time Series, Second Edition
FULLER · Measurement Error Models

*Now available in a lower priced paperback edition in the Wiley Classics Library.

GALLANT · Nonlinear Statistical Models.
GEISSER · Modes of Parametric Statistical Inference
GELMAN and MENG (editors) · Applied Bayesian Modeling and Casual Inference from Incomplete-data Perspectives
GEWEKE · Contemporary Bayesian Econometrics and Statistics
GHOSH, MUKHOPADHYAY and SEN · Sequential Estimation
GIESBRECHT and GUMPERTZ · Planning, Construction and Statistical Analysis of Comparative Experiments
GIFI · Nonlinear Multivariate Analysis
GIVENS and HOETING · Computational Statistics
GLASSERMAN and YAO · Monotone Structure in Discrete-Event Systems
GNANADESIKAN · Methods for Statistical Data Analysis of Multivariate Observations, Second Edition
GOLDSTEIN and WOLFF · Bayes Linear Statistics, Theory & Methods
GOLDSTEIN and LEWIS · Assessment: Problems, Development and Statistical Issues
GREENWOOD and NIKULIN · A Guide to Chi-Squared Testing
GROSS, SHORTLE, THOMPSON and HARRIS · Fundamentals of Queueing Theory, fourth Edition
GROSS, SHORTLE, THOMPSON and HARRIS · Solutions Manual to Accompany Fundamentals of Queueing Theory, fourth Edition
* HAHN and SHAPIRO · Statistical Models in Engineering
HAHN and MEEKER · Statistical Intervals: A Guide for Practitioners
HALD · A History of Probability and Statistics and their Applications Before 1750
HALD · A History of Mathematical Statistics from 1750 to 1930
HAMPEL · Robust Statistics: The Approach Based on Influence Functions
HANNAN and DEISTLER · The Statistical Theory of Linear Systems
HARTUNG, KNAPP and SINHA · Statistical Meta-Analysis with Applications
HEIBERGER · Computation for the Analysis of Designed Experiments
HEDAYAT and SINHA · Design and Inference in Finite Population Sampling
HEDEKER and GIBBONS · Longitudinal Data Analysis
HELLER · MACSYMA for Statisticians
HERITIER, CANTONI, COPT and VICTORIA-FESER · Robust Methods in Biostatistics
HINKELMANN and KEMPTHORNE · Design and Analysis of Experiments, Volume 1: Introduction to Experimental Design
HINKELMANN and KEMPTHORNE · Design and analysis of experiments, Volume 2: Advanced Experimental Design
HOAGLIN, MOSTELLER and TUKEY · Exploratory Approach to Analysis of Variance
HOAGLIN, MOSTELLER and TUKEY · Exploring Data Tables, Trends and Shapes
* HOAGLIN, MOSTELLER and TUKEY · Understanding Robust and Exploratory Data Analysis
HOCHBERG and TAMHANE · Multiple Comparison Procedures
HOCKING · Methods and Applications of Linear Models: Regression and the Analysis of Variance, Second Edition
HOEL · Introduction to Mathematical Statistics, Fifth Edition
HOGG and KLUGMAN · Loss Distributions
HOLLANDER and WOLFE · Nonparametric Statistical Methods, Second Edition
HOSMER and LEMESHOW · Applied Logistic Regression, Second Edition
HOSMER and LEMESHOW · Applied Survival Analysis: Regression Modeling of Time to Event Data
HUBER · Robust Statistics
HUBERTY · Applied Discriminant Analysis
HUBERY and OLEJNKI · Applied MANOVA and Discriminant Analysis, 2nd Edition
HUNT and KENNEDY · Financial Derivatives in Theory and Practice, Revised Edition
HURD and MIAMEE · Periodically Correlated Random Sequences: Spectral Theory and Practice
HUSKOVA, BERAN and DUPAC · Collected Works of Jaroslav Hajek – with Commentary

*Now available in a lower priced paperback edition in the Wiley Classics Library.

HUZURBAZAR · Flowgraph Models for Multistate Time-to-Event Data
IMAN and CONOVER · A Modern Approach to Statistics
JACKSON · A User's Guide to Principle Components
JOHN · Statistical Methods in Engineering and Quality Assurance
JOHNSON · Multivariate Statistical Simulation
JOHNSON and BALAKRISHNAN · Advances in the Theory and Practice of Statistics: A Volume in Honor of Samuel Kotz
JOHNSON and BHATTACHARYYA · Statistics: Principles and Methods, Fifth Edition
JOHNSON and KOTZ · Distributions in Statistics
JOHNSON and KOTZ (editors) · Leading Personalities in Statistical Sciences: From the Seventeenth Century to the Present
JOHNSON, KOTZ and BALAKRISHNAN · Continuous Univariate Distributions, Volume 1, Second Edition
JOHNSON, KOTZ and BALAKRISHNAN · Continuous Univariate Distributions, Volume 2, Second Edition
JOHNSON, KOTZ and BALAKRISHNAN · Discrete Multivariate Distributions
JOHNSON, KOTZ and KEMP · Univariate Discrete Distributions, Second Edition
JUDGE, GRIFFITHS, HILL, LU TKEPOHL and LEE · The Theory and Practice of Econometrics, Second Edition
JUREČKOVÁ and SEN · Robust Statistical Procedures: Asymptotics and Interrelations
JUREK and MASON · Operator-Limit Distributions in Probability Theory
KADANE · Bayesian Methods and Ethics in a Clinical Trial Design
KADANE and SCHUM · A Probabilistic Analysis of the Sacco and Vanzetti Evidence
KALBFLEISCH and PRENTICE · The Statistical Analysis of Failure Time Data, Second Edition
KARIYA and KURATA · Generalized Least Squares
KASS and VOS · Geometrical Foundations of Asymptotic Inference
KAUFMAN and ROUSSEEUW · Finding Groups in Data: An Introduction to Cluster Analysis
KEDEM and FOKIANOS · Regression Models for Time Series Analysis
KENDALL, BARDEN, CARNE and LE · Shape and Shape Theory
KHURI · Advanced Calculus with Applications in Statistics, Second Edition
KHURI, MATHEW and SINHA · Statistical Tests for Mixed Linear Models
* KISH · Statistical Design for Research
KLEIBER and KOTZ · Statistical Size Distributions in Economics and Actuarial Sciences
KLUGMAN, PANJER and WILLMOT · Loss Models: From Data to Decisions
KLUGMAN, PANJER and WILLMOT · Solutions Manual to Accompany Loss Models: From Data to Decisions
KOTZ, BALAKRISHNAN and JOHNSON · Continuous Multivariate Distributions, Volume 1, Second Edition
KOTZ and JOHNSON (editors) · Encyclopedia of Statistical Sciences: Volumes 1 to 9 with Index
KOTZ and JOHNSON (editors) · Encyclopedia of Statistical Sciences: Supplement Volume
KOTZ, READ and BANKS (editors) · Encyclopedia of Statistical Sciences: Update Volume 1
KOTZ, READ and BANKS (editors) · Encyclopedia of Statistical Sciences: Update Volume 2
KOVALENKO, KUZNETZOV and PEGG · Mathematical Theory of Reliability of Time-Dependent Systems with Practical Applications
KOWALSI and TU · Modern Applied U-Statistics
KROONENBERG · Applied Multiway Data Analysis
KULINSKAYA, MORGENTHALER and STAUDTE · Meta Analysis: A Guide to Calibrating and Combining Statistical Evidence
KUROWICKA and COOKE · Uncertainty Analysis with High Dimensional Dependence Modelling

*Now available in a lower priced paperback edition in the Wiley – Interscience Paperback Series.

KVAM and VIDAKOVIC · Nonparametric Statistics with Applications to Science and Engineering
LACHIN · Biostatistical Methods: The Assessment of Relative Risks
LAD · Operational Subjective Statistical Methods: A Mathematical, Philosophical and Historical Introduction
LAMPERTI · Probability: A Survey of the Mathematical Theory, Second Edition
LANGE, RYAN, BILLARD, BRILLINGER, CONQUEST and GREENHOUSE · Case Studies in Biometry
LARSON · Introduction to Probability Theory and Statistical Inference, Third Edition
LAWLESS · Statistical Models and Methods for Lifetime Data, Second Edition
LAWSON · Statistical Methods in Spatial Epidemiology, Second Edition
LE · Applied Categorical Data Analysis
LE · Applied Survival Analysis
LEE · Structural Equation Modelling: A Bayesian Approach
LEE and WANG · Statistical Methods for Survival Data Analysis, Third Edition
LEPAGE and BILLARD · Exploring the Limits of Bootstrap
LEYLAND and GOLDSTEIN (editors) · Multilevel Modelling of Health Statistics
LIAO · Statistical Group Comparison
LINDVALL · Lectures on the Coupling Method
LIN · Introductory Stochastic Analysis for Finance and Insurance
LINHART and ZUCCHINI · Model Selection
LITTLE and RUBIN · Statistical Analysis with Missing Data, Second Edition
LLOYD · The Statistical Analysis of Categorical Data
LOWEN and TEICH · Fractal-Based Point Processes
MAGNUS and NEUDECKER · Matrix Differential Calculus with Applications in Statistics and Econometrics, Revised Edition
MALLER and ZHOU · Survival Analysis with Long Term Survivors
MALLOWS · Design, Data and Analysis by Some Friends of Cuthbert Daniel
MANN, SCHAFER and SINGPURWALLA · Methods for Statistical Analysis of Reliability and Life Data
MANTON, WOODBURY and TOLLEY · Statistical Applications Using Fuzzy Sets
MARCHETTE · Random Graphs for Statistical Pattern Recognition
MARKOVICH · Nonparametric Analysis of Univariate Heavy-Tailed Data: Research and practice
MARDIA and JUPP · Directional Statistics
MARKOVICH · Nonparametric Analysis of Univariate Heavy-Tailed Data: Research and Practice
MARONNA, MARTIN and YOHAI · Robust Statistics: Theory and Methods
MASON, GUNST and HESS · Statistical Design and Analysis of Experiments with Applications to Engineering and Science, Second Edition
MCCULLOCH and SERLE · Generalized, Linear and Mixed Models
MCFADDEN · Management of Data in Clinical Trials
MCLACHLAN · Discriminant Analysis and Statistical Pattern Recognition
MCLACHLAN, DO and AMBROISE · Analyzing Microarray Gene Expression Data
MCLACHLAN and KRISHNAN · The EM Algorithm and Extensions
MCLACHLAN and PEEL · Finite Mixture Models
MCNEIL · Epidemiological Research Methods
MEEKER and ESCOBAR · Statistical Methods for Reliability Data
MEERSCHAERT and SCHEFFLER · Limit Distributions for Sums of Independent Random Vectors: Heavy Tails in Theory and Practice
MICKEY, DUNN and CLARK · Applied Statistics: Analysis of Variance and Regression, Third Edition
* MILLER · Survival Analysis, Second Edition
MONTGOMERY, JENNINGS and KULAHCI · Introduction to Time Series Analysis and Forecasting Solutions Set

*Now available in a lower priced paperback edition in the Wiley Classics Library.

MONTGOMERY, PECK and VINING · Introduction to Linear Regression Analysis, Fourth Edition
MORGENTHALER and TUKEY · Configural Polysampling: A Route to Practical Robustness
MUIRHEAD · Aspects of Multivariate Statistical Theory
MULLER and STEWART · Linear Model Theory: Univariate, Multivariate and Mixed Models
MURRAY · X-STAT 2.0 Statistical Experimentation, Design Data Analysis and Nonlinear Optimization
MURTHY, XIE and JIANG · Weibull Models
MYERS and MONTGOMERY · Response Surface Methodology: Process and Product Optimization Using Designed Experiments, Second Edition
MYERS, MONTGOMERY and VINING · Generalized Linear Models. With Applications in Engineering and the Sciences
† NELSON · Accelerated Testing, Statistical Models, Test Plans and Data Analysis
† NELSON · Applied Life Data Analysis
NEWMAN · Biostatistical Methods in Epidemiology
OCHI · Applied Probability and Stochastic Processes in Engineering and Physical Sciences
OKABE, BOOTS, SUGIHARA and CHIU · Spatial Tesselations: Concepts and Applications of Voronoi Diagrams, Second Edition
OLIVER and SMITH · Influence Diagrams, Belief Nets and Decision Analysis
PALTA · Quantitative Methods in Population Health: Extentions of Ordinary Regression
PANJER · Operational Risks: Modeling Analytics
PANKRATZ · Forecasting with Dynamic Regression Models
PANKRATZ · Forecasting with Univariate Box-Jenkins Models: Concepts and Cases
PARDOUX · Markov Processes and Applications: Algorithms, Networks, Genome and Finance
PARMIGIANI and INOUE · Decision Theory: Principles and Approaches
* PARZEN · Modern Probability Theory and Its Applications
PEÑA, TIAO and TSAY · A Course in Time Series Analysis
PIANTADOSI · Clinical Trials: A Methodologic Perspective
PORT · Theoretical Probability for Applications
POURAHMADI · Foundations of Time Series Analysis and Prediction Theory
POWELL · Approximate Dynamic Programming: Solving the Curses of Dimensionality
PRESS · Bayesian Statistics: Principles, Models and Applications
PRESS · Subjective and Objective Bayesian Statistics, Second Edition
PRESS and TANUR · The Subjectivity of Scientists and the Bayesian Approach
PUKELSHEIM · Optimal Experimental Design
PURI, VILAPLANA and WERTZ · New Perspectives in Theoretical and Applied Statistics
PUTERMAN · Markov Decision Processes: Discrete Stochastic Dynamic Programming
QIU · Image Processing and Jump Regression Analysis
RAO · Linear Statistical Inference and its Applications, Second Edition
RAUSAND and HØYLAND · System Reliability Theory: Models, Statistical Methods and Applications, Second Edition
RENCHER · Linear Models in Statistics
RENCHER · Methods of Multivariate Analysis, Second Edition
RENCHER · Multivariate Statistical Inference with Applications
RIPLEY · Spatial Statistics
RIPLEY · Stochastic Simulation
ROBINSON · Practical Strategies for Experimenting
ROHATGI and SALEH · An Introduction to Probability and Statistics, Second Edition
ROLSKI, SCHMIDLI, SCHMIDT and TEUGELS · Stochastic Processes for Insurance and Finance
ROSENBERGER and LACHIN · Randomization in Clinical Trials: Theory and Practice

†Now available in a lower priced paperback edition in the Wiley – Interscience Paperback Series.
*Now available in a lower priced paperback edition in the Wiley Classics Library.

ROSS · Introduction to Probability and Statistics for Engineers and Scientists
ROSSI, ALLENBY and MCCULLOCH · Bayesian Statistics and Marketing
ROUSSEEUW and LEROY · Robust Regression and Outline Detection
ROYSTON and SAUERBREI · Multivariable Model - Building: A Pragmatic Approach to Regression Anaylsis based on Fractional Polynomials for Modelling Continuous Variables
RUBIN · Multiple Imputation for Nonresponse in Surveys
RUBINSTEIN · Simulation and the Monte Carlo Method, Second Edition
RUBINSTEIN and MELAMED · Modern Simulation and Modeling
RYAN · Modern Engineering Statistics
RYAN · Modern Experimental Design
RYAN · Modern Regression Methods
RYAN · Statistical Methods for Quality Improvement, Second Edition
SALEH · Theory of Preliminary Test and Stein-Type Estimation with Applications
SALTELLI, CHAN and SCOTT (editors) · Sensitivity Analysis
* SCHEFFE · The Analysis of Variance
SCHIMEK · Smoothing and Regression: Approaches, Computation and Application
SCHOTT · Matrix Analysis for Statistics
SCHOUTENS · Levy Processes in Finance: Pricing Financial Derivatives
SCHUSS · Theory and Applications of Stochastic Differential Equations
SCOTT · Multivariate Density Estimation: Theory, Practice and Visualization
* SEARLE · Linear Models
SEARLE · Linear Models for Unbalanced Data
SEARLE · Matrix Algebra Useful for Statistics
SEARLE and WILLETT · Matrix Algebra for Applied Economics
SEBER · Multivariate Observations
SEBER and LEE · Linear Regression Analysis, Second Edition
SEBER and WILD · Nonlinear Regression
SENNOTT · Stochastic Dynamic Programming and the Control of Queueing Systems
* SERFLING · Approximation Theorems of Mathematical Statistics
SHAFER and VOVK · Probability and Finance: Its Only a Game!
SILVAPULLE and SEN · Constrained Statistical Inference: Inequality, Order and Shape Restrictions
SINGPURWALLA · Reliability and Risk: A Bayesian Perspective
SMALL and MCLEISH · Hilbert Space Methods in Probability and Statistical Inference
SRIVASTAVA · Methods of Multivariate Statistics
STAPLETON · Linear Statistical Models
STAUDTE and SHEATHER · Robust Estimation and Testing
STOYAN, KENDALL and MECKE · Stochastic Geometry and Its Applications, Second Edition
STOYAN and STOYAN · Fractals, Random and Point Fields: Methods of Geometrical Statistics
STREET and BURGESS · The Construction of Optimal Stated Choice Experiments: Theory and Methods
STYAN · The Collected Papers of T. W. Anderson: 1943–1985
SUTTON, ABRAMS, JONES, SHELDON and SONG · Methods for Meta-Analysis in Medical Research
TAKEZAWA · Introduction to Nonparametric Regression
TAMHANE · Statistical Analysis of Designed Experiments: Theory and Applications
TANAKA · Time Series Analysis: Nonstationary and Noninvertible Distribution Theory
THOMPSON · Empirical Model Building
THOMPSON · Sampling, Second Edition
THOMPSON · Simulation: A Modeler's Approach
THOMPSON and SEBER · Adaptive Sampling
THOMPSON, WILLIAMS and FINDLAY · Models for Investors in Real World Markets
TIAO, BISGAARD, HILL, PEÑA and STIGLER (editors) · Box on Quality and Discovery: with Design, Control and Robustness

*Now available in a lower priced paperback edition in the Wiley Classics Library.

TIERNEY · LISP-STAT: An Object-Oriented Environment for Statistical Computing and Dynamic Graphics
TSAY · Analysis of Financial Time Series
UPTON and FINGLETON · Spatial Data Analysis by Example, Volume II: Categorical and Directional Data
VAN BELLE · Statistical Rules of Thumb
VAN BELLE, FISHER, HEAGERTY and LUMLEY · Biostatistics: A Methodology for the Health Sciences, Second Edition
VESTRUP · The Theory of Measures and Integration
VIDAKOVIC · Statistical Modeling by Wavelets
VINOD and REAGLE · Preparing for the Worst: Incorporating Downside Risk in Stock Market Investments
WALLER and GOTWAY · Applied Spatial Statistics for Public Health Data
WEERAHANDI · Generalized Inference in Repeated Measures: Exact Methods in MANOVA and Mixed Models
WEISBERG · Applied Linear Regression, Second Edition
WELSH · Aspects of Statistical Inference
WESTFALL and YOUNG · Resampling-Based Multiple Testing: Examples and Methods for p-Value Adjustment
WHITTAKER · Graphical Models in Applied Multivariate Statistics
WINKER · Optimization Heuristics in Economics: Applications of Threshold Accepting
WONNACOTT and WONNACOTT · Econometrics, Second Edition
WOODING · Planning Pharmaceutical Clinical Trials: Basic Statistical Principles
WOODWORTH · Biostatistics: A Bayesian Introduction
WOOLSON and CLARKE · Statistical Methods for the Analysis of Biomedical Data, Second Edition
WU and HAMADA · Experiments: Planning, Analysis and Parameter Design Optimization
WU and ZHANG · Nonparametric Regression Methods for Longitudinal Data Analysis: Mixed-Effects Modeling Approaches
YANG · The Construction Theory of Denumerable Markov Processes
YOUNG, VALERO-MORA and FRIENDLY · Visual Statistics: Seeing Data with Dynamic Interactive Graphics
ZACKS · Stage-Wise Adaptive Designs
* ZELLNER · An Introduction to Bayesian Inference in Econometrics
ZELTERMAN · Discrete Distributions: Applications in the Health Sciences
ZHOU, OBUCHOWSKI and McCLISH · Statistical Methods in Diagnostic Medicine

*Now available in a lower priced paperback edition in the Wiley Classics Library.